Dynamics of Ordering Processes in Condensed Matter

Dynamics of Ordering Processes in Condensed Matter

Edited by

S. Komura

Hiroshima University
Hiroshima, Japan

and

H. Furukawa

Yamaguchi University
Yamaguchi, Japan

Plenum Press • New York and London

Library of Congress Cataloging in Publication Data

International Symposium on Dynamics of Ordering Processes in Condensed Matter
(1987: Kyoto, Japan)
  Dynamics of ordering processes in condensed matter / edited by S. Komura and H.
Furukawa.
      p.   cm.
  "Proceedings of the International Symposium on Dynamics of Ordering Processes
in Condensed Matter, held August 27–30, 1987, in Kyoto, Japan"—T.p. verso.
  "Under the auspices of the Physical Society of Japan"—Pref.
  Includes bibliographies and index.
  ISBN-13: 978-1-4612-8295-2   e-ISBN-13: 978-1-4613-1019-8
  DOI: 10.1007/978-1-4613-1019-8
  1. Condensed matter—Congresses. 2. Order-disorder models—Congresses. 3. Phase
transformations (Statistical physics)—Congresses. 4. Polymers and Polymerization—
Congresses. I. Komura, S. II. Furukawa, H. (Hiroshi), date. III. Nihon Butsuri Gakkai.
IV. Title.
QC173.4.C65I46   1987
530.4'—dc19   88-9408
                                                                              CIP

Proceedings of the International Symposium on Dynamics of Ordering
Processes in Condensed Matter, held August 27–30, 1987, in Kyoto, Japan

© 1988 Plenum Press, New York
Softcover reprint of the hardcover 1st edition 1988
A Division of Plenum Publishing Corporation
233 Spring Street, New York, N.Y. 10013

John E. Hilliard
May 14, 1926–April 21, 1987

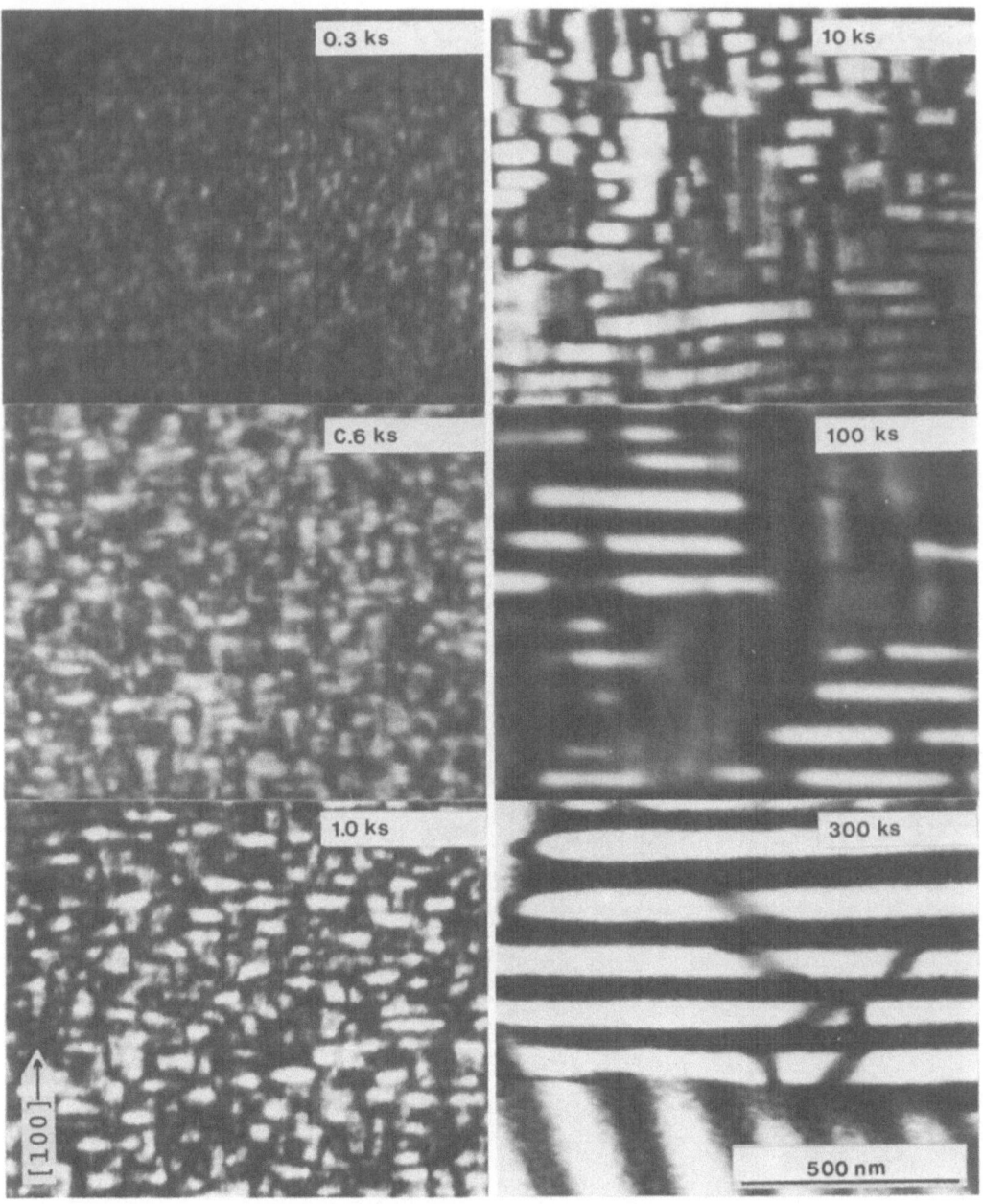

Ordering with phase separation from B2 to (B2 + DO₃) in an Fe-13.8at%Si alloy at 923 K. The photographs of electron microscope images were taken with 111 superlattice reflection from the DO₃ phase. (See page 315.) (Presented by S. Matsumura, H. Oyama, and K. Oki, Department of Materials Science and Technology, Kyushu University.)

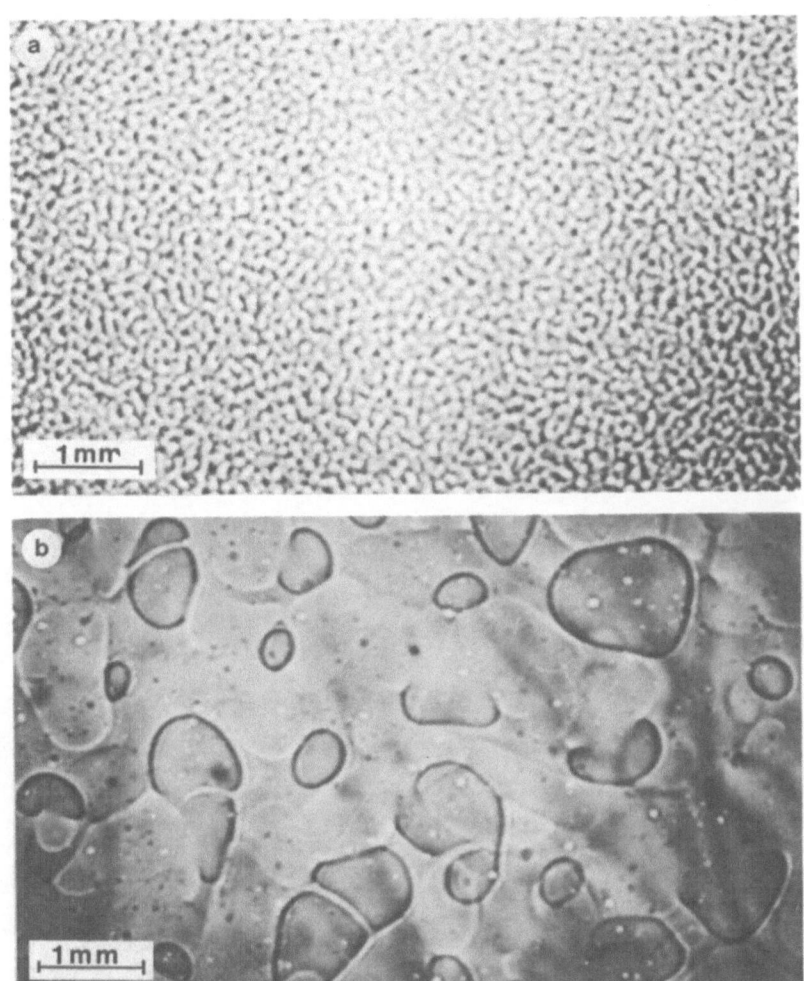

View of phase-separating fluid undergoing spinodal decomposition
in the critical region when the effects of gravity have been
suppressed. Macroscopic interconnected structures can thus be
detected. This has been made possible by performing experiments
under microgravity and/or by using binary fluids with a precise
matching of density, as provided by the mixture methanol +
cyclohexane, the latter partially deuterated (quench of 5 mK
below the critical point, (a) 40 s after the quench, (b) 3 min
after the quech). (See page 373.) (Presented by D. Beysens, P.
Guenoun, and F. Perrot, Service de Physique du Solide et de
Resonance Magnetique, CEN-Saclay.)

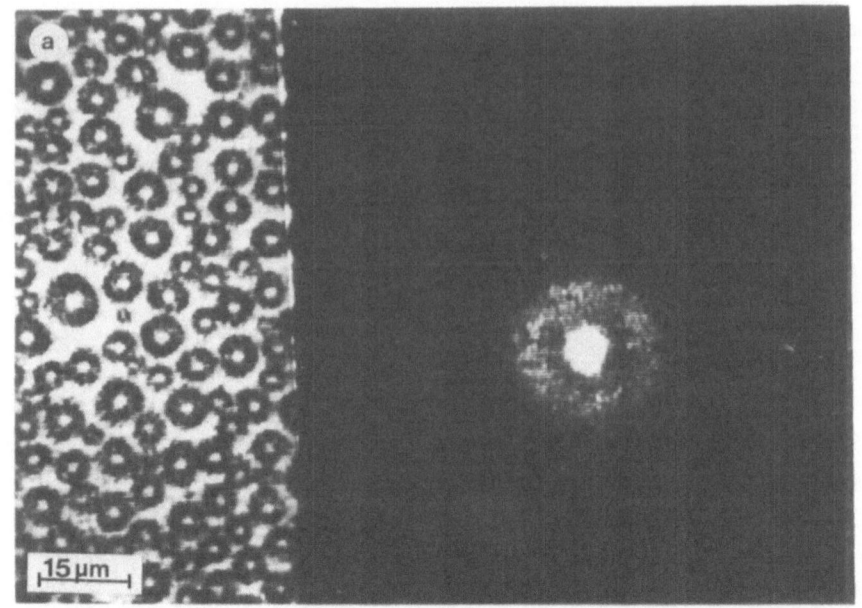

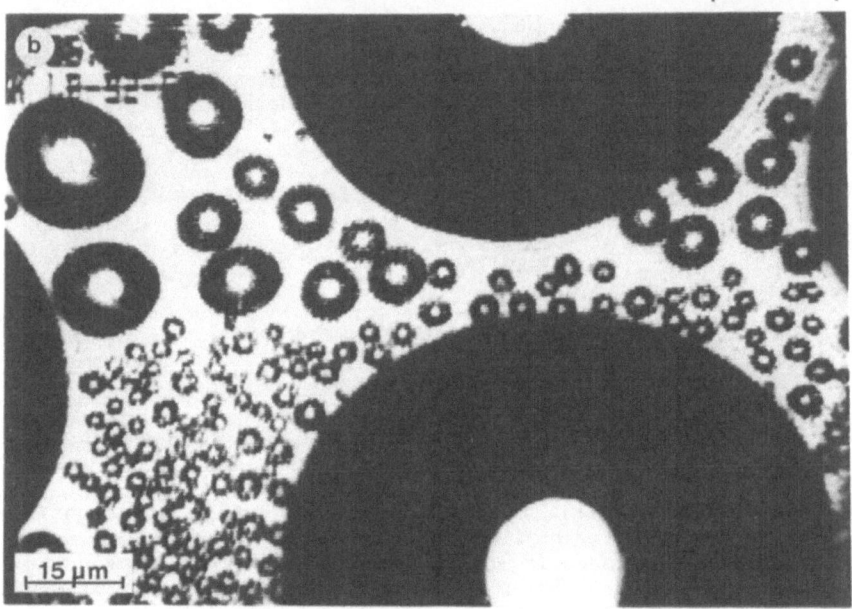

"Breath figures," i.e., the 2-D patterns formed by a fluid (here water vapor) condensing onto a cold surface. The wetting conditions here are such that the droplets are hemispherical. (a) First stages of the growth, where the pattern, evolving by the growth of single droplets and their coalescence, remains remarkably self-similar in time. On the left of the screen is the droplet pattern, on the right the corresponding structure factor as obtained by light scattering. (b) Late stages of the growth, when different "families" of droplets coexist and continue to grow. A new family forms when the separation between droplets is of the order of 100 μm. (See page 403.) (Presented by D. Beysens, Service de Physique du Solide et de Resonance Magnetique, CEN-Saclay.)

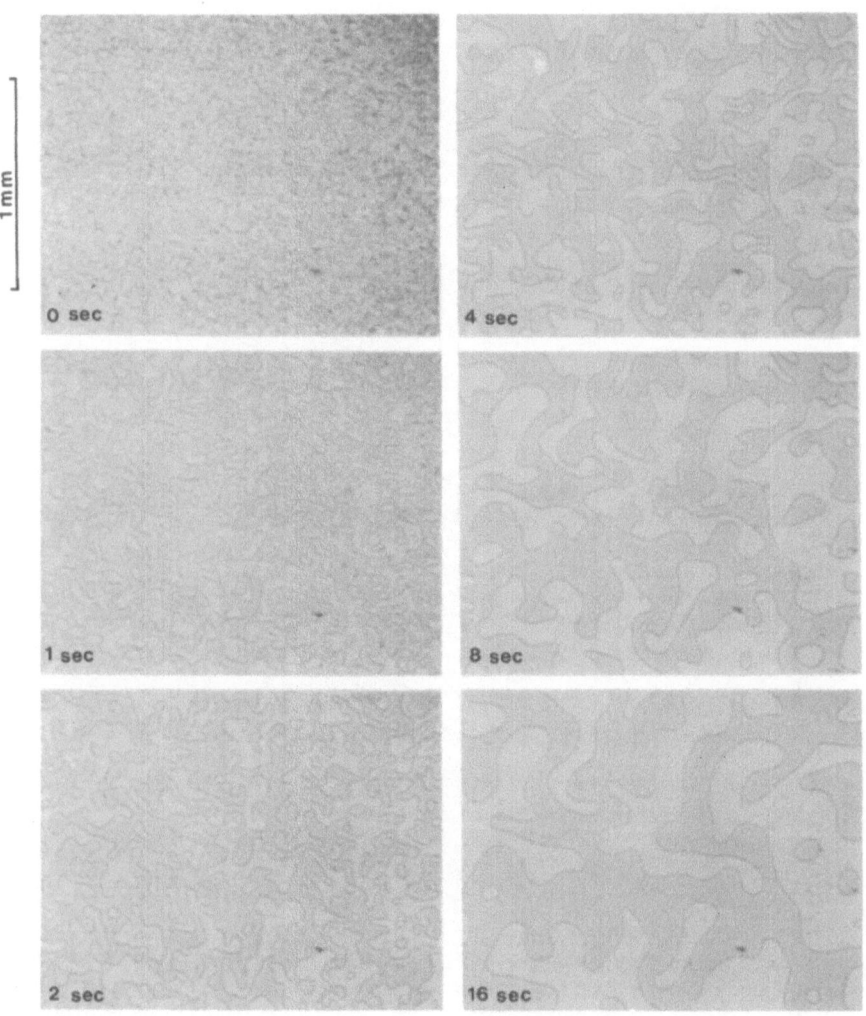

Temporal behavior of optical microscopic images after quenching a twisted nematic liquid crystal. The dark lines are disclinations. (See page 415.) (Presented by H. Orihara and Y. Ishibashi, Faculty of Engineering, Nagoya University.)

CHAIRMAN

Kyozo Kawasaki
Department of Physics
Faculty of Science
Kyushu University
Fukuoka 812, Japan

SECRETARY GENERAL

S. Komura
Faculty of Integrated Arts and Sciences
Hiroshima University
Hiroshima 730, Japan

ORGANIZING COMMITTEE

K. Hirano (Tohoku University)
M. Suzuki (Tokyo University)
Y. Yamada (Tokyo University)
T. Nose (Tokyo Institute of Technology)
T. Miyazaki (Nagoya Institute of Technology)
H. Furukawa (Yamaguchi University)

LOCAL COMMITTEE

M. Furusaka (Tohoku University)
H. Ikeda (Ochanomizu University)
K. Osamura (Kyoto University)
T. Hashimoto (Koyoto University)
K. Kaji (Kyoto University)
T. Takeda (Hiroshima University)
T. Eguchi (Fukuoka University)

INTERNATIONAL ADVISORY COMMITTEE

K. Binder (Mainz)
J.W. Cahn (Washington D.C.)
S. Chikazumi (Tokyo)
P.G. de Gennes (Paris)
W.I. goldburg (Pittsburgh)
J.D. Gunton (Philadelphia)
P. Guyot (Grenoble)
G. Inden (Dusseldorf)
R. Kubo (Tokyo)
J.S. Langer (Santa Barbara)
J. Takamura (Kyoto)

This symposium is financially supported by the following organizations and companies which are gratefully acknowledged.

ORGANIZATIONS

Japan Society for Promotion of Sciences
Commemorative Association for the Japan World Exposition (1970)
Kasima Foundation
Yoshida Foundation for Science and Technology

COMPANIES

Agne Gijutsu Center
Aishin Takaoka, Ltd.
Asahi Chemical Industry Co., Ltd.
Bridgestone Corporation
Daicel chemical Industries, Ltd., Research Center
Daido steel Co., Ltd.
Hatachi Ltd., Central Research Laboratory
Hitachi Metals, Ltd., Magnetic & Electronic Materials Research
    Laboratory
Idemitsu Kosan Co., Ltd., Central Research Laboratory
Idemitsu Petrochemical Co., Ltd., Performance Polymers Research
    Laboratory
JEOL Ltd.
Kanegafuchi Chemical Industry Co., Ltd.
Kansai Paint Co., Ltd.
Kyoritsu Shuppan Co., Ltd.
Matsushita Electric Industrial Co., Ltd., Air-Conditioner Division
Mitsubishi Electric Corp., Central Research Laboratory
Mitsubishi Metal Corporation, Central Research Institute
Mitsubishi Rayon Co., Ltd., Central Research Laboratory
Mitsui Petrochemical Industries Ltd.
Murata Mfg. Co., Ltd.
Nihon Kagaku Engineering Corp.
Nippon Denso Co., Ltd.
Nippon Kokan K.K., Steel Research Center
Nippon Light Metal Company, Ltd.
Nippon Mining
Nippon Steel Corporation, R.& D. Labs.-I
Nippon Steel Corporation, R.& D. Labs.-III
Nisshin Steel Co., Ltd.
Polyplastics Co., Ltd.
Rigaku Corporation
Shoei Chemical Inc.
Showa Denko K.K., Chichibu Research Laboratory
sinku Riko Co., Ltd.
Sony Corporation
Sumitomo Bakelite Co., Ltd.
Sumitomo Chemical Co., Ltd.
Sumitomo Electric Industries, Ltd.
Sumitomo 3M Limited
Toa Nenryo Kogyo K.K.
Tokin Corporation
Tokuyama Soda Co., Ltd.
Toyobo Co., Ltd., Research Center
Toyota Central Research & Development Labs. Inc.
Toyota Motor Corporation
Union Optical Co., Ltd.

PREFACE

The International Symposium on Dynamics of Ordering Processes in Condensed Matter was held at the Kansai Seminar House, Kyoto, for four days, from 27 to 30 August 1987, under the auspices of the Physical Society of Japan. The symposium was financially supported by the four organizations and 45 companies listed on other pages in this volume. We are very grateful to all of them and particularly to the greatest sponsor, the Commemorative Association for the Japan World Exposition 1970. A total of 22 invited lectures and 48 poster presentations were given and 110 participants attended from seven nations.

An objective of the Symposium was to review and extend our present understanding of the dynamics of ordering processes in condensed matters, (for example, alloys, polymers and fluids), that are brought to an unstable state by sudden change of such external parameters as temperature and pressure. A second objective, no less important, was to identify new fields of science that might be investigated by similar, but sometimes more sophisticated, concepts and tactics. An emphasis was laid on those universal aspects of the laws governing the ordering processes which transcended the detailed differences among the substances used. The 71 lectures reproduced in this volume bear witness to the success of the Symposium in meeting amply the first objective and, to a lesser extent, the second.

Discussions were generated by the invited lectures and the poster presentations, supported with the use of video players and personal computers. These discussions were truly interdisciplinary in the sense that the participants included metallurgists, polymer scientists, experimental scattering physicists, and statistical theoretical physicists. The enthusiasm that flavored the discussions meant that the objectives were well satisfied.

We should like to thank all the participants who attended the symposium for their part in making it so lively and rewarding. Our thanks are also due to all those who participated in the organization of the symposium under the general direction of the chairman, Professor Kyozi Kawasaki. These include the organizing committee, the local committee, as well as the international advisory committee, the members of which are listed separately. We are most grateful to Miss N. Terao and Miss A. Samura who lightened our respective burdens as secretary general and editors of this volume.

S. Komura   H. Furukawa

Faculty of Integrated   Faculty of Education
Arts and Sciences   Yamaguchi University
Hiroshima University   Yamaguchi 753, Japan
Hiroshima 730, Japan

CONTENTS

C. Theory for Phase Transition of Polymer

D. Dendrite & Snow

PART II - NUMERICAL SIMULATION

E. Spinodal Decomposition and Order-Disorder

F. Pattern at Phase Transition

G. Other Topics related to Phase Transition

PART III - EXPERIMENTAL: SOLID

H. Spinodal Decomposition

I. Pattern in Spinodal Decomposition

PART IV - EXPERIMENTAL: FLUID

M. Simple Liquid

N. Polymer

PART V -- OTHER TOPICS : THEORETICAL

O. Stochastic Equations

INTRODUCTORY REMARKS

Kyozi Kawasaki

Department of Physics
Faculty of Science
Kyushu University 33
Fukuoka 812, Japan

The study of what happens when a system is suddenly brought into a state which is not thermodynamically stable has a very long history, the origin of which can be traced to the works of Gibbs last century. In the early days the subject was explored in such various corners of sciences as metallurgy, physical chemistry, meteorology, etc. But in more recent years it has established itself as a major field of basic research. This is evidenced, for instance, by the appearance of the very extensive review article written by theoretical physicists such as J.D. Gunton and the coworkers, (Gunton et al, 1983), and also by the fact that this subject was chosen as one of the important topics in the I.U.P.A.P. Statistical Physics Conferences (for instance, Stanley, 1986). Nowadays, the study of ordering processes is no longer limited to the fields of science mentioned above. On the experimental side, the study spreads over many areas dealing with a wide variety of condensed matter systems such as polymers, fluid mixtures, gels, ferroelectrics, membranes, superfluids, superconductors, and the like. The experimental techniques used are equally multi-faceted; some are looking into transient real space patterns that appear during ordering processes and others study fluctuation spectra in wave vector space. On the theoretical side, not only theoretical physicists are involved but also mathematicians and computer scientists are playing important roles. In other words, the study of the dynamics of ordering processes has now developed into a full-fledged interdisciplinary research area of basic science which is relevant to technology.

Under such circumstances it is a natural thing to plan a meeting that emphasizes this interdisciplinary aspect of the subject. In fact, such meetings have already taken place in this country, with the participation, however, of only Japanese scientists. Their impact, nevertheless, has been quite remarkable. For example many theorists were inspired by being exposed to fascinating experimental facts that had been unknown to them and experimentalists received hints crucial to the interpretation their findings. These meetings definitely contributed to progress of the field in this country. This experience engendered the idea of holding an international symposium in the same spirit and the idea materialized through the Kyoto Symposium on Dynamics of Ordering Processes in Condensed Matter. The present volume contains the reports of all the papers presented at the Kyoto Symposium.

On behalf of the organizers I thank all the participants of the Symposium - both overseas and domestic - for their efforts to maintain a high overall scientific quality in the Symposium and its proceedings. It is our sincere wish that the present Symposium will provide a new start for more collaborative research efforts involving scientists of different disciplines on a world-wide scale.

I wish to take this opportunity to thank members of the International Advisory Committee, listed elsewhere, for their very helpful suggestions and encouraging words which were vital to the success of the Symposium.

Last but not least, I thank all the Foundations and the Corporations whose names are listed elsewhere for their generous financial support, without which the Symposium could never have taken off the ground.

References

Gunton, J.D., San Miguel, M. and Sahni, P.S. 1983 "The dynamics of first order phase transitions", in: "Phase Transitions and Critical Phenomena", C. Domb and J.L. Lebowitz, eds., Academic Press, New York.
Stanley, H.E., 1986 "Statistical Physics" North-Holland, Amsterdam.

A HISTORICAL PERSPECTIVE

Dedicated to John E. Hilliard (14 May 1926 - 21 April 1987)

John W. Cahn
National Bureau of Standards
Gaithersburg, MD 20899

The subjects of this conference have a long history. I would like to tell about those that came from metallurgy and those in which my friend, John Hilliard, had a hand.

All important commercial alloys in use at the beginning of this century had been known for two thousand years. Only cast iron was unknown to the Romans. Then in 1906 came the accidental discovery of an aluminum room temperature age hardening alloy by Wilm[1]. This alloy quickly became an important commercial alloy with the trade name Duralumin, but its behavior was judged mysterious and completely different from that of any other alloy. For thirteen years there was a search for other such alloys. Only one other was discovered, but not published. Duralumin was judged to be a unique and curious phenomenon.

Then in 1919 came one of the most important ideas for the field of metallurgy. Three scientists from the then U.S. Bureau of Standards (now NBS) proposed that the age hardening was a room temperature precipitation phenomenon[2]. They suggested that an examination of phase diagrams would reveal not only which other alloys would be candidates for precipitation hardening, but also would suggest compositions, as well as the temperatures of homogenization and precipitation. The importance of this "theoretical" suggestion is clear from the record. A worldwide research effort was launched immediately. Within a decade every metal had been hardened in this way. Most modern alloys derive their properties from a precipitation process. Indeed many of the papers at this conference are about this subject.

Prior to 1919, the altering of properties through phase changes was limited to iron alloys. The discoveries by craftsmen over the millennia, had received a phase diagram explanation that had been confirmed in the second half of the nineteenth century with microscopy. The term "phase change" was limited to cases where a major phase disappears; precipitation was not judged to be a phase change. Indeed in spite of the practical successes of the theory of precipitation hardening, academic skeptics prevailed. Part of the resistance was the result of old misconceptions and definitions about phases. Since the precipitates were not visible, many terms and models were proposed for the hardening agent; "clusters" and "zones" have survived to this day as a relic of the days when a phase was not a phase unless it was large enough to be seen in an optical microscope, and sometimes even had a fixed composition.

Fifty years ago, in 1937 the precipitates were finally seen by diffuse X-ray scattering, and called GP zones in honor of their two independent discoverers, Guinier[3] and Preston[4]. This was a radically new way of seeing, that took decades to gain acceptance. Both the gap in scale and the reciprocal ways of seeing kept the scatterers and the microscopists apart. Only in the late 1950's, when transmission electron microscopy came into full use, were the two ways of seeing merged.

The early studies of phase changes initially were entirely descriptive. Theory impinged in several waves. First came phase diagrams and thermodynamics as the explanation of what was seen; then came nucleation and growth. Even in this age, it should not be hard to understand, that just displaying the iron-carbon phase diagram explained a tremendous amount of craft lore.

The light microscope was the ideal tool for seeing those phase changes in which the nucleation event of a new phase was rare, so that the subsequent growth could be followed over microns or more. The phenomenological theories of Avrami and Johnson and Mehl[5] dealt with the interplay of empirical nucleation and growth rates. Theories of nucleation and of growth rates developed separately and were often tested separately, almost always in systems with long distances between nuclei. Precipitation hardening on the nanometer scale did not fit in, although larger scale precipitation did. It was a very satisfying, almost biological, view of phase changes; phases were born in nucleation, grew, and were consumed in subsequent phase changes.

The scatterers' view of the world was also empirical. Because phase information was lacking, the observed scattering could not be inverted to reveal the real space structures unambiguously. Their effort was on displaying the structures; there was little attempt to propose structures that could arise realistically from known mechanisms. There was much disagreement among proposed structures, in part arising because the various alloys studied had different compositions and histories. For me, one of these proposed structures, by Lipson and Daniel[6], seemed totally unrealistic; there was no way I could imagine how such a compositional modulation could arise from either nucleation and growth or from diffusion. Surely I thought, there must be another, more reasonable structure consistent with the data.

The stability of phases had been considered in detail by Gibbs[7], who had developed a set of necessary conditions. One of these gave a limit to metastability, which we now know as the spinodal, a word coined by van der Waals. For about fifty years the location of the spinodal was considered an important part of phase change lore. It was considered attainable in clean systems, and would result in rapid phase change by barrierless nucleation. Before the spinodal was reached, nucleation was thought always to be heterogeneous. That nucleation could occur homogeneously by thermal fluctuation was first proposed only in 1925[8]. Thereafter, nucleation quickly became the accepted paradigm, supplanting the spinodal almost entirely. One erroneous argument used against the spinodal idea was that it was unnecessary; it neglected surface energy and without surface energy there is no barrier to nucleation and no metastability regardless of the spinodal. One of my earliest pieces of unpublished research was to look experimentally inside what is now known as the chemical spinodal in the gold-nickel system. I found only nucleation and growth.

Surface energy is a key ingredient of nucleation theory. The earliest theory of this, by Young in 1805[9], lead to an inconsistent result, that the surface energy was given by the square of the density or composition difference between the adjoining phases. This was noted by many investigators that such a result makes the surface unstable to spreading, and in 1893 van der Waals produced a diffuse interface theory[10], that was quickly forgotten. Beginning in the 1930's with Becker, and often since then, the forgotten Young result was rederived, and commonly accepted.

In 1956, Mats Hillert[11] proposed a one-dimensional diffusion theory that did indeed predict that a modulated structure would arise within the spinodal. A key ingredient was the theory of Becker for the surface energy. In spite of the fact that nucleation theory is very different in less than two-dimensions, there was a barrier to nucleation that disappeared at the spinodal.

John Hilliard and Mats Hillert had been at MIT together. When Hilliard came to GE, we decided to explore the curious results Hillert had gotten, beginning initially with the problem of the surface energy. This led to our first paper[12]. We next explored the energy of a droplet and nucleation[13], and found to our surprise that the barrier to nucleation disappeared at the spinodal, contrary to my experience with the gold-nickel system. For me this discrepancy was an important barrier to progress, that was only solved with the understanding of the stabilizing influence of composition-induced strains. Once that was solved, the diffusional theory of spinodal decomposition followed immediately[14]. Like Hillert, the theory predicted a modulated structure, but in three dimensions and heavily influenced by the stresses such a structure would generate.

It was John Hilliard's idea to follow the progress of such a process directly by scattering. The diffusion equation predicted directly how the structure factor would evolve. The test of the theory would be on the scattering itself, not on its transform back to a real space structure[15]. It was a revolutionary idea, both for those who followed phase changes with a microscope, and for the others who were reporting empirical structures inferred from scattering. Looking at the scattering curves themselves proved to be a powerful tool, and quickly led to new ideas, such as the need to introduce fluctuations[16], and applications to other phase changes, such as ordering[17]. It is the principal tool for the investigations reported in this conference.

Hilliard also wanted to test the diffusion equation outside of the spinodal. For this he created modulated foils by evaporation to use as diffusion specimens[18]. Not only did he test the modifications of the diffusion equation that had been proposed, and extended the range of diffusion measurements by orders of magnitude into an important technological range, but he also launched a new field. The curious properties of such modulated foils have created much new research[19].

John Hilliard was an extraordinary man. He was a leader in this age of rapid evolution of metallurgical thought. He stayed current, not only in the advances in metallurgy, but also in those aspects of mathematics and physics that he felt would become important to our field. His ability to learn throughout his life and teach himself new topics, that he would then introduce into his work, was all the more remarkable since his formal mathematics education stopped with secondary schooling. He was an inspiring teacher, not only for his students, but also for his colleagues and the wider metallurgical community. His papers on radically new topics were written to teach with clarity, out of respect and understanding for the varied educational background of his readers. He was passionately involved with scientific discovery, and was successful in instilling this in all who were fortunate to have been associated with him.

References

1. A. Wilm Metallurgie 8, 225 (1911)
2. P. D. Merica, R. G. Waltenberg, and H. Scott, Sci. Papers of the U.S. Bureau of Standards #347, Vol. 15, 271 (1919); AIME Bull. 150, 913 (1919).
3. A. Guinier, Compt. Rend. 206, 1641 (1938); Nature 142, 569 (1938).
4. G. D. Preston, Proc. Roy. Soc. A167, 534 (1938); Nature 142, 570 (1938).
5. W. A. Johnson and R. F. Mehl, Trans. AIME 135, 416 (1939).
6. V. Daniel and H. Lipson, Proc. Roy. Soc. A 181, 368 (1943); 182, 378.
7. J. W. Gibbs, Coll. Works, V. 1, Yale Un. Press (1948), pp. 105, 252.
8. D. Turnbull, Solid State Phys. 3, 225 (1956).
9. T. Young, Phil. Trans. Roy. Soc. 95, 65 (1805).
10. J. D. van der Waals, English translation by J. Rowlinson, J. Stat. Phys. 20, 197 (1979).
11. M. Hillert, ScD thesis, MIT (1956); Acta Met 9, 525 (1961).
12. J. W. Cahn and J. E. Hilliard, J. Chem. Phys. 28, 258 (1958).
13. J. W. Cahn and J. E. Hilliard, J. Chem. Phys. 31, 688 (1959).
14. J. W. Cahn, Acta Met. 9, 795 (1961); 10, 179 (1962).
15. J. E. Hilliard, "Spinodal Decomposition" in Phase Transformations, ASM (1970), pp. 497.
16. H. E. Cook and J. E. Hilliard, J. Appl. Phys. 40, 2191 (1969).
17. H. E. Cook, D. de Fontaine and J. E. Hilliard, Acta Met 17, 765 (1969).
18. H. E. Cook, Acta Met. 18, 297 (1970).
19. See for example, W. C. M. Yang, T. Tsakalakos and J. E. Hilliard, J. Appl. Phys. 48, 876 (1977).

ATOM MIGRATION DYNAMICS IN CRYSTALS

Ryoichi Kikuchi

Department of Materials Science and Engineering

FB-10, University of Washington, Seattle, WA 98195

1. INTRODUCTION

The principle of equilibrium statistical mechanics is well established. Backed by the basic physics of quantum mechanics, the mathematics of it is the probability theory. One consequence is that the macroscopic state is the one for the largest probability of appearance.

When we extend this point of view of probability, we can construct the basic principle for atomic migration dynamics in crystals. The basic input of physics still has ambiguity, but the mathematics part is again the probability theory. The state of a system changes towards the direction which is the most probable.

After the basic principles are established, actual implementation of these principles needs a separate consideration. The cluster variation method (CVM) for the equilibrium case and the path probability method (PPM) for the irreversible case will be discussed. Using these methods, important properties which are often misinterpreted in dynamics of phase transitions in crystalline systems will be presented.

2. PROBABILITY BASIS OF EQUILIBRIUM AND NONEQUILIBRIUM STATISTICAL MECHANICS

Quantum mechanics tells us that two different states having the same energy appear with the same probability. We may call this the equal probability postulate.

In statistical mechanics, let us consider a closed system and suppose that a state of a system is specified by a set of parameters α_1, α_2,...(to be written collectively as α). The number of states specified by α is written as $W(\alpha)$. Based on the equal probability postulate, the probability that a state specified by α appears is proportional to $W(\alpha)$. Following Boltzmann,[1] we may define the entropy $S(\alpha)$ corresponding to α as

$$S(\alpha) = k \ln W(\alpha)$$

where k is the Boltzmann constant.

Consider that the system interacts with a heat bath of temperature T. When the energy ΔE flows from the heat bath into the system, the second law of thermodynamics tells us that the entropy of the heat bath changes by the amount

$$(\Delta S)_{h.b.} = -\Delta E/T.$$

When we regard the entirety of the system plus the heat bath as an enlarged closed system, the probability that the original system is in the state α and has the energy ΔE (measured from a certain reference energy state) is proportional to

$$W(\alpha) = \exp[-F(\alpha)\beta],$$

where $\beta = (kT)^{-1}$ and we define

$$F(\alpha) = E(\alpha) - TS(\alpha).$$

In these expressions, we treat ΔE as one of α_i parameters since no confusion is expected. When we plot $W(\alpha)$ against α, we expect a sharp peak when the number of particles is macroscopically large. We may call α the state variable.

When the atomic arrangement (state) in a crystal is changing in time, $\alpha(t)$ is a function of time. Starting from α_0 at t, many states at $t + \Delta t$ are accessible. If we can calculate the probability $P(\alpha_0, t; \alpha, t + \Delta t)$ of going from a state α_0 at t to a state α at $t + \Delta t$, P is expected to have a sharp peak, as was the case with the equilibrium statistical mechanics. The details of the discussions are presented in Ref. 2.

3. CLUSTER VARIATION METHOD (CVM) FOR EQUILIBRIUM

In Section 2 the basic principle of equilibrium statistical mechanics is stated as a minimization of free energy. The actual implementation of the principle, however, is not so easy. The CVM is considered the best available answer to the need, and has been used in progressive frequency in recent years. It is not a single method, but is a hierarchy of methods. In general, more reliable results are obtained when a larger cluster is used and hence a longer time is spent by computation.

Usually the energy of the system is written as a sum of pairwise potentials:

$$E(\alpha) = (Nz/2) \sum_{i,j} \varepsilon_{ij} y_{ij},$$

where N is the total number of lattice points in a system, z is the coordination number, ε_{ij} is the pairwise potential of the i-j pair, and the variable y_{ij} is the fraction of i-j pairs. Although strictly speaking ε_{ij} is to be carefully estimated either by quantum mechanics or by comparing with experiments, in most of phase diagram computations, ε_{ij} is treated as given.

The entropy expression $S(\alpha)$ has been the source of the CVM hierarchical structure. It can be written as a modification of the Boltzmann expression as follows:[3]

$$\exp[S(\alpha)/k] = W_1 \times W_2 \times W_3 \times \dots .$$

The first factor is for a point:

$$W_1 = \frac{N!}{\prod_i (Nx_i)!} = \frac{N!}{\{o\}_N} .$$

where N is the total number of lattice points, x_i is the fraction of the i species, and the $\{o\}_N$ is the notation used in the CVM literature for the sake of brevity. Using similar notation, the rest of the factors are

$$W_2 = \prod_{[2\ pt.]} \frac{\{a\}_N \{b\}_N}{\{a-b\}_N N!} ,$$

$$W_3 = \prod_{[3\ pt.]} \frac{\{ _b/^a \}_N \{b - c\}_N \{ ^a \searrow _c \}_N N!}{\{ _b/^a \searrow _c \}_N \{a\}_N \{b\}_N \{c\}_N} ,$$

where [2 pt.] means to multiply all possible pair shapes (to be included in the designed treatment), while [3 pt.] means to multiply all possible shapes and sizes of three-point clusters.

Usually the accuracy of the CVM formulations improved when a larger cluster is chosen. One of the importnat findings in the CVM series is that[4] in a two-dimensional lattice, one-dimensionally long clusters improve the accuracy. A. Schlijper[5] proved that as the number of squares becomes infinite in the 2-D series, the CVM treatment converges to the exact result.

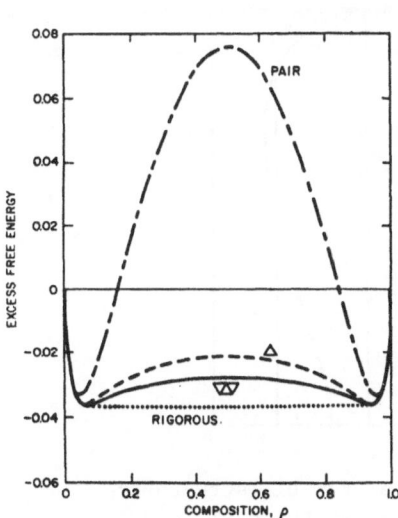

Fig. 1 Excess free energy for a binary alloy, Ref. 6 for $\exp(-4\beta\varepsilon')=0.3$.

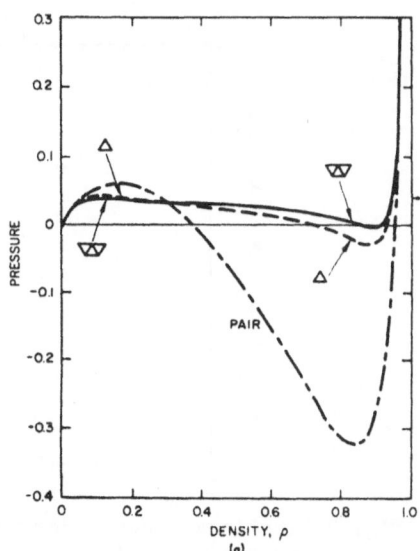

Fig. 2 Equation of state for the lattice-gas-liquid system, Ref. 6 for $\exp(-\beta\varepsilon)=0.3$.

An important consequence of the convergence properties of 2-D series is shown in Fig. 1. This plots the Helmholtz free energy F of a binary AB alloy against the density ρ of B for different approximations (different clusters). The lattice is a two-D triangular lattice. It is clearly seen that as the approximation improves the central hump of F decreases, showing the tendency that in the limit of the rigorous statistical mechanics treatment the F curve has a flat bottom. The same mathematics is used in the gas-liquid isotherms of the lattice gas model in Fig. 2, where we see the clear trend that the van der Waals-type loop is an artifact due to approximation and the exact statistical mechanics must show a flat pressure isotherm.

An important consequence of Fig. 1 is that the spinodal point cannot be calculated from equilibrium statistical mechanics. This clearly shows that the spinodal decomposition[7] is a kinetic process and thus depends on the path of the previous heat treatment.

Another consequence of Fig. 1 which causes more serious trouble than the spinodal decomposition is that it jeopardizes the foundation of the boundary structure calculations. It does not seem completely clarified why the boundary structure calculations, which the present author also participated actively[9,10], give reasonably acceptable results, in spite of the apparent contradiction stated above. However, the boundary structure formulation is not of our primary concern in the present paper, and will not be discussed further.

As an example of the ordering process in condensed matter (the theme subject of the present symposium), the solid curve in Fig. 3 is appropriate. The curve[11] is for the long-range order parameters for L1$_0$ system calculated by the tetrahedron approximation of the CVM.

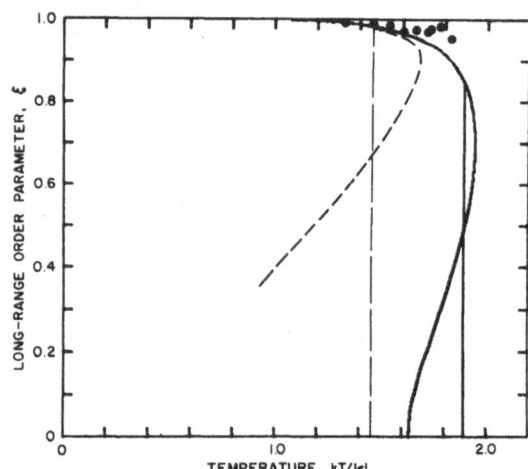

Fig. 3 The long-range order parameter for L1$_0$ structure calculated using a tetrahedron as the basic cluster of the CVM. Dots are experimental results. See Ref. 11.

4. PATH PROBABILITY METHOD (PPM) FOR IRREVERSIBLE PROCESSES

The CVM can treat not only homogeneous systems but inhomogeneous cases as well, including for example boundaries. In the latter, the state variables depend on where the cluster is located with respect to the boundary center. A state variable may contain configurations both at the position m and m+1 as in Fig. 4(a). In a PPM variable (which we call a path variable) a space configuration at t and that at t + Δt appear as in Fig. 4(b). We may say in short that the PPM treats the time as the fourth space axis.

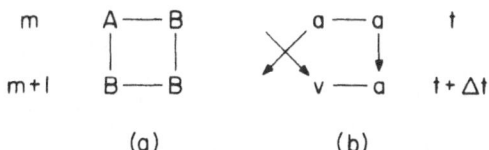

(a)   (b)

Fig. 4   (a) An example of a state
variable in an AB binary
system with a boundary.
(b) An example of a path
variable connecting time t and
t + Δt. A crossed arrow
indicates that an atom jumps
away and a vacancy moves in.

The path probability that a certain path appears in Δt is written in terms of the path variables as an extension of the $\exp[-F(\alpha)\beta]$ function for the equilibrium case. Two basic assumptions are made in writing the path probability function. One of them is the vacancy mechanism; namely, except for special purposes we assume that an atom jumps into an available vacancy located in a nearby lattice point. The second is the Markovian property: after a jump, the atom spends at the new lattice point a long enough time that when it makes the next jump, the memory of the previous jump direction is completely lost.

Usually an atomic jump is helped by thermal fluctuation which is caused by the thermal energy supplied from the heat bath, and can help the atom to overcome the potential barrier for a jump. After the jump, the atom drops to a lower potential energy level and the activation energy dissipates in the form of the kinetic energy, i.e., the excess thermal energy. The exact procedures of treating the activation energy has not been settled.

In writing the path probability function, no concept of the state entropy or the free energy is used. Nevertheless, some results of the PPM can be interpreted using equilibrium thermodynamic concepts. In Fig. 5, a_1 and a_2 are the long-range order and the short-range order parameters[12] for the bcc ordered structure, B2. The free energy contours for $kT/\varepsilon = 4.5$ are shown and E is the equilibrium point. (The order-disorder critical point is $kT_c/\varepsilon = 6.95$). Isothermal relaxations were calculated at this temperature using the PPM (the initial points being at A, B, etc.). For the isothermal annealing, the PPM path from the initial point D or F goes along the dotted curve to the final equilibrium point E, the path being obviously different from the steepest descent direction.

It is worth pointing out here that the decisive advantage of the PPM is that it is closely tied to its equilibrium counterpart, the CVM. For isothermal annealing calculations, the d/dt = 0 limit of the PPM is identical with the CVM result.

As was discussed in Section 2, and in Ref. 2 in detail, starting from a state at t, the system can go to one of many possible states. When the system is left by itself under the imposed constraints, it will change into the state at $t + \Delta t$ corresponding to the maximum probability. This is the most probable change and thus we call it the natural path. One of the most important properties of the natural path for isothermal annealing is that the free energy $F(\alpha)$ always decreases monotonically in time.[13]

In contrast to the natural path, starting from the initial state α_0 at t_0, the probability that the system finds itself at α_1 at a finite later time t_1 is called the two-gate probability $P[\alpha_0, t_0; \alpha_1, t_1]$. As a special case of this, the probability of fluctuating from the equilibrium state at $t_0 \longrightarrow -\infty$ to a state α_1 at t_1 is shown[2] as

$$P[\alpha_{eq}, -\infty; \alpha_1, t_1] = \exp\{-[F(\alpha_1) - F(\alpha_{eq})]\beta\}$$

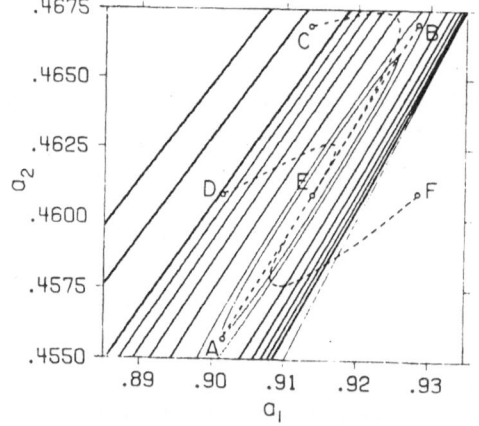

Fig. 5   The isothermal relaxation paths of the PPM superposed on the free energy surface of the CVM. The temperature is at $kT/\varepsilon = 4.5$ (c.f. $kT_c/\varepsilon = 6.95$).

Before going to applications of the PPM, it is important to point out that the PPM of the pair approximation leads exactly the same results[14] as the corresponding approximation of the master equation treatment. When a triangle is used as the basic cluster the two methods are also the same[15] except for a slight difference.

5. NUCLEATION

The "pair" free energy curve in Fig. 1 was sometimes used as the basis of the growth kinetics argument of a new phase. It was argued that if the initial composition was between the free energy minimum and the inflexion point in Fig. 1 on the left-hand side, the new phase near the right end is initiated by the nucleation process.

Although the discussion in Section 3 rules out the validity of the pair-shape curve in Fig. 1 in equilibrium, the nature of the nucleation process is still shrouded by nebulous clouds. One of the reasons for confusion may be the well-known curve for the excess free energy ΔF for an emerging droplet of nucleus r: ΔF starts as r^2, goes through a maximum at r* before it comes down. In the nucleation process of a new phase, the free energy barrier at r* must be overcome, and thus in analogy we may say that starting from the initial state near the left end of Fig. 1, nucleation process proceeds by going over a free energy barrier, in contrast to the spinodal decomposition process which does not go over the barrier. Note that the free energy of a metastable state (not an unstable state) of Fig. 1 is locally minimum in a hyperspace.

We can present a similar question based on Fig. 2. In Fig. 2, gas and liquid phases coexist at the pressure around $p_{gl} = 0.03$ in the scale of the ordinate. The maximum of the pair curve is around $p_{sp} = 0.06$. The question is the following: if the pressure of the gas is kept at 0.05 between p_{gl} and p_{sp}, does the gas condense into a liquid? Physics says it does condense. However, if the PPM based on the pair approximation is used, how can the transformation be described? Does the nucleation process go over a free energy barrier? If so, does the part of the process follow a fluctuation path?

In order to answer these questions, we formulate the PPM for nucleation of liquid droplets from a vapor phase.[16] A droplet made of condensing n atoms is called an n-mer. The number of n-mers in a system is $N(t)x_n(t)$ where $N(t)$ is the total number of clusters in a system and $x_n(t)$ is the state variable (n = 1,2,...). For the change of state, $N(t)X_{n,m}(t, t + \Delta t)$ is the number of n-mers at t which changes into m-mers at t + Δt (n $\geq$ 2, m = n-1, n, n+1; n=1, m=2); $X_{n,m}$ is a path variable.

Without specifying the detailed mechanism, we write the unit probabilities for $X_{n,n+1}$ and $X_{n+1,n}$ processes as $\Theta_{n,n+1}$ and $\Theta_{n+1,n}$, respectively. The path probability function is written using the standard techniques of the PPM. We keep the state at t fixed and minimize the path probability function with respect to $X_{n,m}$ (n≠m) to obtain the relations for the most probable path. The results are

$$X_{1,2} = 1/2 \; x_1(t)^2 \; \Theta_{1,2}\Delta t \; ,$$

$$X_{2,1} = x_2(t) \; \Theta_{2,1}\Delta t \; ,$$

$$X_{n,n+1} = x_1(t)x_n(t) \; \Theta_{n,n+1}\Delta t; \quad n \geq 2 \; ,$$

$$X_{n+1,n} = x_{n+1}(t) \; \Theta_{n+1,n}\Delta t; \quad n \geq 2 \; .$$

By inspection we can tell that these results are identical to the kinetic expressions written down based on physical reasoning in conventional nucleation theory. Thus we can conclude that the conventional kinetic differential equations for the cluster nucleation and growth are derived using the PPM as a natural path. This conclusion further implies that the free energy of the entire system always decreases as the nucleation proceeds.

A corollary from this conclusion is that if we start with the pressure 0.05 in Fig. 2 and keep the same temperature, we can never arrive at the liquid phase. The reason is that between p_{g1} and p_{sp}, the state is metastable and hence the free energy for the entire system is locally a minimum. In order to move away from this local minimum, the system is to be fluctuated macroscopically over the barrier and such a fluctuation is not possible.

Let us examine the other first-order process shown in Fig. 3 as an example. This is an order-disorder transition in CuAu calculated using the tetrahedron CVM. The transition occurs at $T_t = 1.89$. Between $T_t = 1.89$ and $T_1 = 1.62$, the ordered state is more stable than the disordered state. Physics of the first-order transition does expect that below T_t the ordered state will nucleate and grow. However, the disordered state is locally minimum and thus the PPM using the tetrahedron cannot describe the transition. The question we face is how we can plan to study nucleation from atomic point of view. We propose the following procedure as the realistic feasible method.

As we have shown that a new phase can nucleate and grow only when the free energy is monotonically decreasing from the initial state to the final state, in order to create the ordered phase in Fig. 3, we have to bring the temperature below T_1. When we quench the system to a temperature T_q very slightly below $T_1(T_q = T_1 - \Delta T)$, the ordered phase nucleates and grows. Since the driving force for nucleation is the derivative of the free energy (with respect to the "reaction path" in the hyperspace), the rate of nucleation is very slow when ΔT is small, and the incubation period shortens as ΔT increases.

6. NUCLEATION AND GROWTH OF A MONOLAYER

The PPM has not developed far enough to demonstrate details of nucleation processes in three-dimensional systems. In this section we show an example of a two-dimensional nucleation using a triangle as the cluster of the PPM. The related work was reported in several papers[15,16,17] and also in unpublished reports.[18]

From a vapor phase, atoms fall on a substrate. Atoms may be captured on the surface or be later reevaporated. The interaction potential with the substrate is w and the atom-atom interaction is ε. A triangle lattice is assumed for atomic arrangement, and we assume only the monolayer coverage. Migration of atoms on the surface was treated in some of the reports,[17-20] but not in this section.

A rate parameter θ_{ip} is the probability that an atom hit a lattice point per unit time, and is used in normalizing the time scale: $\tau = t\theta_{ip}$. Another rate parameter θ_{ev} is the attempt frequency for atomic escape from the surface and appears in the combination:

$$e^{\beta\mu} = \theta_{ip}/(\theta_{ev}e^{-\beta w}).$$

We can prove that μ is the chemical potential of the atoms in the vapor phase, and hence $\exp(\beta\mu)$ is the activity.

When we start with the bare surface at $\tau = 0$, the surface coverage can be calculated using the PPM. The curves depend on $\exp(\beta\mu)$ as shown in Fig. 6. Note that both axes are in the log scale. The curves clearly indicate the incubation period and how the growth stage starts.

The results of the PPM calculation can be demonstrated by visualizing with the simulation method proposed in Ref. 21. The details will be published elsewhere.

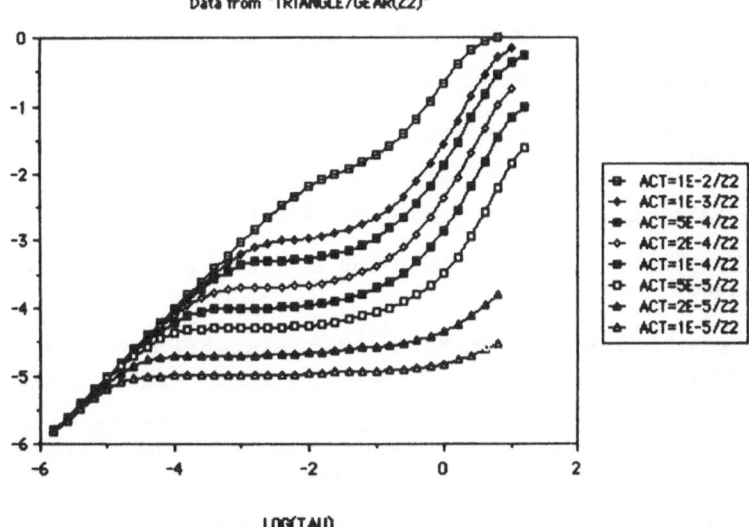

Fig. 6   Nucleation and growth of a monolayer by vapor deposition on a substrate. Ordinate is the log of surface coverage.

7. CONCLUSION

Kinetics in crystalline solids are governed by the requirement that the change of state occurs toward the direction of the most probable path. This requirement is formulated in the path probability method. As the change occurs toward the most probable path, the free energy of the system always decreases.

The nucleation process can be formulated based on the most probable path principle. The free energy of the entire system always decreases as the nucleation process proceeds.

REFERENCES

1. L. Boltzmann, Epitaph on his tombstone, Ehrenfriedhof, Wien, Austria. In Boltmann's epitaph it is written as S = k.log W.

2. R. Kikuchi, Prog. Th. Phys., Suppl. No. 35 (1966) p. 1.

3. R. Kikuchi, "The CVM calculations of phase diagrams" in Computer Modeling of Phase Diagrams (Ed., L.H. Bennett, Met. Soc., Inc., Warrendale, PA) p. 49 (1986).

4. R. Kikuchi and S.G. Brush, J. Chem. Phys. 47, 195 (1967).

5. A. G. Schlijper, Phys. Rev. B27, 6841 (1983).

6. R. Kikuchi, J. Chem. Phys. 47, (1967) 1664.

7. J.W. Cahn, Acta Metall. 10, 907 (1962).

8. J.W. Cahn and J. Hilliard, J. Chem. Phys. 28, 258 (1958).

9. R. Kikuchi and J.W. Cahn, Acta Met. 27, 1337 (1978).

10. R. Kikuchi and J.W. Cahn, Phys. Rev. B21, 1893 (1980).

11. R. Kikuchi and H. Sato, Acta Met. 22, 1099 (1974).

12. K. Gschwend, H. Sato and R. Kikuchi, J. Chem. Phys. 69, 5006 (1978).

13. R. Kikuchi, Phys. Rev. 124, 1682 (1961).

14. T.Ishikawa, K. Wada, H. Sato and R. Kikuchi, Phys. Rev. A33, 4164 (1986).

15. K. Wada, T. Ishikawa, H. Sato and R. Kikuchi, Phys. Rev. A33, 4171 (1986).

16. R. Kikuchi, "Derivatives of the Cluster Growth Equations Using the Path Probability Method," Hughes Research Laboratories Research Report No. 353 (1966). Unpublished.

17. R. Kikuchi, Crystal Growth (Suppl. to J. Phys. Chem. Solids) p. 605 (1967).

18. R. Kikuchi, J. Chem. Phys. 47, 1646 and 1653 (1967).

19. R. Kikuchi, J. Phys. Soc. Japan 26, Suppl. 136 (1969).

20. R. Kikuchi, Hughes Research Laboratories Research Report NOS. 391-394 (1968).

21. R. Kikuchi, Phys. Rev. B22, 3784 (1980).

THERMODYNAMIC FORMALISM FOR A CRITICAL NUCLEUS

IN CONDENSATION FROM VAPOR

Kazumi Nishioka

Department of Precision Mechanics
The University of Tokushima
Tokushima 770, Japan

INTRODUCTION

Critical nuclei in condensation from the vapor are usually extremely small in size so that the homogeneous properties of the bulk liquid are probably not attained even at the center. Hence, the validity of the thermodynamic method to compute the reversible work to form a critical nucleus becomes questionable. On the other hand, they are not small enough to allow direct evaluation of the partition function with the sufficient accuracy. Thus, those critical nuclei present the characteristic size range where neither the macroscopic thermodynamics nor the statistical mechanics is straightforwardly applicable.

The present work intends to extend the thermodynamic formalism to such extremely small systems by following the idea behind the Gibbs theory of surface tension.[1] In addition, the cluster-size dependence of the surface tension is formulated in such a manner that the results of computer simulations for microclusters[2] and for the bulk liquid[3] may be employed. Complication due to translation and rotation of a microcluster in vapor is also discussed in relation to the estimated size dependence of the surface tension.

THERMODYNAMIC FORMALISM

Consider a critical nucleus of the liquid phase in a supersaturated vapor, which is supposed to consist of a pure substance for simplicity. The system is defined here by a mathematical boundary having a conic shape intersected by a sphere as shown by the heavy solid line in Fig.1. The center of the sphere coincides with the 'center' of the nucleus, and its radius R_g is taken to be so large that it passes through the homogeneous vapor phase. Complication due to translation and rotation of the nucleus will be discussed later.

Consider next the thermodynamic variables to define a state of the system. The spherical boundary passes through the homogeneous vapor phase, hence the effect of the interaction between the molecules in each side of the boundary on internal energy and entropy may be considered as shared equally between the two sides of the boundary. Thus, the internal energy $E_{4\pi}$ and entropy $S_{4\pi}$ for the entire sphere are well defined, where the subscript 4π indicates the

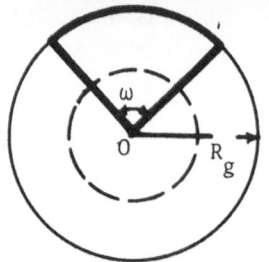

Fig. 1. System defined by a mathematical boundary.

solid angle for the entire sphere. Since the thermodynamic properties are homogeneous along the direction perpendicular to the boundary defining the cone, the internal energy E and the entropy S of the system are given by

$$E = E_{4\pi}\omega/4\pi, \quad S = S_{4\pi}\omega/4\pi, \tag{1}$$

where ω denotes the solid angle of the cone. The number of molecules within the system may fluctuate, but only the average value is treated in thermodynamics. Thus,

$$N = N_{4\pi}\omega/4\pi, \tag{2}$$

where N and $N_{4\pi}$ denote the number of molecules in the system and in the entire sphere, respectively.

The temperature T and the chemical potential μ of a supersaturated vapor determine the state of the critical nucleus as well as the intensive properties of the supersaturated vapor. Hence, the state of the entire sphere in Fig. 1 is determined by $S_{4\pi}$, $N_{4\pi}$ and R_g, thus the state of the system is determined by S, N, R_g and ω.

The fundamental equation for the system is given by

$$E = E(S,N,R_g,\omega), \tag{3}$$

which may be rewritten as

$$E = E_{4\pi}(S_{4\pi},N_{4\pi},R_g) \cdot (\omega/4\pi). \tag{4}$$

Employing the chain rule in differentiation, the differential form of Eq.(3) is obtained as

$$dE = (\partial E_{4\pi}/\partial S_{4\pi})dS + (\partial E_{4\pi}/\partial N_{4\pi})dN + (\omega/4\pi) \cdot (\partial E_{4\pi}/\partial R_g)dR_g$$
$$+ (1/\omega) \cdot [E-(\partial E_{4\pi}/\partial S_{4\pi})S-(\partial E_{4\pi}/\partial N_{4\pi})N]d\omega. \tag{5}$$

Note that the system is supposed to remain in equilibrium both before and after an infinitesimal change in state. Employing the relations

$$(\partial E_{4\pi}/\partial S_{4\pi}) = T, \quad (\partial E_{4\pi}/\partial N_{4\pi}) = \mu, \quad (\partial E_{4\pi}/\partial R_g) = -p4\pi R_g^2 \tag{6}$$

in Eq.(5), we get

$$dE = TdS + \mu dN - p\omega R_g^2 dR_g + \sigma d\omega, \tag{7}$$

where p denotes the pressure of the supersaturated vapor and σ represents the following:

$$\sigma = [E-TS-\mu N]/\omega. \tag{8}$$

Thus, the fundamental equation of the system in differential form is shown to be expressible in terms of the intensive variables T, μ and p of the supersaturated vapor.

To consider the possibility of classifying the variables into the extensive and the intensive ones, let us suppose two systems both of which are specified by (S,N,R_g,ω) and combine them together to form a single system. The system thus formed is specified by $(2S,2N,R_g,2\omega)$ due to Eqs.(1) and (2), and Eq.(4) leads to

$$E(2S,2N,R_g,2\omega) = 2E(S,N,R_g,\omega). \tag{9}$$

Hence, S, N, ω and E may be considered as extensive variables under a constant value for R_g, and E is a homogeneous function of the first degree in S, N and ω. It follows from Euler's theorem that

$$(\partial E/\partial S)S + (\partial E/\partial N)N + (\partial E/\partial \omega)\omega = E, \tag{10}$$

which is equivalent to Eq.(8) since $(\partial E/\partial S)$ and $(\partial E/\partial N)$ have been identified as T and μ of the supersaturated vapor and $(\partial E/\partial \omega)$ is denoted as σ. Under a constant value for R_g, the variables T, μ and σ are homogeneous functions of the zeroth degree in the variables S, N and ω, hence they may be considered as intensive variables in the present formalism. Taking differential of Eq.(10) under a constant value for R_g and comparing the result with Eq.(7), we get the Gibbs-Duhem relation as

$$SdT + Nd\mu + \omega d\sigma = 0. \tag{11}$$

Eq.(11) plays an indispensable role in obtaining the cluster-size dependence of the surface tension. Note that the present definition of the system having a conic shape is the clue in classifying the variables into the extensive and the intensive ones and in obtaining the Gibbs-Duhem relation.

DIVIDING SURFACES AND THE SURFACE TENSION

It is shown in the previous section that the thermodynamic formalism can be established even for an extremely small system containing a critical nucleus. However, for the practical usefulness of the formalism, thermodynamic variables must be either experimentally measurable or must be explicitly related to macroscopic variables. In this section the term $\sigma d\omega$ will be represented in terms of experimentally measurable quantities.

Following Gibbs,[1] we employ a dividing surface and introduce the hypothetical system[4] by filling each side of it with a reference bulk liquid and a bulk vapor. For a given state of the system described by S, N, ω and R_g, the radius R and the area A of the dividing surface are determined once the dividing surface condition is specified. We introduce the work terms $\gamma dA + \rho dR$ associated with the dividing surface as

$$\gamma dA + \rho dR = \sigma d\omega - [-p_\ell dV_\ell - pdV_g'], \tag{12}$$

where the subscript ℓ and g denote the quantities of the hypothetical liquid phase and the hypothetical vapor, respectively, and dV_g' the volume change with R_g kept invariant. Eq.(12) may be interpreted as that the unmeasurable work term $\sigma d\omega$ of the real system is represented by the work terms of the hypothetical system and that their difference is ascribed to the work terms of the abstract dividing surface. Necessity of the term ρdR may be understood in connection with the rotation of the mathematical boundaries defining the system.[5]

Integrating Eq.(12) by keeping R_g and the nature of the system invariant, we obtain

$$\sigma\omega = -p_\ell V_\ell - pV_g + \gamma A. \tag{13}$$

Substituting Eq.(13) into Eq.(10), one gets

$$E = TS + \mu N - p_\ell V_\ell - pV_g + \gamma A. \tag{14}$$

Differentiating Eq.(13) under a constant value for R_g and employing Eqs.(11) and (12), we get the following form for the Gibbs-Duhem relation:

$$S^{ex}dT + N^{ex}d\mu - N_\ell(d\mu^o - d\mu) + Ad\gamma - \rho dR = 0, \tag{15}$$

where S^{ex} and N^{ex} denote the excess entropy and the excess number of molecules, respectively, and N_ℓ the number of molecules in the hypothetical nucleus. Anticipating an extension of the Gibbs method, we allowed in Eq.(15) a possibility that the chemical potential μ^o of the reference bulk liquid may be different from that of the real system.

It is intended[4] that the results of computer simulation[2] for liquid microclusters are utilized to derive the cluster-size dependence of the surface tension. For this purpose we need to employ the equimolecular dividing surface, because the simulation results are obtained in terms of the number of atoms contained in the clusters. However, the term ρdR remains in Eq.(12) under this choice of the dividing surface. To avoid this inconvenience, the present author[4] chose the pressure p_ℓ^o for the reference bulk liquid so that ρdR vanishes in Eq.(12). Accordingly, the chemical potential μ^o of the reference bulk liquid will be different from that of the real system in this case. Under Nishioka's choice[4] of the dividing surface and the reference bulk liquid, Eq.(15) becomes

$$S^{ex}dT - N_\ell(d\mu^o - d\mu) + Ad\gamma = 0. \tag{16}$$

It follows from the condition of thermodynamic equilibrium that

$$p_\ell^o - p = 2\gamma/R. \tag{17}$$

Eliminating γ from Eqs.(14) and (17), one obtains

$$\omega(p_\ell^o - p) = 6(E - TS - \mu N + p\omega R_g^3/3)/R^3. \tag{18}$$

The condition $N^{ex}=0$ for the equimolecular dividing surface is practically equivalent to

$$4\pi R^3/3 = nv_\ell, \tag{19}$$

where n denotes the number of molecules contained in a critical nucleus and v_ℓ the molecular volume of the reference bulk liquid. For a given state of the system, R and p_ℓ^o are determined by Eqs.(18) and (19). However, since bulk liquid is nearly incompressible, pressure dependence of v_ℓ may be neglected and Eq.(19) alone determines R. Then, p_ℓ^o is determined, in principle, by Eq.(18).

Let us consider the physical meaning of the surface tension introduced here. For the entire sphere Eq.(14) may be rewritten as:

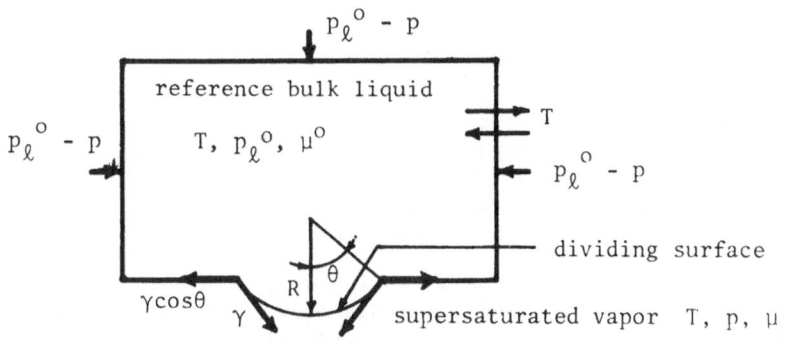

Fig. 2. Reversible thought process to form a critical nucleus
from the reference bulk liquid.

$$\gamma A = E^{ex} - TS^{ex} - (\mu - \mu^{o})n, \tag{20}$$

where E^{ex} denotes the excess internal energy. Consider a thought process of
forming a critical nucleus from the reference bulk liquid as shown in Fig.2.
The reference bulk liquid is enclosed by an envelope to maintain, with the
help of a reversible work source, the difference in the pressure and the chemical potential. Suppose that a small aperture is opened and then closed in
the envelope so as to extrude a critical nucleus of n molecules into the
vapor. During this process, the interface between the nucleus and the vapor
is formed. But, as long as the mechanical work term is concerned, it may be
regarded as replaced by the reference bulk phases and the flexible membrane
which is located at the dividing surface and possesses the mechanical tension
γ. Thus, to keep the process reversible, the reversible work source must provide the force $\gamma\cos\theta$ at the edge of the aperture. Amount of the work done at
the edge is given by

$$\int_0^R \gamma\cos\theta\cdot 2\pi R\sin\theta\, d(R\cos\theta) + \int_R^0 \gamma\sin\theta\cdot 2\pi R\cos\theta\, d(-R\cos\theta), \tag{21}$$

in which the first term represents the work done during the opening process
and the second term during the closing process. Eq.(21) results in $4\pi R^2\gamma/3$.
In addition, amount of the work $4\pi R^3(p_\ell^{o}-p)/3$, which may be rewritten as
$8\pi R^2\gamma/3$ due to Eq.(17), is done to push the envelope inward during the extrusion process. Summation of those two amounts of the reversible work results
in γA.

Besides this mechanical work, the reversible chemical work $(\mu-\mu_\ell^{o})n$ is
to be done during the extrusion process because of the difference in the chemical potentials between the nucleus and the reference bulk liquid. Thus, the
total reversible work in the extrusion process is

$$W^{rev} = \gamma A + (\mu - \mu_\ell^{o})n. \tag{22}$$

Since the extrusion process may be considered as that to form the real system
from the hypothetical system, W^{rev} can be identified with $E^{ex}-TS^{ex}$. Hence,
Eq.(22) is equivalent to Eq.(20) as it should, and γA is ascribed to the reversible mechanical work during the extrusion process shown in Fig.2. Thus,
it becomes clear that γ in the present formalism possesses basically the same
physical meaning as that of the surface tension in the Gibbs formalism.

DISCUSSION

The cluster-size dependence of the surface tension in Nishioka's formalism is governed by Eq.(16), from which the following expression is derived at

a constant temperature:[4]

$$\gamma(n) = - (Rn/v_\ell) [\partial(F(n)/n)/\partial n],\tag{23}$$

where $F(n)$ denotes the Helmholtz free energy of a liquid microcluster containing n molecules. $\gamma(n)$ was computed for argon[4] by employing the results of computer simulation due to Lee et al.[2] Difficulty caused by translation and rotation of a microcluster in vapor was avoided by assigning the internal free energy $F^{int}(n)$ to $F(n)$. Let us consider this point a little further.

Canonical partition function of a liquid microcluster in vapor may be separated into the translational, the rotational and the internal parts by introducing the coordinate frame so that its origin coincides with the center of mass of the microcluster and there is no angular momentum relative to it.[6] Hence, the center of the sphere in Fig.1 may be regarded as the center of mass of a microcluster, and each of $E_{4\pi}$ and $S_{4\pi}$ may be considered as the sum of the internal part for the microcluster and the value for the surrounding vapor. Similarly, the center of the hypothetical microcluster may be considered as its center of mass, and its internal energy and entropy as their internal parts. Following these considerations, the translational and the rotational degrees of freedom are supposed to be excluded from both the hypothetical and the real microclusters in the extrusion process shown in Fig.2. This view corresponds to the Lothe-Pound theory.[7]

Thus, S^{ex} in Eq.(16) represents the difference in the internal parts of of the entropies between the real and the hypothetical microclusters. In addition to Eq.(16), the following rewritten form of Eq.(14) is employed to derive Eq.(23):[4]

$$\gamma A/3 = [(E - TS) - (E_g - TS_g)] + pV_\ell - n\mu.\tag{24}$$

The first term in Eq.(24) represents the Helmholtz free energy of the real microcluster, which is denoted as $F(n)$ and appears in Eq.(23). It follows from the foregoing discussion that the first term in Eq.(24) actually represents the internal free energy $F^{int}(n)$, hence the assignment[4] of $F^{int}(n)$ to $F(n)$ in Eq.(23) is shown to be reasonable. Furthermore, both γ in Nishioka's formalism and the Gibbs surface tension represent the reversible work of the same nature as clarified in the preceding section. Thus, the cluster-size dependence of the Gibbs surface tension is expected to possess the behavior similar to that of γ for argon investigated earlier.[4]

REFERENCES

1. J.W.Gibbs,"The Scientific Papers of J.W.Gibbs", Dover, New York(1961), Vol.I, pp.219-258.
2. J.K.Lee, J.A.Barker and F.F.Abraham, J.Chem.Phys.58:3166(1973).
3. J.-P.Hansen and L.Verlet, Phys.Rev.184:151(1969).
4. K.Nishioka, Phys.Rev.A16:2143(1977).
5. K.Nishioka, to be published.
6. K.Nishioka, G.M.Pound and F.F.Abraham, Phys.Rev.A1:1542(1970).
7. J.Lothe and G.M.Pound, J.Chem.Phys.36:2080(1962).

SCALING AND CAM THEORY IN FAR-FROM-EQUILIBRIUM SYSTEMS

Masuo Suzuki

Department of Physics, Faculty of Science
University of Tokyo, Hongo, Bunkyo-ku
Tokyo 113, Japan

INTRODUCTION

This paper consists of the following three parts, namely (i) to review the scaling theory[1-4] of transient phenomena near the instability point, which was proposed by the present author[1-3] and which has been developed for these ten years,[3,4] and to propose a new general scaling function to describe ordering processes of systems with a conservation law, (ii) to explain the basic idea of the coherent-anomaly method (CAM) by the present author[5] which has been applied to spin systems[6-8], and (iii) to discuss some possible applications of the CAM theory to the dynamics of ordering processes in condensed matter.

PART I: SCALING THEORY OF TRANSIENT PHENOMENA AND ORDERING PROCESSES

In 1976, the present author[1,2] proposed the following general scaling theory of the relaxation and fluctuation in non-equilibrium systems: The fluctuation of the order parameter x at the time t obeys the following scaling form

$$\langle x^2 \rangle_t \simeq \langle x^2 \rangle_{st} f^{(sc)}(\tau), \tag{1}$$

where τ denotes a generalized scaling variable of the form

$$\tau = S(t, \varepsilon, \delta, \cdots). \tag{2}$$

Here ε denotes an appropriate smallness parameter such as the inverse system size $1/\Omega$ (or the strength of noise) for uniform systems described by a macrovariable and the wave-number k for non-uniform systems. For the former case, τ takes the form[1-3]

$$\tau = \varepsilon e^{2\gamma t} \tag{3}$$

with the growing rate γ, and for the latter case, τ takes[1]

$$\tau = kt^\phi \text{ (or } k^m t) \tag{4}$$

with some scaling exponent ϕ (or $m = 1/\phi$).

The strategy of the present author to confirm or to derive (1) in some specific model was[1-4] to make use of perturbational expansions of the relevant physical quantity $\langle Q \rangle_t$ with respect to the smallness parameter ε and to sum up all the most dominant terms for large t and for τ fixed. The last condition that τ be fixed is the key point of our scaling theory, namely τ is invariant under the scale transformation $t \to t'$, $\varepsilon \to \varepsilon'$, $\cdots$, that $\tau (t', \varepsilon', \cdots) = \tau(t, \varepsilon, \cdots)$, and consequently the normalized physical quantity $\langle Q \rangle_t / \langle Q \rangle (t, 0, \cdots)$ is also scale-invariant.[1-3] Thus, the above limit $t \to \infty$ for τ fixed is called the scaling limit in non-equilibrium systems. The present author[2] demonstrated explicitly the existence of such a scaling limit and scaling function $f^{(sc)}(x)$ in a specific model of the form

$$\frac{d}{dt} x = \gamma x - g x^3 + \eta(t) \tag{5}$$

with the gaussian-white noise $\eta(t)$ satisfying $\langle \eta(t)\eta(t') \rangle = 2\varepsilon\delta(t-t')$, or equivalently in the following Fokker-Planck equation

$$\frac{\partial}{\partial t} P(x,t) = [-\frac{\partial}{\partial x}(\gamma x - g x^3) + \varepsilon\frac{\partial^2}{\partial x^2}] P(x,t) \tag{6}$$

for $\gamma > 0$, which corresponds to the situation that the system is unstable. As is well known[1-3], the scaling solution of (5) or (6) for the initial condition that $x = 0$ (or $P(x, 0) = \delta(x)$) is given by

$$\langle x^2 \rangle_t = \langle x^2 \rangle_{st} \frac{1}{\sqrt{2\pi}} \int_{-\infty}^{\infty} e^{-\xi^2/2} \frac{\xi^2 \tau}{1+\xi^2 \tau} d\xi, \tag{7}$$

where $\langle x^2 \rangle_{st} = \gamma/g$ and $\tau = (g\varepsilon/\gamma^2) \{\exp(2\gamma t)-1\} \simeq (g\varepsilon/\gamma^2)\exp(2\gamma t)$ for large t. This result is very instructive in considering more general problems of ordering processes in unstable systems.

In fact, Kawasaki et al.[9] extended this scaling idea of non-equilibrium systems to the TDGL model without conservation law, and obtained a scaling variable of the form (4), or more explicitly $\phi = 1/2$. Recently Hurukawa[10] extended this idea to the conserved TDGL system and proposed the following scaling form

$$S(k,t) \simeq t^{-\Delta} f_H^{(sc)}(x); \quad x = kt^{\phi} \tag{8}$$

for the structure function $S(k,t)$, where $f_H^{(sc)}(x) = 3x^2/(2 + x^6)$.

The scaling function $f^{(sc)}(x)$ may, however, be obtained more generally in the form

$$f^{(sc)}(x) = \frac{dx^2}{a+bx^2+cx^4+x^6}, \tag{9}$$

where we have used the evenness of $f(x)$ and also have used the asymptotic conditions that $f(x) \propto x^2$ for small x and that $f(x) \propto x^{-4}$ for large x. If we impose the normalization condition that $f(x)$ takes the maximum $f_{max} = 1$ at $x = 1$ as in Hurukawa's argument, then we obtain $c =$

a-2 and d = 2a+b-1, and consequently we have

$$f^{(sc)}(x) = \frac{(2a+b-1)x^2}{a+bx^2+(a-2)x^4+x^6}.$$ (10)

Here, we have the restriction that

$$0 < a < 8 \qquad \text{and} \qquad 2a+b > 1,$$ (11)

which comes from the condition that f(x) has a single peak and from the positivity of f(x) for all positive x. Hurukawa's scaling function corresponds to the case that a = 2 and b = 0. The parameters a and b may depend on the quenched temperature T and quenching mechanism.

In other words, the two parameters a and b are characterized by the coefficients of the following asymptotic behavior

$$f^{(sc)}(x) \simeq Ax^2 = \frac{2a+b-1}{a}x^2 \qquad \text{for } x \ll 1,$$ (12)

and

$$f^{(sc)}(x) \simeq Bx^{-4} = (2a+b-1)x^{-4} \qquad \text{for } x \gg 1,$$

namely we have

$$a = \frac{B}{A} \text{ and } b = B - \frac{2B}{A} +1.$$ (13)

Equivalently to (10), the scaling function $f^{(sc)}(x)$ is also parameterized as

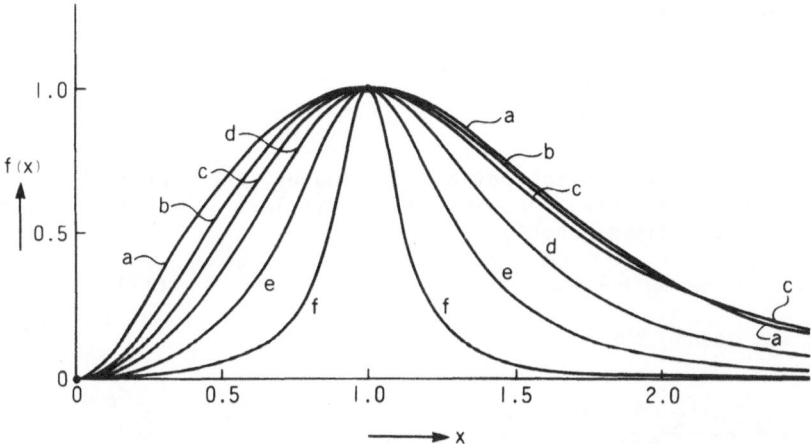

Fig. 1   The dependence of the scaling function f(x) on the two parameters a and b: (a) a=1, b=5; (b) a=2, b=4; (c) a=4, b=2; (d) a=2, b=0; (e) a=1.5, b=-1; (f) a=0.5, b=0.1.

$$f^{(sc)}(x) = \frac{ABx^2}{B+(A-2B+AB)x^2+(B-2A)x^4+Ax^6} \qquad (14)$$

with $0 < B < 8A$. It will be quite interesting to fit experimental data[11] with our scaling function (10) by taking a and b as fitting parameters, or by using the relation (13). A much better fitting must be obtained owing to the free choice of the two parameters. For convenience of comparison, some scaling functions are shown explicitly in Fig. 1 for a few typical values of the parameters a and b. Some explicit applications of (10) will be published elsewhere.[12]

A possibility to derive the above scaling function as well as the scaling exponents ϕ and ψ will be discussed in PART III on the basis of the scaling theory.

PART II: COHERENT-ANOMALY METHOD (CAM)

The basic idea of the coherent-anomaly method (CAM)[5] is to consider systematic self-consistent mean-field approximations for each phenomenon and to extract a common feature inherent among them, by making an analytic continuation of the degree of approximation. The key point of the CAM theory is to note[5] the appearance of the coherent anomalies in the coefficients or amplitudes of the classical singularities obtained in the generalized systematic mean-field (or effective-field) approximations.

For simplicity, we consider here the magnetic susceptibility $\chi_0(T)$ of ferromagnets which is expected to show the following fractional singularity

$$\chi_0(T) \sim \frac{1}{(T-T_c^*)^\gamma} \qquad (15)$$

near the true critical point T_c^* with the fractional critical exponent γ. As is well known, even generalized mean-field approximations yield the Curie-Weiss law

$$\chi_0(T) \simeq \frac{\bar{\chi}(T_c)}{\varepsilon}, \quad \varepsilon = \frac{T-T_c}{T_c} \qquad (16)$$

near each mean-field critical point T_c. However, the mean-field critical coefficient $\bar{\chi}(T_c)$ becomes anomalously large[5-8] as the degree of approximation increases, namely

$$\bar{\chi}(T_c) \to \infty \quad \text{as} \quad T_c \to T_c^*. \qquad (17)$$

Thus, we may assume that

$$\bar{\chi}(T_c) \simeq \frac{f_\chi}{(T_c-T_c^*)^\psi} \qquad (18)$$

near $T_c = T_c^*$, namely for $\delta(T_c) \equiv (T_c-T_c^*)/T_c^* \ll 1$. This coherent-anomaly exponent ψ can be easily estimated from the mean-field critical data on $\bar{\chi}(T_c)$ as a function of T_c obtained in cluster-mean-field approximations.[5-8] The non-classical critical exponent γ can be estimated through the coherent-anomaly relation[5,7]

$$\gamma = 1 + \psi, \tag{19}$$

which has been derived in several methods.[5,7] That is, the deviation of γ from the classical value $\gamma_0 = 1$ comes from the intrinsic part of fluctuation which is expressed as the coherent anomaly (18). This concept of the coherent-anomaly is quite general and can be applied to many other cooperative phenomena[5] such as spin glasses and percolation.[13]

The above CAM theory has the following merits, namely (i) this is a convergent theory[7], (ii) its convergence is very rapid[6-8], and (iii) it is applicable to any kind of phase transition by constructing "super-effective-field approximations", which will be published in the near future.[14]

PART III: SOME POSSIBLE APPLICATIONS OF THE CAM THEORY TO THE DYNAMICS OF ORDERING PROCESSES

It will be extremely interesting to apply the CAM theory to the dynamics of ordering processes in condensed matter. As has been discussed in PART II, we have to construct first some mean-field (or effective-field) approximations to describe the relevant ordering process with classical exponents ψ_0 and Δ_0. Then, the true scaling exponent ψ and Δ in (8) can be estimated as

$$\phi = \phi_0 + \psi_\phi \text{ and } \Delta = \Delta_0 + \psi_\Delta \tag{20}$$

through the coherent-anomaly exponents[5] ψ_ϕ and ψ_Δ. More explicitly, we study first, in some generalized mean-field approximations, the maximum wave-number $k_m(t)$ at which $S(k,t)$ becomes maximum with respect to k, namely we may have

$$k_m^{(mf)}(t) \simeq \bar{k}(L)t^{\phi_0}, \tag{21}$$

where L denotes the degree of approximation, that is, the cluster size in cluster approximations. Thus, we may have the following coherent anomaly

$$\bar{k}(L) \simeq L^\psi. \tag{22}$$

A generalized scaling argument yields[5] the following relation

$$\phi = \phi_0/(1+\psi). \tag{23}$$

Therefore, by estimating the coherent-anomaly exponent ψ in (22), we can estimate the non-classical critical exponent ϕ through (23). There is also a possibility to find the scaling function $f^{(sc)}(x)$, which will be reported elsewhere.

The present work is partially financed by the Research Fund of the Ministry of Education, Science and Culture.

REFERENCES

1. M. Suzuki, Phys. Rev. 56A: 71 (1976).
2. M. Suzuki, Prog. Theor. Phys. 56: 77, 477 (1976). J. Stat. Phys. 16: 11, 477 (1977).
3. M. Suzuki, Adv. Chem. Phys. 46: 195 (1981); Prog. Theor. Phys. Suppl. 79: 125 (1984).
4. M. Suzuki, Y. Liu and T. Tsuno, Physica 138A: 433 (1986).
5. M. Suzuki, J. Phys. Soc. Jpn. 55: 4205 (1986). See also M. Suzuki, Phys. lett. 116A: 375 (1986), and Quantum Field Theory (Proc. Int. Symp. Positano, Salerno, Italy, June 5-7, 1985) ed. F. Mancini (North- Holland, Amsterdam, 1986).
6. M. Suzuki and M. Katori, J. Phys. Soc. Jpn. 55: 1 (1986).
7. M. Suzuki, M. Katori and X. Hu, J. Phys. Soc. Jpn. 56: 3092 (1987).
8. M. Katori and M. Suzuki, J. Phys. Soc. Jpn. 56: 3113 (1987). X. Hu, M. Katori and M. Suzuki, J. Phys. Soc. Jpn. 56: No. 11 (1987).
9. K. Kawasaki, M. C. Yalabik and J. D. Gunton, Phys. Rev. 17: 455 (1978).
10. H. Hurukawa, Physica 123A: 497 (1984).
11. J. Marro, J. L. Lebowitz and M. H. Kalos, Phys. Rev. Lett. 43: 282 (1979). S. Komura, K. Osamura, H. Fujita and T. Takeda, Phys. Rev. 31B: 1278 (1985).
12. M. Suzuki, in preparation.
13. M. Suzuki, to be published in Phys. Lett. A.
14. M. Suzuki, to be submitted to J. Phys. Soc. Jpn.

THE CRYSTAL GROWTH KINETICS IN THE CLUSTER APPROXIMATION

K. Wada, H. Tsuchinaga and T. Uchida

Department of Physics, Faculty of Science
Hokkaido University, Sapporo 060, Japan

INTRODUCTION

Many aspects of the equilibrium and growth kinetics of crystals have been studied on the solid on solid (SOS) model. In order to set up the kinetic equations of crystal growth the cluster approximation has been often used. As the approximation proceeds from the Bragg-Williams (point)[1] approximation to the pair approximation[2], the region of unphysical meta-stable state at low temperatures for small driving force $\Delta\mu/kT$ is considerably reduced. Here $\Delta\mu$ is the chemical potential difference between vapor and solid and T is the absolute temperature. However, there can be seen still a wide non-growth region compared with the Monte Carlo simulation result[3].

In this paper, we will apply the triangle approximation of the path probability method (PPM)[4] to the adsorption and evaporation processes and analyze the growth kinetics from the vapor of the hexagonal |001| face of a crystal. The PPM is known as a dynamical version of the cluster variation method (CVM)[5]. The results show that the non-growth region for small driving force is further reduced and are in good quantitative agreement with the Monte Carlo simulation. We also examine the effect of diffusion processes and we see the enhancement of the crystal growth velocity.

CRYSTAL GROWTH EQUATION

The present SOS model is defined as a hexagonal lattice which can be filled with atoms, but such that in the normal direction to a hexagonal face |001| a solid atom can occupy only a site on the other solid atom. The Hamiltonian of the system is written as

$$H = -2J \sum_{\langle ij \rangle} c_i c_j - \Delta\mu_0 \sum_i c_i + V([c_i]) \qquad (1)$$

where the first sum runs over all neighboring pairs on each |001| face, c_i takes 1 or 0 according as the i-th site is occupied or empty, $\Delta\mu_0$ the bare chemical potential difference between vapor and solid, and $V([c_i])$ denotes a potential which is infinite for a forbidden configuration and zero otherwise. Further, the crystal is divided into monatomic layers, parallel to a |001| face, and each layer is assigned a number $n(-\infty < n < \infty)$. We apply the triangle approximation of the PPM to the relaxation processes of adsorption and evaporation.

Before we write down the kinetic equations, we give a brief outline of the PPM. The PPM is formulated as follows. Let us consider the change of state of an ensemble of M equivalent systems within a small interval of time Δt. In addition to the number of systems $MP_1^{(s)}([c],t)$ taking a configuration $[c]$ at t, we introduce the number of systems $MP_2^{(s)}([c],t;[c'],t+\Delta t)$ which is defined as the number of systems taking a configuration $[c]$ at t and $[c']$ at $t+\Delta t$. Here $[c]$ denotes a configuration $(c_1,c_2,\cdots)$. From definitions, $P_1^{(s)}$ and $P_2^{(s)}$ represent the one time and two time probabilities and are connected with each other by

$$P_1^{(s)}([c],t)= \operatorname*{Tr}_{[c']} P_2^{(s)}([c],t;[c'],t+\Delta t), \qquad (2)$$

where Tr denotes a trace over all configurations. Under the condition that a state $[MP_1^{(s)}([c],t)]$ of the ensemble at time t is given, the transition probability of the ensemble in a short time interval Δt through a path $[MP_2^{(s)}([c],t;[c'],t+\Delta t)]$ is given by

$$T(t,t+\Delta t) = \frac{\displaystyle\prod_{[c]} (MP_1^{(s)}([c],t))!}{\displaystyle\prod_{[c]}\prod_{[c']} (MP_2^{(s)}([c],t;[c'],t+\Delta t)!}$$

$$\times \prod_{[c]}\prod_{[c']} (W^{(s)}([c]|[c'],\Delta t))^{MP_2^{(s)}([c],t;[c'],t+\Delta t)}$$

$$(3)$$

where $W^{(s)}([c]|[c'],\Delta t)$ is the transition probability of a system changing from $[c]$ to $[c']$ in Δt. When $[P_2^{(s)}([c],t;[c'],t+\Delta t)]$ is varied starting from a state $[P_1^{(s)}([c],t)]$, the maximization requirement of eq.(3) gives a master equation of the system in the limit of $\Delta t \rightarrow 0$. Thus eq.(3) is considered a variational function determining the evolution of the system. From eq.(3) we can proceed to the cluster approximation, which treats information of the whole system in terms of finite number of clusters (points, pairs, triangles, etc.) of lattice points. Recently we have shown the following facts[6]: (A) Eq.(3) can be written down systematically in terms of the one time and two time probabilities of cluster corresponding to a chosen approximation. (B) By differentiating eq.(3) thus constructed with respect to two time probabilities of cluster keeping one time probabilities fixed, we have the master equation of the chosen cluster, that is, the rate equation of order parameters of the system.

In the present triangle approximations we choose triangles in the n-th layer as the basic clusters. Then the order parameters representing the system can be taken as the concentration of solid atom $x_1^{(n)}(t)$, the pair correlation $y_{12}^{(n)}(t)$ and the three site correlation $s^{(n)}(t)=z_{112}^{(n)}(t)-z_{122}^{(n)}(t)$. (Fig.1) After some calculation of the PPM the rate equations of the order parameters in the relaxation processes are given by

$$\frac{dx_1^{(n)}(t)}{dt} = \frac{1}{\tau_R}(\rho_{n-1}(t)e^L - \rho_n(t)\lambda_n^z)$$

$$\frac{dy_{12}^{(n)}(t)}{dt} = \frac{1}{\tau_R}(\rho_{n-1}(t)e^{L\frac{y_{22}^{(n)}-y_{12}^{(n)}}{x_2^{(n)}}} - \rho_n(t)\frac{\lambda_n^z}{D_n}(\frac{z_{122}^{(n)}e^k}{y_{12}^{(n)}} - \frac{z_{111}^{(n)}e^{-k}}{y_{11}^{(n)}}))$$

$$\frac{ds^{(n)}(t)}{dt} = \frac{1}{\tau_R}(\rho_{n-1}(t)e^L(z\frac{z_{122}^{(n)}}{z_2^{(n)}}-1)-\rho_n(t)\lambda_n^z(z\frac{z_{112}^{(n)2}}{y_{11}^{(n)}y_{12}^{(n)}\lambda_n D_n}-1))$$

$$(4)$$

where $\rho_n(t)=x_1^{(n)}(t)-x_1^{(n+1)}(t)$ is the concentration of surface atom in the n-th layer, τ_R a characteristic time constant of relaxation processes, $z=6$ the coordination number, $L=(\Delta\mu_0+zJ)/kT$ is the driving force, $K=J/kT$, and λ_n represents an environmental effect per bond when an n-th layer atom evaporates from the crystal and is given by

$$\lambda_n = \frac{1}{2}\left(\frac{z_{111}^{(n)}e^{-k}}{y_{11}^{(n)}} + \frac{z_{122}^{(n)}e^{k}}{y_{12}^{(n)}} + D_n\right)$$

$$D_n = \left\{\left(\frac{z_{111}^{(n)}e^{-k}}{y_{11}^{(n)}} - \frac{z_{122}^{(n)}e^{k}}{y_{12}^{(n)}}\right)^2 + \frac{4z_{112}^{(n)}2}{y_{11}^{(n)}y_{12}^{(n)}}\right\}^{\frac{1}{2}}$$

In the SOS model all processes occur on the surface of the crystal and in each of eq.(4) the first term is connected with adsorption processes and the second with evaporation processes. With the help of the CVM we have a following free energy F corresponding to these kinetic equations;

$$F/NkT = \sum_n\left[zKy_{12}^{(n)} - Lx_1^{(n)} + S(\rho_n) - 3\sum_{\langle ij\rangle} S(y_{ij}^{(n)}) + 2\sum_{ijk} S(z_{ijk}^{(n)})\right],$$

$$(5)$$

where N is the number of sites on a layer and $S(x)=x(\ln x-1)$. It can be shown that a configuration of a minimum free energy satisfies the stationary condition of eq.(4).

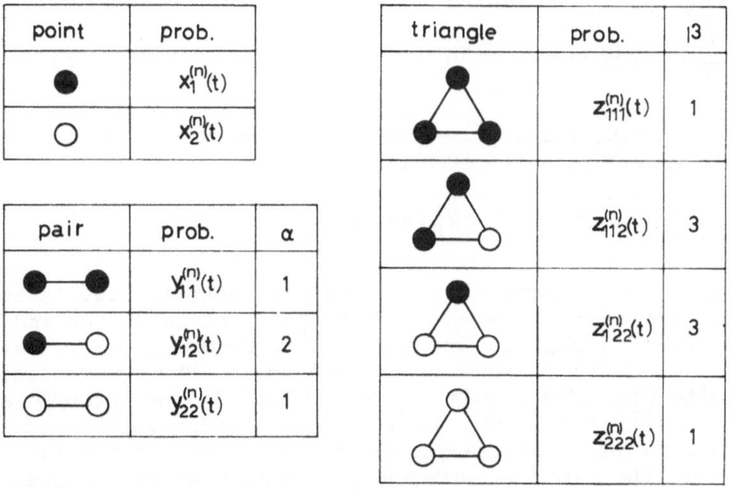

point	prob.
●	$x_1^{(n)}(t)$
○	$x_2^{(n)}(t)$

pair	prob.	α
●—●	$y_{11}^{(n)}(t)$	1
●—○	$y_{12}^{(n)}(t)$	2
○—○	$y_{22}^{(n)}(t)$	1

triangle	prob.	β
▲ (●●●)	$z_{111}^{(n)}(t)$	1
▲ (●●○)	$z_{112}^{(n)}(t)$	3
▲ (●○○)	$z_{122}^{(n)}(t)$	3
▲ (○○○)	$z_{222}^{(n)}(t)$	1

Fig.1 One time probabilities of configurations of clusters in the n-th layer in the triangle approximation. ●(○) denotes that a site is occupied (empty) and the suffix 1 (2) is used, respectively. α and β indicate the number of configurations having the same probabilities. Since there are geometrical relations among one time probabilities, $x_1^{(n)}(t)$, $y_{12}^{(n)}(t)$, and $s^{(n)}(t)=z_{112}^{(n)}(t)-z_{122}^{(n)}(t)$ may be chosen as independent variables.

We can also have the equations for inter- and intra-layer diffusion processes

$$\frac{dx_1^{(n)}(t)}{dt_D} = \frac{z}{\tau_D}(\rho_{n-1}(t)e^{L_0} - \rho_n(t)\lambda_n^z)$$

$$\frac{dy_{12}^{(n)}(t)}{dt_D} = \frac{z}{\tau_D}(\rho_{n-1}(t)e^{L_0}\frac{y_{22}^{(n)}-y_{12}^{(n)}}{x_2^{(n)}} - \rho_n(t)\frac{\lambda_n^z}{D_n}(\frac{z_{122}^{(n)}e^k}{y_{12}^{(n)}} - \frac{z_{111}^{(n)}e^{-k}}{y_{11}^{(n)}}))$$

$$\frac{ds^{(n)}(t)}{dt_D} = \frac{z}{\tau_D}(\rho_{n-1}(t)e^{L_0}(z\frac{z_{122}^{(n)}}{z_2^{(n)}} - 1) - \rho_n(t)\lambda_n^z(z\frac{z_{112}^{(n)2}}{y_{11}^{(n)}y_{12}^{(n)}\lambda_n D_n} - 1)),$$

$$(6)$$

where $\exp(L_0)$ is given by

$$e^{L_0} = \sum_{n=-\infty}^{\infty} \rho_n(t)\lambda_n^z$$

and τ_D is the characteristic time constant of diffusion process in which the diffusion length X_s is defined by $X_s = \sqrt{z\tau_R/\tau_D}$. It should be noted that the above equations can be obtained if in eq.(4) we replace $1/\tau_R$ by z/τ_D and $\exp(L)$ by $\exp(L_0)$, respectively. Though the present treatment takes care of correlation within the triangle in the layer, the particle exchanges between the layers are treated in the point like approximation.

NUMERICAL RESULTS

The growth rate of the crystal is defined as the combined process of relaxation and diffusion:

$$R = \frac{d}{dt}\sum_n x_1^{(n)}(t) \qquad\qquad (7)$$

where $d/dt = d/dt_R + d/dt_D$. The Wilson-Frenkel growth rate denoting the maximum growth rate is given by

$$R_{WF} = k_0^+(1-e^{-L}) \qquad (k_0^+ = e^L/\tau_R) . \qquad\qquad (8)$$

The results in the present approximation are shown for two typical temperatures together with the results of the pair approximation and the Monte Carlo simulation in Fig 2. For low temperatures and weak driving forces the growth rate in the present approximation is improved considerably compared with that of the pair approximation. In the triangle approximation, there exists fluctuation of triangle clusters which helps the crystal growth even in low temperatures. However, for the intermediate driving forces the relative abundance of atoms with few nearest neighbor atoms leads to a little lower growth rate than those of point and pair approximation. Further, as is expected, we see that the growth rate increases with increase of diffusion length and goes near Wilson-Frenkel growth rate. However, it should be noticed that the diffusion processes do not change the critical field L_c from which the crystal growth begins. The present results show rather good quantitative agreements with the Monte Carlo calculations.

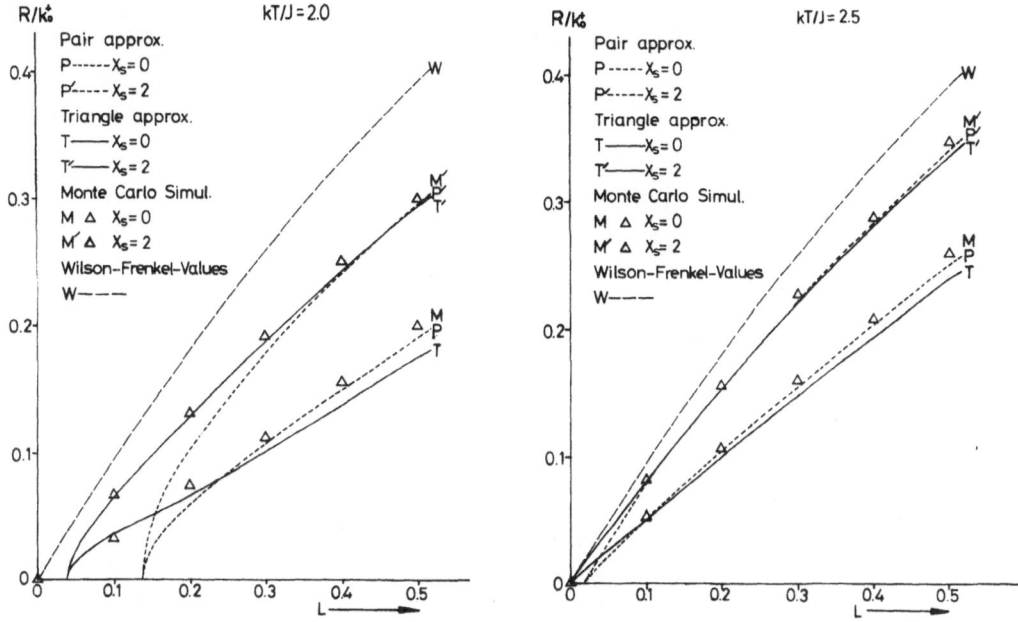

Fig.2 The average growth rate R vs driving force L at temperatures
kT/J=2.0 and 2.5. The results of triangle approximation (T) with and
without diffusion are shown with those of pair approximation (P) and
Monte Carlo simulation (M).

References

1) D.E.Temkin, Sov.Phys.Crystallogr. **14**(1969)344
2) Y.Saito and H.Müller-Krumbhaar,
    J.Chem.Phys.**70**(1979)1079
3) G.H.Gilmer and P.Bennema, J.Appl.Phys.**43**(1972)1349
4) R.Kikuchi, Prog.Theor.Phys.(Kyoto) Suppl.**35**(1966)1
    K.Wada, T.Ishikawa and H.Tsuchinaga, Physica **142A**(1987)38
5) R.Kikuchi, Phys.Rev.**81**(1951)988
6) K.Wada, M.Kaburagi and T.Uchida, submitted to Physica A

SOME PROBLEMS RELATED TO SCALING

IN SYSTEMS UNDERGOING PHASE SEPARATIONS

Hiroshi Furukawa

Institute of Physical Sciences
Faculty of Education
Yamaguchi University
Yamaguchi 753, Japan

INTRODUCTION

We consider the dynamics of phase separation where coarsening of droplets or domains proceeds continously [1-3]. We omit the nucleation and growth. In the late stage, however, no clear distinction exists between the nucleation and growth and the continous spinodal decomposition. The dynamics of phase separation of a material such as an alloy may be classified into three stages, i.e., the early, intermediate and late stages. When the average domain size becomes large enough, the interfacial thickness, s, of domains or droplets is set vanishingly small compared with the average domain size, R. This stage is called the late stage of the phase separation. In this stage the length scale of the system is only the average domain size, R. Then the dynamical scaling assumption is introduced. The process of phase separation is invariant under the rescalings of distance r and time t. Let $F(r,t)$ be a function of r and t. Let t be assumed to be rescaled as $b^{1/a}t$, then r is rescaled as br and

also F is rescaled as $b^{-c}F$. Then $F(br,b^{1/a}t)=b^{-c}F(r,t)$. By setting $b^{1/a}t=1$, we find that

$$F(r,t) = t^{ac}F(rt^{-a},1) = [R(t)]^{c}F(r/R) , \qquad (1.1)$$

where

$$R \propto t^{a} \qquad (1.2)$$

The Fourier coefficient $F_k(t)$ of $F(r,t)$ is scaled as

$$F_k(t) = R^{d+c} F(kR) , \qquad (1.3)$$

where d is the spatial dimension and k is the wave number. For the structure function $S_k(t)$ we find c=0 because the amplitude of the two point correlation function of particle density is invariant under the rescalings of length due to the saturation of the order parameter. The true scaling state is believed to be attained in a long time limit:

$$t \longrightarrow \infty \text{ or } R \longrightarrow \infty .$$

However, there are several problems to be discussed in such a limit. Also approximate scaling stste are realized in a finite time, where however, scaling exponents are in general different from those of the true scaling state.

GROWTH LAWS

We first consider a dissipative system with a conserved order parameter $Y(r,t)$. The equation of motion for the fourier component $Y_k(t)$ is written as

$$\frac{d}{dt} Y_k(t) = - M_k k^2 \mu_k(t) + f_k(t) , \qquad (2.1)$$

where $M_k(t)$, $\mu_k(t)$ and $f_k(t)$ are the mobility, the thermodynamic force and the fluctuating force. The mobility and the fluctuating force are related to each other by the fluctuation-dissipation theorem:

$$<f_k(t) f_{-k}(t')> = 2k_B M_k k^2 \delta(t-t') . \qquad (2.2)$$

The equation of motion for the structure function, which is defined by $S_k(t)=<Y_k(t)Y_{-k}(t)>$ is given from (2.1) and (2.2) as

$$\frac{d}{dt} S_k(t) = 2k^2 M_k(t)[k_B T - H_k(t)] , \qquad (2.3)$$

where $H_k(t)=<\mu_k(t)Y_{-k}(t)>$ has a dimension of energy and may be called an energy function. Now the scalings of $S_k(t)$, $M_k(t)$ and $H_k(t)$ are written as

$$S_k(t) = R^d S(kR), \quad M_k(t) = R^{-\zeta} M(kR), \quad H_k(t) = R^h H(kR). \qquad (2.4)$$

By substituting (2.4) into (2.3) we obtain the kinetic exponent a [1]:

$$a = 1/(d+2+\zeta-h) . \qquad (2.5)$$

There are several possibilities of choosing ζ [4] and h [1], and accordingly the kinetic exponent a takes various values. $\zeta = 1$ for the surface mobility which comes from the individual motion of atoms on the domain surfaces, $\zeta = 0$ for the bulk mobility which comes from the individual motions of atomes inside or outside domains. For fluid, the materials is conveyed directly by the transport of droplets. In such a case the mobility is given by the Kawasaki-Stokes law and we have $\zeta = -2$. h=0 if thermal fluctuation is effective and h=d-1 if the surface tension is effective. There are some cases where both the thermal fluctuation and the surface tension are ineffective, for instance, as in one-dimensional systems at low temperatures [5]. In such a case h=-∞, which gives a=0 and this is a logarithmic growth of droplet $R \propto \ln t$. In a system where the local order parameter is not conserved, the kinetic exponent a is given by $a=1/(d+\zeta-h)$. Many kinetic exponents thus obtained were derived by several authors. a=1/3 for solid (by the surface tension and the bulk mobility) is given by Lifshitz and slyozov [6] and Wagner [7], and a=1 for liquid is given by McMaster [8] and Siggia [9]. These exponents are the largest ones and are generally believed to be obtained in the late stages of the phase separations of solid and liquid, respectively. Actually, however, they are not always obtained. For instance, at low temperatures in solid the bulk mobility becomes ineffective due to the disappearence of individual motions of atoms. In such a case the surface mobility becomes effective and the kinetic exponent a=1/4 may become observable [1]. However, only at early time (at low temperatures) the exponent is 1/4, which crosses over at late times to 1/3 [10]. For a system with nonconserved order parameter the surface mobility is effective at high temperatures or in very early stage of the phase separation. Hence we usually observe a=1/2, which was first suggested by Lifshitz [11].

The application of (2.5) is not fully investigated, yet. An interesting case is that the surface mobility is not by individual atoms but by convective motions. This situation occurs when surfaces melt but interiors do not (surfaces are fluids but interios are solids). Let the portion of the melted part be C. Then the dominant mobility is the contribution from liquid (surface) parts. Therefore, M is proportional to R^2 (Stokes-Kawasaki) multiplied by C, $M \propto CR^2$. For the completely melted situation we set C=1, which gives a=1 with the surface tension as a driving force. But for the surface melt we have $C \propto R^{-1}$, and we have a=1/2 with surface tension as a driving force. The surface melt is not a normal phenomenon. But the same effect might be observed in a polymer blend if the molecular weights of two polymers are very different from each other. Interiors of domains with larger molecular weight have less fluidity, but polymers near surfaces may recover the fluidity due to the small molecular weight of polymers of the other phase. a=1/2 is really observed in a polymer blend [12]. We note a different mechanism is proposed by Kawasaki andSekimoto for this experiment [13]. The present mechanism does not contradict with the structure function scaling as shown experimentally [12]. If the thermal fluctuation is effective the exponent 1/2 is replaced by 1/(d+1). For the grain growth one may also expect an edge melt under the high pressure as in the interior of the earth. If melted edges percolate flows along edges are responsible to the coarsening of grains. In this case, however, we observe that the mobility is of the order R^0, which has the same R-dependence as the solid bulk mobility.

TURBULENT GROWTH OF DROPLET

So far coarsening of droplets are described by dissipative equations. However, for a simple liquid mixture in a very late stage the Reynolds number increses indefinitely as droplets grow. Then the dissipation by a usual friction does not work. (2.3) is a balance equation between force (the right hand side) and the friction (the left hand side). In a simple liquid with a large Reynolds number the inertial friction become important. The equation (2.8) is then replaced by [14]

$$\frac{d}{dt} S_k(t) + 2I_k(t) = 2k^2 M_k(t)[k_B T - H_k(t)] , \qquad (3.1)$$

where $I_k(t) = mMRe < ik.(Du_k/Dt)]Y_{-k}(t) >$ with D/Dt being the time derivative along the fluid motion, m is a quantity with a mass dimension and u is the velocity field. $I_k(t)$ corresponds to the inertial friction and therefore becomes important for large Reynolds numbers. A dimensional analysis of (3.1) gives four typical kinetic exponents a=1/d, 1, 2/(d+2) and 2/3 [14], according as what force and what friction are effective. The last two are for large Reynolds numbers. We note that the inertia is always effective for a two dimensional simple liquid, and therefore a=1 can not be observed for two dimensional simple liquid. This is consistent with the numerical simulation [15]. However, for two dimensional viscous liquids the kinetic exponent a=1 must be observed, though an experiment exhibits somewhat peculear growth [16].

Now we notice that the basic equation has a time reversal symmetry, if the dissipative friction is neglected for large Reynolds numbers. Due to the surface tension droplets coalesce into larger ones. Thus the energy from the surface tension is transfered into the fluid kinetic energy without a loss. Then the droplet is separated into many droplets because of its kinetic energy. Is this picture correct, and does the coarsening stop at large Reynolds numbers? The answer to this question is "no". The kinetic energy transfered from the surface tension is transfered into thermal fluctuation by the turbulence eddies. One large turbulence eddy separates into a few eddies in a time interval of the order of the eddy turn over time. The size of separated eddies decreases exponentially, and the turn over time also decreases exponentially. Therefore, the time interval for which the kinetic energy of the droplet is transfered into the thermal fluctuation is of the same order as the eddy turn over time. This eddy turn over time is of the same order as the time for which droplets coalesce into a larger one. Therefore, for the turbulent growth the energy is dissipated in a time interval of the same order as that for the droplet growth [17] (see Fig.1). Thus the growth laws derived only by the dimensional analysis of (3.1) hold. In the long time limit the growth law R∝ $t^{2/3}$ might be observed and hence this must be for the true scaling state of a simple liquid. The turbulent growth law is not yet observed experimentally. To observe this growth law a special technique [18] must be needed. This growth law may be verified in somewhat different way as Onuki suggested for phase separation under shear [19].

DROPLET GROWTH (TIME --->)

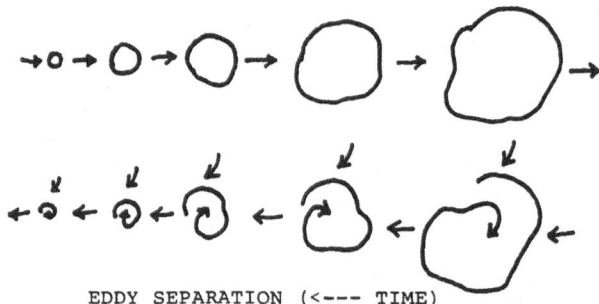

EDDY SEPARATION (<--- TIME)

Fig.1   Schematics of the turbulent growth of droplets in fluid. Droplets grow and the kinetic energies are transfered by turbulent eddies into thermal energies. The characteristic time of the droplet growth is of the same order as the turbulence eddy separation. Thus a grown droplet looses its kinetic energy in its characteristic growth time and its size does not decay, though the basic kinetic equation has a time reversal symmetry.

There are many elementary growth mechanisms as seen in the above. The growth mechanism of domains which is observed depends on how it is effective at the observed temporal region. Therefore, usually, the dominant kinetic exponent changes as time proceeds. What growth law is singled out in the long time limit? Three types of crossovers can be considered [20]. Consider the situation where two growth mechanisms coorporate. Assume that the average domain radius increases by the amount $\Delta R = \Delta R_1 + \Delta R_2$ in a time interval Δt. Here ΔR_1 and ΔR_2 are increments by the growth mechanisms 1 and 2, respectively. Then $dR/dt \stackrel{=}{=} \Delta R/\Delta t = \Delta R_1/\Delta t + \Delta R_2/\Delta t$. If the two increments occur in a parallel way, then $\Delta R_1/\Delta t$ and $\Delta R_2/\Delta t$ are determined simply by the average radius R as $\Delta R_i/\Delta t = C_i' R^{1-1/a_i}$ ($i \stackrel{=}{=} 1$, or 2), where C_i are constants and a_i are the kinetic exponents. Thus we have

$$\frac{dR}{dt} = C_1 R^{1-1/a_1} + C_2 R^{1-1/a_2} . \qquad (4.1)$$

This equation gives $R \propto t^{a_1}$ initially and $R \propto t^{a_2}$ finally, where $a_1 < a_2$. In such a case we find that the largest growth mechanism dominates the late stage. This is the first prototype of the crossover, and this behavior of the change in the kinetic exponent is usually observed. If the two growth mechanisms occur in a series way or exclusively, then the average radius increases by the amount ΔR in a time interval $\Delta t_1 + \Delta t_2$. Here $\Delta R/\Delta t_i = C_i' R^{1-1/a_i}$ (i=1 or 2). Therefore, we have [20]

$$[\frac{dR}{dt}]^{-1} = [C_1' R^{1-1/a_1}]^{-1} + [C_2' R^{1-1/a_2}]^{-1} \qquad (4.2)$$

This equation gives $R \propto t^{a_2}$ initially and $R \propto t^{a_1}$ finally. This is the second prototypes of the change of kinetic exponents. The third prototype is the mixture of the above two crossovers. The droplet growth is descrived by an effective exponent a_I, ($a_1 < a_I < a_2$) (intermittent state).

Although such reductions of the kinetic exponents as the third prototype are reported in grain growths or in systems with many degeneracies [21], more elaborated studies are needed. This is because the effect of the surface tension is not strong in such systems and the crossover region becomes wide. Thus the effective kinetic exponent becomes small for a relatively short time interval. We show a result of numerical situation on a modified Potts model. We simulated using a method like the Glauber dynamics, but the Monte Carlo sampling was done in a somewhat different way from that of Ref. [21]. First, a site is randomly chosen. Let the spin state of this site be Z. Next, one of its neighboring sites is randomly chosen. Let the spin state of this site be Z'. Then the Monte Carlo try is done under the restriction that the state of the first spin can be changed only into Z' with the state of the second spin fixed. This model is suitable for the dynamics of domain walls. In fact a spin state changes only at the interfaces. The advantage of this model is that the coarsening time is effectively larger than that of usual kinetic Potts model [21]. One Monte Carlo step of this model corresponds to P Monte Carlo steps of the usual model with P degeneracies. It should be noted that the present model does not have a critical point, i.e., the model is intrinsically irreversible. In Fig.2a we show a situation on the 200×200 sized triangular lattice with nearest neighbor interactions for infinite degeneracies of a spin state at a low temperature quenched from infinite temperature. Let us remember that two growth mechanisms with a=1/2 (due to the surface tension) and a=1/3 (due to the thermal fluctuation) cooperate. In this figure a dotted curve represents (4.1) with a_1=1/3 and a_2=1/2. We can see that the reduction of the kinetic exponent in this model is strongly suggested to be by a simple crossover of the first type. On the square lattice at low temperatures the situation becomes somewhat different. The kinetic exponent a=0 also cooperates. Even in this case we find that the kinetic exponent crosses over to a=1/2 for time longer than 10000 MCS (see Fig.2b). Recently, Grest et al. [22] have made a simulation on a larger lattice (1000 by 1000) and have found the crossover of the exponent to 1/2 in late times. The degeneracy of their model is still small due to the employment of the usual Glauber dynamics, however.

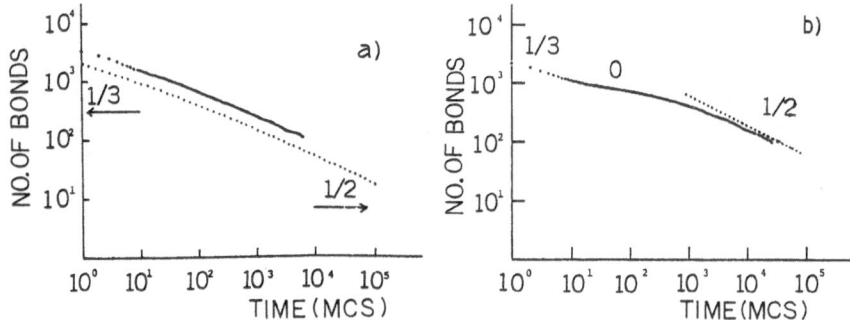

Fig.2   a) The temporal evolution of the number of bonds between different spin states on a triangular lattice at temperature $4J/3k_B$, where J is the strength of a nearest neighbor interaction energy. The number of bonds is regarded as the total length of the interface and is proportional to the energy of the system, and is considered to be proportional to the inverse average grain radius R. The dotted curve represent (4.1) with a_1=1/3 and a_2=1/2.
b) Similar as a) on a square lattice at temperature $J/5k_B$. The dotted line indicates slope -1/2.

Concerning the above problem we here show that in the coarsening of the two dimensional soap froth the kinetic exponent is a=1/2, which is given by a simple dimensioanl analysis of the kinetic equation. Sides (membranes) always meet with two other ones with the angle 120, due to the balance among surface tensions of three sides at vertex. Also due to the dispersion of bubble sizes sides are almost always curved. The curvature of a side is proportional to the pressure difference between two opposite bubbles. The total vapor flow through a side is proportional to the side length multiplied by its curvature, which does not depend on the side of the bubble. The total vapor flow flowing into a bubble is easily evaluated and this is in proportion to n-6 with n being the number of the sides. The time rate of the change of the area of the bubble, A is thus given by von Neumann's equation:

$$\frac{dA}{dt} = K(n-6) .$$

(4.3)

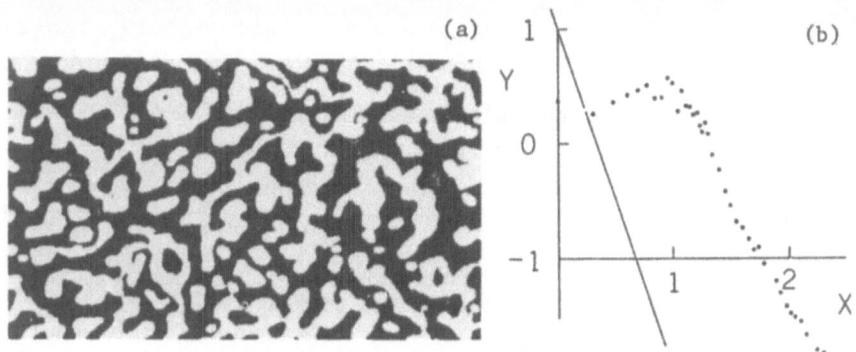

Fig.3   a) Transmission-electron-microscope image of Au film
            (from R.B. Laibowitz and Y. Grefen, Phys. Rev. Lett. 53,
            380 (1984). b) Its one dimensional power spectrum.
            X indicates kL/2π with L being the one dimensional length
            of the system, and Y indicates the one dimensional power
            spectrum. The straight line indicates a slope -3, which
            corresponds to the slope -4 in two dimensions (see
            H. Furukawa, Phys. Rev. A, 35, 3961 (1987)).

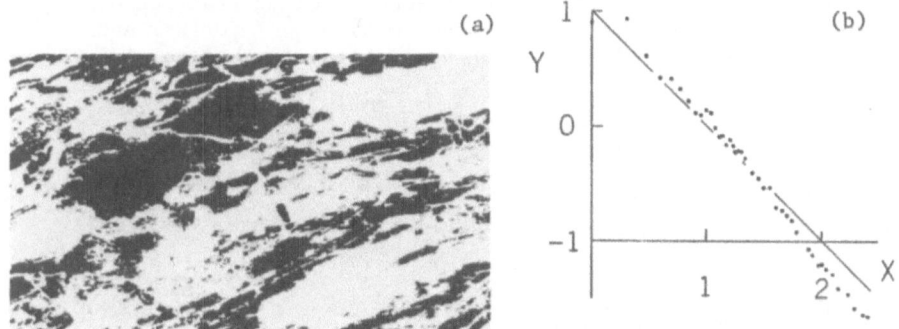

Fig.4   a) Microscope image of a rock (Schistosity) (from
            K. Miyagi, Ganseki Gairon (Kyoritsu, 1983)).
            b) Its one dimensional power spectrum. The straight
            line indicates a slope -1, which corresponds to the
            slope -2 in two dimensions (see H. Furukawa,
            Phys. Rev. A, 35, 3961 (1987)).

where K is a constant. In this equation a bubble with six vertexes does
not change its area. But this does not mean that this bubble stops growing
or shrinking. Due to zero net of incoming and outgoing vapors the total
area is invariant for a 6-sided bubble, but its shape is changing with the

same characteristic time as others and in this time the number of sides changes into five or seven. This is the reason why the kinetic exponent for the soap froth is a=1/2 as acertained numerically [23] or experimentally [24].

SCALED STRUCTURE FUNCTIONS

Finally we consider the explicit scaled structure function $S(x)$ for the system with conserved order parameter. Due to the conservation of matter the scaled structure function $S(x)$ has x^2 asymptotic form at small wave numbers. This is because droplets move randomly obeying the diffusion equation. In most cases interfaces of domains or droplets are smooth and sharp. Then the scaled structure function has x^{-d-1} tail at large wave numbers. Besides, droplets or domains have similar sizes in everywhere and therefore the system is nearly periodic. Thus the structure function has a peak near $k=R^{-1}$ with R being the average radius of the droplet. The simplest structure function satisfying the above three conditions is given by [25]

$$S(x) = (1+\gamma/2)x^2[\gamma/2+x^{2+\gamma}]^{-1} \ , \quad (\gamma=d+1) \ . \tag{5.1}$$

This simple structure function is shown to agree with experimental ones at the volume fraction 0.1 [26], where droplets do not percolate. The assumptions for (3.1) are not valid when the volume fraction of the minority phase exceeds the percolation threshold volume fraction 0.15. More generally the asymptotic forms are set $S(x) \propto x^{2+z}$ for x<1 and $x^{-d-1-z'}$ for x>1. Notice that d-1-z' is the surface fractal dimensionality [27].

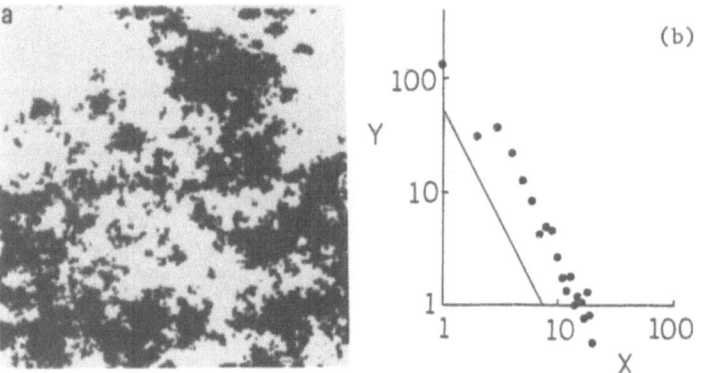

Fig.5   a) Pattern at time 2400 MCS of an Ising spin system on a 200×200 sized squared lattice with nearest neighbor interaction at infinte temperature. The spin is not conserved, but the spin flip-flop is assumed to be possible only at interfaces.
b) Its two dimensional power spectrum (structure function). X indicates $kL/2\pi$ with L being the one dimensional length of the system, and Y is the structure function. The straight line indicates a slope -2. This case corresponds to the case of Fig.4.

Therefore, in a usual situation z' may vary from -1 to d-1. For smooth interfaces the two points correlation function has a distance dependence as r at small distance r's. On the other hand for entangled interfaces it may have r^d dependence, which gives z'=d-1 corresponding to setting $\gamma=2d$. Here we show in figure two examples of the surface conditions. Fig.3 corresponds to the approximate surface fractal dimension d-1-z'=0, whereas Fig.4 corresponds to d-1-z'=d. The surface fractal dimension 0 can be really realized by the model introduced above to study the crossover. We consider the Ising spin with nonconserved order parameter where the spin flip-flop is allowed only at interfaces of domains (see the above). We simulated this on 200×200 sized square lattice with nearest neighbor interaction at infinite temperature. The pattern at time 2400 is shown in Fig.5a, and the two dimensional structure function is shown in Fig.5b. The structure function has a k^{-2} singularity. No surface tension plays a role in this model at infinite temperature even in the long time limit.

The structure function (5.1) with $\gamma=6$ is compared with the scattering function in a polymer system [28]. The value $\gamma= 6$ (=2d) was first given by considering that the entanglement of the interface changes the probability of finding the other phase from one of phases within a given distance [29]. It is to be noted that $\gamma=6$ is only for the intermediate values of the wave number: At larger values of the wave number the exponent recovers the value 4 (=d+1). It is also to be noted that $\gamma=6$ seems not to have a firm theoretical basis, but was qualitatively derived so as to understand how the structure function becomes steeper due to the change in morphologies of droplets. Recent experimental observations indicate a continuous change in z' as a function of time [30]. Fratzl [31], et al found that z is not zero for deep quench but z=1. This z=1 may be a result of the strong correlation among droplets due to the percolation of droplets.

The most serious shortcoming of (5.1) is that it does not contain volume fraction dependence. For instance the actual S(x) remains finite at x=0 for the zero volume fraction, since S(x) in this case is the structure function of a single droplet. There are several attempt to obtain the volume fraction dependent structure function [32]. At this moment almost attempts are for small volume fractions where droplets are assumed to be compact. The most theories predict such a volume fraction dependence as $S(x) \propto (v^{-1/2}Rk)^2$ at small x, where v is the volume fraction of a minority phase.

REFERENCES

[1] H. Furukawa, Adv. Phys. 34, 703 (1985).
[2] K. Binder and D.W. Heermann, Scaling Phenomena in Disordered Systems, edited by R. Pynn and T. Skjeltorp (New York:Plenum press, 1985).
[3] J.D. Gunton, M. San Miguel, and P.S. Sahni, in Phase Transition and Critical Phenomena, Vol.8 edited by C. Domb and J.L. Lebowitz (new York, Academic Press, 1983), p267.
[4] K. Binder, Phys. Rev. B 15, 4424 (1977).
[5] K. Kawasaki and T. Nagai, Physica, A 121, 175 (1983): H. Ikeda, 1983 J. Phys. Soc. Jpn. Suppl., 52, 33 (1983).
[6] I.M. Lifshitz and V.V. Slyozov, J. Phys. Chem. Solids, 19, 35 (1961).
[7] C. Wagner, Z. Electrochem., 65, 581 (1961).
[8] L.P. McMaster, Adv. Chem. Ser. 142, 43 (1975).
[9] E.D. Siggia, Phys. Rev. A 20, 595 (1979).
[10] Y. Oono and S. Puri, Phys. Rev. Lett. 58, 836 (1987); and preprint. See J.D. Gunton, in Proceedings of this Symposium.
[11] I.M. Lifshitz, Zh. sksp. teor. Fiz., 42, 1354 (1962) (Soviet Physics JETP, 15, 939 (1962)).

S.M. Allen and J.W. Cahn, Acta Metall., $\underline{27}$, 1089 (1979).

[12] T. Izumitani, M. Takenaka, T. Hashimoto, Polymer Prepr. Jpn., Soc. Polym. Sci. Jpn., $\underline{35}$, 2978 (1986)., See also T. Hashimoto, in Proceedings of this Symposium.

[13] K. Kawasaki and K. Sekimoto, Physica A, $\underline{147}$, 349 (1987).

[14] H. Furukawa, Phys. Rev. A $\underline{31}$, 1102 (1985).

[15] F.F. Abraham, S.W. Koch and R.C. Desai, Phys. Rev. Lett., $\underline{49}$, 923 (1982).

[16] A. Nakai, T. Shiwaku, H. Hasegawa, and T. Hashimoto, Macromolecules, $\underline{19}$, 3008 (1986).

[17] H. Furukawa, Phys. Rev. A$\underline{36}$, 2288 (1987).

[18] C. Houessou, P. Guenoun, R. Gastaud, F. Perrot and D. Beysens, Phys. Rev. A$\underline{32}$, 1818 (1985).

[19] A. Onuki, Phys. Rev. A. $\underline{34}$, 3528 (1986).

[20] H. Furukawa, Phys. Rev. A $\underline{30}$, 1052 (1984): $\underline{29}$, 2160 (1984).

[21] P.S. Sahni, D.J. Srolovitz, G.S. Grest, M.P. Anderson and S.A. Safran, Phys. Rev. B$\underline{28}$, 2705 (1983).

[22] G. Grest, M.P. Anderson and D.J. Srolovitz, in Proceedings of Nato Advanced Study Institute on Time Dependent Effects in Disordered Materials, Geilo, Norway 1987, ed. R. Pynn (plenum).

[23] C.W. Beenakker, preprint.

[24] J. Wejchert, D.Weaire, J.P. Kermode, Phylosophical Magazine, B$\underline{53}$, 15 (1986).

[25] H. Furukawa, Physica, A $\underline{123}$, 497 (1984).

[26] S. Komura, K. Osamura, H. Fujii and T. Takeda, Phys. Rev. B$\underline{31}$, 1278 (1985). See also, M. Takahashi, H. Horiuchi, S. Kinoshita, Y. Ohyama and T. Nose, J. Phys. Soc. Jpn., $\underline{55}$, 2687 (1986).

[27] H.D. Bale and P.W. Schmidt, Phys. Rev. Lett., $\underline{53}$, 596 (1984).

[28] T. Hashimoto, M. Itakura, H. Hasegawa, J. Chem. Phys. $\underline{85}$, 6118 (1986).

[29] H. F urukawa, Phys. Rev. A $\underline{23}$, 1535 (1981).

[30] S. Fujikawa, M. Furusaka, M. Sakauchi and K. Hirano, Proc. 4th Int. Conf. on Aluminum-Lithium Alloys, (Paris, 1987). See also S. Katano and M. Iizumi, Phys. Rev. Lett. $\underline{52}$, 835 (1984), and M. Furusaka, Y. Ishikawa and M. Mera, Phys. Rev. Lett. $\underline{54}$, 2611 (1985).
C.G. Windsor and R. M. Barron, in Proceedings of this Symposium.
M. Furusaka, Y. Ishikawa, S. Fujikawa, M. Sakauchi and K. Hirano. in Proceedings of this Symposium.

[31] P. Frazl, J. L. Lewowitz, J. Marro and M.H. Kalos, Acta metall. $\underline{31}$, 1849 (1983). See also S. Spooner, in Kinetics of Aggregation and Gelation, eds. F. Family and D.P. Landau (North-Holland 1984).

[32] P.A. Rikvold and J.D. Gunton, Phys. Rev. Lett. $\underline{49}$, 286 (1982).
G.F. Mazenko and O.T. Valls., Phys. Rev. Lett., $\underline{51}$, 2044 (1983).
T. Ohta, Ann. Phys. $\underline{158}$, 31 (1984); Prog. Theor. Phys. $\underline{71}$, 1409 (1984).
H. Tomita, Prog. Theor.Phys. $\underline{71}$, 1405 (1984).
H. Furukawa, Prog. Theor. Phys. $\underline{74}$, 174 (1985).
M. Tokuyama, Y. Enomoto and K. Kawasaki, Physica A $\underline{143}$, 183 (1987).

GROWTH KINETICS PROBLEMS AND THE RENORMALIZATION GROUP

Z.W. Lai and Gene F. Mazenko

The James Franck Institute and Department of Physics
The University of Chicago
Chicago, IL 60637
Oriol T. Valls

School of Physics and Astronomy
University of Minnesota
Minneapolis, MN 55455

Abstract

We discuss the structure of the renormalization group and the determination of universality classes for growth kinetics problems. Our analysis is based on a differential renormalization group equation of the Callen-Symanzik type. We find that many growth kinetics problems can be classified into four basic groups characterized by different low temperature behavior.

I. Introduction

One of the driving forces in the increasing interest in growth kinetics problems is the existence [1] of scaling behavior. One finds rather generally that order parameter, $\phi(\vec{R})$, correlation functions

$$C(\vec{R},t) = <\phi(\vec{R})\phi(\vec{0})>_t \quad , \tag{1.1}$$

evaluated at a time t after a temperature quench from an initial disordered state, shows a scaling behavior

$$C(\vec{R},t) = F(\vec{R}/L(t)) \tag{1.2}$$

where L(t) is a characteristic domain or droplet size. This form, or the associated Fourier transform, has been verified [1] for a variety of systems. It seems immediately apparent that this scaling should be associated with some deeper set of invariances in the theory and, in direct analogy with critical phenomena, one begins to look for a renormalization group (RG) description for these problems. Many workers [2] have implicitly assumed that such an RG description exists, but there has been rather little work on this problem outside our own group [3] and that of Gunton. [4]

A key aspect of the RG approach is the identification of universality classes and those variables which are relevant and irrelevant for the determination of those classes. From a practical point of view it is also very important to establish the existence of those variables which are marginal or irrelevant but, which as slow transients, strongly influence the access to the scaling regime. Another important concept in developing a RG approach and identification of classes is the notion of a fixed point. This concept has been used rather loosely in this field and we would like to develop it a bit more carefully here. Let us consider a particular example of the evolution of a binary alloy subjected to a deep temperature quench. One realization of this situation is shown in Fig. 1. This is an electron micrograph taken from the work of Eguchi et al. [5] It shows the development of the ordered phases as phase separation progresses in time. From a fundamental point of view, scaling and fixed points result from a self-similarity of this system under simultaneous rescaling of space and time:

$$C(\vec{R}, t) = C(\vec{R}/b, \Delta(b)t) \tag{1.3}$$

where the spatial rescaling factor $b > 1$ and the time rescaling factor $\Delta(b) < 1$. We demonstrate this scaling behavior in Fig. 1 where we have taken a section out of an early time t' portion and blown it up by a factor b. We then compare it with the system at a later time $t = t'/\Delta$. The scaling property seems roughly correct.

One may appropriately ask at this point, what about temperature? In critical phenomena this is the most important variable in the problem. Consider the implementation of scaling in that case. Any change in a length, $l \rightarrow l' = l/b$, can be compensated for by changing the temperature and in turn the correlation length $\xi(T)$ so that

$$\xi(T'(b)) = \xi(T)/b \tag{1.4}$$

can be satisfied and the ratio l/ξ is invariant under RG transformations.

How does temperature enter into growth kinetics problems? If we return to Fig. 1, we see that the blown up earlier time segment has rougher edges than the later time element which has a relatively smooth interface. We have not quite achieved self-similarity. We must also include temperature renormalization if we are to have complete self-similarity. Practically this means that we should have run the earlier time system at a lower temperature which would have produced sharper edges. Thus the transformation which should lead to the *fixed point* is given by

$$\vec{R} \rightarrow \vec{R}' = \vec{R}/b \tag{1.5a}$$

$$t \rightarrow t' = \Delta(b,T)t < t \tag{1.5b}$$

$$T \rightarrow T' = T'(T,b) < T . \tag{1.5c}$$

In the appropriate long time *scaling* region the system is invariant under such a transformation and one has a growth kinetics *fixed point*. This treatment of the temperature is sensible since under rescaling the effective temperature is lowered and interfaces appear sharper - which also occurs through spatial rescaling which weakens ones resolution. It is this observation which leads people to conclude that growth kinetics correspond to a *zero temperature fixed point*. We have verified [6,3] that this picture holds at low temperatures for the simplest systems, as has Gunton [4] and his group. However there are some surprises when one goes away from the simplest systems.

The question is whether this scaling behavior under temperature renormalization is generally satisfied. If we have a characteristic length, L(t,T), does it generally satisfy the scaling relation:

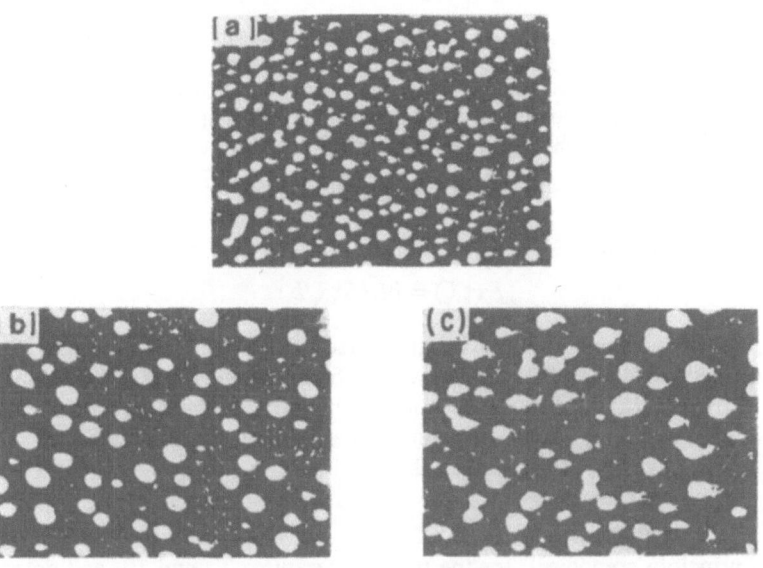

Fig. 1 Domain structures imaged (Ref. 5) with B_2 superlattice reflection in 23.0 at .% Al in Fe alloy. The samples were quenched from $630°$ C and annealed at $570°$ C. In panel (a) the system was annealed for 100 min. while in panel (b) it was annealed for 1000 min. In panel (c) we have taken a portion of panel (a) with approximately the same number of domains as in panel (b) and then blown this portion up to the same size as panels (a) and (b). Comparison of panels (b) and (c) illustrates the self-similar simultaneous rescalings of space and time.

$$L(t,T) = bL(\Delta t, T'(T,b)) \ ? \tag{1.6}$$

In our recent work [7] we have been able to draw some conclusions about the answer to this question for a variety of systems.

II. Renormalization Group Formulation

Let us return to the scaling form (1.2) and start again in a more general fashion. Implicit in writing (1.2) is that L(t,T) is large compared to any other length in this system. Let us assume that there exists another length [8] in the problem, $\zeta(T)$, which may be large enough to compete with L(t,T) over some *intermediate* time regime. For temperatures below but near the critical temperature the equilibrium correlation length $\xi(T)$ will be large and one can identify $\zeta(T) = \xi(T)$. In the presence of the additional length we must modify (1.2) to read

$$C(\vec{R},\tau,\zeta) = F(\vec{R}/L , \zeta/L) \quad . \tag{2.1}$$

In writing (2.1) we have also introduced a scaling time $\tau = \tau(t,T)$ which may differ from the "natural" time t. We showed in earlier work [3] that a scaling form of this type follows if the correlation functions satisfy a self-similarity relation

$$C(\vec{R},\tau,\zeta) = C(\vec{R}/b,\Delta\tau,\zeta/b) \tag{2.2}$$

and the length L satisfies

$$L(\tau,\zeta) = bL(\Delta\tau,\zeta/b) \quad . \tag{2.3}$$

How can this phenomenology be quantified? In Ref. 7 we discuss in more detail how to systematically obtain the time rescaling factor Δ and the scaling results of the type given by (2.2) and (2.3). Here we simplify things somewhat by considering only the characteristic length $L(\tau,\zeta)$. If L is a monotonically increasing function of τ then it is sensible to **define** a quantity $\tau'(\tau,\zeta,b)$ such that

$$L(\tau,\zeta) = bL(\tau',\zeta/b) \quad . \tag{2.4}$$

One can in principle compute L for the two temperatures T_1 and T_2, and determine b via

$$b = \zeta(T_1)/\zeta(T_2) \tag{2.5}$$

where $T_1 > T_2$, and then match $L(\tau,\zeta_1)$ and $bL(\tau',\zeta_2)$ to determine $\tau'(\tau,\zeta,b)$. This is what is done in standard [9] real space renormalization group type calculations.

We present here a local Callen-Symanzik type formulation which avoids many of the matching problems encountered in the method described above. The RG equation is obtained by differentiating both sides of (2.4) with respect to τ, ζ and b. We have

$$(\frac{\partial L}{\partial \tau})_\zeta = b(\frac{\partial L'}{\partial \tau'})_{\zeta'} (\frac{\partial \tau'}{\partial \tau})_{\zeta'} \tag{2.6a}$$

$$(\frac{\partial L}{\partial \zeta})_\tau = b(\frac{\partial L'}{\partial \tau'})_{\zeta'} (\frac{\partial \tau'}{\partial \zeta})_\tau + b(\frac{\partial L'}{\partial \zeta'})_{\tau'} (\frac{\partial \zeta'}{\partial \zeta})_\tau \tag{2.6b}$$

$$0 = L' + b(\frac{\partial L'}{\partial \tau'})_{\zeta'} (\frac{\partial \tau'}{\partial b})_{\tau,\zeta} + b(\frac{\partial L'}{\partial \zeta'})_{\tau'}(\frac{\partial \zeta'}{\partial b})_{\tau,\zeta} \tag{2.6c}$$

where $L' = L(\tau',\zeta/b)$ and $\zeta' = \zeta/b$. We can use (2.6a) and (2.6b) to eliminate the partial derivatives of L' with respect to ζ' and τ' in (2.6c) to obtain

$$[b\frac{\partial}{\partial b} + \zeta\frac{\partial}{\partial \zeta}+D\tau\frac{\partial}{\partial \tau}]\tau'(\tau,\zeta,b) = 0 \tag{2.7}$$

where

$$D = \frac{1-n_\zeta}{n_\tau} \tag{2.8a}$$

and n_ζ and n_τ are the logrithmic derivatives

$$n_\zeta = \frac{\zeta}{L}\frac{\partial L}{\partial \zeta} \tag{2.8b}$$

$$n_\tau = \frac{\tau}{L} \frac{\partial L}{\partial \tau} \quad . \tag{2.8c}$$

The boundary condition associated with (2.7) is

$$\tau'(\tau, \zeta, b = 1) = \tau \quad . \tag{2.9}$$

It is convenient to introduce the time rescaling parameter Δ defined in the usual way:

$$\tau'(\tau, \zeta, b) = \Delta(\tau, \zeta, b)\tau \quad . \tag{2.10}$$

Eqs. (2.7) and (2.9) then become:

$$[b\frac{\partial}{\partial b} + \zeta\frac{\partial}{\partial \zeta} + D\tau\frac{\partial}{\partial \tau} + D]\Delta(\tau, \zeta, b) = 0 \tag{2.11a}$$

and

$$\Delta(\tau, \zeta, b = 1) = 1 \quad . \tag{2.11b}$$

Eqs. (2.11) are our fundamental RG equations. They are of the same standard RG form as found in Ref. 10. The information specific to the sytstem is contained in $D(\tau, \zeta)$ and the problem reduces to a determination of D. As discussed in Ref. 3, we find a fixed point [11,12] if $\lim_{\tau \to \infty} D(\tau, \zeta) = D(\zeta)$ and $D(\zeta) > 0$ for all ζ. Once D is known the time rescaling parameter Δ is obtained by solving (2.11), and the low temperature form for L(t) follows then from a solution of (2.4).

III. Class 1 Systems

Class 1 systems are very simple since the natural variables (t,T) can be taken to be the scaling variables (τ, ζ) for low temperatures. In this case we expect that ζ can be chosen to be the equilibrium correlation length, ξ, for all $T < T_c$ and for low temperatures $\zeta \sim \xi \sim T$. Consequently the fixed point is simple since

$$\lim_{t \to \infty} D(t, T) = D* \tag{3.1}$$

for all $T < T_c$ and D* is independent of T. It is easy to show that if D* is independent of T, that (2.11) has the solution:

$$\Delta = b^{-D*} \tag{3.2}$$

and (2.4) leads to

$$L(t) = Bt^{1/D*} \tag{3.3}$$

Let us summarize the procedure. First compute D(t,T) using any available means (Monte Carlo, numerical simulation, etc.) and look at the long-time limit for fixed T. If this limit is positive and independent of T then one has what we have called a class 1 growth kinetics system. As shown in Fig. 2, the spin flip kinetic Ising model on a square lattice is such a system with $D* = 2$.

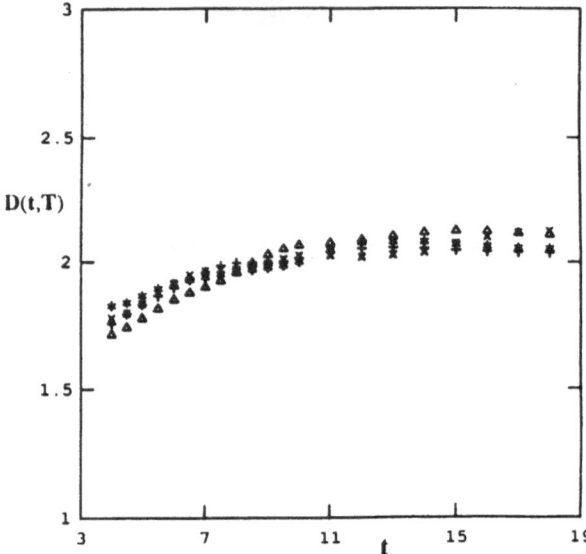

Fig. 2. $D(t,T)$ vs. t for quenches to y = 0 (plusses), 0.01
(asterisks), 0.02 (crosses), and 0.03 (triangles) for the SFKI
model on a square lattice. See Ref. 3 for details.

IV. Class 2 Systems

What other possibilities are there? Let us consider the spin flip kinetic Ising model
on a hexagonal lattice. The major difference between this model and the same model on
a square lattice is that this model freezes after some finite time upon quenching to zero
temperature. There is therefore an associated freezing length

$$\lim_{t \to \infty} L(t,0) = L_0 \quad . \tag{4.1}$$

What happens in this case if we assume that scaling and the RG hold in exactly the same
form as on the square lattice (where the growth on quenching to T=0 is qualitatively the
same as quenches to T > 0) ? In Fig. 3 we show the results for D(t,T) for quenches to
various temperatures. We note immediately that D does not approach an acceptable fixed
point value for long enough times. It approaches a constant value as $t \to \infty$ but this
value is negative and unphysical since it would correspond to scaling to later times
($t' > t$). We find that we can represent our results for $\tilde{D} = \lim_{t \to \infty} D(t,T)$ as a function of
temperature in the form

$$\tilde{D} = D^* - \frac{E_0}{T} \quad . \tag{4.2}$$

We find that we can not achieve a fixed point in the theory as formulated. However,
scaling can be achieved if we change to a new time variable

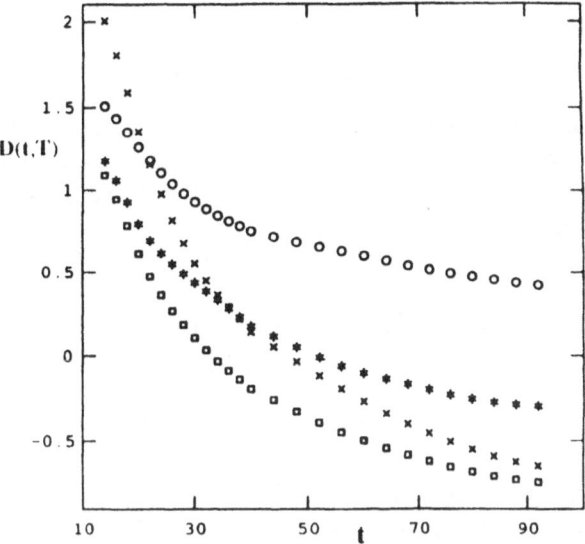

Fig. 3. $D(t,T)$ vs. t for $y_H = 0.02$ (crosses), 0.03 (squares), 0.04 (asterisks), and 0.055 (circles) for the SFKI model on a hexagonal lattice.

$$\tau = t/\tau_0 \qquad (4.3)$$

where τ_0 is a function of T and to be determined. We can relate the D computed with the natural variables t and T to that computed using the scaling variables τ and T using the chain-rule for differentiation to obtain

$$D(t,T) = D(\tau,T) + \frac{T}{\tau_0}\frac{\partial \tau_0}{\partial T} \qquad . \qquad (4.4)$$

If we identify

$$\frac{T}{\tau_0}\frac{\partial \tau_0}{\partial T} = -\frac{E_0}{T} \qquad , \qquad (4.5)$$

we can easily solve for τ_0 to obtain

$$\tau_0 = \tau_1 e^{E_0/T} \qquad (4.6)$$

where, here and below, τ_1 has a weak temperature dependence as $T \rightarrow 0$. Note that τ_0 has a very strong activated temperature dependence. We can then also identify the long time value of $D(\tau,T)$ as D^* and we have again found a fixed point.

The form taken by the growth law in this case can be written

$$L(t,T) = L_0 + B(t/\tau_0)^n \qquad (4.7)$$

where

$$n = 1/D^* \quad . \tag{4.8}$$

Systems of this type which freeze upon quenching to T = 0 but which grow with a power law in time for quenches to T > 0 have been classified by us as class 2 systems.

If we choose our normalization of τ_0 such that B = 1 and subtract off the *slow transient L_0* from L(t,T), then

$$\tilde{L} = L - L_0 = \tau^{1/2} \tag{4.9}$$

as shown in Fig. 4. The temperature dependence of τ_0 is found to be given by

$$\tau_0^{-1} = ay_H (1 + by_H) \tag{4.10}$$

where $y_H = e^{-2J/k_B T}$ where J is the exchange energy and a = 14.5 and b = −1.5. The low temperature behavior of L_0 is given by $L_0(T) = 3.5 - 2.4 k_B T/J$.

Contrary to our previous [3] assertions, the spin exchange or Kawasaki model on a square lattice is also a class 2 system. It freezes for quenches to T = 0, but for T > 0

$$\tilde{L} = \tau^{1/3} \tag{4.11}$$

as shown in Fig. 5, with

$$\tau_0^{-1} = ay^2(1 + by) \tag{4.12}$$

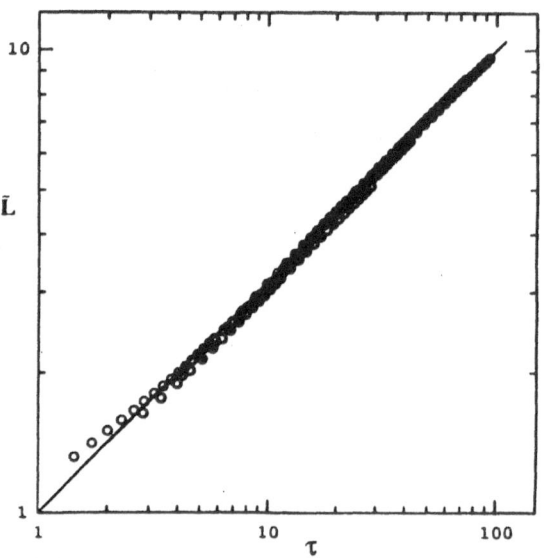

Fig. 4. ln − ln plot of $\tilde{L}$ vs. $\tau = t/\tau_o$. Circles are the data points for six different temperatures y_H = 0.01, 0.02, 0.03, 0.04, 0.055, 0.0718, and the solid line is the function $\ln\tilde{L} = \frac{1}{2}\ln\tau$.

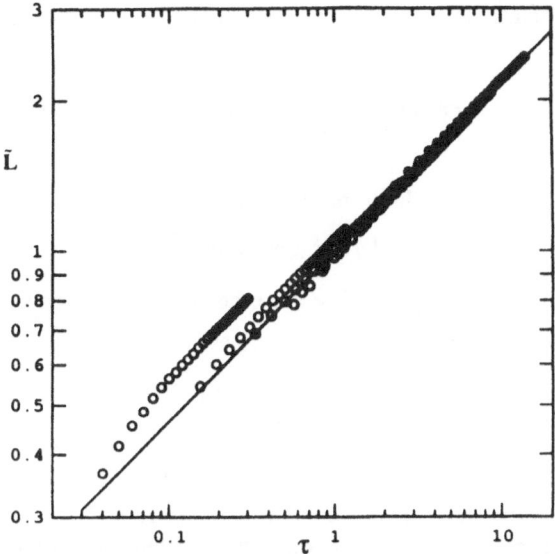

Fig. 5. ln − ln plot of $\tilde{L}$ vs. τ for seven different temperatures (y = 0.005, 0.01, 0.015, 0.02, 0.025, 0.03, 0.04) for the SEKI model on a square lattice. The solid line is the function $\ln\tilde{L} = \frac{1}{3}\ln\tau$.

where $y = e^{-4J/k_B T}$, $a = 4.2$ and $b = -7.5$. The freezing length is given by $L_o = 1.9 - 9.2y$ for low temperatures. The 1/3 exponent is in agreement with the theory of Lifshitz and Slyozov [13] and the recent large scale Monte Carlo simulations of Amar, et al. [14] While we find agreement with Huse's [12] prediction that $\tau_0^{-1} \sim y^2$, we find disagreement with his prediction that L_0 blow up exponentially as $T \to 0$. We find that L_0 does increase as $T \to 0$, but it saturates at the values controlled by the zero temperature freezing. For the two temperatures studied by Amar, et al. we obtain a ratio $L_0(0.3T_c)/L_0(0.5T_c) = 1.15$ in very good agreement with their result of 1.18 and in contrast with the result 5.33 proposed by Huse.

V. Random Systems: Class 3 and 4 Behavior

Let us consider the rather different growth kinetics of systems with quenched impurities. Again systems in this class freeze when quenched to zero temperature. However for quenches to T > 0 this class is characterized by a D(t,T) which differs from the behavior described for class 1 and class 2 systems. The basic physics is simple. There exist energy or free energy barriers due to the randomness in the system which depend on the characteristic length L:

$$E = E(L) \quad . \tag{5.1}$$

If the associated characteristic times are activated then

$$t = \tau_1 e^{E(L)/T} \quad . \tag{5.2}$$

It has been asserted [15] in the case of the random field Ising model (RFIM) that

$$E(L) \approx L \tag{5.3}$$

so one can easily invert (5.2) to obtain the logrithmic behavior given by

$$L(t,T) = L_0 + AT \ln(t/\tau_1) \tag{5.4}$$

where A has a weak T dependence. There are other random systems where

$$E(L) \approx L^{1/m} \tag{5.5}$$

with $m \neq 1$, as has been suggested for the case of diluted ferromagnets [16,17] and spin glasses. [18] In these cases one expects a growth law

$$L(t,T) = L_0 + [AT \ln(t/\tau_1)]^m \quad . \tag{5.6}$$

The RG treatment of these systems depends on whether m = 1, or not. Consider first the case m = 1 which we call a class 3 system. Then, given the growth law (5.4), one can compute D(t,T) and find that D(t,T) has a definite long time limit, but that limit is strongly dependent on temperatures for low temperatures:

$$\lim_{t \to \infty} D(t,T) = D^*(T) = \frac{L_0}{AT} \quad . \tag{5.7}$$

This behavior leads to a perfectly acceptable growth kinetics fixed point with a time rescaling factor which is strongly temperature dependent

$$\Delta = e^{-(b-1)D^*(T)} \quad . \tag{5.8}$$

and the growth law (5.4) follows from (2.4) and (5.8).

For $m \neq 1$, class 4 systems with L given by (5.6), one obtains

$$D(t,T) = \frac{(1-m)L + mL_0}{mAT(L-L_0)^{(m-1)/m}} \quad , \tag{5.9}$$

which has the long time behavior

$$D(t,T) = \frac{1-m}{m} \ln(t/\tau_1) \tag{5.10}$$

which, for m > 1, is negative and does not reach a fixed point. Again, as for class 2 systems, using the natural variables (t,T) does lead to a fixed point. One must choose (τ,ζ) such that one does find a fixed point. Looking at the equation for the growth law one easily sees that scaling is achieved with the choice of variables

$$\ln\tau = [\ln(t/\tau_1)]^m \tag{5.11}$$

and

$$\zeta = (AT)^m \tag{5.12}$$

In this case, in terms of these variables, the case $m \neq 1$ is mapped onto the case m = 1 and D, computed in terms of the scaling variables τ and ζ, takes the form

$$\lim_{\tau \to \infty} D(\tau,\zeta) = \frac{L_0}{\zeta} \tag{5.13}$$

which is essentially of the same form as (5.7) for m = 1.

The scaling form for L for class 3 and 4 systems is

$$\tilde{L} = L_0 + \zeta \ln \tau \tag{5.14}$$

which clearly satisfies (2.13) with $\Delta = e^{-(b-1)L_0/\zeta}$. There is, as yet, only [19] theoretical evidence for these class 3 and 4 systems.

Acknowledgements

This work was supported in part by the National Science Foundation Grant No. DMR-84-12901, the Central Computer Facility of the NSF Materials Research Laboratory at the University of Chicago and by the Microelectronics and Information Sciences Center and by the Graduate School at the University of Minnesota.

References

1. See the review in: J.D. Gunton, M. San Miguel and P.S. Sahni, in *Phase Transitions and Critical Phenomena*, Vol. 8, p. 267 (1983), ed. by C. Domb and J. Lebowitz (New York, Academic Press).

2. For example, in very recent work on Cell Dynamical Models by Y. Oono and S. Puri (preprint) it is stated that they believe their "models are in the same universality class as the Cahn-Hilliard equation".

3. G.F. Mazenko, O.T. Valls and F. Zhang, Phys. Rev. B **31**, 4453 (1985).

4. J. Viñals, M. Grant, M. San Miguel, J.D. Gunton and E.T. Gawlinski, Phys. Rev. Lett. **54**, 1264 (1985); S. Kumar, J. Viñals and J. D. Gunton, Phys. Rev. B **34**, 1908 (1986).

5. K. Oki, H. Sagana, and T. Eguchi, J. Phys. (Paris) **C7**, 414 (1977).

6. See G.F. Mazenko and O.T. Valls, Phys. Rev. B **27**, 6811 (1983) for a discussion and earlier references.

7. Z.W. Lai, G.F. Mazenko and O.T. Valls, Preprint.

8. In this paper we are primarily concerned with low temperatures where, for class 1 and 2 systems, $\zeta \sim \xi \sim T$. For class 3 and 4 systems the identification of ζ is determined by the collective activated nature of the droplets pinned by the impurities and not governed by the width of interfaces ($\sim \xi(T)$) as in class 1 and 2 systems.

9. S. Ma, Phys. Rev. Lett. **37**, 461 (1976).

10. D.J. Amit, *Field Theory, the Renormalization Group and Critical Phenomena* (McGraw-Hill, New York, 1978) Chapter 8.

11. It is suggested in Ref. 12 that we did not recover the Lifshitz-Slyosov result for the SEKI model in Ref. 3 because we only allowed for a T=0 fixed point in our analysis. This suggestion is incorrect since the T=0 fixed point associated with this problem corresponds to a high temperature disordered fixed point on scales large compared to the freezing length L_0.

12. D. Huse, Phys. Rev. B **34**, 7845 (1986).

13. I.M. Lifshitz and V.V. Slyozov, J. Phys. Chem. Solids **19**, 35 (1961).

14. J. Amar, F. Sullivan and R. Mountain, to be published.

15. G. Grinstein and J.F. Fernandez, Phys. Rev. B **29**, 6389 (1984), and J. Villain Phys. Rev. Lett. **52**, 1543 (1984).

16. D. Huse and C. Henley, Phys. Rev. Lett. **54**, 2708 (1985).

17. G. Grest and D. Srolovitz, Phys. Rev. B **32**, 3014 (1985). See also D. Chowdhury, M. Grant and J.D. Gunton, Phys. Rev. B **35**, 6792 (1987).

18. D. Fisher and D. Huse, Phys. Rev. Lett. **56**, 1601 (1986).

19. There is numerical evidence in Ref. 17.

GROWTH PROCESS OF ORDER PARAMETER

--- BLOCK SPIN ANALYSIS

Macoto Kikuchi and Yutaka Okabe[*]

Department of Physics, Osaka University
[*]Toyonaka 560, Japan
Department of Physics, Tohoku University
Sendai 980, Japan

INTRODUCTION

In this paper we study the ordering process of the three-dimensional Ising model of the non-conserving order parameter after a rapid quench from $T=\infty$ to a thermodynamically unstable state by means of Monte Carlo simulation. The temporal evolution of the probability distribution function (PDF) of the magnetization per spin, $P(m,t)$ is investigated. We discuss the qualitative fearures of PDF, the finite-size scaling and the effective potential which governs the evolution of PDF.

There exist two characteristic length scales in the present problem, that is, the correlation length ξ of the thermal fluctuation and the average radius $r(t)$ of the growing domain. It is believed from the curvature-driven dynamics that $r(t)$ grows as $t^{1/2}$ once the domains are well established. [1,2] From the renormalization-group standpoint, ordering process is related to two stable fixed points, that is, $T=0$ and $T=\infty$. Thus ξ is an irrelevant length and the growth of order is essentially temperature independent; temperature dependence is of secondary interest. The block-spin transformation (BST) is an effective method to extract the temperature independent fixed point behavior. Some attempts have been made to apply BST to the study of the ordering process. [3,4]

The temporal evolution of PDF is intrinsically system-size dependent. If the only relevant length scale is $r(t)$, the linear size of the system L should be scaled by $r(t)$. Then PDF for the finite-system is expected to satisfy a following dynamical finite-size scaling form:

$$P(m,t;L) = \tilde{P}(m,tL^{-2}).$$

At finite temperature, the finite thermal correlation length ξ gives rise to some corrections to the above leading scaling form; a most trivial one comes from the fact that the equilibrium magnetization is less than unity. If ξ is much smaller than L, BST makes the system be quenched to the zero-temperature fixed point, and then the above scaling form is valid. Here we restrict our attention mainly to the temperature independent (fixed point) behavior.

METHOD OF SIMULATION

We treat the simple-cubic spin-flip Ising model of the nearest-neighbour interaction. The whole lattice is divided into eight sublattices for the vectorization. All the spins on each sublattice are updated simultaneously with the Metropolis' transition probability. We expect that the choice of an update scheme and that of a transition probability affect only a time-scale. BST is defined by the majority rule of the block size $2\times2\times2$ with a random tie-breaker. Two of the eight possible choices for the location of a block are employed at each transformation. Therefore at the n-th blocking level, 2^n block-spin systems are created; the block-spin magnetization is defined by the average magnetization over them. That procedure makes the magnetization a quasi-continuous variable even for high-level block-spin systems. Initial configurations are prepared to be completely random. We make simulations on 9000 samples for each quench condition; all the samples have different initial configurations and are run with different random number sequences.

RESULTS

We show the evolution of $P(m,t)$ for L=64 at $T=0.74T_c$ in Fig. 1(a).

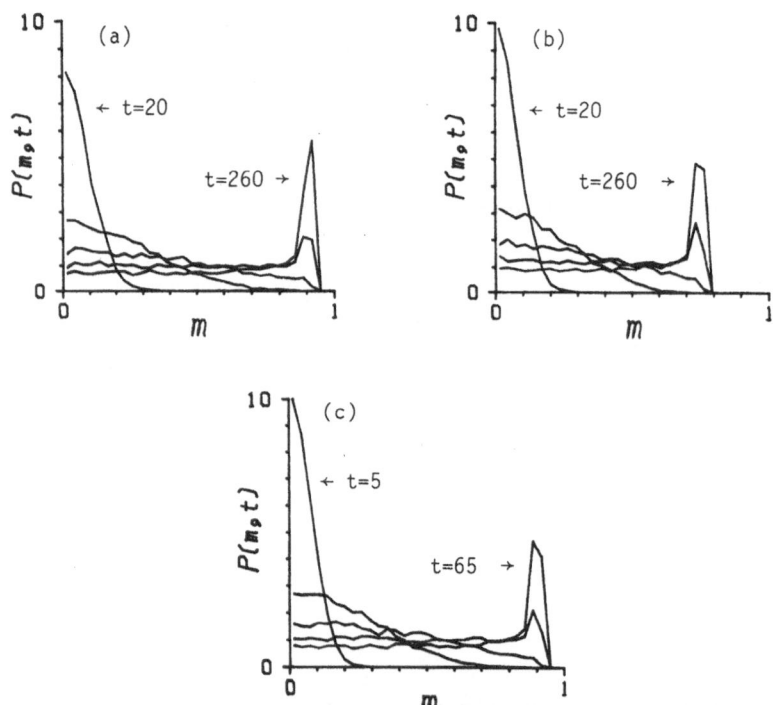

Fig. 1. The distribution function $P(m,t)$ versus m for L=64 and $T=0.74T_c$ (a), for L=64 and $T=0.89T_c$ (b), and for L=32 and $T=0.74T_c$ (c). The data for 20, 80, 140, 200 and 260 MCS/spin are shown in (a) and (b), and those for 5, 20, 35, 50 and 65 MCS/spin are shown in (c).

Data for 20, 80, 140, 200 and 260 Monte Carlo steps (MCS) per site are
presented. From the ±m symmetry, we take signs of magnetization for all
the data as positive. The whole range of magnetization ($0 \leq m \leq 1$) are divided
into 32 meshes. Initially, PDF has a a Gaussian form with the width of
$O(L^{-3/2})$ because the spins are assigned ±1 randomly. We see from the
figure that P(m,t) at t=20 MCS/spin still keeps the Gaussian feature,
although the width becomes macroscopic. The width broadens with time, and
then a peak at equilibrium magnetization appears. This feature agrees
qualitatively with Suzuki's scaling solution for the one-parameter system
with the Landau-type quartic potential. [5] Some differences, however, are
found. For example, the peak position of m moves with time in the scaling
solution, but no appreciable movement is found in the present result. This
fact indicates that the effective potential which governs the dynamics is
of different form from a quartic function. Figures 1(b) and 1(c) are the
similar plots for L=64, T=0.89T_c and L=32, T=0.74T_c respectively. In
Fig. 1(c) we show the data for 5, 20, 35, 50 and 65 MCS/spin. Figures 1(a)
and 1(b) have the similar qualitative feature; the difference in the quench
temperature appears as the difference in the equilibrium magnetization.
Similarity seen in Fig 1(a) and 1(c) suggests that the finite-size scaling
indeed holds. In order to see the fixed point behavior, we show PDF for
the systems after two-times BST in Figs. 2(a)-2(c). Parameters are the
same as in Figs. 1(a)-1(c). We see that the three figures are almost

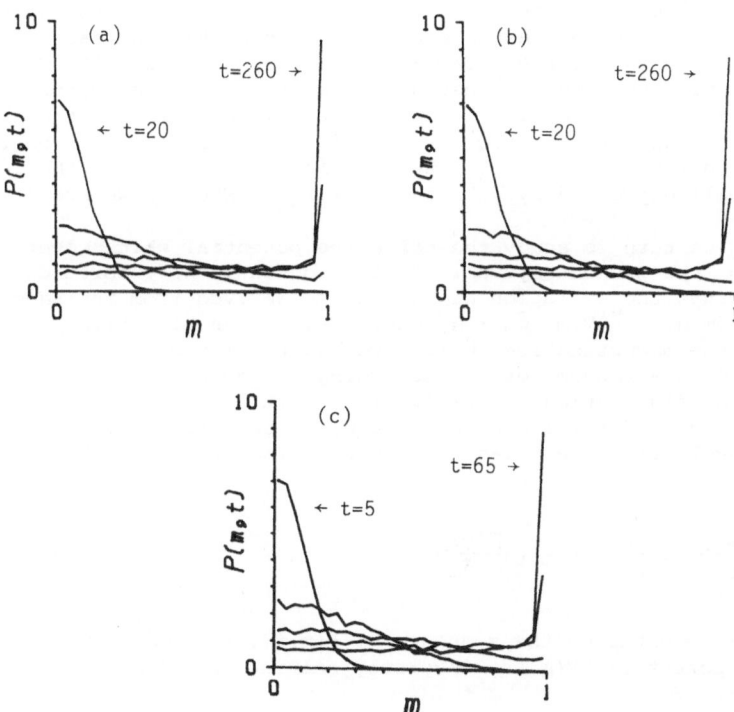

Fig. 2. The disribution function P(m,t) versus m for the block-spin system
with the blocking level n=2. The same parameters are used as in
Fig. 1.

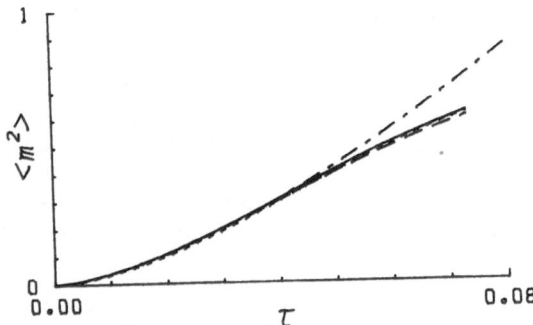

Fig. 3. The temporal evolution of the second moment $\langle m^2 \rangle$ against the scaling variable $\tau \equiv tL^{-2}$. The solid, dashed and dotted lines show the data for L=64 and T=0.74T$_c$, for L=64 and T=0.89T$_c$, and for L=32 and T=0.74T$_c$, respectively. The curve $\sim \tau^{3/2}$ is represented by the dash-dotted line

idendical. Thus P(m,t) for the block-spin systems are independent of the quench condition.

In order to see the finite-size effect more explicitly, we plot the second moment of PDF, $\langle m^2 \rangle$ for the two-times transformed system against the scaling variable $\tau \equiv tL^{-2}$ in Fig. 3. This quantity expresses the fluctuation around m=0. The solid, dashed and dotted lines show the data for L=64, T=0.74T$_c$, the data for L=64, T=0.89T$_c$ and the data for L=32, T=0.74T$_c$ respectively. Those three lines agree well; thus the asymptotic finite-size scaling is confirmed. The dash-dotted line expresses the curve $\sim \tau^{3/2}$, which is the expected behavior of $\langle m^2 \rangle$ from the random distribution of the domains with the radius $\sim t^{1/2}$. We find in the figure that the initial stage is well expressed by $\tau^{3/2}$ for as large system as we are treating.

Next, we turn to study the effective potential which governs the dynamics of the present system. As the conventional theoretical treatment, we assume that the evolution of PDF can be derived from the Langevin equation, $dm/dt = \Gamma \delta V(m)/\delta m + \eta$, where V(m) is an effective potential acting on the magnetization, Γ is a constant which determines the time scale and η is a random force. According to the Suzuki's scaling theory, [5] when the fluctuation, initially is of $O(L^{-3/2})$, grows to $O(1)$, the effect of random force can be neglected; in that time region (scaling region), evolution of PDF can well be described by a deterministic drift equation,

$$\frac{dP(m,t)}{dt} = \frac{\delta}{\delta m} \left(P(m,t) \frac{\Gamma \delta V(m)}{\delta m} \right) .$$

Since we have obtained the sequence of P(m,t), we can calculate the effective potential $\Gamma V(m)$ by inversely solving the above drift equation:

$$\Gamma V(m) = \int^m dm' \frac{1}{P(m',t)} \int^{m'} dm'' \frac{dP(m'',t)}{dt} .$$

After checking that the obtained potential is almost independent of time in the scaling region, time average is taken. We divide the range $0 \leq m \leq 1$ into 64 meshes for the numerical integration.

In Figs. 4(a)-4(c) we plot the potentials $\Gamma V(m)$ obtained through the above procedure for the block-spin systems of the blocking-level $n \leq 2$. The same parameters are used as in Figs. 1(a)-1(c). The scale of potentials for Fig. 4(c) is four times as large as those for Figs. 4(a) and 4(b). We cannot calculate the potentials for m larger than the potential minima, because we use the information of the relaxation from the disordered state. We find that the potential converges to a single function as n increases. All the potentials for n=2 coincide with each other. Therefore they give the fixed-point potential, which is independent of temperature and is proportional to L^{-2}.

It is convenient to plot the derivative of the potential $\Gamma \delta V(m)/\delta m$ for investigating the detailed functional form of the potential. In Fig. 5 we show $\Gamma \delta V(m)/\delta m$ for L=64, T=0.89T_c and n=0, for example. The dotted and dashed lines are the functions $am+bm^3$ and $am+cm^5$ respectively, which are fitted so that their derivatives at m=0 and the position of the maxima coincide with those of $\Gamma \delta V(m)/\delta m$. We find that the fitting of the latter to $\Gamma \delta V(m)/\delta m$ is better than the former in the whole range of m. Therefore the function Am^2+Cm^6 gives a good approximation for the effective potential compared with the Landau-type function. Beyond the inflection point of the potential, however, the higher terms than m^6 are required for a much better fitting.

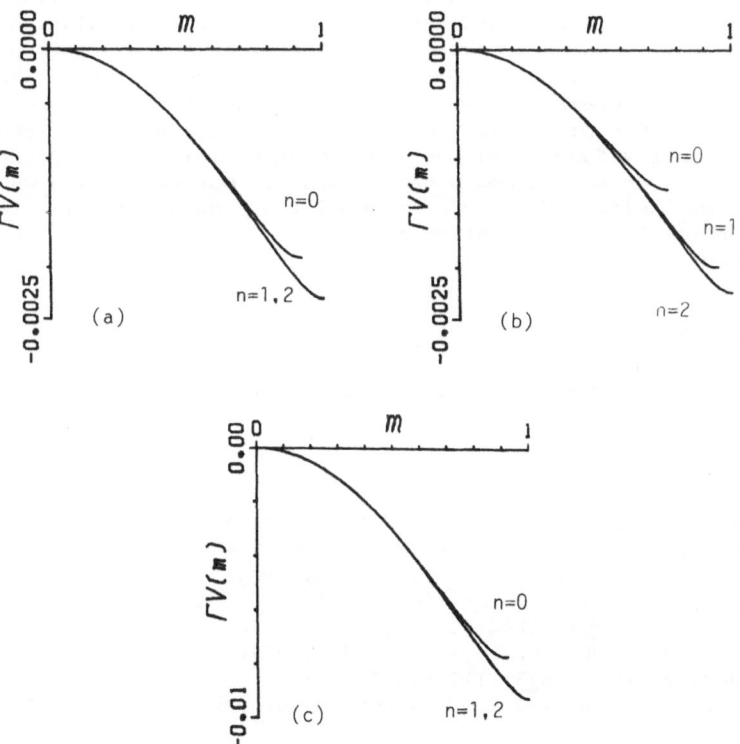

Fig. 4. The effective potentials $\Gamma V(m)$ for the block-spin systems of blocking level $n \leq 2$. The same parameters are used as in Fig. 1.

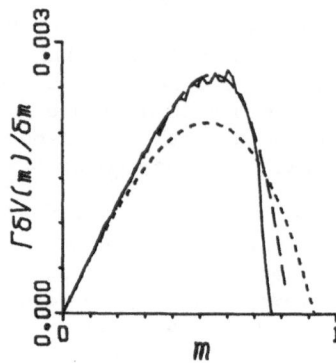

Fig. 5. The derivative of the effective potential, $\Gamma \delta V(m)/\delta m$, for L=64, T=0.89$T_c$ and n=0. The dotted and dashed lines are the functions $am+bm^3$ and $am+cm^5$ respectively, which are fitted so that their derivatives at m=0 and the position of the maxima coincide with those of $\Gamma \delta V(m)/\delta m$.

SUMMARY AND DISCUSSION

We have studied the temporal evolution of the probability distribution function of the order parameter in the ordering process by the Monte Carlo simulation. The dynamical finite-size scaling has been shown to hold for the evolution of PDF. Consequently we have confirmed the scaling by the domain size r(t). A new method has been proposed for calculating the effective potential which governs the growth process. Apart from the boundary condition, our potential V(m) can be regarded as the single-site part of the coarse-grained free energy functional whose coarse-graining cell size is L. The probability distribution function and the coarse-grained free energy functional was studied at equilibrium by Binder [6] and Kaski et al. [7] using the Monte Carlo method. Our study is a dynamical counterpart of theirs. It should be noted that the potential $\Gamma V(m)$ serves as a good device for investigating the growth law of r(t); the observed L^{-2} dependence of $\Gamma V(m)$ is a clear evidence for $t^{1/2}$ law.

ACKNOWLEDGEMENT

We are grateful to Dr. Seiji Miyashita for valuable discussions.

REFERENCES

1. S.M. Allen and J.W. Cahn, Acta. Metall. 27, 1085 (1979).
2. K. Kawasaki and T. Ohta, Prog. Theor. Phys. 67, 147 (1982).
3. G.F. Mazenko, O.T. Valls and F.C. Zhang, Phys. Rev. B31, 4453 (1985).
4. J. Vinals, M. Grant, M. San Miguuel, J.D. Gunton and E.T. Gawlinski, Phys. Rev. Lett. 54, 1264 (1985).
5. M. Suzuki, Prog. Theor. Phys. 56, 77 (1976); 56, 477 (1976).
6. K. Binder, Z. Phys. B43, 119 (1981).
7. K. Kaski, K. Binder and J.D. Gunton, Phys. Rev. B29, 3996 (1984).

STATISTICAL PHYSICS THEORY

OF OSTWALD RIPENING

Michio Tokuyama

General Education
Tohwa University
Fukuoka 815, Japan

INTRODUCTION

We review recent theoretical developments in our understanding of the late-stage processes of phase separation in binary alloys. When a system is quenched into a metastable state, phase separation occurs by nucleation and growth[1,2], In the late stage, known as Ostwald ripening[3], the minority phase takes the form of spherical droplets whose growth and dissolution proceeds by an evaporation-condensation mechanism. There are two theoretical aspects in understanding of the dynamics of such a phase separation, depending on what processes we are interested in. The first is to study the causal motion which is described by a single droplet size distribution function $f(R,t)$ with radius R, and which is experimentally observable by an electron microscope. This was first done in the monumental works by Lifshitz and Slyozov[4] and independently by Wagner[5]. They found the celebrated scaling law $f(R,t) = [n(t)/R(t)]p_0(R/R(t))$, where the average droplet radius grows as $R(t) \sim t^{1/3}$ and the number density of droplets decays as $n(t) \sim t^{-1}$. The relative droplet size distribution function $p_0(\rho)$ is a time-independent function of ρ. Although their works were the origin of later theoretical studies of Ostwald ripening, their results were valid only in the limit of zero volume fraction of the minority phase and did not agree with experimental observations where the volume fraction is not infinitely zero. After their works, many attempts[6-10] to extend their theory to the case of finite volume fraction have been proposed. The second is to explore the fluctuations around the causal motion. Although they are small as compared to the causal motion, they are still important since they are observable as a structure function $S_k(t)$ by small-angle scattering experiments. The structure function has been found to satisfy a scaling law $s_k(t) \sim R(t)^3 S(kR(t))$ in a computer simulation based on a kinetic Ising model[11] and also in real systems such as binary alloys[12-16]. Although there have been several phenomenological theories to study the scaled structure function $S(x)$ based on the droplet dynamics[2,17-20], there does not exist a first-principles theory which enables us to derive an equation of motion for $S_k(t)$ and thus to determine an explicit, analitic form for $S(x)$.

In the following sections, we discuss our recent advances in both theoretical aspects mentioned above from a new unifying viewpoint based on the statistical physics theory of Ostwald ripening by the present author and Kawasaki[21].

SPHERICAL DROPLET MODEL

Here we describe a derivation of a model equation used to study Ostwald ripening. We consider a three dimensional classical system which consists of spherical droplets of the minority phase and a supersaturated solution of the majority phase which is characterized by a concentration field $C(r,t)$. The droplets are sufficiently large so that the distribution of their positions can be assumed to be stationary. Such a system has two kinds of characteristic lengths; the average droplet radius $R(t)$ and the correlation length $L(t)=1/[4\pi R(t)n(t)]^{1/2}$ within which two droplets have correlations[21]. We assume that the initial total supersaturation Q is small so that $R(t)/L(t)=(3\phi(t))^{1/2}\sim Q^{1/2}\ll 1$, where $\phi(t)=4\pi n(t)R(t)^3/3$.

The driving force for Ostwald ripening is the interfacial surface energy of droplets. Due to such surface free energy contributions, the solubility of droplets, $C_{eq}(R)$, is decreasing function of size R which is given by the Gibbs-Thompson relationship

$$C_{eq}(R) = C_\infty(1 + a/R), \tag{1}$$

where C_∞ is the equilibrium solubility of a droplet with infinite radius, and $a=2\sigma v/k_B T$, σ being the surface tension and v the atomic volume of solute. Because of the difference in solubility, diffusion gradients are set up in the supersaturated solution between droplets. Thus, the growth and dissolution of droplets is limited by the diffusion through the solution from dissolving droplets to growing droplets. Then, the concentration $C(r,t)$ is described by the diffusion equation[7]

$$(\partial/\partial t)C(r,t) = D\nabla^2 C(r,t) \tag{2}$$

with the boundary condition

$$C(r=r_i,t) = C_{eq}(R_i(t)), \tag{3}$$

where D is the diffusion coefficient and $R_i(t)=|R_i(t)|$ the radius of the ith droplet. r_i denotes the position vector from the origin to a point on the surface of the ith droplet and is given by $r_i=X_i+R_i(t)$, where X_i is the position vector of the center of the ith droplet and $R_i(t)$ is the vector from the center of the ith droplet to a point on its surface.

By adding an appropriate source term, the boundary condition (3) can be eliminated from eq.(2). Thus, on the length scale of order L, the diffusion process is described by[10,22]

$$(\partial/\partial t)C(r,t) = D\nabla^2 C(r,t) + (4\pi\alpha D/v)\sum_{i=1}^{N}\delta(r-X_i)M_i(t), \tag{4}$$

where $M_i(t)$ denotes the strength of the interaction between the concentration and the ith droplet located at position X_i, and $\alpha=avC_\infty$ the capillary length. Since the characteristic relaxation time of the diffusion field, $\tau_0=L(t)^2/D=R(t)^2/3D\phi(t)$, is much shorter than the coarsening time $\tau=R(t)^3/\alpha D$ in which the droplet has an appreciable change in size, on the time scale of order τ, the l.h.s. of eq.(4) is assumed equal to zero, leading to the quasi-static diffusion equation. This approximation is valid as long as $\phi(t)\gg\alpha/3R(t)$. Then, solving eq.(4), we have the formal solution

$$C(r,t) = C_0(r,t) + (\alpha/v) \sum_{i=1}^{N} M_i(t)/|r-X_i|, \tag{5}$$

where $C_0(r,t)$ is a bare concentration field in the absence of droplets. Solving eq.(5) for $M_i(t)$ and using eqs.(1) and (3), on the length scale of order L, we thus find

$$M_i(t) = 1 - R_i(t)/R_0(X_i,t) - R_i(t)\sum_{j\neq i} M_j(t)/|X_i-X_j|, \tag{6}$$

where $R_0(r,t)$ is the bare critical radius and is given by

$$R_0(r,t) = \alpha/v[C_0(r,t) - C_\infty]. \tag{7}$$

Next we derive the equation of motion for the radius of the ith droplet, $R_i(t)$. Since conservation of mass holds for each droplets, the time evolution of the mass of the ith droplet is described by

$$(d/dt)[(4\pi/3)R_i(t)^3] = -vDR_i(t)^2\int(n_i\nabla C)d\Omega_i, \tag{8}$$

$$= 4\pi\alpha DM_i(t), \tag{9}$$

where Ω_i denotes the orientation of the vector R_i, and n_i the unit normal vector. In order to derive eq.(9) from eq.(8), we have used eq.(5). Since the total mass is also conserved, eq.(9) must be suppemented by the conservation law. Let us introduce a homogeneous concentration by $C(t)= V^{-1}\int dr C(r,t)$ where V is the total volume of the system. Using eqs.(4) and (9), we then find the conservation law

$$q(t) + \Delta(t) = Q \tag{10}$$

with the volume fraction of the minority phase

$$q(t) = V^{-1}\int dr\int dR(4\pi/3)R^3 N(R,r;t) = v[C_{in} - C(t)], \tag{11}$$

and the supersaturation $\Delta(t)=v[C(t)-C_\infty]$, where C_{in} is the concentration before nucleation has occured, and $N(R,r;t)$ is the number density of droplets with radius R and at the position r and is given by

$$N(R,r;t) = \sum_{i=1}^{N} \delta(R-R_i(t))\delta(r-X_i(0)). \tag{12}$$

Eq.(9) with eq.(6) is a starting equation for studying the late stage processes[7,23]. This equation consists of two terms. The first term, $1-R_i/R_0$, is the mean-field term and predicts that the droplet grows if $R_i>R$, and it dissolves if $R_i<R$,. The second term represents the spatial interactions between droplets separated by the distance of order L and contributes to a renormalization of the first term. Because of the Coulomb-like long-range interaction, it is beyond our capacity to deal with eq.(9) analytically, although it was recently solved numerically by computer simulations[11,24,25]. Therefore, we should first reduce eq.(9) to obtain macroscopic equations, which we can reasonably analyze. Recently, we have derived the system of kinetic equations for the distribution functions from eq.(9), which enable us to describe not only the causal motion of the droplet coarsening but also the fluctuations around it[21]. This reduction was done by using a systematic expansion in powers of $\phi^{1/2}$, namely, the ratio of the average droplet radius and the correlation length, since the spatial interaction term is of order $\phi^{1/2}$.

The single droplet distribution function is given by

$$F(i;t) = \overline{N(i;t)}, \tag{13}$$

where i denotes the set of R_i and r_i, and the bar means the average over the initial ensemble $\rho_0(\{R_i(0), X_i(0)\})$. Differentiating $F(1;t)$ with respect to t and using eq.(9) leads to the BBGKY-like hierarchy equations for the distribution functions. The systematic expansion in powers of $\phi^{1/2}$ is then achieved to truncate such hierarchy equations[21]. In the late stage where $t > \tau(3\phi)^{-1/2}$ and $r >> L$, the distribution function $F(R,r;t)$ becomes homogeneous in space, leading to $f(R,t)$. Thus, we obtain the Fokker-Planck type kinetic equation for $f(R,t)$[21,26]

$$(\partial/\partial t)f(R,t) = \alpha D(\partial/\partial R)R^{-2}[1 - \rho + (3\phi(t))^{1/2}\{v(\rho) + C(\rho)\}]f \tag{14}$$

with the drift term

$$v(\rho) = \rho(m_2 - \rho), \tag{15}$$

and the soft-collision operator

$$C(\rho) = c(\rho) + (d/d\rho)E(\rho), \tag{16}$$

where ρ is a relative droplet radius defined by $\rho = R/R(t)$, $R(t)$ being the average droplet radius given by $R(t) = \int Rf(R,t)dR/n(t)$, and the explicit forms of the soft-collision terms $c(\rho)$ and $E(\rho)$ are given by eqs.(2.19) and (2.30) of Ref.26, respectively. m_i denotes the ith momont defined by $m_i = \int \rho(t)^i f(\rho,t)dR/n(t)$, where $n(t) = \int f(R,t)dR$.

Eq.(14) is a generalization of the Lifshitz-Slyozov equation to the case of finite volume fraction and is different from those proposed by the other theories[6-10] in several important points. Firstly, to order ϕ^0, the spatial interaction term is shown to renormalize the bare critical radius $R_0(t)$ so as to be $R(t)$. Secondly, to order $\phi^{1/2}$, such an interaction term enters both the collisionless drift term and the soft-collision term. The soft-collision term represents a collision between droplets separated by the distance of order L due to the diffusive long-range interactions and is different from the hard collision between two touching droplets discussed by Lifshitz and Slyozov. Although this collision has been studied by none of the previous authors, it plays an important role in the late stage. Thirdly, eq.(14) ensures to each order in $\phi^{1/2}$ that the volume fraction $q(t)$ becomes constant in the late stage. In fact, using the readily verifiable facts that

$$\int_0^\infty (1-\rho)f(R,t)dR = \int_0^\infty v(\rho)f(R,t)dR = \int_0^\infty C(\rho)f(R,t)dR = 0, \tag{17}$$

we find, from eqs.(11) and (14), the equation $(d/dt)q(t) = 0$. This is combined with eq.(10) to obtain $\Delta(t) = 0$ and $q(t) = Q$.

For long time t the distribution function $f(R,t)$ satisfies scaling

$$f(R,t) = [n(t)/R(t)]p(R/R(t); Q^{1/2}) \tag{18}$$

where $p(\rho)$ is a time-independent function of ρ. In fact, this has been confirmed recently by solving eq.(14) under various initial conditions[27]. By using eqs.(14) and (18), we then find the temporal power laws

$$n(t) = (3Q/4\pi m_3)R(t)^{-3}, \quad R(t)^3 = K(Q)t, \quad L(t) = (m_3/3Q)^{1/2}R(t), \tag{19}$$

where the coarsening rate $K(Q)$ is given by

$$K(Q) = \alpha D \lim_{\rho \to 0} p(\rho;Q^{1/2})/\rho^2. \tag{20}$$

The relative droplet size distribution function $p(\rho)$ thus obeys a second-order differential equation[26]

$$[4+\rho(d/d\rho)p(\rho)]=-(3\alpha D/K)(d/d\rho)\rho^{-2}[1-\rho+(3Q/m_3)^{1/2}\{v(\rho)+C(\rho)\}]p \tag{21}$$

with the boundary conditions

$$\int_0^\infty p(\rho)d\rho = 1, \quad <\rho> = \int_0^\infty \rho p(\rho)d\rho = 1, \tag{22}$$

where m_3 is time independent, and the angular brackets denote the average over $p(\rho)$.

We should mention here that since eq.(21) does not contain any adjustable parameters, $p(\rho)$ can be determined by solving eq.(21) self-consistently so as to satisfy eqs.(20) and (22). In fact, because of the second-derivative term, eq.(21) has non-vanishing solution at all finite ρ, leading to asymptotic forms $p(\rho)\sim\rho^2$ for small ρ and $p(\rho)\sim\rho\exp(-A\rho^4)$ for large ρ, where A is a constant[28]. This situation is quite different from the other theories[6-10], where the soft-collision process is left out and hence $p(\rho)$ obeys a first-order differential equation which has a solution cut off at some finite value of ρ. Because of the soft-collision term, the present theory leads to the higher coarsening rate $K(Q)$ and the broader and flatter distribution $p(\rho)$ than the others do at the same value of Q^{28}. Thus the role of the soft-collision process in coarsening turns out to be quite different from that of the collisionless drift process. This will be seen more clearly in fluctuations.

FLUCTUATIONS

The fluctuations around the causal motion, $\delta N(1;t)=N(1;t)-F(1;t)$, are small as compared to the mean value $F(1;t)$; $|\delta N/F|\sim Q^{1/4}<<1^{21}$. However, they are still important since they are observable as a structure function by small-angle scattering experiments. The structure function $S_k(t)$ is defined by

$$S_k(t) = \int_0^\infty dR_1\int_0^\infty dR_2(4\pi R_1^3/3)(4\pi R_2^3/3)\Psi(kR_1)\Psi(kR_2)\chi_k(R_1,R_2;t) \tag{23}$$

with a structure factor of a single droplet $\Psi(x)=3(\sin x-x\cos x)/x^3$, where $\chi_k(R_1,R_2;t)$ is a spatial Fourier transform of the variance

$$\chi(1,2;t) = \delta N(1;t)\delta N(2;t). \tag{24}$$

In order to obtain an equation of motion for $S_k(t)$, we first derive the equation of motion for the variance. Similarly to the case of the causal motion, differentiating eq.(24) with respect to t and using eq.(9) also leads to the hierarchy equations for distribution functions. Since $|\delta N/F|<<1$, such hierarchy equations can be truncated by employing the same procedure as that used in the derivation of eq.(14). After the Fourier transformation is taken, in the late stage we thus obtain the following linear variance equation for $\chi_k(R_1,R_2;t)$[21,29]:

$$(\partial/\partial t)\chi_k(R_1,R_2;t) = (1+e_{12})\alpha D(\partial/\partial R)R^{-2}[\int d\rho_3 H(y,\rho_2,\rho_3)\chi_k(R_3,R_1;t)$$

$$+ (3Q/4\pi m_3)(3Q/m_3)^{1/2}R(t)^{-5}I(y,\rho_2,\rho_1)] \tag{25}$$

with

$$H(y,\rho_2,\rho_3)=[1-\rho_3+(3Q/m_3)^{1/2}\{v(\rho_3)+C'(\rho_3)\}][\delta(\rho_3-\rho_2)-\rho_2 P(\rho_2)/(y^2+1), \tag{26}$$

where $y=kL(t)$, $C'(x)\delta(x-z)=C(z)\delta(z-x)$ and e_{12} denotes the exchange operator between 1 and 2. The term I is the source term generated by the soft-collision process[29]. Thus there are two kinds of macroscopic fluctuations. The first originates from the thermal fluctuations already existing at the beginning, which are described by $\chi_k(\rho_1,\rho_2;t=0)$. The second is the non-thermal fluctuations generated by the soft-collision process, leading to the source term I. Similarly to eq.(18), the variance $\chi_k(\rho_1,\rho_2;t)$ satisfies scaling

$$\chi_k(\rho_1,\rho_2;t) = (3Q/4\pi m_3)R(t)^{-5}\chi(y,\rho_2,\rho_1;Q^{1/2}), \tag{27}$$

where the temporal power laws for $R(t)$ and $L(t)$ are given by eq.(19). Differentiating eq.(27) with respect to t and inserting it into eq.(25), one can also easily find a second-order differential equation for $\chi_k(\rho_1,\rho_2;t)$.

The equation of motion for $S_k(t)$ is now derived from the linear variance equation (25). In fact, by applying the projection operator method and making a Markoffian approximation, we obtain the following new linear equation of motion for $S_k(t)$[29]:

$$(d/dt)S_k(t) = 2(\alpha D/R(t)^3)[h(x,y) - \gamma(x,y)]S_k(t) + 2\alpha D(4\pi Q/3m_3)\Gamma(x,y) \tag{28}$$

with the instantaneous term

$$h(x,y) = 3\int_0^\infty d\rho'<\rho^3 Y(\rho x)H(y,\rho',\rho)>\mu(\rho'x)/w(x), \tag{29}$$

and the damping term

$$\gamma(x,y) = \int_0^\infty d\rho'<\rho^3 Y(\rho x)H(y,\rho',\rho>\rho'^{-2}(d/d\rho')\eta(x,y,\rho')/w(x), \tag{30}$$

where $x=kR(t)$, $w(x)=<\rho^6 Y(\rho x)^2>$, $\mu(x)=-\sin x/x$, and $\eta(x,y,\rho')$ a fluctuating force given by eq.(4.9) of Ref.29. The source term $\Gamma(x,y)$ is given by

$$\Gamma(x,y) = w(x)\gamma(x,y) + 3(3Q/m_3)^{1/2}\int_0^\infty d\rho\int_0^\infty d\rho'\rho^3\Psi(\rho x)\mu(\rho'x)I(y,\rho',\rho). \tag{31}$$

Thus the source term I consists of two types of terms related to the two kinds of fluctuations discussed above. The first term $w\gamma$ is related to the initial fluctuations and the second term to the fluctuations generated by the soft-collision process.

Use of eqs.(23) and (27) leads to the scaling form

$$S_k(t) = (4\pi Q/3m_3)R(t)^3 S(x,y;Q^{1/2}), \tag{32}$$

where $S(x,y)$ is a time-independent function of x and y. Thus, the structure function $S_k(t)$ is scaled by two kinds of characteristic lengths, $R(t)$ and $L(t)$ which satisfies eq.(19). We should remark here that $S(x,y)$ depends on Q not only through $L(t)$ but also through the drift and collision terms. This situation is quite different from the other theories[18-20], where it depends only through the correlation length. Differentiating eq.(32) with respect to t and using eq.(28), we then find

$$[3 + x(d/dx)]S(x,y) = (6\alpha D/K)[\{h(x,y) - \gamma(x,y)\}S(x,y) + \Gamma(x,y)]. \tag{33}$$

Since this equation does not contain any adjustable parameters, the scaled structure function $S(x,y)$ can completely be determined by solving it so as to satisfy the boundary condition $S(x=0,y=0)=0$, which comes from the fact that total droplet volume is conserved[29].

As mentioned before, the soft-collision process plays a different role in coarsening from the collisionless drift process. To see this, let us discuss the asymptotic form of $S(x,y)$ in two regions; $y>>x>>1$ and $1>>y>>x$. At small x the fluctuations are governed by the process with the chracteristic length $L(t)$ and the role of the collision process is not qulitatively different from that of drift process. Both processes lead to the asymptotic solution

$$S(x,y) \sim y^2, \quad \text{for } 1>>y>>x, \tag{34}$$

which ensures volume conservation. At large x, however, the fluctuations are dominated by the process with the characteristic length $R(t)$ and hence the role of the collision process is quite different from the other. In fact, the collision term makes h, γ and Γ x^2 times larger than the drift term does. This is due to the collision operator $C(\rho)$, which averages out the short-wavelength components of the fluctuations. Thus the collision process leads to the following asymptotic solution at smaller values of x than the drift process does:

$$S(x,y) \sim x^{-4}, \quad \text{for } y>>x>>1, \tag{35}$$

where x^{-4} tail results from a well-defined smooth interface for a droplet and is known as Porod's law.

CONCLUSION

In this article we have discussed the systematic theory recently developed for the late stage processes of phase separation. There are three important features in comparing it with others. First, it enables us to derive not only the kinetic equation for the single droplet distribution function $f(R,t)$ but also the equation of motion for the structure function $S_k(t)$ from a new unifying point of view. Secondly, it is a first-principles theory which yields the explicit equations for the scaling functions $p(\rho)$ and $S(x,y)$. Thirdly, it stresses that the role of the soft-collision process in the late stage is quite different from that of the collisionless drift process. In fact, the former makes $p(\rho)$ broader and flatter and $S(x,y)$ narrower and higher than the latter does at the same value of Q[28,29]. The present results for $f(R,t)$ are in better agreement with recent computer simulations[24,25] and experiments[27,30-32] than those of the other theories, where the soft-collision process is left out. Those for $S_k(t)$ are also expected to be in better agreement with experiments[16,33]. This confirms the importance of soft-collision process. More experimental studies on this subject may be encouraged.

REFERENCES

1. J.D. Gunton, M. San Miguel and P.S. Shani, in: "Phase Transition and Critical Phenomena," vol.8, C. Domb and J.L. Lebowitz, eds. Academic press, London,(1983).
2. K. Binder and D. Stauffer, Phys. Rev. Lett. 33:1006 (1974).
3. W. Ostwald, Z. Phys. Chem. 37:385 (1901).
4. I.M. Lifshitz and V.V. Slyozov, J. Phys. Chem. Solids 19:35 (1961)
5. C. Wagner, Z. Electrochem. 65:581 (1961).
6. A.J. Ardell, Acta. Met. 20:61 (1972).
7. J.J. Wein and J.W. Cahn, Materals Research 6:151 (1973).
8. A.D. Brailsford and P. Wynblatt, Acta. Met. 27:489 (1979).
9. P.W. Voorhees and M.E. Glicksman. Acta. Met. 32:2001, 2013 (1984).
10. J.A. Marqusee and J. Ross, J. Chem. Phys. 80:536 (1984).
11. J.L. Lebowitz, J. Marro and M.H. Kalos, Acta. Met. 30:297 (1982).
12. M. Hennion, D. Ronzaud and P. Guyot, Acta. Met. 30:599 (1982).
13. P. Fratzl, J.L. Lebowitz, J. Marro and M.H. Kalos, Acta. Met. 31:1849 (1983).
14. S. Katano and M. Iizumi, Phys. Rev. Lett. 52:835 (1984).
15. S. Komura, K. Osamura, H. Fujii and T. Takeda, Phys. Rev. B31:1278 (1985).
16. K. Osamura, H. Okuda and S. Ochiai, Preprint.
17. H. Furukawa, Prog. Theor. Phys. 59:1072 (1978).
18. P.A. Rikvold and J.D. Gunton, Phys. Rev. Lett. 49:286 (1982).
19. T. Ohta, Ann. of Phys. 158:31 (1984).
20. H. Tomita, Prog. Theor. Phys. 71:1405 (1984).
21. M. Tokuyama and K. Kawasaki, Physica 123A:386 (1984).
22. M. Tokuyama and R.I. Cukier, J. Chem. Phys. 76:6202 (1982).
23. K. Kawasaki and T. Ohta, Physica, 118A:175 (1983).
24. C.W.J. Beenakker, Preprint.
25. Y. Enomoto, K. Kawasaki and M. Tokuyama, Acta. Met. 35:907 (1987).
26. M. Tokuyama, K. Kawasaki and Y. Enomoto, Physica, 134A:323(1986).
27. Y. Enomoto, K. Kawasaki and M. Tokuyama, Acta. Met. 35:915 (1987).
28. Y. Enomoto, M. Tokuyama and K. Kawasaki, Acta. Met. 34:2119(1986).
29. M. Tokuyama, Y. Enomoto and K. Kawasaki, Physica 143A:183 (1987).
30. P.K. Rastogi and A.J. Ardell, Acta. Met. 19:321 (1971).
31. T. Hirata and D.H. Kirkwood, Acta. Met. 25:1425 (1977).
32. T. Eguchi, Y. Tomokiyo and S. Matsumura, Preprint.
33. K. Osamura, these proceedings.

ON VOLUME PHASE TRANSITION IN GELS

Akira Onuki

Research Institute for Fundamental Physics
Kyoto University, Kyoto 606

1. Introduction

Gels can undergo a volume phase change with varying some parameter such as the temperature T, the degree of ionization, or the pH.[1] The phase transition has been observed under the condition of zero osmotic pressure. It is mostly of first order and the critical point can be approached by appropriately adjusting external conditions. The equilibrium aspect can be well understood by Flory's theory[2] as shown by Tanaka. The aim here is to present preliminary analysis on (i) the phase transition occurring in anisotropically deformed gels and on (ii) the critical dynamics.

Within the Flory scheme the mixing free energy ΔF_M is of the form

$$\Delta F_M / k_B T = a^{-3} \int d\mathbf{r} [(1-\phi)\ln(1-\phi) + \chi\phi(1-\phi) + \frac{1}{36\phi} a^2 |\nabla\phi|^2] , \qquad (1)$$

where a is the monomer size, $\phi(\mathbf{r})$ is the polymer volume fraction, χ is the polymer-solvent interaction parameter dependent on T, and the last term accounts for density inhomogeneities.[3] The network structure gives rise to the elastic free energy. Flory's expression is[2]

$$\Delta F_{e\ell} / k_B T = \sum_{i=1}^{\nu_e} [\frac{1}{2}(\alpha_{i1}^2 + \alpha_{i2}^2 + \alpha_{i3}^2) - \frac{3}{2} - \frac{1}{2}\ln(\alpha_{i1}\alpha_{i2}\alpha_{i3})] , \qquad (2)$$

where i denotes the i-th chain and ν_e is the total effective chain number. The α_{i1}, α_{i2} and α_{i3} are the elongation ratios of the i-th chain along the three principal axes of the deformation under consideration. In the isotropic case we have $\alpha_{i1} = \alpha_{i2} = \alpha_{i3} = (\phi_0/\phi)^{1/3}$ and

$$\Delta F_{e\ell} / k_B T = \frac{1}{2}\nu_0 \int d\mathbf{r}_0 [3(\phi_0/\phi)^{2/3} - 3 + \ln(\phi/\phi_0)] . \qquad (3)$$

Here, $\nu_0 \equiv \nu_e/V_0$, V_0 and ϕ_0 are the volume and the volume fraction of the relaxed network, respectively, and $\mathbf{r}_0$ is the coordinate fixed to the relaxed system. In the expanded state the position vector $\mathbf{r}$ is related to $\mathbf{r}_0$ by $\mathbf{r} = (\phi_0/\phi)^{1/3}\mathbf{r}_0$. Thus,

$$\Delta F_{e\ell} / k_B T = \frac{1}{2}\nu_0 \int d\mathbf{r} [3(\frac{\phi}{\phi_0})^{1/3} - 3\frac{\phi}{\phi_0} + \frac{\phi}{\phi_0}\ln(\frac{\phi}{\phi_0})] . \qquad (4)$$

Furthermore, when the gel is ionized, the mixing free energy of mobile ions is given by[2,4]

$$\Delta F_{ion}/k_B T = \frac{1}{2}\nu_0 f_i \int d\mathbf{r} \; \frac{\phi}{\phi_0} \ell n(\frac{\phi}{\phi_0}) \; , \tag{5}$$

where f_i is the number of dissociated hydrogen ions per chain.

Now the statics is determined by the free energy

$$F = \Delta F_M + \Delta F_{e\ell} + \Delta F_{ion} \tag{6}$$

$$= k_B T \int d\mathbf{r} [f_0(\phi) + \frac{1}{36\phi} a^{-1} |\nabla\phi|^2] \tag{7}$$

The second line is the definition of $f_0(\phi)$. Remarkably there exists a critical line of second order phase transition in the χ-f_i-ϕ space.[4]

If the gel is homogeneous, the osmotic pressure is given by

$$\pi = k_B T [\phi \frac{\partial}{\partial\phi} f_0 - f_0] \; . \tag{8}$$

This determines the average volume fraction for a given π in homogeneous, isotropic states.

2. Phase Transition in Anisotropically Deformed States

In the isotropic case $I \equiv \alpha_1^2 + \alpha_2^2 + \alpha_3^2$ in (2) has been set equal to $3(\phi_0/\phi)^{2/3}$. For general deformations the following is the simplest approximation[5,6]:

$$I = \sum_{\ell,m} (\frac{\partial X_\ell}{\partial x_m^0})(\frac{\partial X_\ell}{\partial x_m^0}) \; , \tag{9}$$

where x_m^0 and X_ℓ are the coordinates before and after the deformation, respectively. Here we are interested in the fluctuations around an anisotropically deformed state $\{x_\ell^s\}$ which is maintained by external forces acting on the boundary. Then the displacement vector $\{u_\ell\}$ from the new reference state is defined by

$$X_\ell = x_\ell^s + u_\ell \; . \tag{10}$$

We assume that $\partial x_\ell^s/\partial x_m^0$ are constant in space. It is convenient to choose Cartesian coordinates such that

$$\sum_\ell \frac{\partial x_m^s}{\partial x_\ell^0} \frac{\partial x_n^s}{\partial x_\ell^0} = \mu_m \delta_{mn} \; . \tag{11}$$

In particular, when the gel is uniaxially stretched, we have

$$\mu_x = \alpha_x^2 \; , \; \mu_y = \mu_z = \alpha_y^2 = \alpha_z^2 \equiv \alpha_\perp^2 \; . \tag{12}$$

The u_ℓ will be regarded as functions of x_m^s and, as a new reference frame, x_m^s will be simply written as x_m. Then,

$$I = \sum_\ell \mu_\ell + 2\sum_\ell \mu_\ell \frac{\partial u_\ell}{\partial x_\ell} + \sum_{\ell,m} \mu_m \frac{\partial u_\ell}{\partial x_m} \frac{\partial u_\ell}{\partial x_m} \; . \tag{13}$$

On the other hand, the volume change is expressed as

$$\phi_0/\phi = \text{Det}\{\frac{\partial X_\ell}{\partial x_m^0}\} = (\phi_0/\phi_s)\text{Det}\{\delta_{\ell m} + \frac{\partial u_\ell}{\partial x_m}\} \ , \tag{14}$$

where $\phi_0/\phi_s=(\mu_1\mu_2\mu_3)^{1/2}$ and Det $\{\cdots\}$ denotes the determinant of the matrix. Thus, to first order in the deviations, we find

$$\delta\phi/\phi_s = -\sum_\ell \frac{\partial}{\partial x_\ell} u_\ell + \cdots \ . \tag{15}$$

Now by setting $\nu_s=\nu_0\phi_s/\phi_0$ the free energy can be expressed as

$$F/k_B T = \int d\mathbf{r}[f(\phi) + \frac{1}{36\phi} a^{-1}|\nabla\phi|^2 + \frac{1}{2}\nu_s \sum_{\ell,m} \mu_m \frac{\partial u_\ell}{\partial x_m} \frac{\partial u_\ell}{\partial x_m}] \tag{16}$$

Let us then calculate the variance $\chi_\mathbf{k}=<|\phi_\mathbf{k}|^2>$ in the mean field theory, $\phi_\mathbf{k}$ being the Fourier component of $\phi(\mathbf{r})$. Note that the transverse displacement $\mathbf{u}_{\perp\mathbf{k}}=\mathbf{u}_\mathbf{k}-k^{-2}(\mathbf{k}\cdot\mathbf{u}_\mathbf{k})\mathbf{k}$ is decoupled from $\phi_\mathbf{k}$ in (16). Thus,[†]

$$\chi_\mathbf{k} = 1/[\tau + K_s k^2 + \nu_s\sum_\ell \mu_\ell \hat{k}_\ell^2] \ , \tag{17}$$

where τ is a constant arising from $f(\phi)$ in (16), $K_s \equiv 1/(18a\phi_s)$, and $\hat{k}_\ell \equiv k^{-1}k_\ell$. Strikingly $\chi_\mathbf{k}$ depends on the direction $\hat{\mathbf{k}}$ even in the limit $k\to0$. Particularly, in the uniaxial case (12), we have

$$\chi_\mathbf{k} = 1/[\bar{\tau} + K_s k^2 + \nu_s(\alpha_x^2-\alpha_\perp^2)\hat{k}_x^2] \ , \tag{18}$$

where $\bar{\tau}=\tau+\nu_s \alpha_\perp^2$. Notice that (18) coincides with the variance of Ising spin systems with dipolar interaction.[8] In the shear deformation case, $x=x_0+\gamma y_0$, $y=y_0$, $z=z_0$, we find

$$\chi_\mathbf{k} = 1/[\bar{\tau} + K_s k^2 + \nu_s(2\gamma\hat{k}_x\hat{k}_y + \gamma^2\hat{k}_x^2)] \ , \tag{19}$$

where $\bar{\tau}=\tau_0+\nu_s$. Therefore we expect strong anisotropy of light scattering near the instability (spinodal) point where $\bar{\tau}$ in (18) or (19) is very small.

Next we give a general expression for the stress tensor consistent with (16) and including the fluctuation contributions. It consists of two parts, $\Pi_{ij}=\Pi_{ij}^{(1)}+\Pi_{ij}^{(2)}$. The first part is determined by ϕ as

$$\Pi_{ij}^{(1)}/k_B T = [P_0 + \phi\frac{\partial}{\partial\phi}f - f - K\phi\nabla^2\phi]\delta_{ij} + K(\frac{\partial}{\partial x_i}\phi)(\frac{\partial}{\partial x_j}\phi) \ , \tag{20}$$

where P_0 is a constant, $f(\phi)$ is defined in (16), and $K\equiv1/(18a\phi)$. The second part is the elastic contribution.[5] Because its general form is somewhat complicated, we only give an expression valid for the uniaxial case,

---

[†] The term $K_s k^2$ in (17) should also be changed to a form dependent on $\hat{\mathbf{k}}$ at small k as claimed by Daoudi.[7] We are neglecting such an effect here.

$$\Pi_{ij}^{(2)}/k_BT = -\nu_0(\frac{\phi}{\phi_o})[\mu_i\delta_{ij} + \mu_i\frac{\partial u_j}{\partial x_i} + \mu_j\frac{\partial u_i}{\partial x_j} + \sum_\ell\mu_\ell\frac{\partial u_i}{\partial x_\ell}\frac{\partial u_j}{\partial x_\ell}] \ , \quad (21)$$

The above results are consistent with those in the theory of rubber elasticity[2] (provided that the fluctuation terms are omitted). An expression valid for shear will be given in another paper.

Summary

(i) The phase transition can be greatly altered by homogeneous deformations. The light scattering intensity and mechanical properties have very aniso-tropic structures especially near the spinodal. Furthermore we should encounter unique phase separation processes under such anisotropic situa-tions. The domain structure emerging due to the thermodynamic instability will be nearly lamellar in the case of strong uniaxial compression and nearly cylindrical in the cases of strong uniaxial stretching and strong shear.

(ii) The study in this direction also seems to be indispensable to under-stand pattern formation on the gel surface.[1] Upon swelling or shrinking of a gel, a layer near the surface must be uniaxially stretched or compressed in the course of the process.

3. Construction of a Dynamic Model

For simplicity we assume that the system is isotropic and near the critical point. The order parameter ϕ fluctuates around an average ϕ_s. The free energy is then given by (7) plus the elastic contribution due to the transverse displacement $u_\perp$:

$$F = k_BT\int d\mathbf{r}[f_0(\phi) + \frac{1}{2}K_s|\nabla\phi|^2 + \frac{1}{2}\nu_s\sum_{\ell,m}(\frac{\partial}{\partial x_m}u_{\ell\perp})^2] \ , \quad (22)$$

where K_s and ν_s are constants. To set up dynamic equations we follow the ideas in Refs.[9] and [10]: When the gel velocity $\mathbf{v}_g$ and the solvent velocity $\mathbf{v}_f$ are different, the gel should be acted on by a frictional force of the form $\zeta(\mathbf{v}_g-\mathbf{v}_f)$ per unit volume, $\zeta(\sim6\pi\eta_0a^{-2}\phi)$ being the friction constant. This force is then balanced with the force due to gel deformations, $-\nabla\cdot\Pi$, Π being given by (20) and (21). Thus, $\mathbf{v}_g-\mathbf{v}_f\simeq-\zeta^{-1}\nabla\cdot\Pi$.

After some manipulations we arrive at the following equations

$$\frac{\partial}{\partial t}\phi = -\nabla\cdot(\phi\mathbf{v}_f) + \lambda_0\nabla^2\frac{\delta F}{\delta\phi} \ , \quad (23)$$

$$\rho_0\frac{\partial}{\partial t}\mathbf{v}_f = -[\phi\nabla\frac{\delta F}{\delta\phi} + \frac{\delta F}{\delta u_\perp}]_\perp + \eta_0\nabla^2\mathbf{v}_f \ , \quad (24)$$

$$\frac{\partial}{\partial t}u_\perp = \mathbf{v}_f - (\lambda_0/\phi_s)\frac{\delta F}{\delta u_\perp} \ . \quad (25)$$

where $\nabla\cdot\mathbf{v}_f=\nabla\cdot\mathbf{u}_\perp=0$, $\lambda_0(=\phi/\zeta)$ and η_0 are the kinetic coefficients, ρ_0 is the average mass density of the solvent, and $[\cdots]_\perp$ in (24) denotes taking the transverse part of the vector. The first equation (23) has been derived from $\partial\phi/\partial t=-\nabla\cdot(\phi\mathbf{v}_f)-\nabla\cdot\phi(\mathbf{v}_g-\mathbf{v}_f)$. We may add random source terms to the right hand sides of (23)$\sim$(25) which are related to the kinetic coefficients via the fluctuation-dissipation relations. Then the equilibrium is surely given by $\exp[-F/k_B T - \frac{1}{2}\int dr\rho_0 \mathbf{v}_f^2/k_B T]$.

We readily notice the following :

(i) If the convection term $-\nabla \cdot (\phi\mathbf{v}_f)$ in (23) is neglected, the decay rate of ϕ is

$$\Gamma_q^0 = (k_B T)^{-1}\lambda_0 \left(\frac{\partial \pi}{\partial \phi}\right)q^2(1 + q^2\xi^2) , \tag{26}$$

where ξ is the correlation length and $(\partial\pi/\partial\phi)\propto\xi^{-2}$.

(ii) The transverse parts, $\mathbf{u}_\perp$ and $\mathbf{v}_f$, are strongly coupled at long wavelengths, $q<q_c\equiv(\rho_0\mu)^{1/2}/\eta_0$, $\mu=k_B T\nu_s$ being the shear modulus. There, the gel and the solvent move in phase and we have two shear sound modes oscillating as $\exp(\Omega_q t)$ with

$$\Omega_q \cong \pm i(\mu/\rho_0)^{1/2}q - \frac{1}{2}(\eta_0/\rho_0)q^2 . \tag{27}$$

(iii) The dynamics of ϕ is crucially dependent on the magnitude of the shear modulus μ. If μ is large, $\mathbf{v}_f$ must be small since $\mathbf{v}_f\cong\mathbf{v}_g$ and the convective term in (23) will be negligible. However, if μ is very small, the kinetic coefficient λ_0 will be increased considerably by the mode coupling effect. This seems to explain Tanaka's finding[11] of an apparent critical divergence of λ_0. The oscillating behavior (27) indicates that λ_0 does not diverge at the critical point.

(iv) Also in the spinodal decomposition process $\mathbf{u}_\perp$ and $\mathbf{v}_f$ can be important if μ is very small. But details are not well understood at present.

Remarks

In this short note I present only starting equations. Analysis of the critical dynamics and the spinodal decomposition will be reported in the near future. It is also of great interest to investigate such dynamical processes in anisotropically deformed cases. The pattern formation on the surface is of course very intriguing. Indeed there remain a rich variety of unexplored phenomena in gels.

REFERENCES

1. T. Tanaka, Physica 140A (1986) 261.
2. P.J. Flory, *Principles of Polymer Chemistry* (Cornell Univ. Press, Ithaca, N.Y., 1953).
3. P.G. de Gennes, *Scaling Concepts in Polymer Physics* (Cornell Univ. Press, Ithaca, N.Y., 1980).
4. T. Tanaka, D. Fillmore, S-T. Sun, I. Nishino, G. Swislow and A. Shah, Phys. Rev. Lett. 45 (1980) 636.
5. R.S. Rivlin, *Rheology*, vol.1 (edited by F. Eirich, Academic Press, N.Y., 1956).
6. R. Kubo, J. Phys. Soc. Japan 3 (1948) 312.
7. S. Daoudi, J. de Physique 38 (1977) 1301.
8. A.I. Larkin and D.E. Khemelnitskii, Sov. Phys. J.E.T.P. 29 (1969) 1123.
9. T. Tanaka, L.O. Hocker and G.B. Benedek, J. Chem. Phys. 59 (1973) 5151.
10. P.G. de Gennes, Macromolecules 9 (1976) 594.
11. T. Tanaka, S. Ishiwata and C. Ishimoto, Phys. Rev. Lett. 38 (1977) 771.

MORPHOLOGY DYNAMICS OF POLYMER

BLENDS AND BLOCK COPOLYMER MELTS

Kyozi Kawasaki and Ken Sekimoto

Department of Physics, Faculty of
Science, Kyushu University 33
Fukuoka 812, Japan

SUMMARY OF OUR PREVIOUS WORKS

Polymer melt systems present some novel features for theorists and experimentalists concerned with dynamics of ordering processes[1,2]. First of all, slowness of the process permits studies of the entire stage of kinetics including the early linear stage. Secondly, many different kinds of polymer systems like blends of homopolymers, block copolymers of all sorts, liquid crystal polymers etc. show a variety of ordered phase structures and hence ordering kinetics. Finally, complexity of chain dynamics presents challenging theoretical problems which still remain to be solved.

Recently we have been attempting to apply the reptation model to morphology dynamics of polymer melt systems. The reptation model developed by de Gennes, Doi, Edwards and others[3,4] has been successful mostly for explaining viscoelastic properties as well as the self-diffusion of homopolymer melts, and its extension to polymer systems with mutually incompatible heterogeneous components is by no means straightforward and without doubt. Nevertheless, in view of absence of any other dynamical model of polymer melts with the same degree of success, consequences of the reptation model on ordering kinetics should be of considerable interests. Since the detailed accounts of our works at the time of writing this article have been or will be published elsewhere[5,6,7] we will be rather brief in describing these works except for some new results for block copolymers to be discussed at the end.

We consider binary systems and start from the following equations for the density variables $\rho_K(r)$ where K=A,B denotes monomer species[5,6,7]:

$$\frac{\partial}{\partial t} \rho_K(r) = -L \int dr' \sum_{K'} \Lambda^{KK'}(rr') \frac{\delta H\{\rho_A,\rho_B\}}{\delta \rho_{K'}(r')} \tag{1}$$

where L is a kinetic coefficient, H the system free energy functional and $\Lambda^{KK'}(rr')$ are certain functions related to the joint probability distributions of chain configurations in the local equilibrium state with given density variables. For instance, for the AB diblock copolymer system we have

$$\Lambda^{AA}(rr') = [p_A(r)+p_J(r)]\delta(r-r')-P_{AJ}(rr')-P_{JA}(rr') \qquad (2)$$

$$\Lambda^{AB}(rr') = P_{AJ}(rr')+P_{JB}(rr')-P_{AB}(rr')-p_J(r)\delta(r-r') \qquad (3)$$

and similarly for $\Lambda^{BA}(rr')$ and $\Lambda^{BB}(rr')$. Here the P's are the properly normalized joint probability distribution functions where A,B and J denote polymer ends with species A and B and AB junction point, respectively. The microscopic chain dynamics underlying (1) is the reptation dynamics biased by the mean field arizing from short-range segment interactions.

The density equations of the form (1) supplemented with the incompressibility condition and the thermal noise term can be used to describe various stages of (micro) phase separation kinetics. In particular, in the late stage where A-rich and B-rich domains are well separated, (1) can be reduced to the equation of motion of domain walls. For polymer blends[5] we predicted existence of a regime in the late stage where the average linear domain size grows as $t^{1/2}$. For block copolymers, however, the reptation dynamics alone may not be sufficient to obtain the domain wall equations[6].

RECOVERY OF DEFORMATION OF LAMELLAR STRUCTURE

We now turn to a new result on dynamics of weak and gentle deformations of lamellar structure of the micro phase separated diblock copolymers. Then there are two types of deformations: slow variations of the lamellar spacing (or repeat unit ℓ) and gentle bending of the lamellae. These are characterized by the wave vectors k and k_n which are parallel and perpendicular to the lamellae, respectively, whose magnitudes are much smaller than ℓ^{-1}. Such deformations are familiar in many layered structures and have been thoroughly studied, for instance, in the Rayleigh-Benard convection[8].

In the following we obtain the decay rate $\Omega(k,k_n)$ of the infinitesimal deformations with k and k_n near the micro phase separation point with weak segregation[6,7,9,10]. In doing so we heavily quote from our previous works to be referred to as I, K, OK and KO, respectively. In I we obtained the expression

$$\Omega(k,k_n) = (L\ell^2/S^2)\Lambda(k,k_n)B(k,k_n) \qquad (4)$$

Here, denoting by n the coordinate normal to the lamellae and the differentiation by a prime,

$$S \equiv \int_0^\ell \psi'(n)^2\,dn \qquad (5)$$

where $\psi(n)$ is the order parameter profile in the undeformed lamellar state and

$$\Lambda(k,k_n) = \frac{1}{\ell^2}\int_0^\ell dn \int_0^\ell dn'\,\Lambda(k,k_n;nn')\psi'(n)\psi'(n') \qquad (6)$$

where $\Lambda(k,k_n;nn')$ is ℓ-periodic in n and n' and is related to the function $\Lambda(rr')$ similar to (2) and (3) which is appropriate for the order parameter. $B(k,k_n)$ is the coefficient of the expansion of H in Fourier components of the deformation field $u(k,k_n)$:

$$H = \frac{1}{2} \frac{1}{(2\pi)^3} \int dk \int dk_n \, B(k,k_n) |u(k,k_n)|^2 + \ldots \tag{7}$$

For weak segregation $\psi'(n)$ is small and is [9]

$$\psi'(n) = -(2\pi \, \psi_e /\ell)\sin(2\pi n/\ell) \tag{8}$$

The $\Lambda(kk_n;nn')$ in (6) can be obtained for the homogeneous phase and we find

$$\Lambda(kk_n;nn') = \sum_{m=-\infty}^{\infty} \Lambda(k,k_n + \frac{2\pi m}{\ell}) \exp[i \, \frac{2\pi m}{\ell} (n-n')] \tag{9}$$

where $\Lambda_{1/2}$ is the Fourier transform [7] of $\Lambda(r,r')=\Lambda(r-r')$. Denoting $q=(k^2+k_z^2)^{1/2}$ we find

$$\Lambda(k,k_z) = \lambda(q) \equiv 2\eta_q(N_A)\eta_q(N_B)[1- \frac{1}{4} \eta_q(N_A)\eta_q(N_B)]/\eta_q(N_A+N_B) \tag{10}$$

with

$$\eta_q(N) \equiv 1 - \exp(- \frac{1}{6} Nb^2 q^2) \tag{11}$$

where N_A and N_B are the polymerization indices of the two blocks of a copolymer chain and b the segment length. On the other hand $B(k,k_n)$ for small $|k|$ and $|k_n|$ is obtained in KO which takes the following form if we supplement with proper coefficients in front by consulting with the results of OK:

$$B(k,k_n) = B_n k_n^2 + B_\perp k^4 \tag{12}$$

with

$$B_n \equiv \frac{T}{4f(1-f)} \frac{b^2}{\rho_0} 2(2\pi \, \psi_e /\ell)^2 \tag{13}$$

$$B_\perp \equiv \frac{T}{4f(1-f)} \frac{b^2}{\rho_0} \frac{1}{2} \psi_e^2 \tag{14}$$

where $f \equiv N_A/(N_A+N_B)$, ρ_0 the total constant monomer number density and T the Boltzmann constant times the absolute temperature. In view of (8) only the terms with $m=\pm1$ of (9) survives in (6), which permits us to take $k=k_n=0$ in (6). Finally, we note that L is proportional to the chain diffusion constant along the tube, which is inversely proportional to $N=N_A+N_B$, and also to the chain number density ρ_0/N. Hence we can put, taking dimensionality also into account,

$$L = D_1\rho_0/T(Nb)^2 \tag{15}$$

where D_1 is the diffusion constant of a single segment.

Assembling all these findings together we finally obtain

$$\Omega(k,k_n) = D_n k_n^2 + D_\perp k^4 \tag{16}$$

where

$$D_n \equiv \frac{D_1}{N^2} \Delta_n(f) \tag{17}$$

$$D_\perp \equiv \frac{D_1 b^2}{4N} \Delta_\perp(f) \tag{18}$$

and

$$\Delta_n(f) \equiv \frac{\lambda_1}{f(1-f)} \ , \tag{19}$$

$$\Delta_\perp(f) \equiv (\ell/2\pi b)^2 \ N^{-1} \ \Delta_n(f) \tag{20}$$

with λ_1 the dimensionless number defined with (10) by

$$\lambda_1 \equiv \lambda(2\pi/\ell) \tag{21}$$

We have evaluated the functions $\Delta_n(f)$ and $\Delta_\perp(f)$ which are shown in Fig.1 where[11] the value of ℓ as a function of f was taken from Fig.2 in Leibler's work[8]. Here D_n is the phase diffusion coefficient and is to be compared with the self-diffusion constant D_G. The latter is given by the form similar to (17) with Δ_n replaced by 1/3, which is considerably smaller than Δ_n .

The first and second terms of (16) describe, respectively, the decay rate of dilation (or compression) type and bend type deformations of lamellar structure, respectively. In this weak segregation case we do not have difficulties encountered in the case of strong segregation[6]. We strongly encourage experimental test of our theoretical results obtained here, which will shed light on applicability of reptation ideas to morphology dynamics of block copolymer melts.

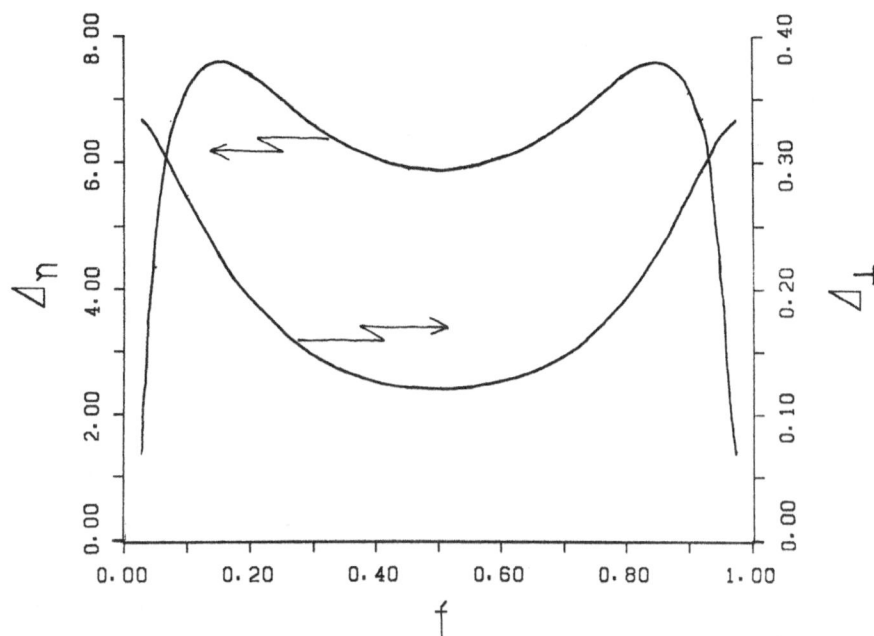

Fig.1 The behaviers of Δ , (19), and
Δ , (20), as functions of f.

Reference

1. T. Hashimoto, Structure formation in polymer mixtures by spinodal
   decomposition, in : "Current Topics in Polymer Science, Vol.II", R. M.
   Ottenbrite, L.A. Utracki and S. Inoue, eds. Hanser Publishers, Munich
   (1987), and also this proccedings.
2. T. Hashimoto, "Time-resolved small angle X-ray scattering studies on
   kinetics and molecular dynamics of order-disorder transition of block
   polymers, in : "Physical Optics of Dynamic Phenomena and Processes in
   Macromolecular Systems", B. Sedlacek, ed., Walter de Gruyter & Co.,
   Berlin (1985).
3. P.G. de Gennes, "Scaling Concepts in Polymer Physics", Cornell
   University Press, Ithaca (1979).
4. M. Doi and S.F. Edwards, "Theory of Polymer Dynamics", Oxford
   University Press, Oxford (1986).
5. K. Kawasaki and K. Sekimoto, Dynamical theory of polymer melt
   morphology, Physica A, 143: 349 (1987).
6. K. Kawasaki and K. Sekimoto, Morphology dynamics of block copolymer
   systems, Physica A (submitted)
7. K. Kawasaki, Ordering kinetics in quasi-one-dimensional systems and
   polymer melts, in : "Competing Interactions and Microstructures:
   Statics and Dynamics", A. Bishop, R. Heffner and R. LeSar, eds.,
   Springer, Heidelberg (1987).
8. J.E. Wesfreid and S. Zaleski "Cellular Structures in Instabilities"
   Springer, Heidelberg (1984).
9. T. Ohta and K. Kawasaki, Equilibrium morphology of block copolymer
   melts, Macromolecules, 19: 2621 (1986).
10. K. Kawasaki and T. Ohta, Phase Hamiltonian in periodically modulated
    systems, Physica A, 139: 223 (1986).
11. L. Leibler, Theory of microphase separation in block copolymers,
    Macromolecules, 13: 1602 (1980).

TOWARDS A THEORY OF INTERFACIAL PATTERN FORMATION

David A. Kessler* and Herbert Levine**

*Department of Physics, University of Michigan
Ann Arbor, MI 48109

**Schlumberger-Doll Research, Old Quarry Road
Ridgefield, CT 06877

INTRODUCTION

An understanding of the origin of snowflake shapes, and other such non-equilibrium growth patterns[1], has long eluded physicists. Although the governing differential equations and boundary conditions have been known for some time[2,3], at least to a working approximation, the nonlinearities and instabilities present therein have made it difficult to understand even qualitatively how these shapes arise. In a variety of physical systems including solidification, multiphase fluid flow, electrochemical deposition etc. one finds interfacial patterns that in one or another aspect vary between compact and branched, between symmetric and irregular, and between stable and unstable. Recent experimental and theoretical developments have revealed unexpected similarities among these ostensibly different physical systems and viable calculational approaches to the elucidation of these issues. This article aims to review these developments, in their current incomplete state, placing particular emphasis on dendritic crystal growth.

DENDRITIC CRYSTAL GROWTH

In this section, we will focus on the simplest set of assumptions that one can make in trying to describe the growth of a dendritic crystal such as that shown in Fig. 1. We imagine the crystal immersed in a supercooled melt, which we take to be completely free of any chemical impurities. Furthermore, we assume that the rate at which atoms can attach to the interface is sufficiently rapid as to ensure local thermodynamic equilibrium. This idealized situation can be approximately reached by using the organic liquid succinonitrile[4] which solidifies into a plastic crystal; for this substance attachment kinetics effects have been measured to be small.

For the above system, the rate of crystal growth is controlled simply by the rate of heat diffusion. At the interface, the first order phase transition between liquid and solid leads to the emission of heat. This heat must be conducted away to a distant cold bath (at $T = T_m - \Delta$) in order for the growth to proceed. This leads to the conservation law condition

$$D\nabla^2 T - \dot{T} = \frac{L}{c_p} \int ds' v_n(s')\delta(\vec{x} - \vec{x}'(s'))$$ (1)

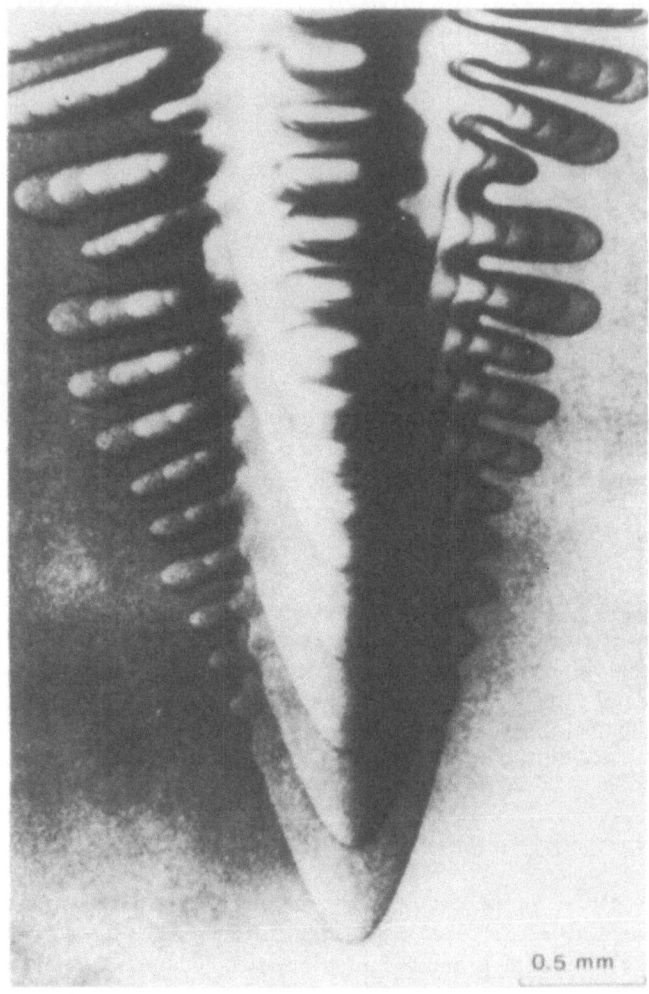

Fig. 1. Time-exposure sequence of the solidification of succinonitrile (Courtesy M. E. Glicksman, Rensselear Polytechnic
Institute).

where L is the latent heat, c_p the specific heat (both assumed equal for solid and liquid), v_n is the normal growth velocity of the interface $\vec{x}(s)$, and D the thermal diffusion constant. The final equation is the local equilibrium condition, which in two dimensions takes the form

$$T(\vec{x}(s)) = T_m \left(1 - \frac{\gamma(\theta)\kappa}{L}\right) \tag{2}$$

with κ the curvature and $\gamma(\theta)$ the surface tension, which explicitly depends on θ, the angle between the interface normal and the crystal symmetry axes. We will take the form

$$\gamma(\theta) = \gamma(1 - \epsilon \cos 4\theta) \tag{3}$$

for fourfold anisotropy of strength ϵ. Note that the crystallinity of the material enters the growth law only through the surface energy correction to the melting temperature.

The first step in treating the pattern evolution is the derivation of an integro-differential equation of motion for the interface. To do this, we solve eq. (1) for the temperature via a Green's function technique. We will use a quasistatic approximation which amounts to considering growth rates such that the velocity remains fairly constant (equal to v_0 in the $\hat{y}$ direction) over typical diffusion times. Then, defining $u = \frac{T-T_m+\Delta}{L/c_p}$ we find

$$u(\vec{x}, t) = \frac{1}{\pi} \int ds' v_n(s', t) K_0 \left(\frac{v_0}{2D} \mid \vec{x} - \vec{x}'(s') \mid\right) e^{-\frac{v_0}{2D}(y-y'(s'))}. \tag{4}$$

Finally, we evaluate u on the interface and rewrite eq. (2) as $u = \tilde{\Delta} - d_0(\theta)\kappa$, with $\tilde{\Delta} = c_p\Delta/L$, and $d_0 = \gamma c_p T_m/L^2$. The normal velocity is determined implicitly as a function of the interface position via these equations.

Before proceeding to dendrites, it is worthwhile recalling the reason that diffusion limited crystal growth gives rise to patterns. Imagine a planar interface moving at velocity v_0. It is easy to check by direct substitution into the above equation that this will be a solution if and only if $\tilde{\Delta} = 1$. Now we perturb the interface by assuming a small deformation $\delta_k \cos kx e^{\omega t}$. A simple computation[5] leads to the dispersion relation

$$\omega = \frac{v_0}{l} \left[A^{1/2}(1 - ld_0 k^2) - 1\right] \tag{5}$$

with $A \equiv 1 + (kl)^2$. where $l \equiv 2D/v_0$ is the diffusion length. For wavelength short compared to the diffusion length, this approximately becomes

$$\omega = v_0 \mid k \mid (1 - ld_0 k^2). \tag{6}$$

We see that there is a band of unstable wavenumbers where the growth rate is positive, and a critical length scale defined by the wavelength of the fastest growing mode

$$k_c^{-1} = \sqrt{3d_0 l}. \tag{7}$$

We therefore expect the system to develop some type of pattern in this range of length scales. Note too that in the absence of surface tension, $\tilde{d}_0 = 0$, the growth rate increases indefinitely for shorter wavelengths, and the system is wildly unstable. In this case, corresponding to the classical Stefan problem in applied mathematics, the physical problem is ill-posed.

Now, let us turn to the problem at hand, that of finding dendritic solutions. First, let us study the Ivantsov[6] limit, $d_0 = 0$. Then, there exist steady-state parabolic solutions $y(x,t) = y(x) + vt$, with $y(x) = -x^2/2\rho$. Substituting this ansatz into eq. (4), we derive the Ivantsov relation

$$\tilde{\Delta} = e^p \sqrt{\pi p} \; \mathrm{erfc}(\sqrt{p}) \tag{8}$$

with Peclet number p equal to $\rho v/2D$. This relationship (actually its three-dimensional generalization) is in good agreement with the measured properties of dendrites. There is however the outstanding deficiency that the undercooling only selects the *product* of tip radius and velocity, whereas in experiment one finds that both are unique and reproducible. This failure is not entirely surprising, as in the absence of surface tension the problem has only one length scale and there is no way to select a unique shape. Furthermore, as already noted the problem is not well-defined without some short distance cutoff. We therefore need some way to properly incorporate the effect of surface tension and thereby select the correct pattern.

Let us describe a simple method[7] (initially suggested by Langer[8], with some modification by Shraiman[9]) for doing this; a more rigorous treatment is also available in the work of Ben-Amar and Pomeau[10]. Consider the steady-state equation in the schematic form

$$F[y(x), \alpha] = 0 \tag{9}$$

where α is an abbreviation for the parameters entering into the equation and boundary conditions (in the present case d_0, ϵ and pe). Since the problem is translation invariant, if $y(x)$ is a solution so is $y(x) + y_0$ for any constant y_0. If we then expand $F[y(x) + y_0, \alpha]$ for infinitesimal y_0, we find $Ly_0 = 0$, where the operator $L \equiv \delta F/\delta y$. Now consider varying the parameters α infinitesimally along some trajectory; we expect a translation-invariant finger solution will continue to exist, which requires the perturbed operator $L + \delta L$ to continue to have a constant zero mode. In lowest order perturbation theory this requires

$$\hat{y}_0^\dagger \, \delta L \, y_0 = 0 \tag{10}$$

where the adjoint mode $\hat{y}_0^\dagger$ obeys $L^\dagger \hat{y}_0 = 0$.

To obtain the adjoint equation, we substitute $y(x) + \delta(x)$ and expanding to first order in δ:

$$\tilde{v} \frac{\delta''(x)}{(1 + x^2/p^2)^{3/2}} (1 - \epsilon \cos 4\theta) = \int dx' \frac{\partial G}{\partial y'} [\delta(x') - \delta(x)] \tag{11}$$

where G is the product of K_0 and the exponential in eq. (4), and $\tilde{v} = vd_0/2D$. We have dropped terms involving fewer derivatives of δ, as these will be irrelevant in the small $\tilde{v}$ limit. The adjoint equation to the same level of approximation just involves changing the sign of the last term of the integrand on the right hand side. In the limit $\tilde{v} \to 0$ we look for rapidly oscillating solutions of the adjoint equation of the form

$$\delta(x) \sim e^{i\psi(x)/\sqrt{\tilde{v}}} \tag{12}$$

Substituting this *ansatz* into (11), we can evaluate the resulting integrals by residues. The final result is the WKB equation

$$\frac{\tilde{v}\delta''(x)}{(1 + x^2/p^2)^{3/2}}(1 - \epsilon \cos 4\theta) = -\frac{(1 - ix/p))\delta(x)}{1 + x^2/p^2} \tag{13}$$

Substituting in the previous *ansatz* and keeping the leading term as $\tilde{v} \to 0$, we obtain

$$\psi(x) = -i \int_0^x dx' \frac{(1 - ix'/p)^{3/4}(1 + ix'/p)^{1/4}}{(1 - \epsilon \cos 4\theta)^{1/2}} \tag{14}$$

where the normalization has been chosen so that $\delta(0) = 1$. The solvability condition (10) takes the explicit form

$$I \equiv \int dx f(x) e^{i\psi(x)/\sqrt{\tilde{v}}} = 0 \tag{15}$$

where $f(x)$ is a slowly varying function.

The scaling behavior of the solution follows from the singularities in the last expression for ψ. Note the saddle point in the integral expression (14) at $x/p = -i$; if $\epsilon = 0$, the denominator has no singularities and the integration contour can be simply deformed to pass through this saddle point, leading to an asymptotic estimate for I of the form

$$I \sim A\tilde{v}^{-\alpha}e^{-B/\sqrt{\tilde{v}}}$$

for some constants α, A and B, and I never vanishes. There are no dendritic solutions without crystal anisotropy!

On the other hand, assume now a non-zero value of ϵ. There is a new singularity which occurs when the anisotropy factor goes to zero. The deformed contour will pass around this singularity and I will oscillate. Using $\tan\theta = x/p$, one can show[7,10,11] that the vanishing of integral requires the scaling

$$\tilde{v} \sim \sigma_0(\epsilon)p^2$$

with $\sigma_0 \sim \epsilon^{7/4}$. The constant factor in the last expression can be found by either a more sophisticated analysis or by numerical solution.

There are a number of general remarks worth noting. First, one can extend the analysis to show that the steady-state solution with the largest $\tilde{v}$ is the only linearly stable solution[12]; hence it determines the unique pattern. Next, it was pointed out by Pelce and Pomeau[13] that the scaling $\tilde{v} \sim p^2$ follows from simple dimensional considerations once one assumes that the relevant length scale for forming a velocity comes from the surface energy. Finally, the fact that crystallinity is absolutely necessary for stable dendritic growth meshes nicely with recent experiments[14,15] where the reduction of effective anisotropy by growth on a rough surface causes a transition to more disorderly patterns. The non-perturbative solvability condition, with its sensitivity to small changes in the surface energy can directly account for an otherwise surprising dependence on what used to be considered an unimportant detail of the growth process.

Let us briefly discuss the extension of these ideas to three dimensions. The problem of three-dimensional dendrites is technically difficult because the existence of steady-state solutions (parabolic needle crystals) again requires non-isotropic surface tension[7,16,17]. Any physical anisotropy (such as that arising, say, from an underlying cubic symmetry) will give rise to a non-axisymmetric form for the shape. For such shapes, the Gibbs-Thomson condition eq. (2) is more complicated than that given above; also, the solvability condition must now be extended to require smoothness of all azimuthal Fourier components of the shape. Nevertheless, results to date[18] indicate that the same qualitative picture, a single stable steady-state solution in the presence of crystal anisotropy, remains valid. Comparison to experiments then require sufficiently precise data on the surface energy as a function of angle.

The last point to be discussed is that of sidebranching. Since the steady-state solution is linearly stable, sidebranching must involve a non-linear effect. The current approach to the question of determining the sidebranch wavelength starts by assuming a broadband external noise near the tip[19]. This noise is selectively amplified[20], leading eventually to a non-linear transition to the sidebranching pattern as one moves away from the tip. The selectivity of the amplification gives rise to a peak in the spectrum, with the immediate result that ratio of wavelength to tip radius is independent of p and scales with anisotropy as

$$\lambda/\rho \sim \epsilon^{7/8}.$$

It might be possible to estimate ϵ via this formula for materials whose surface energy has not been adequately measured (by comparing this ratio to the same ratio is succinonitrile with known ϵ) and thereby test the validity of the solvability approach.

OTHER SYSTEMS

After a detailed look at the problem of dendritic crystals, we now turn to a general survey of where we stand in the study of interfacial patterns. From the discussion thus far, we can abstract a framework which subsumes many other cases of interface evolution. The basic mechanism in all these systems is a growth instability brought about by enhanced field gradients near protruding tips. For pure crystals, temperature is the relevant field; for directional solidification, concentration plays this role, for multiphase fluid flow, pressure etc. Coupled to the initial instability mechanism is a restabilization of short wavelength fluctuations due to surface energy effects. Finally, there are additional factors governing the pattern, including microscopic symmetries, externally imposed macroscopic gradients, and boundary conditions.

To illustrate this way of thinking about pattern forming systems, we will describe a case of multiphase fluid flow, air displacing viscous liquid in a Hele-Shaw cell[21]. In this confined geometry, the pressure approximately obeys Laplace's equation

$$\nabla^2 p = 0.$$

At the fluid interface, $v_n \sim \hat{n} \cdot \nabla p$, giving rise to the macroscopic instability. Similarly, $p = -\gamma\kappa$ for surface tension γ and curvature κ, suppressing short wavelength deformations of the air-fluid interface. This system can give rise to stable steady-state fingers, dendrites or disordered structures, depending on the injection method (channel flow vs. radial cells), the surface energy (immiscible vs. miscible displacements), the type of liquid (simple, liquid crystalline, polymeric) and whether there is any imposed flow anisotropy, say by adding grooves.

The selection of the velocity of the simple Saffman-Taylor finger[22] is now well understood. Again, there is a continuum of allowed shapes in the absence of surface tension; this continuum breaks down to a discrete set due to the same solvability mechanism outlined above for dendrites. Here, a direct comparison to experiments can be made, with the result[23] that the theory (once modified for the presence of thin films) accurately predicts the observed structure.

To illustrate some of the questions currently being addressed, we focus on the simple fluid channel geometry with the addition of a either a thin wire or small air bubble[24] near the tip of the moving finger. Without such a perturbation, the finger undergoes a non-linear tip-splitting instability at small surface tension. The wire or bubble somehow stabilizes this mode, and can therefore give rise to stable narrow fingers. Under certain circumstances, the bubble can oscillate, generating a nice sidebranching train down the side of the finger. In the absence of this natural oscillation, forcing the system with an external oscillator generates a linear response curve with a sharp peak at some frequency.

Using our current ideas, we can understand some of these observations. As already mentioned, there exists a complete theory of finger width selection without the bubble. Hong and Langer[25] have shown how the extreme sensitivity of the solvability mechanism to small changes in the boundary conditions near the tip can be used to explain the width changing effect of these perturbations. Also, the linear response of the finger can be computed in the same way[26] that was used to find the sidebranching response to external noise. This calculation predicts a frequency peak with a sharp dropoff ($\sim \omega^3$) past the maximum. Unfortunately, we do not have a non-linear theory of pattern stability. So, although there are some heuristic ideas regarding the onset of tip-splitting[27], there is currently no explanation as to why the wire or bubble (or an imposed anisotropy) manages to suppress this behavior. And, a full treatment of the dynamics which would enable us to derive a Hopf bifurcation to the oscillatory state is still unavailable.

Another system of tremendous interest is directional solidification. This system consists of a crystal growing under concentration-diffusion control, with the addition of an externally imposed thermal gradient[3]. The main difference between this system and the free space dendrite is that now there are two macroscopic length scales, the diffusion length and the thermal gradient scale. A preliminary computation by Karma[28] indicated that there would be a solvability mechanism at work to determine the wavelength of the celluar pattern. Recently however, Dombre and Hakim[29] and Ben-Amar[30] have suggested that Karma may have missed an additional continuum of steady-state solutions. If this conclusion is valid, we may need a separate mechanism for wavelength selection.

In our work on solidification from the melt, we have always assumed that the system operates close to local thermodynamic equilibrium. In electrochemical deposition, however, it is well known that there are important effects due to non-trivial kinetics of the charge exchange reaction that occurs upon adding an ion to the growing deposit. In general, we can expect a non-linear relationship between the growth velocity and the conditions right at the interface, which will have to be handled for quantitative, if not qualitative, understanding of the structure. For example, the classic aggregation study of Ball and Brady[31] on copper was mapped onto the Diffusion-Limited -Aggregation (DLA) model of Witten and Sander[32] by using a boundary condition that the concentration $c = 0$ at the interface; this is obviously not a local equilibrium condition but instead an assumption of complete irreversibility. This condition completely neglects surface tension and, more importantly, crystal anisotropy; the validity of this approach seems to be related to the observation that DLA aggregates are not single crystals but rather are polyscrystalline[33]. How this works in detail is an open issue.

Most of our to date studies have focused on local properties such as tip shape, initial sidebranching and tip-splitting. There are also fascinating global issues in systems such as electrochemical deposition. First, there appear to be two alternatives open to a system undergoing local tip-splitting; it can either form a fractal or a dense radial structure[33,34] (see Fig. 2); the difference may be related to the diffusion length and/or to voltage drops in the aggregate itself[35]. Another experimental observation[36] is a frequency peak in the AC current even when the system is grown under constant voltage conditions. This peak appears to be related to sidebranching and may represent frequency locking of the sidebranch oscillations on differing dendritic arms. This result opens up the possibility

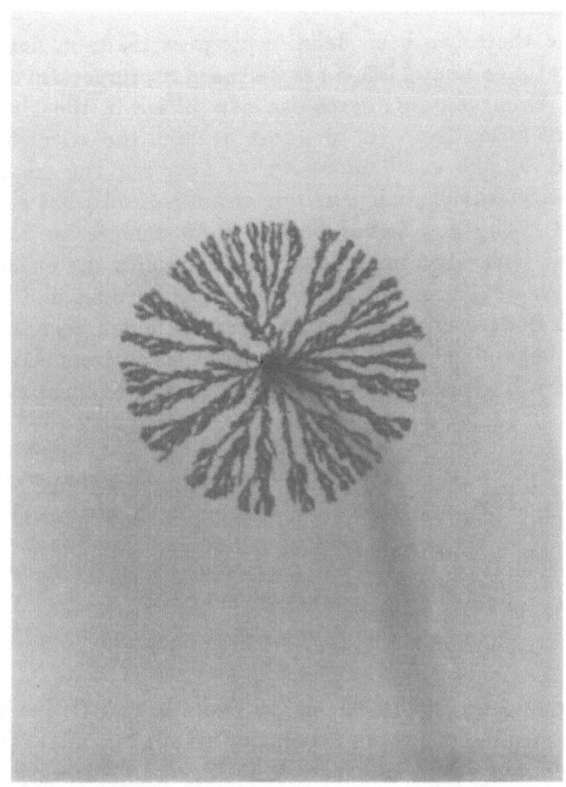

Fig. 2. Dense branching morphology observed in the electrodeposition of $ZnSO_4$ (D. Grier, Univ of Michigan).

of using electrical measurements to study morphology changes involved in going from dendrite to tip-splitting to fractal.

Many other examples that could be given of interfacial patterns and current attempts to understand them; we specifically have in mind spiral crystals in phospholipid monolayers[37], lamellar eutectics[38], and rising bubbles[39]. We would also like to note in passing that much interesting work has been done in the field of chemical reaction fronts including the propagation of flames[40]; here too there seems to room for both orderly finger-like structures and disorderly fractal interfaces, as the governing parameters are varied. This system is more complex inasmuch as the bulk transport equation is non-linear (as opposed to the linear diffusive behavior we have been considering till now) but this may not alter the physics very much if the important instabilities driving pattern formation are due to the interface.

So at this stage there has been definite progress made in understanding pattern formation, but many fundamental issues (as well as many important applications) remain to be worked out. Eventually, we foresee a new phase in the history of the subject of interfacial pattern formation. For the past decade, the scientific focus of research in this field was set by the excellent review of Langer[3]; the emphasis was on what determines the velocity, why there is a unique structure and what aspects of the system are most important. The new issues beginning to emerge are how does the global pattern relate to the local structure, how can we predict the effects of perturbations on either the governing equations or the boundary conditions and finally, whether the mechanisms at work in determining interfacial structure can play any role in furthering our understanding of complex behavior in other systems far from equilibrium. Hopefully, these new questions will generate as many surprises and exciting new ideas as did the previous ones.

The work of one of us (D.K.) was supported by U.S. Department of Energy Grant No. DE-FG-02-85ER54189.

REFERENCES

1. For recent reviews, see D. Kessler, J. Koplik, and H. Levine, *Pattern Selection in Fingered Growth Phenomena*, to appear in Advances in Physics; *On Growth and Form*, ed. Stanley, H. E. and Ostrowsky, N. (Dordrecht: Martinus Nijhoff; G. M. Homsy, *Ann. Rev. Fluid Mech. 19*, 271 (1987).

2. D. P Woodruff, *The Solid-Liquid Interface* (Cambridge University Press, 1973).

3. J. S. Langer, *Rev. Mod. Phys. 52*, 1 (1981).

4. M. E. Glicksman *Matl. Sci. and Engr. 65*, 45 (1984) and references therein.

5. W. W. Mullins and R. F. Sekerka, *J. Appl. Phys 34*, 323 (1963).

6. G. P. Ivantsov, *Dokl. Akad. Nauk SSSR 58*, 567 (1947).

7. The application of the WKB method to free dendrites appears in B. Caroli, C. Caroli, B. Roulet, and J. S. Langer, *Phys. Rev. A 33*, 442 (1986); D. Hong and J. Langer, preprint; D. Kessler, J. Koplik, H. Levine, *"Pattern Formation Far From Equilibrium: The Free Space Dendritic Crystal"*, *Patterns, Defects, and Microstructure* NATO ASI Series E, Vol. 121 D. Walgraef, ed. (Nijhoff, 1987); A. Barbieri, D. Hong, and J. S. Langer, *Phys. Rev. A 35*,1802 (1986).

8. J. S. Langer, *Phys. Rev. A 33*, 435 (1986).

9. B. I. Shraiman, *Phys. Rev. Lett. 56*, 2028 (1986).

10. M. Ben-Amar and Y. Pomeau, *Europhys. Lett. 2*, 307 (1986).

11. B. Caroli, C. Caroli, C. Misbah, and B. Roulet, B. 1987 (to be published).

12. D. A. Kessler and H. Levine, *Phys. Rev. Lett. 57*, 3069 (1986). D. Bensimon, P. Pelcé, and B. Shraiman, 1987 (to be published).

13. P.Pelce and Y. Pomeau, *Studies in Appl. Math. 74*, 245 (1986).

14. H. Honjo, S. Ohta, and M. Matsushita, M. 1986 *J. Phys. Soc. Japan 55*, 2487.

15. For the reverse experiment of adding anisotropy to an otherwise isotropic system, see E. Ben-Jacob, R. Godbey, N. D. Goldenfeld, J. Koplik, H. Levine, T. Mueller, and L. M. Sander, *Phys. Rev. Lett.* **55**, 1315 (1985).

16. M. Ben-amar and Y. Pomeau, 1986 *Physiochem. Hydro.* to appear.

17. D. A. Kessler, J. Koplik, H. Levine, *Phys. Rev. A* **33**, 3352 (1986).

18. D. Kessler and H. Levine, 1987 preprint.

19. R. Pieters and J. S. Langer, *Phys. Rev. Lett.* **56**, 1948 (1986).

20. D. A. Kessler and H. Levine, *Europhysics Lett.* to appear; M. Barber, A. Barbieri, and J. S. Langer, 1987 preprint; B. Caroli, C. Caroli, and B. Roulet, 1987 (to be published); for the same idea in a different context, see R. J. Deissler, *J. Stat. Phys.* **40**, 371 (1985).

21. H. J. S. Hele-Shaw, *Nature* **58**, 334 (1898).

22. P. G. Saffman and G. I. Taylor, *Proc. Roy. Soc. A* **245**, 312 (1958).

23. P. Tabeling, G. Zocchi, and A. Libchaber *J. Fluid Mech.* to appear.

24. Y. Couder, O. Cardoso, D. Dupuy, P. Tavernier, *Europhys. Lett.* **2**, 437 (1986); Y. Couder, N. Gérard, and M. Rabaud, *Phys. Rev. A* **34**, 5775 (1986).

25. D. Hong and J. Langer, (to be published).

26. D. Kessler, J. Koplik and H. Levine, ref. 1.

27. D. Bensimon, *Phys. Rev. A* **33**, 1302 (1986).

28. A. Karma, *Phys. Rev. Lett.* **57**, 858 (1986).

29. T. Dombre, T. and V. Hakim, 1987 (to be published).

30. M. Ben-Amar, private communication.

31. R. M. Brady, and R. Ball, *Nature* **309**, 225 (1984).

32. T. Witten and L. Sander, *Phys. Rev. B* **27**, 5696 (1983).

33. D.Grier, E. Ben-Jacob, R. Clarke, and L. M. Sander, *Phys. Rev. Lett.* **56**, 1264 (1986).

34. Y. Sawada, A. Dougherty, and J. P. Gollub, *Phys. Rev. Lett.* **56**, 1260 (1986).

35. D. Grier and L. Sander, private communication.

36. R. Suter and P-Z. Wong, in preparation.

37. R. M. Weis and H. M. McConnell, *Nature* **310**, 47 (1984).

38. W. Kurz and D. J. Fisher, *Fundamentals of Solidification* (Aedermannsdorf, Switzerland: Trans Tech Publications, 1984).

39. J.-M. Vanden-Broeck, *Phys. Fluids* **27**, 1090,2604 (1984).

40. G. I. Sivashinsky, *Ann. Rev. Fluid Mech.* **165**, 179 (1983).

PATTERN FORMATION OF SNOW CRYSTALS

-SIMULATION OF DEVELOPMENT OF FACETS BY MEANS OF BOUNDARY ELEMENT METHOD

E. Yokoyama and T. Kuroda

Institute of Low Temperature Science
Hokkaido University
Sapporo 060, Japan

Introduction

A spherical single ice crystal with radius of 1-10μm is first formed by freezing of a supercooled water droplet in cloud, then it grows into a hexagonal prism bounded by two basal {0001}-and six prism {10$\bar{1}$0}-faces by consuming supersaturated water vapor and develops into various forms, e.g. plate, column, dendrite, needle and so on, according to the growth conditions such as temperature and supersaturation[1,2].

In this paper, we simulate the development of prism facets starting from a spherical crystal by taking into account i) the surface kinetic process for incorporating the molecules into a crystal and ii) the diffusion process for supplying the molecules to the surface, and investigate a role of each process in pattern formation under various growth conditions. The moving boundary problem in which two processes are coupled is extremely difficult to solve. Therefore, we treat two-dimensional crystals perpendicular to c-axis of a hexagonal ice crystal even though the diffusion process is three dimensional in the actual growth of ice crystals, and we solve the problem by means of the boundary element method. A morphological instability of a polyhedral crystal is also discussed briefly.

It has recently found that the complicated habit change depending on temperature is attributed to surface melting of basal and prism faces. For details, the reader may refer to the papers 3 and 4 in references.

Growth Mechanism and Simulation Method

Fig.1 shows a part of a growing surface. The growth rate R_k determined by surface kinetic process in the direction normal to a singular surface, i.e. molecularly smooth surface is given by

$$R_k = b(\theta)\sigma_s . \qquad ---(1)$$

In eq.(1) σ_s is the local supersaturation at the surface and b(θ) is the kinetic coefficient which represents the activity of the surface for incorporating the molecules into crystal and depends on the local slope s=tan θ of the surface against the singular surface and also on σ_s for singular surface. That is[5]

$$b(\theta) = b_{max} \cdot \frac{s}{s_1} \tanh \frac{s_1}{s} \cdot \frac{1}{\cos\theta} , \qquad ---(2)$$

$$s \quad = \tan\theta = d/\lambda, \quad s_1 = d/2x_s, \qquad ---(3)$$

$$b_{max} = \alpha v_c p_e / \sqrt{2\pi mkT} , \qquad ---(4)$$

where λ is the mean step distance, d the step height, x_s the mean surface diffusion distance of a molecule on the surface, α the condensation coefficient, v_c the volume of a molecule in the crystal, m the mass of a molecule, k the Boltzmann constant, T the absolute temperature, p_e the equilibrium vapor pressure of ice.

Thus, the growth rate V_k in the direction normal to the surface determined by surface kinetic process is given by

$$V_k = R_k \cos\theta = \beta(\theta)\sigma_s, \qquad ---(5)$$

$$\beta(\theta) = b(\theta)\cos\theta , \qquad ---(6)$$

where $\beta(\theta)$ is also the kinetic coefficient. The dependence of $\beta(\theta)$ on θ which is rotation angle about c-axis is schematically shown in Fig.2. It possesses six minima at $0°$, $\pm 60°$, $\pm 120°$, $180°$ corresponding to six prism faces, i.e. singular surfaces. Although there is no step on the singular surface in equilibrium state, steps are actually generated by two-dimensional nucleation or with the aid of screw dislocation in non-equilibrium state. Since the frequency of step generation increases with increasing σ_s, the slope θ_0 of growth hillock formed by the steps on the prism faces as well as the kinetic coefficient $\beta_0 = \beta(\theta_0)$ of the prism face increases with increasing σ_s. In this paper, we consider only the spiral growth to occur with the aid of a screw dislocation emerging at the center of each prism face. Therefore the value of β_0 of the prism face is determined by the slope $s_0 = \tan\theta_0(\propto \sigma_s)$ of a spiral growth hillock on it and increases with increasing σ_s at the position where the screw dislocation emerges (see arrows in Fig.2). It should be noticed that the anisotropy of $\beta(\theta)$ decreases with increasing β_0. Namely, the anisotropy increases with decreasing σ_s.

On the other hand, the growth rate V_d determined by volume diffusion process is given by

$$V_d = \frac{v_c p_e}{kT} D q_s , \qquad ---(7)$$

where D is the diffusion constant, $q_s=(\partial\sigma/\partial n)_{surface}$ is the normal gradient of supersaturation at a position on the surface, which is unknown parameter given as a boundary condition for diffusion equation.

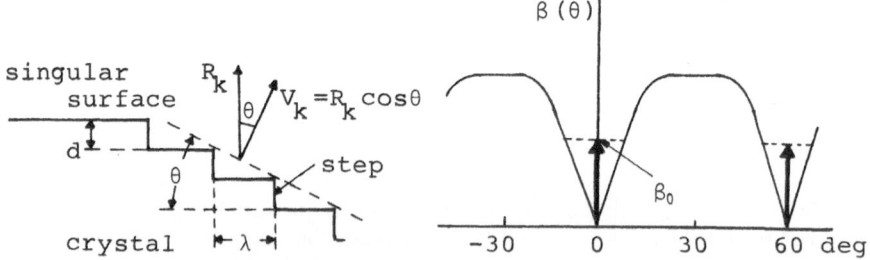

Fig.1　Schematic representation
　　　 of growing surface

Fig.2　Dependence of $\beta(\theta)$ on
　　　 $\theta = \tan^{-1}s$

The growth rate V_k determined by surface kinetic process should be equal to the growth rate V_d determined by volume diffusion process under steady state conditions[6]:

$$V = V_k = V_d . \qquad ---(8)$$

By inserting eqs.(5) and (7) into eq.(8), we obtain

$$q_s = \beta(\theta)kT\sigma_s/v_c P_e D , \qquad ---(9)$$

This is an important relation for determining self-consistently the growth rate V controlled both by diffusion process and by surface kinetic process.

The supersaturation σ surrounding a crystal in the region Ω (Fig.3) is governed by the quasi-static diffusion equation

$$\Delta\sigma = 0 , \qquad ---(10)$$

subject to the boundary conditions: $\sigma = \sigma_\infty$ on the boundary Γ_1 which is apart from the center of the crystal by the distance R and normal gradient $(\partial\sigma/\partial n)_{surface} = q_s$ on the boundary Γ_2 representing the crystal surface. We solve the problem by means of the boundary element method as follows. At first, we obtain the boundary integral equation from eq.(10) using Green's theorem and mass conservation conditions eq.(9):

$$\sigma_{si}/2 + \int_{along\Gamma_2} (q* - \sigma*\beta(\theta)kT/v_c P_e D)\sigma_s d\Gamma = \int_{along\Gamma_1} (q_\infty\sigma* - \sigma_\infty q*)d\Gamma , \qquad ---(11)$$

where σ_{si} is the supersaturation at a point i on Γ_2, q_∞ the normal gradient of supersaturation on Γ_1, $\sigma*$ satisfies $\sigma* + \delta_i = 0$ (δ_i: the Dirac delta function), $q*$ is $\partial\sigma*/\partial n$. Then, we change the eq.(11) to the algebraic equations with respect to the surface supersaturation σ_s. We approximately replace q_∞ with following equation under the assumption that Γ_1 is far from Γ_2.

$$q_\infty \simeq \sigma_\infty/R \ln(R/r_c) , \qquad ---(12)$$

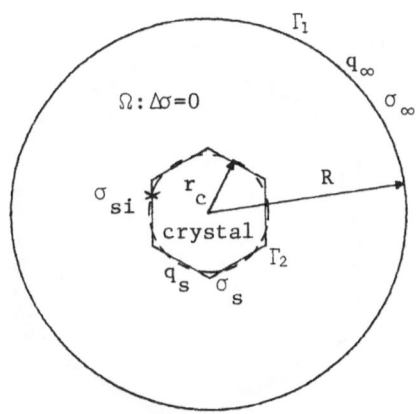

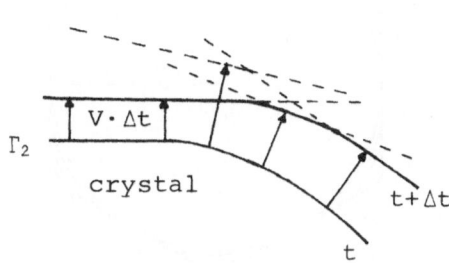

Fig.3 Diffusion filed surrounding a crystal

Fig.4 A procedure for drawing a growth form

which is obtained for diffusion field surrounding a circular crystal with a radius r_c which is half of mean length along a-axis and b-axis and with perfect sink ($\sigma_s=0$).

By solving the algebraic equations, we obtain σ_s, consequently the growth rate V at each position on the surface at a certain moment and then determine a shape of the growing surface after Δt second. The surface growing faster is cut by neighboring surfaces growing slower, while the surface growing slowest develops its area largest. Therefore, a growth form is determined by inner envelope of the lines perpendicular to each vector $\vec{V} \cdot \Delta t$ at its point(Fig.4). By repeating this procedure we simulate the development of facet.

Results

The numerical values used for the simulation are as follows: T=253.15K(=-20°C), k=1.38×10^{-16}erg/deg, m=3×10^{-23}g, d=4.5×10^{-8}cm, x_s=400d, v_c=3.25×10^{-23}cm^3, p_e=1.03×10^{3}dyn/cm^2, R/r_c=1000, α =0.1[7,8] and initial radius of circular crystal r_0 is 10^{-3}cm.

(a)   (b)

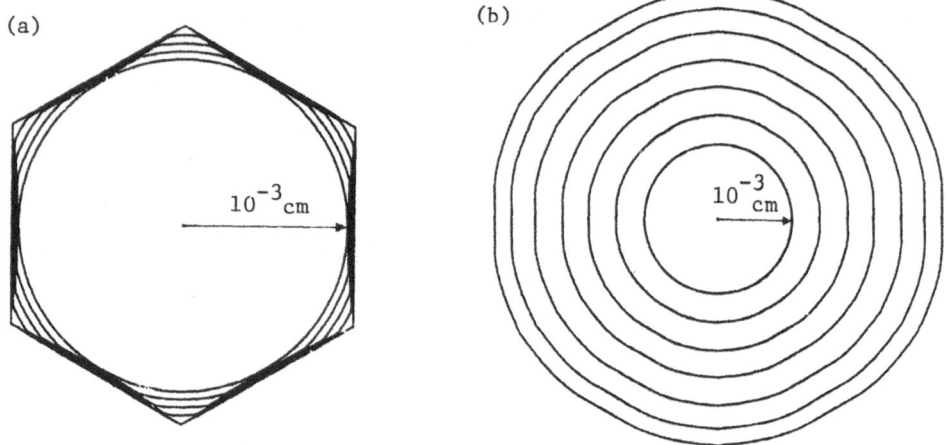

Fig.5   The development of prism faces for D=447cm^2/sec and (a) σ_∞= 1% at times 0, 6, 11, 17, 23 sec (b) σ_∞= 15% at times 0, 3.1, 6.1, 9, 12, 14.7, 16.7 sec.

(a)   (b)

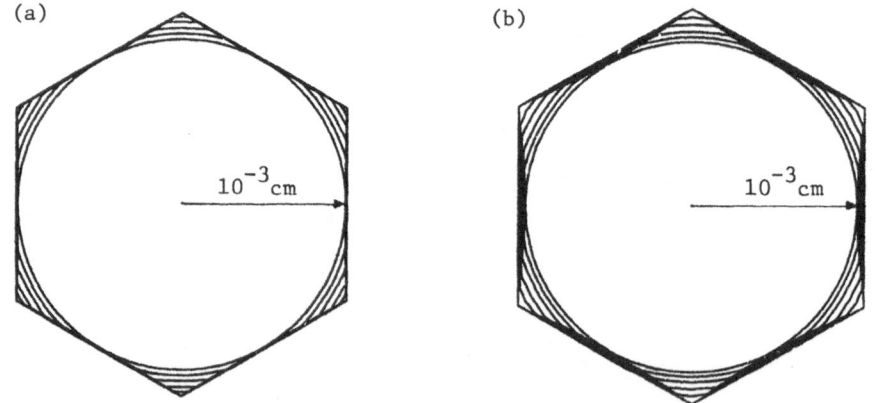

Fig.6   The development of prism faces for D=0.22cm^2/sec and (a) σ_∞= 1% at times 0, 233, 401, 575, 703, 750 sec (b) σ_∞= 15% at times 0, 13, 27, 44, 60, 84 sec.

In Fig.5 we present two examples of development of six prism faces simulated for larger diffusion constant $D=447 cm^2/sec$ (corresponding to 0.3Torr of air pressure at which the role of the surface kinetic process is more important than that of the diffusion process). In the case of lower supersaturation $\sigma_\infty=1\%$ (Fig.5(a)), anisotropy of growth rate is so large that prism faces growing with smallest rate develop their area quickly and an initial circular crystal becomes a perfect hexagon within 23 seconds. On the other hand, in the case of $\sigma_\infty=15\%$(Fig.5(b)), anisotropy of growth rate becomes so weak that prism faces can hardly develop, even though the crystal size increases to three times initial size during growth for 16.7 seconds. It should be noted that the supersaturation $\sigma_\infty=15\%$ is the maximum supersaturation at $-20°C$ which is realized when water vapor is equilibrated with supercooled water droplets in cloud.

Fig.6 shows two examples for smaller $D=0.22 cm^2/sec$ (corresponding to 1 atm of air pressure) at (a) $\sigma_\infty=1\%$ and (b) $\sigma_\infty=15\%$. Since the surface supersaturation σ_s largely drops from σ_∞ because of insufficient supply of molecules by diffusion, growth rate is much smaller than that for $D=447 cm^2/sec$ (compare growth times in Fig.6 with that in Fig.5). Especially in the case of $\sigma_\infty=1\%$ (Fig.6(a)), prism faces can scarcely grow because of too small σ_s so that a perfect hexagon which is nearly circumscribed about the initial circular crystal is formed. It is to be noticed that an initial circular crystal can develop into a perfect hexagon even at higher supersaturation $\sigma_\infty=15\%$ in case of lower D (Fig.6(b)).

Discussions

In this section we discuss the conditions for development of facets. We have shown in the former section that an initial circular crystal develops into a hexagon because of strong anisotropy of growth rate, i.e. that of the kinetic coefficient $\beta(\theta)$. Since the value of β_0 for singular surfaces decreases with decreasing surface supersaturation σ_s, anisotropy of $\beta(\theta)$ increases with decreasing surface supersaturation σ_s, which is determined by both the diffusion process and the surface kinetic process.

Let us now consider the role of each process as rate determining process under various growth conditions. The growth rate under steady state conditions is formally expressed as[9]:

$$V = \sigma_\infty / (\Omega_k + \Omega_d), \qquad ---(13)$$

where Ω_k is the resistance of the surface kinetic process given by

$$\Omega_k = 1/\beta(\theta), \qquad ---(14)$$

Ω_d the resistance of the volume diffusion given by

$$\Omega_d = kT \delta / v_c P_e D. \qquad ---(15)$$

Here δ is the thickness of the diffusion boundary layer which is defined by $(\partial\sigma/\partial n)_{surface} \delta = \sigma_\infty - \sigma_s$ and nearly equal to the local radius of curvature.

In the case of higher diffusion constant $D=447 cm^2/sec$ (Fig.5), the volume diffusion is so easy that the surface supersaturation σ_s does not drop largely from the supersaturation σ_∞ at Γ_1 (For example, $\Omega_k/\Omega_d=1500$ and $\sigma_s=0.66\%$ in Fig.5(a), $\Omega_k/\Omega_d=340$ and $\sigma_s=9.7\%$ in Fig.5(b)). Since σ_s increases with increasing σ_∞, the anisotropy of growth rate decreases because of an increase in growth rate of prism faces. Namely, the development of facets becomes more difficult with increasing σ_∞ for higher D. It is a so-called kinetic roughening, if the anisotropy completely vanishes.

As the diffusion constant D decreases, the growth rate decreases because of an increase in Ω_d. If Ω_d became much larger than Ω_k, the growth rate would be expressed as $p_e v_c D \sigma_\infty / kT \, \delta$ from eqs.(13) and (15) and prism faces would not appear. However, prism faces well develop, in fact,(see Fig.6), since σ_s decreases because of difficulty of the volume diffusion under the conditions of lower D. It means that not only Ω_d but also Ω_k increase with decreasing D because of the decrease in σ_s (eq.(14)) (For example, $\Omega_k / \Omega_d \simeq 4$ in Fig.6(a) and $\Omega_k / \Omega_d \simeq 2.8$ in Fig.6(b)). Therefore, the development of facets becomes easier with decreasing D in spite of larger σ_∞(Fig.6(b)). It should be noticed that the hexagonal symmetry of snow crystals is caused by the anisotropy of $\beta(\theta)$.

It is also interesting to consider the transition from a hexagon to a hexagonal dendrite. When a polyhedral crystal grows preserving its flat surface, the surface supersaturation σ_s is not uniform over the surface, i.e. it is largest at corners and smallest at the center of the surface. For example, σ_s is 0.56% at the corner and σ_s is 0.17% at the center in Fig.6(b). With increasing σ_s at corners, the rate of two-dimensional nucleation exponentially increases at corners and when its growth rate becomes lager than that of center, instability of a polyhedral crystal occurs. This is onset of dendritic growth of polyhedral snow crystals. Here we should distinguish this morphological instability of polyhedral crystal from the Mullins-Sekerka instability[10] for the rough interface like crystal/melt interface. In the latter problem, $\beta(\theta)$ is extraordinarily large for any surface orientation so that only the diffusion process is important.

ACKNOWLEDGMENT

The authors would like to thank Mr.T.Irisawa for his kind suggestions for computer programming.

REFERENCES

1. T.Kobayashi, The growth of snow crystals at low supersaturation, Phil. Mag., 6:1363 (1961).
2. T.Kuroda, Recent developments in theory and experiment of growth kinetics of ice crystals from the vapour phase and their growth forms, J.Crystal Growth, 65:27 (1983).
3. T.Kuroda and R.Lacmann, Growth kinetics of ice from the vapour phase and its growth forms, J.Crystal Growth, 56:189 (1982).
4. Y.Furukawa, M.Yamamoto and T.Kuroda, Ellipsometric study of the transition layer on the surface of an ice crystal, J.Crystal Growth, 82:665 (1987).
5. W.K.Burton, N.Cabrera and F.C.Frank, The growth of crystals and the equilibrium structure of their surface, Phil.Trans.Roy.Soc., A243:299 (1951).
6. T.Kuroda, T.Irisawa and A.Ookawa, Growth of polyhedral crystal from solution and its morphological stability, J.Crystal Growth, 42:41 (1977).
7. W.Beckmann and R.Lacmann, Interface kinetics of the growth and evaporation of ice single crystals from the vapour phase, II.Measurements in a pure water vapour environment, J.Crystal Growth,58:433 (1982).
8. T.Kuroda and T.Gonda, Rate determining processes of growth of ice crystals from the vapour phase, II.Investigation of surface kinetic process, J. Meteor. Soc. Japan, 62:563 (1984).
9. T.Kuroda, Rate determining processes of growth of ice crystals from the vapour phase, I.Theoretical consideration, J.Meteor.Soc.Japan, 62:552 (1984).
10. W.W.Mullins and R.F.Sekerka, Morphological stability of a particle growing by diffusion or heat flow, J.Appl.Phys., 34:323 (1963).

NUMERICAL SIMULATION STUDIES OF THE KINETICS OF FIRST ORDER PHASE TRANSITIONS

J. D. Gunton and E. T. Gawlinski

Department of Physics and
Center for Advanced Computational Science
Temple University
Philadelphia, Pa. 19122

K. Kaski

Department of Electrical Engineering
Tampere University of Technology
Tampere, Finland

INTRODUCTION

Numerical simulation methods have a rich history in the study of the kinetics of first order phase transitions[1]. These include Monte Carlo and molecular dynamics simulations and numerical integration of stochastic differential equations, some of which will be discussed in this article. It has been recognized for many years that such studies are useful in obtaining qualitative insights, such as the role of conservation laws, vertices, dimensionality, etc. in ordering processes. However, it is quite difficult to determine asymptotic domain growth laws for bulk systems by such methods, due both to finite size effects and related finite time limitations imposed by current computer capabilities[2]. It is particularly difficult to determine growth laws and scaling functions for problems in which one has no theoretical prediction to test. Similar difficulties exist for experimentalists, of course! Recent theoretical developments, however, have led to renewed interest in determining the growth law for systems which undergo spinodal decomposition and coarsening, such as occurs following a quench below a critical point at a critical value of the order parameter. We will primarily focus on this issue here, given its importance in understanding phase-separation dynamics in two and three dimensional systems. Huse[3] has recently given a heuristic argument that for such a quench the asymptotic domain growth behavior is described by a Lifshitz-Slyozov exponent, i.e. the characteristic length, $L(t)$, behaves like

$L(t) \sim t^X$, with x = 1/3. However, this behavior is modified by an important correction term arising from surface diffusion along interfaces of interconnected structures. Mazenko and Valls, on the other hand, have used renormalization group ideas together with Monte Carlo and numerical integration techniques to study the kinetic Ising model[4] and continuum Langevin model,[5] respectively, in two dimensions. They have predicted a logarithmic growth law for the kinetic Ising model and an exponent of x = 1/4 for the Langevin model. If true, this means that the two models belong to different dynamic universality classes, which would be presumably due to the discrete nature of the conserved order parameter. Although the Mazenko-Valls approach is interesting, it is based on assumptions whose validity we find difficult to determine. Thus it is noteworthy that recent Monte Carlo studies[6,7] of the kinetic Ising model, including work which we summarize here, agree with the Huse prediction and show no evidence of a logarithmic behavior. It is also obviously of interest to examine the Langevin model, where one can also test whether the Huse or Mazenko-Valls prediction is valid. We will summarize new results for the Langevin equation in this article. Details will be published elsewhere.

Before discussing the two dimensional kinetic Ising and Langevin model results, we summarize our findings. In both our Monte Carlo study[7] and one carried out by Amar et al.[6], the domain growth law is found to satisfy $R(t) = A + B \, t^{1/3}$, for a variety of different definitions of the characteristic length. Thus the prediction by Huse seems to describe the data for R(t) very well. There is no evidence of the logarithmic behavior predicted by Mazenko and Valls in the extensive time regime considered in these Monte Carlo studies. In addition, the work of reference 7 shows that the order parameter order parameter correlation function g(r,t) satisfies a scaling law of the form g(r,t) = F(r/R(t)) after some early time regime. As well, the structure factor S(k,t) was found in both studies to satisfy an analogous scaling behavior. The results for the Langevin model[8] are remarkably similar. The only possible difference is that there appears to be a small deviation from scaling for both g(r,t) and S(k,t) in the time regime studied. Whether this is real or a numerical artifact remains to be seen. In any event, the domain growth law for both the kinetic Ising and Langevin models is in good agreement with the prediction by Huse. There is no evidence to indicate that these models belong to different dynamical universality classes. There is clearly a need for further theoretical investigation of this problem, particularly as regards predicting the scaling functions. As well, it would be desirable to extend the simulation studies to longer times and bigger systems and, hopefully, to three dimen-

sions. It of course would be most useful to compare these results with
experimental studies.

2. Langevin Model

In this section we discuss recent results of a numerical integration
of the Langevin equation for a conserved order parameter in two dimensions[8]. This is a model of adatom ordering in chemisorbed systems or of
phase separation in binary alloys (in three dimensions). Since the details of this simulation will be reported elsewhere[8], we will summarize
only a few major results. The study was carried out for quenches that
correspond to relatively low temperatures, in that a small value was
chosen for the Ginzburg parameter which occurs in the scaled version[9] of
the Langevin equation. (In this section results are given in terms of
scaled distances and times as defined in reference 9, for a value of the
Ginzburg parameter ε = 0.1 given in that reference.) Thus critical fluctuations are not important, so that mean field theory should provide a
first approximation to the equilibrium properties of the model.

The integration was carried out for a discretized version of the
scaled Langevin equation, using a square lattice of (128) x (128) points,
with the order parameter $\psi(\vec{r}_i)$ given at each of these lattice points.
This order-parameter would be the local concentration field for a binary
alloy, say. The dimensionless mesh size and time step were chosen as Δr =
1.00 and $\delta\tau$ = 0.01 respectively. Averages were carried out over the initial conditions and the noise term in the Langevin equation. A large
number of runs (solutions of the equation) was carried out, ranging from
75 - 200 to ensure good statistics, with the maximum dimensionless time
studied being τ = 1000.

Three quantities in particular were computed: the order parameter
correlation function, $g(r,\tau)$, the structure factor $S(k,\tau)$, and the probability distribution function $P(c,\tau)$ describing the distribution of curvatures of the interfaces separating domains of opposite phase. We believe
that this is the first calculation of $P(c)$ carried out in studies of spinodal decomposition. It is hoped that this will lead to a better morphological characterization of the interconnected structure associated with
spinodal decomposition and coarsening. The $g(r,\tau)$ is a circular average
of $g(\vec{r},\tau)$ = < $\delta\psi(\vec{r},\tau)$ $\delta\psi(\vec{0},\tau)$ >. From this a characteristic length scale,
R_g, was computed, where $R_g(\tau)$ is defined as the smallest value of r for
which $g(r,\tau)$ = 0. The first moment of $P(c)$ defines a second length scale,
R_c. We consider the calculation of $P(c)$ to be at best a first approximation to the interface morphology, so that R_g is a much more reliable measure of a characteristic length than R_c. Nevertheless, as shown in Fig. 1

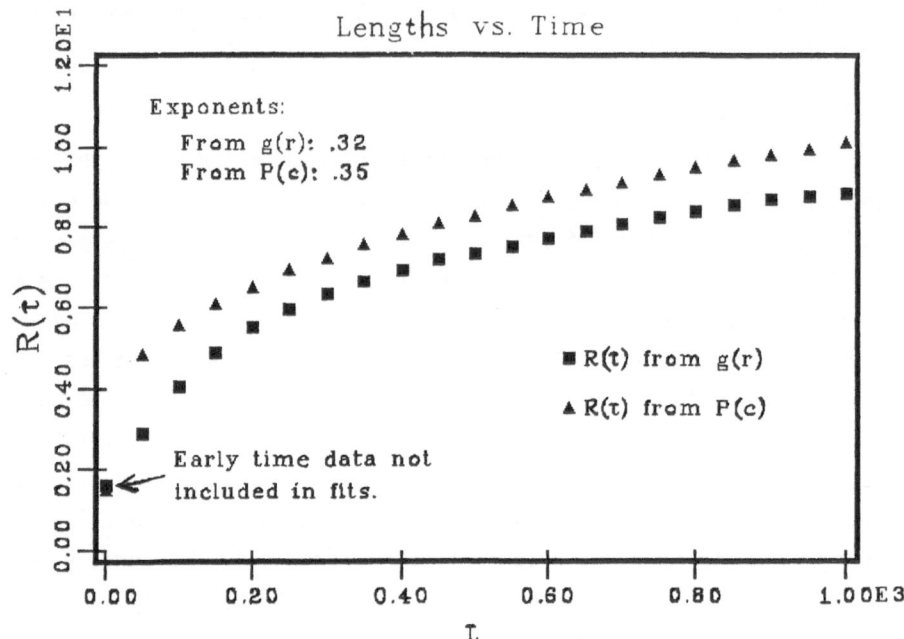

Figure 1. Behavior of two characteristic lengths, R_g (solid squares) and R_c (solid diamonds), as a function of the dimensionless time τ. R_g is calculated from the order-parameter order-parameter correlation function, while R_c is determined from the probability distribution function for the curvature of the interfaces.

the behavior of these two lengths as a function of time is quite similar. We defer a discussion of the exponent which characterizes R_g to the end of this section.

First we address the scaling ansatz, which states that the order parameter-order parameter correlation function satisfies $g(r,\tau) = F(r/L(\tau))$ beyond some initial "transient" time. In Fig. 2 we show the results obtained from plotting our data for $g(r,\tau)$ vs. $r/R_g(\tau)$ for $\tau = 50 - 950$. (We have omitted many other sets of data points obtained in this time interval in this plot, since they fall on the same curve.) It seems clear that to a first approximation the scaling regime corresponds to $\tau \gtrsim 150$, although small deviations might be present even out to $\tau = 1000$. It does seem clear, however, that the data at $\tau = 50$ (which lies above the other data in the vicinity of the first minimum of $g(r)$) is <u>not</u> in the scaling regime.

We have also calculated $S(k,t)$, using 200 runs, and find that it also satisfies a scaling form, although deviations from scaling seem to be present. The approximate scaling function is similar to that found in Monte Carlo studies. These results will be published elsewhere.

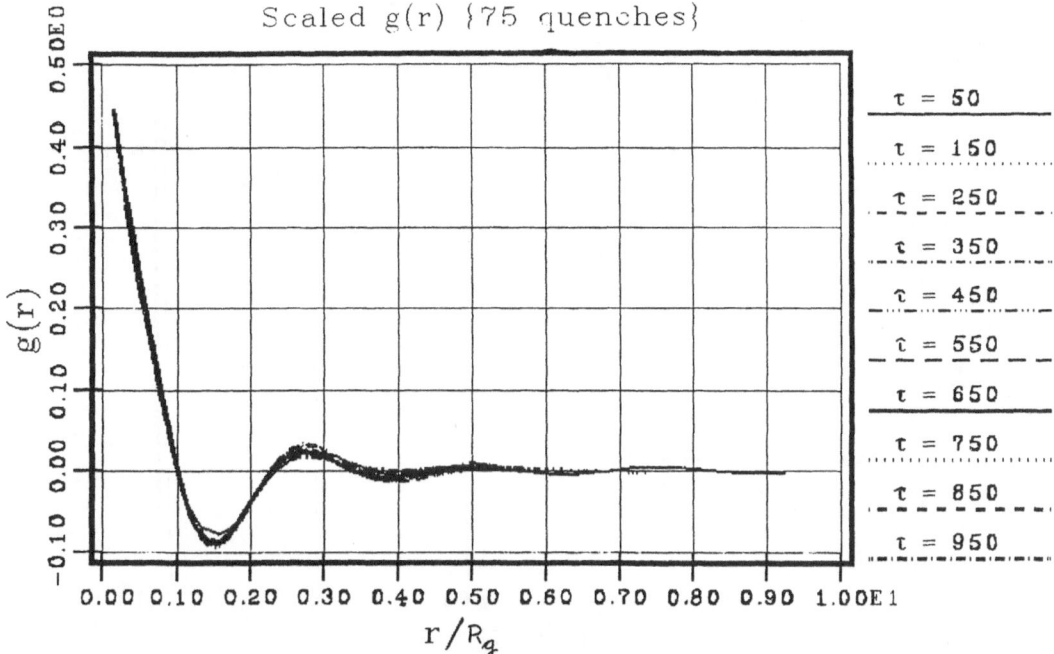

Figure 2. The order-parameter order-parameter correlation function, $g(r,\tau)$, plotted with respect to r/R_g, to test the scaling ansatz.

Finally, we turn to the difficult question of the domain growth law obeyed by R_g. To test between the two theories, we have fit our data to the form $R_g = A + B\tau^x$. We find that an excellent fit is given, with $x = 0.32 \pm 0.04$. We have further plotted our data vs. $\tau^{1/3}$ and $\tau^{1/4}$, respectively, as shown in Fig. 3. In such a plot the data should lie on a straight line corresponding to the correct exponent. As can be seen, the data is extremely consistent with $\tau^{1/3}$ and inconsistent with $\tau^{1/4}$. Thus our data fit the Huse prediction[3] and are inconsistent with the recent renormalization group argument[5] that $x = 1/4$.

3. Kinetic Ising Model

We next summarize the results of a recent Monte Carlo simulation[7] of domain growth in the two-dimensional spin-exchange kinetic Ising model. This work is in agreement with a similar study by Amar et al.[6] and in addition yields the scaling function for the order parameter correlation function (the spin - spin correlation function in this case). Both Monte Carlo studies were carried out at the critical value of the magnetization (i.e. $M = 0$), with equal amounts of up and down spins present. As is well known, this model of a conserved order parameter can be used as a microscopic model of a binary alloy, chemisorbed system, etc. Indeed, the continuum

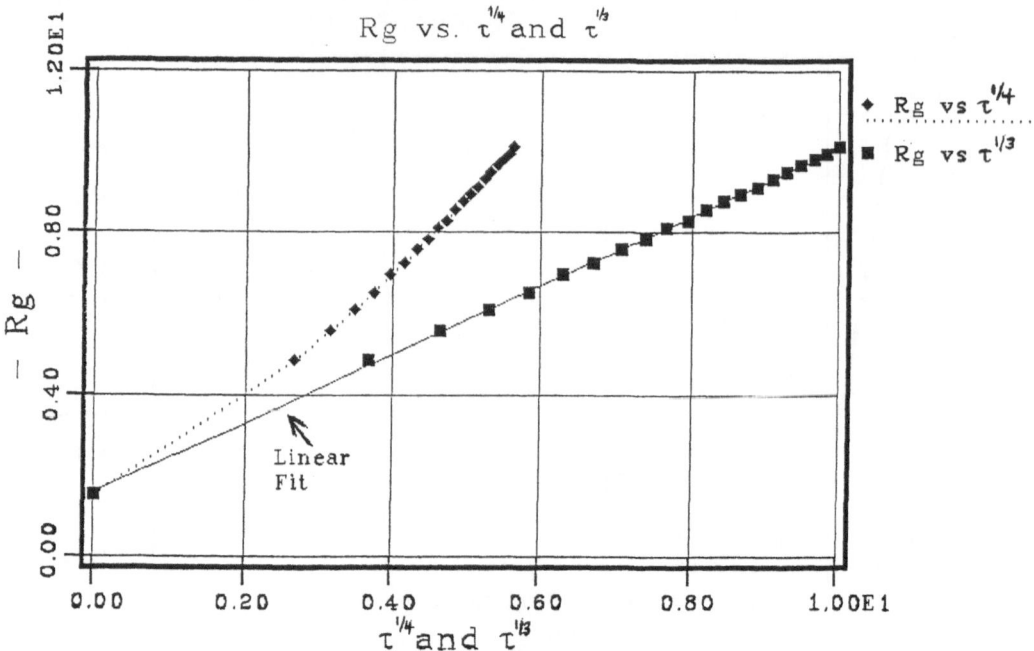

Figure 3. Plots of the data for R_g vs. $\tau^{1/3}$ and $\tau^{1/4}$, respectively. As can be seen, the data is well fit by the $\tau^{1/3}$ form.

Langevin model discussed in section 2 can be "derived" from the kinetic Ising model by a semi-macroscopic, coarse graining procedure. Since this is not an exact derivation, it is of interest to see if these models belong to the same dynamical universality class. Our work, plus that of Amar et al., strongly suggests both models do belong to the same universality class, as we will see below.

Since the details of this work will be published elsewhere[7] and the procedure in the Monte Carlo simulation is standard, we again restrict ourselves to a discussion of the main results. Several quantities were computed, the spin-spin correlation function $G(r,t)$, the energy and the circularly averaged structure factor. The calculations were carried out for two different quench temperatures, $T_F = 0.6\ T_c$ and $T_F = 0.9\ T_c$. We only discuss the former case here. The calculations were carried out for an Ising model with nearest-neighbor interactions only, on lattices of size 100 x 100 and 200 x 200 respectively. In the former case averages were computed over 200 independent runs, while in the latter case averages were computed over 100 runs. The dynamical properties were computed out to a maximum time of 150,000 Monte Carlo steps per spin. The work of Amar et al.[6] considered a larger system of 512 x 512 spins. The bulk of the work dealt with a quench to $T_F = 0.5\ T_c$, with 100 runs out to 100,000 Monte

Carlo steps per spin. One long run out to 10^6 MCS/spin was also carried out. The results of these two studies are extremely similar.

In Fig. 4 we show the results[7] of plotting the correlation function $G(r,t)$ vs. r/R_g. Note that a very good fit to a scaling form $F(r/R_g)$ is obtained over the wide range of 1,500 - 150,000 Monte Carlo steps per spin. The behavior is similar to that shown in Fig. 2. One could also argue that small deviations from scaling might be present in Fig. 4, although

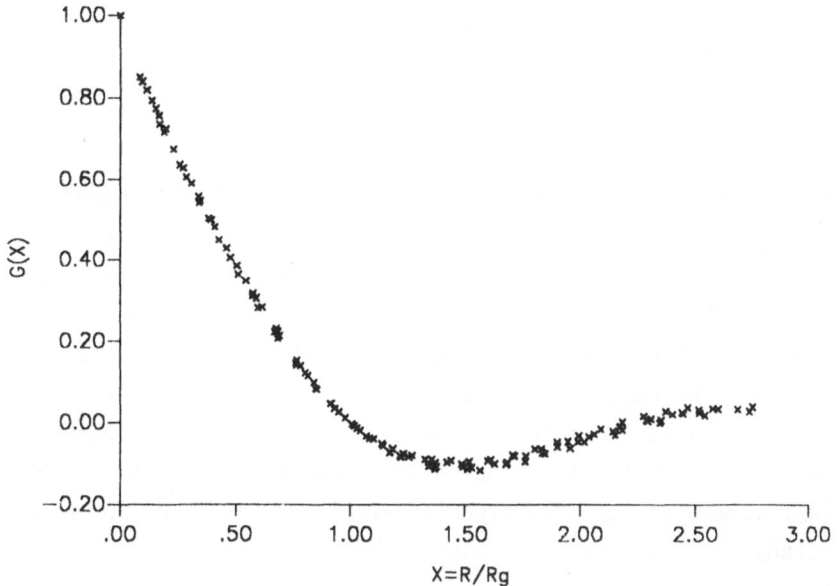

Figure 4. The spin-spin correlation function vs. r/R_g, to check the scaling ansatz. The data is for the 200 x 200 system at $T = 0.6 \, T_c$, averaged over 100 runs. Data points are shown for nine different times, ranging from 1500 - 150,000 Monte Carlo steps per spin.

as in Fig. 2 this is difficult to determine. Similar scaling results are obtained if we plot $G(r,t)$ vs. r/R_E, where R_E is a length scale determined from the energy per spin, $E(t)$, with

$$R_E(t) = 2J/(E(t) - E_0) \tag{1}$$

In (1) J denotes the coupling constant between the spins while E_0 denotes the equilibrium energy per spin at the quench temperature T. An analogous quantity has been computed for the Langevin model discussed in section 2. Note that R_E is an intuitively appealing length scale, since it is a measure of the inverse perimeter density. The circularly averaged structure factor also satisfies a scaling form, as shown for example in reference 6.

Finally, we consider the domain growth law for the kinetic Ising model. There are two equivalent ways to express the domain growth law predicted by Huse. The first is the form used in fitting the data for the Langevin model in section 2, namely

$$R(t) = A + Bt^{1/3} \tag{2}$$

An alternative form is to define an effective exponent $n_{eff}(t)$ as

$$n_{eff} = d(\log R(t))/d\log t \quad . \tag{3}$$

Eq. (2) implies

$$n_{eff}(t) = \frac{1}{3}\left(1 - \frac{A}{R(t)}\right) \tag{4}$$

Amar et al. found that their data agrees with (2) (Within 0.3%). They also computed n_{eff} for various choices of the length $R(t)$. The extrapolated value of n_{eff} as $t \to \infty$ was found to be 1/3, within the statistical error of the data. Essentially the same results have been found in reference 7, at $T = 0.6\ T_c$. For example, in Fig. 5 we plot the effective exponent versus $1/R_g$. The extrapolated value is 0.322. Doing a similar plot using R_E as the length scale yields a slightly higher value, but both estimates are consistent with x = 1/3. As in the study by Amar et al.[6], no evidence of a logarithmic growth predicted in reference 3 is found in this Monte Carlo study[7].

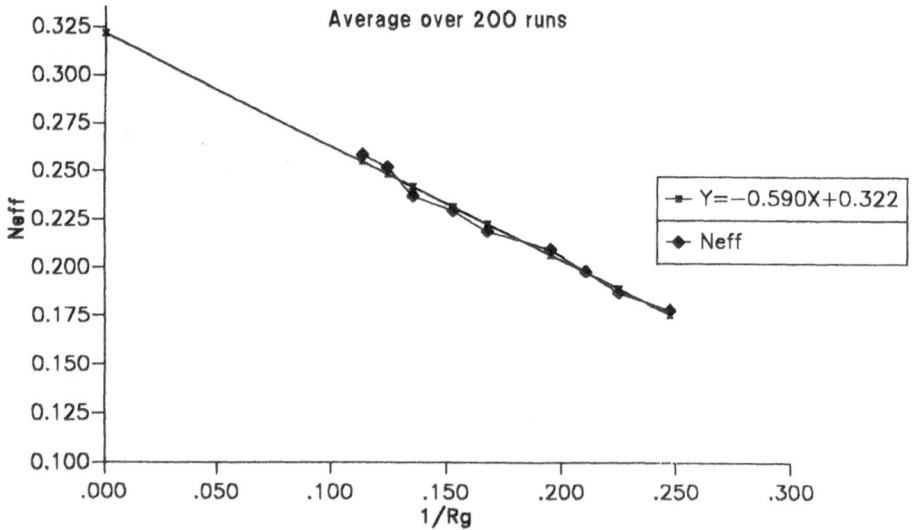

Figure 5. n_{eff} vs. $1/R_g$. Extrapolated value is 0.322

4. Conclusion

We have concentrated in this article on a rather narrow issue, namely the domain growth law and scaling function for two dimensional systems with a conserved, scalar order parameter, following a quench below the critical point at a critical value of the order parameter. Nevertheless, two important points should be stressed concerning the behavior of three dimensional systems. The first is that if $R = A + Bt^{1/3}$ does describe the growth law for critical quenches in two dimensions, this law should also be valid for analogous three dimensional systems. The second concerns the issue of a mathematical characterization of the interconnected structure associated with spinodal decomposition and coarsening. This would seem to be a fundamental issue in the entire field of pattern formation, which includes many disciplines. The question is easier to answer for the regime of nucleation and growth, where one has well defined, relatively isolated droplets. In this case the morphological structure is characterized by specifying the droplet distribution function, $f(R,t)$, which is the number of droplets per unit volume with radius R, at time t. Quite extensive theoretical predictions for $f(R,t)$, as well as $S(k,t)$, now exist for three dimensional systems, for relatively small volume fractions of these droplets[10-12]. More experimental work is necessary to test the validity of these theories, in particular to test the predictions for the dependence of $f(R,t)$ and $S(k,t)$ on the volume fraction of droplets. In the interconnected regime, the analogue of the droplet distribution function would seem to be the time dependent curvature distribution function, $P(c,t)$, describing the curvature of the interfaces which separate domains of opposite phase. This seems to be a more difficult quantity to compute. The work on the Langevin model[8] is at best a first step toward characterizing the morphology of the interconnected structure. Nevertheless, it is gratifying that reasonable results are obtained for $R_c(t)$. Much more theoretical attention should be given to this problem, as well as to calculating scaling functions in the interconnected regime.

We conclude with a brief remark concerning the problem of determining growth laws, etc., which arises in both numerical simulation and experimental studies. It is obvious that one must obtain accurate data over several decades in time before a definitive comparison with theoretical predictions can be made. Critical phenomena matured as a discipline when high quality data over several decades of reduced temperature were obtained. Although there are several such examples of analogous experimental studies in the kinetics of first order phase transitions, more of this seems necessary in both the numerical simulation and experimental areas.

This should provide a strong challenge to the many scientists in this field. In the same sense, it would seem very worthwhile to explore theoretical models which attempt to focus on the asymptotic behavior of ordering systems. These include interface models[13], cell dynamics[14] and renormalization group approaches[4,5,15].

Acknowledgements

This work was supported by NSF Grant #DMR8612609 and the computing facilities at Temple University, Tampere University at the IBM Bergen Scientific Centre. We wish to thank Dr. Amit Chakrabarti, Dr. Joseph Collins, Dr. Debashish Chowdhury, Professor Martin Grant and Dr. Jorge Vinals for many helpful conversations concerning the subject of this paper.

References

1. Reviews of the field are given in J. D. Gunton, M. San Miguel and P. S. Sahni, in Phase Transitions and Critical Phenomena, Vol. 8, ed. C. Domb and J. L. Lebowitz (1983); K. Binder in Condensed Matter Research using Neutrons, S. W. Lovesey and R. Scherm, eds. (Plenum, 1984); J. D. Gunton in Proceedings of NATO Advanced Study Institute on Time Dependent Effects in Disordered Materials, Geilo, Norway 1987, ed. Roger Pynn, (Plenum), to be published.
2. See, for example, P. Fratzl, J. L. Lebowitz, J. Marro and M. H. Kalos, Acta Metall. $\underline{31}$, 1849 (1983) and references therein; K. Binder and D. W. Heermann, in Scaling Phenomena in Disordered Systems, ed. by R. Pynn and A. Skjeltorp (Plenum, 1985); A. Milchev, K. Binder and D. W. Heermann, Z. Phys. B$\underline{63}$, 521 (1986).
3. D. Huse, Phys. Rev. B$\underline{34}$, 7845 (1986).
4. G. F. Mazenko, O, T. Valls and F. C. Zhang, Phys. Rev. B$\underline{32}$, 5807 (1985).
5. G. F. Mazenko and O. T. Valls, Phys. Rev. Lett. $\underline{59}$, 680 (1987).
6. J. G. Amar, F. E. Sullivan and R. D. Mountain, to be published.
7. K. Kaski, J. D. Gunton and D. Chowdhury, submitted for publication.
8. E. T. Gawlinski, J. D. Gunton and J. Vinals, submitted for publication.
9. M. Grant, M. San Miguel, J. Vinals and J. D. Gunton, Phys. Rev. B$\underline{31}$, 3027 (1985).
10. M. Tokuyama, K. Kawasaki and Y. Enomoto, Physica 134A, 323 (1986).
11. M. Tokuyama, Y. Enomoto and K. Kawasaki, Physica 143A, 183 (1987).
12. Y. Enomoto, M. Tokuyama and K. Kawasaki, Acta Met. 34, 2119 (1986).
13. K. Kawasaki and T. Ohta, Prog. Theor. Phys. $\underline{68}$, 129 (1982); K. Kawasaki, Ann. Phys. (N. Y.) $\underline{154}$, 319 (1984).
14. Y. Oono and S. Puri, Phys. Rev. Lett. $\underline{58}$, 836 (1987).
15. J. Vinals, M. Grant, M. San Miguel, J. D. Gunton and E. T. Gawlinski, Phys. Rev. Lett. $\underline{54}$, 1264 (1985); J. Vinals and J. D. Gunton, Phys. Rev. B$\underline{33}$, 7795 (1986).

MONTE CARLO SIMULATION OF MODULATED PHASES

D. J. Srolovitz[#,*], G. N. Hassold[*], and J. Gayda[+]

[#]Los Alamos National Laboratory   [+]NASA Lewis Research Center
Los Alamos, NM 87545 USA   21000 Brookpark Road
Cleveland, OH 44135 USA

[*]Department of Materials Science and Engineering
University of Michigan
Ann Arbor, MI 48109 USA

INTRODUCTION

Modulated phases exist over a wide range of material classes. In general, the term 'modulated phases' refers to the existence of an ordered array of multiple constituent elements of the material. These elements are typically either atoms of different Z or blocks of different phases. A good example of the former are intermetallics, such as the Ll_2 structure Ni_3Al which consists of four interpenetrating simple cubic lattices (one occupied by Al and the other three by Ni). If we ignore chemical type, this structure is simply face centered cubic. Materials such as SiC are also ordered, but unlike Ni_3Al it exhibits a wide range of phases, some with unit cells as large as 1200 nm on an edge[1]. The existence of multiple ordered structures of the same material is known as polytypism and has been observed for such disparate materials as metal iodides, micas, chalcogenides, opal, and graphite[1]. Different polytypes are often related to each other by the presence of periodic arrays of planar defects such as stacking faults or antiphase boundaries. Recent work[2] has suggested that spin models, such as the Axial Next Nearest Neighbor Ising (ANNNI) model provide a reasonable description of polytypism. In two dimensions, this model exhibits[3] both ferromagnetic and ordered (alternating two spin wide stripes - ↑↑↓↓↑↑↓↓↑↑↓↓↑↑↓↓) phases at low temperatures. In addition to a low temperature ferromagnetic phase, the three dimensional ANNNI model has a large number of different ordered phases at different temperatures and magnitudes of the frustration parameter.

In addition to exhibiting a number of different ordered crystal structures, the Ni-Al system also shows modulation of the microstructure in the two phase regime. This modulation consists of a relatively regular array of cuboidal, Ll_2 Ni_3Al particles, separated by regions of Ni with Al in solution. Application of a uniaxial external stress can convert these arrays of cuboidal particles to arrays of needle-like or plate-like particles[4]. This change in the morphology on application of an external stress is due to the elastic anisotropy of the particles versus the matrix and occurs by the diffusional linking of the cuboidal particles. Since the modulation, in this case, is due to the presence of elastic interactions between precipitates, no spin model has yet been applied to this type of modulation.

In this paper, we present Monte Carlo simulation results for the formation of modulated phases in the framework of the two dimensional ANNNI model with a nonconserved order parameter[5]. This work complements the earlier studies of Kaski, et al.[6] by examining a different, wider area of parameter space and temperature. Like Kaski, et al., we find[5] that for certain temperatures and values of the frustration parameter, κ, ordered domains form quickly and the correlation length grows as the square root of time. However, there exists a range of κ for which a quench from high to low temperature results in the formation of a metastable glassy phase. In addition to the ANNNI model study, preliminary results are presented on a newly developed model which exhibits phase modulation due to the presence of elastic interactions between the different phases and with an externally applied stress.

THE 2-d ANNNI MODEL

Simulation Procedure

In two dimensions, the ANNNI Hamiltonian may be written as:

$$H = -(1/2) \sum_{i,j} J_0 \, S_{i,j} S_{i\pm1,j\pm1} + J_1 S_{i,j} S_{i\pm2,j} \qquad (1)$$

where $S_{i,j}$ is the spin orientation (±1) on site i,j, and J_0 and J_1 are constants. It is convenient to normalize the ANNNI Hamiltonian by the nearest neighbor coupling constant, J_0, such that the site energy, E_{ij}, is:

$$E_{ij} = -S_{ij}[S_{i\pm1,j\pm1} - \kappa \, S_{i\pm2,j}]/2 \qquad (2)$$

where $\kappa=-J_1/J_0$. The temperature, T, has been normalized by J_0/k_B, where k_B is Boltzman's constant. The phase diagram for the 2-d ANNNI model is well established[3,7] and is indicated[7] in Fig. 1. The ferromagnetic phase is denoted by F, the paramagnetic phase by P, the incommensurate phase by I, and the modulated phase by <2>, indicating that it corresponds to alternating stripes of two up and two down spins oriented perpendicular to the axial direction (i.e. the i direction).

In the present study, we perform[5] Monte Carlo simulation of the evolution of the ANNNI model quenched from $T \gg T_c$ to a finite temperature, $0.02 \le T \le 0.4$, for $0.5 \le k \le 20.0$. Simulations were also performed for $\kappa < 0.5$, however those results are presented elsewhere[5]. The simulations were performed with nonconserved dynamics with a random updating scheme. The transition probability employed was: $W = [1-\tanh(\delta E/2T)]/2$, where δE is the change in energy of the system due to an attempted spin flip. All of the simulations were performed on $N = 200 \times 200 = 40,000$ site square lattices and the data was averaged over at least 5 simulations.

The simulations were analyzed by monitoring the magnitude of the structure factor in the axial (i) and perpendicular (j) directions. Since the excess energy of the system, ΔE, (i.e., the instantaneous energy minus the ground equilibrium energy) scales inversely with the correlation length[8], the time dependence of the excess energy was also monitored. The equilibrium energy was determined as a function of κ from 2000 MCS (1 MCS is defined as N microtrials) simulations starting from the appropriate T=0 ground state.

112

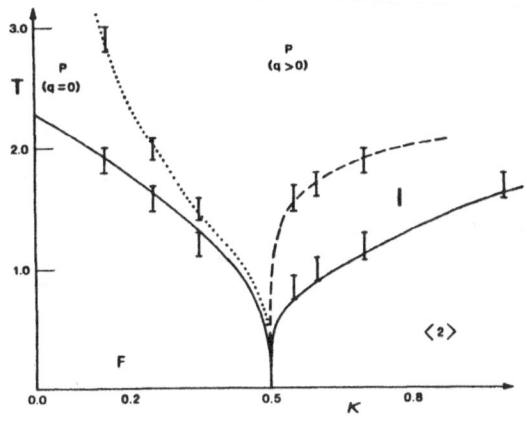

Fig. 1. The 2-d ANNNI phase diagram, after Beale, et al.[7]

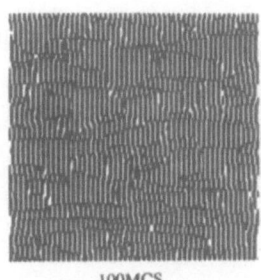
100MCS

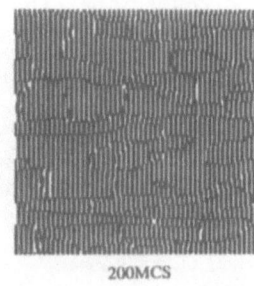
200MCS

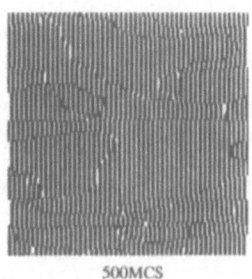
500MCS

Fig. 2. The temporal evolution of the ANNNI microstructure following a quench from T≫T_c to T=0.02 for κ=20.0. The dark and light regions correspond to spin up and down, respectively.

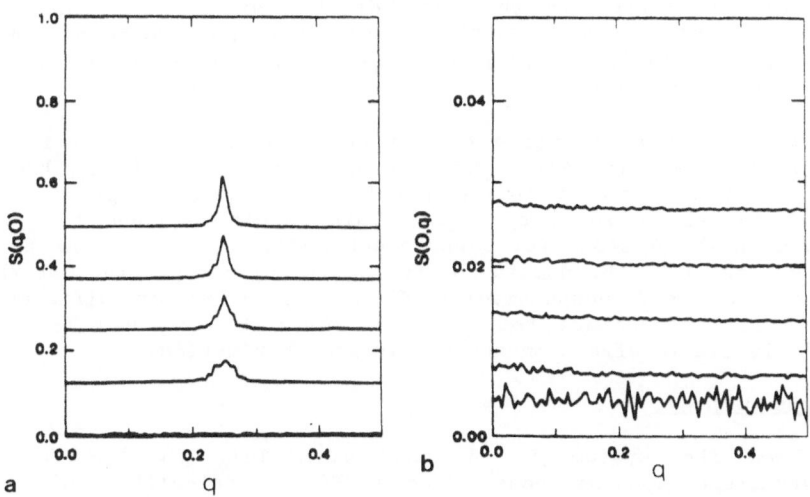

a

b

Fig. 3. The time dependence of the structure factor in the direction (a) parallel and (b) perpendicular to the aixal direction for the microstructure of Fig. 2. The curves correspond to 400, 300, 200, 100, and 0 MCS, in order from top to bottom.

Quenches With $\kappa > 1$

The temporal evolution of the ANNNI microstructure following a quench from $T \gg T_c$ to $T=0.02$ for $\kappa=20$ is shown in Fig. 2. The <2> phase, which consists of two spin wide stripes, forms rapidly and the resultant domain structure coarsens with time. Unlike in the 2-fold degenerate Ising model, the <2> phase in the ANNNI is 4-fold degenerate and hence has domain wall vertices. The orientation of the domain walls in the <2> phase region depends on the relative energies of the different types of domain walls and their relative orientation. For $\kappa > 1$, the energies of the domain walls are such that the domain walls lying in the axial direction are lower in energy than those in the perpendicular direction. This asymmetry is manifested in the microstructures of Fig. 2, where the domain walls are seen to lie preferentially in the low energy, axial direction.

Figure 2 shows that the model has both 3- and 4- fold vertices, like the 4-fold degenerate Potts and clock models, respectively. The 4-fold vertices are such that as one circumscribes the vertex, the 4-spin pattern of the <2> phase is phase shifted by one spin at each domain wall. This leads to two types of 4-fold vertices having either a clockwise or counter-clockwise vorticity. It is convenient to describe the properties of the vertex in the framework of dislocation theory in the sense that a Burger's circuit around the vertex yields a closure failure which is associated with an extra half-stripe either pointing up or down (depending on the sign of the vorticity). This extra half-stripe is akin to the extra half-plane of atoms associated with an edge dislocation. 3-fold vertices are equivalent to partial dislocations separated by stacking faults.

The asymmetry between the axial and perpendicular directions is seen clearly in Fig. 3 where we show the time dependence of the structure factor. In the axial direction, a peak at $q=1/4$ forms at early times and then slowly sharpens. The width of the peak indicates the <2> phase domain size. On the other hand, one should expect a peak at $q=0$ in the the structure factor in the perpendicular direction. The lack of a pronounced peak at $q=0$ is attributed to the preponderance of domain walls in the axial direction which tend to destroy long range order in the perpendicular direction.

The temporal evolution of the correlation length is indicated in Fig. 4a where we plot ΔE vs time for quenches to $T=0.02$ at $1 \le \kappa \le 20$. The slopes of these curves indicate that the correlation length is increasing as $t^{1/2}$ for all κ in this range. This result is consistent with that observed in the 4 state Potts and clock models. However, as pointed out by Kaski, et al.[6], the domain coarsening is anisotropic in that the growth in the axial and perpendicular directions occur at different rates. Varying the temperature to which the model is quenched (Fig. 4b) has relatively little effect on the domain growth kinetics.

Quenches With $\kappa < 1$

When the system is quenched form $T \gg T_c$ to $T \approx 0$ for $1/2 \le \kappa \le 1$, microstructures such as those shown in Fig. 5 typically result. Although the temperature and κ for the simulation shown in this figure are well within the ordered <2> phase region, the characteristic two spin wide strip pattern clearly does not form. Instead we observe a disordered phase which does not appear to be evolving toward the equilibrium <2> state. Comparison of the spin configurations at early and late times show little difference. There does, however, appear to be an extremely slow

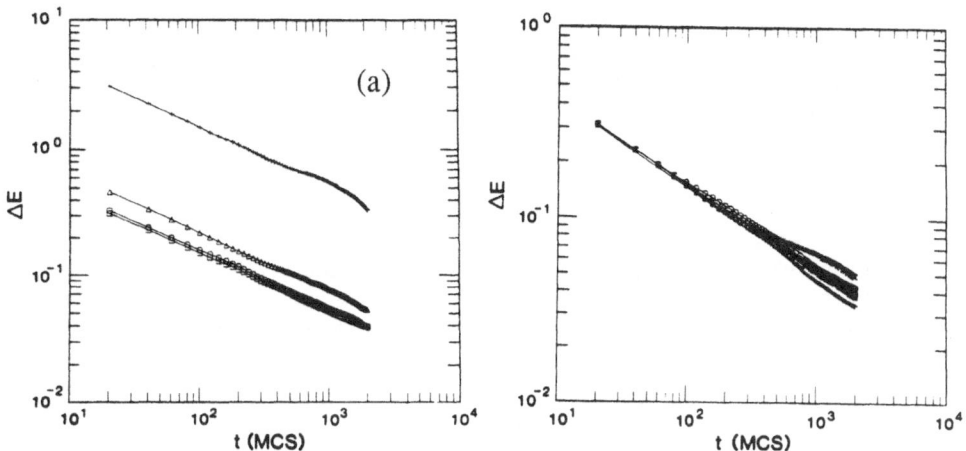

Fig. 4. The time dependence of the excess energy, ΔE, following a quench from $T \gg T_c$. (a) Quench to T=0.02 with $\kappa = 1$ (squares), 1.2 (circles), 2.0 (triangles) and 20 (+'s). (b) Quench to T=0.02 (squares), 0.1 (circles), 0.2 (triangles), 0.3 (+'s), and 0.4 (x's) at $\kappa=1$.

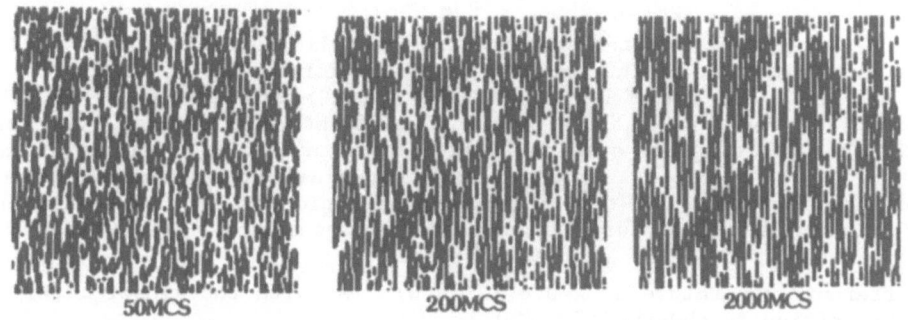

Fig. 5. The temporal evolution of the ANNNI microstructure following a quench from $T \gg T_c$ to T=0.02 for $\kappa=0.8$.

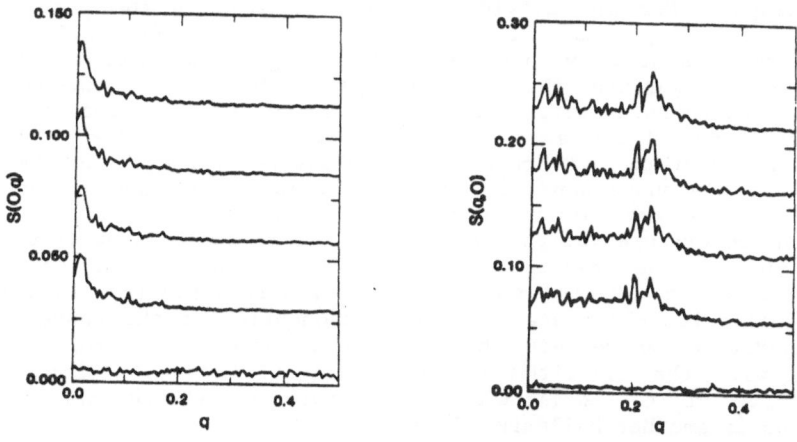

Fig. 6. The time dependence of the structure factor in the (a) axial and (b) perpendicular direction for a quench to T=0.02 at $\kappa=0.6$. The curves in each figure correspond to 2000, 1500, 1000, and 500 MCS, in order from top to bottom.

increase in the correlation length with time. Although the equilibrium <2> phase does not appear, short range order clearly exists. The presence of short range order with no accompanying long range order is a common feature of glassy systems. The microstructures quenched into this region of the phase diagram are nearly indistinguishable from those quenched to low temperature for $0 \leq k \leq 1/2$, where the equilibrium state is ferromagnetic.

The time dependence of the structure factor for a quench from $T \gg T_c$ to $T=0.02$ at $\kappa=0.6$ is shown in Fig. 6. The structure factor in the axial direction shows only a very small amplitude for $q>1/4$ and a larger amplitude in the range $0<q<1/4$. While the amplitude is largest around $q=1/4$ (as is expected for the <2> state), it is interesting to note that an increased amplitude is also present near $q=0$. This shows the remarkably strong competition between the F and the <2> phases, even when κ is well within the <2> state phase field. In the perpendicular direction, the model clearly exhibits ferromagnetic tendencies, as indicated by the peak near $q=0$. The amplitude of this peak appears to increase slowly with time, although no clear <2> or F ordering is apparent in the microstructure. The temporal evolution of the correlation length is indicated in Fig. 7a for quenches with $0 \leq \kappa \leq 1$. In the range of κ between 1/2 and 1, the growth is sub-power law, indicative of a kinetically frozen system.

Increasing the temperature to which the quench was performed leads to the formation of the equilibrium phase for $1/2 < \kappa < 1$ (see Fi. 7b). For $T>0.2$ at $\kappa=0.6$ and $T>0.1$ at $\kappa=0.8$, the <2> state appears to form and exhibits power law kinetics within the 2000 MCS simulations. However, the <2> state does appear even at lower temperature given sufficient time (see Fig. 7c). As for all true glasses, the glass transition temperature is only defined once the time scale of observation is set. The existence of a transition temperature is also seen by consideration of the structure factor after a 2000 MCS aging of the system following quenches to different temperature at $\kappa=0.6$ (Fig. 8). As with the kinetics (Fig. 7b), a transition is observed for $T>0.2$.

The transition from defected <2> phase formation and growth to glass formation that occurs at $\kappa=1$ can be understood in terms of the growth of a single stripe. For an individual stripe to grow in the perpendicular direction, its tip must be able to advance through all possible environments presented by the quench from high temperature. The most unfavorable configuration that the tip must be able to grow through occurs when, say, the up spin strip encounters a region of down spins at its tip. The energy required for such growth is proportional to $1-\kappa$. Therefore only for $\kappa>1$ can the <2> phase form following a quench to $T=0$. For $1/2<\kappa<1$, the <2> phase can form only by thermally activated growth. The required thermal activation increases with decreasing κ. The glass transition temperature is, therefore, expected to scale linearly with $1-\kappa$. The magnitude of the transition temperature also depends on the duration of the observation. Recognizing that the time required for the formation of the equilibrium phase scales as the exponential of the energy barrier over temperature, we expect the glass transition temperature to scale inversely with the logarithm of the observation time. This relative insensitivity of the transition temperature to the duration of the observation is another hallmark of a glass.

MODULATED PHASES - ELASTIC INTERACTIONS

Simulation Procedure

The model employed in simulating a two phase system with elastic

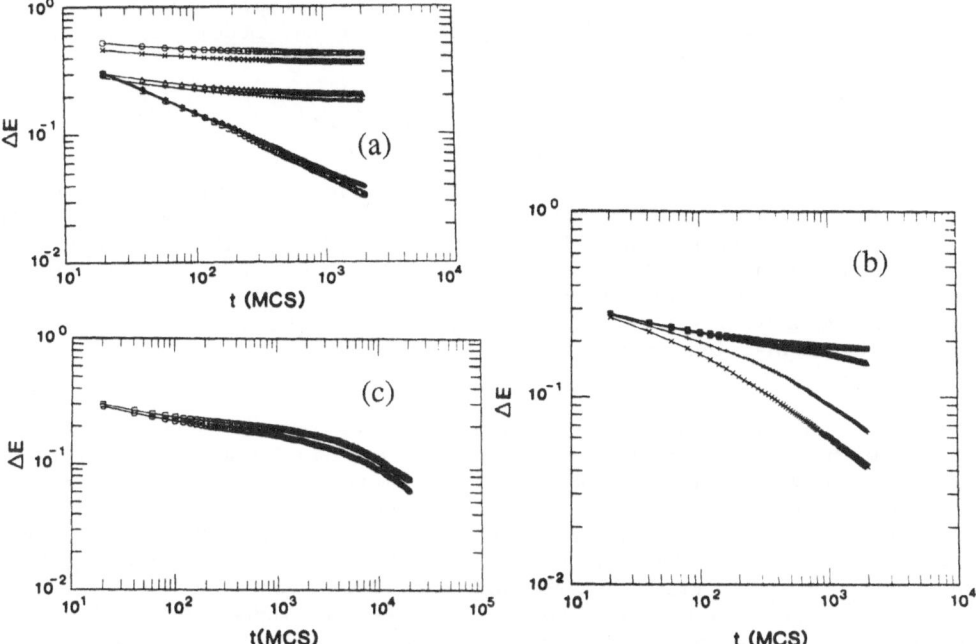

Fig. 7. The time dependence of the excess energy following a quench from $T > T_c$. (a) The curves correspond to a quench to T=0.02 at $\kappa = 0$ (squares), 0.2 (circles), 0.4 (triangles), 0.6 (+'s), 0.8 (x's), and 1.0 (diamonds). (b) Quenches at κ=0.4, and T=0.02 (squares), 0.1 (circles), 0.2 (triangles), 0.3 (+'s), amd 0.4 (x's). (c) Same as (a) but for 20,000 MCS for k = 0.4 (upper) and 0.6 (lower).

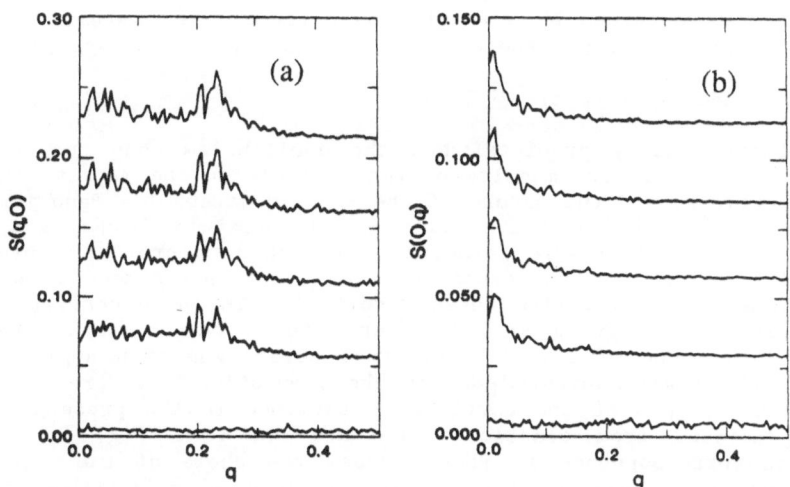

Fig. 8. The time dependence of the structure factor in the (a) axial and (b) perpendicular directions for a quench to T=0.02 at κ=0.6. The differenct curves in each figure correspond to 2000, 1500, 1000, and 500 MCS, in order from top to bottom.

interactions is based upon an Ising phase description with a finite
element elastic analysis. The Hamiltonian for the model may be written
as:

$$H = \sum_{i,j} -(J_0/2) \, S_{i,j} S_{i\pm1,j\pm1} + E_{el}(S_{i,j}, S, \sigma_{ext}) \qquad (3)$$

where J_0 is a positive constant, $S_{i,j}=\pm1$ indicates the phase of site i,j,
and $E_{el}(S_{i,j}, S, \sigma_{ext})$ is the elastic energy of the system associated with
site i,j, which depends on all of the other spins in the system (S) and
the externally applied stress σ_{ext}. The first term in Eq. 3 accounts for
the energy of the phase 1/phase -1 interface. The simulations were
performed on square lattices using spin-exchange (Kawasaki) dynamics with
the same transition probability as employed in the ANNNI model
simulations.

In performing the elastic analysis we allow one phase to have a
misfit with respect to the other and the two phases may have different
elastic constants (cubic symmetry is assumed). Since a finite element
calculation must be performed for each spin exchange attempt, it is
impractical to evaluate the elastic energy over the entire Monte Carlo
grid. Instead, once the two nearest neighbor sites to be exchanged are
identified, a small grid (between 6x6 and 18x18) centered on these sites
is chosen on which a finite element analysis is performed before and after
the attempted exchange. The nodes of the finite element grid are the same
as the lattice sites of the Monte Carlo grid. The finite element
calculations were performed using simplex grid elements and the finite
element code adopted was that described in Ref. 9. Although the elastic
energy depended sensitively on grid size, the change in energy due to a
spin exchange showed less than a 1% variation on going from the 6x6 to
18x18 finite element grid size. Since each finite element calculation was
performed on a sub-grid, the elastic effect of the material outside the
sub-grid was included by applying stresses to the surface of the sub-grid
equal to that obtained by averaging over the entire system. In effect,
these stresses are simply the external stresses (i.e., $<\sigma>=\sigma_{ext}$).

Stress Dependence of Particle Shape

Elastic theory predicts that the equilibrium shape of a particle
depends on the relative modulus of the particle to the matrix, the misfit
of the particle, and the nature of the applied stress[10]. Ignoring surface
energy Pineau[11] has produced a map of stable particle shapes as a function
of the appropriate elastic parameters. In order to provide a preliminary
test of the simulation technique, we have performed a series of
simulations to compare with these results[11]. The spin configuration was
initialized by assigning S=-1 to four sites arranged in a square in a
uniform background of S=1. An external stress was then applied and the
particle shape was monitored during the simulation. A direct comparison
between the shapes of the particles determined in the present study and
those predicted from elastic theory are shown in Fig. 9. Clearly,
excellent correspondence is found across the whole of the map, thereby
indicating that the approximate finite element elastic calculations are
reproducing the exact (within linear elasticity) theoretical results.

The interaction between particles can modify the particle shapes
determined in the single particle simulations. In order to investigate
this effect we initialized the spin configuration to consist of two
particles (4-spin, square particles; $E_p/E_m=0.8$, where E is Young's modulus
of the particle and matrix, respectively) in an otherwise uniform matrix.
An external stress was then applied and the system was allowed to
equilibrate at T=0 (again with zero surface tension). The resultant

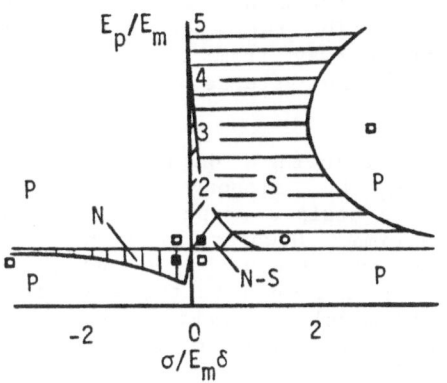

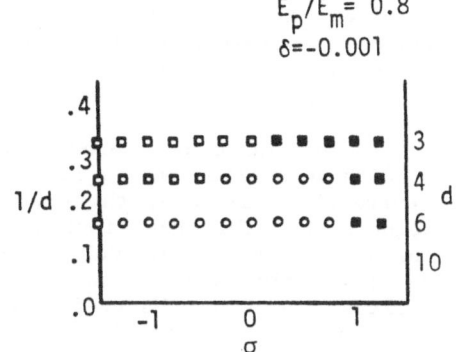

Fig. 9. Equilibrium particle shape in the absence of surface tension after Ref. 11. The circle, open and closed squares correspond to spheres (S), plates(P) and needles(N) determined by simulation. σ=stress and δ=misfit.

Fig. 10. Particle shape dependence on spacing, d, and stress, σ. The circle, open and closed squares correspond to spheres, plates and needles. E_p/E_m=modulus ratio of precipitate vs. matrix.

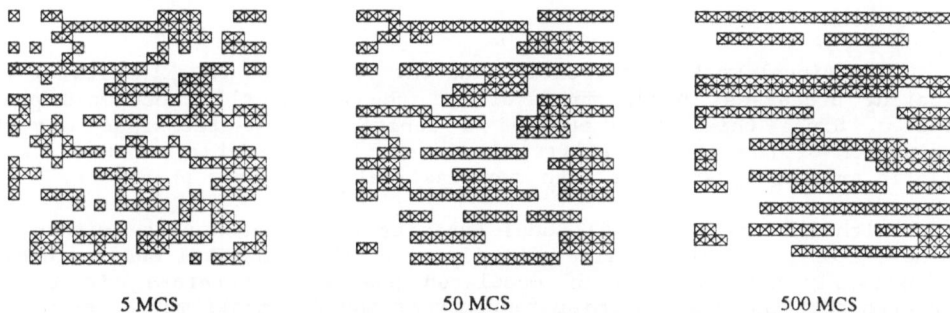

5 MCS   50 MCS   500 MCS

Fig. 11. The temporal evolution of the microstructure following a quench from $T \gg T_c$ to $T \approx 0$ with E_p/E_m=1.2, δ=.0005, and a negative stress.

Fig. 12. Stress dependence of the microstructure starting from a regular array of particles (center) with E_p/E_m=1.2 and δ=.0005.

particle shapes as a function of applied stress and the interparticle separation, d, are indicated in Fig. 10. The influence of the second particle is clearly to bias the particle away from the equiaxed and towards elongated shapes. This effect becomes stronger as the particle separation decreases.

The evolution of the microstructure following a quench from $T \gg T_c$ to $T \approx 0$, with an applied stress is shown in Fig. 11. Although no kinetic data is currently available, the characteristic striped morphology is clearly developing and the order increases with time. However, only short range order is visible and no well defined domain structure is evident. This type of ordering appears to be similar to that seen in the evolution of the glass-like phase observed in the ANNNI model. In making a comparison, it is important to note that in this model the evolution is controlled by spin-exchange dynamics, instead of the spin-flip dynamics employed in the ANNNI study.

Finally, we note that even when the structure is completely ordered at $T \ll T_c$, application of an external stress creates a rafted (elongated) structure (see Fig. 12). Such a change in order may be surprising in light of recent nonconserved order parameter studies where highly ordered metastable states are found to be relatively stable[12]. However, in the present case interactions are relatively long ranged compared to the interactions assumed in those nonconserved order parameter studies.

FUTURE WORK

Additional work is currently underway to examine the kinetics of the ordering processes in the framework of the elastically modulated phase model. Since this model employs a finite element technique for the evaluation of the elastic energy, it is easily extendible to the case of elastic anisotropy and more than two elastically distinct phases.

Although the 2-d ANNNI model exhibits a modulated phase within the framework of a relatively simple Hamiltonian, the 3-d ANNNI model exhibits an extremely wide variety of modulated phases. Therefore, it may be expected that the glass forming tendency of the 2-d model may be even more pronounced in 3-d. Future work aimed at providing a clearer understanding of this metastable, glass-like phase and a search for similar phases in other simple spin models is currently planned.

REFERENCES

1. G. C. Trigunayat and G. K. Chadha, Phys. Stat. Sol. a 4, 9 (1971).
2. G. D. Price and J. Yeomans, Acta Cryst. B 40, 448 (1984).
3. J. Villain and P. Bak, J. Phys. (Paris) 42, 657 (1981).
4. J. K. Tien and S. M. Copely, Met. Trans. 2, 215 (1971).
5. G. N. Hassold and D. J. Srolovitz, Phys. Rev. B, to be published.
6. K. Kaski, T. Ala-Nissila, and J. D. Gunton, Phys. Rev. B 31, 310 (1985).
7. P. Beale, P. M. Duxbury, and J. Yeomans, Phys. Rev. B 31, 7166 (1985).
8. J. Vinals, M. Grant, M. San Miguel, J. D. Gunton, and E. T. Gawlinski, Phys. Rev. Lett. 54, 1264 (1985); G. S. Grest and D. J. Srolovitz, Phys. Rev. B 32, 3014 (1985).
9. L. J. Segerlind, "Applied Finite Element Analysis", (Wiley, New York, 1976).
10. J. D. Eshelby, Prod. Roy. Soc. (London) A241, 376 (1957).
11. A. Pineau, Acta Metall. 24, 559 (1976).
12. S. A. Safran, Phys. Rev. Lett. 46, 1581 (1981).

VERTEX DYNAMICS OF TWO DIMENSIONAL DOMAIN GROWTH

Tatsuzo Nagai,[*] Kyozi Kawasaki,[+] and Katsuhiro Nakamura[#]

[*] Department of General Education, Kyushu Kyoritsu University
Kitakyushu 807, Japan
[+] Department of Physics, Faculty of Science, Kyushu University
33, Fukuoka 812, Japan
[#] Department of Physics, Fukuoka Institute of Technology
Fukuoka 811-02, Japan

INTRODUCTION

Many domain structures are seen in foams, grain aggregates, magnetic systems, and so on. These systems are characterized by compact domains and non-uniform domain sizes and shapes. Our interest is to describe such features in the non-equilibrium case.

The problem of domain growth has two aspects, time-variation of the domain size and statistics of domain shapes. Many computer simulations have been carried out on the first aspect.[1] However, the problem on the asymptotic growth law of domain size still remains unsettled, because of the difficulty of obtaining large scale systems in computer simulations. On the second aspect, Rivier gave a statistical mechanical theory of the stationary states,[2] but no dynamical theory has been given so far.

In order to find definitive asymptotic laws for these properties by computer simulation, we need large scale systems. For this purpose, we have adopted a coarse-grained, simplified model and have found the asymptotic behavior for the two-dimensional domain system.

DISSIPATIVE VERTEX MODEL

Our model for domain growth is as follows. The system consists of many vertices in two dimensions which are connected to each other by straight domain walls. Three domain walls emerge from each vertex. Domain states are P-degenerate and are specified by integers $\alpha = 1, 2, \ldots, P$. The equation of motion for the i-th vertex with its position vector $\underset{\sim}{r}_i$ at time t is given by

$$\eta \frac{d\underset{\sim}{r}_i}{dt} = -\underset{\sim}{\nabla}_i V(\underset{\sim}{r}_1, \underset{\sim}{r}_2, \ldots, \underset{\sim}{r}_N) \ , \ i = 1, 2, \ldots, N, \tag{1}$$

where η denotes the friction coefficient which may generally depend on t, N the number of vertices, and V the potential energy. The following three elementary processes occur when two or three vertices come within a short distance Δ (vertex size).

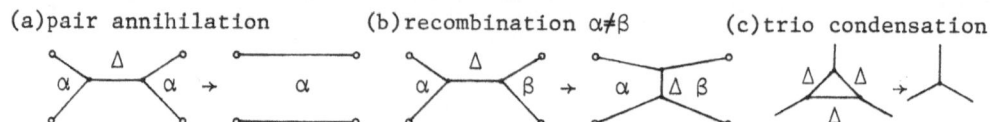

(a)pair annihilation   (b)recombination $\alpha \neq \beta$ (c)trio condensation

The equation of motion, (1), can be derived from the more general equation of motion for curved domain wall in the spirit of mean field theory. (Details will be presented elsewhere). Although the above model is thus quite general we only consider the simplest case with constant η, $P=\infty$, and the line tension energy $V=\sigma \Sigma_{<ij>}|r_i - r_j|$, where σ is a positive constant and the sum is taken over all vertex pairs (i,j) directly connected on the domain wall network. Equation (1) then becomes

$$\frac{dr_i}{dt} = - \sum_j{}' \frac{r_i - r_j}{|r_i - r_j|} \quad , \quad i=1,2,\ldots,N, \tag{2}$$

where the sum should be taken over three vertices, j, bonded to i on the network, and units of length and time have been suitably chosen to make them dimensionless. Note that the velocity of a vertex only depends on the relative angles between the three domain walls emerging out of the vertex but not on their lengths. Since we have $P=\infty$ all domains are always in different states from each other, which may be the case in foams and grain aggregates.

COMPUTER SIMULATION

We numerically solve the equations of motion for vertices, (2), by using the Runge-Kutta-Gill method with periodic boundary conditions.

The initial structure is constructed from a regular hexagonal network where some randomly chosen hexagons were divided into two rectangles and a smaller hexagon in random fashion and all of the vertex positions were shifted at random by a small amount. Thus we have a random but somewhat special initial structure which contains $N_0 \equiv 4800$ vertices and 3600 polygons with even numbers of edges (n=4,6,8,10,12). We take the time step $\Delta t=0.01$ and the vertex size $\Delta=0.01$. The initial average domain size is of order one.

In Fig.1 we show our results : Fig.1(a),(b),(c) and (d) are snapshots of the domain structure at t=2,10,20 and 30, respectively. At t=2 the domain structure is still undergoing a transient process. After t=10 universal behavior independent of the initial state can be observed, i.e. algebraic growth of the average domain size R(t) and self-similar domain network patterns. In fact, the patterns of Fig.1(b),(c) and (d) seem to be the same except for differing length scales. This suggests the scale invariance of the appropriately scaled domain size distribution function and hence the invariance of the edge number distribution function in time, as will be discussed below. Furthermore, the average domain size R(t) can be evaluated by the relation $R(t)/R_0 = \sqrt{N_0/N(t)}$ where R_0 denotes the initial average domain size and the result is shown in Fig.2. From this result we obtain the linear growth law $R(t) \sim t$ for t>10. The deviations of our data from the linear behavior are due to statistical fluctuations. We have carried out many simulations by changing N_0 and the initial domain distribution and have reached the same conclusion as mentioned above (Details will be presented elsewhere).

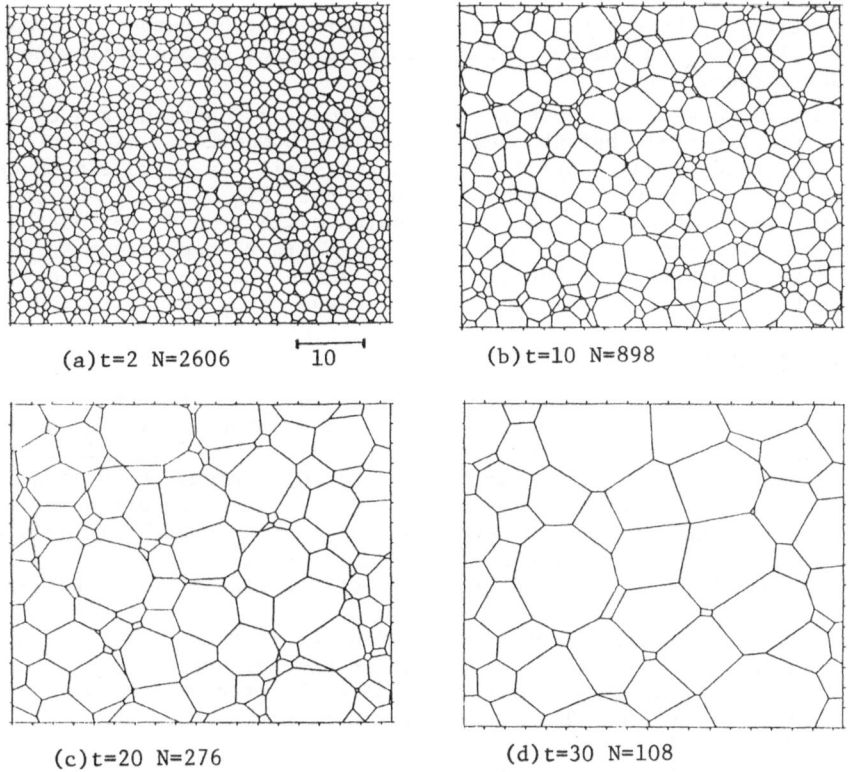

(a) t=2 N=2606 |―――| 10   (b) t=10 N=898

(c) t=20 N=276   (d) t=30 N=108

Fig.1. Time evolution of the domain structure for P = ∞.

The most elementary topological property of the domain system is the edge number distribution function f(n,t) where n denotes the number of edges of a domain.   This function is normalized as

$$\sum_{n=3}^{\infty} f(n,t) = 1 \ .$$
(3)

In our model we have the following topological identity for the average number of edges of a domain :

$$\overline{n} = \sum_{n=3}^{\infty} nf(n,t) = 6 \ .$$
(4)

Fig.2. Time dependence of the average domain size divided by its initial value (solid line).

123

This identity can be derived in the limit $N_D \to \infty$ (N_D: the number of domains) from Euler's relation and our assumption that three domain walls emerge from each vertex.[2]

In Fig.3 we show the edge number distribution functions $f(n,t)$ for the cases of Fig.1(b)($t=10, N_D=405$) and Fig.1(c)($t=20, N_D=124$), while for the case of Fig.1(d)($t=30$) we have $N_D=46$ which is too small for us to construct a reliable $f(n,t)$. At the transient stage of $t=2$, Fig.1(a), we obtain a more symmetric $f(n,t)$ which has the maximum value 0.413 at $n=6$, reflecting the initial distribution. Figure 3 suggests that $f(n,t)$ approaches a stationary function as time goes on. The dotted line shows the edge number distribution function found theoretically by Rivier and takes the following form :[2]

$$f(n) = C(n-n_1)e^{-\gamma n} , \qquad (5)$$

where we have only one adjustable parameter because of the two constraints (Eqs.(3) and (4)) to find $C=3.60$, $n_1=2.97$ and $\gamma=0.680$ for a stationary form for $t \to \infty$ which is expected from our results at $t=10$ and $t=20$. In fact n_1 was so chosen as to reproduce our result $f(3,20)$ approximately at $n=3$. However overall shape of $f(n)$ is insensitive to small variations of n_1 as long as it does not exceed 3. Our results seem to approach Rivier's one which is characteristic of the matured domain system. Moreover, our data give rise to $\bar{n}=5.93$ at $t=10$ and $\bar{n}=5.95$ at $t=20$ which are close to $\bar{n}=6$, (4).

The second elementary topological quantity is the topological correlations described by $m(n)$, the average number of sides of the nearest neighbor of an n-sided domain. Figure 4 shows $m(n)$ calculated using Fig.1(c) at $t=20$ as a function of $1/n$. The straight line is given by

$$m(n) = K_1 + \frac{K_2}{n} , \qquad (6)$$

where $K_1=5.04\pm0.13$ and $K_2=8.66\pm0.63$. Equation (6) has been obtained by applying the method of least squares to the seven data points ($n=3-9$) and well represents our simulation result. This means that there exist topological correlations in our domain system. Equation (6) is known as the Aboav-Weaire law in grain aggregates.[3] This law was first discovered by Aboav for polycrystalline magnesium oxide, for which the values $K_1=5$ and $K_2=8$ were obtained within an error of 5%.[4] The agreement between these values and ours is remarkable.

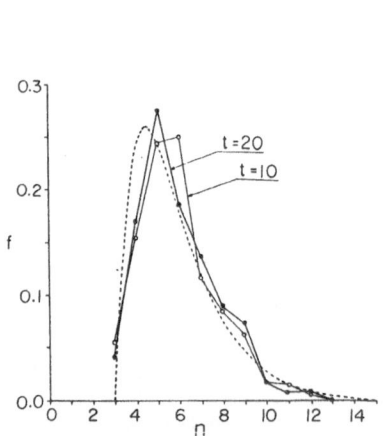

Fig.3. Edge number distribution functions of the domain.

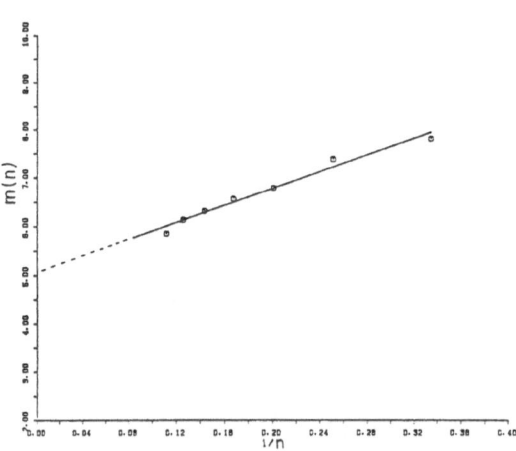

Fig.4. Average number of sides of the nearest neighbor of an n-sided domain.

DISCUSSION AND CONCLUSION

In this paper we have obtained the linear growth law $R \sim t$ for the simple case where η was constant, $P=\infty$, and Eq.(2) governed the motion of the vertices. This linear growth law is expected based on the dimensional analysis of Eq.(2) if one-parameter dynamical scaling is assumed. As mentioned before, Eq.(1) can be derived from the more general equation of motion appropriate for grain aggregates and then the average friction coefficient takes the form $\eta=\gamma R(t)$ with γ a positive constant. Changing the time variable from t to τ given by $d\tau=\sigma\gamma^{-1}R^{-1}dt$ in Eq.(1), we again obtain the same equation as Eq.(2) except for the change of the time variable for the case of the line tension energy V. The above simulation result leads to $R \sim \tau$ in this case and in turn to $\tau^2 \sim t$. As a result we obtain the square root growth law $R \sim t^{1/2}$. The exponent $1/2$ can be again expected by the dimensional analysis of the equation of motion, Eq.(1) with $\eta=\gamma R(t)$ and V the line tension energy, directly. Therefore we can conclude that the exponent of the growth law is determined by the form of the equation of motion for the vertices and not by the topological property of the domain structure for the case $P=\infty$.

One can refine Eq.(1) on the basis of the more general equation above mentioned for the case of grain aggregates where local fluctuation effects on the effective vertex friction coefficients are taken into account. Simulation results for the refined model look very similar to our results for the case $\eta=\gamma R(t)$ (Y. Enomoto, private communication).

Our results suggest that the edge number distribution approaches the stationary distribution in the long time regime which was given by Rivier. Furthermore we conclude that there exist topological correlations between neighboring domains in our domain system which are well described by the Aboav-Weaire law. These results on topological properties seem to depend neither on the specific forms of the equation of motion for vertex and nor on the initial distribution of domains but do depend on whether the domain structure is fully developed or not.

We would like to thank Drs. T. Ohta and K. Sekimoto for valuable discussions, and Kyushu University Computer Center for the use of their facilities.

REFERENCES

1. See, for example, J. Wejchert, D. Weaire and J. P. Kermode, Phil. Mag. B53: 15(1986) for references for foams and G. S. Grest, M. P. Anderson, and D. J. Srolovitz, in Proceedings of NATO Conference on "Time-Dependent Effects in Disordered Materials" (Geilo, Norway, 1987) for references for grain aggregates.
2. N. Rivier, Phil. Mag. B52: 795(1985).
3. C. J. Lambert and D. L. Weaire, Metallogr. 14: 307(1981).
4. D. A. Aboav, Metallogr. 3: 383(1970).

ORDERING PROCESS IN A QUENCHED TRICRITICAL SYSTEM

Takao Ohta[*], Yoshihisa Enomoto[+], Kyozi Kawasaki[#], and Akinori Sato[#]

[*] Department of Physics, Faculty of Science, Ochanomizu University
Tokyo 112, Japan
[+] Department of Applied Physics, Faculty of Engineering
Nagoya University, Nagoya 464, Japan
[#] Department of Physics, Faculty of Science, Kyushu University
Fukuoka 812, Japan

INTRODUCTION

Among the various ordering dynamics a tricritical system provides us with several interesting characteristic features.[1-3] To be specific we consider a binary alloy consisting of A and B atoms, which exhibits a phase diagram as shown in Fig.1. At high temperatures the system is in a disordered phase. The broken line in Fig.1 indicates the second order order-disorder transition line below which the system undergoes a macroscopic order. Note that the concentration X is defined such that the numbers of A and B atoms are equal at X=0. For simplicity we assume that the system is symmetric under the interchange of A and B atoms. Below the tricritical temperature T_t the order-disorder transition line separates into two coexistence lines as indicated by the full lines in Fig.1. Thus we note that at least three different types of temperature quench are possible. The arrows show such quenches.

The fundamental quantities of the ordering dynamics are the local order parameter S(r,t) and the local concentration difference X(r,t). Thus the interplay between the conserved and nonconserved quantities X(r,t) and S(r,t) respectively can be realized. For instance in the quench DO S(r,t), which is linearly unstable, governs the ordering dynamics. Since the ordered state is doubly degenerate antiphase boundaries are formed as the ordering proceeds. When X=0 the excess concentration is adsorbed at the antiphase boundaries.[3] In the quench DC S(r,t) is again a linearly unstable mode initially. However as the amplitude of X(r,t) increases via the nonlinear coupling between S and X the time-evolution will eventually be governed by X(r,t) because of its conservation. Here even when the initial volume fraction is not close to 0.5 interconnected thin domains are expected to emerge in an intermediate stage of the ordering process. This is a striking difference from a droplet pattern observed in the ordinary phase separation in an off-critical quench.

In this paper we describe a cell dynamic approach[4,5] to the quenched tricritical system to simulate the ordering dynamics. Since visualization of the growing domains is to be published separately[6] we here focus our attention mainly on the nonequilibrium scattering functions and explore the possible scaling law[7,8] expected at the late stage of the ordering process.

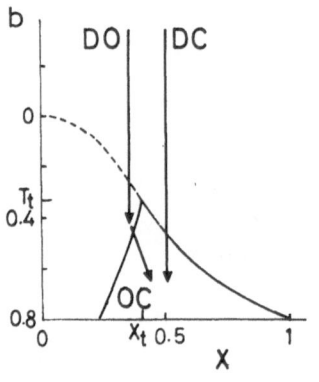

Fig. 1   Phase diagram in T-X space for g=2(1-b). The temperature is assumed to be proportional to b whose origin is the order-disorder point at X=0.

MODEL AND SIMULATION

Following the method proposed by Oono and Puri[4] for an ordinary phase separation kinetics we construct a space-time discrete model suitable for the tricritical system[6] (See also Ref.9.) We divide the space into small cells arranged in the square lattice. Within each cell S and X are constant in space. The time evolution is described by a map of these quantities. This map must have two stable fixed points corresponding to the equilibrium values $S=\pm S_e$ in the ordered phase and three stable fixed points inside the coexistence curve reflecting the triple degeneracy of the state. The coupling among cells is introduced through the local spatial averaging[4]

$$\ll S(n,t)\gg = \frac{1}{4(1+\varepsilon)} \{\sum_{nn} S(n-i,t) + \varepsilon \sum_{nnn} S(n-i,t)\} \tag{1}$$

where $S(n,t)$ stands for the value of S at the n-th cell. nn(nnn) means the summation over the four (next) nearest neighbour cells. The constant ε is here chosen as $\varepsilon=0.5$. The system should be invariant under the interchange of S and -S. Bearing this in mind and taking into account the assumed symmetry and the conservation law for X the model equation reads

$$S(n,t+1)=S(n,t)- M\{F(S(n,t),X(n,t))+ D_s[\ll S(n,t)\gg -S(n,t)]\} \tag{2}$$

$$X(n,t+1)=X(n,t)+ L[\ll G(S(n,t),X(n,t))\gg -G(S(n,t),X(n,t))] \tag{3}$$

where

$$G(S,X)=-D_x[\ll X\gg -X]+H(S,X) \tag{4}$$

and L, M, D_x and D_s are positive constants. The functions F and H are defined by

$$F(S,X)=\frac{\partial W(S,X)}{\partial S} \tag{5}$$

$$H(S,X)=\frac{\partial W(S,X)}{\partial X} \tag{6}$$

where $W(S,X)$ is given by

$$W(S,X)=\frac{1}{2}X^2-\frac{b}{2}S^2+\frac{g}{4}S^4+\frac{f}{2}S^2X^2-\mu X \qquad (7)$$

with b, g, f and μ positive constants. It is readily verified that (2) and (3) have three stable uniform fixed points for $g \leq 2bf$. Thus the condition

$$g=2bf \qquad (8)$$

specifies the tricritical point. By rescaling we may put b=1 and f=2 without loss of generality. Hereafter we use this unit of scales.

The discrete model (2) and (3) with (5)~(7) is not unique. Oono and Puri[4,5] in the study of spinodal decomposition in a system with one relevant variable have employed another form of the nonlinear coupling where the state converges more rapidly to the fixed points and which inherently exclude a chaotic behavior. However we do not use such a form in the present two-variable system simply because it complicates the task of obtaining the phase diagram such as the one in Fig. 1. It is noted that in a continuum limit (2) and (3) agree with the partial differential equations proposed previously for this system.[10,11]

RESULTS AND DISCUSSION

We have solved numerically the equations (2) and (3) on a 100x100 square lattice with a periodic boundary condition. In most of the simulations we have chosen the parameters as $D_s = D_x =1$. Initially X and S on each cell are chosen to be Gaussian random numbers with the averages $\langle S \rangle = S_i$ and $\langle X \rangle = X_i$ and with the variances $\langle(S-\langle S \rangle)^2\rangle^{1/2}$=0.04 and $\langle(X-\langle X \rangle)^2\rangle^{1/2}$ =0.02. Other parameters and S_i and X_i for each quench are summalized in Table 1.

A typical domain growth in the quench DC is displayed in Fig. 2. In this case the volume fraction of the disordered state (black area) is 0.443. As mentioned before the interconnected thin domains are formed and their thickness gradually increases. A droplet pattern is prohibited since there must be a disordered state between the regions where $S=S_e$ and $-S_e$.

We have evaluated the normalized scattering function for $S(n,t)$ defined by

$$I_s(k,t)=\langle S_k(t)S_{-k}(t)\rangle/\sum_k\langle S_k(t)S_{-k}(t)\rangle \qquad (9)$$

and similarly for $I_x(k,t)$. $S_k(t)$ is the Fourier components of $S(n,t)$. The cross correlation between S and X is also finite although we do not discuss this quantity here. If there is only one characteristic length scale $\ell(t)$ in the late stage of the ordering process the scattering function is

Table 1. The parameters used in the simulations

type of quench	L	M	g	X_i	S_i	result
DC	0.4	0.25	1	0.5477	0	Fig. 2
			2.5	0.6428	0	Fig. 3
DO	0.2	0.05	2.7	0.5477	0	Fig. 4
OC	0.2	0.05	1	0.5477	0.3849	Fig. 5

Fig.2   Growing patterns of X(n,t) for DC at t=250(left) and 1000(right).

expected to obey the scaling law[7],[8] i.e.

$$I_s(k,t)=\ell(t)^d f(k\ell(t))$$ (10)

and a similar relation for $I_x(k,t)$ where $\ell(t)$ increases in time as

$$\ell(t)=\ell_0 t^{\phi}$$ (11)

with ϕ a positive constant. d is the dimensionality of space.

Figures 3∿5 display the time evolution of $I_s(k,t)$ and $I_x(k,t)$ for the respective quenches. These are averaged over seven independent runs respectively. In the quench DC with $X_i=0.6187$, where the volume fraction of the two coexistence phases is 0.5, there appear three kinds of domains i.e. those characterized by $S=\pm S_e$ and $X=X_1$ and by $S=0$ and $X=X_2$. Thus as far as we are concerned with $I_x(k,t)$ it looks like the ordinary spinodal decomposition. However because of the degeneracy of S $I_s(k,t)$ behaves as that in the ordering kinetics in a system with a nonconserved order parameter. The degeneracy of S can be masked if we consider the correlation $\langle(S(n,t)S(m,t))^2\rangle$. In fact the Fourier transform of this correlation is found to be almost the same as $I_x(k,t)$. The exponent ϕ defined in (12) was evaluated from $[k^2]$ for $I_s(k,t)$ and $[k]$ for $I_x(k,t)$ where

$$[k^n]\equiv \int_0^\infty dk\ k^n I(k,t)/\int_0^\infty dk I(k,t)$$ (12)

From the last three time steps in Fig. 3 we obtain $\phi_s\simeq0.25$ (0.32) from $I_s(k,t)$ and $\phi_x\simeq0.30$ (0.24) from $I_x(k,t)$. The values in the parentheses were obtained from the peak heights assuming scaling. In a Monte Carlo simulation of a tricritical spinodal decomposition in a two-dimensional metamagnet[12] the corresponding exponents are estimated as $\phi_s\simeq0.35$ and $\phi_x\simeq0.21$. These are compared with the above values although both results are only approximate estimates due to the large fluctuations in simulations. In the quench DO we obtain $\phi_s\simeq0.33$ and $\phi_x\simeq0.49$. Thus the coarsening in the quench DC is slower than that in DO. We have not estimated the exponent ϕ for the quench OC (which can be realized by a pressure change in a binary fluid) since the time evolution is so slow that the power law behavior (12) was not seen within the time steps of the present simulation. In the plot of $I_s(k,t)$ we have excluded the time-dependent k=0 component. The similarity between $I_s(k,t)$ and $I_x(k,t)$ is due to the fact that there are

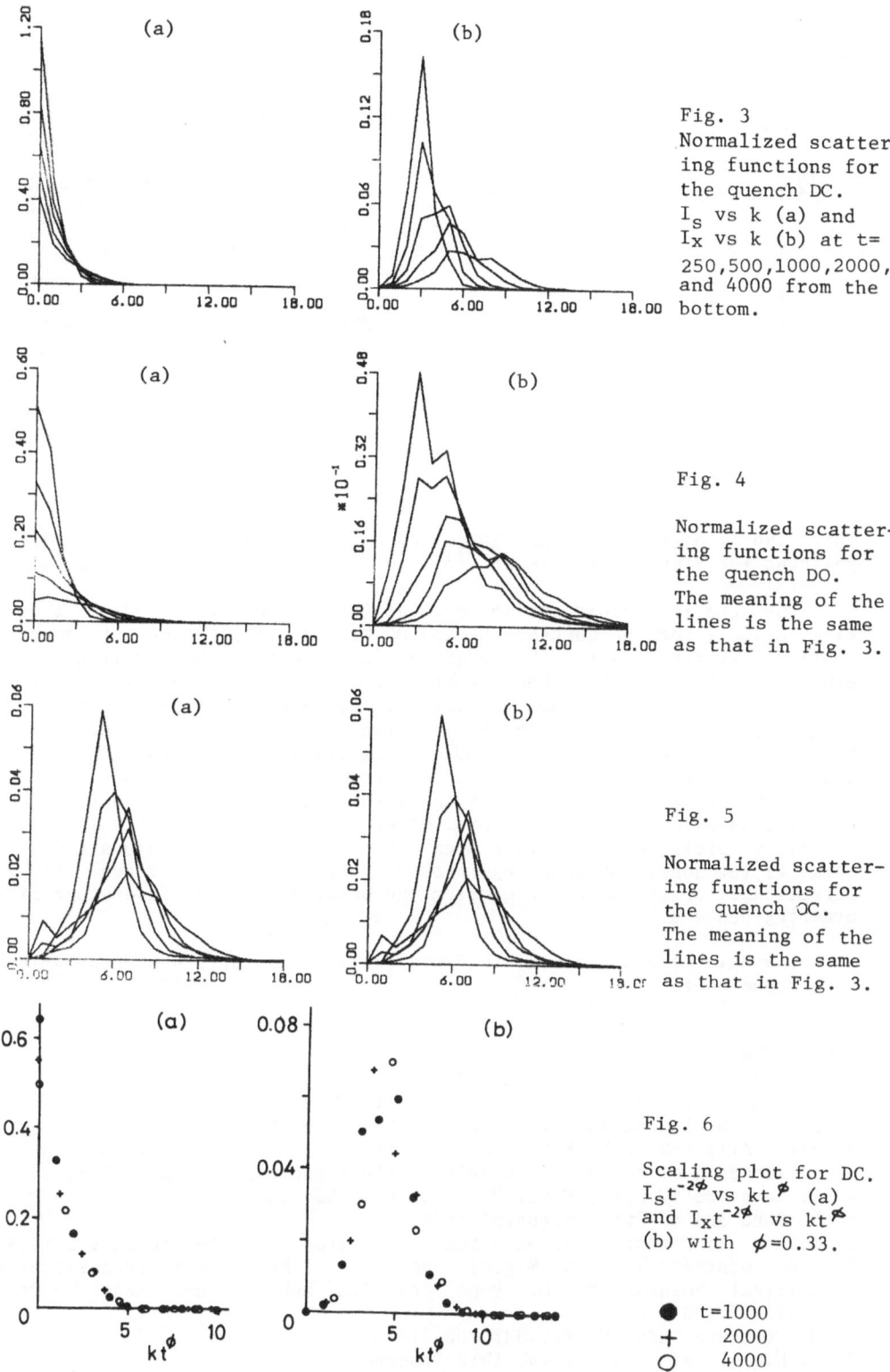

Fig. 3
Normalized scattering functions for the quench DC. I_s vs k (a) and I_x vs k (b) at t= 250,500,1000,2000, and 4000 from the bottom.

Fig. 4

Normalized scattering functions for the quench DO. The meaning of the lines is the same as that in Fig. 3.

Fig. 5

Normalized scattering functions for the quench OC. The meaning of the lines is the same as that in Fig. 3.

Fig. 6

Scaling plot for DC. $I_s t^{-2\phi}$ vs kt^{ϕ} (a) and $I_x t^{-2\phi}$ vs kt^{ϕ} (b) with ϕ=0.33.

● t=1000
+ 2000
○ 4000

131

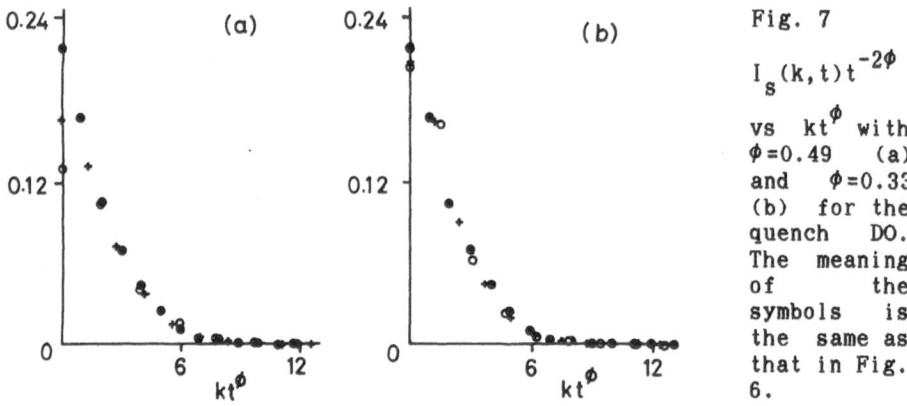

Fig. 7

$I_s(k,t)t^{-2\phi}$

vs kt^ϕ with $\phi=0.49$ (a) and $\phi=0.33$ (b) for the quench DO. The meaning of the symbols is the same as that in Fig. 6.

only two states $S=S_e$ and $X=X_1$ and $S=0$ and $X=X_2$. In fact the growing patterns for S and X are almost identical with each other in the quench OC.

The scaling plots for the data at t=1000, 2000 and 4000 are shown in Figs. 6 and 7 for the quenches DC and DO respectively. We do not expect the scaling law for $I_x(k,t)$ for the quench DO since the concentration does not exhibit a macroscopic phase separation. Actually the values of X(n,t) within the interfaces does not saturate but changes gradually in time. In other cases as shown in the figures the scaling seems to hold. Comparing Figs. 7a and 7b we see that $\phi=0.49$ is more favorable than $\phi=0.33$ except for the data at k=0. Theoretically we expect $\phi=1/3$ for DC and $\phi=1/2$ for DO in the interfacial approach. Thus the values estimated from the figures are not inconsistent with the theory. However the present results for scaling are very preliminary. Further study by changing the parameters and/or the form of the nonlinear coupling in such a way that the interface thickness becomes smaller is needed to confirm the behaviors in the late stage of the ordering process.

The authors are grateful to Professor Y. Oono for a number of valuable correspondences.

REFERENCES

1. S.M. Allen and J.W. Cahn, Acta. Metallurgica 24: 42 (1976).
2. K. Oki, H. Sagane and T. Eguchi, J. Physique 38: C7-414 (1977).
3. S.M. Allen and J.E. Krzanowski, "Solid-Solid Phase Transformations", H. I. Aaronson et al eds., The Metallurgical Society of AIME, (1982).
4. Y. Oono and S. Puri, Phys. Rev. Letters, 58: 863 (1987).
5. Y. Oono and S. Puri, preprint (1987).
6. T.Ohta, K. Kawasaki, A. Sato and Y. Enomoto, submitted to Phys.Lett.A.
7. J.D. Gunton, M. San Miguel and P.S. Sahni, "Phase Transitions and Critical Phenomena", C. Domb and J.L. Lebowitz ed., Academic Press, London (1983).
8. H. Furukawa, Adv. Phys., 34: 703 (1985).
9. T. Eguchi and H. Ninomiya, this Proceeding.
10. K. Kawasaki, unpublished work (1978).
11. T. Eguchi, K. Oki and S. Matsumura, Mat. Res. Soc. Symp. Proc., 21: 589 (1984).
12. P.S. Sahni and J.D. Gunton, Phys. Rev. Letters, 45: 369 (1980).

TEMPERATURE-DEPENDENCE AND CROSSOVER EFFECTS IN DOMAIN-GROWTH KINETICS

Ole G. Mouritsen

Department of Structural Properties of Materials
Technical University of Denmark, Building 307
DK-2800 Lyngby, Denmark

Hans C. Fogedby

Institute of Physics, Aarhus University
DK-8000 Aarhus C, Denmark

Eigil Præstgaard

Department of Chemistry
Roskilde University
DK-4000 Roskilde, Denmark

I. INTRODUCTION

The domain-growth kinetics in systems undergoing ordering processes after thermal quenching is believed to be determined by a few relevant properties of the systems.[1] Among the candidates for relevant properties are the ordering degeneracy, p, and the nature of the conservation laws in effect. It has also been suggested that at zero temperature the domain-wall softness may be relevant.[2,3] A major part of our current knowledge on domain-growth kinetics stems from computer-simulation studies of microscopic interaction models.[1,4,5] A quantitative analysis of the computer-simulation data, as well as of real experimental data in terms of dynamical scaling and a growth law, is severely hampered by the lack of appropriate theories, in particular for p>2. Moreover, the analysis is made difficult by a poor understanding of crossover phenomena in time and as a function of thermodynamic potentials, such as temperature, and various non-universal model parameters.

In this paper we discuss such crossover phenomena in systems with non-conserved order parameter. Our discussion is based on computer-simulation studies of two-dimensional kinetic lattice-spin models with different ordering degeneracies. A kinetic lattice model is specified by a Hamiltonian and a kinetic process, e.g. Glauber or Kawasaki dynamics.[1] Crossover can therefore be considered in terms of (i) temperature, (ii) parameters of the Hamiltonian (including lattice topology), and (iii) control parameters of the kinetic process. Specifically, we shall discuss the effects of the ratio of nearest- and next-nearest-neighbour interaction strengths, the range of Kawasaki spin exchange, and the degree of softness of the domain walls.

The coherent picture which emerges from the computer-simulation studies is that, with the exception of possible crossover effects to a special zero-temperature behaviour, the algebraic growth law

$$R(t) \sim t^n \qquad (1)$$

(R is the time-dependent average linear domain size) holds with the classical Lifshitz-Allen-Cahn[6,7] exponent n = ½ for all models studied with non-conserved order parameter. The exponent is independent of p, independent of the domain-wall softness, and independent of whether or not the density is conserved.

II. ZERO-TEMPERATURE KINETICS: CROSSOVER IN HAMILTONIAN PARAMETERS

We shall here be concerned with two-dimensional lattice models with sub-lattice ordering of an antiferromagnetic type. Several of the models describe ordering in physisorbed and chemisorbed overlayers on solid surfaces. For these models neither Glauber nor Kawasaki dynamics conserve the order parameter. However, Kawasaki dynamics conserve the density (the bulk magnetization) whereas Glauber dynamics do not. Hence we study the dynamical models A and C in the classification scheme of Hohenberg and Halperin.[8] The quenches are performed from infinite temperature to some temperature, T_f, below the transition temperature, T_c, using Monte Carlo computer-simulation techniques.[5]

A key lattice model is the antiferromagnetic square-lattice Ising model

$$H = J\left(\overset{nn}{\underset{i>j}{\sum}} \sigma_i \sigma_j + \alpha \overset{nnn}{\underset{i>j}{\sum}} \sigma_i \sigma_j \right); \; J>0, \; \sigma_i = \pm 1, \qquad (2)$$

with ratio α of next-nearest-neighbour (nnn) and nearest-neighbour (nn) interaction strengths. The domain-growth kinetics in the case of $\alpha = 0$, leading to p = 2 (1x1)-antiferromagnetic ordering, is well-known as being described by the Lifshitz-Allen-Cahn law, independent of whether or not the density is conserved.[1] However, at zero temperature with Glauber dynamics the model freezes into a metastable glassy structure and growth ceases. Conversely, in the case of $\alpha = 1$, leading to the p = 4 (2x1)-anti-ferromagnetic ordering, the model grows with Glauber dynamics and n ≃ ½.[9,10] A systematic study[10] as a function of α, cf. Fig. 1, reveals that $\alpha = 1$ is a border case below which there is always freezing-in at zero temperature. For $\alpha > 1$, Lifshitz-Allen-Cahn kinetics apply. This very abrupt crossover may be understood[10] by considering the energetics of the various types of domain walls: For $\alpha < 1$, kink-formation in the walls is an activated process and the domain pattern becomes pinned to the square lattice.

A similar type of crossover has been observed in p-state ferromagnetic Potts models (p > 2) with nn and nnn interactions.[11] For these models (with Glauber dynamics), zero-temperature growth is persistent (with n ≃ 0.50) for $\alpha = 1$ only; otherwise the system freezes-in. This behaviour has been related to the finding of two fixed points of the dynamical process, one being a freezing fixed point (also identified by Mazenko et al.[12]) which is stable only at T = 0 and another being an equilibrium fixed point which is stable only for T > 0. Recently, Viñals and Grant[13] related the existence of the fixed points to the frustration of the thermodynamic driving force and to topological properties of the local-equilibrium domain shapes. In particular, they introduced the concept of *tiling*: If differently-sized domains of the same local-equilibrium shape (e.g. squares on a square lattice) can tile the system, growth is limited to activated processes. These ideas also apply to the Ising problem discussed above.

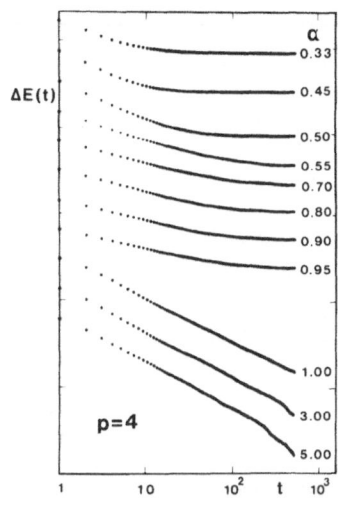

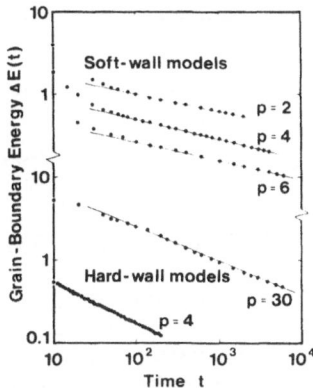

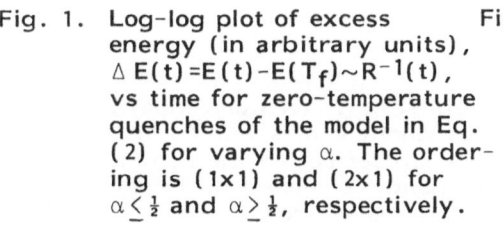

Fig. 1. Log-log plot of excess
energy (in arbitrary units),
$\Delta E(t) = E(t) - E(T_f) \sim R^{-1}(t)$,
vs time for zero-temperature
quenches of the model in Eq.
(2) for varying α. The ordering is (1x1) and (2x1) for
$\alpha \leq \frac{1}{2}$ and $\alpha > \frac{1}{2}$, respectively.

Fig. 2. Log-log plot of excess energy
vs time for zero-temperature
quenches of a variety of soft-
wall and hard-wall models
with different ordering degeneracy, p. The soft-wall
and hard-wall kinetic exponents are n≃0.25 and 0.50,
respectively.

Crossover in terms of p in large-p Potts models with Glauber dynamics has
been discussed fervently since it was found[14] that n varied smoothly from 0.50
at p = 2 to 0.41 for p>30. It has recently been shown[15] that this crossover is
an artifact of short-time simulations and that n ≃ ½ for all p.

Ising and standard Potts models give rise to sharp (hard) domain walls.
The finding, cf. Fig. 2, of a kinetic exponent of n ≃ 0.25 in zero-temperature
quenches of several soft-wall models[3,16] with Glauber dynamics (XY-type
spin models, p = 2, 4, 6, and high-p-state Potts models with anisotropic
grain-boundary energies) has led to the proposal of the existence of a separate zero-temperature universality class. Crossover to this special zero-
temperature behaviour is expected when the domain walls possess the capacity of softening (widening) locally in response to high curvature.[17] It is
noteworthy that the soft-wall exponent is found to be independent of p (similar to the hard-wall models), thus suggesting that the low exponent value is
not likely to signal a crossover to zero-temperature freezing-in. So far, no
explanation has been found for this behaviour.

III. ZERO-TEMPERATURE KINETICS: CROSSOVER IN CONTROL PARAMETERS
OF THE KINETIC PROCESS

In an extensive computer-simulation study of the domain-growth kinetics
of Eq. (2) with α = 1 and conserved density (nn Kawasaki dynamics), Sadiq
and Binder[9] found zero-temperature freezing-in and finite-temperature crossover to n ≃ 1/3 Lifshitz-Slyozov-like kinetics.[18] This is a very interesting
result since it indicates that for p > 2, the kinetics depends on whether or
not a quantity (density) other than the order parameter is conserved. It
was argued[9] that the excess-wall density of p > 2 models would imply that the
wall dynamics are controlled by long-range diffusion and hence obey Lifshitz-
Slyozov kinetics. However, it has recently been discovered that,[19] allowing
for nnn spin exchange as well as nn exchange (described by a frequency
ratio δ), cf. Fig. 3, the zero-temperature freezing-in only occurs for δ = 0

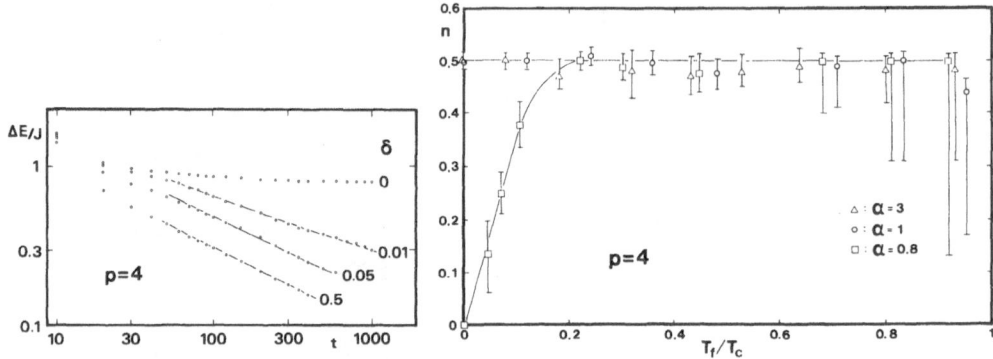

Fig. 3. Log-log plot of excess energy vs time for zero-temperature quenches of the model in Eq. (2) with variable ratio, δ, of nnn and nn Kawasaki exchange frequencies.

Fig. 4. Kinetic exponent for the model in Eq. (2) as a function of α and quench temperature.

and that there is a distinct crossover to Lifshitz-Allen-Cahn kinetics for $\delta > 0$. For small values of δ, effective exponent values $0 < n < \frac{1}{2}$ are consequences of this crossover. The dramatic effect on the zero-temperature kinetics of allowing for nnn pair exchange is related to the increased diffusion range which spans the region with excess density.

IV. CROSSOVER IN TEMPERATURE

When the quenching temperature is non-zero, activated processes come into play and the freezing-in at $\delta = 0$ seen in Fig. 4 is released, the frozen-in domain pattern gets unpinned, and growth resumes. Sadiq and Binder[9] found for $\delta = 0$ that n varies with temperature, and argued that a plateau around $n \simeq 0.35$ is attained before crossover to the critical region is encountered. However, for $\delta \geq 0.02$ it is found[19] that the slow kinetics at $T = 0$ quickly crosses over to $n = \frac{1}{2}$ kinetics at low finite temperatures. The crossover is slower, the lower the temperature and the lower the value of δ. These results (as well as other results on models with excess-wall density[20]) suggest a reinterpretation of the $\delta = 0$ kinetics[9] as being influenced by the $T = 0$ freezing-in behaviour for all temperatures. The asymptotic growth region is not reached within the available time span.

A quite analogous crossover in temperature is observed in the square-lattice Potts model[11,13] with non-conserved density. For finite temperatures, the freezing fixed point becomes unstable and there is a crossover to the equilibrium fixed point characterized by a growth law with $n \simeq 0.50$.

Figures 5 and 6 show that the finding of a special exponent value of $n \simeq 0.25$ for models with soft domain walls is a special zero-temperature phenomenon.[17,21] For $T > 0$, there is a crossover to Lifshitz-Allen-Cahn kinetics for the soft-wall models with p = 2, 4, and 6. It would be interesting to see if the same crossover takes place in the high-p Potts models with anisotropic grain-boundary energy.[3]

V. CONCLUSIONS

The large body of computer-simulation studies referred to in this paper of a great variety of different two-dimensional kinetic lattice models with different ordering degeneracies and non-conserved order paramter is con-

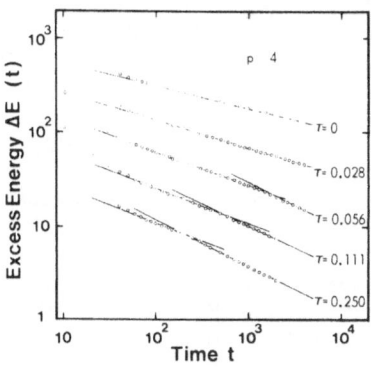

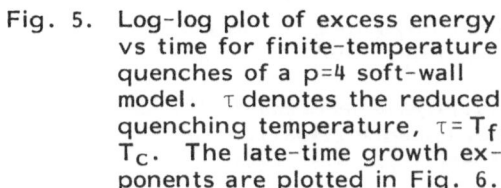

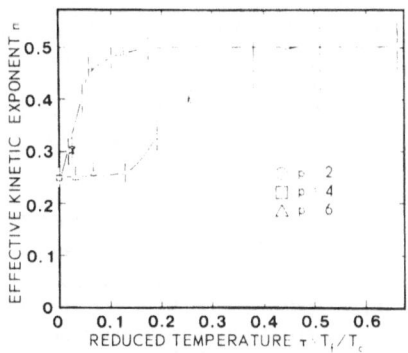

Fig. 5. Log-log plot of excess energy vs time for finite-temperature quenches of a p=4 soft-wall model. τ denotes the reduced quenching temperature, $\tau = T_f / T_c$. The late-time growth exponents are plotted in Fig. 6.

Fig. 6. Plot of effective late-time growth exponents vs temperature for p = 2, 4, and 6 soft-wall models.

sistent with Lifshitz-Allen-Cahn kinetics, Eq. (1), with $n \simeq \frac{1}{2}$. At zero temperature, some models exhibit freezing-in behaviour and other models have a different kinetics, $n \simeq 1/4$, which seems to be related to their capacity of locally widening the domain wall in response to curvature. This universality in finite-temperature domain-growth kinetics is very remarkable.

The true asymptotic growth behaviour is, however, often veiled by crossover effects. Our understanding of these crossover phenomena is still very limited, although renormalization group results and tiling concepts have provided some insight.[11,13] A description of the various types of crossover, of which some have been discussed here, is important for the interpretation of experimental data of growth kinetics which often yield exponent values different from those predicted theoretically.[22]

Acknowledgements: This work was supported by the Danish Natural Science Research Council under grants J.nr. 5.21.99.72 and J.nr. 11-5593.

REFERENCES

1. For a review, see J.D. Gunton, M. San Miguel, and P.S. Sahni, in *Phase Transitions and Critical Phenomena*, edited by C. Domb and J.L. Lebowitz (Academic, New York) 1983, Vol. 8, p. 269.
2. O.G. Mouritsen, Phys.Rev. B 28, 3150 (1983); Phys.Rev. B 31, 2613 (1985); Phys.Rev. B 32, 1632 (1985); Phys.Rev.Lett. 56, 850 (1986).
3. G.S. Grest, D.J. Srolovitz, and M.P. Anderson, Phys.Rev.Lett. 52, 1321 (1984).
4. J.D. Gunton, in *Time-Dependent Effects in Disordered Materials*, edited by R. Pynn (Plenum Publ. Co., New York, 1987), in press.
5. O.G. Mouritsen, *Computer Studies of Phase Transitions and Critical Phenomena* (Springer Verlag, New York, 1984), Chap. 5.4.
6. I.M. Lifshitz, Zh.Eksp.Teor.Fiz. 42, 1354 (1962) [Sov.Phys. JETP 15, 939 (1962)].
7. S.E. Allen and J.W. Cahn, Acta Metall. 27, 1085 (1979).
8. P.C. Hohenberg and B.'. Halperin, Rev.Mod.Phys. 49, 435 (1977).
9. A. Sadiq and K. Binder, J.Stat.Phys. 35, 517 (1984).
10. A. Høst-Madsen, P.J. Shah, T.V. Hansen, and O.G. Mouritsen, Phys. Rev. B 36, 2333 (1987).

11. J. Viñals and J.D. Gunton, Phys.Rev. B $\underline{33}$, 7795 (1986).
12. G.F. Mazenko, O.T. Valls, and F.C. Zhang, Phys.Rev. B $\underline{31}$, 4453 (1985).
13. J. Viñals and M. Grant, preprint (1987).
14. P.S. Sahni, D.J. Srolovitz, and G.S. Grest, Phys.Rev. B $\underline{28}$, 2705 (1983).
15. G.S. Grest, M.P. Anderson, and D.J. Srolovitz, in *Time-Dependent Effects in Disordered Materials*, edited by R. Pynn (Plenum Publ. Co., New York, 1987) in press.
16. O.G. Mouritsen, in *Annealing Processes – Recovery, Recrystallization and Grain Growth,* edited by N. Hansen et al. Proceedings of the Seventh Risø International Symposium on Metallurgy and Materials Science, Risø 1986, p. 457.
17. O.G. Mouritsen and E. Præstgaard, Phys.Rev. B, in press (1987); see also W. van Saarloos and M. Grant, Phys.Rev. B, in press (1987) for a critique of this point of view.
18. I.M. Lifshitz and V.V. Slyozov, J.Phys.Chem. Solids $\underline{19}$, 35 (1961).
19. H.C. Fogedby and O.G. Mouritsen (preprint 1987).
20. J. Viñals and J.D. Gunton, Surf.Sci. $\underline{157}$, 473 (1985).
21. O.G. Mouritsen and E. Præstgaard (preprint 1987).
22. M.C. Tringides, P.K. Wu, and M.G. Lagally, Phys.Rev.Lett., in press (1987).

ORDER-DISORDER TRANSITIONS IN HEXAGONAL BINARY ALLOYS

- A MONTE CARLO ANALYSIS

Sabine Crusius and Gerhard Inden

Max-Planck-Institut für Eisenforschung GmbH
Düsseldorf, Fed. Rep. Germany

INTRODUCTION

Ordering reactions in alloys have been studied most extensively and successfully in the cubic crystal structures bcc and fcc. In contrast to this only little work has been devoted so far to alloys with the hexagonal structure. This is to a good part due to a complication which arises even in the simplest approach of a pairwise interaction model, that is an anisotropy of the interactions. Consequently a correct treatment of the ordering reactions should take into account this anisotropy of correlations between atom pairs and higher order clusters. This is a difficult task with analytical models.

Therefore, in the present study the Monte Carlo (MC) simulation technique has been adopted to study order/disorder transformations in the hexagonal alloys. This method is a computer experiment which is performed on an alloy crystal stored in a computer. The advantage of this method is its flexibility with respect to the "experimental conditions": any crystal structure can be studied, the range of interactions and correlations is only limited by the crystal size, spatial asymmetry can be introduced without difficulty, any superstructure compatible with the boundary conditions may develop, and the equilibrium state can be reached from any starting state.

The hexagonal close packed structure with $c/a=1.633$ is a special case which is topologically similar to the fcc lattice differing only in the stacking sequence of otherwise identical {111}-planes. In a pairwise interaction model with first and second nearest neighbours the two structures are also energetically identical. Therefore, there exists a correspondence between the ordering reactions and phase diagrams of both structures. The latest phase diagram for fcc alloys with nearest-neighbour interactions was determined by Ackermann et al. [1].

In real alloys c/a will rarely take the ideal value. The distance of nearest neighbours within the basal plane and between adjacent planes will not be the same and the corresponding interaction energies will thus take different values. In the treatment of ordering reactions the so-called interchange energies $W^k = -2V_{AB}^k + V_{AA}^k + V_{BB}^k$ between k-th neighbours are generally introduced for convenience. The interchange energies take positive values for an ordering tendency between the corresponding neighbours (atomic ordering is sometimes formulated in an Ising model formalism; the interactions J^k used there correspond to $-W^k/4$). In the present instance interchange energies between nearest neighbours (k=1) within and between adjacent basal planes need to be distinguished. This is done by adding a second index taking the value 1 and 2 for the two cases, i.e. W^{11} and W^{12}, respectively. Only positive values will be considered in this study. Three cases will be studied in full detail: (i) $W^{12}/W^{11}=1$ which is the isotropic case corresponding to the hcp structure, (ii) $W^{12}/W^{11}=0$ which corresponds to a total decoupling of neighbouring {001}-planes (this case should thus be equivalent to the two-dimensional triangular lattice), (iii) $W^{12}/W^{11}=0.8$ which is between the preceding limiting cases and corresponds to the general hexagonal case. Results of further studied cases $W^{12}/W^{11}=$ 0.7, 0.6 will be mentioned.

GROUND STATES

The ground states of the three-dimensional (3-dim.) lattice have not been analyzed so far. For the 2-dim. hexagonal lattice a full analysis has been made by Kudo and Katsura [2] including up to third-nearest neighbour interactions J^1, J^2, J^3. Fortunately, there exists a correspondence between the 3-dim. hexagonal structure with anisotropic first-neighbour interactions and the planar lattice with third-neighbour interactions. The results from [2] can thus be immediately transposed to the 3-dim. case. This is visualized in Fig. 1 showing that the projection of two adjacent basal planes of the 3-dim. lattice, Fig. 1b, is identical to the planar hexagonal lattice, Fig. 1c. Atoms linked by the interaction J^2 of the planar lattice correspond to the atoms within the basal plane of the 3-dim. lattice linked by W^{11}, those linked by J^1 correspond to those linked by W^{12} between adjacent basal planes. This correspondence holds up to second-nearest neighbours of the 3-dim. lattice.

The correspondence between the interchange energies J^k of the planar lattice and the W^k of the 3-dim. lattice is given by

$$J^1 = 2W^{12}, \quad J^2 = W^{11}, \quad J^3 = 2W^2$$

The factor 2 takes account of the difference in coordination number between the equivalent bonds of the two lattices.

Fig. 2 shows the ground state diagram which has been derived from [2]. The lines indicate, for a given interchange energy ratio, the values

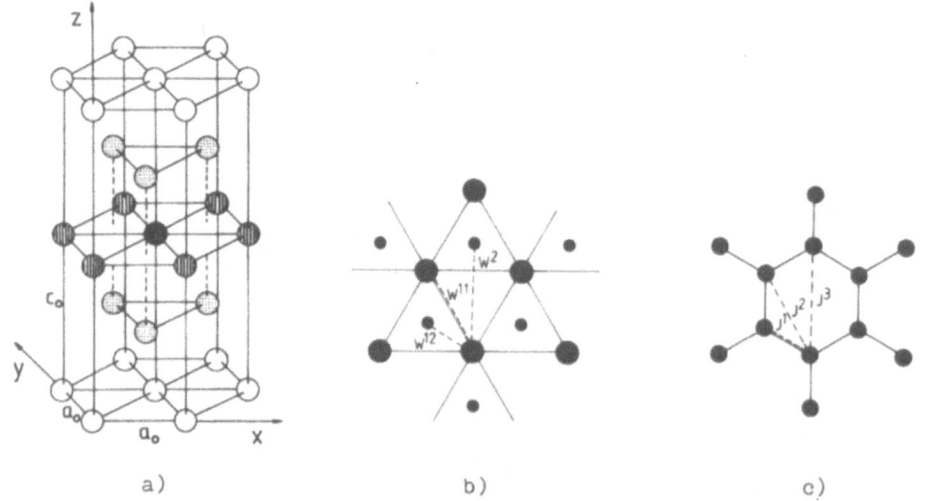

a)   b)   c)

Fig. 1 Correspondence between the 3-dim. and 2-dim. hexagonal lattices a) ● central atom, ◍ nearest neighbours within the basal plane (interchange energy W^{11}), ⊕ nearest neighbours between basal planes (interchange energy W^{12}). b) projection of two neighbouring basal planes of Fig. 1a) (z=0 large symbols, z=$c_0/2$ small symbols); interactions up to second-nearest neighbours are indicated. c) 2-dim. hexagonal lattice with interactions up to third neighbours.

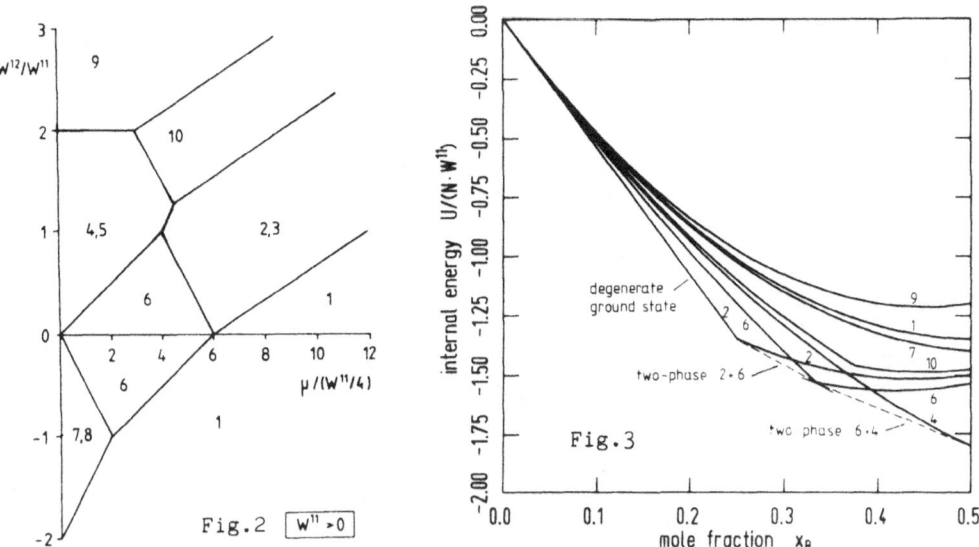

Fig. 2 Ground state diagram of the 3-dim. hexagonal lattice with anisotropic interactions. The ranges of stability of the various superstructures are given as a function of chemical potential and ratio W^{12}/W^{11}. The
superstructures are crystallographically identified in Table I.

Fig. 3 Ground state energy (T=0 K) versus mole fraction of the various
superstructures for $W^{12}/W^{11}=0.8$. The ground states are degenerate (linear
variation of U for off-stoichiometric compositions) or two-phase.

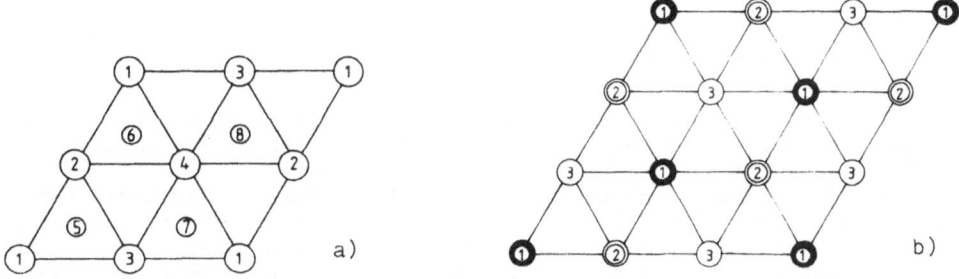

Fig. 4 Sublattices describing the superstructures DO_{19} and B19 (4a) and
the planar structure A_2B (4b). Small circles refer to positions in
adjacent planes.

of the chemical potential $\mu = (\mu_A - \mu_B)/2$ for which two-phase equilibria
exist between the phases labeled on either side. The phases labeled from 1
to 10 are listed in Table I with their crystallographic identification.
The structures (2,3), (4,5) and (7,8) are pairs of energetically
equivalent structures differing by periodic antiphase boundaries (APB)
which do not contribute to the energy. From Fig. 2 it follows that for the
values $0 \leq W^{12}/W^{11} \leq 1$ considered in this study only three superstructures
will be relevant: DO_{19}, B19 and A_2B (2-dim. ordering in the basal planes).
The two variants of DO_{19} and B19 with periodic APB (3 and 5 in Table I)
should also be considered. No constraint was imposed in the Monte Carlo
simulations as to their occurrence. Yet they have never been observed.
Therefore they will not be considered any further in this paper. Fig. 3
completes the ground state discussion showing the internal energy at T=0 K
versus mole fraction. The most stable state is either degenerate (linear

variation of U for offstoichiometric ordered alloys) or two-phase. The slope of the lines corresponds to the values of μ which can be read from Fig. 2.

In order to describe the observed ordered states sublattices can be introduced in an enlarged unit cell with the basis $(2a_0, 2a_0, c_0)$. Eight sublattices are required to describe all superstructures. Fig. 4a shows the arrangement of the sublattices within the enlarged hexagonal unit cell. The planar structure A_2B requires an even larger basis $(3a_0, 3a_0, c_0)$ and three sublattices within the basal plane, Fig. 4b.

MONTE CARLO PROCEDURE

General Aspects

The computer crystal was composed of 24x24x12 hexagonal unit cells, i.e. $24^3 = 13824$ lattice positions. Periodic boundary conditions in the directions x, y, z, see Fig. 1, were imposed to avoid surface effects. Both canonical and grandcanonical calculations were made.

In the canonical calculation the composition of the computer crystal is fixed and the equilibrium state at a selected temperature is obtained by randomly selecting two positions with different kinds of atoms and performing an interchange of these atoms according to the probability

$$p = \exp(-\Delta U/k_B T) \quad \text{if } \Delta U > 0$$
$$= 1 \quad \text{if } \Delta U < 0$$

where $\Delta U = U_2 - U_1$, i.e. the internal energy difference between the states after and before the atomic interchange.

In the grandcanonical calculation the chemical potential difference μ is fixed, here defined as $\mu = (\mu_A - \mu_B)/2$ for convenience (μ_A, μ_B with reference to the pure components in the hexagonal structure), and the equilibrium state at a given temperature is obtained by randomly selecting one position and changing the atom on it into the other component A or B with the probability

$$p = \exp(-[\Delta U - \mu\Delta(N_A - N_B)]/k_B T) \quad \text{if } [\ldots] > 0$$
$$= 1 \quad \text{if } [\ldots] < 0$$

where N_A, N_B are the number of A and B atoms of the system, $\Delta U = U_2 - U_1$. U_1, U_2 are the internal energies before and after the transformation of A into B (or vice versa).

The starting state can be any configuration. The two limiting cases, i.e. the most perfect long-range ordered (lro) state and the random state, are most convenient. Starting with a most perfect lro state avoids the problem of antiphase boundaries. Starting from the random state introduces antiphase boundaries when the equilibration temperature is below the critical temperature of lro.

The total number of successful Monte Carlo steps (MCS), i.e. performed interchanges or atom changes, and not the number of attempts is relevant for the equilibration process. In the present work 70 successful MCS per site were performed for equilibration. Depending on temperature and the state (below, at or above a transition temperature) the number of attempted MCS per site varied between 500 and 5000. The equilibrium state was analyzed by taking averages over 60 consecutive equidistant configurations created by further 13 successful MCS per site. In order to decide about the stability this averaging procedure was repeated. States called lateron "stable lro" showed no variation in the consecutive average values. The contrary was found only in a limited temperature and composition range of the short-range ordered state.

Critical temperatures of order-disorder transformations are generally detectable by the change in the properties, e.g. lro parameters, as a function of temperature or chemical potential. A problem of determination of the critical temperature or chemical potential arises in the case of first order transformations which show a hysteresis when the critical temperature is crossed in both ways. This hysteresis was removed by taking a mixture of the two phases as the starting state. In this instance the number of successful MCS per site in the equilibration process was increased to 140.

States of LRO

The state of lro is fully determined by the fraction p_A^L of A-atoms on each of the eight sublattices. Since the p_A^L obey the relation $\Sigma_L p_A^L = 8x_A$ only seven parameters are independent. Those are the lro parameters defining the lro state with respect to type and degree of ordering: the type of ordering derives from the sets of sublattices which are equally occupied, the degree of lro derives from the difference in occupation between these sets.

If only the DO_{19} and B19 structures are considered most of the sublattices are equally occupied and the number of independent parameters reduces to one. Then it is convenient to define the degree of lro by a parameter S which varies from 0 to 1 between the random and fully ordered state. For the DO_{19} structure such a parameter is given by

$$S = \left|3(p_A^4 + p_A^5) - \sum_{i \neq 4,5} p_A^i\right|/S_0$$

$$\text{with } S_0 = 24x_B \quad \text{for } 0.75 \leq x_A \leq 1.0$$
$$= 8x_A \quad \text{"} \quad 0.50 \leq x_A \leq 0.75$$

if the sublattices 4,5 are those which are preferentially occupied with the minority component B as assumed in Table I. In a Monte Carlo simulation this constraint cannot be imposed and other equivalent settings, 1,8 or 2,7 or 3,6 are possible leading to different values in the above formula for S. This problem can be overcome by taking the average over all possible settings:

$$DO_{19}: \quad S = \sum_{(i,j)} \left|3(p_A^i + p_A^j) - \sum_{k \neq i,j} p_A^k\right|/2S_0$$
$$(i,j)=(1,8), \ (2,7), \ (3,6), \ (4,5)$$

Similarly one gets for

$$B19: \quad S = \sum_{(i,j,k,l)} \left|p_A^i + p_A^j + p_A^k + p_A^l - \sum_{n \neq i,j,k,l} p_A^n\right|/16x_B$$

$$(i,j,k,l)=(1,2,7,8), \ (1,3,6,8), \ (1,4,5,8), \ (2,3,6,7), \ (2,4,5,7), \ (3,4,5,6)$$

The A_2B structure consists of an ordering within the basal plane with no correlation between these planes. In this instance three sublattices are required and the lro parameter is

$$A_2B: \quad S = \sum_{(i,j,k)} \left|p_A^i - (p_A^j + p_A^k)/2\right|/S_0$$

$$(i,j,k) = (1,2,3), \ (2,3,1), \ (3,1,2)$$
$$S_0 = 6x_B \quad \text{for } 2/3 \leq x_A \leq 1.0$$
$$= 3x_A \quad \text{"} \quad 0.5 \leq x_A \leq 2/3$$

States of SRO

Short-range order (sro) is determined by the pair correlation functions ξ_k between k-th neighbours. They are defined by (e.g. [3])

$$\xi_k = (1/z^{(k)}N)\sum_{ij} \sigma^i \sigma^j$$

the summations going here over all N lattice points i and j being k-th neighbours, $z^{(k)}$ is the coordination number, $\sigma^i = 1$ or -1 if an A or a B atom is on position i. It is immediately evident that the ξ_k can be evaluated very easily from the computer crystal. Instead of ξ_k the Cowley-Warren sro parameters are more often used. They are given by

$$\alpha_k = [\xi_k - (x_A - x_B)^2]/4x_A x_B$$

The α_k will be used in the following since they give immediately the

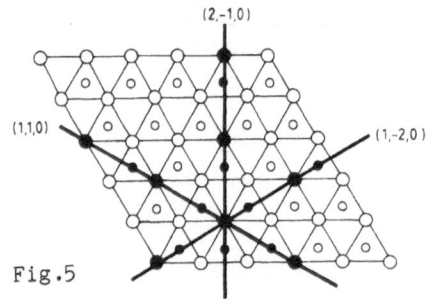

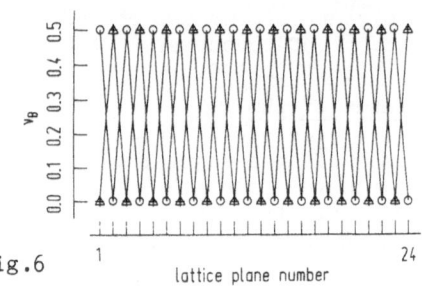

Fig.5   Fig.6

Fig. 5 Traces of the three types of lattice planes analyzed for their composition ν_B as a function of their position within the computer crystal.

Fig. 6 Composition ν_B of the lattice planes + (110), o ($1\bar{2}0$), Δ ($2\bar{1}0$) as a function of their position in the computer crystal for the stoichiometric compositions A_3B (DO_{19}) and perfect long range order.

deviation from the random state for which $\alpha_k = 0$. They have been analyzed in this work up to the 10-th neighbours. It must be mentioned, however, that the 1-st and 4-th neighbour α_k have to be split into the neighbours within the basal plane and out of it (in these cases k represents the double indices 11, 12, 41, 42).

In fcc alloys the sro state showed large fluctuations close to the triple point of the phases A1, $L1_2$ and $L1_0$ [1]. This was detected by observing a variation of the composition of lattice planes as a function of position in the computer crystal and time. The same kind of analysis will be applied here. Fig. 5 shows the traces of three lattice planes which give the clearest identification of the superstructures DO_{19} and B19. E.g. Fig. 6 shows the periodic variation of the composition ν_B of the lattice planes as a function of their position within the computer crystal which is typical for the perfectly ordered DO_{19} superstructure at the stoichiometric composition. The more complicated cases with off-stoichiometry and imperfect ordering will be discussed later.

NUMERICAL RESULTS

$W^{12}/W^{11} = 1.0$

The calculated phase diagram is shown in Fig. 7. As expected for this case of isotropic nearest neighbour interactions this diagram is topologically the same as that obtained in [1] for fcc alloys. Of course, the superstructures are now the hexagonal equivalents. In the sro state close to the triple point strong fluctuations are observed. This range called "fluctuating lro" in [1] is shown as shaded area. More details are given for $W^{12}/W^{11}=0.8$.

$W^{12}/W^{11} = 0.$

In this instance the basal planes are energetically decoupled. Correlations only exist within these planes and only the 2-dim. superstructure A_2B forms. Any kind of correlation between these planes which might come up in the 3-dim. simulation is purely accidental.

The phase diagram is shown in Fig. 8. No two-phase equilibria, no discontinuity and no hysteresis effects have been observed on crossing the critical temperature of lro. The transition thus exhibits all the characteristics of a 2nd. order transformation. The hatching of the critical temperature line in Fig. 8 indicates this continuous transformation. Numerically the critical temperature has been defined by plotting S^2 vs. temperature or chemical potential and extrapolating linearly to S=0. It is worth mentioning that in the A_2B structure any APB contributes to the energy. Therefore APB are not stable within this structure. If they

144

are built up casually they will be removed in the continuation of the simulation.

At $\mu=0$ (i.e. $x_B=0.5$) no long range ordering has been found down to temperatures $\tau = k_BT/(W^{11}/4) = 0.6$. A pronounced sro has been observed as shown by the sro paramters α_1, α_2, α_3 in Fig. 9. At $\tau=0.6$ they have already reached their limiting values for $\tau=0$ which are known from a

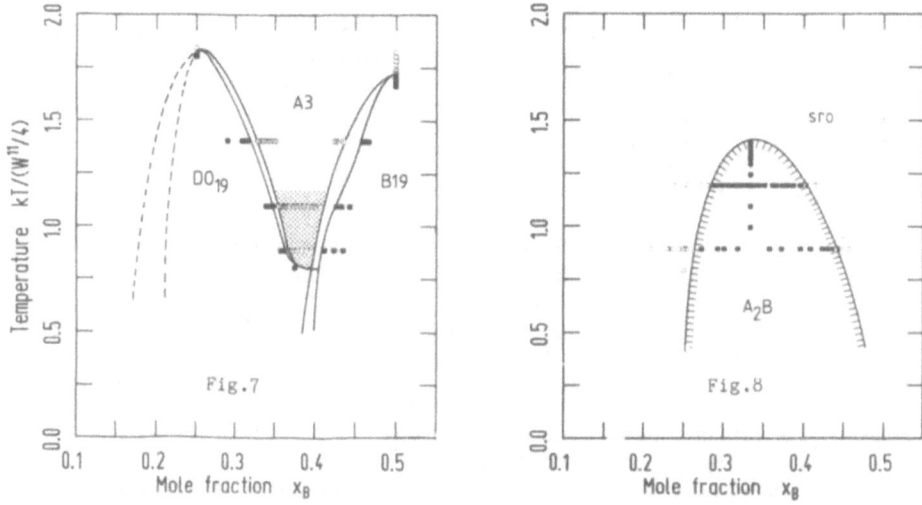

Fig. 7 Phase diagram of hexagonal alloys with isotropic nearest neighbour interactions, $W^{12}/W^{11} = 1$. This situation is energetically identical and topologically equivalent to the fcc situation. Therefore the phase diagram is also topologically the same as for fcc alloys [1]. For compositions $x_B < 0.25$ the phase boundaries have thus been taken from the fcc calculation [9] (broken lines). The shaded area indicates the range of strong fluctuations. The points indicate steps of grandcanonical and canonical calculations.

Fig. 8 Phase diagram of hexagonal alloys with $W^{12}/W^{11}=0$, i.e. with complete decoupling of the basal planes. Only 2-dim. long-range order exists. The hatching shall indicate the 2nd order character of the transformation. Symbols as in Fig. 7.

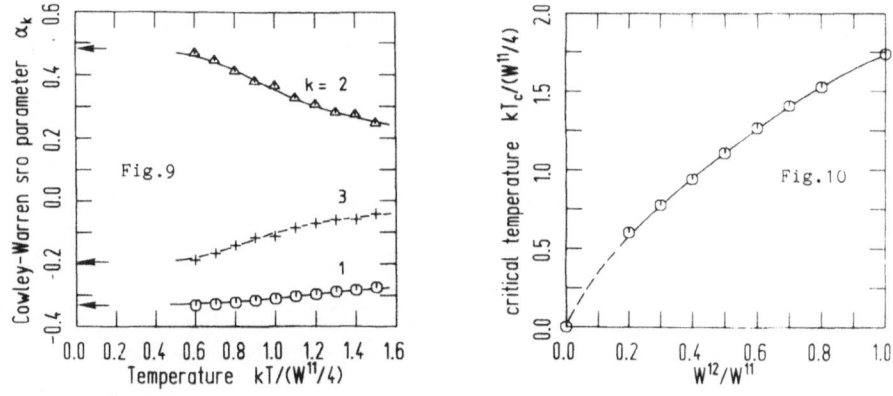

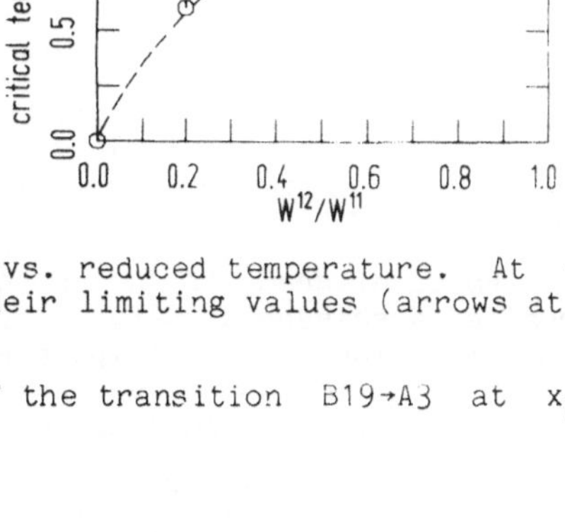

Fig. 9 Cowley-Warren sro parameters α_k vs. reduced temperature. At $\tau=0.6$ α_1, α_2, α_3 have already reached their limiting values (arrows at $\tau=0$) without any lro.

Fig. 10 Reduced critical temperature of the transition $B19 \rightarrow A3$ at $x_B=0.5$ vs. W^{12}/W^{11}.

ground-state analysis of the 2-dim. triangular lattice by Stephenson [4]. This means that it is topologically possible to produce the maximum number of unlike bonds without the necessity of producing lro correlations and the critical temperature of lro for this composition will be zero. This is also suggested from a plot of the critical temperature of the transition B19→A3 vs. W^{12}/W^{11} for $x_B=0.5$ which extrapolates to $\tau<0.2$ for $W^{12}=0$, see Fig. 10.

$\underline{W^{12}/W^{11} = 0.8}$

The calculated phase diagram is shown in Fig. 11. All three superstructures DO_{19}, B19 and A_2B appear. The shaded area indicates the range where strong fluctuations in the sro state have been observed. This is visualized in Fig. 12 and 13 where the average composition ν_B of lattice planes is shown versus their position in the computer crystal. In Fig. 12 three subsequent calculations are shown which are performed at the same temperature within the domain of stable lro (12a and b) and of sro (c) according to the three selected steps of μ-values. The three calculations are separated by 70 successful MCS per site. In the stable lro state the fraction ν_B of B-atoms on the three previously selected lattice planes varies periodically with the position in the crystal as it should be for the DO_{19} structure (compare Fig. 6), and these composition waves are stable in time (12a and b). In the sro state, Fig. 12c, the amplitude of the waves ν_B is almost zero, and ν_B fluctuates around the constant value corresponding to the random solution. In Fig. 13 an analogous calculation has been done within the shaded area of Fig. 11. Here the amplitude ν_B of the composition waves is comparable to those of the lro state. However, the waves do not go through the whole crystal. The amplitude is handed over, e.g. from a wave on $(1\bar{2}0)$ planes to another one on (110) planes. These are not stable in time as shown by comparing the two subsequent equilibrium calculations shown in Fig. 13 a, b which are separated by 70 successful MCS per site for otherwise identical parameters of temperature and chemical potential.

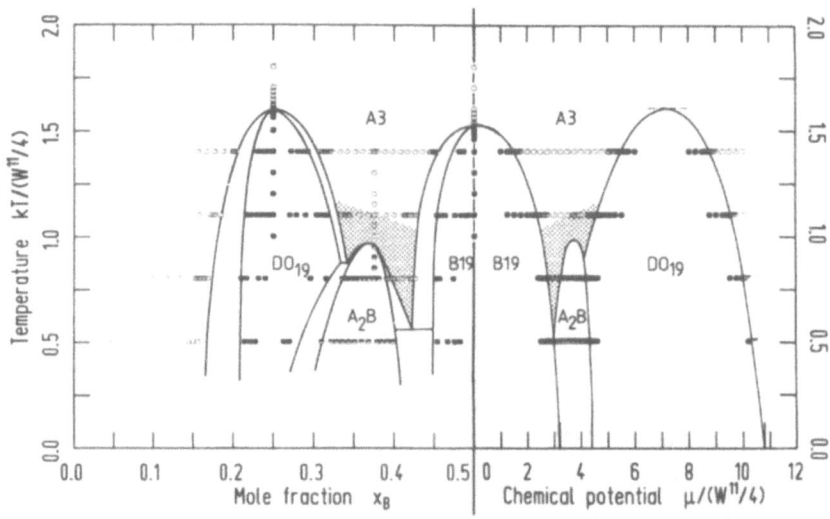

Fig. 11 Phase diagram of hexagonal alloys with $W^{12}/W^{11}=0.8$ with two choices of abscissae as obtained from grandcanonical and canonical calculations. The lines in the temperature/potential diagram extrapolate to the values at T=0 given in the ground-state diagram. Symbols as in Fig. 7. The shaded area indicates the range of strong fluctuations in the sro state.

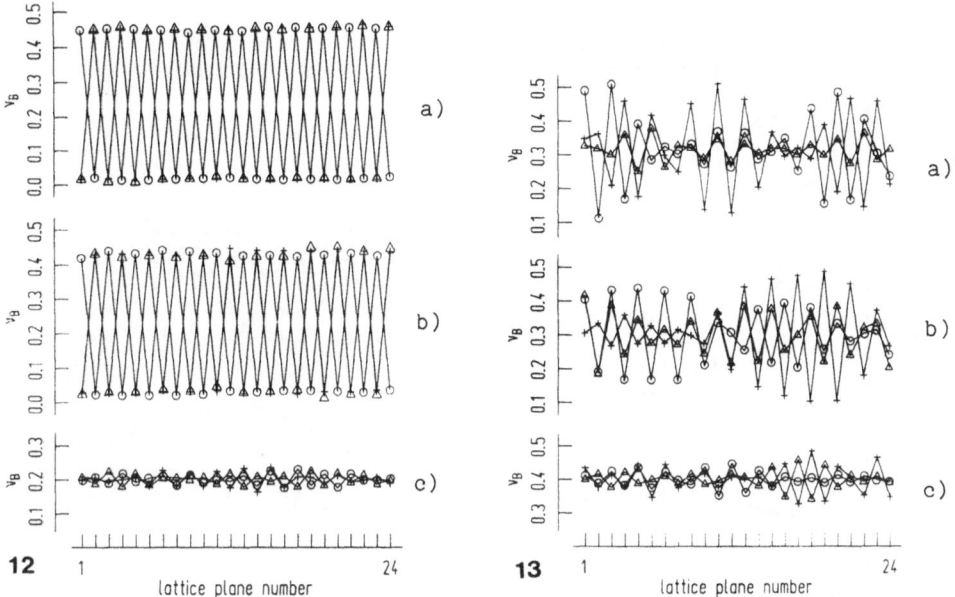

Fig. 12 Stable lro (a and b) and sro (c) visualized with the fraction ν_B of B-atoms on the lattice planes (110)=+, (1$\bar{2}$0)=o, (2$\bar{1}$0)=Δ versus position in the computer crystal. Reduced temperature τ=1.4 and W^{12}/W^{11}=0.8. (a) μ=8.75, stable lro DO_{19}; (b) μ=8.85, stable lro DO_{19}; (c) μ=8.95, sro. The states (a), (b), (c) are each separated by 70 successful MCS per site.

Fig. 13 Fluctuations in the sro state observed in the shaded area of the phase diagram in Fig. 11. They are visualized with the fraction ν_B of B-atoms on lattice planes (as in Fig. 12) vs. position in the computer crystal for two subsequent equilibrium calculations (a), (b) separated by 70 successful MCS per site. Reduced temperature τ=1.1 and W^{12}/W^{11}=0.8. (a) and (b), μ=4.8; (c), μ=3.0, sro outside the shaded area of Fig. 11.

Fig. 14 shows the Cowley-Warren sro parameters as a function of chemical potential. It is remarkable that the sro parameters of fourth neighbours, particularly α_{41}, change sign on passing from lro to sro. In this instance the sro state exhibits qualitatively different features from the lro state. This is typical for the fcc and hexagonal structures.

With $W^{12} \neq 0$ there exists an energetic coupling between adjacent basal planes. This coupling is, however, not effective in the perfectly ordered stoichiometric structure A_2B, since this structure is still degenerate with respect to translations of basal planes against each other. At off-stoichiometric compositions or non-perfect lro this coupling becomes effective. Fig. 15 shows the atomic arrangement in a basal plane of an off-stoichiometric A_2B-type lro structure as obtained by the MC equilibration calculation. Large domains with stoichiometric A_2B structure have been formed and the excess atoms are concentrated in the boundaries. The next basal plane exhibits an arrangement which is correlated to the previous one: the nodes where three boundary lines meet are generally on top of each other and the boundary lines turn by 60° from one plane to the next. Due to this correlation the microstructures of two overnext basal planes are generally very similar.

W^{12}/W^{11} = 0.7, 0.6

In these instances the phase diagrams are similar to Fig. 11 but the maxima of the transformation temperatures change. The variation of the maximum transformation temperature B19→A3 at x_B=0.5 with the strength of W^{12} has already been given in Fig. 10. The corresponding curve for the

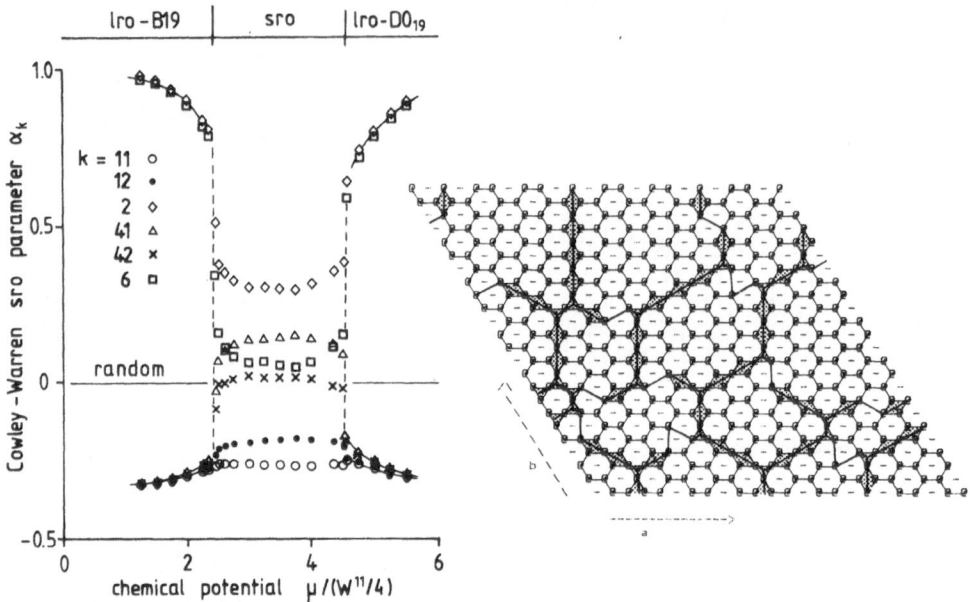

Fig. 14 Cowley-Warren sro parameters calculated for $W^{12}/W^{11}=0.8$ at $\tau=1.1$ as a function of μ (labels $k=11$, 12, 41, 42 mean nearest and fourth neighbours within a basal plane and between two basal planes, respectively). Within the state of sro the parameter α_{41}, α_{42} change their sign compared to the lro state.

Fig. 15 Lro state of type A_2B (- : B-atom, @ : A-atom) at off-stoichiometric composition calculated for $W^{12}/W^{11}=0.8$, $\tau=0.5$, $\mu=4.35$ (i.e. $x_B=0.321_5$). The shaded range represents an A-rich boundary, a single line represents a B-rich boundary.

transformation $DO_{19} \rightarrow A3$ at $x_B=0.25$ is parallel to the one in Fig. 10, and the difference between both is small (see Fig. 11). The corresponding temperature for $A_2B \rightarrow A3$ varies opposite to the previous ones and the maximum position shifts in composition with varying W^{12}.

DISCUSSION

The present 3-dim. MC analysis can be compared only with a recent Cluster Variation calculation by Kikuchi and Cahn [5] who calculated the critical temperature of lro for the DO_{19} structure as a function of W^{12}. At the stoichiometric composition A_3B their critical temperature is slightly higher than the results of this work. This is to be expected since the tetrahedron cluster they have used takes into account only tetrahedron correlations. The discrepancy is more striking for compositions closer to AB at which the CV method yields a high critical temperature. This seems to be an artefact of the approximation since tetrahedra describing the B19 structure (which is stable at composition AB) were not included in the CV treatment.

The present results for the limiting case $W^{12} = 0$ agree with MC calculations for the 2-dim. triangular lattice by Metcalf [6]. They confirm the theoretical result of the exact 2-dim. solution for the composition AB saying that there exists no finite critical temperature of lro [7]. Also for this 2-dim. case the CV method with the triangle as basic cluster yields a finite critical temperature of lro at AB [8] in conflict with the exact solution.

Table I Hexagonal superstructures. Basis and origin relative to hexagonal coordinate system (x,y,z) in Fig.1a. Designation: Strukturbericht

No.	Design. Compos.	Spacegrp	Basis	Origin	Positions	Equivalent Sublattices
1	A	A3 $P6_3/mmc$ hexagonal	$\underline{a}=a_0[100]$ $\underline{b}=a_0[010]$ $\underline{c}=c_0[001]$	$a_0/3[\overline{2}10]+c_0/4[00\overline{1}]$	2d (1/3 2/3 3/4)	1=2=3=...=8
2	A_3B	DO_{19} $P6_3/mmc$ hexagonal	$\underline{a}=2a_0[100]$ $\underline{b}=2a_0[010]$ $\underline{c}=c_0[001]$	$a_0/3[\overline{2}10]+c_0/4[00\overline{1}]$	6h (5/6 2/3 1/4) 2c (1/3 2/3 1/4)	1=2=3=6=7=8 4=5
3	A_3B	DO_a Pmmn orthorhombic	$\underline{a}=2a_0[\overline{1}\overline{1}0]$ $\underline{b}=c_0[00\overline{1}]$ $\underline{c}=a_0[1\overline{1}0]$	$a_0/3[120]$	4f (1/4 0 1/6) 2b (0 1/2 1/3) 2a (0 0 2/3)	
		-->No. 2 with periodic APB: every fourth {$\overline{1}$10} plane, vector: $a_0[\overline{1}\overline{1}0]$				
4	AB	B19 Pmma orthorhombic	$\underline{a}=c_0[00\overline{1}]$ $\underline{b}=a_0[010]$ $\underline{c}=a_0[210]$	$a_0/3[\overline{2}10]+c_0/4[00\overline{1}]$	2f (1/4 1/2 5/6) 2e (1/4 0 1/3)	2=3=6=7 1=4=5=8
5	AB	Pnma orthorhombic	$\underline{a}=a_0[\overline{1}20]$ $\underline{b}=c_0[00\overline{1}]$ $\underline{c}=2a_0[100]$	$a_0/6[\overline{1}10]+c_0/4[00\overline{1}]$	4c (1/12 1/4 5/8) 4c (1/12 1/4 1/8)	
		-->No. 4 with periodic APB: every fourth {010} plane, vector: $a_0[100]$				
6	A_2B 2-dim, hexagonal	p6m	$\underline{a}=a_0[1\overline{1}]$ $\underline{b}=a_0[12]$	(00)	2b (1/3 2/3) 1a (0 0)	2=3 1
	A_2B 3-dim, monoclinic	$P2_1/m$	$\underline{a}=a_0[1\overline{1}0]$ $\underline{b}=a_0[120]$ $\underline{c}=c_0[001]$	$a_0/6[210]+c_0/4[00\overline{1}]$	2e (1/2 1/6 1/4) 2e (1/6 1/2 1/4) 2e (5/6 5/6 1/4)	2 3 2=3 1
7	AB	Pmmn orthorhombic	$\underline{a}=a_0[010]$ $\underline{b}=c_0[00\overline{1}]$ $\underline{c}=a_0[\overline{2}10]$	$5a_0/12[210]$	2b (0 1/2 1/12) 2a (0 0 5/12)	2=3=5=8 1=4=6=7
8	AB	$P2_12_12$ orthorhombic	$\underline{a}=a_0[120]$ $\underline{b}=c_0[001]$ $\underline{c}=2a_0[100]$	$a_0/12[1\overline{4}0]+c_0/4[00\overline{1}]$	4c (1/6 1/4 3/8) 4c (2/3 1/4 1/8)	
		-->No. 7 with periodic APB: every fourth {010} plane, vector: $a_0[100]$				
9	AB	B_h $P\overline{6}m2$ hexagonal	$\underline{a}=a_0[100]$ $\underline{b}=a_0[010]$ $\underline{c}=c_0[001]$	$a_0/3[210]+c_0/2[001]$	1a (0 0 0) 1d (1/3 2/3 1/2)	5=6=7=8 1=2=3=4
10	A_5B_3	$P6_3/mmc$ hexagonal	$\underline{a}=4a_0[100]$ $\underline{b}=4a_0[010]$ $\underline{c}=c_0[001]$	$a_0/3[\overline{2}10]+c_0/4[00\overline{1}]$	2d (2/3 1/3 1/4) 6h (11/12 5/6 1/4) 12j(5/12 1/12 1/4) 6h (1/6 1/3 1/4) 6h (5/12 5/6 1/4)	

Acknowledgement: The financial support by Deutsche Forschungsgemeinschaft is gratefully acknowledged.

REFERENCES

[1] H. Ackermann, S. Crusius and G. Inden, Acta Met. 34 (1986) 2311
[2] T. Kudo and S. Katsura, Progr. Theor. Physics 56 (1976) 435
[3] J.M. Sanchez, D. de Fontaine, Phys. Rev. B17 (1978), 2926
[4] J. Stephenson, J. Math. Phys. 5 (1964) 1009
[5] R. Kikuchi, J.W. Cahn in 'User Application of Phase Diagrams', L. Kaufman (Ed.), ASM 1987
[6] B.D. Metcalf, Phys. Letters 45A (1973) 1
[7] R.M.F. Houtappel, Physica 16 (1950) 425
[8] D.M. Burley, Proc. Phys. Soc. 85 (1965) 1163
[9] K. Binder, J.L. Lebowitz, M.K. Phani and M.H. Kalos, Acta Met. 29 (1981) 1655

PC-VISUALIZATION OF PHASE TRANSITIONS IN ALLOYS

···KINETICS OF ORDERING WITH PHASE SEPARATION

Tetsuo Eguchi and Hiroshi Ninomiya

Department of Applied Physics, Fukuoka University
Fukuoka 814-01, Japan

1. Introduction

If an alloy, which was originally in the state of a uniform and disordered solid solution, is brought into a mixed phase field of the phase diagram,
then the process of phase separation takes place either by nucleation and
growth of precipitates or by spinodal decomposition. On the other hand when
an alloy is aged in an ordered phase field, the process of ordering develops
in the alloy, until ordered domains of opposite phases grow into an antiphase ordered domain structure.   Furthermore, in some alloys, such as Fe-Al
and Fe-Si, the process of ordering sometimes accompany a phase separation
resulting in a two-phase state of order and disorder[1-3].   These processes
are caused by migration of solute atoms to create a new phase of inhomogeni-
ety, which is thermodynamically more stable than the original homogeneous
phase, and are regarded as examples of relaxation from a nonequilibrium
state to the one of thermal equilibrium.

One of the present authors and his former associates have already developed a theoretical analysis of the case, in which the ordering and phase
separation occur simultaneously enhancing each other, using the method of
Ginzburg and Landau for irreversible processes in nonequilibrium thermodynamics[4], as applied to a system of mean field continuum, which is assumed to
represent the alloy under consideration. Comparing the result of our theory
with our experimental observation in Fe-based alloys, we obtained a certain
theoretical interpretation as to the process of ordering with phase separation[2,3]. It was pointed out there that under certain circumstances a wave-
like variation in the local degree of order is picked up by the local concentration of solute atoms, resulting in a spinodal-like decomposition into
two phases with different values of composition and degree of order.

Such an investigation is of interest not only for the interpretation of
microstructures of alloys, but also for the fundamental analysis of the dynamics of phase transitions and the pattern formation in alloys.   The equations of motion obtained there, however, are those of a coupled nonlinear
type, which include correlations between the local concentration and the
local degree of order in such a complicated way, that it was hard to obtain
any physically direct insight of the phenomena by mere inspections of the
equations.

Personal computers (PC), on the other hand, are now rather powerful and are distributed widely not only in offices but also in home studies, and are available for various research purposes. In the present report PC's are used for the analyses of the kinetics of phase transformations in alloys, especially in the cases of (i):nonlinear spinodal decomposition in disordered alloys, (ii):formation of antiphase ordered domain structure in a uniform solid solution, and (iii):ordering coupled with phase separation. The equations of motion governing these processes are derived from the first principles of thermodynamics of irreversible processes, and are solved numerically in one and two dimensions. The time evolution of the local concentration wave and/or the local degree of order is displayed in colour on the monitor screen of a PC. This helps us understand the processes more profoundly and vividly than we could imagine otherwise.

2. PROCEDURE

We consider an alloy of the composition $A_{(1-X)/2}B_{(1+X)/2}$ whose free energy is given by $\phi(X, S)$, where X and S are the concentration of solute atoms and the degree of long range order. In the state of thermal equilibrium the free energy must satisfy the following conditions:

$$\partial \phi / \partial X = \mu \quad \cdots\cdots\cdots (1), \quad \text{and} \quad \partial \phi / \partial S = 0 \quad \cdots\cdots\cdots (2),$$

where μ is the chemical potential. These conditions must represent the equations of state as well as the exact phase relations. In the case when the composition and the degree of order depend upon the position and time, the thermodynamical potential governing the behaviour of the system is given as a functional of the local composition $x(r,t)$ and the local degree of order $s(r,t)$ as

$$\Phi [\{x(r,t)\}, \{s(r,t)\}]$$

$$= \int \{\phi(x,s)+(h/2)\cdot(\nabla x)^2+(k/2)\cdot(\nabla s)^2\} \, d^3r \quad \cdots\cdots(3).$$

In the above expression $\phi(x,s)$ is the bulk free energy per unit volume and h and k are the interface energies per unit length, when x and s vary with position r. When the system is in a state of thermal equilibrium, the thermodynamical potential $\Phi [x,s]$ must satisfy the conditions:

$$\delta \Phi / \delta x = \mu \quad \cdots\cdots\cdots (4), \quad \text{and} \quad \delta \Phi / \delta s = 0 \quad \cdots\cdots\cdots (5).$$

When the system is not in thermal equilibrium, then the relaxation takes place in such a way that $x(r,t)$ and $s(r,t)$ change to approach conditions (4) and (5). The equations of motion in this case are given by the following, where L and M are the rate parameters for the reaction, which depend upon temperature.

$$\partial x / \partial t = L \nabla^2 (\delta \Phi / \delta x) \quad \cdots\cdots\cdots\cdots (6),$$
and
$$\partial s / \partial t = -M(\delta \Phi / \delta s) \quad \cdots\cdots\cdots\cdots (7).$$

The difference in the forms of the equations for x and s is due to the fact that $x(r,t)$ is conserved in the sense that

$$\int x(r,t) \, d^3r = V \cdot X \quad \cdots\cdots\cdots\cdots\cdots (8),$$

(V and X are the volume and the average composition of the alloy, respectively), whereas $s(r,t)$ is not conserved.

In applying the above formulas to the actual cases, we have to choose the bulk free energy, such that it reproduces the equations of state by (1) and (2), especially the phase relations of the system, and then try to solve the kinetic equations (6) and (7) simultaneously. The purpose of the present investigation, however, is to visualize the general features of the kinetics of phase transformations, and consequently we have chosen the free energy ϕ as simple as possible but to include the essential terms, which are relevant for the phenomena under consideration. The method of scaling also has been used extensively in order to remove the unknown parameters from the kinetic equations. All calculations are carried out at constant temperature, which corresponds to the processes under isothermal aging.

The model free energies we chose for our investigation and the corresponding kinetic equations are given below.

(i) For a nonlinear spinodal decomposition:

$$\phi(x) = a - b X_0^2 x^2 / 2 + b x^4 / 4 \cdots\cdots (9),$$

with X_0, a and b positive constants depending upon the temperature, and

$$\dot{u} = -\partial^4 u - (1 - u^2)\partial^2 u + 2u(\partial u)^2 \cdots\cdots (10),$$

where $u = \sqrt{3} x / X_0$.

(ii) For a process of formation of the antiphase ordered structure:

$$\phi(s) = a + b X^2 - b X_0^2 s^2 / 2 + b X_1^2 s^4 / 4 + b X^2 s^2 / 2 \cdots\cdots (11),$$

where X is the composition and X_0 and X_1 are constant parameters satisfying the inequalities $0 < X_1 < X_0 < 1$, $X_0^2 + (1 - X_1)^2 < 1$, $X < X_0$ and

$$\dot{v} = v(1 - v^2) + \partial^2 v \cdots\cdots\cdots\cdots (12),$$

where $v = X_1 s / (X_0^2 - X^2)^{1/2}$.

(iii) For a process of ordering accompanying phase separation:

$$\phi(x, s) = a + b x^2 - b X_0^2 s^2 / 2 + b X_1^2 s^4 / 4 + b x^2 s^2 / 2 \cdots (13),$$

and

$$\dot{x} = -\partial^4 x + 2\partial^2 x + \partial^2(x s^2) \cdots\cdots\cdots (14),$$

with

$$\dot{s} = \alpha s(X_0^2 - x^2 - X_1^2 s^2) + \beta \partial^2 s \cdots\cdots (15),$$

where α and β are the positive parameters. In the equations of motion (10), (12), (14) and (15) the dots over the variables represent the time derivative, and ∂ the gradient operator, both in the scaled space-time units.

In the actual calculation to obtain solutions for the above equations, these were replaced by the appropriate difference equations, and were solved either in one dimension along straight lines of 600 to 630 mesh, or in two dimensional matrices of 100·100 mesh, each under cyclic conditions. A series of random numbers were put into the equations as the starting noise distributions. The time evolution of wave-like solutions on isothermal aging was displayed in colour on the monitor screen of PC.

3. RESULTS AND DISCUSSIONS

Since the printing of this paper cannot reproduce the colour of the display, the results of our visualization are given here only in black and white. In Figure 1a to 1d the results are given of our calculation for case (i) of the preceeding section. Figure 1a gives an example of the solutions of Equation (10) in one dimension, for a nonlinear spinodal decomposition from its early to late stages. It is seen in the figure that the equation itself picks up the proper wavelength from a random distribution of the initial noise. In contrast to the almost distinct wavelength from the beginning the amplitude of the waves has a long distance of modulation, indicating the necessity of taking a mesh of fairly large number of units. In the late stage, however, the amplitude and wavelength become almost uniform with the exception of one or two odd wave still remaining. Figure 1b is a demonstration of the Fourier analyses of the patterns, in which the process of "coarsening" is noticed as the principal wavelength growing up from about 8, which corresponds to the proper wavelength of the linear approximation, to nearly 12, its nonlinear value. Such a change of the wavelength in the course of spinodal decomposition has also been predicted by Langer[5] and Tsakalakos[6] in their more elaborate analyses in one dimension.

It should be noted here that in our analysis the process of spinodal decomposition terminates at the stage when the segregation reached the spinodal points where the diffusion constant is nil. Figure 1c shows the enlargement of the last solution of Fig. 1a, and indicates that the phase separation is slowly approaching the spinodal points on both sides of the composition but never goes beyond them. In order for the process of phase separation actually to terminate at the point of stability, it is necessary for us to consider a constantly acting thermal noise, under which the system would not be able to rest at the spinodal points. Another point of interest is that almost any case of our calculation included one or two odd waves, as shown in Figs. 1a, 1b and 1c, which we attibuted as due to the mismatching of the proper wavelength with the period of cyclic condition. In order for the system to accomodate a complete set of the proper waves in the final stage, the total length must be an integral multiple of the wavelength, which is proper to the almost stationary solution of the equation. In Figure 1d an example of our solution is shown, in which the length and the mesh unit were adjusted so that a large number of waves were included in the solution without any exceptionally odd ones.

Figure 2 shows an example for the result of our calculation for case (ii) of the preceeding section. In contrast to the case of spinodal decomposition of case (i), the process of ordering starts out with an irregular sequence of waves without definite proper wavelength. The process terminates in an antiphase ordered domain structure with large and small sizes of domains. Each domain wall has the same gradient of degree of order, the value of which can be predicted from the stationray solution of Equation (12). In the one dimensional case, the minimum energy of the stationary state is a single domain without antiphase boundaries, and the next lowest energy state is the one with two boundaries, and so forth. Thus even in the very late stage of ordering small domains with inferior boundaries tend to vanish, as is seen in Figure 2.

Now the main problem of our interest lies in case (iii), the ordering with phase separation. Figure 3a shows the phase relation of the system at a certain temperature, which can be obtained from the free energy (13) under the equilibrium conditions (1) and (2), with the appropriate values of the parameters X_0 and X_1. As can be seen in the figure there exists a phase field consisting of mixed order and disorder. The phase boundary between the ordered and disordered phases is given by X_0, whereas the boundaries between the ordered, the order+disorder and the disordered phases by X_1 and X_2, respectively. In the figure the free energies of the equilibrium phases are given as functions of the concentration X for the ordered and disordered states, which have a common tangent between $X = X_1$ and X_2. In this case the spinodal point X_s is one-sided, and at the point $X = X_0$ the curvature of the free energy for stable phase changes discontinuously from negative for the ordered state to positive for the ordered one, as was shown by Allen and Cahn[7]. Thus it is customary to define a spinodal phase field between X_s and X_0. Once the values for X_0 and X_1 are given, then the phase boundaries X_s and X_2 are obtained by the following expressions.

$$X_s{}^2 = (X_0{}^2 + 2X_1{}^2)/3 \quad \cdots\cdots\cdots\cdots\cdots \quad (16),$$

and

$$X_2 = (X_0{}^2 + X_1{}^2)/2X_1 \quad \cdots\cdots\cdots\cdots\cdots \quad (17).$$

Figure 3b shows two examples of our calculation of the coupled changes of local concentration and local order parameter. The upper set of the curves show the time development of the concentration(c-wave) and order(s-wave), in which the disturbance in the local degree of order in the vicinity of $s = 0$ is picked up instantly by the local concentration around $x = .35$ just above $X_s = .33$. The "resonance" in this case, however, is of a nonlinear nature, and the wavelength of the c-wave is one half of that of the s-wave, as was predicted by analytical manipulation of Equations (14) and (15)[2]. On the other hand, in the lower case, in spite of the larger disturbance in the concentration around $x = .47$ (just below $X_0 = .5$) than in the degree of order around $s = .05$ (of the slightly ordered state),the c-wave calms down at first and then starts to pick up slowly the influence of the s-wave. In this case the development itself is much slower than the other one, showing a nonresonant nature of this case. It should be noted that the two waves (s-and c-wave) have almost the same form but with opposite phase, as was predicted also by an analytical approximation. These examples demonstrate the physical situation in which the coupled motions of the conserved c-wave and the nonconserved s-wave proceed either resonantly or nonresonantly, and seem to confirm our interpretation of the patterns formed in the ordering Fe-Al and Fe-Si alloys[1,2].

We should now proceed to our two dimensional calculation of the cases (i) to (iii) of the forgoing section. Figure 4 is an example of our PC calculation of the nonlinear spinodal decomposition in two dimensions in its early stage, where the formation of the pattern has already started. In Figure 5 the process of formation of antiphase ordered structure is demonstrated in its late stage, where the main pattern has already been established, but small domains of inferior sizes and degrees of order are still disappearing. Such a process of diminishing small antiphase domains was actually observed by Allen et al.[8] in their in situ electron microscopic observation. Figure 6 is a reproduction of the display of a set of solutions for case (iii) of the ordering with phase separation in two dimensions in its intermediate stage. The process of ordering, which has developed in the early stage, has been inducing that of the concentration waves, and it is seen in the figure that the latter is already forming the domain structure, which resembles that of the spinodal decomposition in a disordered alloy. All through in these figures the time flows from bottom to top and then from left to right. It has become clear by our analysis that in the process of ordering with phase separation such as in Fe-Al and Fe-Si local degree of order does not necessarily follow the equilibrium value corresponding to the local concentration, although in the final stage these two parameters in the same domain should coincide with the equilibrium.

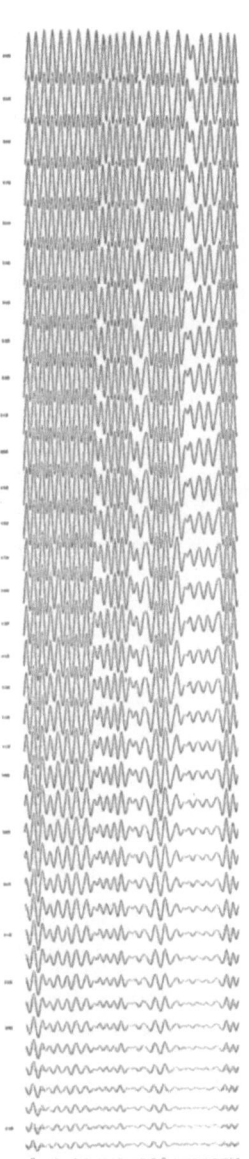

Fig. 1. Time sequence of the 1-dimensional spinodal decomposition. The time goes from the bottom to the top of the figure.

Fig. 1b (right). The development of the concentration wave (right) and the corresponding spectra of Fourier components (left). Note that the largest component shifts on aging from 8.3 to 11 in wavelength, showing the "coarsening" effect.

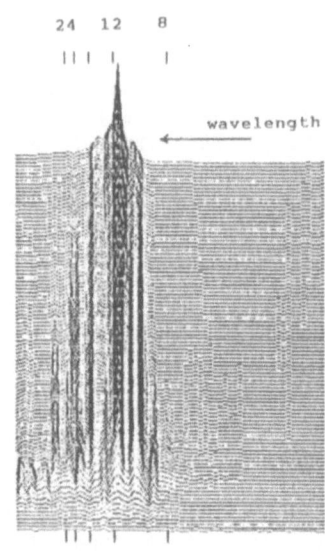

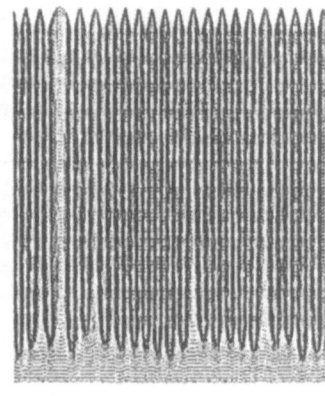

Fig. 1c (below). The wave profile of the last stage of Fig. 1a. The wave tops approach the spinodal point X_s but do not reach the saturation limit X_o. Also note the existence of an odd wave.

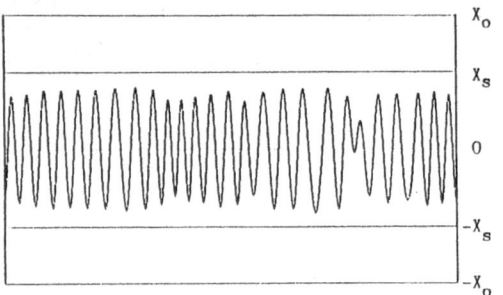

Fig. 1d (below). An example of a nonlinear spinodal decomposition in one dimension with a large number of waves accomodated in the linear mesh. The unit length was so adjusted that all the waves are safely accomodated with little distortion.

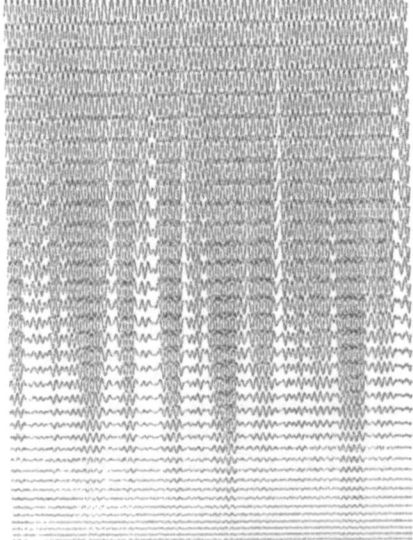

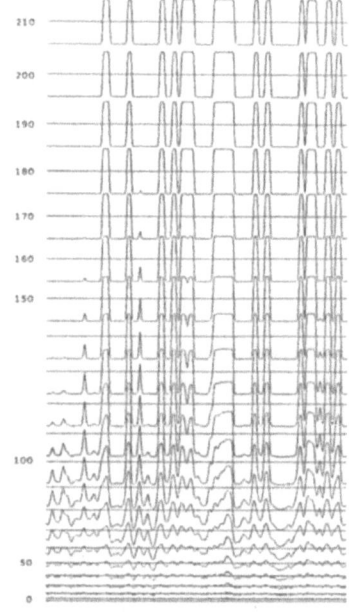

Fig. 2. (above right). An example of the process of formation of ordered antiphase domain structure in a uniform alloy. The wave of the local degree of order starts out with a irregular shape, picked up from the initial noise, and ends up in a series of square-like waves forming the antiphase structure. Note that the gradient of each domain boundary is the same, the value of which can be obtained from the limiting stationary solution of Eq. (12).

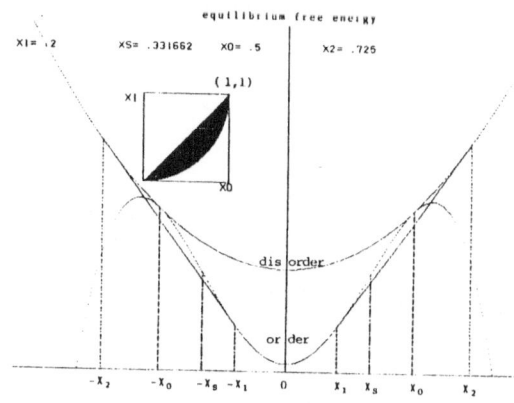

Fig. 3a (Left). The equilibrium free energy obtained from (2) and (13), with $X_0 = .5$ and $X_1 = .2$. The ordered single phase state is stable for $x^2 < X_1^2$, the mixed phase of order and disorder for $X_1^2 < x^2 < X_2^2$, and the disordered phase for $x^2 > X_2^2$.

Fig. 3b (below). Two examples of the process of ordering with phase separation in one dimension; one is the "resonant" and the other the "nonresonant" case.

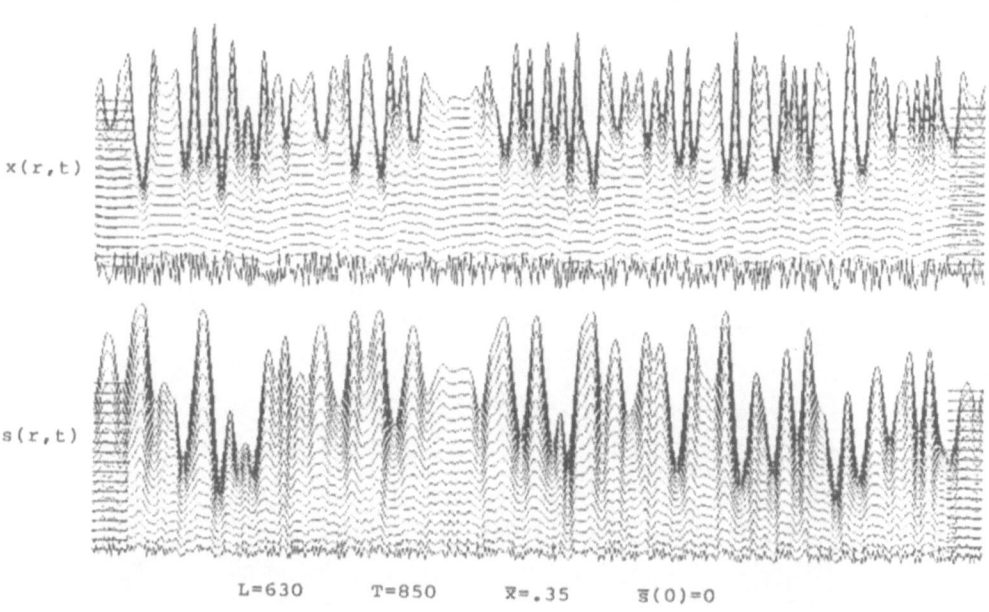

$L=630 \qquad T=850 \qquad \bar{x}=.35 \qquad \bar{s}(0)=0$

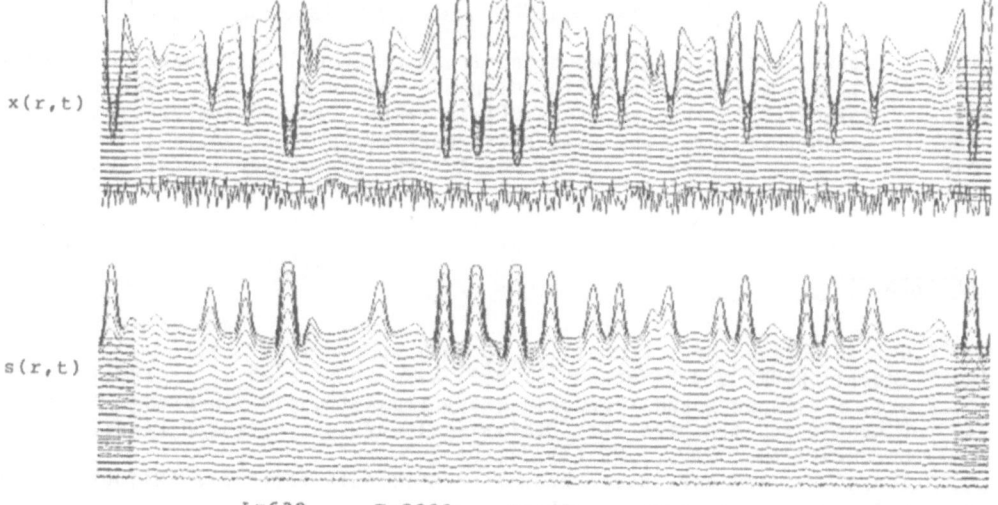

$L=630 \qquad T=3000 \qquad \bar{x}=.47 \qquad \bar{s}(0)=.05$

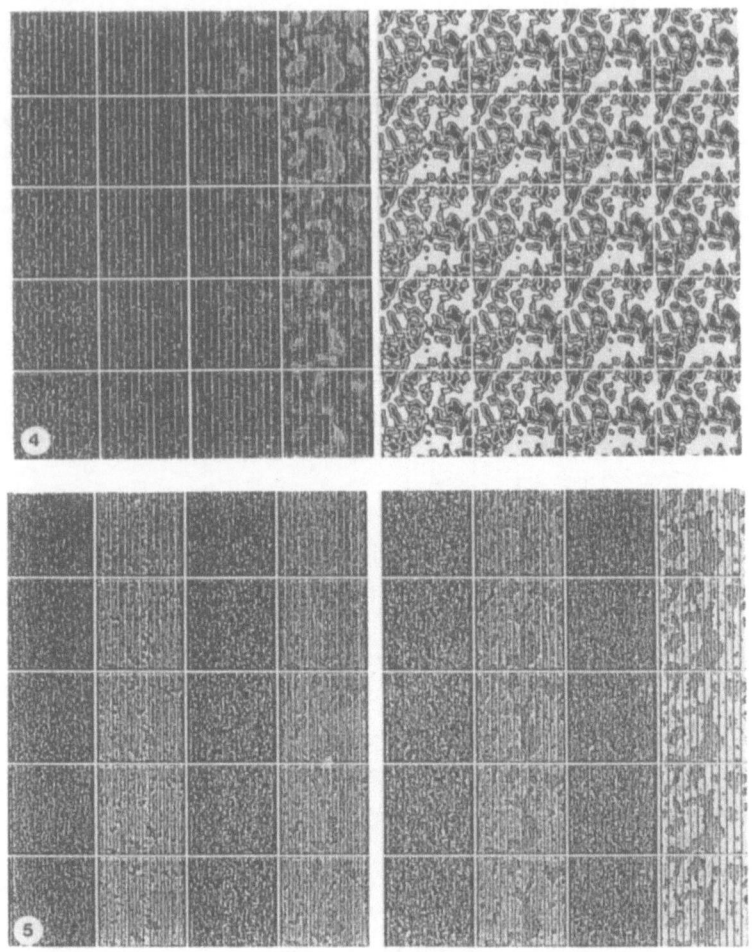

Fig. 4 (above). Examples of two dimensional cases of the spinodal decomposition (left) and the ordering with antiphase domain structure (right).   Time flows from bottom to top and from left to right. Fig. 5 (below). Example of the two dimensional ordering with phase separation in an early to intermediate stage of the process.   The s- and c-waves are paired in left and right.

REFERENCES

1. K.Oki, H.Sagane and T.Eguchi: J. de Phys., C-7 (1977), 414
2. T.Eguchi, H.Oki and S.Matsumura: Mat. Res. Symp. Proc., 21 (1984), 589
3. K.Oki, S.Matsumura and T.Eguchi: Phase Transitions B, in press
4. L.D.Landau and E.M.Lifshitz: Statistical Physics, Pergamon Press, 1959
5. J.S.Langer: Ann. Phys., 65 (1977), 53
6. T.Tsakalakos and M.P.Dugan: Modulated Structure Materials, Nijhoff, 1984
7. S.M.Allen and J.W.Cahn: Acta Met., 24 (1976), 425
8. W.Park and S.M.Allen: Mat. Res. Symp. Proc., 62 (1986), 303

TRANSIENT REVERSIBLE GROWTH AND

PERCOLATION DURING PHASE SEPARATION

Dieter W. Heermann

Institut für Physik
Johannes-Gutenberg Universität
Staudinger Weg 7
6500 Mainz
West Germany

Abstract

Binary mixtures when quenched into the two-phase region exhibit transient percolation phenomena. These transient percolation phenomena and the underlying mechanism of transient reversible growth are investigated. In particular, one of the possible dynamical percolation lines between the dynamical spinodal and the line of macroscopic percolation is traced out. Analyzing the finite size effects with the usual scaling theory one finds exponents which seem to be inconsistent with the universality class of percolation. However, at zero temperature, where the growth is non-reversible and the transition of a sol-gel type, the exponents are consistent with those of random percolation.

What we consider in this talk is a two component system. It may be a metallic alloy, a polymer blend, or a lattice gas. The central question which we address is: What happens when we vary the concentration at a fixed temperature?

At infinite temperature there is no correlation between the atoms. Atoms and B-atoms occupy the sites on the lattice randomly, to use the lattice gas language. The above question is then the random percolation problem [1,2].

In the random percolation problem one finds finite clusters below a certain concentration threshold or percolation threshold $c_p(T=\infty)$. Clusters are only considered for the minority species. A cluster is here to be understood as a collection of like atoms connected by nearest neighbour bonds [3]. Above the percolation threshold there is a cluster which spans the system. In other words, the probability for a given atom to be part of the largest cluster (order parameter P_∞) is zero below the threshold and increase to one above the threshold. At the percolation threshold there is a geometric phase transition.

Upon lowering the temperature one introduces correlations between the atoms. Let us assume that the system is a lattice gas with the hamiltonian

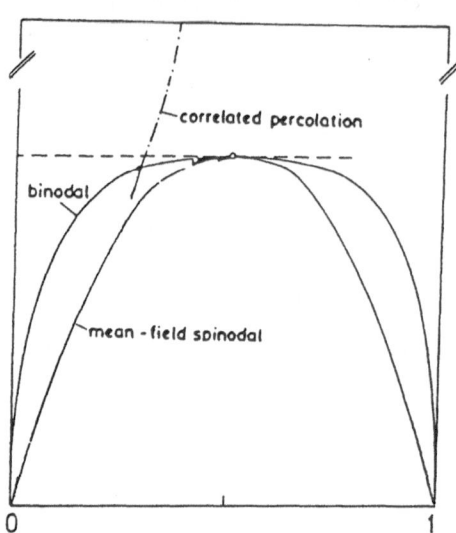

Figure 1. Shown is a schematic phase diagram for a lattice gas. The dashed line indicates the percolation thresholds c_p^{corr} (T) as a function of temperature.

$$\mathcal{H}_{Ising} = -J \sum_{<i,j>} s_i s_j \tag{1}$$

Here J is the exchange coupling between nearest neighbours, denoted by the symbol $<i,j>$, on a lattice. For simplicity the lattice is taken to be $N = L^d$, with d the space dimension. In the lattice gas interpretation the s_i's represent atoms of two types with $s_i = \pm 1$ attached to a lattice site labeled i. A schematic phase diagram is shown in Fig 1.

We can now investigate if there is a percolation threshold, ie. a geometric phase transition for the correlated problem. For this correlated percolation problem [4-6] one finds a line of percolation thresholds c_p^{eq} (T) as indicated schematically by the dashed line in Fig 1. The line of correlated percolation threshold starts at the threshold concentration of the random percolation problem. The line ends on the coexistence curve at a temperature $T/T_c = 0.96$[7,8]. All along this line one finds the same critical exponents as for the random percolation problem [9].

Does this line have a continuation into the two-phase region? In the two-phase region the extra dimension of time is added to the problem. Whereas states in the one-phase region are thermodynamically stable, and as such time independent, the states inside the two-phase region are not thermodynamically stable and time dependent.

What we consider are quenches, at a fixed concentration, from a thermodynamically stable state at high temperature, to a state below the coexistence curve [10-12]. After surch a quench the system will evolve through relaxation patterns, eventually reaching a stable equilibrium state on the coexistence curve.

What we analyze are the relaxation patterns, in particular the clustering properties during the approach towards equilibrium. Let us follow the evolution of the mass of the largest cluster in the system after a quench [13-16]. For illustrative purposes two typical patterns are shown for the two-dimensional case in Fig 2. The figures are time ordered from left to right. The patterns show that atoms rapidly aggregate to form a spanning cluster.

Quantitatively the evolution of the mass in the largest cluster is shown in Fig 3 for the three dimensional case, which we will discuss from now on.

The results were obtained using the non-equilibrium Monte Carlo method (NMC) [18-20]. Initially there is a steep increase in the mass of the largest cluster at fixed concentration after a quench. It should be noted that the concentration studied are far away from the random percolation threshold. After some time the cluster mass decreases. There is a transient, reversible growth phenomenon in the two-phase region.

Figure 2. Shown is a typical relaxation pattern for a two-dimensional system after a quench from the one-phase region into the two-phase region. The largest cluster is highlighted by the full squares.

At fixed temperature and certain concentrations there is an initial rapid aggregation of atoms after a quench from the one-phase into the two-phase region. The aggregation structure is not fixed. Only for zero temperature the aggregation structure freezes. At non-zero temperature the structure can rearrange, unlike the diffusion-limited-growth (DLA) [17] where also particles aggregate to form branched structures. Such branched structures are also observed during the transient reversible growth of the largest cluster. Eventually the structure even looses material and shrinks quite substantialy.

We define now the dynamic percolation as follows [16]. Let the concentration be fixed and select a time t_p. We ask: What is the probability for a given atom to belong to the largest cluster at the time t_p for the concentration c after a quench from $T = \infty$ to some $T < T^{coex}$? This is the order parameter which depends on the concentration, temperature and time t_p.

In Fig 4 are shown the order parameter for several temperatures ranging from $T/T =0.75$ to $T =0$. These curves show the same finite size behaviour as the random and the correlated problem. Indeed, they can be analyzed by the same finite size scaling method [15, 16].

Recall that for the order parameter and the order parameter susceptibility one has the scaling relations

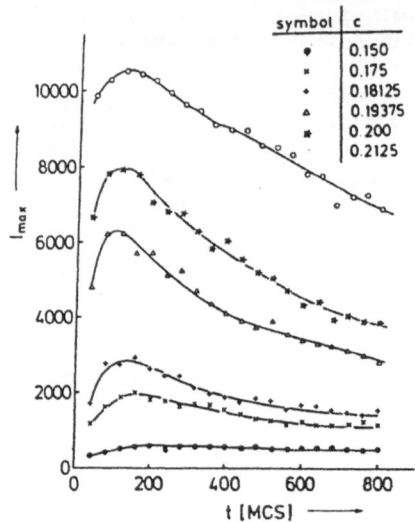

symbol	c
•	0.150
×	0.175
▴	0.18125
△	0.19375
✳	0.200
	0.2125

Figure 3. Time evolution of the size of the largest cluster in a system of linear dimension of L =40. The temperature was T/T_c =0.3 and the various concentrations are indicated in the figure.

$$P_\infty(L,c) = L^{-\beta/\nu}\bar{P}(\delta c L^{1/\nu}) \tag{2}$$

$$\chi(L,c) = L^{\gamma/\nu}\bar{\chi}(\delta c L^{1/\nu}) \tag{3}$$

where we have used the abbreviation $\delta c = c - c_p^{corr}$. The exponent ν is the correlation length exponent, β describes how the order parameter P_∞ vanishes and γ the divergence of the order parameter susceptibility X. The functions $\bar{P}$ and $\bar{X}$ are the scaling functions.

To use these scaling relations, the percolation thresholds were first obtained by the intersection between the order parameters for different system sizes. The resulting line of dynamic percolation points is depicted in Fig 5.

Analyzing the finite size behaviour using the scaling relations and the percolation thresholds one finds the exponents. The analysis thus far made [16] indicates that the exponents β, γ etc are not those of the random or correlated percolation problem. Only for T – 0 one finds agreement with these exponents. For T $\neq$ 0 the exponets do not seem to belong to the universality class of the random and correlated problem.

For the random and the correlated problem one finds

$$\beta = 0.45 \quad random \quad percolation \tag{4}$$

$$\gamma = 0.88 \tag{5}$$

while from the scaling plots (cf. Fig. 6) one finds

$$\beta = 0.48 \quad non-eq \quad percolation \quad T \neq 0 \tag{6}$$

$$\nu = 0.7 \tag{7}$$

Of course, much better statistics is needed to determine the exponents more precisely.

This is only one of the possible dynamic percolation lines. There is a range of percolation lines bordered by the dynamical spinodal [14] on the metastable side and the line of macroscopic percolation on the spin-

162

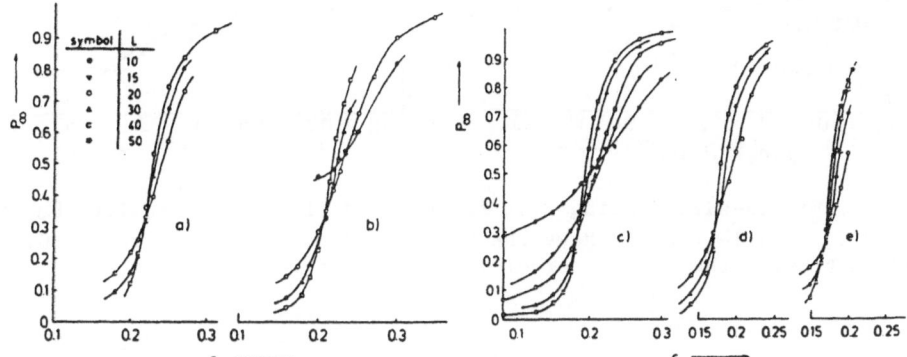

Figure 4. Percolation probability P_∞ plotted vs. the concentration in a lattice gas model for the temperatures $T/T_c=0.75$ (a), $T/T_c=0.6$ (b), $T/T_c=0.3$ (c), $T/T_c=0.15$ (d), $T/T_c=0$ (e).

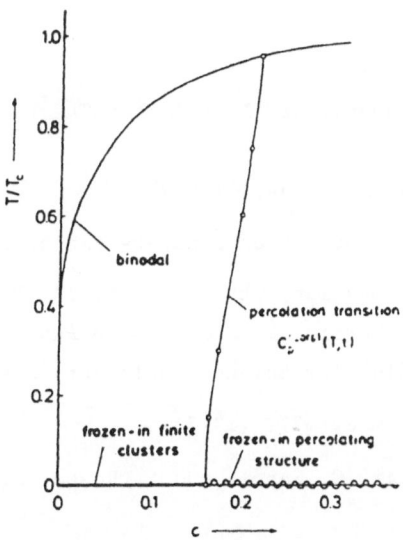

Figure 5. Phase diagram for the three-dimensional lattice gas model including the line of dynamic percolation.

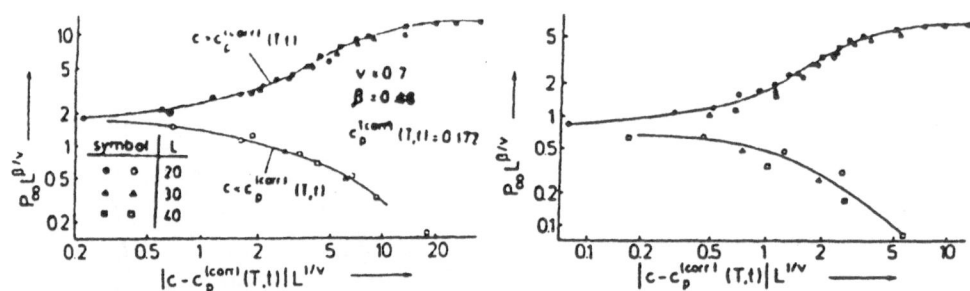

Figure 6. Finite-size scaling plot for the percolation probability P_∞ at $T/T_c =0.3$. The left part shows the "best fit" values $\nu =0.7$ and $\beta =0.48$; right part using the random percolation exponents.

odal decomposition side. Changing the time t_p where we require percolation to appear we can tune to different dynamic percolation lines.

In summary, it was shown that the line of percolation has a continuation from the one-phase into the two-phase region. This continuation is not unique. There is a range of percolation lines. The particular line depends on the value of t_p where one requires the system to exhibit percolation. The percolation phenomenon is brought about by a transient reversible growth of one cluster after a quench.

ACKNOWLEDGMENT   I would like to thank K. Binder and S. Hayward for the many stimulating discussions. Also partial support from the Sonderforschungsbereich *41* is gratefully acknowledged.

REFERENCES

1. D. Stauffer, *An Introduction to Percolation Theory*, Taylor and Francis, London, 1985

2. J.W. Essam, *Rep. Prog. Phys.* 54,1 (1979)

3. Algorithms for the recognition of clusters can be found in

    - J. Hoshen and R. Kopelman, *Phys. Rev.* B. 14, 3428 (1976)
    - R. Dewar and C.K. Harris, in *Fractals in Physics*, eds. L. Pietronero and E. Tosatti, Elsevier Science Publishers Amsterdam, 1986

4. A. Coniglio, *J. Phys.* A 8, 1773 (1975)

5. A. Coniglio, F. Peruggi, C. Nappi, and L. Russo, *J. Phys.* A 10, 205 (1977)

6. A. Coniglio and W. Klein, *J. Phys.* A 13,2775 (1980)

7. H. Müller-Krumbhaar, *Phys. Lett.* A 50,27 (1974)

8. D.W. Heermann and D. Stauffer, *Z. Phys.* B 44,339 (1981)

9. D. Stauffer, A. Coniglio and M. Adam, *Adv. Poly. Scie.* 44,103 (1981)

10. J.D. Gunton, M. San Miguel, and P.S. Sahni, in *Phase Transitions and Critical Phenomena*, vol 8, eds C. Domb and J.L. Lebowitz, Academic Press, New York, 1983

11. K. Binder and D.W. Heermann, in *Scaling Phenomena in Disordered Systems*, eds P. Pynn and A. Skeltrop, Plenum, 1984

12. K. Binder, in *Condensed Matter Research using Neutrons*, eds S.W. Lovesey and R. Scherm, Plenum Press, New York, 1984

13. K. Binder, *Solid State Comm.* 34,191 (1980)

14. D.W. Heermann, *Z. Phys.* B 55,309 (1984)

15. D.W. Heermann, K. Binder, and S. Hayward, in *Proceedings of the 2nd Bar-Ilan Conference* ed. C. Domb, *Phil, Mag.* (1987)

16. S. Hayward, D.W. Heermann, and K. Binder, *J. Statis. Phys.* to appear

17. P. Meakin, *Phys. Rev. Lett.* 51, 1119 (1983)

18. K. Binder, ed. *Applications of the Monte Carlo Method in Statistical Physics*, Springer Verlag, Berlin, 1984

19. D.W. Heermann, *Introduction to the Computer Simulation Methods of Theoretical Physics*, Springer Verlag, Heidelberg, 1986

20. M.H. Kalos and P.A. Whitlock, *Monte Carlo Methods* Vol. 1, Wiley, New York, 1986

STATISTICS OF RANDOM PATTERN --- CURVATURE, PERCOLATION AND OTHERS

Hiroyuki Tomita and Chikara Murakami[*]

Department of physics, College of General Education
Kyoto University, Kyoto 606

[*]Department of Physics, Faculty of Science
Kyoto University, Kyoto 606

[Abstract: We consider topological natures of random patterns generated by excursion set of Gaussian, random fields in arbitrary dimensional Euclidean spaces. The Euler characteristic of the excursion manifold is calculated as a function of the volume fraction. The positions of zeros of it are shown to be unchanged by any coarse-graining or scaling transformations, and then the lowest one is identified with the percolation limit. Others are transition points where the effective dimension of the backbone of percolating manifold changes. A phase diagram to classify the percolating manifold according to its connectivity is obtained by an analytic continuation using non-integer dimension.]

The significant feature of the late stage of phase separation process is the development of smooth, random interfaces. Some general aspects of the scaling form of the structure function $S(q)$ result from this picture: The singular profile of the interface causes, as the lowest order in q^{-1}, the long tail $S(q) \sim A/q^{d+1}$, i.e. the well known Porod law, where d is the dimension of the system and A is essentially the interface area density. As the first order, it can be shown that the smoothness condition requires the following sum-rule relation[1,2]

$$\int_0^\infty [\, q^{d+1}\, S(q) - A\,]\, dq = 0 \, , \tag{1}$$

which should be examined at the same time when the validity of the Porod law is considered in the experimental or theoretical studies of the scaling function. The second order term is related to a curvature invariant averaged over the interface.[1] The purpose of the present paper is to discuss the statistical geometry of random interface system by calculating the curvature invariants, particularly, the Euler characteristic (EC) which is a useful quantity to investigate the geometrical connectivity of the interface or cluster itself. The statistics of the interface is incorporated by using the notion of excursion set of random field.[3] That is, the excursion set and its boundary set correspond to the 'sea' and the 'coast' with a given level of water surface in a random topography.[4] We restrict the problem within tractable Gaussian models. Many works on dynamic theory of phase separation process are in this frame,[5,6,7,8,9] i.e. the field is approximated by a linear, Gaussian one and the non-linearity is incorporated in a separate way. Note that the non-linear transformation of the field parameter [5,6,7] is just identical with the notion of excursion set.

Let $\{u(r)\}$ be a Gaussian, random field with mean $\langle u\rangle=0$ and variance $\langle u^2\rangle=1$ in d-dimensional Euclidean space R^d, and define the excursion set U and its boundary set ∂U by

$$U = \{\, r \mid u(r) < U \,\},$$

and

$$\partial U = \{\, r \mid u(r) = U \,\}. \tag{2}$$

Here let us restrict the Gaussian field to isotropic, homogeneous and non-erratic one, i.e. we assume that the spatial correlation function has the following expansion form:

$$\langle u(0)u(r)\rangle = \sigma(r)$$

$$= 1 - |\sigma''(0)| \, r^2/2! + \sigma^{(4)}(0) \, r^4/4! + \cdots . \tag{3}$$

In this case the excursion set U constructs a well-defined manifold with smooth surface ∂U, and EC $\chi(U)$ of it is given by a surface integral of the d-dimensional total curvature K over ∂U, or by the number of critical points(with index) of $\{u(r)\}$ in U.[3] Because U is an infinite manifold, we use the density of EC defined by

$$\bar{\chi}_d(U) = \chi(U)/V , \tag{4}$$

where V is the volume of the total system. By replacing the volume integral over V by ensemble average using a multivariate Gaussian distribution function, we find the following analytic expression using the Hermite polynomials $H_n(x)$[10]:

$$\bar{\chi}_d(U) = \lambda^{-d} \, (d/dU)^d \, \phi(U) \tag{5}$$

$$= \lambda^{-d}(2\pi)^{-1/2} \, H_{d-1}(-U) \, \exp(-U^2/2) ,$$

where

$$\phi(U) = (2\pi)^{-1/2} \int_{-\infty}^{U} \exp(-u^2/2) \, du = \langle \theta(U-u(r))\rangle , \tag{6}$$

which is used as the volume fraction ϕ of the 'sea' in the following. λ is the correlation length defined by

$$\lambda = (2\pi/|\sigma''(0)|)^{1/2} . \tag{7}$$

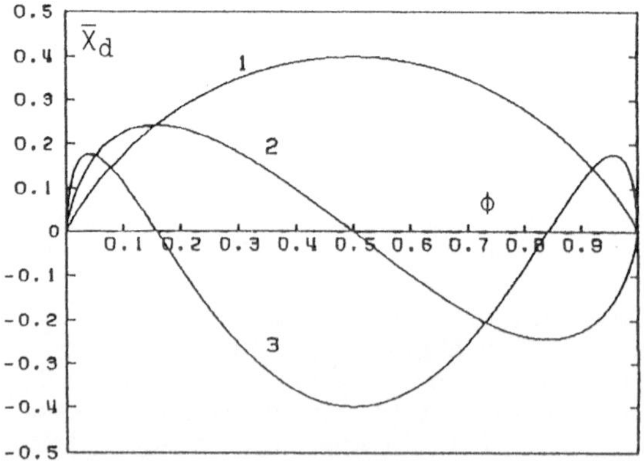

Fig.1 The EC density $\bar{\chi}_d$ for d=1,2,3 with unit $\lambda=1$.

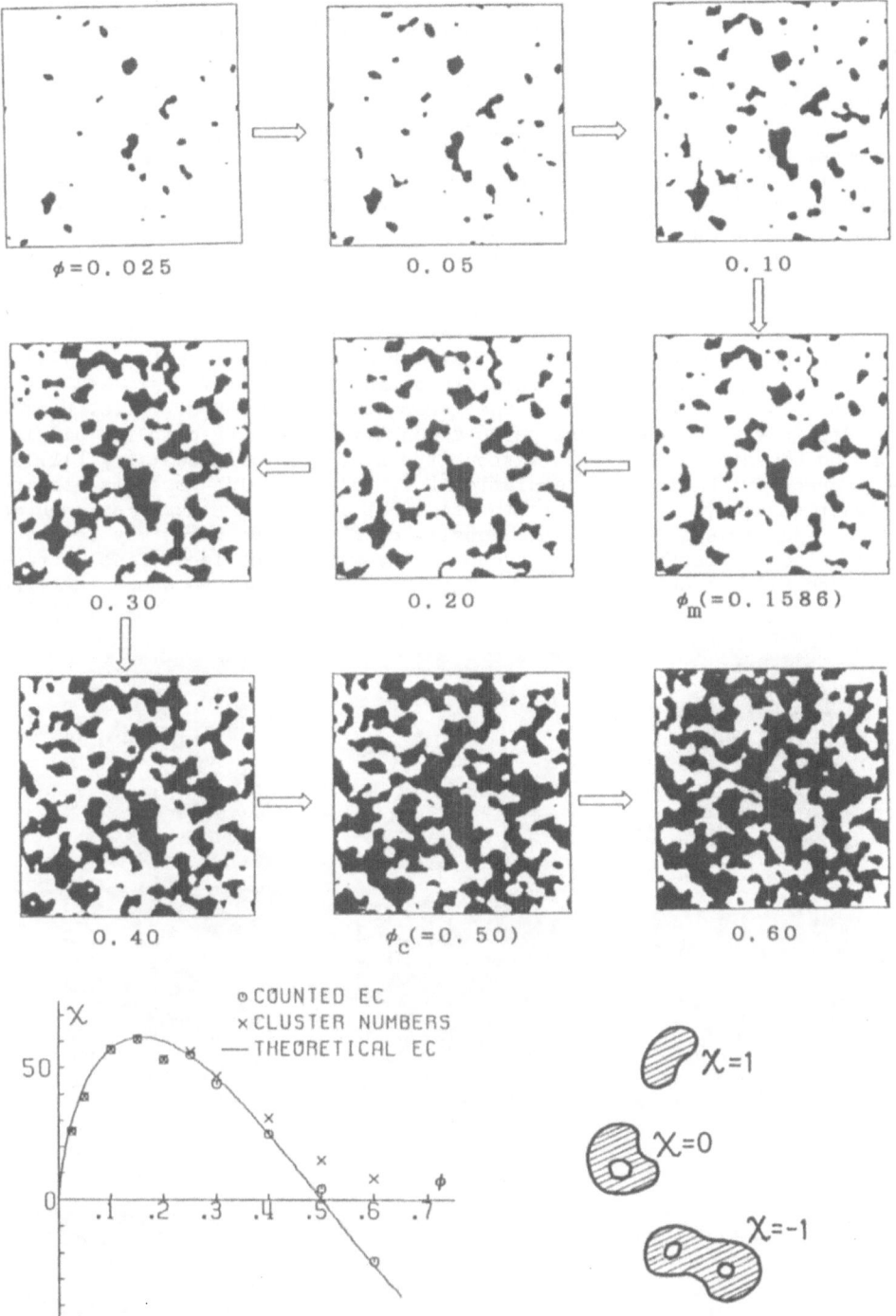

Fig.2　Excursion set in a sample field on　200×200　square and counted
EC.　Correlation function is $\sigma(r) = \exp(-r^2/50)$ in this scale
and $\lambda = 12.533$.

Note that the expression of the EC density given by Eq.(5) is universal, i.e. does not depend on the individuality of the system except for this parameter λ. Results are shown in Fig.1 for d=1,2,3 using units λ^{-d}. Excursion patterns in a sample field of d=2 system and counted EC using this single sample are shown in Fig.2. When the volume fraction ϕ is sufficiently small, the excursion set U is composed of isolated 'ponds' or 'lakes' and $\bar{\chi}_d(U)$ is equivalent to the number density of them, and it increases with ϕ till it reaches a maximum at ϕ_m. For $\phi > \phi_m$, $\bar{\chi}_d$ decreases because of the coalescence of the lakes. Note that when a lake coalesces by itself, it changes into an annulus (or solid torus in d=3), which causes the difference between EC and cluster number. At a certain value ϕ_c the sign of $\bar{\chi}_d(U)$ changes from plus to minus, i.e. the mean topology changes there.

Apparently there are d-1 non-trivial zeros of $\bar{\chi}_d(U)$. The values $\phi_c=1$ for d=1 and $\phi_c=0.5$ for d=2 just agree with the exact (trivial?) values of the percolation limit of continuum system of d=1 and d=2, respectively.[11] The value of the lower zero $\phi_c=0.15866...$(i.e. $U_c=-1$) for d=3 is very close to the percolation limit found by computer simulation on Gaussian fields with various types of co-variance $\sigma(r)$,[12,13] and agrees with Ziman's speculation.[4] This coincidence seems to be somewhat strange, if one calls to mind that any local structures such as 'lakes' and 'islands' affect the density of EC and the zeros may be moved by a kind of coarse-graining procedure by which such local fine structures are painted out.(See Fig.3) This is not. It can be shown that the

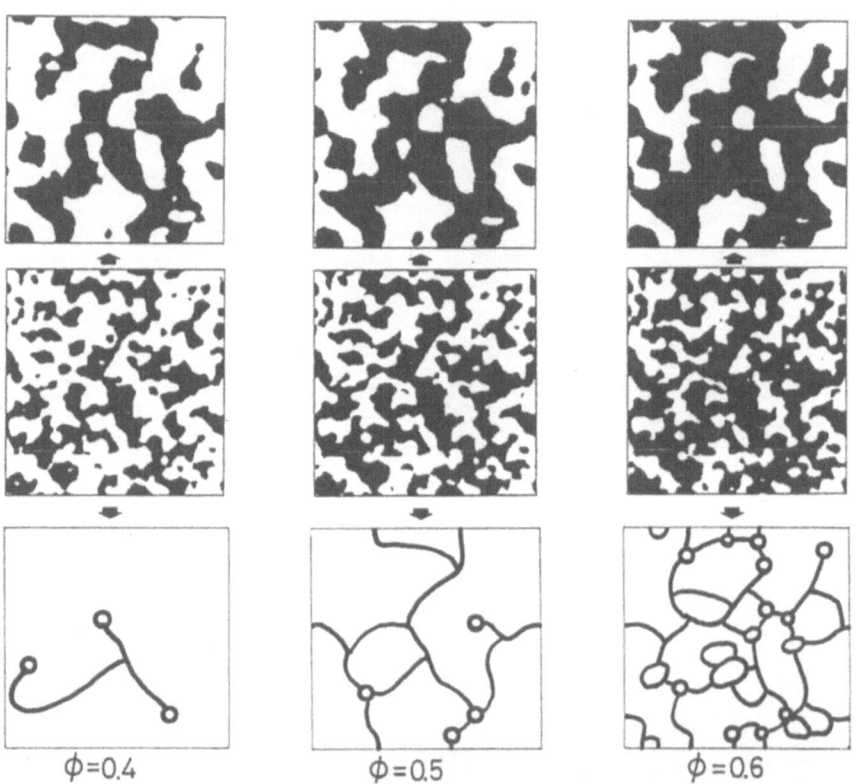

$\phi=0.4$ $\phi=0.5$ $\phi=0.6$

Fig.3 Coarse-graining (upword) and retraction (downword). The same samples in Fig.2 are used. Coarse-graining is performed using an equal weight kernel in 21×21 cell. Retracted backbone of only the largest cluster is shown.

positions of the zeros are never changed by any kind of coarse-graining transformations: Suppose that coarse-graining or smoothing of the Gaussian field is performed with use of a proper kernel by

$$\tilde{u}(r) = \int K(r-r') \, u(r') \, dr' \; . \tag{8}$$

This new field $\{\tilde{u}(r)\}$ is also Gaussian. Let the variance $\langle \tilde{u}^2 \rangle = 1$ be conserved. Then, the form of the EC density $\overline{\chi}_d(U)$ is not changed because of the universal expression of it as is mentioned below Eq.(7). Only the amplitude of it decreases, because the correlation length λ defined by Eq.(7) is increased by smoothing. Thus the positions of zeros are not changed by coarse-graining, and then does not depend on the local structures. This means that a kind of self-similarity exists at $\phi = \phi_c$. As is well known in the lattice percolation problem, the system at the percolation limit has a self-similar structure.[14] If one may assume the same nature of the continuum system, at least the mean topology should be invariant under scale transformation at the percolation limit. We can find none of such invariant points other than the zeros of $\overline{\chi}_d(U)$.

Now let us discuss on the other zeros for $d \geq 3$. According to the symmetry of the field, the upper zero $\phi_c = 0.84133\ldots$ for $d=3$ is the limit of 'land' percolation. In the range $0.15866\ldots < \phi < 0.84133\ldots$, both 'sea' and 'land' percolate mutually. This range cannot be found in $d=2$ system. [11] This is the simplest understanding. Suppose 'land' part is vacuum. It is easy to imagine a $d=3$ manifold U with many bubbles in it for $\phi \simeq 1$. Then the upper zero, which is also scale invariant, is expected to be a transition point where the geometry of connectivity of the percolating cluster changes. This can be understood more clearly, if the spatial dimension d in Eq.(5) is extended to continuous one by using the fractional Riemann-Liouville integral form[15] as follows:

$$\overline{\chi}_{d-\nu}(U) = \Gamma(\nu)^{-1} \int_{-\infty}^{U} \overline{\chi}_d(t) (U-t)^{-(1-\nu)} \, dt \; , \tag{9}$$

where $\Gamma(\nu)$ is the Gamma function. One may restrict ν as $0 < \nu \leq 1$ to avoid the ambiguity of singular integral. This is essentially a unique analytic continuation into continuous d. (Another choice is to replace the range of integration by (U, ∞).) In Fig.4(a), branches of computed zeros

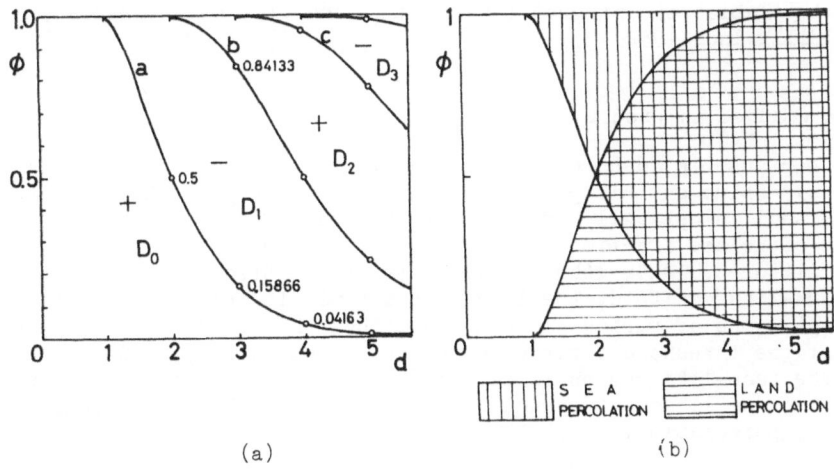

(a)   (b)

Fig.4   Phase diagrams for mean topology of percolating manifold.
Lines a,b,c,.. and the sign +,- in (a) show the branches
of zeros and the sign of the EC density.

and signs of $\bar{\chi}_d(U)$ are shown. This is a phase diagram for the mean topology of the excursion set U. In the region D_0, percolating manifold does not exist. The lowest branch a is the ordinary percolation limit. Note that the next branch b, which starts at d=2, passes $\phi=0.5$ at d=4. This is a reflection of the symmetry of the field, just as the branch a passes $\phi=0.5$ at d=2. Thus this branch b is interpreted as a transition line where the effective dimension of the percolating manifold changes from 1 to 2(=4/2). To relate this fact clearly with the connectivity, it is more suitable to imagine the retraction of clusters. Note that EC is homotopy invariant. Examples for d=2 are shown in Fig.3. After retraction, in D_1 the percolating manifold may become a net-work 'backbone' with negative EC and d=1 geometry, in D_2 a foam-like 'skeleton' with positive EC and d=2 geometry, and so on. And on the boundary line between these zones, the geometry of percolating manifold should be fractal.[15] This is a reasonable understanding of the phase diagram, though the physical nature* of corresponding transitions and the meaning of the non-integer dimension are left as interesting future problems.

References

[1] H.Tomita, Prog.Theor.Phys.72(1984),656.
[2] H.Tomita, Prog.Theor.Phys.75(1986),482.
[3] R.J.Adler, 'The Geometry of Random Field'(John-Wiley and Sons, 1980)
[4] J.M.Ziman, 'Models of Disorder',(Cambridge Univ. Press,1979)
[5] K.Kawasaki, M.C.Yalabik and J.D.Gunton, Phys.Rev.B17(1978),455.
[6] T.Ohta, D.Jasnow and K.Kawasaki, Phys.Rev.Lett.49(1982),1223.
[7] T.Ohta, Ann. of Phys.158(1984),31.
[8] J.S.Langer, M.Bar-on and H.D.Miller, Phys.Rev.A11(1975),1417.
[9] H.Tomita, Prog.Theor.Phys.59(1978),1116.
[10] H.Tomita, Prog.Theor.Phys.76(1986),952.
[11] R.Zallen and H.Scher, Phys.Rev.B4(1971),4471.
[12] A.S.Skal, B.I.Shklovskii and A.L.Efros, JETP.Lett.17(1973),377.
[13] R.Zallen, 'The Physics of Amorphous Solids',(John Wiely and Sons, 1983)
[14] For example, D.Stauffer, Physics Reports 54(1979),1.
[15] For example, B.B.Mandelbrot, 'The Fractal Geometry of Nature', (Freeman, San Francisco,1982). The formula used in Eq.(9) is found in 'Tables of Integral Transforms', chap.13, ed. A.Erdélyi,(MacGraw-Hill, Inc. 1954).

*) Note added in proof.

A plausible idea we have got by discussions with the participants of the symposium on this point is as follows: Suppose that the medium of the 'sea' is optically transparent and that of the 'land' is not. What value of volume fraction is the threshold for macroscopic transparency of this mixtures, when coherent light is transmitted from one side? In the network-like manifold, the light waves randomly interfere after traveling through paths of various length, and then the ordinary percolation limit is not the threshold for the macroscopic transparency. On the contrary, in the foam-like region it is possible to find coherent paths on it as many as one likes. Consequently one may call the upper transition the optical percolation.

MOLECULAR DYNAMICS OF VORTEX-POINTS

IN THE GROWTH PROCESS OF A QUENCHED COMPLEX FIELD

Hiroyasu Toyoki

Physics Institute, Faculty of Education and Liberal Arts
Yamanashi University, Kofu 400, Japan

1. Introduction

Many recent studies on dynamics of systems quenched below their critical temperatures have shown that most of these ordering processes exhibit scaling behaviors. It is a fundamental problem how these scaling behaviors can be divided into universality classes. Main ones of relevant factors in the classification will be the symmetry of ordered state (the component N in vector order parameters) and the spatial dimension D. These factors determine the nature of defects such as kinks, vortex strings, domain walls and so on ,which play important roles in the growth processes. Equations for defects in various systems have been derived from time-dependent Ginzburg-Landau(TDGL)-type field equations mainly by Kawasaki and his collaborators[1]. However, there are not so many cases of which statistical behaviors are known.

For systems with a continuous symmetry, the large-N limit case have been investigated by Mazenko and Zannetti, and de Pasquale and Tartaglia[2]. These analyses would be generally applicable to systems having no defects. However, the dynamics of systems containing defects will differ from them. Recently we have considered the ordering process of a system with a nonconserved complex order parameter in three dimension[3]. Our picture is as follows. In the early time stage after quenching, vortex strings form densely, which means that the order parameter is fully developed locally without orientational correlation. Subsequently the strings shrink gradually, so that the macroscopic order grows. We have obtained the two-point correlation function of both the string density and the gradient of the order parameter within an approximated model derived by the localized induction approximation for the strings. They can be written as scaling forms with the characteristic length varying as $t^{1/2}$.

In this paper we study a two-dimensional system with the same symmetry, where the interaction among vortices plays a direct role as the driving force. The order of this system grows through the pair annihilations of vortices and antivortices.

2. Model

The energy for a configuration of vortex points $\{r_i\}$ is written as

$$H = - \sum_{\langle ij \rangle} (-1)^{\sigma_i + \sigma_j} \log |\mathbf{r}_i - \mathbf{r}_j| \, , \tag{1}$$

where σ_i takes 0 for a vortex and +1 for an antivortex. The equation of motion for a vortex[*] is given by

$$\frac{d}{dt}\mathbf{r}_i = -\frac{\partial H}{\partial \mathbf{r}_i}$$

$$= \sum_j (-1)^{\sigma_i + \sigma_j} \frac{\mathbf{r}_i - \mathbf{r}_j}{|\mathbf{r}_i - \mathbf{r}_j|^2} \, . \tag{2}$$

This equation does not contain any nondissipative terms and thermal noises because they will have less effects on the ordering process than the deterministic dissipative term. The total number of vortices changes solely by pair annihilations, i.e., a pair of a vortex and an antivortex is annihilated when the pair comes near within a distance ξ of the order of the core radius. Consequently the number of vortices decreases with time. Exactly speaking, this model corresponds to the case quenched to zero temperature because the pair creations can take place at finite temperature in two dimension. It is, however, applicable to the deeply

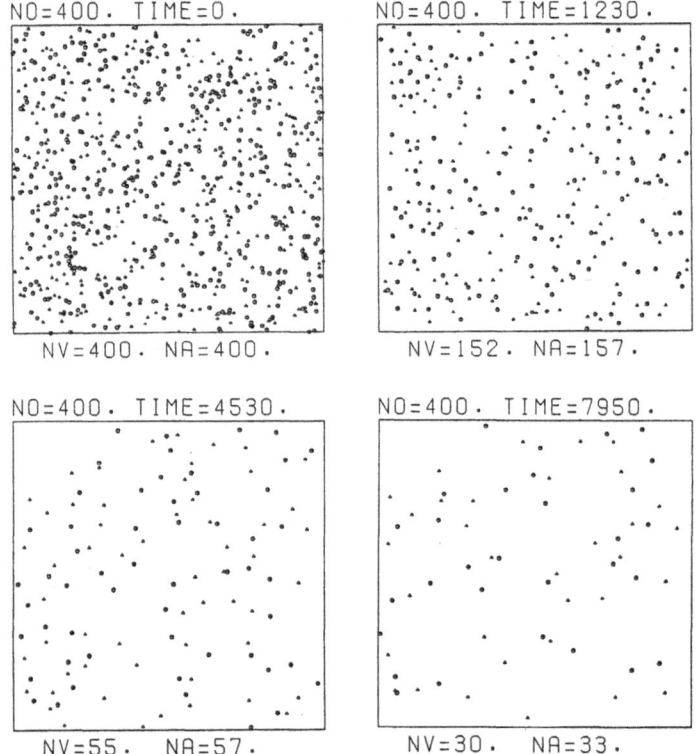

NO=400. TIME=0. NV=400. NA=400.
NO=400. TIME=1230. NV=152. NA=157.
NO=400. TIME=4530. NV=55. NA=57.
NO=400. TIME=7950. NV=30. NA=33.

Fig.1 The evolution of a vortex configuration for $N_0 = 400$, $\Delta t = 2 \times 10^{-7}$ and $\xi = 5 \times 10^{-3}$. The marks ⊙ and ▲ indicate vortex and anti-vortex respectively.

----

[*] We often use the word "vortex" as the generic name for vortex and antivortex.

quenched case in which the number of vortices will exceed sufficiently its equilibrium value.

3. Method of Simulation and Results

We have studied numerically the above model initially consisting of N_0 vortices and N_0 antivortices in a unit square. We have calculated its evolution by Euler's method, which is simple but expected to give a correct result for dissipative cases. The initial configuration is given by a set of random numbers. The boundary is taken to be an absorption-type one with mirror-image effect for vortices. This is appropriate for superfluid [4]He in which the stream lines produced by vortices run along the boundary near it. We consider the eight first-order image vortices for a vortex, though there are infinite number of images.

We show in Fig.1 an example of the temporal change of a configuration for $N_0=400$, the time step $\Delta t=2\times10^{-8}$ and $\xi=5\times10^{-3}$. We have computed the behavior of the number of vortices and the mean distance of nearest neighbor(n.n.) vortex-vortex pairs d_{++} and the one of n.n. vortex-antivortex pairs d_{+-} defined by

$$d_{++}^{+-} = \frac{1}{N_+} \sum_i \text{Min.}(\{ |r_i^{(+)} - r_j^{(\pm)}|, 1 \le j \le N_+ \}) \quad , \tag{3}$$

where the superscript (+) and (−) denote vortex and antivortex, and their numbers are indicated by N_+ and N_- respectively. The results for $N_0=500$ are shown in Fig.2. After the early transient stage, the average number

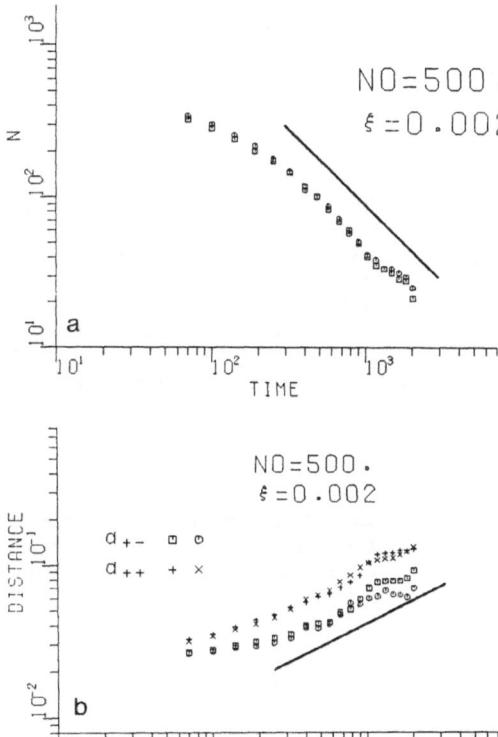

NO=500.
$\xi=0.002$

NO=500.
$\xi=0.002$

d_{+-} □ ○
d_{++} + ×

Fig.2

(a) The change of the total number N(t) for $N_0=500$, $\Delta t=10^{-6}$ and $\xi=2\times10^{-3}$. Two runs with different initial configurations are plotted. The line has a slope of −1.

(b) The changes of the mean distances of n.n. vortex-vortex pairs d_{++} and the one of vortex-antivortex pairs d_{+-} for the same runs with (a). The line has a slope of 1/2.

$N(t)$ of $N_+(t)$ and $N_-(t)$ decreases with time as t^{-1} and the both n.n. distances increase as $t^{1/2}$. These indicate the existence of only one characteristic length in the ordering process.

In order to perform the simulation for a larger system, we use the following cell method. Dividing a system into cells having n_c vortices in an average, we consider the interaction of pairs in the same cell and in the n.n. cells. The size of a cell is adjusted step by step so that n_c may remain constant. Employing this method, we have tested the dependence on the initial configuration and the core radius ξ. We show in Fig.3 the results for $N_0=2000$ and $n_c=15$ with two different ξ's. For each ξ, we have averaged five runs with different initial configurations. The $t^{1/2}$-behavior same as in Fig.2 is seen over a wider range of time, irrespective of the value of ξ. This result shows that the interactions among vortices much further than d_{+-} have little effect by canceling one another.

From the results shown in Fig.2(b) and Fig.3(b) we obtain $d_{++}/d_{+-}=1.6\pm0.1$, which is a characteristic of this system distinguished from the random configuration of vortices and antivortices because this ratio should be unity for the latter case.

4. A Kinetic Idea

The above power-law behavior can be derived by assuming the self-similarity, i.e., the n.n. distances of the vortices d_{++} and d_{+-} have the

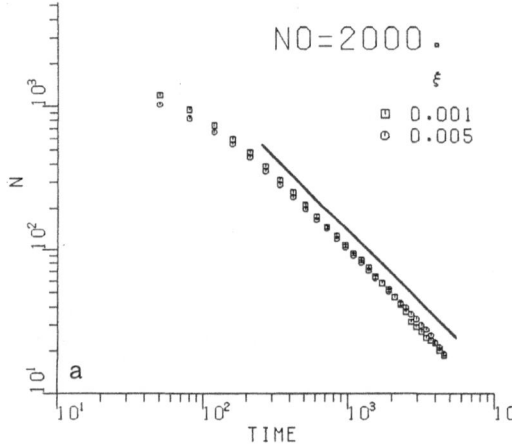

The results by the cell method
(a) The change of $N(t)$ for $\xi=5\times10^{-3}$ and 10^{-3}, which represented by $\circ$ and $\square$ respectively. The parameters are both taken to be $N_0=2000$ and $\Delta t=5\times10^{-7}$. The line has a slope of -1.

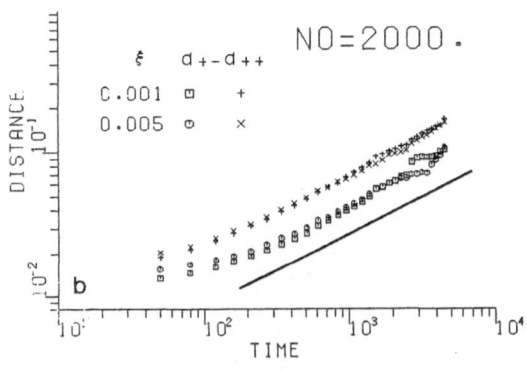

(b) The change of the n.n. distances for the same runs with (a). The has a slope of $1/2$.

same time-dependence and the vortices (not including the antivortices) continue to be randomly distributed in a scale larger than d_{++}. Then we can write the rate equation as

$$\frac{d}{dt}N(t) = -\alpha(t)N(t) \qquad (4)$$

The coefficient $\alpha(t)$ is proportional to the inverse of the mean collision time t_0 which is defined by the time till a vortex-antivortex pair initially separated by $d_{+-}(t)$ is annihilated. When the equation for a pair distance $z(t)$ is given by $dz(t)/dt = -z(t)^{-\beta}$, we obtain

$$\alpha(t)^{-1} \sim t_0 \sim d_{+-}(t)^{\beta+1} . \qquad (5)$$

The number of vortices $N(t)$ is related to $d_{+-}(t)$ as $N(t)^{-1/D} \sim d_{++} \sim d_{+-}$. As a result, we obtain

$$\frac{d}{dt}N(t) \sim -N^{(\beta+1)/D+1} . \qquad (6)$$

Equation (6) immediately leads to $N(t) \sim t^{-D/(\beta+1)}$. Putting $D=2$ and $\beta=1$ we obtain $N(t) \sim t^{-1}$ and $d_{++} \sim d_{+-} \sim t^{1/2}$.

5. Discussions

Our numerical experiment leads us to conclude that the ordering process in the N=2, D=2 system has a characteristic length $\ell(t)$ varying as $\ell(t) \sim t^a$ with a=1/2. Since the mean distances of n.n. pairs represent the range over which the orientation of the order parameter is correlated, the structure factor will also be scaled by $t^{1/2}$, though it has not been investigated so far.

In Table 1, we summarize the growth laws of systems with nonconserved order parameter. The systems have the same exponent a=1/2 except for the N=1, D=1 system. For systems with defects, this fact is related to the common nature that the strength of force is proportional to the inverse of length. However, it is not clear why this exponent coincides with the one of systems containing no defects.

Another interesting problem is whether the power relation derived in §4 holds generally in the annihilation processes of points with different β's, apart from the dynamics of quenched systems. Recently we have found numerically for two-dimensional systems with $\beta \gtrsim 2$ that the number of points decreases more slowly than the results in §4. Furthermore, the exponents of d_{+-} is generally different from the one of d_{++}, though d_{++} is proportional to $N^{-1/2}$ as well as the case $\beta=1$. This means that the system with $\beta=1$ is a special case exhibiting the self-similar evolution. The details on this point will be presented elsewhere.

Acknowledgment

The author would like to thank Dr. K. Honda and Prof. H. Kimura for valuable discussions. Numerical computations were executed at Yamanashi University Computer Center.

Table 1. Defects and Growth Laws in Nonconserved Systems

D\N	1	2	3 - - - - - - ∞
1	kinks $\log t$ [4]		
2	interfaces $t^{1/2}$ [5]	vortex points[*] $t^{1/2}$	none $t^{1/2}$ [2]
3		vortex strings $t^{1/2}$ [3]	

[*]this work

References

[1] K.Kawasaki, Ann.Phys. (N.Y.) **154**(1984),319.
  K.Kawasaki and H.R.Brand, Ann.Phys. (N.Y.) **160**(1985),420.
  K.Kawasaki, Phys.Rev. **A31**(1985),3880 and references therein.
[2] G.F.Mazenko and M.Zannetti, Phys.Rev. **B32**(1985),4565.
  F.de Pasquale and P.Tartaglia, Phys.Rev. **B33**(1986),2081.
[3] H.Toyoki and K.Honda, Prog.Theor.Phys. **78**(1987),No.2.
[4] T.Nagai and K.Kawasaki, Physica **120A**(1983),587;
  K.Kawasaki and T.Nagai, Physica **121A**(1983),175.
[5] T.Ohta, D.Jasnow and K.Kawasaki, Phys.Rev.Lett. **49**(1982),1223.

LATTICE SIMULATION OF INTERFACE

DYNAMICS IN EUTECTIC SOLIDIFICATION

Alain Karma

Theoretical Physics Department 405-47
California Institute of Technology
Pasadena, CA 91125

ABSTRACT

A two dimensional lattice model with random walkers is used to investigate eutectic solidification of a thin film. The instabilities and dynamical mechanisms underlying the formation of a wide variety of eutectic patterns ranging from orderly to chaotic are discussed.

INTRODUCTION

Problems in pattern formation during crystal growth have attracted a large amount of theoretical and experimental interest over the past few years[1,2]. An important problem in this area which has so far received less attention than the solidification of a single phase is eutectic growth. This problem deserves special attention at this point for at least two reasons. Firstly, it is of considerable technological importance since eutectic growth provides a practical method to produce composite materials which are particularly strong and durable. Hopefully, a deeper theoretical understanding of the mechanisms governing the formation of eutectic microstructures could eventually help in the design of new materials of desired physical properties. Secondly, in this system, the presence of more than one solid phase gives rise to complex spatiotemporal phenomena far richer than in mono-phase crystallization. A detailed study of these phenomena can contribute to a more general understanding of spontaneous pattern formation in nature.

Fig. 1 shows schematically a simple phase diagram for a binary mixture consisting of A and B molecules. The eutectic point in this phase diagram corresponds to the point at which two solid phases, one rich in A molecules (α phase), the other in B molecules (β phase), and a liquid mixture of A and B at eutectic concentration C_E are all in thermal equilibrium. When the liquid phase is prepared at a concentration C_∞ near C_E and directionally solidified in the presence of a temperature gradient a eutectic pattern is formed. In the simplest case where the solidification front advances in steady-state the pattern obtained consists of a periodic array of juxtaposed α and β lamellae as shown in Fig. 2. (The numerical algorithm used to generate Fig. 2 will be discussed further and corresponding experimental photographs can be found inf Ref. 3.)

On the theoretical side, steady-state growth is by far the best understood aspect of eutectic solidification. The pioneering work of Jackson and Hunt[3] has shown that for a given thermal undercooling $\Delta T = T_E - T_O$, where T_O is the isothermal temperature of the interface, the lamellar spacing λ is not uniquely determined by the steady-state equations but can take on a continuous range of values. The growth velocity of the interface ν and λ are related by

$$ \nu = \nu_m \left\{ \frac{2}{\Lambda} - \frac{1}{\Lambda^2} \right\}, \quad \Lambda = \frac{\lambda}{\lambda_m} \tag{1} $$

where $\lambda_m \sim (\Delta T)^{-1}$ is the spacing at which the growth velocity reaches its maximum value $\nu_m \sim (\Delta T)^2$. The theoretical model of Jackson and Hunt permits steady-state lamellar eutectic growth to take place at all melt composition C_∞ between the two solidus lines of the binary eutectic phase diagram shown in Fig. 1. However, in practice, steady-state growth is limited to a small range of melt composition in the vicinity of the eutectic composition. Outside this range the growth of one solid phase becomes sufficiently enhanced relative to the growth of the other phase to cause instabilities to develop along the eutectic solidification front[4,5]. The wide variety of non-steady-state modes of growth which occur above threshold of these instabilities and the resulting patterns have remained largely unexplored. Probably the best documented non-steady-state behavior observed experimentally is one in which dendrites of one solid phase and lamellar eutectics both coexist[6]. There is also experimental evidence for dynamical modes of growth in which dendrites are absent[7]. One of them is an oscillatory mode on twice the lamellar spacing which causes lamellae from the minor phase (the phase of smaller volume fraction) to follow sinusoidal solidification paths, as opposed to the vertical paths followed during steady-state growth. Other modes, which seem to involve more convoluted solidification paths of these lamellae, generate complex eutectic patterns ranging from orderly to chaotic.

The main purpose of this short communication is to demonstrate that random walk computational methods can provide a powerful tool to investigate the formation of these intricate eutectic microstructures. In a recent paper[8] I have shown that the random walk model developed by Kadanoff[9], Tang[10], and Liang[11] for viscous flow in a Hel Shaw cell can be suitably extended to simulate eutectic growth in two dimensions and used to generate complex eutectic patterns similar to the ones which have been observed experimentally. In addition the results of simulations have provided a more complete understanding of the diffusional instabilities underlying the formation of these patterns.

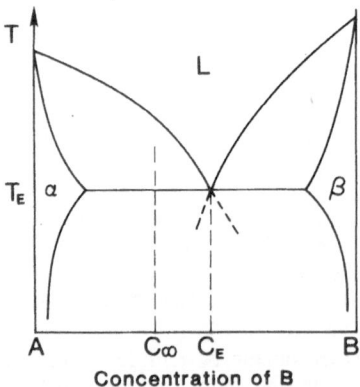

Fig. 1 Simple binary eutectic
phase diagram.

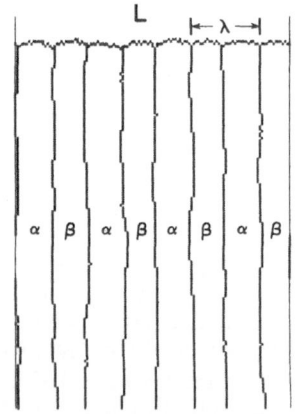

Fig. 2 Simulated lamellar
eutectic pattern.

THE ALGORITHM

The random-walk algorithm is best suited to describe eutectic growth in the limit of slow growth rate where the diffusion equation can be approximated by the Laplacian. In this limit the motion of the eutectic interface is governed by,

$$\nabla^2 u = 0 \tag{2}$$

conservation of mass at both solid-liquid interfaces,

$$V_n = -\hat{n} \cdot \vec{\nabla} u \qquad \alpha\text{-}phase \tag{3a}$$

$$Q V_n = \hat{n} \cdot \vec{\nabla} u \qquad \beta\text{-}phase \tag{3b}$$

local thermodynamic equilibrium at the interface (Gibbs-Thomson relations),

$$u(\alpha) = \Delta T\, m_\alpha^{-1} - d_\alpha \kappa \qquad \alpha\text{-}phase \tag{4a}$$

$$u(\beta) = -\Delta T\, m_\beta^{-1} + d_\beta \kappa \qquad \beta\text{-}phase \tag{4b}$$

and the constraint of mechanical equilibrium which requires the sum of surface tensions to vanish at a triple point

$$\sigma_{L\alpha}\, \hat{\imath}_{L\alpha} + \sigma_{L\beta}\, \hat{\imath}_{L\beta} + \sigma_{\alpha\beta}\, \hat{\imath}_{\alpha\beta} = 0 \tag{5}$$

where $\hat{\imath}_{\gamma\nu}$ is a unit vector which is tangent to the γ-ν interface at a triple point and points away from this point. Here $\sigma_{\alpha\beta}$ is the α-β surface tension, σ_{LS} the L-S surface tension, m_S is the liquidus slope of each phase dT/du defined to be positive, $d_S = \sigma_{LS}\, T_E/(m_S\, L_S)$ and L_S are respectively the capillary length and latent heat per unit volume of each phase ($s = \alpha, \beta$), κ is the local interfacial curvature, $u \equiv (C - C_E)/\Delta C$ is a dimensionless composition field where C denotes the number of B molecules per unit volume, C_S is the concentration of each solid phase, and $\Delta C \equiv C_\beta - C_\alpha > 0$ is the miscibility gap. In addition we have defined $V_n = (-u_\alpha)\, v_n/D$ where v_n is the local normal velocity of the interface, D the coefficient of solute diffusivity, and $Q \equiv u_\beta/(-u_\alpha)$ with $u_S \equiv (C_S - C_E)/\Delta C$ ($u_\alpha < 0, u_\beta > 0$). Finally, the exponentially decaying part of u in the upward z direction is translated, in the Laplacian limit of the diffusion equation, into a linear gradient boundary condition on u far ahead of the interface where, in steady state, $(\partial_z u)_\infty = (v/D)\, u_\infty$ with $C_\infty = C_\alpha \eta + C_\beta (1 - \eta)$; $\eta \equiv \lambda_\alpha/\lambda$ is the volume fraction of the α phase, λ_S ($s = \alpha, \beta$) the lamellar width of each phase, $\lambda = \lambda_\alpha + \lambda_\beta$ the lamellar spacing and $u_\infty \equiv [C_\infty - C_E]/\Delta C$.

Simulations take place on a two dimensional square lattice where sites are divided into into three categories: α-sites and β-sites (s-sites; $s = \alpha, \beta$) represent sites occupied by the solid α and solid β phases respectively and empty sites (e-sites) those occupied by the liquid phase. The lattice spacing is set equal to unity and W measures the lateral width of the system where periodic BC are imposed at the endpoints. The interface, from which walkers are released, is composed of all s-sites which have at least one bond connected to an e-site (i.e. the solid-liquid interface). The essence of the algorithm then consists in solving Laplace equation (eqn. 2) by releasing random walkers from the interface and moving the interface according to specific rules which are consistent with the mass conservation relations (eqns. 3a-3b). The Gibbs-Thomson relations (eqns. 4a-4b) are automatically satisfied by choosing the escape probability of a random walker from site s as

$$P(s) = |u(s)|/\max\{|u|\} \tag{6}$$

181

and by assigning to this walker a flux $f(s)$ equal to the sign of $u(s)$ $(f(s) = \pm 1)$, where max $\{|u|\}$ is the maximum value of $|u|$ on the interface. The intrinsic noise in the algorithm which comes from the random walks is diminished by moving the interface only when a bond connecting and s-site with an e-site has been visited a certain number of times (M or QM times, depending whether the s-site is of α or β type). Finally, the constraint of mechanical equilibrium (eqn. 5) is incorporated into the algorithm by using in eqn. 6 a form of the Gibbs-Thomson relation which takes into account the interaction energy between the three phases at triple points. Away from these points this form reduces to eqns. 4a-b. In their vicinity, that is within a distance r of triple points where r measures the range of the interaction energy between different phases, this form causes the composition field $u(s)$ to become large when slight deviations from the constraint of mechanical equilibrium occur. This increase in composition in turn induces large normal composition gradients which cause interface motion to smooth out these deviations in a time much shorter than the time it takes for triple points to move a distance r.

Three types of walks can be distinguished and each type given clear physical interpretation. Walks that start from an s-site and end on an s-site of the same phase ($\alpha-\alpha$ and $\beta-\beta$ walks) do not contribute to net interface motion, since they conserve flux and only displace material. They simulate the effect of capillary forces at the solid-liquid interface. On the contrary $\alpha-\beta$ and $\beta-\alpha$ walks correspond to diffusion between neighboring lamellae of rejected A and B molecules and contribute to the net growth of both solid phases. Finally walkers are released from infinity at off-eutectic compositions to simulate the gradient BC on u. These walks enhance the growth of one solid phase preferentially.

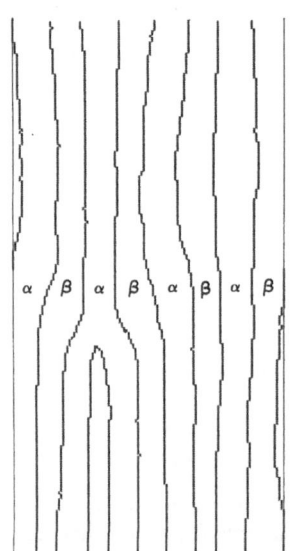

Fig. 3   Simulation exhibiting the termination of one lamella.

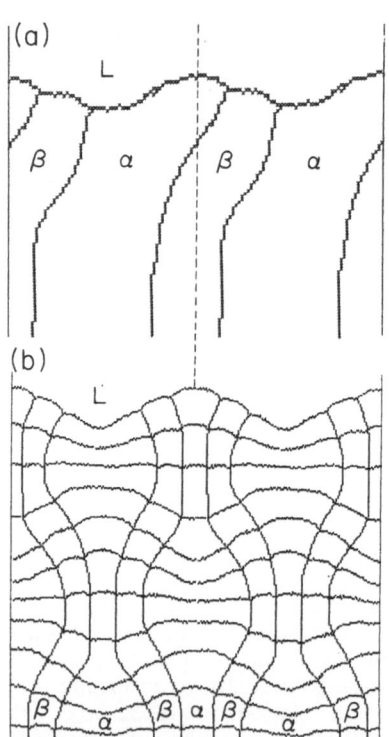

Fig. 4   Interface deformations associated with (a) tilting instability and (b) oscillatory instability.

LIMITATIONS OF THE ALGORITHM

It should be emphasized that this random-walk model simulates the deterministic continuum eqns. 2-5 only when both limits $M \gg 1$ (deterministic) and $1 \ll r \ll \lambda$ (continuum) one satisfied simultaneously. In simulations values of M in a range between 10 and 20 are typically sufficient for the interface evolution to be independent of the sequence of random numbers used to generate the walks, apart from noise induced global symmetry breaking (i.e. left-right symmetry of the "tilting mode", displayed in Fig. 4a), and therefore deterministic for all practical purposes. Because of limited computation time most simulations were performed with values of r and λ ($r \simeq 5, \lambda \simeq 60$) which introduce corrections to the continuum limit. Our hope is that these corrections only have quantitative effects and do not change the qualitative dynamical behavior of the model. This was found to be true in all cases where a simulation was repeated with larger values of λ and r.

INSTABILITIES

In this study we have restricted our attention to a symmetrical binary eutectic phase diagram with $m_\alpha = m_\beta \equiv m$, $Q = 1$, and have chosen equal surface tensions $a_{\alpha L} = a_{\beta L} = a_{\alpha\beta}$. Three different instabilities were observed: the classic long-wavelength instability leading to termination of lamellae (Fig. 3), the oscillatory instability at twice λ predicted successfully by the discrete stability analysis of Dayte and Langer (Fig. 4b), and a new tilting instability on the lamellar spacing which forces lamellae from the phase of smaller volume fraction (β-phase here) to bend coherently on one side of the vertical growth axis (Fig. 4a). At fixed Λ the tilting instability first occurs at a more off-eutectic composition (larger value of $|u_\infty|$) than the oscillatory one, and at fixed composition it occurs at a larger value of Λ. Here, both oscillatory and bending instabilities are driven by the composition gradient ahead of the interface present at off-eutectic compositions.

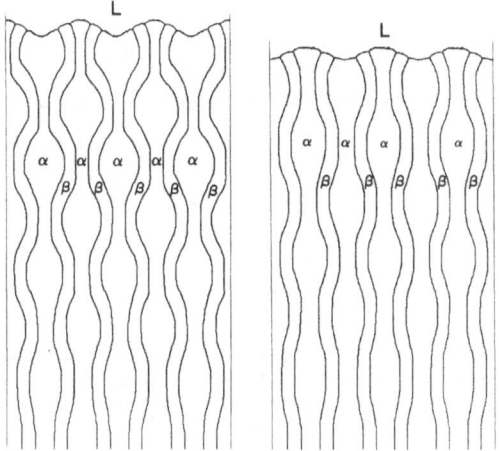

Fig. 5 Oscillatory states.

FROM ORDER TO CHAOS

Above onset of short wavelength instabilities we find that a rich dynamical behavior can take place in between steady-states and dendrites within some range of spacings and compositions. The general understanding of this behavior emerging so far from our simulations is that when either instability (oscillatory or tilting) dominates the dynamics coherent structures are formed (Figs. 5 and 6a), while in regions where both instabilities compete the motion of lamellae can become chaotic and generate disordered eutectic patterns (Fig. 7). Let us note that for the values of the parameters we have studied the tilting instability was always accompanied by the oscillatory one. Therefore Fig. 6a is only ordered because the oscillatory instability has been quenched artificially by choosing the width of the system equal to the lamellar spacing ($W = \lambda$). When the simulation leading to Fig. 6a is repeated with $W = 4\lambda$ (thereby allowing the oscillatory instability to develop) a disordered pattern similar to the one displayed in Fig. 7 is generated.

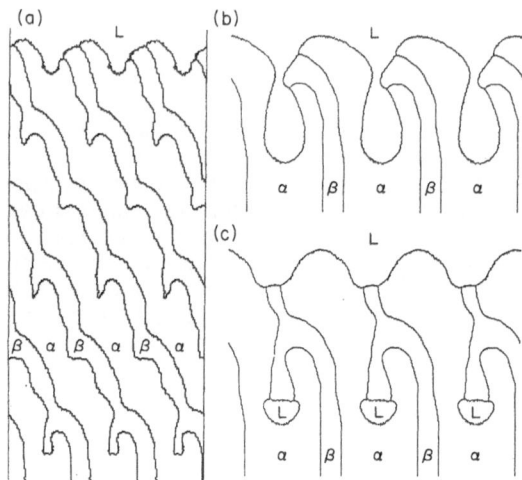

Fig. 6 (a) Exotic eutectic texture; $u_\infty = -1/6$, $W = \lambda = 61$, $M = 15$, $r = 5$ and $\Lambda = 3$. This run took about 30 hrs. of cpu time on a Ridge computer. (b) and (c); time sequence exhibiting the formation of one convolution $W = \lambda = 121$, $\Lambda = 2$, $u_\infty = -1/4$, $M = 10$, and $r = 5$.

FUTURE PROSPECTS

We have only considered so far eutectic solidification of a thin film in the limit of slow velocity. Of course, much more work is now needed to fully understand this system and in particular the eutectic-dendrite transition. Probably the most promising outlook for the near future is to extend lattice simulations to three dimensions where questions of pattern selection involving the motion of defects in nearly regular eutectic patterns remain poorly understood. It is also conceivable that the model discussed in this paper could be modified to investigate the even vaster class of eutectic microstructures which form when the interface of either one or both solid phases is faceted[6].

Acknowledgements

I wish to thank M.C. Cross for many helpful discussions. This research was supported by the California Institute of Technology through a Weingart Fellowship and through the Program in Advanced Technologies which is funded by GM, GTE, TRW and Aerojet.

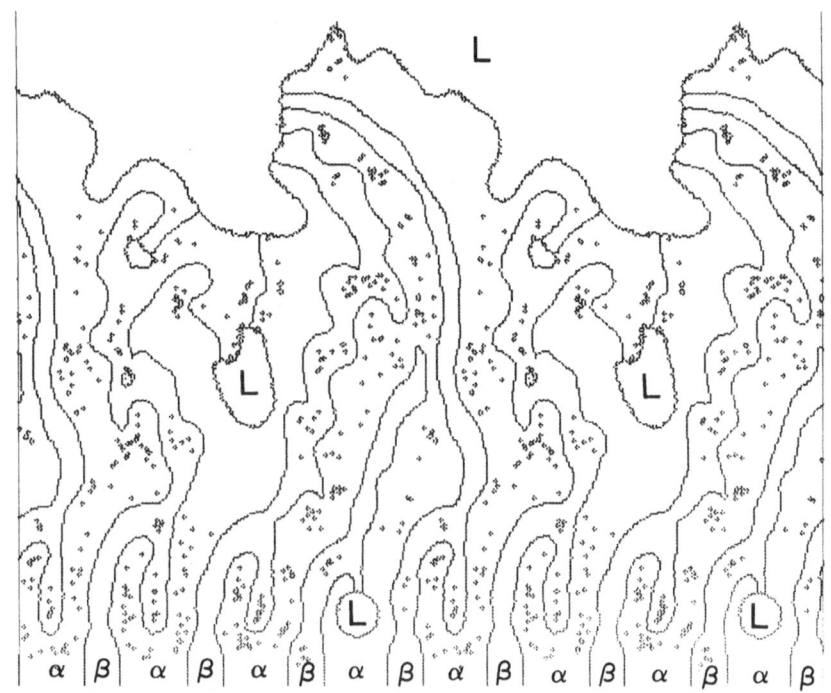

Fig. 7 Disordered eutectic pattern. $W = 4\lambda = 241$. Liquid islands are left behind in the solid but do not affect the growth of the open solid-liquid interface.

References

1. J.S. Langer, Rev. Mod. Phys. 52, 1 (1980).
2. For more recent reviews see J.S. Langer, Les Houches Summer School of 1986 - Lecture Notes; and D. Kessler, J. Koplik, H. Levine (preprint).
3. K.A. Jackson, J.D. Hunt, Trans. Metall. Soc. AIME 236, 1129 (1966); and J.D. Hunt and K.A. Jackson, Trans. Metall. Soc. AIME 236, 843 (1966).
4. D.T.J. Hurle and E. Jakeman, J. Cryst. Growth 3,4, 574 (1968).
5. V. Dayte, J.S. Langer, Phys. Rev. B24, 4155 (1981).
6. D.P. Woodruff, The Solid-Liquid Interface (Cambridge University Press, 1973).
7. J. Van Suchtelin (unpublished).
8. A. Karma, Phys. Rev. Lett. 59, 71 (1987).
9. L.P. Kadanoff, J. Stat. Phys. 39, 267 (1985).
10. C. Tang, Phys. Rev. A31, 1977 (1985).
11. S. Liang, Phys. Rev. A33, 2663 (1986).

PHASE DECOMPOSITION IN AL-ZN ALLOYS

Kozo Osamura

Department of Metallurgy, Kyoto University
Sakyo-ku, Kyoto 606, Japan

INTRODUCTION

Behaviour of phase decomposition in the Al-Zn binary system has been extensively investigated as a model case in the statistical thermodynamic study as well as a base alloy in the industrial materials. In the alloys with relatively low solute concentration, it is knwon[1] that spherical G.P. zones are formed after the alternative unmixing process by the nucleation and growth or the spinodal decomposition at its early stage. However the experimental information is still insufficient to understand the whole process of phase decomposition. In the theoretical viewpoint, the mechanism of phase decomposition in this alloy system has been not yet correctly solved, even though several attempts[2~5] have been reported. At present, it is important to re-examine the structure change during phase decomposition as presicely as possible and to investigate the scaling properties for comparing with the recently progressing theories as mentioned in the present paper.

In the present study, the structure change during phase decomposition has been investigated by means of small-angle scattering (SAS) techniques. The scaling property of decomposed structure has been analysed in two aspects. Firstly the time scaling has been concerned. After the decomposition process was divided into three stages, the structural and kinetic feature of each stage has been discussed by comparing with several recent theories.

EXPERIMENTAL PROCEDURE

The samples investigated here were mainly three Al-X at%Zn alloys (X= 4.0, 6.7, and 10) and other alloys with different solute concentration. Two types of X-ray SAS measurements have been done. One is the so-called conventional measurements using the laboratory apparatus[6]. Another is the in-situ SAS measurments using synchrotron radiation which were done at BL-15A in Photon Factory, National Laboratory for High-Energy Physics, KEK. The details were mentioned elsewhere[7,8]. The meaured intensity was transformed into the coherent scattering intensity $I(k)$ after corrections with the incident beam intensity, sample thickness, absorption, background and fluorescence.

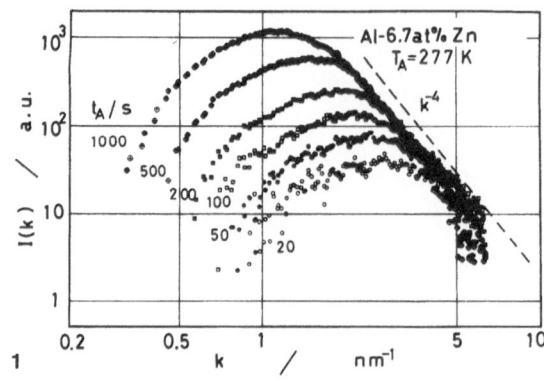

Fig.1 The k-dependence of SAS intensity for Al-6.7 at%Zn alloy aged at 277 K. t_A is the aging time.

EXPERIMENTAL RESULTS

Figure 1 shows the SAS intensities as a function of scattering vector for the Al-Zn alloy during aging at 277 K. The characteristics of intensity profiles are as follows. The intensity maximum shifts towards lower scattering vector region with increasing aging time. At the higher scattering vector region, the intensity seems to decrease obeying the power law of k^{-n}. It is clear that n becomes 4 for the data aged for longer times. However, it is difficult to judge the power law for the data for short time aging. Fig. 2 shows the diffuse scattering intensity observed at 573 K for the same alloy. The intensity decreases gradually at the higher scattering vector region. When the composition fluctuation exists in the solid solution, the following Ornstein -Zernike formula for the scattering intensity might be expected[9];

$$I(k) = I(0)k_c^2/(k_c^2 + k^2), \qquad (1)$$

where k_c is equal to the inverse of correlation length. The above equation could be well fitted with the experimental data as shown by the solid curve in Fig. 2, where I_L indicates the level of Laue monotomic scattering.

In order to make clear the power law at the early stage, the inverse scattering intensity was plotted against the square of scattering vector as shown in Fig. 3. When the k^{-2} dependence holds, we can look a straight line in this figure. Of course, the data obtained at 573 K shows this linear relation. On the other hand, the data aged at 277 K for 135 s and 1010 s give the k^{-4} dependence, which is indicated by the respective solid curve. For the data of short time aging for 20 s, it is difficult to ensure either k^{-2} or k^{-4} dependence. As discussed later, the k-dependence can be described in terms of the zones with diffuse interface.

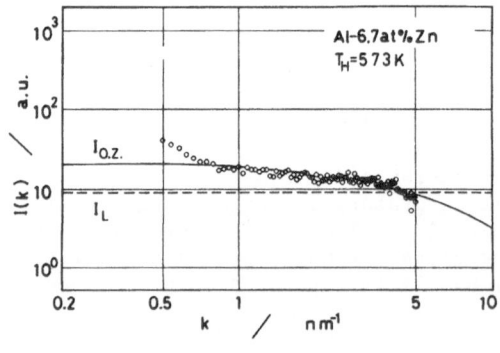

Fig.2 Diffuse scattering intensity as a function of scattering vector, when the Al-6.7at%Zn alloy was held at 573 K.

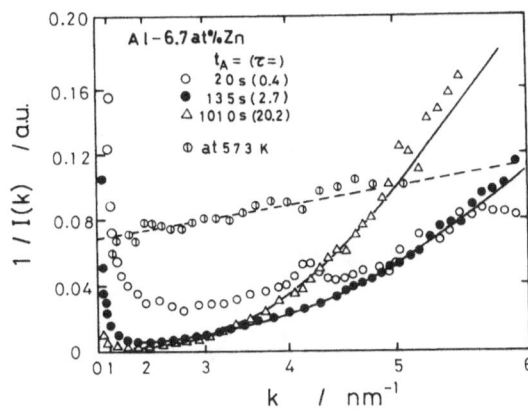

Fig.3 The k-dependence of inverse scattering intensity for the Al-6.7at%Zn alloy aged at 277 K. t_A and τ are the aging time and its specific value.

The integrated intensity is a measure of the degree of phase decomposition. It was calculated by the equation,

$$Q = \int_0^{k_p} 4\pi k^2 I(k) \ dk \ + \ 4\pi k_p^3 I(k_p), \tag{2}$$

where Porod law was assumed for the integration over the entire scattering vector region. In the present case, the integration was performed numerically up to $k_p = 5$ nm^{-1} and beyond this point it was analytically evaluated by the second term of the above equation. The evaluation using Eq. (2) is not accurate at the very early stage, where Porod law does not hold. Its error is estimated within 50 % from the reason mentioned below. As shown in Fig. 4, the normalized integrated intensity, Q/Q^∞ increases with increasing a specific aging time and saturates to a fixed value. The saturated value is defined as Q^∞, of which level for each alloy is proportional to that calculated from the phase diagram[10]. Being based on the half-completion time, $t_{1/2}$ when the ratio Q/Q^∞ reaches 1/2 during isothermal aging, the specific aging time was defined by the relation,

$$\tau = t \ / \ t_{1/2}. \tag{3}$$

Fig. 4 suggests that regardless of the alloy composition, the temporal evolution of the normalized integrated intensity can be scaled against the specific ageing time.

Figure 5 shows the alloy-composition dependence of half-completion time. When the alloy composition increases, the time decreases greatly changing in magnitude. Also it differs depending on the experimental technique. In the conventional SAS experiments[11], the specimen was quenched into brine and stored in liquid nitrogen. The measurements were performed using the cryostat cooled at liquid nitrogen temperature and then the aging treatment was done. The repeat of experiments would introduce a complicated thermal history. On the other hand, the in-situ experiments give a simple situation. When the specimen was quenched from different temperatures and aged at the same temperature, the decomposition rate differed each other as shown in Fig. 6, where the Al-6.7 at% Zn alloy was quenched from the temperature T_Q and aged at 293 K. It is found that the integrated intensities are well scaled as a function of specific aging time. As the respective half-completion time is indicated in the figure, the time is shorter for the higher quenching temperature. This is attributed to the quantity of quenched-in vacancies, because the mobility of solute atoms is primarily proportional to the vacancy concentration.

The decomposition rate has been discussed by Langer[12] and Binder and Stauffer[13]. The alloy composition dependence is connected with the supersaturation, which is defined as

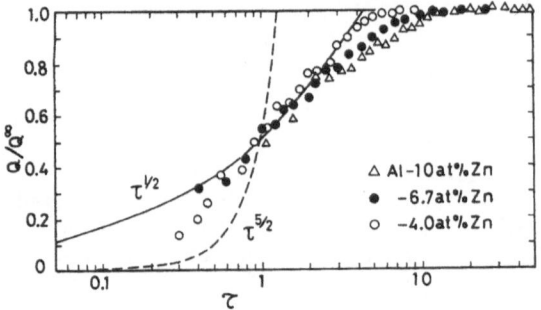

Fig.4 The specific aging time dependence of the normalized integrated intesity for the alloys aged at 277 K.

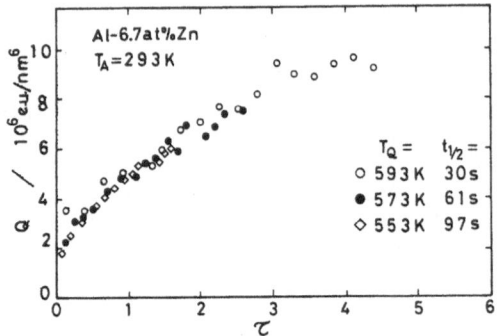

Fig.6 Change of the integrated intensity as a function of specific aging time for Al-6.7at%Zn alloy aged 293 K after quenching from various temperatures T_Q.

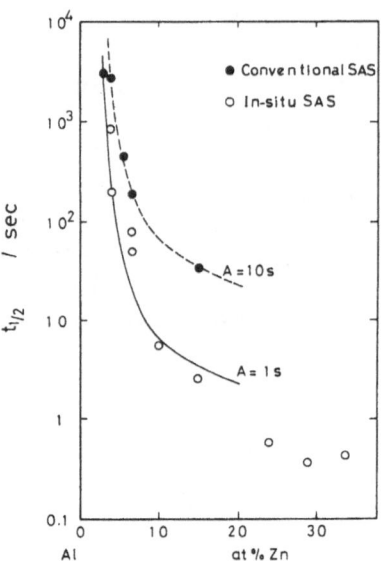

Fig.5 The half-completion time as a function of alloy composition, when the alloy was quenched from 573 K and aged at about 277 K (ref.8) Both solid and dotted curves are the thoretical ones given by Eq.(6).

$$y_s = (T_{coex} - T_A) / (T_c - T_A), \qquad (4)$$

where T_c is the critical temperature and T_{coex} is the value on the coexistence curve corresponding to the alloy composition. The above two temperatures can be determined by using the excess free energy of mixing reported by Lasek[10]. According to the recent nucleation theory by Langer[12], the half completion time is expressed as

$$t_{1/2} = [(32/15) \pi \, \beta^{1/2} \, D^{3/2} \, X_1^{1/2} \, J(X_1)]^{-2/5}, \qquad (5)$$

where β is the critical exponent ,of which value is 0.35 by Stauffer and Kiang[14], D is the diffusion coefficient, X_1 is equal to y_s/β, and $J(X_1)$ is the nucleation rate. For the present experimental condition, Eq.(5) is simply expressed as

$$t_{1/2} = A^* \, y_s^{-13/5} \, \exp(0.248 \, y_s^{-2}), \qquad (6)$$

where A^* is constant including the diffusion coefficient. Equation 6 was fitted with the experimental data as shown by the solid curve in Fig. 5. Then the constant A^* was estimated to be 10 sec for the data due to conventional SAXS and 1 sec for the in-situ SAS. Consequently, the Langer's expression seems to fit with the present experimental results. However, as discussed later, it should be noted that there exist some discrepancies.

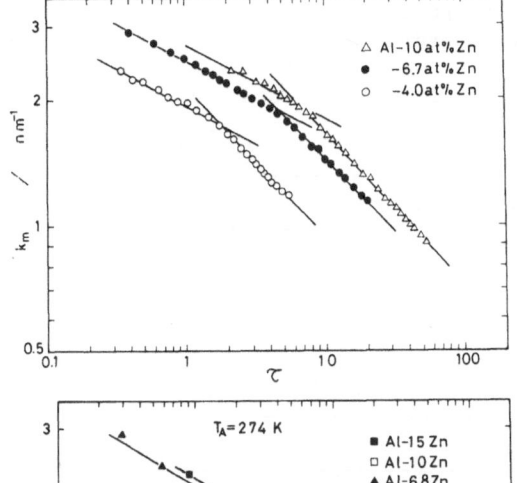

Fig.7 Change of k_m as a function of specific aging time for the Al-Zn alloys aged at 277 K (ref. 8).

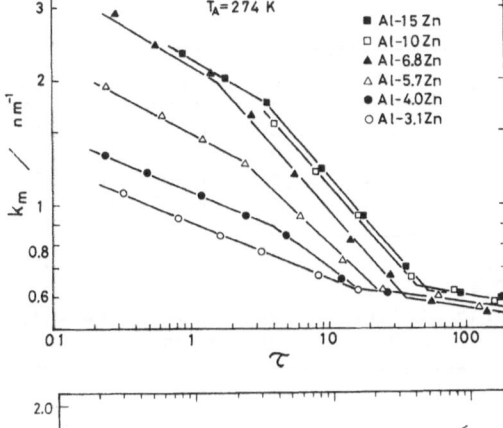

Fig.8 Change of k_m as a function of specific aging time for the Al-Zn alloys aged at 274 K (ref. 11).

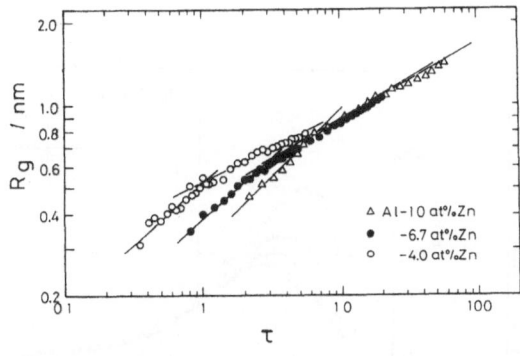

Fig.9 Change of radius of gyration as a function of specific aging time for the Al-Zn alloys aged at 277 K (ref.8).

The specific aging time dependence of k_m, which is the position of intensity maximum, I_{max}, is displayed in Fig. 7 for the data obtained by the in-situ SR-SAS measurements and in Fig. 8 for the data by the conventional SAS. It is clear that a single power law does not hold over the whole region of aging time. The slope decreases with increasing time and three types of slope can be seen depending on the time duration, which is terminated as listed in Table 1. At stage I, k_m shows a weak time dependence. In stage II, k_m decreases rapidly with increasing time. Finally k_m tended to saturate at a constant value. The time dependence of I_{max} has been also examined, which could be correspondingly divided into three stages. The exponents of the following power laws for each stage are summarized in Table 2, in $k_m \propto t^{-n}k$, $I_{max} \propto t^{n}I$ and $R_g \propto t^{n}R$. Fig. 9 shows the change of radius of gyration as a function of the specific aging time. The radius increases gradually in stage I, and further increases in stage II, but at stage II, the growing rate slowed down. The time exponents at each stage were determined as listed in Table 2. The similar study has been performed previously[15].

Table 1   Definition of phase decomposition process in Al-Zn alloy system.

Stage I	Stage II	Stage III
$\tau < \sim 5$	$\sim 5 < \tau < 50$	$50 < \tau$
$Q/Q^\infty < 0.9$	$0.9 < Q/Q^\infty$	$Q/Q^\infty = 1$
$-n_k < n_R$	$-n_k = n_R$	$-n_k = n_R = 0$
increasing V_f	constant V_f	constant V_f

Table 2   Power law for the time evolution of structure parameters in Al-Zn alloy system.

Alloy Comp. at%Zn	Aging Temp. K	Stage I			Stage II			Remarks
		n_I	$-n_k$	n_R	n_I	$-n_k$	n_R	
4.0	277	0.94	0.18	0.46	1.13	0.34	0.27	in-situ
6.7	277	0.83	0.17	0.44	0.95	0.32	0.31	SAS
6.7	277	0.78	0.19	0.48	0.98	0.34	0.30	(ref.8)
10	277	0.71	0.14	0.50	1.00	0.34	0.29	
4.0	274	0.62	0.15	0.45	–	0.33	–	convent.
5.7	274	–	0.19	0.44	1.04	0.36	0.35	SAS
6.8	274	0.63	0.18	0.45	1.06	0.38	0.36	(ref.11)
10	274	–	–	–	0.98	0.36	0.35	

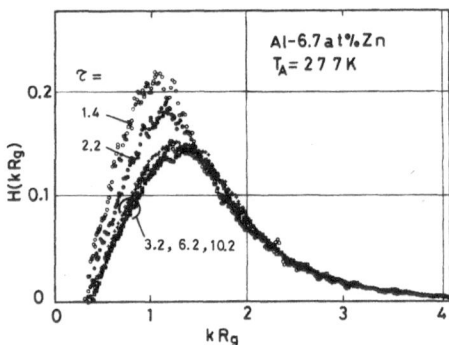

Fig.10 Change of the normalized intensity H(x) as a function of reduced scattering vector x during aging at 277 K.

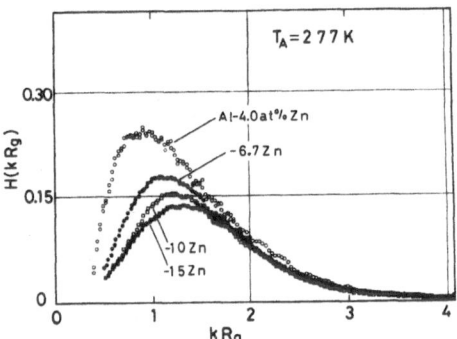

Fig.11 Alloy composition dependence of the normalized intensity at stage II for the Al-Zn alloys aged at 277 K.

Many studies on scaling behaviour have been reported for the phase decomposition. Two structure parameters are needed for this purpose. Usually the following set of two parameters are used; Guinier radius and the integrated intensity[1], or the k_m and the integrated intensity[5], or the first moment k_1 and the integrated intensity[16]. We adopted here the scaling using the integrated intensity and the radius of gyration. We define the following normalized intensity,

$$H(x) = I(x)/(QR_g^3), \tag{7}$$

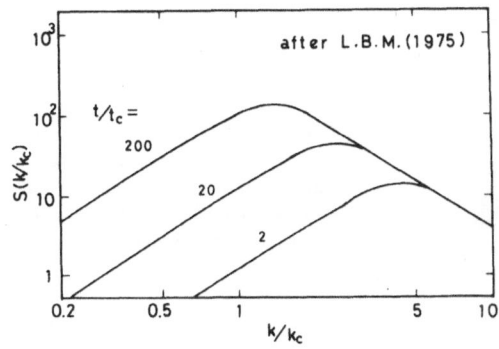

Fig.12 Theoretical scaled struc-
function as a function of reduced
scattering vector after Langer,
Bar-on and Miller (1975).

plotted as a function of kR_g as shown in Fig.10. The half-completion time,
$t_{1/2}$ for this alloy is about 50 sec. The function H(x) decreases with
increasing time and reaches to a saturated value beyond twice the
half-completion time. For the other alloys with different composition, the
same time dependency was observed. The alloy composition dependence of the
scaled structure function at Stage II is shown in Fig. 11. The maximum
value of this function becomes smaller for the alloy with the higher solute
concentration, but the half-width tends to decrease.

DISCUSSION

The structure change during phase decomposition in Al-Zn alloys has
been investigated by monitoring the time evolution of representative
structure parameters. Clearly the time dependence of structure parameters
is divided into three stages. The structural and kinetic feature of stages
I and II will be discussed below in details.

Stage I

As a typical data of scattering intensity at stage I is shown in Fig.
3, it does not obey Porod law and also Ornstein-Zernike formular does not
hold. At the beginning of stage I, the zone size is very small, which is
comparable with the coherence length ξ, which is estimated[10] to be 0.25 nm at
277 K for Al-6.8at%Zn alloy using the thermodynamic data. In order to
understand the structural feature at this very early stage, there exist
essentially two approaches; one is the cooperative model proposed by
Langer, Bar-on and Miller (LBM)[17] and another a simple cluster model.

According to LBM's scaling analysis, the dimensionless structure
function is expressed by using the reduced wave number, q and the
dimensionless time variable, τ' as

$$\partial S/\partial \tau' = -q^2(q^2 - \mu)S + q^2, \qquad (8)$$

where μ indicates the derivatives of free energy of the system. When μ is
a slowly varying function of time, or q^2 is much larger than μ, the above
kinetic equation has an approximate solution of the form,

$$S(\tau') = \left\{ S(\tau'=0) - 1/(q^2 - \mu)\right\} \exp[-q^2(q^2-\mu)\tau'] + 1/(q^2 - \mu). \quad (9)$$

For simplicity, it is assumed that $S(\tau' = 0)$ is very small. Then the
structure function has a maximum in the condition of $q_m^4 \tau' = 1.2564$. And
the power law is deduced as $q_m \propto \tau'^{-1/4}$ and $S(q_m) \propto \tau'^{1/2}$. The q
dependence of the solution of Eq.(9) is displayed in Fig. 12. The structure
function is characterized by an envelope proportional to q^{-2}. Comparing
Fig. 12 with Fig. 13, it seems that the experimental data can be explained

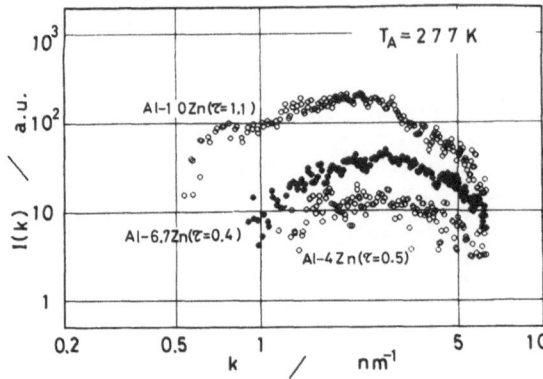

Fig.13 SAS intensity as a function of scattering vector for various Al-Zn alloys aged for very short time as indicated by τ less than unity.

by the q^{-2} rule at the high scattering vector region. This behaviour was pointed out firstly by Furusaka[18]. However, as examined here, the same scattering vector dependence has been observed for the alloys with different composition from 4.0 at%Zn to 10 at%Zn as shown in Fig. 13. At least, one of their alloys locates outside the spinodal curve. Therefore, the explanation due to LBM's nonlinear theory is not valid.

In the cluster kinetics regime, the Gibbs heterophase concentration fluctuation is suggested to be present at the early stage. As mentioned above, the diffuse interface can be expected for these small zones. The composition distribution around zones is proposed to be described by the exponential function as follows;

$$\rho(r) = \rho_o \exp[-(r/R)^m], \qquad (10)$$

where r is the radial distance measured from the center of each zone and ρ_o is the amplitude of composition. When m is assumed to be 2, then the structure function for a single zone is given by

$$G(kR)^2 = v_{eff}^2 \rho_o^2 \exp[-(kR)^2/2], \qquad (11)$$

where v_{eff} is the effective volume of zone. Considering the spatial distribution of zones, the apparent scattering intensity is expressed by the equation,

$$I(k) = N G(kR)^2[1 - \psi(kL)], \qquad (12)$$

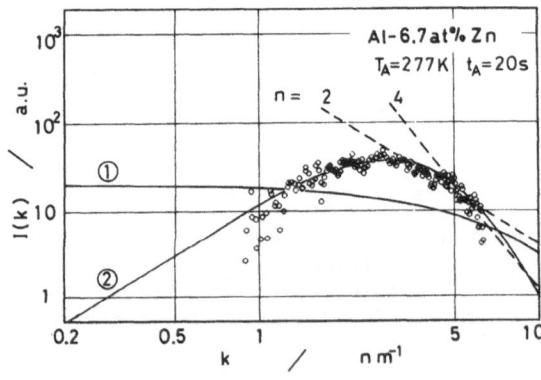

Fig.14 Experimental and theoretical SAS intensity as a function of scattering vector. Open circles are the data for Al-6.7at%Zn alloy aged for 20 s at 277 K. The solid curve ② is theoretical one due to the cluster model.

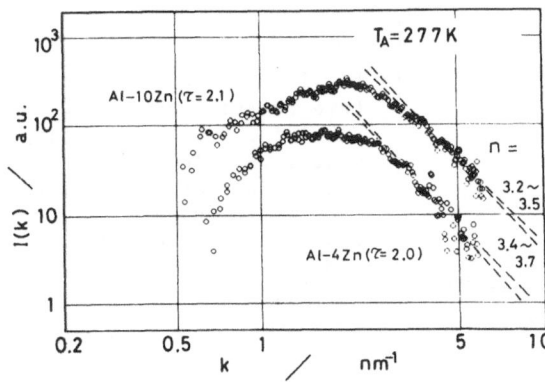

Fig.15 SAS intensity as a
function of scattering
vector for two Al-Zn alloys
aged at 277 K up to $\tau = 2$.

where N is the number density of zones. $\psi(kL)$ is the interparticle
interference function and is given by the equation,

$$\psi(kL) = (4\pi L^3/3v^*)[3(\sin kL - kL\cos kL)/(kL)^3], \qquad (13)$$

where the pre-term $4\pi L^3/3v^*$ is assumed to be unity. Putting that the
radius of gyration is 0.4 nm and the interparticle distance is 3 nm as
obtained experimentally for the early stage, Eq. (13) is evaluated as
shown by the solid curve (②) in Fig. 14, where the absolute value was
fitted with the data. The function can explain well the experimental
scattering vector dependence of the intensity, especially at the high
scattering vector region. This fact indicates that the interface of zone
is very diffuse at the early stage. On the other hand, the scattering
intensity with t_A = 135 s shown in Fig. 3 corresponds to the later period
of stage I, of which scattering vector dependence obeys Porod law.
Fig. 15 shows the k dependence of scattering intensity for the data aged up
to $\tau = 2$, which corresponds to the intermadiate period of stage I. It is
found that the apparent exponent takes a value between 2 and 4. Therefore
it is suggested that the interface of zone becomes gradually sharp during
the process of stage I. In the present study, the integrated intensity has
been obtained by using Eq.(2). The error becomes large at the early stage,
where the Porod law does not hold. Supposing the cluster regime, however,
it is proved that the maximum error does not exceed 50 % beyond the time
scale of $\tau = 0.4$.

As seen in Fig. 4, the volume fraction is suggested to increase at
stage I. According to Langer[12], its rate is expected to be proportional to
$\tau^{5/2}$ as displayed by the dotted curve in Fig. 4. Obviously the theory does
not fit with the experimental data. As mentioned previously[8], the half-
completion time has a linear dependence on the diffusion coefficient.
Therefore the expression of Eq. (5) should be modified in a more rigorous
manner.

The time exponents at stage I as listed in Table 2 are consistent
with those obtained from the computer experiments based on three
dimensional kinetic Ising model. M. Rao, Kalos, Lewowitz and Marro[19]
reported that the time exponents, n_I, $-n_k$ and n_R are 0.7, 0.2 and
0.63~0.36, respectively at 0.6 T_c. Stage I can be characterized as the
process of cluster diffusion and coalescence, because the average zone size
increased, but the number density decreased with increasing time as seen in
Figs. 7, 8 and 9. Here the number density is assumed to be proportinal to
k_m^3 as confirmed previously[20,21]. However it seems at present that there
is no theory to explain quantitatively this stage.

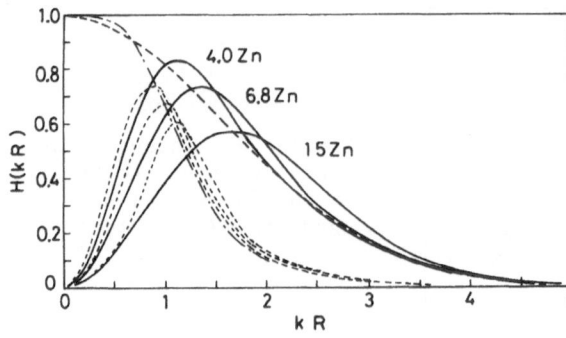

Fig.16 The scaling structure function as a function of reduced scattering vector kR. The solid curves indicate the present simple two phase model and the dotted curves are deduced from the Furukawa's theory.

Stage II

In this stage, the zones have sharp interface, because Porod law holds for scattering vector dependence of intensity as shown in Figs. 1 and 3. As a first approximation, it is possible that the structure of decomposed alloy at stage II is described by a simple cluster model, which is essentially similar with that proposed by Rikvold and Gunton[22]. The zones distribute almost randomly in space, but other zones do not come in a region around each zone. This exclusive volume is given by $4\pi L_p^3/3$. The characteristic length L_p is connected with the interparticle distance L as $L_p = L - R$. Beyond this volume, the probability finding other clusters is equal to the average number density of clusters. Under the condition of constant volume fraction, the next solute atom conservation holds,

$$L = R (1.81V_f^{-1/3} - \sqrt{2}).\tag{14}$$

Then the normalized function H(x) is expressed by the equation[11],

$$H(x) = 1/(1 - V_f) \, \Phi(x)^2 \, \left[1 - \psi(kL)\right].\tag{15}$$

When kR is less than 4, the structure factor is approximated by

$$\Phi(kR)^2 = \exp\left[-(kR)^2/3\right].\tag{16}$$

In the multiparticle system, the average radius is regarded as the radius of gyration. Eq.(15) can be calculated as a function of x=kR when the volume fraction is given. The calculated results shown in Fig. 16 might be compared with the experimental data as shown in Fig. 10. It is seen that the calculated curves can explain qualitatively the experimental data. The volume fraction dependence of $k_m R_g$ (= $(kR)_m$) is shown in Fig. 17. It should be noted that the assessment of the average zone radius is rather arbitrary and here the radius of gyration was used. The half width of intensity profile divided by k_m is shown as a function of V_f in Fig. 18. Here the present result is indicated by solid curve together with the more exact results proposed by Tokuyama and Kawasaki[23] and Furukawa[24].

At stage II, the volume fraction is constant, and the time exponents, n_I, $-n_k$ and n_R are typically represented to be 1.0, 0.33 and 0.29, respectively as listed in Table 2. These experimental results suggest that Ostwald ripening takes place at stage II. Several theories[23-26] predict the similar time exponents. Lifshitz and Slyozov[25] and Wagner[26] (LSW) concerned the evaporation and condensation of solute atoms from small cluster to larger one, but their theory is valid in the limit of infinitesimal volume fraction. Tokuyama and Kawasaki[23] (TK) took into account about the volume fraction dependence and pointed out the important role of soft-collision process among clusters. They led the same temporal

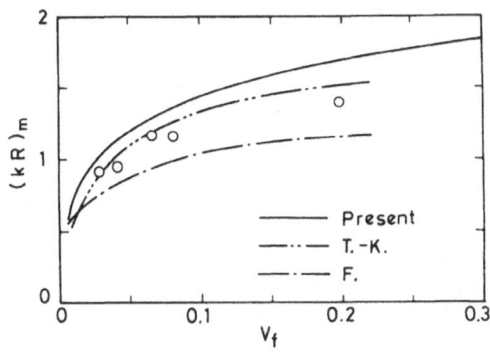

Fig.17 Volume fraction dependence of $(kR)_m$. Open circles indicate the experimental data of k_m multiplied by R_g, and three curves are the theoretical results; present,T.-K. (23) and F.(24).

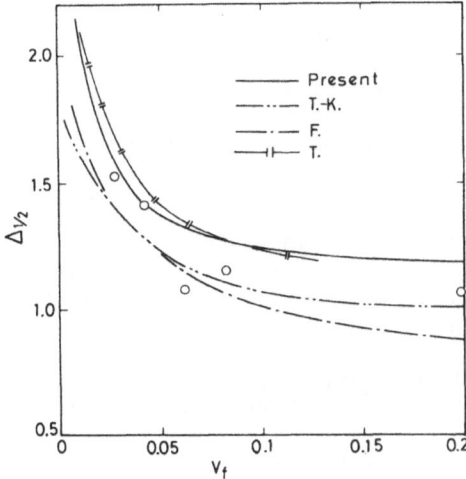

Fig.18 Volume fraction dependence of the half-width of the scaled structure function. The details are the same as in Fig. 17. T.;(27)

power law as LSW, where the time exponents are expressed as $n_I = 1.0$, $-n_k = 1/3$ and $n_R = 1/3$. At present, it seems that TK theory explains reanonably the structure change of stage II.

CONCLUSION

Phase decomposition process in the Al-Zn alloy system has been investigated by means of SAS techniques. The present conclusions are as follows:
1) The composition dependence of half-completion time can be explained by Langer's theory (1980).
2) Using a specific time normalized by the half-completion time, the temporal evolution of decomposition process is divided into three stages with nearly the same time scale, independently of alloy composition and quenching condition.
3) Stage I is a growing process of clusters, where cluster size increases, but the number density decreases. The clusters have diffuse interface. Kinetics of this stage seems to be described by the process given by the kinetic Ising model.
4) Stage II is so called Ostwald ripening process. The dynamical structure change is explained well by Tokuyama and Kawasaki's theory.

Acknowledgements

The authors acknowledge the partial financial support of Light Metal Foundation Inc. The present work was fulfilled by a Grant in Aid for

Scientific Research (Project No. 60460202) of the Ministry of Education, Science and Culture of Japan.

REFERENCE

1. R.Baur and V.Gerold, Acta Metall., 10:637 (1962); W.Merz and V. Gerold, Z. Metallkde., 57:607 (1966); V.Gerold,: J. Appl. Cryst., 10:25 (1977).
2. K.B.Rundman and J.E.Hilliard, Acta Metall., 13:1025 (1967).
3. J.S.Langer, Acta Metall., 21:1649 (1973).
4. R.Acuna and A.Bonfiglioli, Acta Metall., 22:399 (1974).
5. S.Komura, K.Osamura, H.Fujii and T.Takeda, Phys. Rev., B11:1278 (1985).
6. K.Osamura and Y.Murakami, Jap. J. Metal. Soc., 43:537 (1979).
7. K.Osamura, H.Okuda, H.Hashizume and Y.Amemiya, Acta Metall., 33:2199 (1985).
8. K.Osamura, H.Okuda, H.Hashizume and Y.Amemiya, Metall.Trans., submitted (1987).
9. D.Schwahn and W.Schmatz, Acta Metall., 26:1571 (1978).
10. J.Lasek, Czechoslov. J. Phys., 15:848 (1965).
11. K.Osamura, H.Okuda and S.Ochiai, Materials Science Forum, 13/14:57 (1987).
12. J.S.Langer, System Far From Equilibrium, ed. L. Carrido, Springer Verlag, 1980, p. 12.
13. K.Binder and D.Stauffer, Adv. Phys., 25:343 (1976).
14. D.Stauffer and C.S.Kiang, Adv. Coll. and Interf. Sci., 7:103 (1977).
15. M.Hennion, D.Ronzaud and P.Guyot, Acta Metall., 30:599 (1982).
16. J.P.Simon and P.Guyot, Phil. Mag., A49:151 (1984).
17. J.S.Langer, M.Bar-on and H.D.Miller, Phys. Rev., A11:1417 (1975).
18. M.Furusaka, Y.Ishikawa and M.Mera, Phys. Rev. Lett., 54:2611 (1985).
19. M.Řao, M.H.Kalos, J.L.Lebowitz and J.Marro, Phys. Rev., B13:4328 (1974).
20. K.Osamura, H.Okuda and S.Ochiai, Scripta Metall., 19:1379 (1985).
21. V.Syneček, J. de Phys., 23:828 (1962).
22. P.A.Rikvold and J.D.Gunton, Phys. Rev. Lett., 49:286 (1982).
23. M.Tokuyama and K.Kawasaki, Physica. A123:386 (1984).
24. H.Furukawa: Prog. Theor. Phys., 74:174 (1985).
25. I.M.Lifshitz and V.V.Slyozov, J. Phys. Chem. Solids, 19:35 (1961) ; C.Wagner,:Z. Elektrochem., 65:581 (1961).
26. K.Binder and D.Stauffer, Phys. Rev. Lett., 33:1006 (1974).
27. H.Tomita, Prog. Theor. Phys., 71:1405 (1984).

EXPERIMENTAL STUDIES OF ORDERING AND DECOMPOSITION PROCESSES IN ALLOYS

Gernot Kostorz

Institut für Angewandte Physik
ETH Zürich
CH-8093 Zürich, Switzerland

INTRODUCTION

As the local atomic arrangement in alloys is responsible for many macroscopic properties, local ordering and decomposition processes have always received particular attention in physical metallurgy (see, e.g. Cahn and Haasen, 1983). Often, starting from a supersaturated solid solution, ordering and phase separation are intimately connected as one of the decomposition products may be an ordered intermetallic phase. In Ni-rich solid solutions, ordered structures frequently form at intermediate temperatures. The initial stages of ordering and (coherent) precipitation and the shape and arrangement of inhomogeneities are of interest not only for fundamental reasons, but also for an understanding of the outstanding mechanical properties of multi-component technical "superalloys". Four important binary systems, Ni-Al, -Ti, -Cr, and -Mo, are therefore currently studied by combining several advanced techniques, i.e. diffuse and small-angle scattering with X-rays and neutrons, weak-beam and high-resolution transmission electron microscopy as microstructural tools, with physical property measurements, e.g. of electrical resistivity and mechanical strength. The results obtained to date are briefly summarized, and first details on short-range order in α-Cu-Zn, obtained from diffuse neutron scattering, are also presented.

NICKEL-RICH NI-AL AND NI-TI

Al and Ti are the two most important alloying elements in Ni for the formation of the ordered coherent γ' phase which is at the origin of an important part of the high-temperature strength of superalloys. Fig.1 shows the Ni-rich sides of the two relevant phase diagrams. While γ' (Ni_3Al, $L1_2$ structure) is the stable phase occurring in Ni-Al with more than roughly 10 at.% Al, it appears as a metastable phase in binary Ni-Ti where the stable Ni_3Ti phase is hexagonal. The formation of γ' can be studied between about 500 and 750°C in samples quenched from the γ-field of the phase diagram. As the solid-solubility line is very steep, small variations in the initial concentration and slightly different quenching conditions may impair the quality of any study involving more than one sample, as became evident from first survey measurements of small-angle neutron scattering (SANS) on Ni-Al (Beddoe et al., 1984) and Ni-Ti (Kostorz, 1985; Cerri et al., 1987). In-beam aging of individual samples

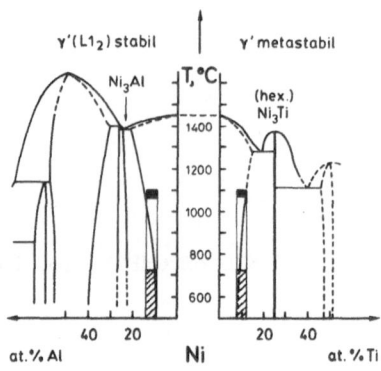

Fig. 1. Ni-rich sections of the Ni-Ti and Ni-Al phase diagrams. The temperature and concentration ranges of γ'-formation studies are indicated.
(After Hansen and Anderko, 1958)

yields more reliable results on the decomposition kinetics. The question whether the γ matrix itself is homogeneous may be studied by diffuse scattering techniques with alloy concentrations below the solid solubility limit. For these measurements, single crystals are mandatory. Single crystals were also grown from the decomposing alloys, as SANS and plastic deformation are thus not further complicated by additional effects (heterogeneous nucleation, hardening) due to grain boundaries.

Decomposition

Early stages of decomposition have been studied by SANS in Ni-(12-14) at.% Al single crystals (Beddoe et al., 1984) with (110) faces perpendicular to the incident beam. The samples were slices 2 to 3 mm thick, cut (by spark erosion) from cylindrical crystals (60 mm long, 10 mm in diameter) grown in alumina crucibles using the Bridgman technique. In a first series, several samples were solution treated in argon for at least 4 h at temperatures above 1150°C and quenched into iced brine. Individual samples were aged for 10 min to ~ 200 h in salt baths at 550, 575 and 625°C, quenched and electropolished at room temperature. SANS measurements were performed on the D11 instrument of the ILL, Grenoble. Scattering vectors Q (= $4\pi\sin\Theta/\lambda$ where Θ is half the scattering angle and λ is the wavelength of the incident neutrons) ranging from 0.06 nm^{-1} to 1.7 nm^{-1} were accessible by measuring with the two-dimensional position-sensitive detector at two positions (2 and 10 m from the samples). The coherently scattered intensity is proportional to the macroscopic differential cross-section

$$\frac{d\Sigma}{d\Omega} = \frac{1}{V_s} \left| \int_{V_s} \{\rho(\underline{r}) - \bar{\rho}\} \exp(i\underline{Q} \cdot \underline{r}) d^3\underline{r} \right|^2 \tag{1}$$

where V_s is the sample volume, $\rho(\underline{r})$ is the locally averaged scattering length density, and $\bar{\rho}$ is the macroscopically averaged scattering length density (see e.g. Kostorz, 1983, for details).

The scattering patterns for aged Ni-Al crystals showed no radial symmetry (see Fig.2, where two early-stage patterns for Ni-Al and Ni-Ti can be compared), and SANS intensities were evaluated along the main symmetry directions, <111>, <110> and <100>, in the detector plane (± 15° sectors). Absolute cross-sections were obtained by calibrating the high-Q range with the incoherent scattering from a vanadium single crystal. Fig.3 shows an example. According to atom-probe field ion microscopy results (AP-FIM, Wendt and Haasen, 1983), decomposition of Ni-Al in this concentration and temperature range follows a nucleation-and-growth path

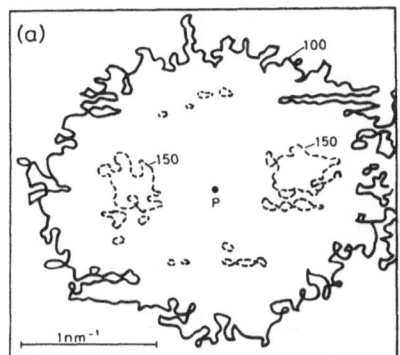 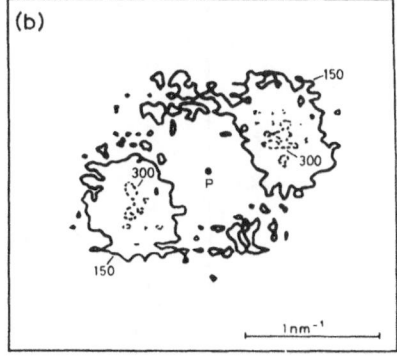

Fig. 2. Lines of equal intensity (number of counts) on the two-dimensional
position-sensitive detector illustrating the small-angle scattering from
(a) a Ni-12.8 at.% Al single crystal aged at 575°C for 10 min,
(b) a Ni-10 at.% Ti single crystal aged at 540°C for 15 min.
A <110> direction is parallel to the incident beam. The peaks in
scattering intensity appear along <100>. (P denotes the primary
beam, absorbed by a beam-stop.)

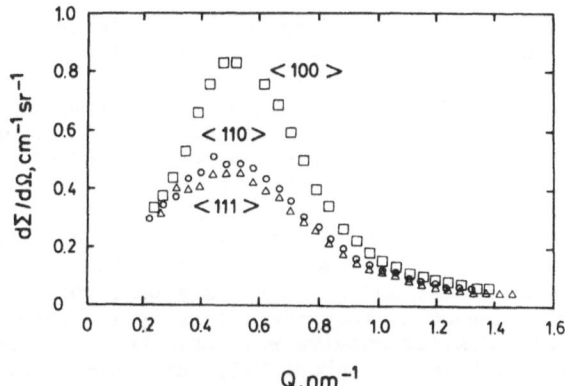

Fig. 3. SANS curves for a Ni-13 at.% Al single crystal with <110> parallel
to the incident beam after 18 h of aging at 575°C. Cross-sections
for scattering vector Q parallel to <100> (□), <110> (o) and <111>
(Δ). The mean radius of the γ' precipitates is ~ 3 nm.

(see also Kampmann and Wagner, 1984). The evolution of γ'-particle sizes
at 550°C and 575°C as obtained from in-beam SANS studies (and AP-FIM for
Ni-14 at.% Al at 550°C) is shown in Fig.4. (A critical radius of ~ 0.4
nm for Ni-14 at.% Al at 550°C is suggested by the model calculation of
Kampmann and Wagner, 1984.) Even for short aging times, the SANS measurements show a preferential alignment of the precipitates along <100> directions (see Fig.2a). A detailed model that roughly accounts for all the
SANS features requires spherical particles at the beginning but a prolate
rotational ellipsoid contribution at later stages, a size distribution and
an average distance parameter with an adjustable distance distribution
function (orientation dependent). The orientation of the ellipsoidal component is mainly <100> but with some variation as scetched in Fig.5, where
a rotation around the other two <100> axes is indicated by the function
$G(\alpha)$. This model accounts for particle elongation along <100> and "an-
gling" (as seen in TEM for much larger particle sizes and volume fractions, cuboidal γ' precipitates form closely-spaced rows along <100> and

201

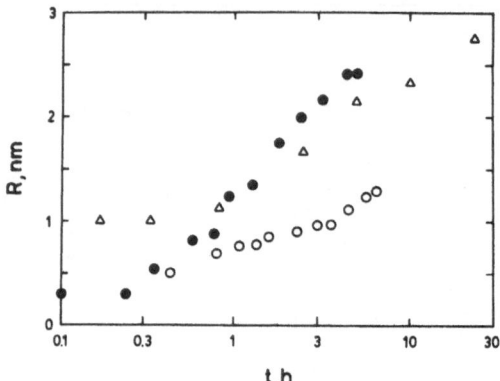

Fig. 4. Radius of γ'-precipitates in Ni-Al from SANS measurements during in-beam aging as a function of aging time at 550°C (o, 13.24 at.% Al) and at 575°C (•, 13.43 at.% Al). FIM-results at 550°C for Ni-14 at.% Al are included for comparison (Δ, from Wendt and Haasen, 1983).

sometimes L-shaped arrangement). Without $G(\alpha)$ the SANS curves for <110> and <111> would be more separated than measured (cf. Fig.3).

The decomposition of Ni-Ti, despite the similarity of the SANS patterns (Fig.2), follows different paths. As the neutron scattering contrast of Ni and Ti is about four times larger than for Ni-Al, compositional inhomogeneities can be detected with a higher sensitivity. i.e. earlier stages of decomposition become accessible. Also, the homogeneity of the initial, as-quenched state of a sample can be verified with good precision, and the quenching conditions were considerably improved until the quenched samples showed a flat scattering curve within a few percent of the expected incoherent and Laue scattering. Fig.6 shows that, for Ni-12 at.% Ti aged in-beam at 540°C, a very broad SANS peak starts developing during the first few minutes and grows and sharpens while its position remains essentially unchanged. The peak intensity is proportional to t^{γ} with $\gamma = 0.24 \pm 0.03$. Similar results were obtained for a sample aged at 500°C where initially, γ is only about 0.1, but after 10 h, a similar exponent as at 540°C seems to control the further increase in intensity. The position of the peak is practically the same as at 500°C ($Q_m \simeq 0.9$ nm^{-1}). As reported by Grüne and Haasen (1986), AP-FIM shows that in this alloy, aging up to 256 h at 550°C is still not sufficient to achieve the composition of {100} planes as required for the γ'-phase (i.e. pure Ni planes alternating with Ni-50 at.% Ti planes). This suggests that the SANS results should be interpreted in terms of compositional waves, with a wavelength of ~ 7 nm. Indications of ordering are, however, found even for the as-quenched (and, according to SANS, "homogeneous") samples. In electron and X-ray diffraction, a peak is observed at the L1$_2$ superstructure Bragg positions. It is initially a broad short-range order peak that sharpens considerably after more than ten minutes of aging. The summed-up intensity at superstructure positions also follows a power law as a function of aging time, but with a somewhat smaller exponent (0.18 at 540°C). High-resolution electron microscopy (HREM) for these alloys, aged after a good quench, revealed signs of ordering (superstructure peaks in laser diffraction patterns taken from multiple-beam images) only after about 10 h of aging. A minimum amount of order must apparently be reached before it can be observed by this technique, but the local Ti concentration is still far below 25%. The ordered regions observed in HREM images of Ni-9.3 and 14.7 at.% Ti (Yoshida et al., 1986a) aged at 570°C (the quench-rate of these samples was lower, and some compositional fluctuations were already

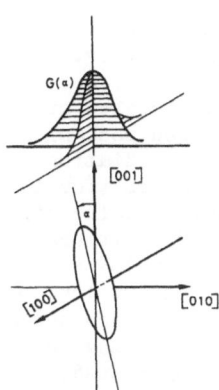

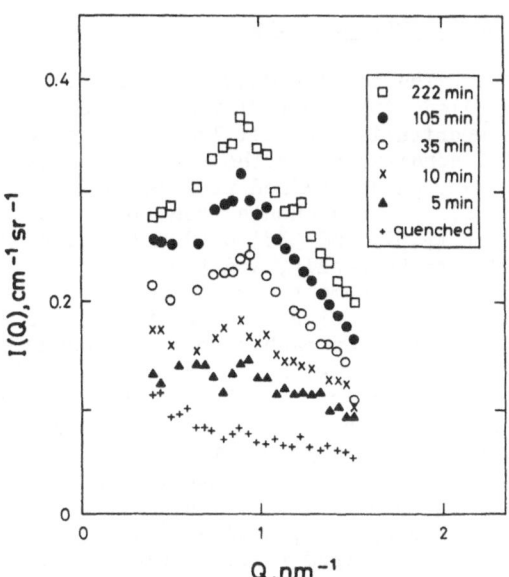

Fig. 5. Orientation and
orientation distri-
bution of ellipsoids
used to model the
Ni-Al SANS patterns
patterns (from Beddoe
et al., 1984).

Fig. 6. SANS intensity along <100> for
a Ni-12 at.% Ti single crystal
during in-beam aging at 540°C
(from Cerri et al., 1987).

present before aging) are compatible with local order on an $L1_2$ lattice
with the Ti atoms occupying the correct sublattice, $Ni_3 (Ni_{1-x}Ti_x)$,
as shown in Fig.7 (Yoshida et al., 1986b). This partial ordering is initially the faster process. Via the coupled lattice parameter changes, it
may determine the wavelength of the subsequent compositional changes. The
dependence of the characteristic wavelength on quenching temperature and
Ti concentration has to be studied before a complete picture of this
ordering-decomposition reaction will emerge.

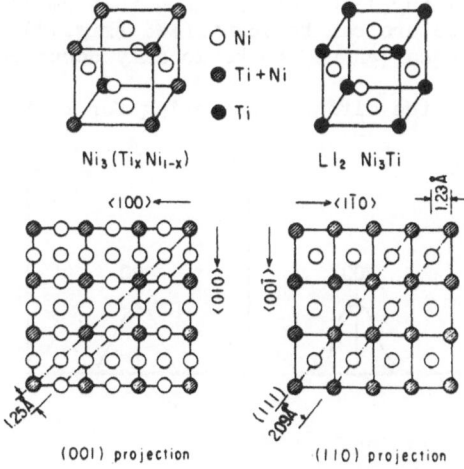

Fig. 7. Unit cells of $Ni_3(Ni_{1-x}Ti_x)$ and Ni_3Ti ($L1_2$) and the
(001) and (110) projections of the atomic arrangements
(from Yoshida et al., 1986a).

Short-range ordering

The local atomic arrangement in Ni-(9.5-9.8) at.% Al has been studied with X-ray diffuse scattering (Klaiber et al., 1987). These Al concentrations do not lead to γ' formation under any accessible experimental conditions, as the solid solubility limit is larger (e.g. 10.8 at.% at 550°C, as determined from magnetization measurements,; see Beddoe et al., 1984). After a quench from 1100°C, samples were heat-treated at 700°C for 48 h (9.5%) and 3354 h (9.8%), water-quenched, polished mechanically and electrochemically, and finally investigated with MoK$_\alpha$ radiation on a four-circle instrument equipped with a high-purity Ge detector (for the elimination of fluorescence scattering and $\lambda/2$ contamination). The diffuse scattering for each sample was measured at about 8000 positions in reciprocal space, with the scattering angle 2Θ ranging from 18° to 90°. Extensive corrections for background, surface roughness, absorption, polarization, Compton scattering and thermal diffuse scattering were necessary before, with appropriate calibration, the diffuse scattering related to short-range order and atomic displacements could be extracted. This scattering intensity can be evaluated following three different schemes (Borie and Sparks, 1971; Williams, 1972; Georgopoulos and Cohen, 1977). The most elaborate method (Georgopoulos and Cohen, 1977) fully includes the variation of atomic scattering factors with Q, and approximates the atomic displacements up to quadratic terms. Thus, at least 25 symmetry related intensities are necessary for the cubic lattice in order to separate the short-range order scattering from 24 functions constituting the various displacement and combined terms (Fig.8 illustrates the three scattering contributions along a cubic axis). In the Borie-Sparks method (Borie and Sparks, 1971), the ratios of scattering amplitudes are assumed to be constant, and only ten different functions are necessary. The method proposed by Williams (1972) is a direct least-squares fit of short-range order and displacement parameters to the data. Fig.9 shows that the three methods give comparable results for the Warren-Cowley short-range order parameters α_{lmn} (l,m,n = site indices with 110 corresponding to the nearest neighbours in the f.c.c. lattice), and the reconstructed short-range order intensities are compared in Fig.10. The diffuse peaks appear at the positions of the L1$_2$ superstructure (100 and 110 type) which is stable for higher Al concentrations, but Fig.9 shows that no long-range order exists. A computer simulation using the first 13 averaged α_{lmn} for a model crystal containing 13104 atoms shows that small regions reminiscent of the L1$_2$ superstructure can be identified (Fig.11), but any of the sublattices has an equal chance to be locally occupied by Al.

In this study, the diffuse scattering for both samples was the same,

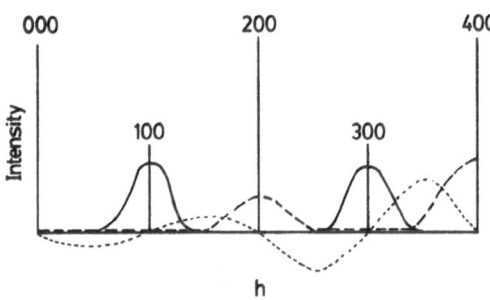

Fig. 8. Schematic of the diffuse scattering intensity components along h00
    in a face-centered cubic crystal (after Schwartz and Cohen, 1977),
    —— short-range order term,
    ····· first-order displacement term,
    – – second-order displacement term.

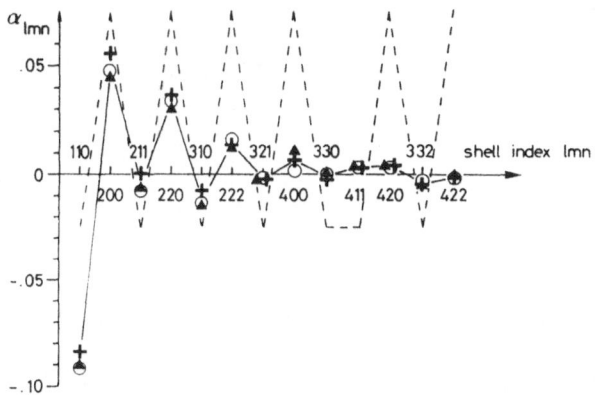

Fig. 9. Warren-Cowley short-range order parameters α_{lmn} for Ni-9.5 at.% Al aged for 48 h at 700°C, obtained with the methods of Georgopoulos and Cohen (o), Borie and Sparks (▲) and Williams (+). The dashed line indicates the sequence of α_{lmn} corresponding to long-range order in the $L1_2$ superstructure at this concentration.

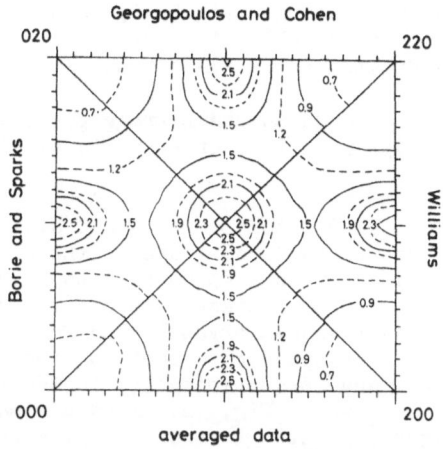

Fig. 10. Iso-intensity plots of $I_{SRO}(\underline{Q})$ in Laue units for Ni-9.5 at.% Al aged for 48 h at 700°C, reconstructed from the short-range order parameters obtained from the three methods and from the averaged data (from Klaiber et al., 1987).

implying that the short-range ordered state was already completely established after 48 h. As even the as-quenched sample shows already some diffuse scattering, the short-range ordering reaction is apparently very fast. Diffuse scattering measurements at temperature are necessary in order to find the equilibrium parameters.

NICKEL-RICH NI-CR AND NI-MO

The solid solubility of Cr and Mo in Ni is larger than for Al and Ti. Resistivity measurements in Ni-(10-20) at.% Cr and electron diffraction in some superalloys point to ordering in this system, but the phase diagram shows an ordered structure only at Ni_2Cr. For Ni-Mo, the stable Ni_4Mo structure can be interpreted as a $\{420\}/5$ compositional wave, whereas the so-called short-range ordered structure extensively studied at

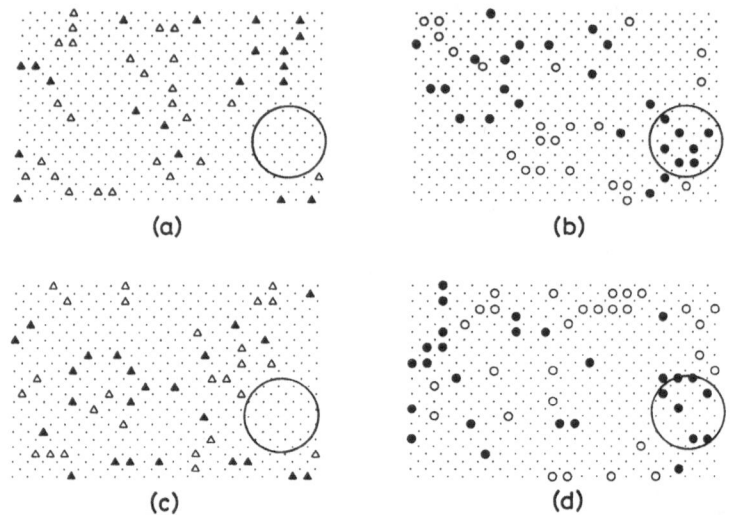

Fig. 11. Sequence (a) to (d) shows consecutive (100) planes of the modelled Ni-9.5 at.% Al sample (Fig.9,10) with ●, o, ▲, △ representing
Al atoms on the four sublattices of the f.c.c. lattice and small
dots for any Ni atom (from Klaiber et al., 1987).

the same composition also leads to sharp Bragg peaks, at 1 1/2 0 positions
corresponding to a {420}/4 wave (see Chakravarty et al., 1974; Mayer and
Urban, 1985).

Single crystals of Ni-20 at.% Cr and Ni-10 at.% Mo are currently
being studied with neutrons and X-rays, and in the electron microscope.
Dislocations in aged and deformed samples of both alloys show planar glide
typical for short-range ordered systems. The elastic diffuse neutron scattering of a Ni-19.4 at.% Cr single crystal aged for 320 h at 555°C has
been measured along the main symmetry directions (Schönfeld et al., 1986).
The reconstructed short-range order intensity in the $(h_1 h_2 0)$ plane, using
the first five α_{lmn}, is shown in Fig. 12. Broad maxima at 1 1/2 0 positions are found, in agreement with the results of a neutron scattering

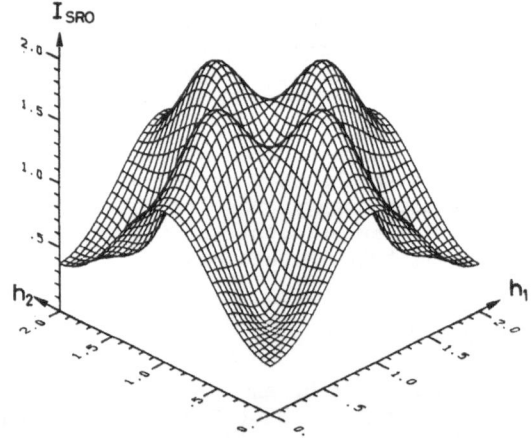

Fig. 12. Short-range order intensity (in Laue units) in the $(h_1 h_2 0)$ reciprocal lattice plane for a Ni-19.4 at.% Cr single crystal aged
for 320 h at 555°C (from Schönfeld et al., 1986).

study by Schweika et al. (1986) who used the Ni-58 isotope in Ni-11 at.%
Cr for an enhancement of the scattering contrast. More detailed measurements of the diffuse intensity on several reciprocal lattice planes and
including different heat treatments are in progress for the Ni-19.4 at.%
Cr alloys (see Fig.13). A distinct decrease in short-range order is found
for a sample aged at 700°C. The inverse Monte-Carlo method (Gerold and
Kern, 1987) is employed to calculate effective pair interaction potentials.

In the search for short-range order in Ni-10 at.% Mo X-rays are currently used, and although some diffuse scattering is observed, the results
to date only indicate that the kinetics of ordering is much slower than in
the electron-irradiation enhanced ordering studies performed by Mayer and
Urban (1985).

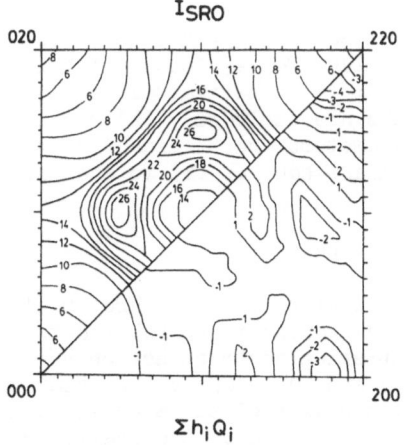

Fig. 13. Short-range order (I_{SRO}) and atomic displacement ($\Sigma h_i Q_i$, i =
1,2,3) scattering in a ($h_1 h_2 0$) reciprocal lattice plane in 0.1
Laue units, reconstructed from α_{lmn} and displacement parameters
(obtained using the Williams method) to fit the experimental data
obtained in (h00) and ($h_1 h_2 0$) for Ni-19.4 at.% Cr aged at 555°C.

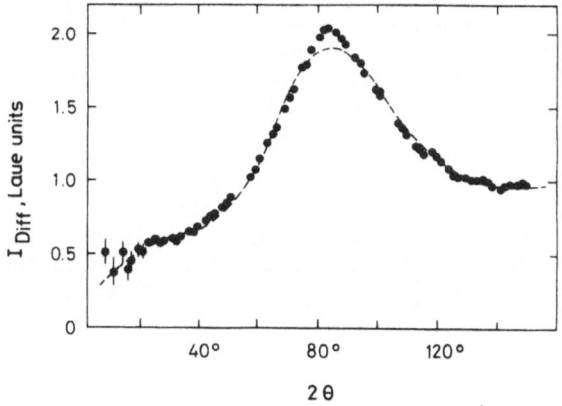

Fig. 14. Diffuse neutron scattering I_{Diff} as a function of scattering
angle 2θ for ^{65}Cu-30 at.% Zn aged for 21 d at 160°C.

207

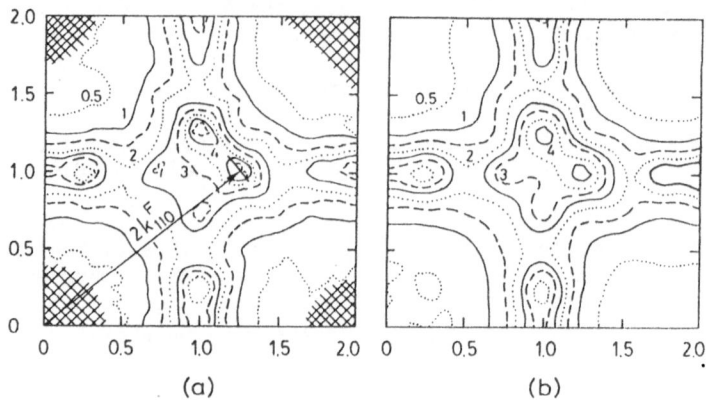

Fig. 15. Diffuse neutron scattering (in Laue units) in the $(h_1 h_2 0)$ reciprocal lattice plane for a ^{65}Cu-31 at.% Zn single crystal aged for 7 d at 200°C; (a) experimental data (cross-hatched regions near Bragg peaks are excluded), (b) reconstructed from 61 short-range order parameters and 34 linear displacement parameters obtained from a least-squares fit to the experimental data.

SHORT-RANGE ORDER IN ALPHA-BRASS

Anomalies in electrical resistivity and specific heat in α-brass have for a long time been assumed to be related to an ordering process upon aging. The low scattering contrast not only in X-ray but also in neutron scattering (with a high incoherent scattering background of Cu) so far prevented a close inspection of the microstructure. An alloy with the Cu-65 isotope was prepared to enhance the scattering contrast for neutrons by a factor of about 5.8. Initial powder measurements of ^{65}Cu-30 at.% Zn, with cold neutrons (λ = 0.48 nm) to avoid multiple Bragg scattering, showed a diffuse scattering peak indicative of short-range order (Fig.14). The position of the diffuse maximum, however, does not coincide exactly with the most common superstructure peaks expected for the f.c.c. lattice (see, e.g. de Fontaine, 1975). Measurements on a ^{65}Cu-31 at.% Zn single crystal confirm that the diffuse peaks are displaced from 100 and 110. As seen in Fig.15a, the diffuse 110 peak is split into a square of four subpeaks that are strongly affected by the antisymmetric linear displacement scattering terms (dlna/dc ≅ 5.7% for Cu-Zn). A least-squares fit according to Williams requires a large number of parameters as the diffuse peaks are rather sharp. Fig.15b shows a reconstructed pattern. The curved line through one of the diffuse peaks in Fig.15a has been drawn with the radius $2k_{110}^F$ (where k_{110}^F is the Fermi wave number in 110-directions). The Fermi surface for Cu-30 at.% Zn has rather flat portions in these directions (see Prasad et al., 1981). The splitting may thus be related to a distortion that will provide more electronic states within the first Brillouin zone (cf. Moss and Walker, 1975). A similar case has been reported by Ohshima et al. (1984) who studied the diffuse X-ray scattering of Ag-15 at.% Mg.

ACHNOWLEDGMENTS

The neutron scattering studies were performed at the Institut Laue-Langevin (ILL), Grenoble, France, and at the Laboratorium für Neutronenstreuung (LNS) der ETH Zürich. The help and interest of G. McIntyre, O. Schärpf and A.F. Wright at ILL and of W. Bührer at LNS are gratefully acknowledged. Useful discussions with M. Arita (ETH) and H. Yoshida (Kyoto University) are appreciated.

The author is grateful to all former and present members of the Institut für Angewandte Physik who were or are involved in the subjects described above, especially to R. Bänninger, A. Blanchard, E. Fischer and P. Studerus for technical support and to B. Schönfeld, A. Cerri, F. Klaiber, L. Reinhard, P. Schwander and U. Stockert (-Zaune) without whose enthusiastic participation the research presented here could not have been performed. A major portion of this work is supported by the "Schweizerischer Nationalfonds zur Förderung der Wissenschaften".

REFERENCES

Beddoe, R.E., Haasen, P., and Kostorz, G., 1984, Early stages of decomposition in Ni-Al single crystals studied by small-angle neutron scattering, in: "Decomposition of Alloys: the Early Stages," P. Haasen, V. Gerold, R. Wagner and M.F. Ashby, eds., Pergamon Press, Oxford.

Borie, B., and Sparks, Jr., C.J., 1971, The interpretation of intensity distributions from disordered binary alloys, Acta Cryst. A, 27:198.

Cahn, R.W., and Haasen, P., eds., 1983, "Physical Metallurgy", 3rd ed., North-Holland, Amsterdam.

Cerri, A., Schmelczer, R., Schwander, P., Kostorz, G., and Wright, A.F., 1987, Combined use of small angle neutron scattering and (conventional and high resolution) electron microscopy for the characterization of early stages of ordering and phase separation in Ni-Ti, in: "Characterization of Defects in Materials", R.W. Siegel, R. Sinclair and J.R. Weertman, eds., Mat. Res. Soc. Symp. Proc., MRS, Pittsburgh, Pa., to be published.

Chakravarty, B., Starke, Jr., E.A., Sparks, C.J., and Williams, R.O., 1974, Short range order and the development of long range order in Ni_4Mo, J. Phys. Chem. Solids, 35:1317.

de Fontaine, D., 1975, K-space symmetry rules for order-disorder reactions, Acta Metall., 23:553.

Georgopoulos, P., and Cohen, J.B., 1977, The determination of short range order and local atomic displacements in disordered binary solid solutions, J. Physique, (Coll.), 12:C7.191.

Gerold, V., and Kern, J., 1987, The determination of atomic interaction energies in solid solutions from short range order coefficients – an inverse Monte-Carlo method, Acta Metall., 35:393.

Grüne, R., and Haasen, P., 1986, Spinodal decomposition of Ni-Ti, J. Physique (Coll.), 47:C2.259.

Hansen, M., and Anderko, K., 1958, "Constitution of Binary Alloys", McGraw-Hill, New York.

Kampmann, R., and Wagner, R., 1984, Kinetics of precipitation in metastable binary alloys – theory and application to Cu-1.9 at.% Ti and Ni-14 at.% Al, in: "Decomposition of Alloys: the Early Stages", P. Haasen, V. Gerold, R. Wagner and M.F. Ashby, eds., Pergamon Press, Oxford.

Klaiber, F., Schönfeld, B. and Kostorz, G., 1987, Investigation of short-range order in Ni-10 at.% Al single crystals by diffuse X-ray scattering, Acta Cryst., A43:525.

Kostorz, G., 1983, X-ray and neutron scattering, in: "Physical Metallurgy", 3rd ed., R.W. Cahn and P. Haasen, eds., North-Holland, Amsterdam.

Kostorz, G., 1985, Recent applications of neutron small angle scattering, in: "Advanced Photon and Particle Techniques for the Characterization of Defects in Solids", J.B. Roberto, R.W. Carpenter and M.C. Wittels, eds., Mat. Res. Soc. Symp. Proc. Vol. 41, MRS, Pittsburgh, Pa.

Mayer, J., Urban, K., 1985, Observation of Ni_8Mo ordered phase in Ni-Mo alloys, Phys. Stat. Sol. (a), 90:469.

Moss, S.C., and Walker, R.H., 1975, Screening singularities and Fermi surface effects in the diffuse scattering from alloys, J. Appl. Cryst., 8:96.

Ohshima, K., and Harada, J., 1986, X-ray diffraction study of short-range-ordered structure in a disordered Ag-15.0 at.% Mg alloy, Acta Cryst. B, 42:436.

Prasad, R., Papadopoulos, S.C., Bansil, A., 1981, Fermi surface properties of alpha-phase alloys of copper with zinc, Phys. Rev. B, 23:2607.

Schönfeld, B., Klaiber, F., Kostorz, G., Zaune, U., and McIntyre, G., 1986, Diffuse neutron scattering and short-range order in Ni-19.4 at.% Cr, Scripta Metall., 20:385.

Schwartz, L.H., and Cohen, J.B., 1977, "Diffraction from Materials", Academic Press, New York.

Schweika, W., and Haubold, H.-G., 1986, Short-range order and atomic interaction in $NiCr_x$, in: "Atomic Transport and Defects in Metals by Neutron Scattering", Proc. IFF-ILL Workshop, C. Janot, W. Petry, D. Richter and T. Springer, eds., Springer, Berlin.

Wendt, H., and Haasen, P., 1983, Nucleation and growth of γ'-precipitates in Ni-14 at.% Al, Acta Metall., 31:1649.

Williams, R.O., 1972, "A Computer Program for the Reduction of Diffuse X-Ray Data from Solid Solutions", Oak Ridge Nat. Lab. Report ORNL-4828, Oak Ridge, Tenn.

Yoshida, H., Arita, M., Cerri, A., and Kostorz, G., 1986a, High resolution electron microscopic studies of Ni-9.3 and 14.7 at.% Ti alloys, Acta Metall., 34:1401.

Yoshida, H., Arita, M., Nissen, H.-U., and Kostorz, G., 1986b, Interpretation of high resolution images of ordered structures in Ni-9.3 and -14.7 at.% Ti alloys, in: "Electron Microscopy 1986", Vol. I, Proc. XIth Int. Cong., T. Imura, S. Maruse and T. Suzuki, eds., Japan. Soc. Electr. Micr., Tokyo, Japan.

MICROSTRUCTURAL PATTERNS AND KINETICS OF PHASE SEPARATION

IN BINARY AND TERNARY METALLIC SYSTEMS

P. Guyot and J.P. Simon

Laboratoire de Thermodynamique et Physico-Chimie Métallurgiques
(UA CNRS N°29). INP Grenoble . ENSEEG B.P.75
38402 Saint Martin d'Hères . France

ABSTRACT

Various aspects of the time evolution of metallic solid solutions quenched under a miscibility gap and evolving towards equilibrium via a phase separation are considered : i) the variety of patterns or microstructures observed at different scales, in correlation with local or long range elastic effects. ii) the different steps of the kinetic evolution towards equilibrium, discussed in terms of the quench depth and theoretical approaches. iii) the impact of anomalous X-ray scattering with synchrotron radiation in elucidating some complex atomic partitioning in ternary systems.

INTRODUCTION

"Quench and aging" is an almost daily process carried out by the metallurgist, who is used to deal with phase transformations starting far from equilibrium, and who is asked to elaborate microstructures of practical interest, whatever they have reached their thermodynamical equilibrium or stay frozen in metastable states of quasi-infinite life time.

At the fundamental level, metallic systems offer several simplifying advantages (short range electronic interactions justifying the use of close neighbours interactions in numerical simulations, generally slow diffusivities, performing characterization techniques with resolution down to the atomic scale ...) but also drawbacks (long range elastic interactions and anisotropy, textures, diffusivity cross-terms in multicomponent alloys...) of intricate complexity.

We develop here what we consider to be essential features of a phase separation occuring in binary and ternary metallic solid solutions, focussing successively on the variety of the patterns or morphologies which form after quench at different steps of the evolution of the system, and on the identification and analysis of the underlying kinetic regimes with respect to theoretical models (coarse grain methods, cluster dynamics, Monte-Carlo simulations of the 3 dim. Ising model). Lastly the possibility to observe complex atomic partioning in ternary alloys, through the determination of partial structure factors, is emphasized.

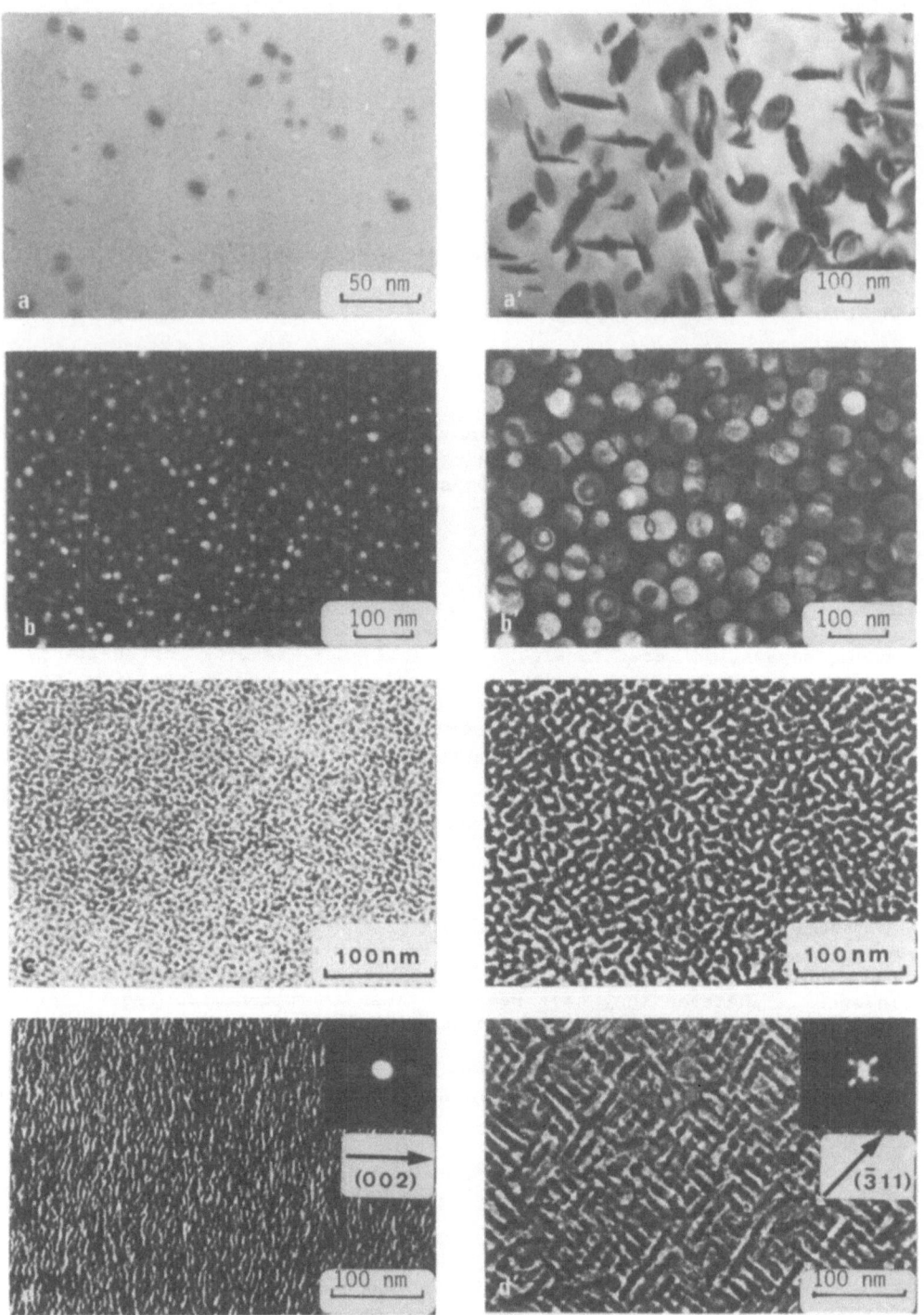

Fig.1. Transmission electron micrographs for various alloys aged at T_a for time t_a.
a)Al-6.8%Zn : T_a = 133°C; t_a = 8h, $t_{a'}$ = 15 days (Courtesy of Laslaz[1]) b)Al-10%Li : T_a =200°C; t_a = 3h, $t_{a'}$ = 48h (Courtesy of Sainfort[2]) c)Fe-33%Cr-20%Co : T_a = 620°C; t_a = 1h, $t_{a'}$ = 8h (Courtesy of Dombre[3]) c)Cu-48%Ni-8%Fe : T_a = 500°C; t_a = 8h, $t_{a'}$ = 65h (Courtesy of Wahi[4])

212

PATTERNS AND MORPHOLOGIES

Snapshots of microstructures evolving at constant temperature, after quench from the homogeneous one phase region in the two phase region, are shown in the transmission electron micrographs (TEM) of Fig.1. Even if the micrographs are 2 dim. projections of 3 dim. structures, some characteristic features may be enhanced :

- two phase patterns with well defined interfaces and discrete space distribution, like AlZn or AlLi.

- eventual shape variation of the growing particles : from spheres to $\{111\}$ ellipsoids in AlZn ; permanently spherical for AlLi, even at high second phase volume fraction ; $<100>$ rods or $\{100\}$ platelets in CuNiFe in the late stages.

- tangled structures of "miscelles" as in FeCrCo, or "tweed patterns" in CuNiFe.

- existence of localized elastic strain fields, (fig.1a').

This description stands at the nanometer scale, the resolution of standard TEM. High resolution TEM may give informations, down to $2\overset{\circ}{A}$, on interface structures and low strain fields. But in order to get rid of projection artefacts, other investigation techniques are necessary. An illuminating example is shown in the field-ion microscopy of Fig.2, where atom imaging on the tip surface evidences percolation of the Cu and (Ni-Fe) species.

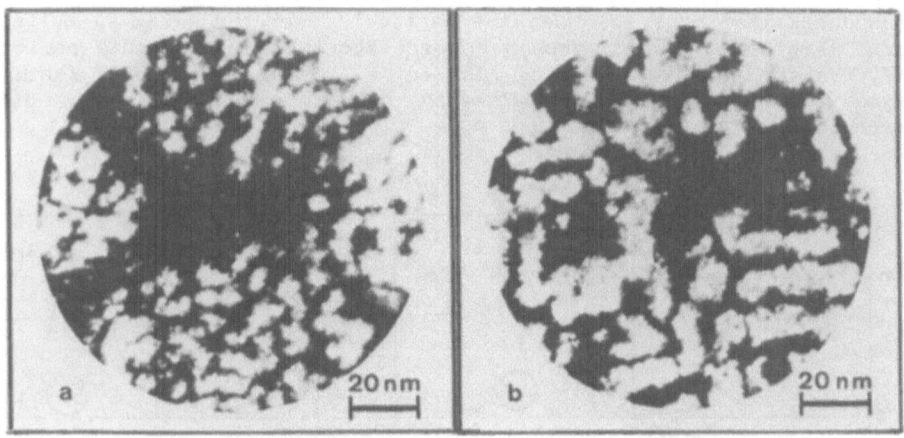

Fig.2. FIM of $Cu_{0.44}Ni_{0.48}Fe_{0.08}$ aged (a) at 500°C for 8h (b) at 450°C for 480h (Courtesy of Piller et al.[5])

Small angle scattering (SAS) has the advantage to give long wave length informations, characteristic of the bulk, and, when bright sources are available, on-line kinetics experiments may be performed. Fig.3 illustrates the reciprocal space neutron SAS signature of the time evolution of decomposed Al-12%Zn single crystals, as recorded on a 2 dim. detector. The informations brought by isointensity maps are two-fold and easily understood in terms of a non connected two-phase modelling of the scattered intensity I(q) :

$$I(q) = Nv^{-1} \Delta f^2 V_p^2 \{ <F_p^2(q)> - <F_p(q)>^2 |1-\tilde{I}(q)| \} \tag{1}$$

where v = average atomic volume, Δf = difference in average scattering length between particle and matrix, V_p = particle volume, N = particle number density.

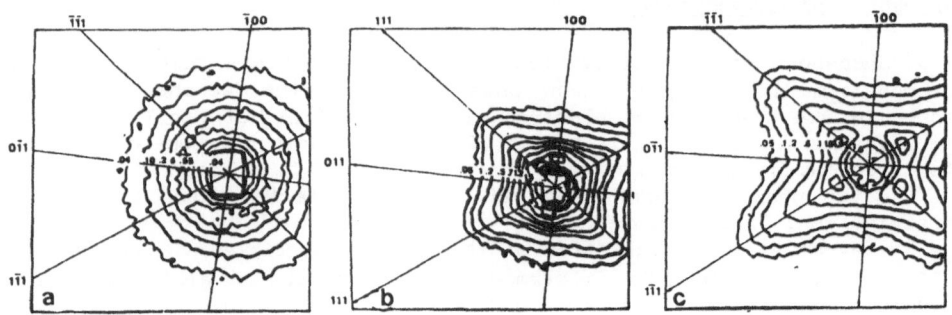

Fig.3 Neutron SAS 2 dimensional patterns of Al-12%Zn single crystals[6].
a)T_a=150°C, t_a=2.3 min ; b)t_a=87 min ; c)T_a=20°C, t_a=972 min.

    i)the large angle scattering comes from the particle form factor $F_p(q)$, and their gradual shape change from spheres to ellipsoids flattened normally to the {111} planes, introduces tails in the <111> directions of patterns initially isotropic.

    ii)the small angle maximum results from the modulation of $F_p(q)$ by the interference term $1-I(q)$, i.e. the Fourier transform of the particle pair correlation function. The appearance of maxima on this interference ring along the <100> matrix directions (Fig.3b), shows that the particules organize themselves on a simple cubic superlattice ; however this effect is weak and killed by $F_p(q)$ when the particule number density is sufficiently large (Fig.3c). Mention should be made about SAS experiments performed on polycrystals, where averaging rubs out all these anisotropic features and gives somewhat universal scattering curves, even for very different morphologies : compare Fig.1 and 4.

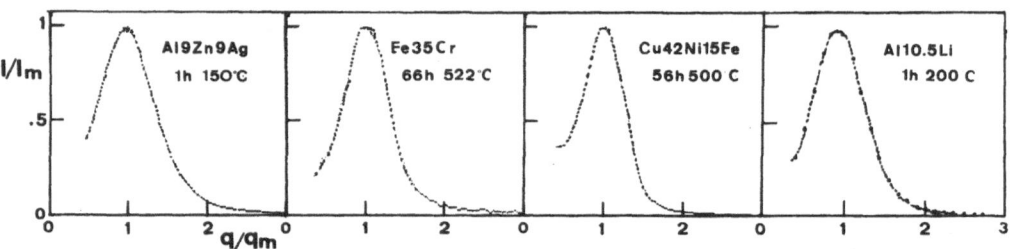

Fig.4. SAS curves for polycrystalline alloys.

    At that level of the morphology description, the understanding of the phase separation dynamics is far to be trivial. Long range elastic effects, atomic mobilities and diffusion paths, interfaces, superimpose to the chemical free energy decrease. When one considers for instance the alloy Al 10%Li, the high volume fraction of the δ' phase (~ 0.3) is expected to lead to percolation ; but the $L1_2$ δ' particles remain spherical, due to the negligible elastic effects,[2] preventing therefore their percolation ; moreover their coagulation would create energetically costly antiphase boundaries. Similarly, "tweed" contrast in TEM has often been considered as a proof of spinodal decomposition, forgetting that pattern regularity is not synonymous of unstable state : intense nucleation current in the metastable region, elastic interactions and proximity effects can act similarly to the fast growth of the λ_m fluctuation wave length. Quantitative analysis in both direct and Fourier spaces are indeed necessary.

For a given quench, at fixed initial supersaturation and aging temperature, the kinetics quantification implies the determination of the two-phase model parameters (size, number density of zones or second phase particles, characteristic lengths for tangled structures, chemical compositions) versus time. The other largely used alternative compares directly SAS profiles with the Fourier transform of the pair correlation function of the concentrations.

Shallow quenches or initial low supersaturated alloys are fairly well understood and relevant to nucleation and growth. Combining TEM and neutron SAS, it has been shown in AlZn[8,9] that the Guinier-Preston zones size and number density vary with time like in the Langer and Schwartz theory[10], where nucleation growth and coarsening occur simultaneously (see Fig.5 for the alloy aged at 133°C). Delay or incubation time for nucleation has also been evidenced and, similarly to fluids, a cloud region may be localized near the miscibility gap[11].

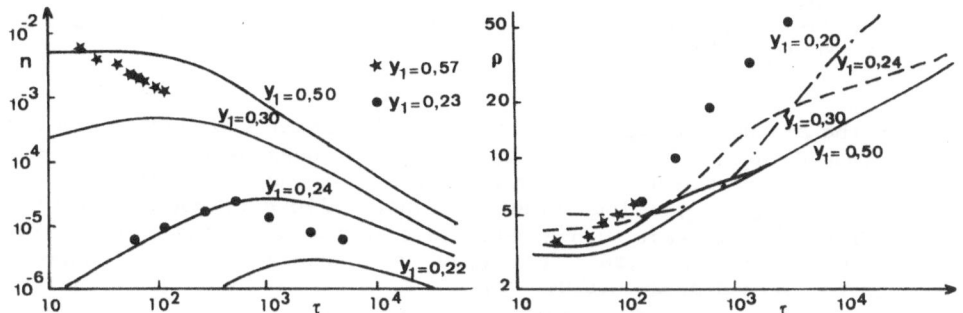

Fig.5. Al-6,8 at%Zn aged at 20°C and 133°C. Number density n and GP zone radius ρ at constant supersaturation y, vs time (all parameters are in reduced units), (from [8,9]).

For deep quenches, the early stages of the kinetics are extremely difficult to catch, for the decomposition of the high temperature random state starts during the quench itself. A successfull experiment, performed "in-situ" with synchrotron radiation[12] is reported in Fig.6a for an alloy Al-12%Zn directly quenched to 139°C. The sequence of spectra recorded every 10s. starts from the homogeneous solid solution, as indicated by the flat t = 0 Laue scattering. The scattering increases first quickly, as does the integrated intensity. The time evolutions of the position q_m of the maximum and its intensity I_m, shown in Fig.6b, follow power laws, with a bimodal regime of cross-over at about 1 min.

In the first domain, I_m varies like $\approx t$, while q_m remains approximately constant. Such features are reminiscent of a linear spinodal regime[13], with a first order development of the exponential. However agings at higher temperatures, 163 and 195°C, Fig.6b, located surely in the metastable region, behave similarly. No singularity of the correlation length is detected, confirming previous claim[9,14,15] that increasing the supersaturation drives only a gradual increase of the second phase nucleation rate. In fact the slope changes of I_m have also been observed by Siebert and Knobler[11] for intermediate quenches (near the spinodal line) of liquid mixtures, and are simply in agreement with the nucleation-growth-coarsening model of Langer and Schwartz (Fig.5).

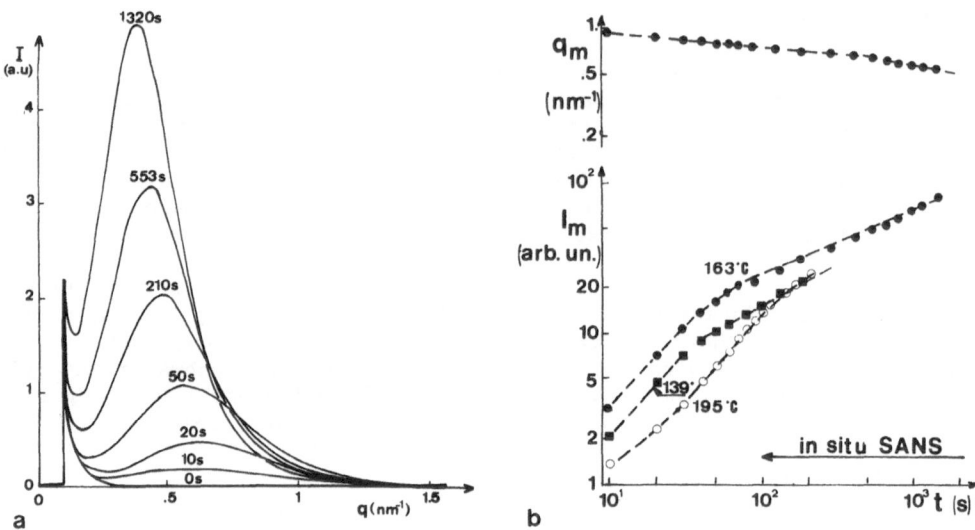

Fig.6. a)Synchrotron radiation SAS of Al-12%Zn aged at 139°C. b)Power laws of the maximum intensity at momentum transfer q_m, for Ta = 139, 163, 195°C (from[12]).

In the second domain, the integrated intensity stabilizes, the growth of I_m slows-down in a $t^{0.5}$ power law and q_m decreases as $t^{-0.15}$. The system is then in the coarsening stage, with constant second phase volume fraction, in a similar way as previous studies of AlZn alloys by neutron SAS[7,8] and CuNiFe alloys[16]. In fact such a late stage is the only stage accessible for large supersaturations (see the aging at 20°C in Fig.5), where the equilibrium second phase volume fraction is reached very quickly. Table 1 gathers some values of the exponents a' ($q_m \propto t^{-a'}$) and a'' ($I_m \propto t^{a''}$) for various AlZn alloys. The relation a'' – 3a' = 0 holds fairly well. a' varies between .15 and.33, as predicted by various models like coarse grain statistic spinodal[17], cluster dynamics[18] and kinetic Ising model[19]. a'=0.33

Table 1. Exponents of the power laws of coarsening (from[12,14])

Alloy	T_a(°C)	a'	a''	Experimental method
Al-5.3%Zn	20	0.20-0.3	0.9-1	NSAS
Al-6.8%Zn	20	0.25-0.3	1	"
	133	0.25-0.3	1	
Al-12.1%Zn	20	0.3	1	NSAS
	130-183	0.15	0.45	XSAS
Al-22%Zn	141-173	0.35	1	XSAS
Al-32%Zn	155	0.24	0.75	XSAS

corresponds to the well known Lifshitz-Slyozov mechanism[20] operating by condensation-evaporation of solute atoms between well separated particles, therefore suitable to dilute systems. The decreases of a' to 0.15-0.20 has been interpreted by Binder[18] in terms of diffusion and coagulation of a high number density of small clusters for systems well inside the two-phase region. It is yet not clear whether second phase percolation should have any effect on the coarsening law in such cases.

---
*$q_m \propto R^{-1}$, R being the size of second phase particles in a 2 phase-model.

To close this part, we briefly outline the scaling of the structure function $S(q,t)$ in the coarsening stage, which is also analyzed in other contributions to this symposium. Suggested by Lebowitz et al[19], the scaled function is defined by :

$$F(qL,t) = L^{-3}(t)S(q,t) / \int q^2 S(q,t)dq \qquad (2)$$

where L is a characteristic length, which can be taken as the size or gyration radius of isolated particles, wave length of percolated systems, or in the reciprocal space q_m^{-1} or q_1^{-1}, q_1 being the first moment of the structure function $S(q)$. It is easy to show that $F(qL,t)$ stabilizes in a function independent on time $F(x=qL)$ as soon as the second phase volume fraction does : see Fig.7 for 2 examples of two-phase modellings of $F(x)$.

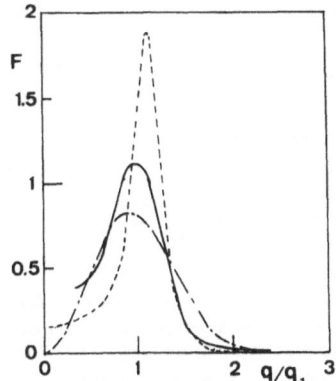

Fig.7. Two-phase hard-spheres modelling of $F(x)$ for a volume fraction of 0.2:—,—,—.—.—from Rikvold and Gunton[21] ; ---- : hard-spheres correlated by an Ashcroft-Lekner correlation function ——Experimental $F(x)$ in CuNiFe[22].

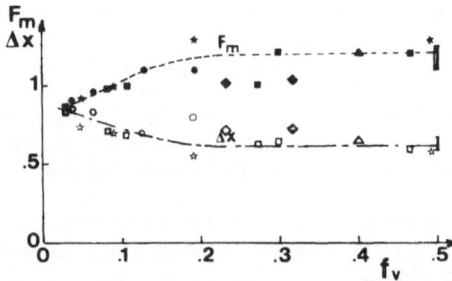

Fig.8. Variations of the scale function with the volume fraction. Full and empty sign correspond respectively to F_m and Δx. stars: Ising model[12,15], squares : AlZn[19], circles : AlZnAg[23-25], rhombi : AlLi[26], triangles FeCr (F. Bley unpublished), vertical bar : CuNiFe[22].

Numerous SAS experiments have proved the validity of such a scaling. Fig.8 show the sensitivity of $F(x)$, expressed in terms of its maximum F_m and half-height width, Δx, versus the second phase volume fraction f.

SOLUTE PARTITIONING IN TERNARY ALLOYS

Kinetic solute partitioning in multi(n)-component systems is complex. The eigenvectors of the diffusion matrix may differ from those of the free energy, leading to decomposition paths deviating from the equilibrium conditions[27]. In the simplest case of ternary alloys, such an eventuality has often been eluded, and the analyses of SAS experiments were conducted with an a-priori simplifying "pseudo-binary" behaviour (CuNiFe, CrFeCo, AlZnAg, $AlMg_2Si$).

The problem can be solved through the determination of the partial structure functions (PSF) $S_{ij}(q)$, i.e. the Fourier transforms of the (ij) pair correlation functions. The scattered intensity writes as a linear combination of PSF :

$$I(q,t) = \sum_{i \neq 0} \sum_{j \neq 0} (f_i - f_o)(f_j - f_o)^* S_{ij}(q,t) \tag{3}$$

where (o) stands for a reference element. In a 2-phase model $S_{ij}(q,t)$ becomes :

$$S_{ij}(q,t) = \Delta c_i(t) \Delta c_j(t) S_{pm}(q,t) \tag{4}$$

where Δc_i is the concentration difference in specie i between the two phases p and m, of own structure factor S_{pm}. Therefore all PSF scale then to S_{pm}, and a pseudo-binary partitioning implies $\Delta c_i / \Delta c_j$ = constant, independently of time, whatever i and j are.

The determination of the $n(n-1)/2$ independent PSF requires at least the same number of scattering experiments with varying atomic contrast $(f_i - f_o)$. This can be achieved by neutron scattering (INS) with isotopic substitution[23,28] or anomalous X-rays scattering (AXS)[22,24,25,29], tuning the wave length near an absorption edge. The inversion of the linear system (3) may amplify the experimental uncertainties, which limits INS to alloys at equilibrium and AXS to alloys with selected contrast[30]. For example, the INS study of the unmixing of AlAgZn[23] partly fails since an identical microstructure at time t can not be exactly achieved in three different samples. On the other hand, although having the advantage to require a single sample, AXS has the drawback in these alloys of a weak contrast variation when working in the vicinity of the Zn edge (Fig.9). Although the three PSF can not been determined precisely, the results for the Guinier-Preston zone formation[24,25] are in agreement with a two phase partitioning, with a fixed kinetic tie-line in agreement with the calculated phase diagram[23] ; the precipitation of the ε' phase produces a rotation of the tie-line which reflects the composition change.

At variance with AlAgZn, CuNiFe is a favourable case for ASAXS study[22]: the intensity rises by a factor 3 when approaching the Fe or Ni edge, Fig.10. The deduced PSF S_{Cu-Cu}, S_{Ni-Ni} and S_{Fe-Fe}, Fig.11a) have all a well defined maximum and, at first glance are homothetic, therefore also presuming a two-phase partitioning. However their relative ratios do not scale with a single S_{pm}. The three S_{ii} are in fact linear combinations of two structure factors, $S_{ii}(q) = x_i S_a(q) + y_i S_b(q)$, with $x_i |S_a(q)| / |y_i S_b(q)|$

218

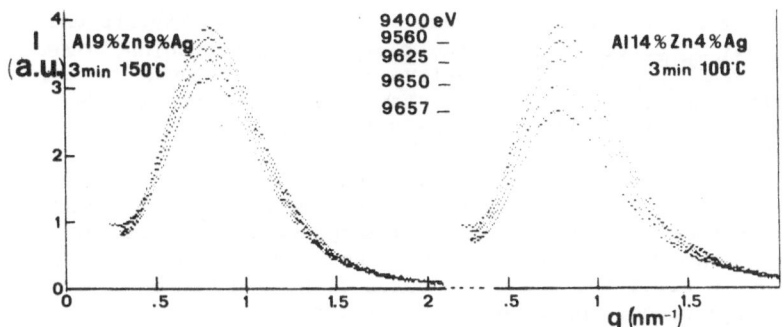

Fig. 9. ASAXS spectra near the Zn edge (9660 eV) for
a poor and a rich Zn, AlZnAg alloy (courtesy of O.
Lyon).

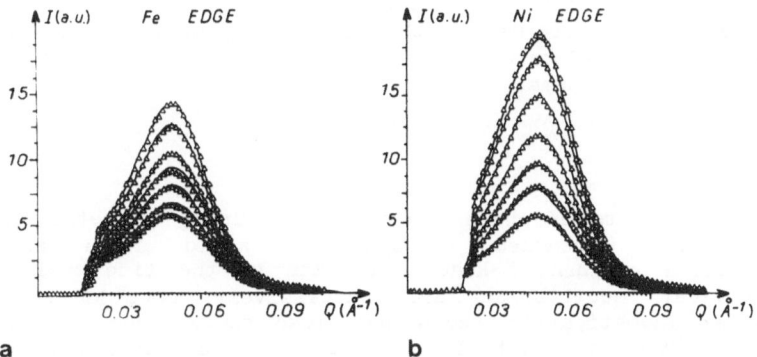

Fig. 10. ASAXS spectra of $Cu_{42}Ni_{43}Fe_{15}$ aged 56 hours at 500°C, near the
Fe and Ni edges, with photon energy increasing upwards. Solid
lines: experimental intensities; triangles : best fit from
calculated PSF of Fig.11 (from [22]).

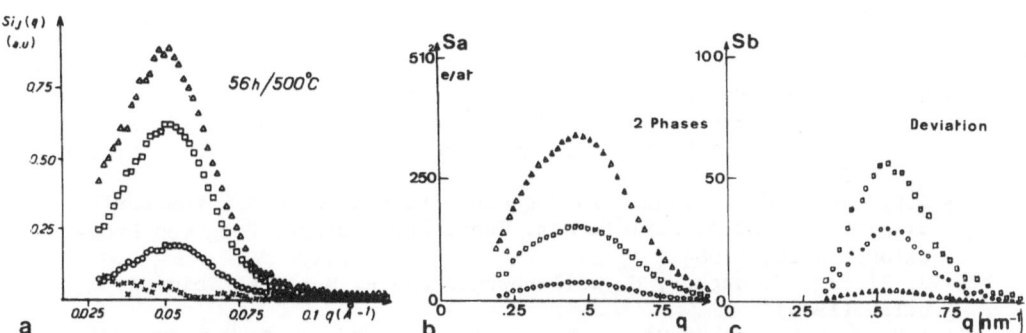

Fig. 11. PSF calculated from data of Fig.10:a) triangles : S_{Cu-Cu}, squares :
S_{Ni-Ni}, circles : S_{Fe-Fe} , crosses : S_{Fe-Ni}); b) and
c) decomposition of the PSF of Fig.a) in the 2 phase term (S_a) and
its deviation (S_b). Sb/Sa scales are in a ratio of 5. (from [31]).

ranging from 2 for i = Fe to about 100 for i = Cu (Fig.11b and 11c). It is shown[31] that S_a corresponds to the two-phase partitioning, already evidenced by TEM[4] and FIM[5] and which tie-line slope corresponds to the equilibrium miscibility gap[32]. S_b, only important for i = Ni and i = Fe represents the deviation from the two-phase partitioning, and indicates composition heterogeneities in the rich (NiFe) phase. Interpretation of $S_b(q)$, accounting for a similar position of the hump of $S_b(q)$ and $S_a(q)$ (same wave length), and larger gyration radius for S_b, has been given[31] in terms of a localization of these heterogeneities at the interface between the (NiFe) and the Cu rich phases. Such interfacial segregation could be correlated to different diffusion rates of Ni and Fe, or percolation effects for such high second phase volume fractions.

Isothermal agings at 460, 500 and 550°C show that the PSF follow a coarsening law, with $I_{ij,m} \alpha t^{.8}$ and $q_{ij,m} \alpha t^{-.26}$, the aging temperature variation inducing only a change in time scaling via the interdiffusion activation energy[31].

Finally the predominence of the 2 phase term S_a in the decomposition of the CuNiFe alloys explains why the ASAXS experiments successfully compare with the previous NSAS results of Aalders et al[33] or Wagner et al[16], or with theory for binary systems.

CONCLUSION

Various aspects of phase separations occuring in metallic binary and ternary systems quenched through the solid state are analyzed. Microstructural patterns, kinetic regimes of the time evolution towards thermodynamical equilibrium are successively considered. Performing experimental investigation techniques lead to a deep insight into the underlying microscopic mechanisms, allowing a discussion with respect to basic concepts like nucleation-growth, spinodal decomposition, coarsening, and complex atomic partitioning.

ACKNOWLEDGEMENTS

The authors are grateful to O. LYON, to authorize them the discussion of not yet published ASAXS results on CuNiFe and to their colleagues F. Bley and F. Livet for unpublished data on AlLi and FeCr.

REFERENCES

1- G. Laslaz and P. Guyot, Acta Met. 25, 277 (1977).
2- P. Sainfort and P. Guyot, Phil. Mag. A, 51, 575 (1985).
3- M. Dombre, C. Allibert, C. Bronner and J. Driole, Mem. Sci. Rev. Met. 75, 605 (1978).
4- R.R. Wahi and J. Stajer, in "Decomposition of Alloys : the early stages", 2nd Acta. scripta Met. Conf. proceedings, Pergamon Press, Oxford, p.165 (1984).
5- J. Piller, W. Wagner, H. Wollenberger and P. Mertens, Ibid. 4, p.156 (1984).
6- A.G. de Salva, J.P. Simon, F. Livet and P. Guyot, Scripta Met., to be published, July (1987).
7- M. Hennion, D. Ronzaud and P. Guyot, Acta Met. 30, 599 (1982).
8- J.P. Simon, P. Guyot and A.G. de Salva, Phil. Mag. A, 49, 151 (1984).
9- P. Guyot and J.P. Simon, J. Chimie Physique 83, 703 (1986).
10- J.S. Langer and A. Schwartz, Phys. Rev. A21, 948 (1980).
11- E.D. Siebert and C.M. Knobler, Phys. Rev. Lett. 54, 819 (1985).

12- J.P. Simon and O. Lyon, "Early stages of decomposition in AlZn" in "Progress in X-ray studies by synchrotron radiation", Strasbourg, 1-4 April (1985), (3-4(A)5).

13- J.W. Cahn and J.E. Hilliard, J. Chem. Phys. 28, 258 (1958); 31, 688 (1959).

14- P. Guyot and J.P. Simon, in "Solid phase transformations", Met. Soc. AIME, p.325 (1982).

15- P. Guyot, Ibid 4 ; p.41 (1984).

16- W. Wagner, R. Poerske, H. Wollenberger, ibid 4; p.170 (1984).

17- J.S. Langer, M. Bar-on and H.D. Miller, Phys. Rev. A, 11, 1417 (1976).

18- K. Binder, C. Billotet and P. Mirold, Z. Physik B, 30, 183 (1978).

19- J.L. Lebowitz, J. Marro and M.H. Kalos, Acta Met. 30, 297 (1982).

20- I.M. Lifshitz and V.V. Slyozov, J. Phys. Chem. Sol. 19, 35 (1961).

21- P.A. Rikvold and J.D. Gunton, Phys. Rev. Lett. 49, 286 (1982).

22- O. Lyon and J.P. Simon, Phys. Rev. B, 35, 5164 (1987).

23- A.G. de Salva, J.P. Simon, P. Guyot and I. Ansara, Acta Met. 31, 1705 (1983).

24- O. Lyon, J.J. Hoyt, R. Pro, B.E.C. Davis, B. Clark, D. de Fontaine and J.P. Simon, J. Appl. Cryst. 18, 480 (1985).

25- O. Lyon and J.P. Simon, Acta Met. 34, 1197 (1986).

26- F. Livet and D. Bloch, Scripta Met. 10, 19 (1985).

27- D. de Fontaine, J. Phys. Chem. Sol. 33, 297 (1972); 34, 1285 (1973).

28- P. Cenedese, F. Bley and S. Lefebvre, Acta Cryst. A, 40, 228 (1984) and F. Bley, P. Guyot and S. Lefebvre, Acta Met. 33, 1235 (1985).

29- P. Goudeau, A. Fontaine, A. Naudon and C.E. Williams, J. Appl. Cryst. 19, 19 (1986).

30- J.P. Simon, O. Lyon and D. de Fontaine, J. Appl. Cryst. 18, 230 (1985).

31- J.P. Simon and O. Lyon, Proceedings of "Phase Transformations 87", Cambridge, to be published.

32- T.G. Chart, D.G. Gohil and X.Z. Shu "Calculated phase equilibria for the Cu-Ni-Fe system", Report DMA (A)54, Teddington (1982).

33- J. Aadlers, C. van Dijk and S. Radelaar, Phys. Rev. B, 30, 1646 (1984).

KINETICS OF PHASE SEPARATION ALONG STABLE TRAJECTORIES

Peter Fratzl[1] and Oskar Blaschko[2]

[1]Institut für Festkörperphysik der Universität Wien
A-1090 Wien, Strudlhofgasse 4
[2]Institut für Experimentalphysik der Universität Wien
A-1090 Wien, Strudlhofgasse 4

The phase separation kinetics of dilute Al-(Zn,Mg) alloys were studied by small-angle and diffuse neutron scattering techniques and the mean cluster radius R as well as the volume fraction V of the solute-rich phase were determined. It could be shown that, within (R,V)-space, the decomposition corresponds to a trajectory, which is stable against small perturbations. Using a simple two-phase model, this trajectory appears as a "valley" of the free energy of the system. When this model is extended to concentrated alloys such a valley still exists in the free energy as function of the different system parameters. During evolution along this valley a time-scaling of the system is obtained. In fact all parameters of the two-phase model, like composition and volume fraction of the phases or the total interphase-surface must be functions of a scaling parameter R, which for dilute alloys, can be identified as the average cluster radius.

INTRODUCTION

The kinetics of alloy decomposition during thermal anneal after a quench from high temperature have been intensively studied for many years as a typical example for non-equilibrium thermodynamics[1]. In particular the time dependence of the structure function I(q), which can be directly measured by neutron or X-ray small-angle scattering, may be described in many cases by a scaling behaviour in the form

$$Q^{-1} I(q) = R^3 F(qR) \qquad (1)$$

Whereas I(q) changes with annealing time, F(x) is a time independent scaling function and the whole time dependence is contained in the scaling parameter R(t). Q is a normalization constant defined by

$$Q = \frac{1}{2\pi^2} \int_o^\infty q^2 I(q) \, dq \qquad (2)$$

Such a scaling behaviour was observed in the case of computer simulations as well as for many real systems[2,3,4]. It is usually interpreted as a stationnary state in which all relevant system parameters (like the total energy, the volume fraction of each phase or the interphase surface, etc.)

are constant during the decomposition process when all length scales are rescaled by the parameter $R(t)$[1,5].

Recently the present authors have shown a first direct experimental evidence for such a relation between the volume fraction V of clusters formed by the decomposition and their average radius R, which in this case is the scaling parameter. In a series of neutron small-angle and diffuse scattering experiments on dilute Al-(Zn,Mg) alloys a time independent relation between R and V was observed in the form[6]

$$\frac{V}{V_\infty} = 1 - \frac{R_0}{R} \tag{3}$$

V_∞ and R_0 are constants, which however depend on temperature and alloy-composition. It could even be shown experimentally that the trajectories defined by (3) are stable against small perturbations[7].

In the present paper, after some introductory remarks about the formulae used for the interpretation of the small-angle scattering data, the results of these experiments are reviewed and discussed within the two-phase model. Within such a model, using a few simple assumptions, trajectories can be obtained for several system parameters as minimum paths, i.e. "valleys", of the free energy. In the case of dilute alloys equation (3) is reproduced, but the model also allows some predictions for concentrated alloys which will be discussed in the last section.

SMALL ANGLE SCATTERING FROM A TWO-PHASE SYSTEM

Throughout this paper we consider a polycrystalline binary A-B alloy, which decomposes into two phases α and β. We call $c(\vec{r})$ the concentration of B-atoms at the position $\vec{r}$ and $\hat{c}$ its average value. We assume a concentration c_1 and c_2 in α and β phase respectively and sharp interfaces between the phases (two-phase model). Depending on the average composition $\hat{c}$ the formation of isolated clusters (Fig. 1a) or of a percolated structure (Fig.1b) are possible. Further we call V the volume fraction of the β-phase and S the total interphase surface per unit volume.

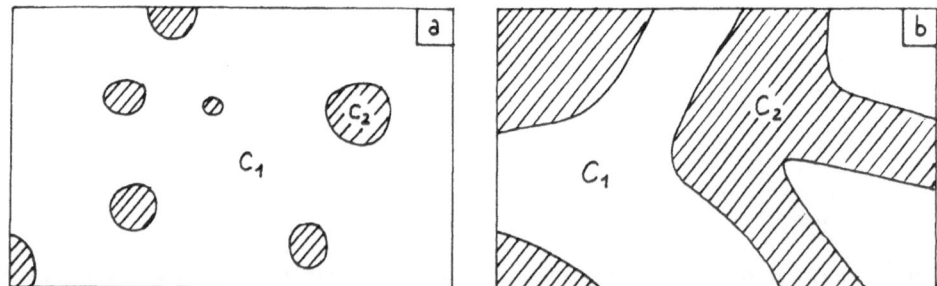

Fig. 1. Possible structure of a two-phase system. In the dilute case (a) there are isolated clusters, and in the concentrated case (b) the structure is percolated.

After some normalization the scattering function (outside the Bragg reflexions) may be written[8]

$$I(q) = \int_0^\infty 4\pi r^2 \, \gamma(r) \, \frac{\sin qr}{qr} \, dr$$

$$\gamma(r) = \langle (c(\vec{r}+\vec{x})-\hat{c}) \, (c(\vec{x})-\hat{c}) \rangle$$

(4)

where $\gamma(r)$ is the spherically averaged correlation function of the concentration fluctuations. Within the two-phase model the concentrations c_1 and c_2 are considered to be constant throughout each phase, which of course cannot be true at the atomic level. Nevertheless, if there is no ordering of the A and B atoms within each phase a random mixture may be assumed, so that the corresponding scattering is constant: $I_0 = V \, c_2 \, (1-c_2) + (1-V) \, c_1 \, (1-c_1)$. We suppose in the following that this small constant has been substracted from intensity.

Then one may calculate the integral intensity[9]

$$Q = \gamma(0) = V \, (1-V) \, (c_2-c_1)^2$$

(5)

In the case of dilute alloys Q is simply proportional to the volume fraction V of the clusters of β-phase. One should however notice here that within the two-phase model Q depends on the structural state of the alloy system (and therefore eventually on time), whereas the total integral intensity would be $Q+I_0 = \hat{c} \, (1-\hat{c})$, which corresponds to the well-known invariant of the system[8].

In addition the correlation function γ may be approximated for small r to the first order as

$$\gamma(r) = Q \, (1- \frac{S}{4V(1-V)} \, r)$$

(6)

leading to the well-known Porod-law for the scattering at high q-values[8]:

$$I(q) = (c_2-c_1)^2 \, \frac{2\pi \, S}{q^4}$$

(7)

Combining this equation with the scaling relation (1) it appears that in the case of dynamical scaling the function $x^4 F(x)$ must be constant for high x. One can choose the function $F(x)$ so that this constant is 6π, then the scaling parameter must be

$$R = 3 \, \frac{V(1-V)}{S}$$

(8)

This is an intrinsic definition of the scaling parameter using the Porod-law. In the case of a dilute alloy ($V \ll 1$) containing a monodisperse distribution of sperical clusters, R represents the cluster radius. This explains why for dilute alloys one can take the scaling parameter as an average cluster radius.

EXPERIMENTS ON Al-(Zn,Mg)

In a series of neutron small-angle and diffuse scattering experiments scattering curves were obtained for dilute pseudobinary Al-(Zn,Mg) alloys annealed after quench from high temperature. The annealing temperatures were chosen to obtain Guinier-Preston zones (25°C to 90°C) or clusters of η'-phase (100°C to 120°C). In all cases the scattering curves showed a scaling behaviour according to eq.(1). This is illustrated for the anneal at room-temperature in Fig.2. No significant change of the scaling function $F(x)$ with temperature could be observed.

Using eq.(1) the parameters R and Q could be determined in all cases. Using the remarks from sect.2, R and Q can respectively be identified as the average cluster radius and the volume fraction V of the clusters respectively (the Guinier-Preston zones are clusters of nearly pure Zn and Mg in this alloy[10], so that we take $c_2-c_1 \simeq 1$). It appears that V does not remain constant, even during the late stages of decomposition, where the scaling relation (1) holds good. During decomposition at room temperature for instance a variation of V of more than a factor 4 was observed. However, a time-independent relation between $V(t)$ and $R(t)$ was found for the decomposition starting with the as-quenched state. The experimental data, as shown in Fig.3, could be well described by eq.(3). The constants R_0 and V_∞ appearing in this equation may be interpreted respectively as the mean radius at the beginning and the volume fraction at the end of decomposition[6]. The values of V_∞ and R_0 obtained for the Al-(Zn,Mg) alloys are summarized in table 1.

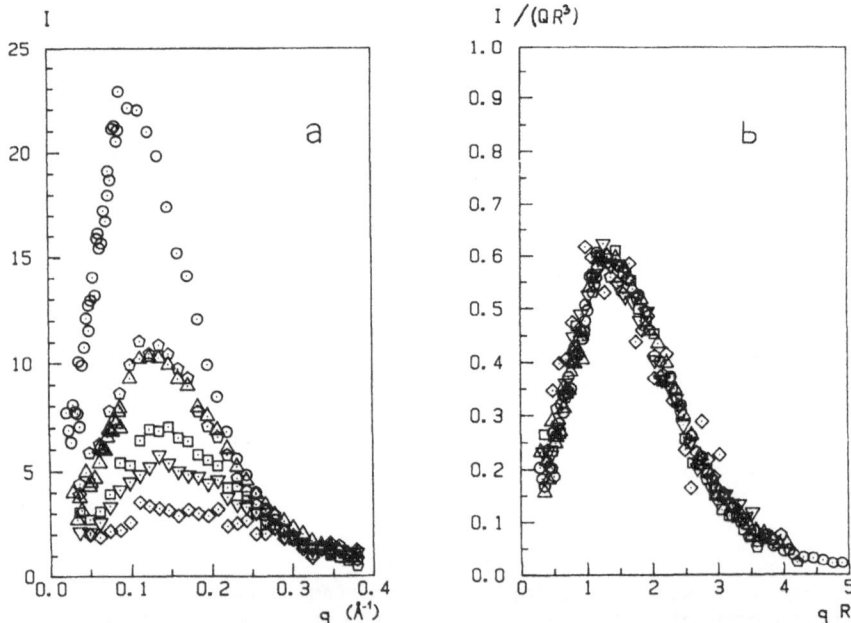

Fig. 2. Neutron scattering curves of Al-2.4at%Zn-1.3at%Mg annealed at room temperature. The time evolution between 24h and 1 year is shown in (a) and the corresponding scaled curves in (b).

Table 1. Constants appearing in equation 3 as determined by
neutron small-angle scattering for Al-(Zn,Mg) alloys

Clustered phase	$\hat{c}(\%)$	$T(K)$	$R_0(\text{Å})$	$V_\infty(\%)$
GP zone	5.4	298	5.1	3.9
GP zone	3.7	298	3.8	2.1
		323	4.4	1.9
		343	5.4	1.6
		363	7.1	1.4
η'	3.7	373	10.8	1.8
		393	17.5	1.3

Equation (3) can be transformed using the lever rule and, if we call $\tilde{c}_1$ the equilibrium concentration of the α-phase, it appears that the product $R(c_1-\tilde{c}_1)$ remains constant during the evolution of the alloy. $1/(c_1-\tilde{c}_1)$ is usually interpreted as the critical radius of the system[11] so that (3) finally expresses the proportionality between mean and critical radius, in agreement with the coarsening theories[11].

The stability of the so obtained trajectories in (R,V) space (eq.3) was studied[7] by introducing perturbations in the following way. During preaging at some temperature a defined point on the (R,V) plane was reached and then after a sudden temperature change (an increase or a decrease) the evolution of the alloy was followed. The results are shown in Fig. 4. After the temperature step the actual point in (R,V) space is outside the trajectory corresponding to the new temperature. However, as shown in the figure, the new trajectory is approached during the subsequent evolution. After the sudden temperature increase this evolution even starts with a partial dissolution of clusters (V decreases) before the new trajectory is reached, where V begins to increase again. The mean cluster size R is growing in all cases.

TRAJECTORIES DURING DECOMPOSITION

The small-angle scattering investigations on Al-(Zn,Mg) have shown that volume fraction and cluster radius in this alloy follow a trajectory in (R,V) space which is stable against small perturbations. This strongly suggests that the relation (3) corresponds to a path of minimum free energy of the system. This possibility has been discussed within a simple model in Ref.7. Here we present an extension of this model, which has the advantage that the results are not limited to the case of dilute alloys only.

The Gibbs free energy per unit volume of the system is written in the form

$$G = \sigma S + V\, g(c_2) + (1-V)\, g(c_1) \qquad (9)$$

(σ is the surface tension and g the local free energy density). During the late stages of decomposition c_2 and c_1 will be close to their equilibrium values $\tilde{c}_2$ and $\tilde{c}_1$. An expansion of G close to its equilibrium value G_e

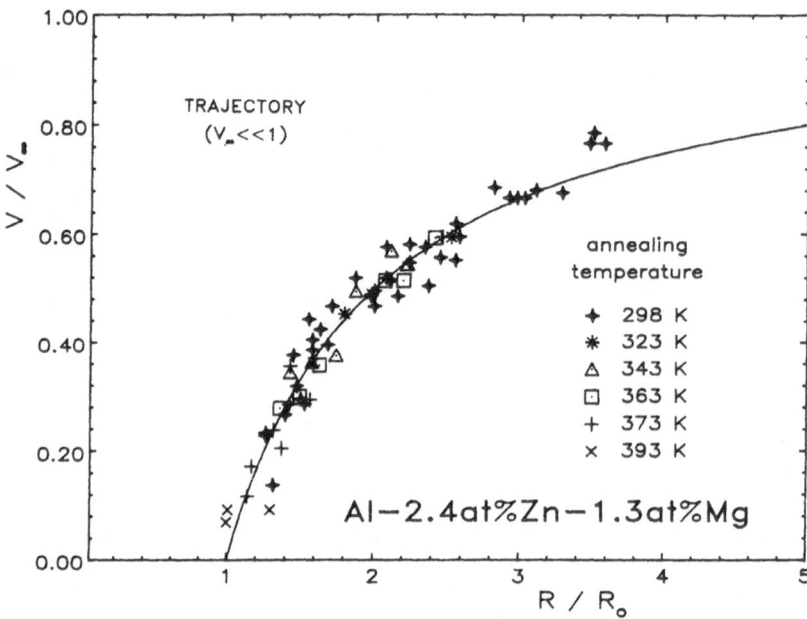

Fig. 3. Mean cluster radius R and volume fraction V in reduced units, as determined by neutron scattering on Al-(Zn,Mg) alloys. The full line corresponds to equation 3.

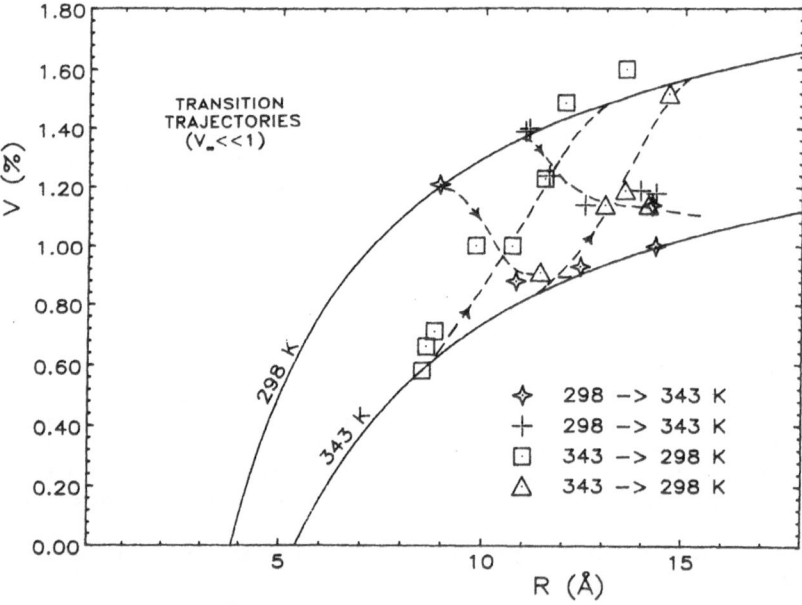

Fig. 4. Transition between trajectories after a temperature step during decomposition in Al-(Zn,Mg). The full lines are trajectories for two temperatures. Broken lines are guides to the eye.

therefore yields:

$$G-G_e = \frac{3\sigma}{R} \ V(1-V) + \frac{1}{2} \ V(c_2-\tilde{c}_2)^2 g_2'' + \frac{1}{2} \ (1-V)(c_1-\tilde{c}_1)^2 g_1'' \qquad (10)$$

S has been replaced in this equation by the parameter R defined in eq.8; g_1'' and g_2'' are the second derivatives of g at $\tilde{c}_1$ and $\tilde{c}_2$ respectively. Using the lever rule, V can also be expanded around its equilibrium value V_∞ as functions of $c_1-\tilde{c}_1$ and $c_2-\tilde{c}_2$, so that we finally conclude from the partial derivatives of eq.(10) that the free energy has a minimum in c_1 and c_2 for each given value of the scaling parameter R, whereas the energy always decreases with increasing R. The free energy G therefore has a valley defined by

$$g_1'' \ (c_1-\tilde{c}_1) \ = \ g_2'' \ (c_2-\tilde{c}_2) \ = \ \frac{1-2V_\infty}{(\tilde{c}_2-\tilde{c}_1)} \ \frac{3\sigma}{R} \qquad (11)$$

Equation (11) defines a stable path for the evolution of the system towards equilibrium. On this path all system parameters (in our model only 2 independent parameters remain besides R) are functions of R, so that by rescaling the system with R the time-dependence can be completely removed. From eq.(11), which is the trajectory of the evolution in the parameter space of the model (R,c_1,c_2), similar relations may be deduced for related parameters like the volume fraction V. Eq.(3) appears then as a consequence of relation (11). The trajectory found experimentally for dilute Al-(Zn,Mg) alloys therefore corresponds within this model to a stable path of minimum free energy[7]. The projection of the trajectory (11) onto the (R,c_1)-plane is shown in Fig.5. The curve for $V_\infty \ll 1$ corresponds to the trajectory shown in Fig.3, the other curves in Fig.5 illustrate the change from a dilute to a concentrated alloy.

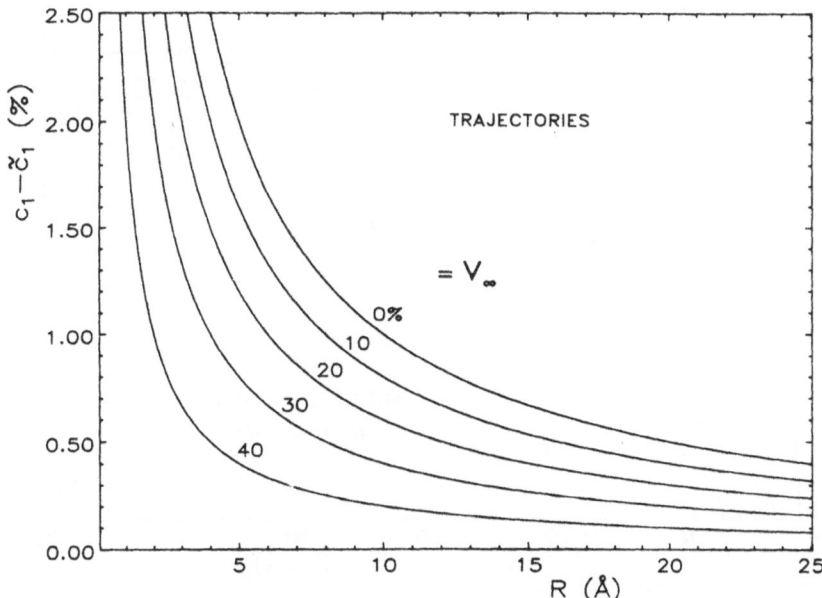

Fig.5. Projection of the trajectory (11) onto the (R,c_1)-plane for different values of V_∞. For the constant $3\sigma/g_1''/(\tilde{c}_2-\tilde{c}_1)$ the value 0.1 Å obtained for Al-(Zn,Mg) alloys has been taken. The trajectory for $V_\infty \ll 1$ is equivalent to eq.(3) via the lever rule.

Unfortunately the integral intensity Q cannot be interpreted directly as volume fraction V in the case of concentrated alloys. Nevertheless a relation can also be derived for Q using eq.(11):

$$Q = Q_\infty \left(1 - \frac{R_1}{R}\right) \tag{12}$$

$$\text{with} \quad R_1 = \frac{A}{V_\infty} \frac{1-2V_\infty}{1-V_\infty} \left(1-V_\infty\left(1 + \frac{g_1''}{g_2''}\right)\right)$$

$$\text{and} \quad Q_\infty = (\tilde{c}_2-\tilde{c}_1)^2 \, V_\infty(1-V_\infty) \quad ; \quad A = \frac{3\sigma}{g_1''(\tilde{c}_2-\tilde{c}_1)^2}$$

This equation also has the form of eq.(3). However, in the case of concentrated alloys, R_1 can be quite small and even negative. The integral intensity Q according to eq.(12) has been drawn in Fig.6 in function of R for several values of V_∞. Only for dilute alloys a significant variation of Q with R can be observed, whereas for concentrated alloys Q remains nearly constant during decomposition. For the parameters chosen (see Fig.6) the highly concentrated alloy even shows a slight decrease of Q with R. Even though well defined trajectories have been obtained also for concentrated alloys in terms of eq.(11), significant variations of the integral intensity Q with R may only be obtained for dilute alloys, in agreement with experimental results[3,4].

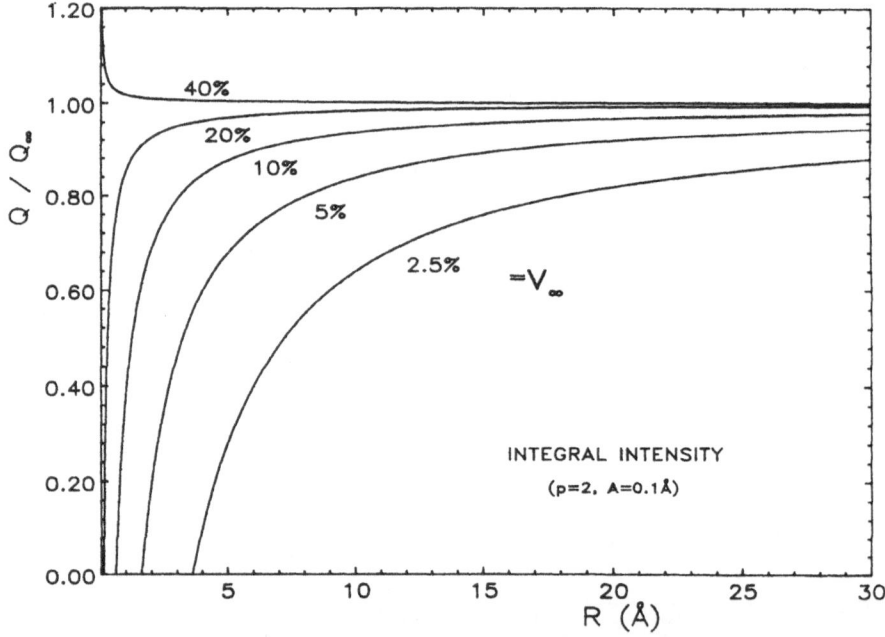

Fig. 6. The integral intensity relative to its equilibrium value is drawn as function of the scaling parameter R for several values of V_∞ according to equ.(12). p is defined as g_1'' /g_2''; the constant A has been taken as 0.1 Å (value obtained experimentally in the case of Al-(Zn,Mg) alloys).

CONCLUSION

As shown by neutron scattering experiments, the decomposition in dilute Al-(Zn,Mg) alloys follows a stable trajectory in (R,V) space defined by eq.(3). Using a simple two-phase model this trajectory can be obtained as a minimum path of the free energy, when the surface tension is supposed to remain constant during decomposition. More generally a trajectory in the (R,c_1,c_2)-space can be obtained, where R is the scaling parameter and c_1, c_2 the concentration of each phase. This trajectory given by eq.(11) holds for dilute as well as for concentrated alloys. The integral intensity Q in a small-angle scattering experiment is proportional to the volume fraction V in the case of dilute alloys, so that eq.(3) can be determined directly. In the case of concentrated alloys, Q cannot be interpreted anymore in this simple way; the R-dependence of Q can however still be obtained from the model, but, unfortunately, it is very small, so that Q will appear constant in a small-angle scattering experiment. Nevertheless it would be possible to check the relation (11) by other experimental methods, for instance using field ion microscopy, where the parameters c_1, c_2 and R may be directly measured.

ACKNOWLEDGEMENT

This work has been supported in part by the "Fonds zur Förderung der wissenschaftlichen Forschung" in Austria.

REFERENCES

1. J. D. Gunton, M. San Miguel and P. S. Sahni, in: "Phase Transitions and Critical Phenomena", C. Domb and J. L. Lebowitz, ed., Academic Press, New York (1983); H. Furukawa, Adv. Phys. 34:703 (1985)
2. J. L. Lebowitz, J. Marro and M. H. Kalos, Acta Metall. 30:297 (1982); P. Fratzl, J. L. Lebowitz, J. Marro and M. H. Kalos, Acta Metall. 31:1849 (1983)
3. S. Komura, K. Osamura, H. Fujii and T.Takeda, Phys.Rev.B 31:1278 (1985)
4. R. Poerschke, W. Wagner, H. Wollenberger and P. Fratzl, J.Phys.F: Met.Phys. 16:1905 (1986)
5. G. F. Mazenko and O. T. Valls, Phys.Rev.B 27:6811 (1983)
6. O. Blaschko and P. Fratzl, Phys.Rev.Lett. 51:288 (1983)
7. P. Fratzl and O. Blaschko, Phys.Rev.B (in press)
8. G. Porod, in: "Small Angle X-ray Scattering", O. Glatter and O. Krakty, ed., Academic Press, London (1982); G. Kostorz in: "Treatise on Materials Science and Technology", Vol. 15:227 Academic Press, New York (1979)
9. V. Gerold, J. E. Epperson and G. Kostorz, J. Appl. Crystallogr. 10:28 (1977)
10. L. F. Mondolfo, Metall. Rev. 153:95 (1971); G. Groma, E. Kovács-Csetényi, I. Kovács, J. Lendvai and T. Ungár, Philos. Mag. A 40:653 (1979)
11. I. M. Lifshitz and V. V. Slyozov, J. Phys. Chem. Solids 19:35 (1961); Y. Enomoto, K. Kawasaki and M. Tokuyama, Acta Metall. 35:907 (1987)

EFFECT OF SPATIAL CORRELATION OF PARTICLES ON OSTWALD RIPENING

Yoshitsugu Tomokiyo*, Kazuo Yahiro **, Syo Matsumura**
Kensuke Oki** and Tetsuo Eguchi***
*HVEM LAB., **Dep. Mat. Sci. & Tech., Kyushu Univ.
and ***Dep. Appl. Phys., Fukuoka Univ, Fukuoka Japan

INTRODUCTION

The study of phase transitions is of interest to scientists concerned with states of aggregation as well as to metallurgists and physicists. The precipitation or phase separation in solids is one of the interesting and useful fields of study for metallurgists and materialists. The precipitation process in alloys generally consists of the initial nucleation and growth, and the late coarsening which is the growth of a larger precipitate particle at the expence of smaller one. The coarsening process of particles, or Ostwald ripening process was theoretically studied by Lifshitz and Slyozov[1] and Wagner[2] (LSW). Their theory is applicable only for the system of infinitesimal volume fraction of precipitates. The modification of the LSW theory has been made by Ardell[3] and many other workers[4,5,6] to take into account the effect of volume fraction. Recently theoretical treatments of coarsening process have been made from the view point of the dynamics of phase transition[7,8,9]. While the $t^{1/3}$ law predicted by the LSW theory is found to hold, the observed size distributions of precipitate particles are frequently different from the one predicted by the theory of LSW or Ardell. We examined the size distribution of particles in Cu-Co and Al-Li alloys and pointed out the importance of the effect of volume fraction on the coarsening[10,11]. As Ostwald ripening is the competitive growth of particles controlled by diffusion of solute atoms, the growth rate of particles is to be influenced by the surrounding particles through the diffusion field. The experimental investigations which paid an attention to the effect of spatial correlations of particles are very scarce. In the present paper Ostwald ripening process of δ' phase in Al-Li alloys was examined by means of transmission electron microscopy to clarify the effect of spatial correlations of particles.

EXPERIMENTAL

An Al-Li alloy is one of the suitable system for the experimental study because it has following characteristics: (i) The δ' phase of the precipitate has the superstructure of Al_3Li (Ll_2 type), and hence we can directly observe the precipitates by an electron microscope using the dark field images with a superstructure reflection. (ii) The lattice misfit between the precipitate and matrix is very small and the effect of strain field around precipitates on coarsening is negligibly small. (iii) So called steady state is attainable on aging[11]. The alloy was prepared from 99.99 wt% Al and 99 wt% Li in an induction furnace at Ar atmosphere. The solution-treatment and subsequent aging were made with caution to avoid oxidation and evaporation of Li. The composition of Li of aged specimen was chemically analyzed to be 1.82 wt%. The thin specimens for electron microscopic observation were prepared by electropolishing with the solution of perchloric acid and ethanol. The electron microscope used was JEM-200B equipped with a heating stage at HVEM LAB., Kyushu University. We attempted to obtain, through stereoscopic and in situ observation, direct information on the microstructure such as the size, shape, volume fraction and spatial distribution of precipitates.

RESULTS AND DISCUSSION

Size distribution of particles

The δ' particles are observed as bright contrast in dark field images taken with a superstructure reflection. One of such examples is shown in Fig. 1. As seen in the figure particles are spherical, their spatial arrangement is metallurgically isotropic and uniform. There is no tendency of directional alignment of particles or change in the shape from a sphere to a cube as observed in Cu-Co alloys[12,13]. Figure 2 shows the size distribution of δ' particles in the alloy aged at 498 K for 18 ks. The observed size distribution function is broader and more symmetric than that of LSW theory, while it is narrower than that of Ardell as apparent from Fig. 2. The size distribution is characterized by parameters such as the volume fraction, ϕ, and the standard deviation, σ, and the skewness, k_s of the distribution function if the size is scaled properly. The measured parameters are listed in Table 1 along with the theoretical ones. The results in Fig. 2 and Table 1 clearly demonstrate that the effect of volume

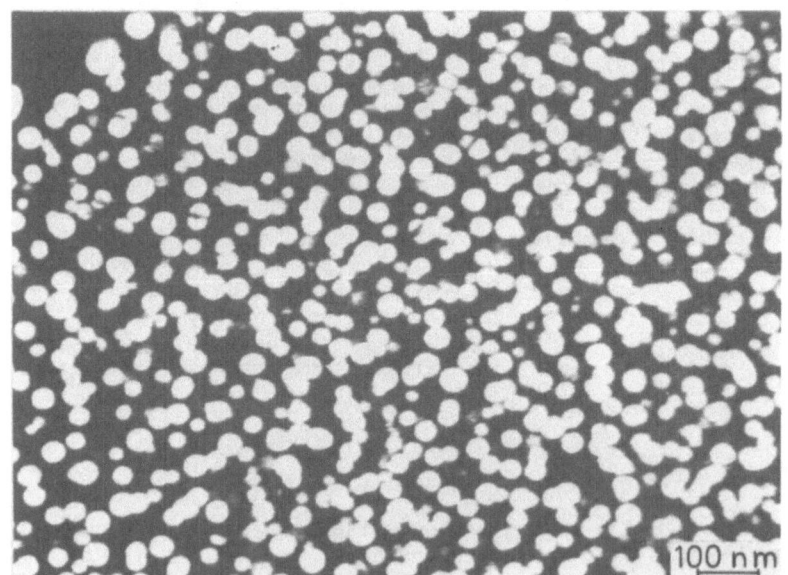

Fig. 1 Electron micrograph of δ' precipitates in an Al-1.82 wt% Li alloy aged at 498 K for 18 ks.
Dark field image taken with 100 superstructure reflection at 200 kV. (One of the images of stereopair.)

fraction plays an important part in the coarsening process but Ardell's theory predicts a far greater effect of volume fraction. Ardell assumed a random distribution of particles which have the same radius, or a mono-disperse assembly of particles. It is suggested that the coarsening process depends on the volume fraction through a spatial correlation of particles. The time evolution of the size distribution of δ' particles at various temperatures has been fully discussed in the previous paper[11], and also compared with the theoretical caluculation by Enomoto et al[14].

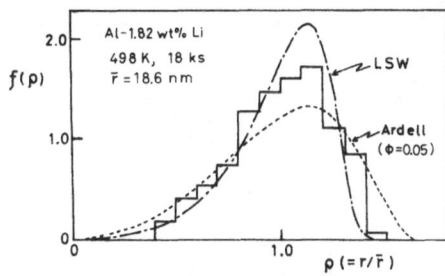

Fig. 2 Comparison of the observed size distribution of particles with those predicted by the theories of LSW and Ardell.

Table 1   Comparison of parameters of particle size distribution.*

	present	LSW[1,2]	Ardell[3]	Kawasaki et al.[14]
Standard deviation, σ	0.23	0.125	0.304	0.2557
Skewness, k_s	-0.37	-0.92	-0.38	-0.623

* The measured volume fraction of ϕ = 0.06 was used.

Spatial distribution of particles

The stereoscopic observation provides us with information on a three dimensional arrangement of particles in an alloy. The arrangement of 600 particles observed in Fig. 1 is shown in three projected figures in Fig.3. The effect of volume fraction is related to diffusion paths and the nearest neighbor interparticle distance, R is important. We measured the values of R for 600 particles shown in Fig. 3. The average value, $\bar{R}$ obtained was 3.3$\bar{r}$ ($\bar{r}$ is the average radius measured), while $\bar{R}$ expected from Ardell's theory for ϕ = 0.06 is 2.8$\bar{r}$. If we expect $\bar{R}$ = 3.3$\bar{r}$ for Ardell's treatment, ϕ should be decreased to 0.03. Thus it is obvious that the effect of volume fraction is overestimated in Ardell's theory. The measured distribution of R is shown in Fig. 4. In order to examine the distribution function F(R), we will derive an analytical expression. If we assume a random distribution of point particles in three dimensions as a simple

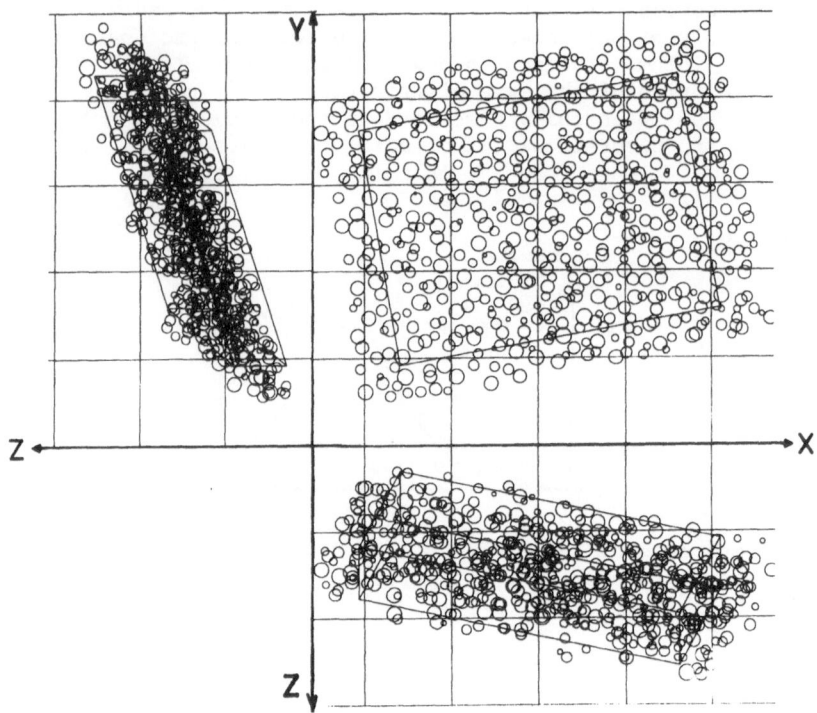

Fig. 3 The spatial arrangement of particles showm in Fig. 1.

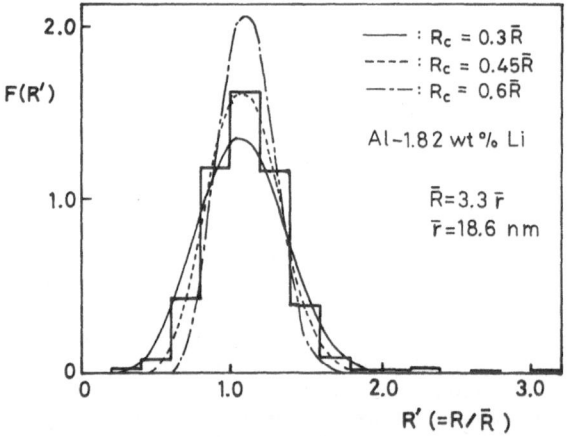

Fig. 4 The distribution of the nearest neighbor interparticle distances.
The histogram is the observed result for particles shown in Fig. 3.
The curves indicate the theoretical predictions by eq. (2).

case, F(R) is given by

$$F(R) = A\,n\,4\,\pi\,R^2\,\exp\left(-\frac{4}{3}\,\pi\,n\,R^3\right)$$ (1)

where, A is the constant and n the number of particles per unit volume. For a practical application, it is necessary to take into accout the size of particles. When the particles have the same radius of $R_c/2$, F(R) is given by

$$F(R) = A\,B\,(R - R_c)^2\,\exp\{-B\,(R - R_c)^3\}$$ (2)

where, A and B are the normalization constants, and R_c the cut off center-to-center distance. The distribution functions calculated by eq.(2) are also shown in Fig. 4 for comparison. As can be seen in the figure the best fitting with the experimental result is obtained when R_c = 0.45 $\bar{R}$ which is converted to 1.5 $\bar{r}$ using R = 3.3 $\bar{r}$. This implies that the observed distribution function is reproduced well by the random spatial distribution of particles which have the same radius of 1.5 $\bar{r}$/2 instead of 1.0 $\bar{r}$. The curve in Fig. 4 for R_c = 0.6 $\bar{R}$ corresponds to the aggregation of particles with the same radius of r. This curve, however, does not fit with the observed one. The disagreement suggests that the size distribution of particles affect the coarsening process.

In situ observation

Observing the same particles in the electron microscope, we pursued the actual competition process. This in situ observation should demonstrate the effect of spatial correlation in the competition. The stereopair of micrographs was taken for the specimen aged at 448 K for 18 ks to know beforehand the spatial arrangement of particles. Then the specimen was reaged on a heating stage at 448 K. An example of the observations is shown in Fig. 5. In the figure the relative depth of particles in the specimen are, for example, 0, −27, 36, −24 and 26 nm for the particles labeled by A, B, C, D and E, correspondingly. Particle B shrinked on account of competition with particle A. Particle C was defeated in the competition with particle D and disappeared within 3.6 ks. On the other

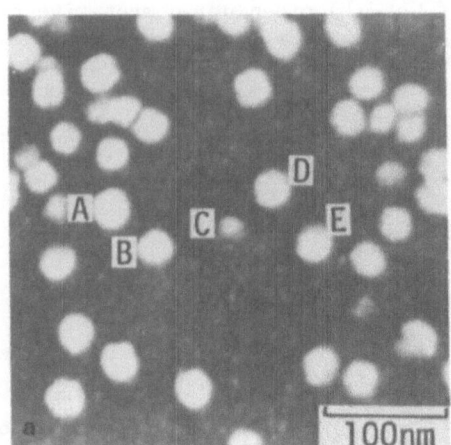 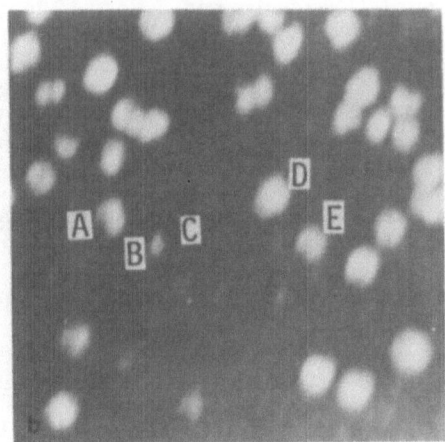

Fig. 5 An example of in situ observation of δ' particles in Al–1.82 wt% Li alloy. (a): The specimen was aged outside the microscope at 448 K for 18 ks, (b): The specimen was reaged at the same temperature for 3.6 ks in the microscope. The micrographs were taken at room temperature.

hand the shrinkage of particle E is not remarkable compared with the case of particle B. This result can be understood, since the nearest neighbor distance R for particles A and B is smaller than that for particles D and E, and furthermore the difference of radius between particles A and B is larger than that between partcles D and E. The result of in situ observation shows that the coarsening proceeds in such a way that a particle closely surrounded by smaller particle(s) can survive and grow, while a particle larger than a certain critical size defined in the LSW or Ardell's theory does not always grow. Especially the relative size and distance between the nearest neighbor particles are important factors to decide whether a particle dissolves or grows.

SUMMARY

It is concluded from present observation that the growth rate of particles strongly depends on the size distribution and spatial distribution of particles. Our results are in good agreement with the predictions by Kawasaki et al.[8,9,14] who took into account the soft-collision process, or the effect of spatial correlations of particles. It is evident from our present investigation that the spatial correlations among particles play an important role in Ostwald ripening in alloys.

Acknowledgement: The authors would like to thank Mr. K. Arita for his assistance in the analysis of electron micrographs.

REFERENCES

1. I. M. Lifshitz and V. V. Slyozov, The kinetics of precipitation from supersaturated solid solution, J. Phys. & Chem. Solids, 19:35 (1961).
2. C. Wagner, Theorie der Alterung von Niederschlägen durch Umlösen, Z. Elektrochemi., 65:581 (1961).
3. A. J. Ardell, The effect of volume fraction on particle coarsening: Theoretical considerations, Acta metall., 20:61 (1972).
4. A. D. Brailsford and P. Wynblatt, The dependence of Ostwald ripening kinetics on particle volume fraction, Acta metall., 27:489 (1978).
5. C.K.L. Davies, P.Nash and R.N. Stevens, The effect of volume fraction of precipitate on Ostwald ripening, Acta metall., 28:179 (1979).
6. J.A.Marqusee and J.Ross, Theory of Ostwald ripening: competitive growth and its dependence on volume fraction, J. Chem. Phys., 80:536 (1984).
7. P. W. Voorhees and M. E. Glicksman, Solution to the multi-particle diffusion problem with applications to Ostwald ripening -I. Theory, II. Computer simulation, Acta metall., 32:2001 (1984).
8. K. Kawasaki and Y. Enomoto, Elementary derivation of kinetic equations for Ostwald ripening, Physica, 135A:426 (1986).
9. Y. Enomoto, M. Tokuyama and K. Kawasaki, Finite volume fraction effects on Ostwald ripening, Acta metall., 34:2119 (1986).
10. T. Eguchi, Y. Tomokiyo, K. Oki and Y. Seno, Coarsening of cobalt precipitates in copper-cobalt alloys, in: "Phase Trasformation in Solids", T. Tsakalakos, ed, North-Holland, Amsterdam, 1984.
11. T. Eguchi, Y. Tomokiyo and S. Matsumura, Electron microscopic observation and its interpretation of Ostwald ripening in precipitation dynamics in alloys, Phase Transitions, 8:213 (1987).
12. S. Matsumura, Y.Seno, Y. Tomokiyo, K. Oki and T. Eguchi, Precipitation behavior in a Cu-4.5 wt% Co alloy, Jpn. J. Appl. Phys., 20:L605 (1981).
13. Y. Tomokiyo, S. Matsumura and M. Toyohara, Strain contrast of coherent precipitates in Cu-Co alloys under exitation of high order reflections, J. Electron Microscopy, 34:338 (1984).
14. Y. Enomoto, K. Kawasaki and M. Tokuyama, The time dependent behavior of the Ostwald ripening for the finite volume fraction, Acta metall., 35:915 (1987).

THE KINETICS OF SPINODAL DECOMPOSITION IN $Mn_{67}Cu_{33}$

Y. Morii[*], B.D. Gaulin[**], and S. Spooner[**]

*Department of Physics, Japan Atomic Energy Research Institute
Tokai, Ibaraki 319-11, Japan
**Solid State Division, Oak Ridge National Laboratory
Oak Ridge, TN 37831, U.S.A.

INTRODUCTION

The kinetics of first order phase transitions[1] has been attracting great interest from physicists studying nonequilibrium phenomena, nucleation and growth, and pattern formation when a rapid change of external variables takes place across the phase boundary. Although there have been many theoretical studies on the kinetics, only a few experiments have been reported on the either order-disorder transitions or the phase separations.[2]

Among possible systems $Mn_{67}Cu_{33}$ alloy is suitable for phase separation studies since the alloy has the following special advantages for neutron scattering experiments. The neutron scattering length of manganese has a negative sign, so the $Mn_{67}Cu_{33}$ composition gives an approximately zero coherent scattering intensity of the disordered phase at any wave vector. The lattice parameter of the alloy at the miscibility gap does not change very much so that the strain between the matrix phase and the precipitate is expected to be small. The composition of $Mn_{67}Cu_{33}$ is located around the central peak of the phase separation curve, so the volume fractions of the manganese rich and manganese deficient phases are almost the same.

Exploiting the advantages of this system, we have pursued[3] a phase separation experiment on a $Mn_{67}Cu_{33}$ alloy after quenching from 800 C to 450 C using the time resolved Small Angle Neutron Scattering (SANS) technique and found that the kinetics can be described by the following three stages. 1. An initial stage where Cahn-Hilliard-Cook (CHC) theory works with a constant peak position, 0.098 1/A, of the structure function. 2. An intermediate stage. 3. A late stage where Furukawa's scaling law is obeyed with the fitting constant γ in his universal scaling function equal to 4.8. It is also found that the measured peak position, $Qmax(t)$, of the structure function, which is inversely proportional to the average domain size of the precipitates, is remarkably well fitted for the entire time region to equation 1 which is a modification from Huse's correction[4] to the asymptotic power behavior of $t^{-1/3}$ in the late stage.

$$\frac{Qmax(t)}{Q_0} = (\frac{t}{\tau})^{-n} \quad ,$$

$$n = \frac{1}{3} - \frac{A}{(\frac{t}{\tau})^{\frac{1}{3}} - A\ln(\frac{t}{\tau})} \qquad (1)$$

Q_0, A, and τ are the constants in equation 1.

Figure 1 shows the experimental results for Qmax(t) and the fitted curve (eq.1) with the constants 0.0995 A^{-1}, 0.293, and 100.2 seconds for Q_0, A, and τ respectively. The three time stages of the evolution of the new phase are also shown in the figure.

The structure factor of a small angle experiment gives information on the polycrystal precipitate in the phase separation. On the other hand the Bragg peak gives information on the crystallites in the precipitate. The intensity of the Bragg peak is proportional to the volume of the crystallites, and the time dependence of the width of the peak is related to the particle size of the crystallites. In this paper we will focus on the kinetics of the crystallites in the precipitate of the new phase in the phase separation of $Mn_{67}Cu_{33}$ alloy by means of a neutron scattering technique.

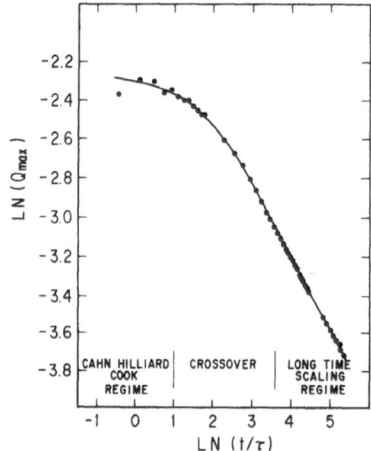

Fig.1 The measured peak positions of the small angle scattering structure function, Qmax (1/A), are shown as a function of reduced time (τ=100.22 sec). The solid line is the result of fitting eq. 1 to the data.

EXPERIMENTAL RESULTS

The Wide Angle Neutron Diffractometer (WAND)[5] is a powerful machine for the study of the kinetics of the phase separation since WAND covers 130 degrees in two theta angle simultaneously with moderately good angle resolution and has a time resolving data acquisition system. A wave length of 1.537 A was chosen. The $Mn_{67}Cu_{33}$ sample, which is the same one as used with the SANS, is placed in the rapid quench furnace[6] and annealed at 800 C for 30 minutes and then quenched to either 450 or 500 C in 30 seconds. As discussed before, the disordered phase gives no coherent scattering at 800 C. The intensity and the Full Width at Half Maximum (FWHM) of the Bragg peaks are studied in situ from immediately after quenching for about 35 hours. The evolution of the Bragg peak indicates the growth of the new phase.

The Bragg peak was broad when it appeared. Since the lattice constants of the two new phases are close to each other, only one peak was observed. The width of the peak decreased as the peak grew. Assuming that the mosaic spread of the small particles of which average size is less than a hundred angstroms is independent of the particle size, the time dependent part of the width is caused mainly by particle size broadening from the crystallites. Since the neutron scattering Bragg peak of the sample is well described by a Gaussian function, the time dependent particle size term of FWHM, $B_{p.s.}(t)$, is related to the observed width, $B_{exp}(t)$, by the following equation 2:

$$B_{exp}^2(t) = B_{instr}^2 + B_{p.s.}^2(t) \tag{2}$$

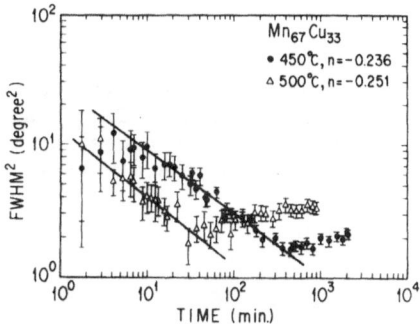

Fig.2 The time dependence of the square of the Full Width at Half Maximum (FWHM) of the (111) Bragg peak of the new phase which displays the particle size broadening. The solid fitting lines are based on a fit of the power law, n, as a function of time to the FWHM.

where B_{instr} is a time independent constant reflecting the instrumental width. B_{instr} involves all kind of extraneous source of broadening, therefore it can be measured using a standard powder sample in which the particle size is large enough to eliminate all particle size broadening. We used Al_2O_3 powder as a standard sample with the same experimental setting and obtained 1.10 degrees for B_{instr} at the (111) Bragg angle.

Figure 2 shows the square of $B_{p.s.}(t)$ for the (111) peak of the crystallite of the precipitate versus time on a logarithmic scale. Data between 10 and 300 minutes at 450 C are well fitted to the power law

$$B_{p.s.}(t) = B_0 t^n \tag{3}$$

where B_0 is an adjustable constant. The fitted power n is evaluated -0.236 for the 450 C.

The $B_{p.s.}(t)$ within 10 minutes is averaged to 2.99 degrees which represents the minimum particle size of the crystallite in this experiment. The constant $B_{p.s.}(t)$ at times later than 350 minutes was probably caused by the instrumental angle resolution of about 1.3 degrees at this two theta angle.

The particle size $L(t)$ is related to the FWHM, $B_{p.s.}(t)$, by the Scherrer equation 4,

$$B_{p.s.}(t) = \frac{\kappa\lambda}{L(t)\cos\theta} \qquad (4)$$

where κ is Scherrer constant for a cubic particle with edges of length $L(t)$, λ the wavelength, and θ the Bragg angle.

Solid circles in figure 3 show the calculated $L(t)$ on a logarithmic time scale. The particle size $L(t)$ is constant at 30 A within 10 minutes after quenching then increases with a power law of $t^{0.236}$ until about 365 minutes when the instrumental angle resolution limited the analysis of the width. The lattice size at 365 minutes is 68 A.

It is shown that the integrated intensity of the (111) Bragg peak, which is proportional to the volume of the decomposed phase, increases with equation 5,

$$I(t) = I_0 t^n \qquad (5)$$

where I_0 is a fitting constant.

A slow but continuous growth with the fitting function of $t^{0.123}$ even at the late times at 450 C was observed in figure 4. The data in the preceding stage has a power law of $t^{0.363}$. The crossover time between the two region is in very good agreement with the starting time of the late stage of SANS data.

DISCUSSION

The importance of the scaling law which is held in the late stage of the present experiment is that the cluster structure in the alloy is characterized by a single parameter $Qmax(t)$. In the cluster regime the average cluster diameter $D(t)$ is given by equation 6.

$$D(t) = 2.571Qmax^{-1}(t) \qquad (6)$$

The calculated $D(t)$ is shown by open circles in figure 3 for comparison with the particle size of the crystallites in the late time regime ($t \geq 72$ min.). The cluster diameter grows from 58 A at 72 minutes to 106 A at 350 minutes by a power function of time, $t^{0.37}$. On the other hand the particle size of the crystallite in the cluster grows from 48 A to 64 A according to the power law $t^{0.236}$ in the same time.

From the data at 500 C, a power n of 0.251 was evaluated. The power $\frac{1}{4}$ might suggest that the growth of the crystallite is driven by the surface mobility and curvature force in the present system.

Although the scaling law does not hold in the initial stage, we extend the calculation of $D(t)$ as shown in figure 3. The cluster size at initial stage turned out to be comparable to the particle size of the crystallites, though the error of the cluster size might be bigger in the stage. One can still say that the new phase clusters are almost single crystal and increase in number but not in size at the early times ($t \leq 10$ min.), then grow faster than the crystallites to the size of about twice that of the crystallites at 350 minutes.

Since the constant FWHM region in the early times was not observed at

242

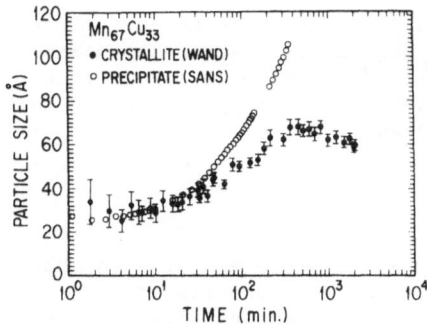

Fig.3 The growth of the averaged particle size of the crystallites (solid circles) and the polycrystal precipitate (open circles) in the phase separation obtained with WAND and SANS respectively.

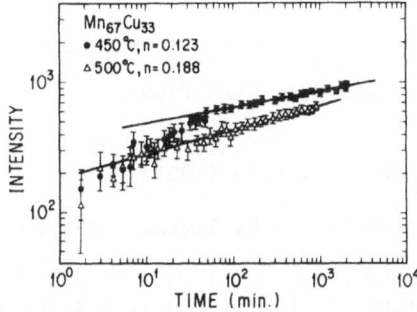

Fig.4 The evolution of the measured intensity of (111) Bragg peak of the new phase. The solid lines are the power law, n, fitted to the data in the scaling regime.

500 C, it suggests that the CHC regime should be within 2 minutes at 500 C. In other words, the process of phase separation goes faster at higher temperatures.

Since no definite crossover time was observed in the intensity data at 500 C in figure 4, it suggests that the scaling regime starts much earlier than that at 450 C, which is consistent with the lack of the constant FWHM region in the early times at 500 C.

The power law n in equation 5 for the later times at 500 C obtained by the fitting is 0.188 which is bigger than that at 450 C. This is probably because that the increase of the number of the crystallite is still effective at 500 C.

YE. Z. Vintaykin et al in reference 2 reported that Bragg peak intensity of the $Mn_{66}Cu_{34}$ alloy reaches its maximum value at about 40 hours in the phase separation by means of a linear time scaled figure. Since the present experiment ended at about 35 hours, the saturation in the peak intensity was not observed.

ACKNOWLEDGMENTS

One of the authors (Y. M.) would like to thank Dr. H. R. Child for his helps in all phases of the WAND experiment.

Research was carried out under U.S.-Japan Cooperative Program on Neutron Scattering, and was supported in part by U.S. DOE under contract DE-AC0584OR21400 with Martin Marietta Energy Systems, Inc.

REFERENCES

1. J. D. Gunton, M. San Miguel, and P. S. Sahni, in "Phase Transition and Critical Phenomena," C. Domb and J. L. Lobowitz (editors) Vol.8, Academic Press, London, (1983)

2. for example on metal: S. Katano and M. Iizumi, Phys. Rev. Lett. **52**, 835 (1984); S. Komura, K. Osamura, H. Fujii, and T. Takeda, Phys. Rev. **B** (RC) **30**, 2944 (1986); YE. Z. Vintaykin, D. F. Litvin, and V. A. Udovenko, Fiz. Metal Metalloved., **37** No.6 1228 (1974)

3. B. D. Gaulin, S. Spooner, and Y. Morii, Phys. Rev. Lett. **59**, 668, (1987)

4. D. A. Huse, Phys. Rev. B**74**, 7845 (1986)

5. M. Iizumi, Proc. of the Int. Conf. on Neutron Scattering, ed. G. H. Lander and R. A. Robinson, p.36 (1985)

6. S. Katano, H. Motohashi, and M. Iizumi, Rev. Sci. Instrum. **57**, 1409 (1986)

7. S. Komura, K. Osamura, H. Fujii, and T. Takeda, Phys. Rev. **B 31**, 1278 (1985)

8. H. Furukawa, Adv. in Phys., **34**, 703 (1986)

DYNAMICAL SCALING IN A Ni-Si ALLOY UNDERGOING PHASE SEPARATION

H. Chen*, S. Polat* and J. E. Epperson**

* Department of Materials Science and Engineering
  University of Illinois
  Urbana, Illinois 61801 USA
** Materials Science Division
  Argonne National Laboratory
  Argonne, Illinois 60439 USA

INTRODUCTION

The dynamics of the decomposition processes in a Ni-12.5 at.% Si alloy has been studied by the small-angle scattering (SANS) method in the temperature range between 400°C and 550°C. The mode of phase separation, as well as the dynamical scaling behavior, was examined in accordance with existing theories and with computer modelling studies.

From previous in-situ x-ray diffraction (XRD) measurements[1-4] it was concluded that the phase separation in this supersaturated Ni-Si alloy begins with nucleation, immediately followed by a growth stage for which the growth of gamma prime precipitates occurs by diffusion of solute from the disordered solid solution matrix, and is completed by Ostwald ripening (coarsening). The nucleation stage is normally too short to be investigated for metallic systems because nuclei of the second phase particles are already present upon quenching from the solid solution regime. Consequently, for practical purposes the kinetics of decomposition may be conveniently divided into three regimes: the growth stage, the crossover or transition regime, and coarsening. The current work was undertaken to examine the self-similarity and dynamical scaling behavior of the microstructural evolution using small-angle neutron scattering techniques. The observations obtained in this experimental study provide a testing ground for the recent theoretical and computer simulation studies on nonequilibrium phenomena. The current results dealing with a decomposition process governed by the classical nucleation and growth mechanism can be viewed in contrast with the spinodal decomposition mechanism occurring in the Mn-Cu system as described in an article by Morii, Spooner and Gaulin[5] in this volume.

EXPERIMENTAL PROCEDURES

A single crystal rod of Ni-12.5 at.% Si was grown by a modified Bridgeman technique. It was homogenized for two days at 1100°C, followed by a water quench. Approximately elliptical shaped samples (10mm in diameter and 1-2mm thick) were cut from the rod and the worked surface layer was removed by standard metallographic techiques, alternated with etching in hot <u>aqua regia</u>. The reference heat treatment before each series of lower temperature measurements was a one hour solution anneal in flowing hydrogen at 1050°C, followed by a rapid quench into iced brine. The experimental heat treatments were carried out in a KNO_3-NaNO_3 salt bath at the indicated temperature.

The SANS measurements were carried out on the small-angle diffractometer (SAD) at the intensed pulsed neutron source (IPNS) of Argonne National Laboratory (ANL). This is a short flight-path, time-of-flight (TOF) instrument which is capable of sampling momentum transfer in the range from 0.005 to 0.35 A^{-1}. Neutrons are generated by pulsed spallation when protons accelerated to 450 MeV strike a depleted uranium target. A spectrum of neutrons suitable for SANS is produced by moderation with solid methane at 20 K, the Maxwellian peak occurring at about 5Å. The scattered neutrons are detected by an area sensitive, gas filled proportional counter of the Kopp-Borkowski type (^{3}He-Xe with 17.8 x 17.8 cm^2 active area) having 64x64 virtual elements. The total flight path is 9m with 1.5m from sample to detector. More information about the TOF technique and the SAD instrument may be found elsewhere[6].

DECOMPOSITION KINETICS

The SANS measurements were performed on quenched specimens annealed at 400, 450, 505 and 550°C for various times[6]. At 505 and 550°C the scattering is characterized by the presence of an interference peak which grows in intensity and shifts to smaller wave vector values as aging time progresses. Such a peak is observed only after 4 hours of annealing at 450°C. For two measurements taken at 400°C, only the one for 200 hours annealing showed a peak. The two dimensional scattering pattern was isotropic for all aging conditions considered. No evidence of preferred alignment of precipitate particles along elastically soft axes was seen[4].

Guinier analysis was performed on the high angle side of the interference peak. The resulting Guinier radii were then plotted as a function of $t^{1/3}$ to test for Ostwald ripening. As shown in Fig. 1, the $t^{1/3}$ law is well obeyed for the 505 and 550°C data. Deviation from linearity is clearly seen for the 450°C data. This observation is consistent with the previous wide-angle XRD results[1], as the decomposition kinetics makes a transition from the growth stage to coarsening after approximately 20 hours of aging at 450°C, whereas above 500°C the growth stage becomes too short to be measureable by the current techniques.

The rate constants obtained from the coarsening data in Fig. 1 can be used to calculate the solute diffusivity according to the Lifshitz-Slyozov-Wagner (LSW) theory provided the interfacial free energy is known. Using the interfacial free energy obtained from the previous XRD work[1], the resulting diffusivity is shown in Fig. 2, together with the XRD values as well as the TEM data from Rastogi and Ardell[7]. Despite the few neutron data points a consistent trend with respect to the XRD results is seen. The activation energy for solute diffusion appears reasonable, and enhanced diffusion due to excess quenched-in vacancies is detected below 500°C. The larger values of the diffusivity obtained from the SANS data probably result from a combination of two factors: the presence of the interference peak and the intrinsic nature of the Guinier approximation, as the Guinier radius gives a weighted average which overemphasizes larger particles in the size distribution.

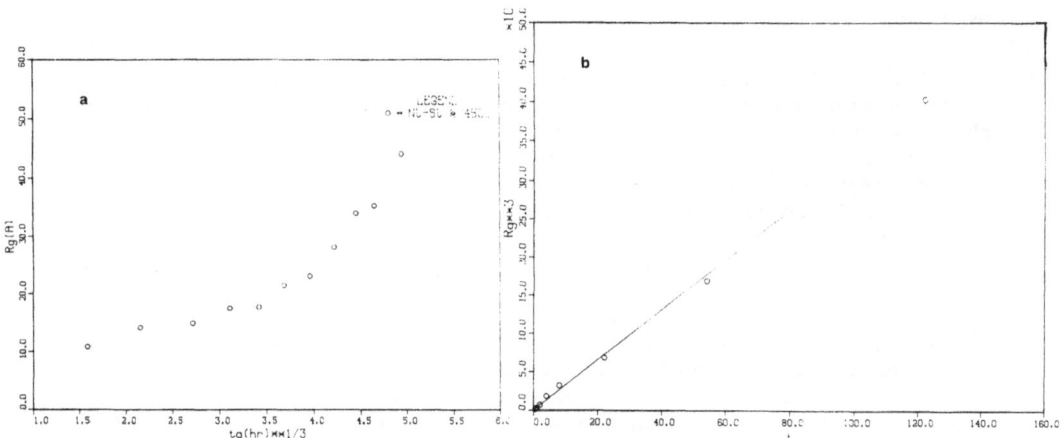

Fig. 1. Guinier radius (Rg) vs. t (hrs)$^{1/3}$ at (a) 450°C and (b) 550°C.

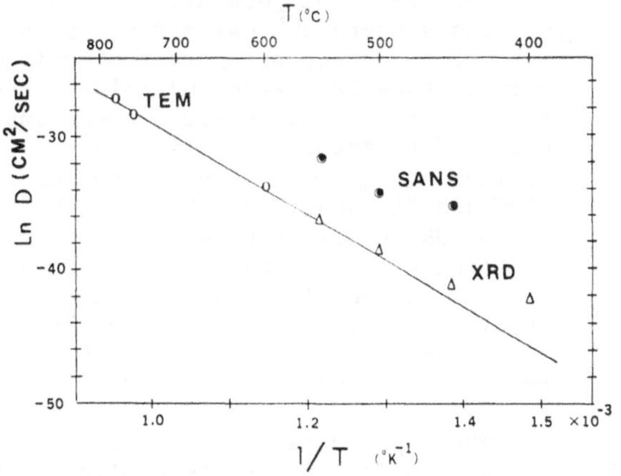

Fig. 2. Comparison of solute
diffusivity obtained
from SANS, XRD and
TEM data.

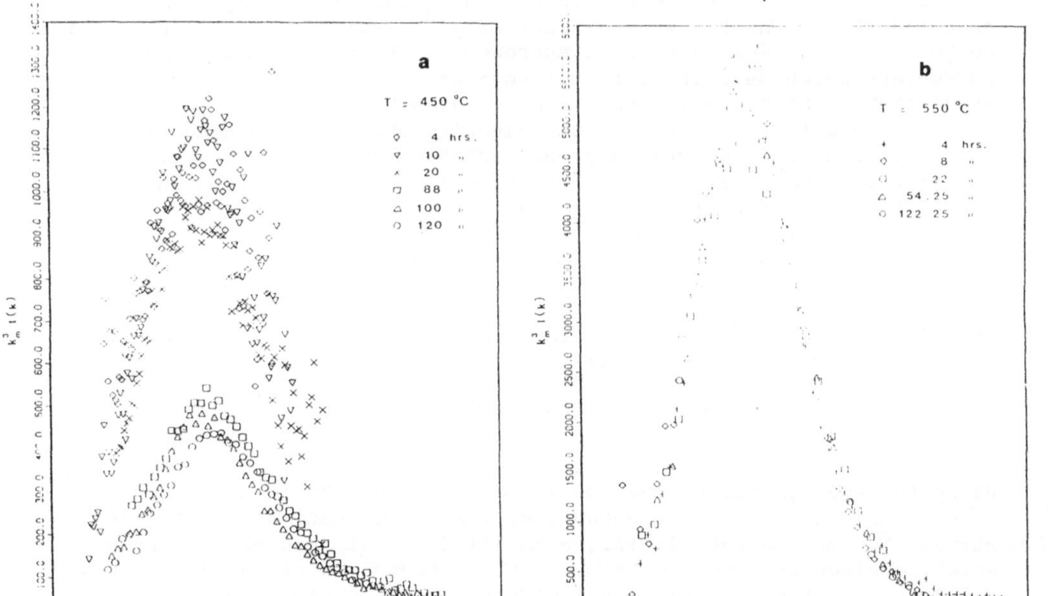

Fig. 3. Furukawa scaling at (a) 450°C and (b) 550°C.

DYNAMICAL SCALING

Dynamical scaling represents a self-similar evolution of decomposition where the configuration of particle distribution remains the same except for a time-dependent characteristic scaling length, $R(t)$, which changes with aging times. In essence the structure function of a decomposing system, which is proportional to the small-angle scattering intensity, may be written as

$$S(k,t) = R(t)^d F(k \cdot R(t)), \qquad (1)$$

where $F(x)$ is the time-independent scaling function and d is the dimensionality. Theoretical and computer simulation approaches[8-10] have been attempted to investigate the dynamical scaling behavior. The proper choice of scaling length and the form of the scaling function have received considerable attention.

Various parameters have been suggested as the scaling length. Furukawa's approach[8] calls for the use of $1/k_m$ where k_m is the wavevector corresponding to the maximum of the interference peak, whereas Lebowitz, Marro and Kalos[9] choose k_1, the first moment of the structure function. Other suggestions include Guinier radius, etc. Testing for dynamical scaling in the Ni-Si alloy can be found in Fig. 3. For long aging times when coarsening is in effect, the scaling is well behaved as it should be, because the scaling law is implicitly contained in the LSW theory of Ostwald ripening. The change in transformation mode from a growth stage to coarsening should also manifest itself in the scaling behavior. It is somewhat surprising, however, to see that dynamical scaling seems to be observed even in the growth stage prior to the crossover regime. The widths of the scaled structure functions remain at a value of 0.74 ($\pm$ 0.01) for all aging conditions considered.

There have been at least three published theoretical predictions[8,10,11] for the explicit form of the time-independent scaled function $F(x)$. The Binder and Stauffer's[10] prediction is too simplified to be accurate. The Furukawa's approach[8] contains many adjustable parameters which make it difficult to judge the significance of the prediction. The phenomenological model of Rikvold and Gunton[11] assumes a gas of spherical droplets of the minority phase, surrounded by depletion zones. It adopts only one variable, the volume fraction (v) of the minority phase; it is thus chosen to fit our SANS data. The explicit scaling function is given as

$$F(x) = C \, \psi^2(Q) \, (1 - \psi(v^{-1/3}Q)), \qquad (2)$$

with $Q = kR$ and $\psi(Q)$ is the Fourier transform of the form factor of a homogeneous sphere of radius R:

$$\psi(Q) = 3 \, (\sin Q - Q\cos Q)/Q^3. \qquad (3)$$

Using the average radius and the volume fraction obtained from our previous XRD data[1], Fig. 4 shows that this model estimates a broader curve. To improve the fitting, a modified formula has been introduced which involves raising the terms in the parenthesis of Eq. 2 to a power P. A much improved fitting is indicated by the dashed line in Fig. 4 when P equals to 1.7. Although this parameter has been adopted before by other researchers[14], justification of this parameter is not at all clear.

248

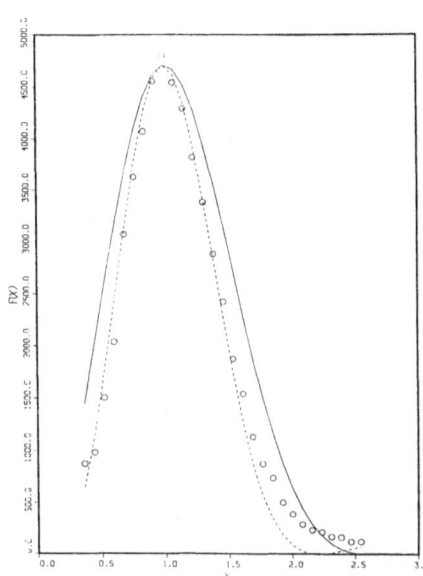

Fig. 4. Rikvold and Gun ton scaling (solid Line) for a sample annealed at 55ºC for 22 h. Dashed line shows modified fitting with P = 1.7.

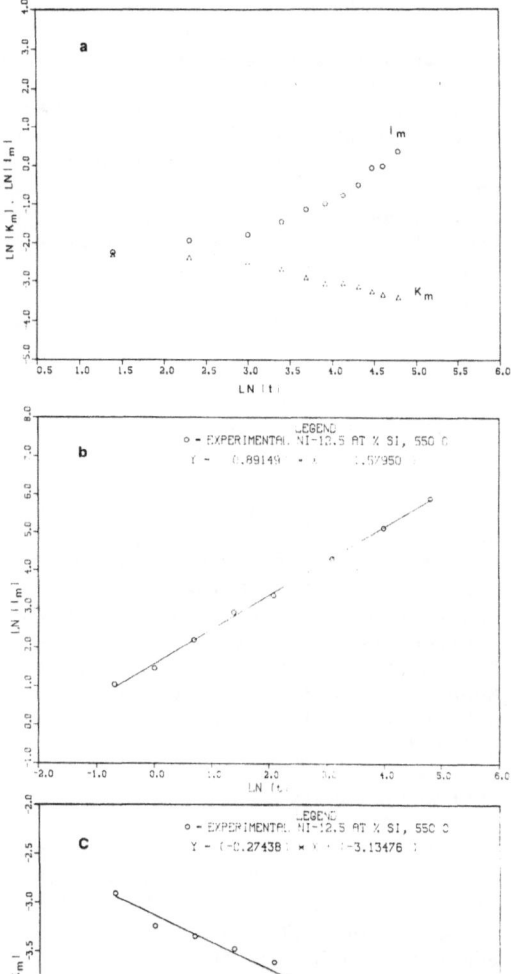

Fig. 4. Log-log plots of SANs peak position and peak intensity vs. time (a) 450ºC and (b),(c) 55ºC.

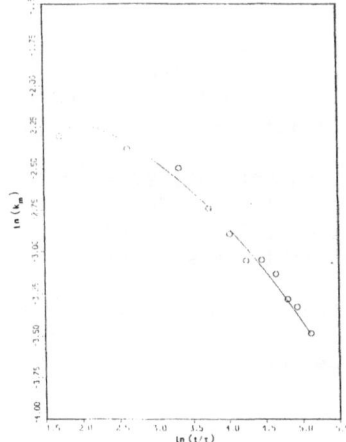

Fig. 6. Fitting to Huse's equation for 450°C data.
$$\ln k_m = -n \ln(t/\tau) + \text{const.}$$
with
$$n = \frac{1}{3} - \frac{A'}{(\frac{t}{\tau})^{1/3} - A' \ln (\frac{t}{\tau})}$$

POWER EXPONENTS

It has been suggested that the growth rate of the characteristic scaling length should follow an asymptotic late time dependence. However, conflicting predictions have recently been put forward concerning the form of this dependence. Furthermore it has been suggested that the intensity maximum of the interference peak should follow a power law at late times. Figure 5 contains log-log plots of k_m as well as $I(k_m)$ versus annealing time. A linear relationship would indicate that power law behavior is followed with the power exponents equal to the slopes as defined in the following equations:

$$k_m \propto t^{-a'} \text{ and } I(k_m) \propto t^{a''}.$$

Such linear relationships are found for the 505 and 550°C data for Ni-12.5 at.% Si, but a break is again seen for the 450°C data in the crossover regime. The least squares fitted values of a' and a" for the late time kinetics can be found in Ref. 12.

The time dependent behavior of k_m for the 450°C data were also analyzed using a modified kinetics formula as shown by Gaulin et al.[14] following the work of Huse[15], which has essentially one adjustable parameter in the fitting function. A good fit can be obtained over the entire range of aging times as shown in Fig. 6. Huse's work has included earlier time corrections to the $t^{1/3}$ coarsening behavior of the LSW theory. It appears such a treatment could generalize the kinetic behavior of a decomposing alloy controlled by nucleation and growth such as this study or by spinodal decomposition such as the work by Gaulin et al.[5,14].

ACKNOWLEDGEMENTS

This research was supported by the Department of Energy under contract No. DE-AC02-76ER01198. The SANS experiments were done at IPNS of ANL under DOE contract No. W-31-109-Eng-38. This manuscript was completed when HC was at ANL as a Faculty Research Participant administered by the ANL-DEP.

REFERENCES

1. S. Polat, M. A. Dvorack and H. Chen, Acta Met. 33:2175 (1985).
2. M. A. Dvorack, J. E. Epperson and H. Chen, Acta Met. 34:117 (1986).
3. M. A. Dvorack, S. Polat and H. Chen, Scripta Met. 18:1395 (1984).
4. S. Polat, C. Marsh, T. Little, C. P. Ju, J. E. Epperson and H. Chen, Scripta Met. 20:1739 (1986).
5. Y. Morii, S. Spooner and H. D. Gaulin, in: "Dynamics of Ordering Process", S. Komura and K. Kawasaki, ed., Plenum, NY (1987).
6. S. Polat, H. Chen and J. E. Epperson, Acta Met. (1987) submitted.
7. P. K. Rastogi and A. J. Ardell, Acta Met. 19:321 (1971).
8. H. Furukawa, Physica, A123:497 (1984).
9. J. L. Lebowitz, J. Marro and M. H. Kalos, Acta Met. 30:297 (1982).
10. K. Binder and D. Stauffer, Phys. Rev. Lett. 33:1006 (1974).
11. P. A. Rikvold and J. D. Gunton, Phys. Rev. Lett. 49:286 (1982).
12. S. Polat, Ph.D. Dissertation, University of Illinois at Urbana-Champaign (1987).
13. A. R. Forouhi, Ph.D. Dissertation, University of California, Berkeley (1982).
14. B. D. Gaulin, S. Spooner and Y. Morii, to be published.
15. D. A. Huse, Phys. Rev. B34:7845 (1986).

VOLUME FRACTION DEPENDENCE OF THE STRUCTURE FUNCTION

IN BINARY SYSTEMS UNDERGOING PHASE SEPARATION

S. Komura and T. Takeda

Faculty of Integrated Arts & Sciences
Hiroshima University
Hiroshima 730, Japan

K. Osamura and K. Okuda

Faculty of Engineering
Kyoto University
Kyoto 606, Japan

INTRODUCTION

Various binary systems, that are initially homogeneous, can be brought by a sudden change of temperature or pressure into an unstable state towards the decomposition of the constituents. The structure function $S(k,t)$, which characterizes the inhomogeneous structure and evolves with real time t, has been found to be expressed by a time-independent scaling function $\widetilde{S}(x)$ in a form

$$S(k,t) \sim k_m^{-d}(t)\, \widetilde{S}(k/\, k_m(t)) \tag{1}$$

in terms of scaling length $k_m^{-1}(t)$ at that time t, where d is the dimensionality of the system[1].

Furukawa has proposed a scaling function[2]

$$\widetilde{S}(x) = 3x^2 /(2 + x^6) \tag{2}$$

for the system with off-critical concentration, i.e. less than 16% for the minority constituent, which forms no percolation but cluster regime. This scaling function has proved to be very successful in interpreting the data of many systems, particularly on Al-Zn systems with concentration around 6-10%[3]. However it has a deficiency that there is no dependence on volume fraction of the constituent, which should certainly be effective for wider range of concentration.

We have undertaken a similar small-angle neutron scattering experiment on Al-Zn binary systems and Al-Zn-Mg pseudo-binary systems as before with volume fraction ϕ of the Zn-rich minority phase ranging from 2.8 to 11.6% and made a cruicial test against the effect of the volume fraction on the scaling function and compared with various existing theoretical predictions.

251

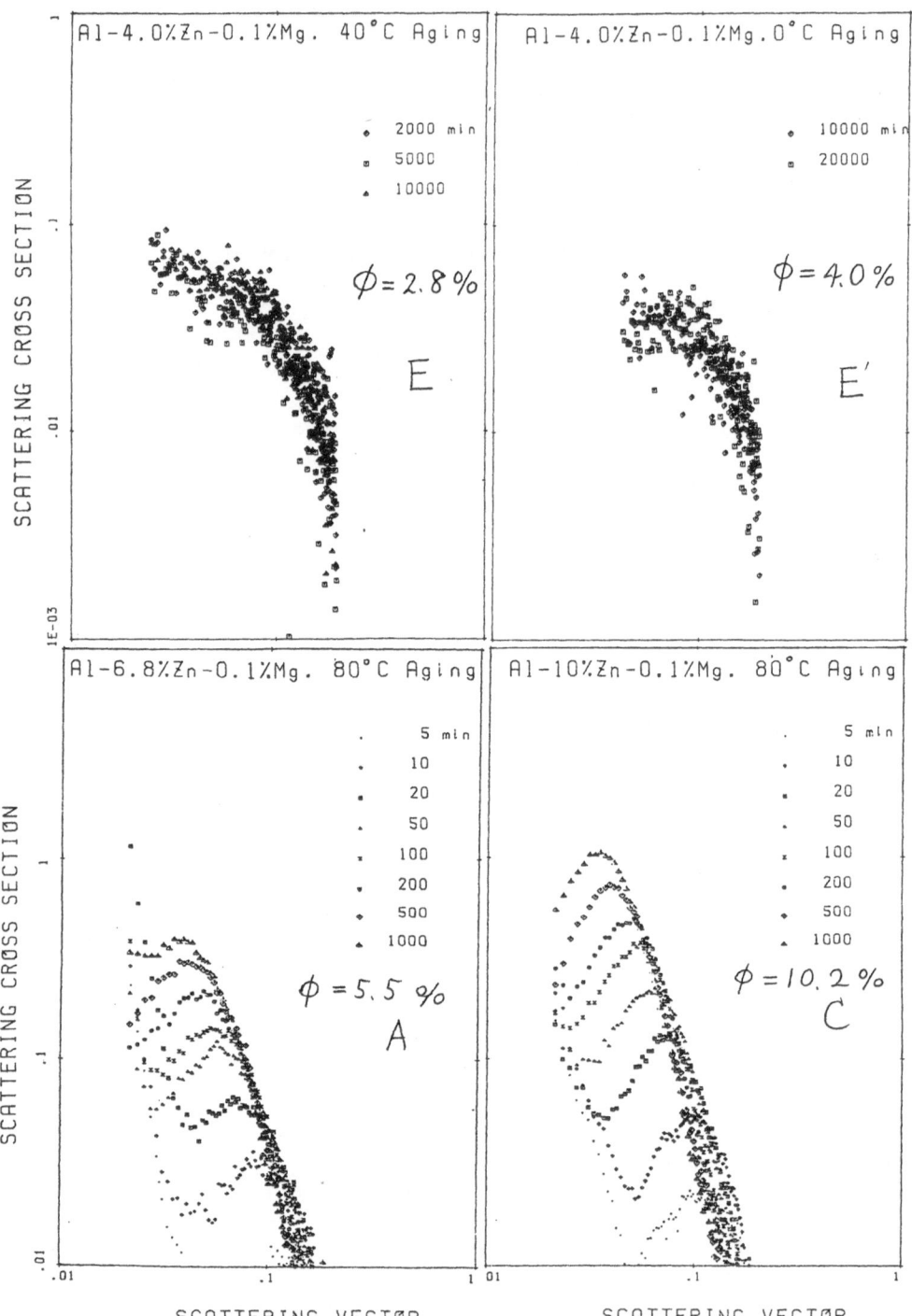

Fig.1. Lograrithmic plot of the scattering cross section $d\Sigma(k,t)/d\Omega$ as a function of the scattering vecter k for ternary Al-Zn-Mg systems with various volume fractions ϕ.

a) sample E (ϕ = 2.8%), b) sample E' (ϕ = 4.0%),
c) sample A (ϕ = 5.5%), d) sample C (ϕ = 10.2%).

EXPERIMENT

The small angle neutron scattering experiment and the sample preparation are similar to the ones described before[3] and are not repeated here. In addition to the eight old samples[3] (A,B,C,D,A',B',C',D'), we have prepared four new samples (E,F,E',F'); these are ternary (pseudo-binary) Al-4 at % Zn-0.1 at % Mg alloy aged at 40° C (E)and 0°C(E') and binary Al-4 at % Zn alloy aged at 40°C(F) and 0°C(F'). For these new samples the aging time t ranged from 2,000 to 20,000 minutes in order to get reasonable structure function S(k,t), which allows us to estimate the relative half width

$$\Delta(t) = \Delta k(t) / k_m(t) \qquad (3)$$

where $\Delta k(t)$ is the full width at half maximum (FWHM) and $k_m(t)$ is the peak position of the structure function S(k,t).

The volume fractions ϕ of the minority phase in these samples were determined by the lever rule in the phase diagram of Al-Zn binary system according to the formula

$$\phi = \frac{C - C_m}{C_p - C_m} \qquad (4)$$

where C is the concentration of the minority atoms in the sample, and C_m and C_p are the equilibrium concentration of the majority and minority phases, respectively, at the aging temperature represented on the binodal line. Therefore the accuracy of the volume fraction depends on the knowledge of the binodal line of the system.

RESULTS

The observed scattering cross sections $d\Sigma(k,t)/d\Omega$ as a function of the scattering vector k are shown for the ternary Al-Zn-Mg alloy system in Fig.1, as expressed in logarithmic scale for both ordinate and abscissa. This way of plotting the data has several advantages; firstly asymptotic laws of S(k,t) for small k and large k are evident; secondly if the scaling holds, the structure functions should be same and coincides with one another by parallel translations; thirdly wide ranges of data are expressible and comparable in a same diagram.

Fig.1 a) represents the data for sample E with ϕ = 2.8%, Fig.1 b) for sample E' with ϕ = 4.0%, Fig.1 c) for sample A with ϕ = 5.5% and Fig.1 d) for sample C with ϕ = 10.2%. It is evident that as the volume fraction increases, the inverse relative width Δ^{-1} grows. The same trends are also observed for the binary Al-Zn alloy system.

From such data we estimated the averages of the inverse relative widths Δ^{-1} for each sample. The results are summarized in Table 1 with errors estimated from standard deviations of various time sequence data for each sample. The inverse relative width Δ^{-1} thus obtained are plotted as a minority phase for both ternary Al-Zn-Mg and binary Al-Zn systems in Fig.2. As the volume fraction decreases, the inverse relative width Δ^{-1} decreases very rapidly below ϕ = 4%.

DISCUSSION

In Fig.2 we have added some other data of the inverse relative width taken from the light scattering in isobutyric acid and water mixtures by

Table 1. Summary of the data on the relative width of the structure function

	Composition Al+		Aging temp. (°C)	Aging time (min.)	Volume fraction ϕ (%)	Inverse relative width $\Delta^{-1}=k_m/\Delta k$
Sample	Zn (at.%)	Mg (at.%)				
E	4.0	0.1	40	2000-10000	2.8	0.433±0.098
E'	4.0	0.1	0	10000-20000	4.0	0.658±0.046
A	6.8	0.1	80	50-500	5.5	0.892±0.005
A'	6.8	0.1	40	50-800	6.8	1.018±0.032
C	10	0.1	80	20-1000	10.2	1.139±0.050
C'	10	0.1	18	10-1000	11.6	1.003±0.097
F	4.0		40	2000	2.8	0.476±0.098
F'	4.0		0	2000-20000	4.0	0.628±0.159
B	6.8		80	5-1000	5.5	0.859±0.005
B'	6.8		40	10-1000	6.8	0.845±0.071
D	10		80	20-1000	10.2	1.065±0.029
D'	10		18	5-1000	11.6	1.080±0.050

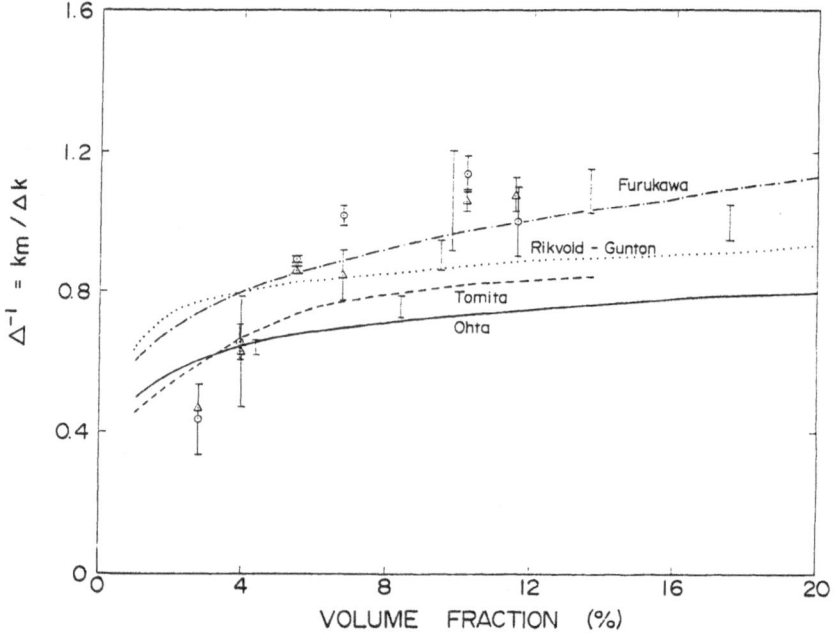

Fig.2. Inverse relative width Δ^{-1} of the structure function in pseudo-binary Al-Zn-Mg system (⊙) and binary Al-Zn system (△) as a function of volume fraction ϕ of the minority phase. Bars without marks are data for isobutyric acid and water system[4] Theoretical predictions by Rikvold-Gunton[5], Ohta[6], Tomita[7] and Furukawa[8] are also included and compared with the experimental results.

Knobler and Wong[4]. These are all in the same trend; the less the volume fraction the less the inverse relative width of the structure function.

In Fig.2 we have included several curves predicted by Rikvold and Gunton[5], Ohta[6], Tomita[7] and Furukawa[8]. All these curves are qualitatively in agreement with the experimental results. However if we look into details there are some systematic deviations of the theoretical curves from the experiment. At large volume fraction (ϕ > 9%) Furukawa's prediction with parameter ϕ_0 = 1/6 fits well the experiment. However at small volume fraction ($\phi \sim 4\%$) Ohta and Tomita's predictions fit well the experiment. At smaller volume fraction (ϕ < 3%) the experimental data decrease much more rapidly than any theories.

CONCLUSION

We have observed the effect of the volume fraction ϕ of the minority phase in pseudo-binary Al-Zn-Mg and binary Al-Zn systems during the phase decomposition to the relative width Δ of the structure function of the systems. The results show that there are general trends that as the volume fraction decreases the relative width grows very rapidly. Qualitatively the results are in agreement with several theoretical predictions. However if we look into details there are systematic deviations of these theories from the experimental results. Particularly the experimental data show more rapid increase of the relative width as the volume fraction decreases below 4%. This fact suggests that at smaller volume fraction the attractive interaction among droplets of minority phase becomes weaker than ever thought; the droplets behaves more or less independently of each other like free molecules in ideal gases.

This experiment was performed at Kyoto University Reactor (KUR). The authors are very grateful to many colleagues at KUR, particularly Professor S. Okamoto, Dr. T. Akiyoshi. The authors wish to thank Professor H. Furukawa at Yamaguchi University for stimulating discussions.

REFERENCES

1. H. Furukawa, Prog. Theor. Phys.,59, 1072 (1978) ; J. L. Lebowitz and M. H. Kalos, Phys. Rev. Lett., 43 282 (1979)
2. H. Furukawa, Physica (Utrecht) 123A, 497 (1984)
3. S. Komura, K. Osamura, H. Fujii and T. Takeda, Phys. Rev. B, 31, 1278 (1985)
4. C. M. Knobler and N. C. Wong, J. Phys. Chem., 85, 1972 (1981)
5. P. A. Rikvold and J. D. Gunton, Phys. Rev. Letters, 49, 286, (1982)
6. T. Ohta, Annals of Phys., 158, 31 (1984)
7. H. Tomita, Progr. Theor. Phys., 71, 1405 (1984)
8. H. Furukawa, Progr. Theor. Phys., 74, 174 (1985)

KINETICS OF ORDERING AND CLUSTER STABILITY :

NON DIFFUSIVE TRANSFORMATION IN A GLASSY CRYSTAL

M. Descamps and C. Caucheteux

Laboratoire de Dynamique des Cristaux Moléculaires
(U.A.801) Université de Lille I
59655 Villeneuve d'Ascq Cedex - France

1 - INTRODUCTION

"Glassy crystals"[1] designate the molecular crystalline solids showing some freezing of the molecular rotations. They are obtained by quenching the equilibrium dynamically disordered phase and investigated as model systems in the study of glasses. Special concerns are the spontaneous relaxations associated to non equilibrium and the mechanism of the glass transition detected by calorimetry at Tg. In spite of active research and newly developed ideas[2,3] these points mostly remain controversal. Since they make it possible to focus mainly on molecular orientations glassy crystals are thus expected to simplify experimental interpretation and lead to reliable theories built on an underlying lattice. A second interesting aspect of these systems that we have developed for the first time on the materials here presented [4,5] – Adamantane derivatives - relates to the kinetics of orientational ordering. Their main feature is that they are of a non conserved order parameter (NCOP). Since they do not involve long range flowing, kinetics of this kind are in general faster than diffusion controlled kinetics[6]. For this reason most of the NCOP systems were studied either in the metastable regime (nucleation and growth) or in the late stage grain growth. As a particular point the early stage of evolution in the regime of unstability is usually too rapid to be carefully followed. In our case the molecular relaxation times after a quench can be very long (minutes to tens of hours) leading to exceptionally slow transformations. In section 2 we present and discuss results of ageing kinetics on a wide interval of temperature showing a very rich behaviour. Around Tg, molecular times are of the same order as any reasonable macroscopic observation. This makes a competition between the rate of reheating and velocity of transformation possible. The result is an extraordinary behaviour which is the structural signature of the glass transition. This is described in section 3.

2 - AGEING KINETICS

We present time-resolved X-ray analysis of single mixed crystals $CN_{1-x}Cl_x$-Adamantane after their quench from room temperature (RT). In all the studied range of compositions (0 < x < .25) initial disordered structures I are isomorphous[7]. They follow the scheme of a f-c-c lattice on which the molecular dipoles randomly tumble among six orientations closely located along the <100> directions. The equilibrium phase diagram is not yet perfectly clarified for values of x slightly larger than 0. For x = 0 the low temperature ordered phase II, stable below 280 K is monoclinic[8]. It does not show simple structural relation with I and the transition is easily by-passed if quenched rapidly enough. Under 180 K the transformation to this phase occurs rapidly and destructively after an incubation time long enough to study ageing kinetics of a metastable phase III. The latter is revealed by the slow growth of superstructure spots at the X boundary points of the Brillouin zone. We were able to follow this with a conventional X-ray source in experimental conditions described previously[5].

At the lower temperatures this transformation implies such infrequent molecular reorientations that it was impossible to detect them by other techniques. Dipoles order locally in an antiparallel way along the four fold axis in a near tetragonal group. The average cubic symmetry is thus achieved by six possible degenerate superlattices along the three<100> directions. When ageing crystals of composition x = .25 no monoclinic recrystallisation is detected whatever the temperature. After low temperature ageing the equilibrium temperature of the first order I-III transition was consistently detected at about T_t = 237K when heating. States related to the X-point superstructure spots could thus be followed in a wider time and temperature range. Below 180K ageing experiments for the different compositions agree and are published in ref.5. Here we present elements of the more extended investigation obtained when using samples of composition x = 0 and x = .25.

Following a deep quench (R.T → T_q < 170 K) the growth of the peak is immediate but very slow and the gradual slowing down of the evolution prevents any reaching of true equilibrium. From the beginning of the evolution, the intensity distribution around the X-position S(q) was found to show dynamical scaling behaviour :

$$S(q,t) = L(t)^z S(q.L(t))$$

Taking the H.W.H.M. of the peak as the inverse of the scaling length gives a universal profile very close to a gaussian. The scaling exponent z is always found to be in a range between 4 and 5. This value which is higher than the dimensionality is consistent with the observed increase of integral intensity and thus of transformed volume. The results show that we are in a regime of unstability which has not reached its late stage. We may retain the picture of ordered microclusters which develop in a disordered matrix. This agrees with an observed bimodal behaviour[9]. Clusters are of small size, typically 40 A in diameter (4 lattice parameters) at the end of the experiments. One of the salient features of the kinetics is their very weak time dependence. The peak parameters follow power-law (t^m) dependences shown in Fig.1, with exponents which are rather insensitive to the ageing temperature. The following values have been found for x = 0 during a 7 day experiment at 157K : $m \cong 0.34$ for the peak intensity and $m' \cong 0.08$ for the typical size of ordered regions. This indicates an increase which is even slower than that expected (m' = .25) when soft walls form[10]. The strong constraints acting against the expansion may have several origins : - elastic interactions between clusters - feedback effect of the orientational freezing induced by the clusters in the disordered matrix. A further effect, coroborated by recent simulations[11] can be that of steric incompatibility between ordered clusters of different orientations. Finally the kinetics follow a very clear Arrhenius dependence on temperature which shows that the influence of the driving force of the transformation is counterbalanced by the molecular mobility : 60 hours are needed to multiply the peak intensity by 3 at 156K whereas 2 hours only are needed at 169K (x = 0).

More shallow quenchs (R.T → 170K < T_q < T_t) are followed by an incubation stage pointing to a metastability. A typical S-shaped evolution of the peak is shown in Fig.2. In opposition to the preceeding regime it cannot be fully described by universal scaling ; this may be due to the variety of occuring regimes. The following stages of evolution may be recognized : at the beginning the incubation period during which the superstructure peaks are non existent or weak but stationary. This period is followed by a net increase of the peak intensity, of the cluster size and the integrated intensity as well. These curves show a distinct last part characterized by very sluggish kinetics where the effective cluster diameter does not change much more and has been found to be an increasing function of the ageing

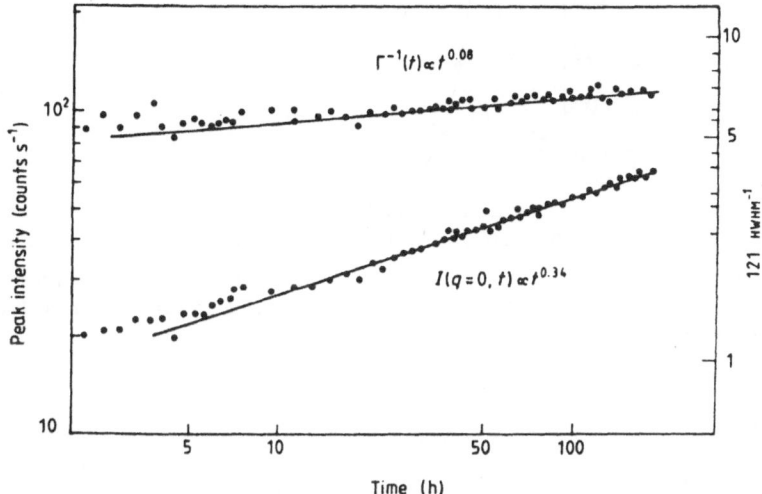

Fig.1 : Log–Log plot of the peak intensity and inverse width Γ^{-1} versus time in the lower temperature regime ($T = 156K$, $x = 0$)

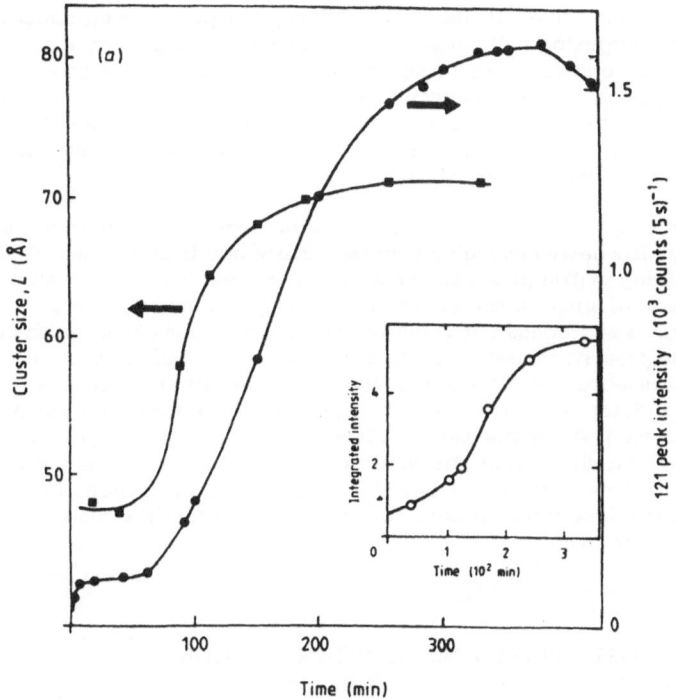

Fig.2 : The typical sigmoïdal behaviour of the growth curve in the higher temperature regime ($T = 177K$, $x = 0$)

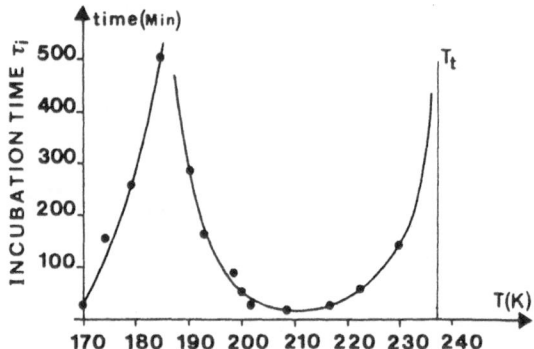

Fig.3 : Variation with temperature of the incubation time τ_i in the higher temperature regime. The left part [170K < T < 185K] marks a gradual transition from
metastability to unstability.

temperature [$\sim$ 55 A at 170K, $\sim$ 100 A at 195K]. One can however notice that the
cluster size reaches this final regime well before the intensity own. This could
indicate that nucleation is predominant and occurs only when clusters have reached
a minimum size. Fig.3 shows the variation with temperature of the incubation time
τ_i for 170K < T < T_t. The most striking result seen in this figure is the existence
of two zones on both sides of about 185K. The right part can be understood as resulting from the competing influences on nucleation of the driving force of the transition and the frequencies of individual molecular rotations. For temperatures lower
than 185K, τ_i decreases with decreasing temperatures. At the same time the growth
curve becomes more and more dissymetric so that it joins the low temperature
behaviour at 170K. This left part of Fig.3 thus apparently defines a transition zone
between metastability and unstability.

The underlying lattice shows a parallel time dependence. It was systematically
monitored by alternatly recording superstructure and Bragg peaks. Ageing induces
a gradual shifting of the peaks towards higher angles. It reveals a slow volume relaxation of the sample on time scales which, under 170K, are of several tens of hours.
This behaviour is analogous with spontaneous relaxations observed in conventional
glasses. In the present model system, it can be clearly associated with the development of local orientational ordering. Another effect of the latter is to progressively
induce lattice distortions. They lead to a peak broadening which increases with
order of reflection. At higher temperatures and in the late stage, we have noticed a
small Q-dependent shifting of the superstructure spots away from exact X-commen-
surate position. This breaking of translational symmetry reveals, in particular, two
different slightly incommensurable environments namely the disordered matrix and
near tetragonal embryos.

3 - CLUSTER UNSTABILITY AND GLASS TRANSITION

During the ordering process, perturbations were applied to the system. They
revealed an extraordinary behaviour which is concerned with cluster stability. The
reactions of the systems appear in the changes of a superstructure peak shown in
Fig.4-a. They are also sketched by the broken path in the T T T diagram of Fig.4-b.
After a pre-ageing stage at low temperature [1] the evolution of the crystal was
followed during a rapid increase of temperature [2]. This increase induces the
complete disappearance of clusters and the recovery of the homogeneous metastable

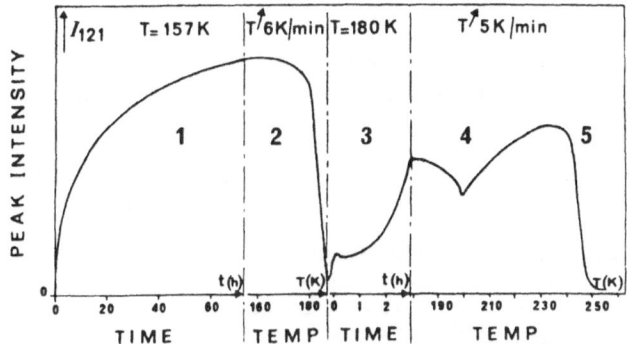

Fig.4-a : Evolutions of the intensity of the superstructure peak 121 during several stages of ageing and reheating (x = .25). The drop of intensity shown in (2) is the structural signature of the glass transition at T_g

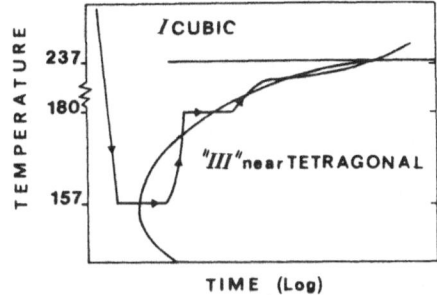

Fig.4-b : TTT sketch of the transformations shown in Fig.4-a

phase I. This reversion typically occurs near 183K, about 50° below the equilibrium transition. It is clearly the structural signature of the glass like endothermic anomaly[11]. After the temperature step and under re-ageing (3), ordering kinetics which correspond to the same superstructure are observed anew. Kinetics are this time preceded by incubation which is standard at this temperature. The temperature of reversion was observed to vary with the previous ageing history. It is apparently an increasing function of the initial cluster size, thus, in practice of the ageing temperature. In this way, upon heating (4), partial reversion could be observed by chance between 190K and 200K. Equilibrium transition is seen by the definitive drop of intensity (5).

This apparent leaving of and returning to metastability poses a lot of questions. The most simple explanation which needs further theoretical studies is that of the competition between the rate of reheating and the kinetic ability of clusters to grow to the higher critical size at the new temperature. The other pressing problem is to more clearly determine the implication of the local orientational ordering in the mechanism of the glass transition.

References

1. SUGA.H (1983) Pure Appl.Chem. 55 427
2. PALMER.R.G., STEIN.D.L., ABRAHAMS.E, ANDERSEN.P.W (1984) Phys.R.Lett 18 214
3. FREDRICKSON.G.H, ANDERSEN.H.C (1985) J.Chem.Phys. 83 11, 1 5822
4. DESCAMPS.M, ODOU.G, CAUCHETEUX.C (1985) J.Phys.Lett 46 L261
5. DESCAMPS.M, CAUCHETEUX.C (1987) J.Phys.C : Solid State Phys. To be published
6. AXE.J.D (1985) Japan.J.Appl.Phys.Suppl 24 46
7. AMOUREUX.J.P, SAUVAJOL.J.L, BEE.M (1981) Acta Crystallogr A 37 97
8. FOULON.M, AMOUREUX.J.P, SAUVAJOL.J.L, CAVROT.J.P, MULLER.M (1984) J.Phys.C : Solid State Phys 17 4213
9. ROLAND.J.P, SAUVAJOL.J.L J.Phys.C : Solid State Phys (1986) 19 3475
10. MOURITSEN.O.G (1986) Phys.Rev.Lett 56 850
11. NAUDTS.J, DESCAMPS.M, To be published
12. FOULON.M, AMOUREUX.J.P, SAUVAJOL.J.L., LEFEBVRE.J., DESCAMPS.M. (1983) J.Phys.C : Solid State Phys 16 L265

CLUSTER SHAPE CHANGES DURING LOW TEMPERATURE ANNEALING

Colin G. Windsor, Ruth M. Barron(a) and John R. Russell.(b)

Materials Physics and Metallurgy Division
Harwell Laboratory, OX11 ORA, UK.
(a) Present address: Merton College, Oxford, UK.
(b) Present address: Imperial College, London, UK.

INTRODUCTION

Recent *in situ* small angle neutron scattering experiments have been performed during the thermal ageing of an evaporated Al–7.5w%Cr–1w%Fe alloy.[1] The sequence of scattering patterns observed suggested that radical changes in cluster shape were occuring during ageing. This paper describes the Monte Carlo simulations on model binary alloys which confirmed this interpretation of the data. An explanation is given in terms of the minimum coordination number of atoms in linear, planar and spherical clusters. New simulations are presented on the formation of $L1_2$ structure clusters in a face centred cubic lattice.

THE SMALL ANGLE SCATTERING EXPERIMENT

Small Angle Neutron Scattering (SANS) permits the size, shape and volume fraction of precipitate clusters to be measured during ageing.[2] The good penetration of neutron beams through most materials enables bulk specimens to be used. It also permits samples to be mounted in a furnace so that precipitate properties can be followed *in situ* during ageing.

Cluster shape may be inferred from the slope of the *log-log* plot of the logarithm of the scattering cross section $d\Sigma/d\Omega$ against the logarithm of the scattering vector $Q = 4\pi\sin\theta/\lambda$, where 2θ is the scattering angle and λ is the neutron wavelength. Figure 1 shows this plot for various ellipsoidal shapes whose cross section may be evaluated analytically.[3] The dotted curve for randomly oriented cigar-shaped ellipsoids shows a –1 slope, the dashed curve for discus-shaped ellipsoids shows a –2 slope, while the full curve for spheres shows the –4 slope of the *Porod law*[2] for the scattering exponent of three dimensional structures at large Q. Note that all shapes show the –4 law at Q values large compared with their inverse size.

Figure 2 shows log-log plots of the small angle neutron scattering results from the evaporated aluminium alloy Al–7.5w%Cr–1w%Fe. Evaporated alloys are unique in that, since they are prepared by condensation of the vapour onto a cooled substrate, they have never been subjected to quenching from high temperatures. The first stages of ageing may therefore be examined. This particular alloy was evaporated on to an aluminium substrate held at 260C.[4] The three different symbols in figure 2 show results at three different ageing temperatures. Their superposition by appropriate scaling of the time gives a unique ageing sequence, covering the equivalent of four decades of real time, to be explained.

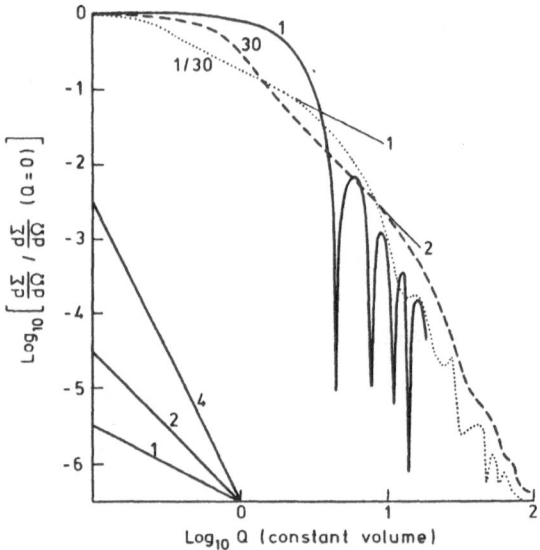

Fig. 1. A log-log representation of the scattering from spheres(solid line), discus-shaped oblate ellipsoids (dashed line) and cigar-shaped prolate ellipsoids (dotted line). The ellipsoids have major axes 30 times the minor axes, and have a constant volume equal to that of a sphere of unit radius.

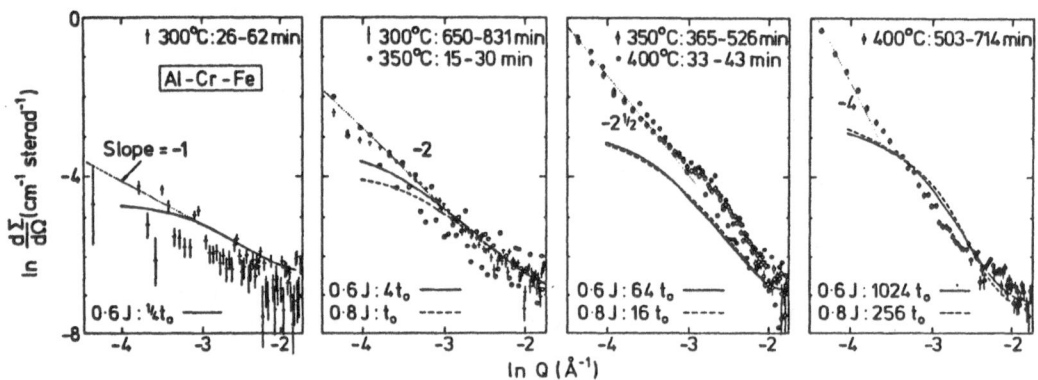

Fig. 2. The small angle neutron scattering macroscopic cross section plotted against the scattering vector Q on logarithmic scales. The points represent measurements at three temperatures over the time intervals shown. The dotted lines indicate the changing exponent of the Q variation. The dashed and full lines represent Monte Carlo computations at two temperatures for the times shown.

THE MODEL ALLOY

The most striking feature to be explained in the experimental results is the change in slope from an initial value close to −1 to a final value close to −4. However these cannot be immediately identified as having been caused by cluster shape changes. Any arbitary distribution, measured over a limited Q range, can always be interpreted in terms of a size distribution of spherical defects.[5] Monte–Carlo computations were therefore undertaken in order to confirm the slope changes as being caused by an evolution of cluster shape.

The initial computations[1] assumed a model binary alloy $A_f B_{1-f}$ with an atomic fraction $f=0.1$ of

A atoms distributed over a simple cubic lattice. Attractive interactions were assumed so that clustering of *A* atoms occured during ageing. The computations reported here have the face centred cubic (fcc) lattice of aluminium, and have repulsive first neighbour interactions, combined with attractive second neighbour interactions, to give the *anti-clustering* $L1_2$ phase.

The Metropolis method for following the approach to thermal equilibrium of a system with defined interactions and temperature is now well established,[6] and its application to alloys widespread.[7] The method used is to choose a random *A* atom within the lattice, then choose a random near neighbour site occupied by a *B* atom, which may be considered for a possible exchange site for the original *A* atom.[8] Having defined the interaction energies between *A* atoms, it is possible to evaluate the energy of the A atom in both its original and its possible new sites, and so calculate the difference ΔE. The exchange proceeds if the $\Delta E < 0$ so that the new energy is lower. It may also proceed if $e^{-\Delta E/kT} \geq R$ where R is a random number between 0 and 1. This allows thermal excitation to proceed towards the Boltzmann distribution. The rate of one trial exchange per atom in the lattice defines the time unit t_0 of the computation. It can be related to the real time through the diffusion constant. The present model is able to treat simple, body centered, and face centered lattices, with interactions to up to four distinct shells of neighbours. All possible exponentials $e^{-\Delta E/kT}$ are evaluated and stored at the start of each run. By using bit addressing and assembler coding on the 68008 processor of a Sinclair QL micro computer it was possible to watch the complete segregation of a $32 \times 16 \times 16$ atom array during a run of two days.[9] During each run the arrays were stored at logarithmic intervals of time, increasing by a factor of two from $\frac{1}{8}t_0$ to $32768t_0$ for analysis. This speed has allowed a large number of computations to be performed investigating the segregation as a function of the important parameters of temperature, time, atomic fraction, lattice structure and interaction strengths. This paper will report mainly the results on the face-centred $L1_2$ structure anti-cluster arrays with a 5% atomic fraction.

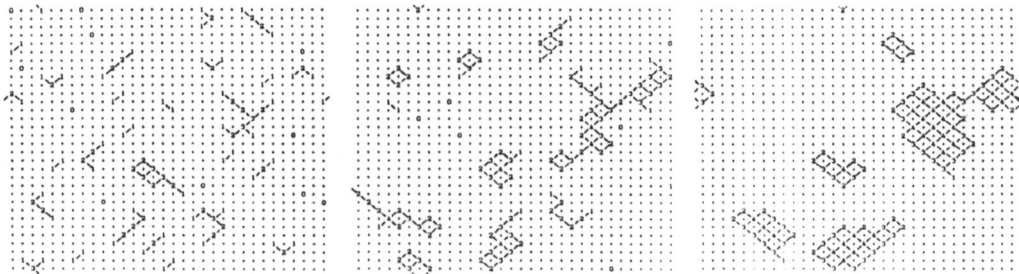

Fig. 3.   The ordering of a two-dimensional array with repulsive first neighbour interactions and attractive 2nd neighbour interaction. The temperature is defined by $kT = -0.25J_1 = 0.25J_2$. Lines join attractive bonds, and define the clusters. The numbers show the coordination number of each atom. The times correspond to $8t_0$, $512t_0$, and $32768t_0$.

ANALYSIS INTO DISTINCT CLUSTERS

Extra insight into cluster growth during ageing is provided by classifying the stored arrays into distinct clusters. In the case of attractive near-neighbour interactions, a cluster is readily defined by the presence of a near-neighbour bond.[6] In the case of arrays with both attractive and repulsive interactions, the best definition for a cluster is less clearcut. We define a distinct cluster as one whose atoms are joined by any non-zero interaction. Figure 3 illustrates this for a 32×32 two-dimensional array with a repulsive first neighbour interaction $2J_1$ and an attractive second neighbour interaction $2J_2$. The lattices are shown at time intervals of a factor of 64. Lines show the attractive bonds which join clusters. The numbers at *A* atom positions show the coordination number of each atom.

This series shows many of the features seen in the three dimensional ageing into anti-clusters.

a) By a time of order $8t_0$, initial clusters containing repulsive bonds are likely to be broken, leaving clusters with essentially only attractive second-neighbour bonds.

b) During an aggregation phase, lasting till around $64t_0$, the mobile non-bonded atoms collect into clusters. At temperatures well below the equilibrium ordering transformation, these clusters often have a diffuse character with many atoms having only a low coordination number.

c) During a compaction phase, the clusters concentrate into a more rounded shape with a higher proportion of atoms having a high coordination number.

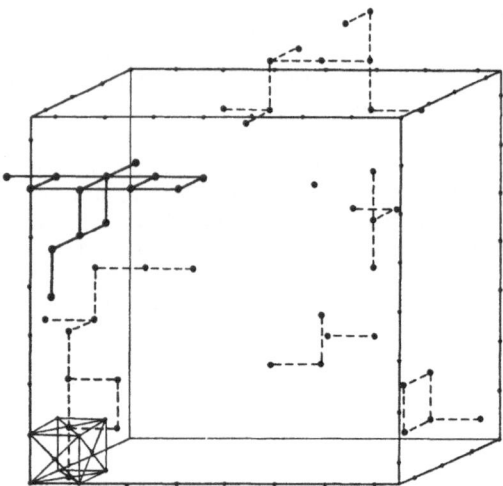

Fig. 4.   Distinct clusters having the $L1_2$ structure growing in a fcc lattice. The lines connect the attractive second neighbour bonds. The small cube shows the underlying fcc structure. The time corresponds to $128t_0$. The growth of the cluster shown with full lines is followed as a function of time in figure 10.

$L1_2$–TYPE ORDERING IN A FACE CENTRED CUBIC ARRAY

Electron, neutron and X–ray diffraction have given some information on the nature of its precipitate phase in the present alloy. At the later stages of ageing, it is certainly the monoclinic θ phase $CrAl_7$. The structure of this phase is noted for the presence of many Al–Cr–Al bonds.[10] At earlier stages the structure is not yet known, but appears to be commensurate with the lattice. In order to obtain an ordering pattern in a fcc lattice with many A-B-A bonds, the model was set up with a repulsive interaction with the 12 nearest neighbours, and an attractive interaction, equal in magnitude, with the 6 second neighbours along the axis of the unit cell. This *ad hoc* choice of the interactions has the merit that at a temperature below about $kT=-0.9J_1=0.9J_2$ the array orders into clusters having A atoms on the corners of the unit cell, with B atoms on the face-centered positions– the $L1_2$ ordering phase. This phase is well known as the γ' phase in the nickel-based superalloys,[11] and as the δ' phase in aluminium lithium alloys. Since there are three face-centred positions to every corner position, the clusters have the formula AB_3. Each face-centered atom also lies on a cubic array, so that the A atoms may lie on any of four interpenetrating cubic lattices. Figure 4 illustrates these clusters in a three dimensional fcc array. The array has a temperature of $kT=-0.2J_1=0.2J_2$, where the clusters show the diffuse structure characteristic of low temperature aggregation.

Having labelled all the clusters in the stored arrays, it is straightforward to evaluate the mean

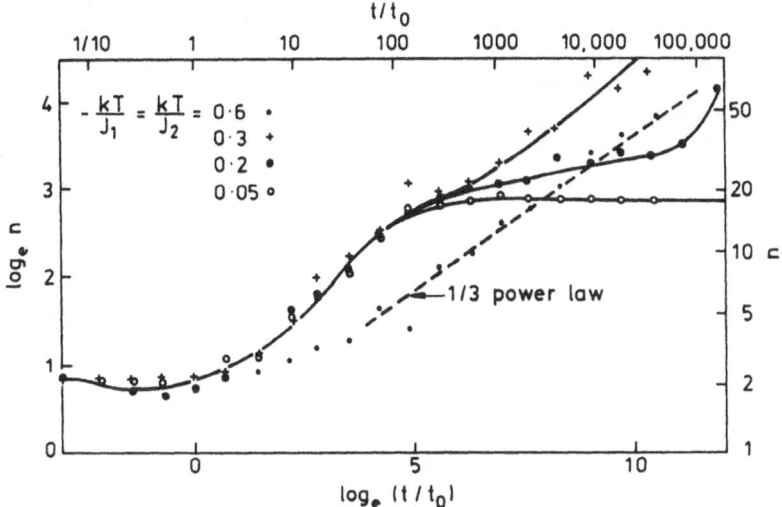

Fig. 5. The root-mean square size of $L1_2$ precipitates in a fcc lattice as a function of time and temperature. Both scales are logarithmic. The full lines are guides to the eye. The dashed line shows the $\frac{1}{3}$ power law of the theory of Lifshitz and Slyozov,[12] which operates at higher temperatures.

cluster size, and cluster size distribution. Figure 5 shows the root-mean-square size of $L1_2$ structure clusters in a face-centred cubic lattice. The points on the y-axis correspond to the initial random array. At temperatures well below the ordering temperature, the probability of evaporation of any atom which has joined a cluster is low. The growth curve during aggregation is therefore almost independent of temperature. This is not true at $kT = -0.6J_1 = 0.6J_2$ when low coordination atoms evaporate readily and nucleation of a more compact cluster must precede growth. Later at lower temperatures the cluster size increases relatively slowly with time.

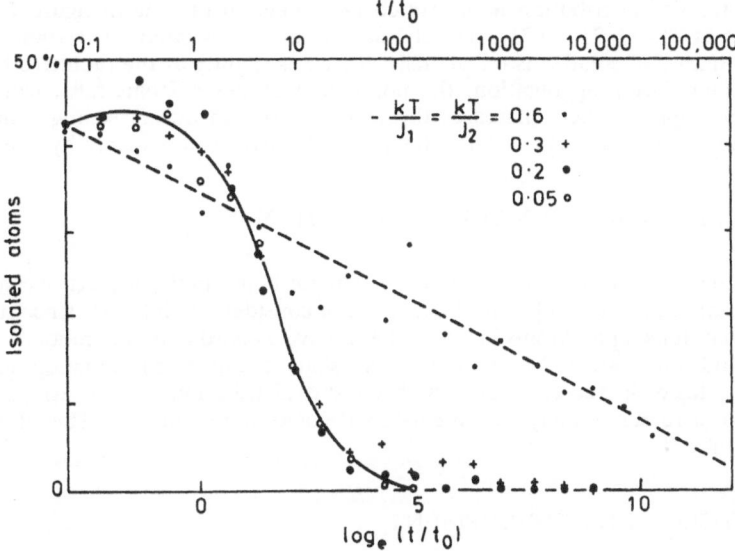

Fig. 6. The percentage of isolated atoms, not bonded to any cluster, as a function of time and temperature during ageing. The full and dashed lines are guides to the eye.

The population of the isolated single atoms is of importance in ageing, since these are the mobile atoms capable of redistributing the cluster size distribution. The temperature dependence of the isolated atom population is shown in figure 6 as a function of time. The initial fall off is very similar for a range of temperatures below $kT=-0.3J_1=0.3J_2$, however the higher temperatures in this range maintain a finite population of isolated atoms at all times. At still higher temperatures the isolated atom population drops much more smoothly, and is always appreciable.

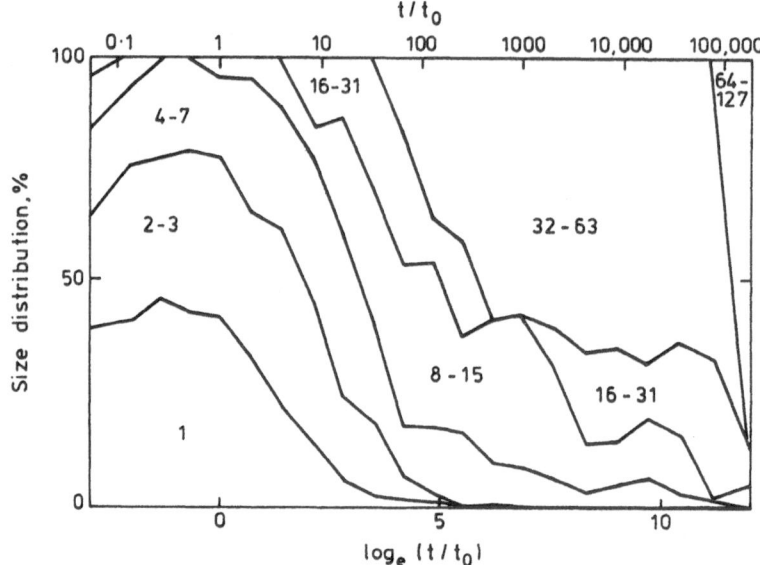

Fig. 7. The size distribution of the $L1_2$ structure precipitates in a fcc lattice as a function of time during ageing. The temperature is defined by $kT=-0.2J_0=0.2J_1$. The lines divide the total number of atoms to groups depending on the number of A atoms in the cluster shown.

The cluster size distribution is illustrated as a function of time in figure 7 for $L1_2$ structure precipitates at $kT=-0.2J_0=0.2J_1$. Initially most atoms are isolated, or in small clusters. At early times the number of small clusters actually decreases slightly as the repulsive bonds break some clusters. Later during aggregation, the population of single atoms falls, while the number of doublet and triplet clusters increases through a maximum during segregation. Larger clusters follow a similar process, but at a later time, until the array is dominated by a few large clusters.

THE COORDINATION NUMBER DISTRIBUTION

The number n of attractive next-nearest neighbour bonds in this fcc lattice can vary from 0 to 6. The distribution of this number in the array is of considerable interest, since it is able to reflect changes in cluster shape. Atoms in linear chains have a coordination number of 1 or 2, planes one of 2 to 4, and cubes one of 3 to 6. Within each shape the mean coordination number rises as the size of the cluster increases. Figure 8 shows this distribution for the arrays of figure 7. The distribution increases steadily as the clusters become more compact. This effect has been seen experimentally.[13]

TRACING CLUSTER DEVELOPMENT

As well as examining the statistical properties of clusters, it is possible, having labelled them, to trace the development of individual clusters. The factor-two time ratio between stored clusters

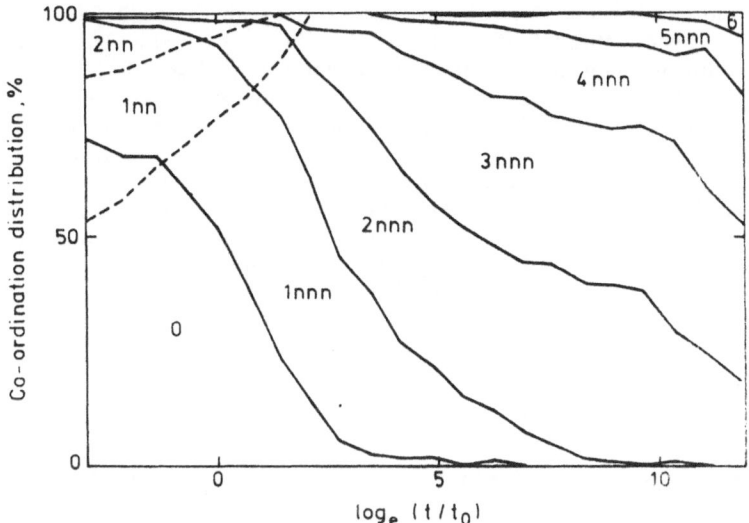

Fig. 8. The distribution of the coordination number of atoms within L1₂ clusters as a function of time during ageing. The arrays are the same as in figure 7. The lines divide the total number of atoms to groups depending on the size range given. Full lines correspond to the attractive second-neighbour interactions (nnn), dashed lines to the repulsive nearest neighbours (nn).

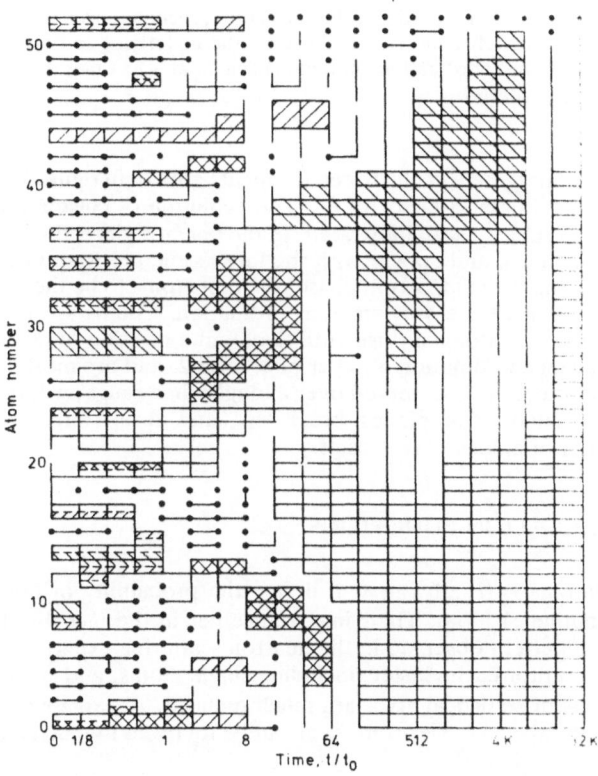

Fig. 9. The development of L1₂ precipitate structure in a fcc lattice as a function of time during ageing. The vertical lines define the cluster sizes. The horizontal lines show the number of atoms common between clusters at adjacent times. The four types of shading define clusters belonging to the four possible sub-lattices.

was chosen with this in mind. A particular cluster can usually be readily identified with its parent and daughter clusters through its centre of gravity in the array. However this method is predictably difficult when clusters of nearly equal size coalesce during a time interval. An automatic procedure for tracing cluster growth was therefore developed, depending on the number of common sites occupied by cluster atoms in stored arrays adjacent in time. Figure 9 shows an example of cluster tracing for a 16×16×8 fcc lattice showing L1$_2$ ordering. The horizontal axis is the logarithm of time, the vertical lines denote the size of each cluster, and the horizontal lines denote the number of overlapping atoms. The vertical ordering of the clusters is arbitary. The early times show the aggegration of the A atoms into numerous small clusters. However the L1$_2$ structure clusters on a fcc lattice may lie on any of 4 sublattices. Clusters on differing sublattices may not merge, so that it is necessary for many clusters to evaporate completely.

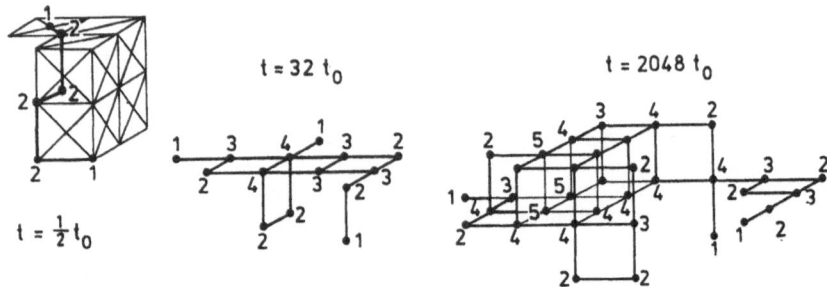

Fig. 10.   The development of a particular L1$_2$ structure precipitate as a function of time during ageing. The cubes in the left hand cluster illustrate the underlying fcc lattice. The numbers at atomic positions give the coordination number of each atom. The times (t) represent the number of trial exchanges per site(t$_0$).

The largest cluster in figure 9 is illustrated during its growth in figure 10. The parts of the figure are again at time intervals of a factor of 64. The evolution of cluster geometry, from linear, to square, and finally to cubic, is well illustrated. In the linear regime, the clusters have not had time to optimise their positions, and appear with the loose structure found by random aggregation of diffusing atoms. During the compaction phase, the first atoms to be thermally excited are those at the ends of chains, which have only single coordination. Chains are thus unstable and tend to break up. The evaporated atoms diffuse until they find a position of higher coordination. Atoms in planar arrays have a coordination number of at least 2, and are most likely to form. At a later stage, atoms of coordination 2, observed over a longer time range, will themselves become likely to be thermally excited. The planes break up, and three dimensional arrays with their coordination number of at least three result.

CLUSTER CORRELATION FUNCTIONS

The pair correlations $g_T(r)$ of a system define the probability of observing an A atom at a distance r from another A atom. Their importance lies in the fact that their Fourier transform gives the diffraction pattern observed in the neutron scattering experiment. By evaluating *cluster* pair correlations $g_C(r)$ for each cluster, including single atoms, and summing over all clusters in the array, the effects of statistical error are much reduced. At large r the cluster correlations go smoothly to zero, and their Fourier transform can be followed to much larger Q values.

Figure 11(a) shows the cluster correlation functions for the fcc arrays as a function of time at a temperature of kT=−0.2J$_1$=0.2J$_2$. The first neighbour correlation starts at the atomic fraction f=0.05, and decays in around a time 8t$_0$. The 2nd neighbour correlation starts at a similar value, and increases rapidly as aggregation occurs. Further neighbour correlations may be divided into

270

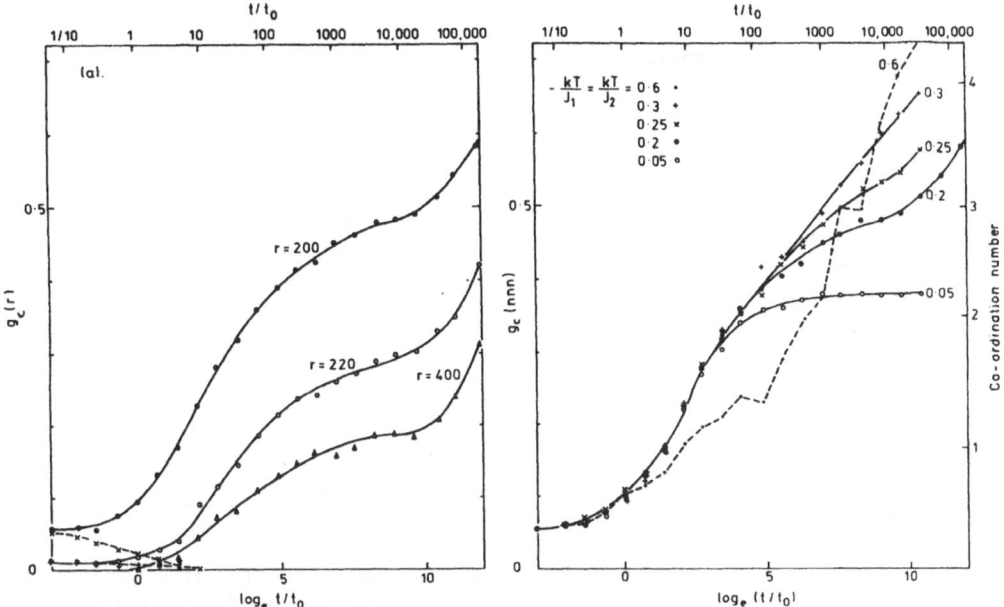

Fig. 11. The cluster correlations for L1₂ stucture clusters in a fcc lattice as a function of time. The left hand figure shows the correlations for several neighbour distances. The right hand figure shows the next nearest neighbour correlation at several temperatures.

those present in the $L1_2$ structure, and those absent. Absent neighbours such as those with Miller indices 110, or 211 decay to zero. Others rise at successively later times, as the size of the cluster increases. The temperature dependence of the cluster correlations is illustrated in figure 11(b) for the case of the attractive 2nd neighbour correlations. The right hand scale expresses this correlation in terms of the mean coordination number. At a temperature low compared with the interactions, say $kT=-0.05J_1 =0.05J_2$, the correlations flatten out, since an atom once joined to a cluster is unlikely ever to be evaporated off. At a temperature $kT=-0.2J_1=0.2J_2$, the clusters aggregate at a similar rate, since evaporation remains unlikely. However thermally activated evaporation eventually becomes significant, and leads to cluster ripening at a rate depending exponentially on the temperature. At temperatures nearer the equilibrium ordering transition, the thermally activated evaporation of atoms with a low coordination number becomes probable, and the more diffuse clusters are unstable. The situation then corresponds to Ostwald ripening. The correlations initially rise more slowly since many small clusters evaporate. Later, due to the statistical emergence of a *seed* cluster, the correlations grow much more rapidly by condensation onto its surface from the continuous flux of diffusing atoms.

SCATTERING SIMULATIONS

By Fourier transforming the cluster correlations, it is possible to obtain the neutron scattering cross section. Interference effects are lost, but none were seen in the experiment. The transforms were shown in figure 2 for attractive near neighbour clusters on a simple cubic lattice. Figure 12 shows a log-log plot of the scattering from the $L1_2$ ordered clusters in a fcc lattice with $kT=-0.3J_1=0.3J_2$. The changing slope is once again clearly seen, as observed in our[1] and earlier experiments,[14] although a precise quantitative fit has yet to be found. We may conclude that the cluster shape changes, and the corresponding changes in slope in the scattering data are not dependent on the details of the computational model.

271

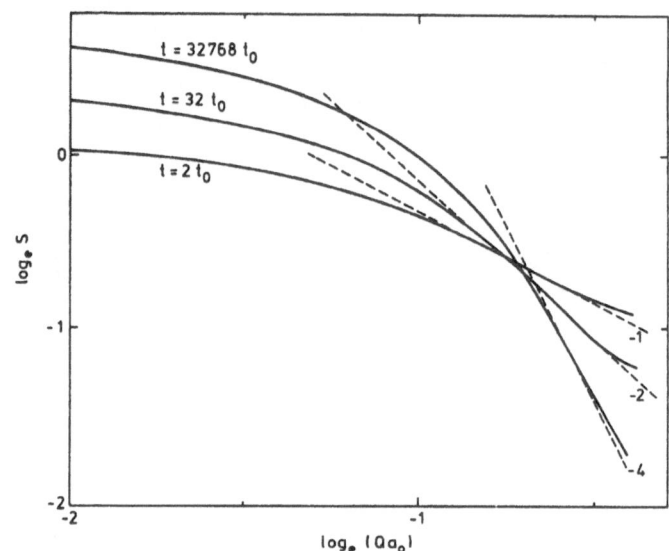

Fig. 12.   The scattering function plotted against the scattering function Q, obtained by Fourier transformation of the cluster correlations. The lines show the predicted scattering from $L1_2$ structure clusters in a fcc lattice at several times. The dotted lines show the slopes indicated. a_0 is the lattice parameter.

ACKNOWLEDGEMENT

This work was undertaken as part of the underlying research programme of the United Kingdom Atomic Energy Authority. We should like to thank Gerhard Inden for his helpful comments.

REFERENCES

1.   C. G. Windsor, R. N. Sinclair, V. S. Rainey, B. Normand, and A. W. Bowen, J. Phys. F. **10,** L229 (1987).
2.   G. Kostortz, in Neutron Scattering, 'Treatise on Materials Science and Technology' Vol 15, Academic Press, New York (1974), p 227.
3.   M. T. Hutchings and C. G. Windsor, Chapter 22, in 'Methods of Experimental Physics', Vol 3, Academic Press, New York, (1987).
4.   V. G. Rivlin and G. V. Raynor, Int. Metals Reviews, No. 4 (1980); R. L. Bickerdike, D. Clark, J. N. Eastabrook, G. Hughes, W. N. Mair, P. G. Partridge and H. C. Ranson Inter. J. of Rapid Solidification **1,** 305 (1985); **2,** 1 (1986); Int. Conf. Rapidly Solidified Alloys, 137 (1986); R. L. Bickerdike, D. Clark, G. Hughes, M.C. McConnell, W. N. Mair, P.G. Partridge and B.W. Viney, Int. Conf. Rapidly Solidified Alloys, 145 (1986).
5.   J. A. Potton, G. J. Daniell and D. Melville, J. Phys D: Applied Physics, **17,** 1567 (1984).
6.   A. Sur, J. L. Lebowitz, J. Marro, and M. H. Kalos, Phys. Rev. B **15,** 3014 (1977).
7.   J. L. Lebowitz, J. Marro, and M. H. Kalos, Comments Solid State Physics, **10,** 201 (1983).
8.   N. Metropolis, A. W. Rosenbluth, M. N. Rosenbluth A. H. Teller and E. Teller, J. Chem. Phys. **21,** 1087 (1953).
9.   C. G. Windsor, Phys. Bull. **38,** 182 (1987).
10.   M. J. Cooper, Acta. Cryst. **13,** 257 (1960).
11.   W. Betteridge and J. Hislop, 'The Nimonic Alloys,' Edward Arnold (London), (1981).
12.   I. M. Lifshitz and V. V. Slyozov, J. Phys. Chem. Solids, **19,** 35 (1961); M. Marder, Phys. Rev. Lett. **55,** 2953·(1985).
13.   O. Blaschko, G. Ernst, P. Fratzl, M. Bernole and P. Auger, Acta. Metall. **30,** 547 (1982).
14.   M. Furusaka, Y. Ishikawa, S. Yamaguchi and Y. Fujino, Phys. Rev. Lett. **54,** 2611 (1985).

DYNAMICAL STRUCTURE CHANGE DURING REVERSION IN AL-ZN ALLOYS

Kozo Osamura[1], Hiroshi Okuda[1]
Yoshiyuki Amemiya[2] and Hiroo Hashizume[3]

1 Department of Metallurgy, Kyoto University
  Sakyo-ku, Kyoto 606, Japan
2 Photon Factory, National Laboratory for High-Energy
  Physics, Tsukuba-gun, Oho-machi, Ibaraki 305, Japan
3 Research Laboratory for Engineering Materials, Tokyo
  Institute of Technology, Nagatsuda, Midori-ku, Yokohama
  227, Japan

INTRODUCTION

It is well known that metastable G.P. zones dissolve partially or completely into matrix when the sample is held at higher temperatures than the pre-aged temperature. This process is called reversion and has been investigated from the metallurgical viewpoint[1]. The kinetic and structural behaviours of this process were discussed by several authors[2~4]. On the reversion in Al-Zn binary system, a few experiments by means of small angle scattering measurments have been perfomed[5~8]. In general, the process has finished so rapidly that the precise structure change is difficult to be detected and therfore there still exsit many discussions to make clear the whole picture of reversion process.

In the present paper, the structure chnage during reversion process for Al-6.8at%Zn and 15at%Zn alloys has been investigated by means of in-situ synchrotron -radiation small-angle scattering(SR-SAS) measurements. An analytical model describing the collapse of zone during reversion has been proposed. The experimental results have been also compared with the numerical features calculated using kinetic Ising model.

EXPERIMENTAL PROCEDURE

The alloys used here were Al-6.8at%Zn and 15at%Zn, prepared in foil shape with about 0.15-0.2 mm in thickness. After the solution treatment at 573 K for 3.6 ks, the specimen was quenched into brine and then pre-aged at 313 K for 60 ks to form metastable G.P. zones. The reversion treatment was done at temperatures as shown in Fig 1, where the temperature range is conventionally divided into two parts by the zone solvus. The measurements were performed using time-resolved SAS apparatus installed at BL-15A in Photon Factory, National Laboratory for High Energy Physics, KEK. The details were described elsewhere[8]. According to the ordinary procedure, the counting rate taken at each scattering vector k (=$4\pi\sin\theta/\lambda$) was transformed into the coherent scattering intensities I(k), after correction with

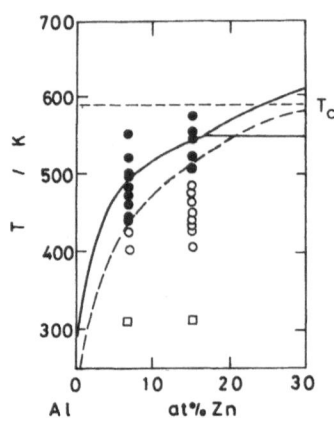

Fig.1 Phase diagram of Al-Zn binary system[10]. Circles indicate the reversion temperatures adopted here.

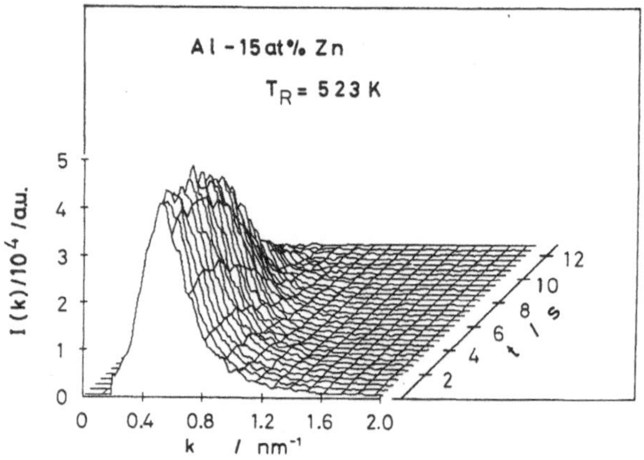

Fig.2 In-situ SR-SAS diagram for Al-15at% Zn alloy during reversion at 523 K after pre-aging at 313 K for 60 ks.

respect to the incident beam intensity, sample thickness, absorption, background and fluorescence.

EXPERIMENTAL RESULTS AND DISCUSSION

The time evolution of coherent scattering intensity profile for Al-15at%Zn alloy reverted at 523 K is shown in Fig. 2, where each intensity profile was the result of integration for 0.2 s. It is seen that the scattering intensity decreases rapidly with time, and vanishes completely within about 12 s, indicating that G.P. zones dissolve completely into the matrix. The behaviour of reversion process is expected to be somewhat different for temperatures above and below the zone solvus. Therefore the experimental results will be separately described as given below.

Reversion Above The Zone Solvus

As shown in Fig. 3, the integrated intensities during reversion for Al-6.8 at%Zn alloy are plotted as a function of time, which was measured from the start pulse of the data colletion system. When the reversion temperature became higher, the integated intensity decreased more rapidly and the dissolution of G.P. zone completed in the shorter time. It should be noted that the reaction period, in which the dissolution is completed, is very short in comparison with the results for the dilute Al-Zn alloy[7]. In order to study the kinetics of dissolution, it is important to estimate the time-lag caused by sample drop into the furnace and the subsequent transient time during which the sample was heated up, which was estimated to be about 0.5 to 2 s for the present experimental condition[9]. Hereafter, the corrected time, which is the time subracted by the time-lag from the real time, is used. The time dependence of the radius of gyration R_g for Al-6.8at%Zn alloy is shown in Fig. 4. It was found that the radius remains constant at first, and after a certain critical time Δt_c, it increases rapidly as indicated by the arrow in the figure. This critical time became shorter with increasing reversion temperature, as shown in Fig. 5. In this figure, the values obtained from the change of other structure parameters are also displayed. One of them is the time, at which the integrated intensity decreased by 20 %. Another is the time deduced from the change of interparticle distance as shown in Fig. 6. From the time depedence of these parameters , the reversion process seems to be divided into two stages as the feature of each stage is listed in Table 1.

Table 1　Change of structure parameters during reversion
above the zone solvus

Parameter	Early Stage	Later Stage
Integrated Intensity, Q	$Q/Q_o > 80$ %	$Q/Q_o < 80$ %
Radius of Gyration, R_g	almost constant	increase
Interparticle Distance, L	almost constant	increase
Electron Density Difference, $\Delta\rho$	decrease	(decrease)
Volume Fraction, V_f	almost constant	(not defined)

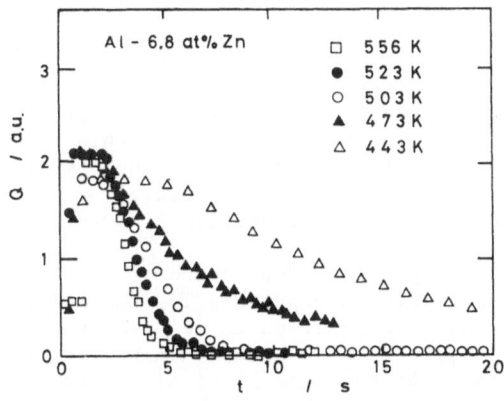

Fig. 3　Change of the integrated
intensity as a function of time.

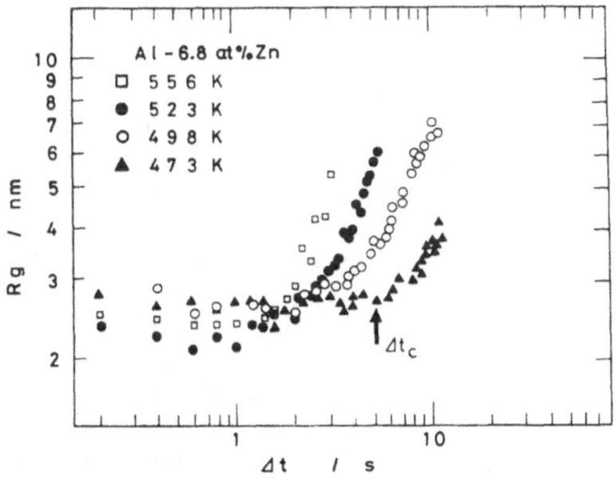

Fig.4　Change of radius of gyration
as a function of the corrected time.

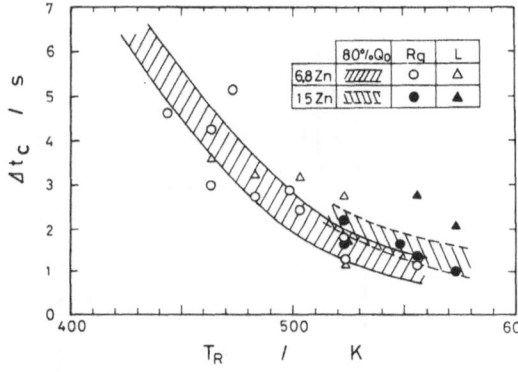

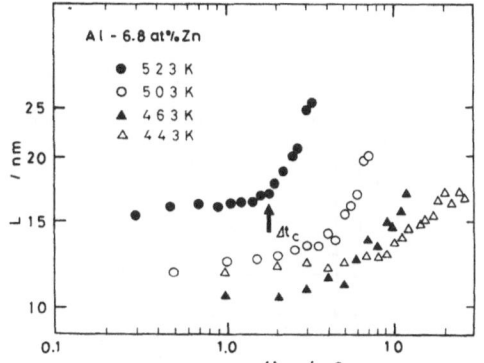

Fig.5　Reversion temperature dependence of the critical time.

Fig.6　Change of interparticle
distance as a function of time.

　　　The structure change at the first stage is characterized as mentioned
below. Supposing the two-phase model, the integrated intensity is
expressed by the equation,

$$Q(t) = \Delta\rho^2 V_f (1 - V_f),\qquad(1)$$

where $\Delta\rho$ is the electron density difference inside and outside the zones
and V_f is the volume fraction of zones. The latter quantity is related
with R_g and L as expressed by the equation,

$$V_f = (4\pi/3)\sqrt{2} (R_g/L)^3.\qquad(2)$$

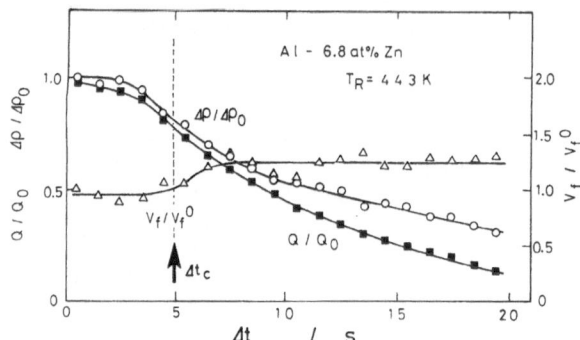

Fig.7 Relative change of structure parameters during reversion.

As the parameters, Q, R_g and L can be experimentally determined, the volume fraction and then the electron density difference are able to be estimated from Eqs. (1) and (2). For instance, the change of structure parameters for Al-6.8at%Zn alloy reverted at 443 K is shown in Fig. 7. Whithin the time of 5 sec, it is found that the volume fraction did not change, while the quantity $\Delta\rho/\Delta\rho_0$ decreased suggesting the decrease of solute concentration inside the zone. This behaviour can be explained as follows.

Fig. 8(a) shows schematically the composition dependence of free energy in this alloy system. At the pre-aging temeprature T_1, the alloy decomposed into two phases, where the zone composition is denoted by $c_p(T_1)$ and the concentration of matrix by $c_m(T_1)$. When the sample is heated up abruptly to a temeprature T_2, which is lower than the critical temperature T_c, the alloy composition c_o comes to lie within the one-phase region. Therefore the zone becomes unstable and tends to dissolve to the matrix.

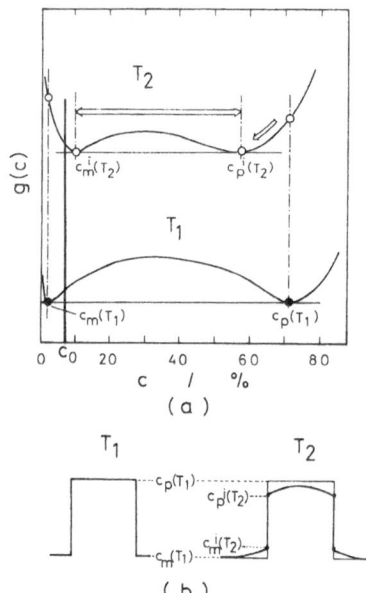

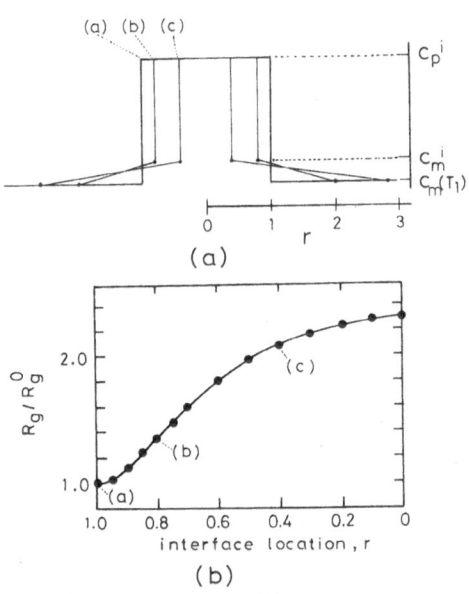

Fig.8(a) The free-energy composition curve at pre-aging temperature T_1 and reversion one T_2. c_o is the alloy composition.
(b) Composition profiles around a zone.

Fig.9(a) Composition profile around a zone. When time proceeds, the location of interface shifts inwards. (b) The corresponding change of radius of gyration as a function of location of interface.

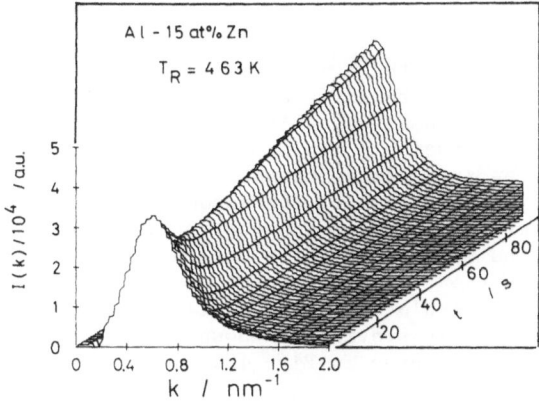

Fig.10 A typical result of SAS intensity during reversion below the zone solvus.

During this process, it is reasonable to assume the local equilibrium, by which both ccompositions of the zone and the matrix at the interface, $c_p^i(T_2)$ and $c_m^i(T_2)$ are fixed from the free energy-composition curve at T_2. For instance, at the reversion temperature of 483 K, the composition difference $c_p^i - c_m^i$ is estimated to be about 50 at%Zn from the coexistence curve, while this difference is 69 at%Zn at the pre-ageing temperature. Consequently the change of concentration inside and around the zone due to the abrupt change of temperature is schematically displayyed in Fig. 8(b). Still at the temperature of T_2, the interface is well defined. This means that the Porod law holds at high temperatures.

In the successive period, that is, the later stage, the observed integrated intensity decreased continuously as shown in Fig.7. Exceeding the time Δt_g, both quantities R_g and L increased rapidly. The former change can be explained in terms of the diffuse tail of concentration profile outside the zone/matrix interface as follows. A model for the concentration profile around a spherical particle is displayed in Fig. 9(a). The radius of gyration for the zone with diffuse concentration profile outside the interface was calculated and its reduced value is shown in Fig. 9(b). When the location of the interface reduces, the radius of gyration is found to increase aapparently. This behaviour explains well the observation shown in Fig. 4. During the process of shrinking zone, the zone with smaller size dissolves faster. Then the average interparticle distance increases, corresponding to the fact as shown in Fig. 6.

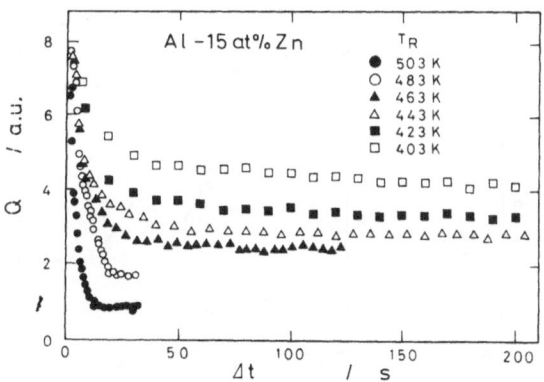

Fig.11 Change of the integrated intensity as a function of time during reversion below the zone solvus.

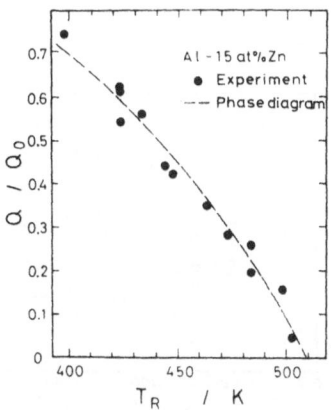

Fig.12 Reversion temperature dependence of the saturated integrated intensity.

277

Reversion Below The Zone Solvus

When the pre-aged specimen is hold at the temperature below the zone solvus, the reversion behaviour becomes quite different. A typical SAS diagram is shown in Fig. 10. During the holding at 463 K, the intensity maximum decreases at first, but after passing a minimum, it begins to increase. The time dependence of the integrated intensity is shown in Fig. 11. The intensity decreases fastly within several ten seconds, and decreases gradually to reach a saturated value at respective temperature. Its saturated value is larger for the specimen reverted at the lower temperature. It is plotted as a function of temperature as shown in Fig. 12. The integrated intensity can be calculated using Eq. (1), when the metastable phase diagram for G.P. zones is previously known[10]. The calculated result is shown by the dotted curve in Fig. 12.

During reversion, R_g does not change at the first period, but increases remarkably beyond a critical time as shown in Fig. 13. As a result, the first decrease of the integrated intensity corresponds to the decrease of the electron density difference as shown in Fig. 14. On the contrary of the result shown in Fig. 7, the integrated intensity does not decrease in the successive period, but both R_g and L increase. Therefore this later period is characterized as a coarsening process of zones survived at the reversion temperature. As a summary, the structure change during reversion below the zone solvus is terminated into three stages as shown in Fig. 15. In the first two stages, the zones formed at the pre-aged temperature are partially dissolved. The behaviour is similar to that for the reversion above the zone solvus. At the later stage, that is, at the third stage, the coarsening of the zones occurs.

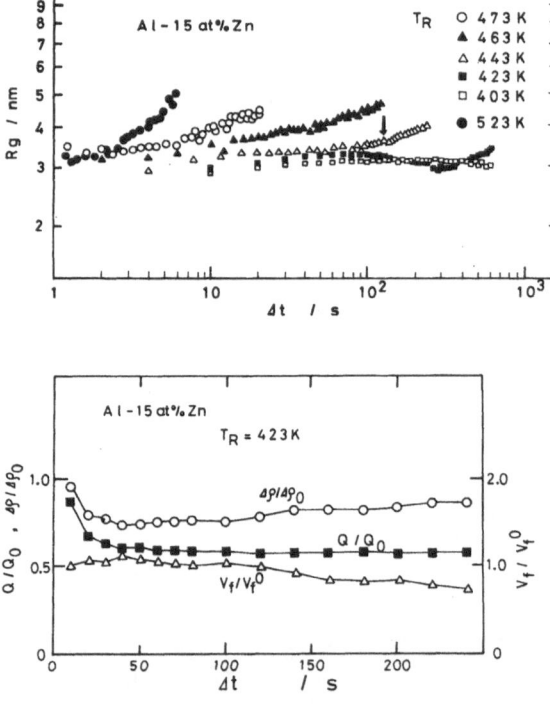

Fig.14 Time dependence of the relative values of structure parameters.

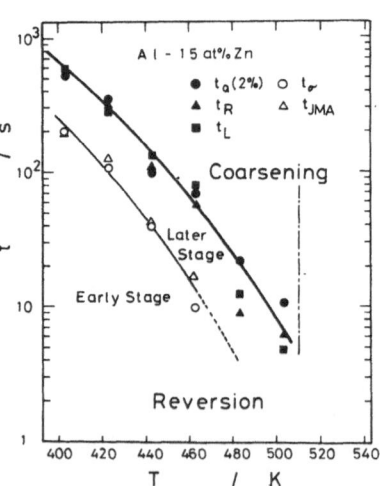

Fig.13 Time dependence of radius of gyration.

Fig.15 Sequence of structure change during reversion below the zone solvus.

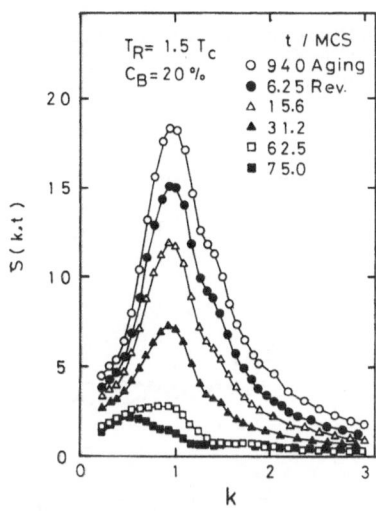

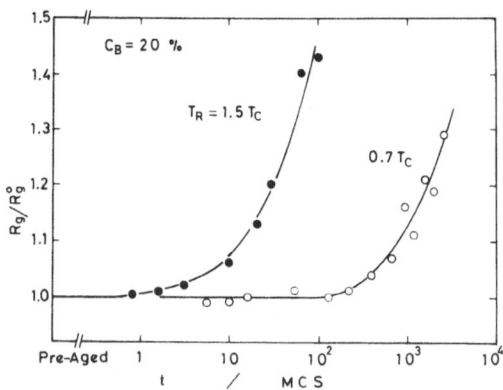

Fig.16 Change of structure
function during reversion in
the Ising system.

Fig.17 Relative change of radius of
gyration as a function of time during
reversion.

Computer Simulation Of Reversion Process

Reversion phenomena was investigated on the FCC Ising lattice. The
simulation procedure was mentioned in details elsewhere[11]. Dimension of
the system was 30 unit cells for each side, and periodic boundary condition
was used. Temperature is measured by the unit of critical temperature T_c,
and length is measured by the unit cell of FCC lattice. The system was
pre-aged for about 1000 MCS at 0.5 T_c. Structure function $S(k,t)$ for the
reversion at 1.5T_c is shown in Fig. 16. It is seen that the structure
function decreases monotonously to vanish. The position of k_m was nearly
constant at first, but shifted to the lower k in the later stage. This
suggests that the average interparticle distance increases, consisting with
the experimental result as shown in Fig. 6. Change of the radius during
reversion at 1.5 T_c and 0.7T_c for the alloy with C_B= 20% is shown in Fig.
17. For the reversion above the phase boundary, the radius remained
constant in the early stage, but exceeding about 80 MCS, it began to
increase. In the case of reversion below the phase boundary, the radius
increased slightly in the first stage and exceeding about 350 MCS, the
radius began to increase rapidly. The onset of this increase of the radius
was coincident with the time when the coarsening process just starts.
These behaviours are very similar to the observations in the reversion
above and below the zone solvus as mentioned here.

In order to make clear the reversion behaviour for the individual
precipitate, the reversion process of a single spherical particle was
examined. In this case, as an initial condition, a spherical particle
with C_p = 100% and radius of 4.5, 6 and 9 was used. Change of the system
with initial radius of R_o = 9 at 0.9T_c is shown as a set of snapshots on
(111) cross section in Fig. 18. Here circles denote the solute atoms and
the solid line denotes the boundary of the particle determined according
to the rule, by which the two third of nearest neighbours are occupied by
solute atoms. In the early stage, the size of the particle was unchanged,
and the concentration inside the particle was gradually decreased. And
after the concentration of the precipitate saturated, it began to shrink.
The relative changes of concentration and the volume of particle during
reversion at 0.8T_c is shown in Fig. 19. This sequence was found to be the
same for all the initial radii examined here. In the later stage where the

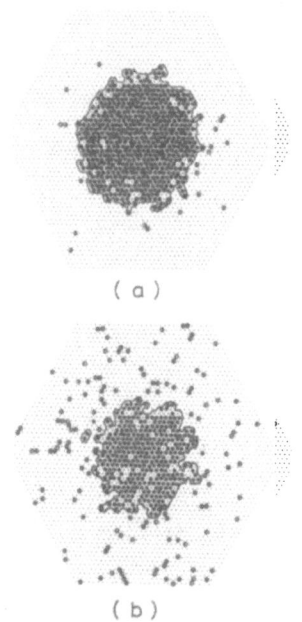

(a)

(b)

Fig.18 Snapshot of single particle on (111) cross section of FCC Ising lattice during reversion. (a) 50 MCS, and (b)2520 MCS.

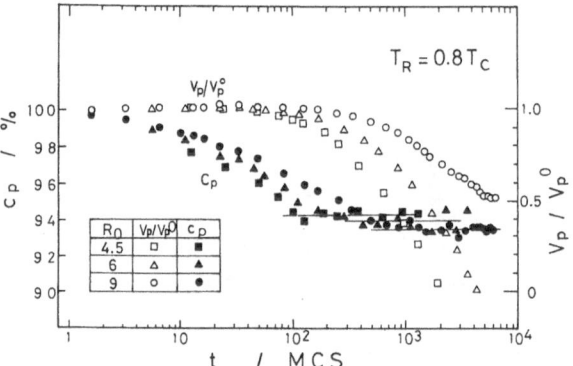

Fig.19 Temporal evolution of concentration and volume of single particle during reversion at $0.8T_c$.

volume of particle decreases, the particle with the smaller initial radius completely disssolved into the matrix and disappeared. In contrast, the particle with the largest radius decreased its volume, but even exceeding 4000 MCS, it did not disappear. From the above computer simulation, the reversion is divided into two stages, regardless of the situation whether the reversion occured inside or outside the phase boundary . The early stage is characterized as the stage of decreasing concentration inside the zone. The later stage is the process of decreasing volume.

CONCLUSION

When the pre-aged Al-Zn alloy was held at high temperatures above the zone solvus, the reversion process could be apparently distinguished into two stages. On the other hand, the coarsning process is followed when it is reverted below the zone solvus. It is made clear that kinetic Ising model can qualitatively well reproduce the structure change during reversion in Al-Zn system.

Acknowledgements: The authors acknowledge the partial financial support of Light Metal Education Foundation Inc. The present work was fulfilled by a Grant in Aid for Scientific Research (Project No. 60460202) of the Ministry of Education, Science and Culture of Japan.

REFERENCES

1. U.Dehlinger and H. Knapp, Z.Metallkde., 43:223 (1952).;V.Merz and V.Gerold, Z.Metallkde., 57:275 (1966).
2. F.V.Nolfi, P.G.Schewmon and J.S.Foster, Trans. AIME, 245:1427 (1969).
3. M.J.Whelan, J. Metal Sci., 3:95 (1969).
4. K.Tsumuraya, Acta Metall., 30:19 (1982).
5. D.Allen, S.Messoloas and R.J.Stewart, J.Appl. Cryst., 11:578 (1978).
6. H.Löffler, R.Kroggel and O.Kabisch, Krist. und Tech., 14:933 (1979).
7. K.Gerstenberg and V.Gerold, Crystal Res. and Technol., 20:79 (1985).
8. K.Osamura,H.Okuda,H.Hashizume and Y.Amemiya, Acta Metall,33:2199(1985)
9. H.Okuda, K.Osamura and Y.Amemiya, Acta Metall., in press (1987).
10. J.Lasĕk, Czechoslov. J. Phys., 15:848 (1965).
11. H.Okuda and K.Osamura, Trans. JIM, submitted (1987).

THE VERY EARLY STAGE OF PHASE SEPARATION PROCESS IN Al-Li ALLOYS STUDIED

BY SMALL-ANGLE NEUTRON SCATTERING

Michihiro Furusaka*, Shin-ichiro Fujikawa**

Masamichi Sakauchi** and Ken-ichi Hirano**

* Department of Physics, ** Department of Materials Science
Tohoku University, Sendai 980, Miyagi, Japan

INTRODUCTION

The phase separation process has been studied extensively, because it is a good example of non-equilibrium phenomena. Recent neutron scattering experiments in Fe-Cr and Al-Zn alloys have revealed a universal behavior that there are two stages in the process.[1,2] The late stage of the phase separation process is defined by the q^{-4} dependences of the scattering function $S(q,t)$ on the higher q side than the peak position. It has been well described by the theory of the kinetics of interfaces between matrix and precipitates. At the late stage, the scaling function proposed by Furukawa[3] holds as follows,

$$F(q/q_1(t)) = q_1^3(t) \cdot \bar{S}(q,t), \cdots \cdots \cdots (1)$$

where $q_1(t)$ is the first moment of $\bar{S}(q,t)$ is a scattering function normalized by second moment of $S(q,t)$ itself. This scaling is supported by many other experimental results.[4,5,6]

The early stage of the phase separation process is also defined by the q^{-2} dependence of the $S(q,t)$ on the higher q side than the peak position. This behavior is well explained by the theory of Langer, Bar-on and Miller[7] which takes into account nonlinearity and thermal fluctuations, but it can not be explained by the linear theory of Cahn.[8] It is still an open question as to what is happening at the very early stage. Some people predict that the linear behavior proposed by Cahn would be found if one could measure an early enough stage. Another question arises as to whether the process described above is the only possible universal behavior.

Recently, Al-Li based alloys have gathered strong interest especially in the aircraft industry because of high strength, high elastic modulus and low density. This alloy is also very suitable for studying the phase separation processes,[9] such as the very early stage and the coarsening stage from the following reasons: i) the atomic diffusion rate of Li can be easily controlled by choosing the ageing temperature, and ii) the main precipitates in this alloy are the metastable δ'(Al$_3$Li) with Ll$_2$ type structure and the coherency strain is very small and isotropic. Therefore, we have decided to study the phase separation process in Al-Li alloys.

EXPERIMENTAL PROCEDURES

The cast ingots of Al–Li alloys containing 9.5 and 11.4 at.% Li were repeatedly hot rolled and finally cold rolled to 2 mm in thickness. These were cut into small specimens with the size of 2 mm x 10 mm x 30 mm, and 3 or 4 pieces of them were used for the scattering experiment. These alloys contained 0.09 mass % Fe, 0.02 mass % Si, 3 ppm Na, 20 ppm Ca and <0.1 ppm K as impurities. Solution treatments were carried out at 773 K for Al–9.5 at.% Li alloy and 823 K for Al–11.4 at.% Li alloy in an argon atmosphere, followed by water-quench. Ageing times at 423 K varied from 60 s to 6.06 x 10^5 s.

Small-angle scattering measurements were carried out SAN[10] at Institute for High Energy Physics (KEK). Scattering path lengths were either 1 m for high momentum transfer region measurements (0.2 < q < 6 nm^{-1}) or 5 m for low q measurement (0.06 < q < 2 nm^{-1}). Since the instrument is installed at pulsed neutron source, by using all of the incident spectra from 0.3 to 0.9 nm, a very wide q range is covered simultaneously. The scattering data were corrected for background and transmission and normalized by scattering from water. All the scattering measurements were carried out at room temperature.

RESULTS AND DISCUSSIONS

Fig. 1 and Fig. 2 show the time evolution of the scattering intensity in Al–9.5 and 11.4 at.% Li alloys aged at 423 K, respectively. Strong small-angle scattering at smaller values of q (< 0.4 nm^{-1}) than the peak position is attributed to the scattering from the precipitates at grain boundaries. The scattering curve from the as-quenched sample of Al–9.5 at.% Li alloy is nearly flat, indicating that the quenching condition is

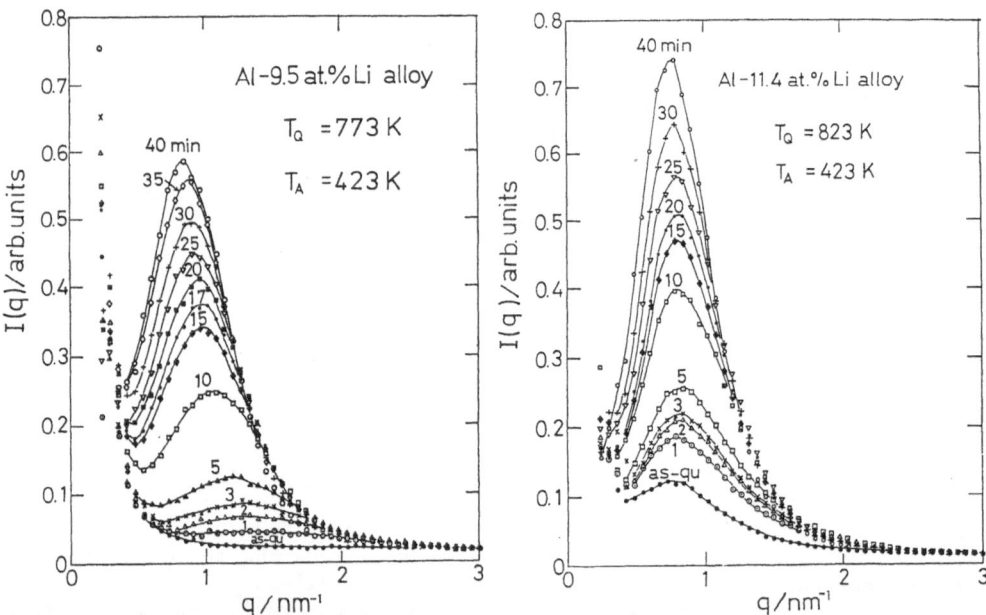

Fig.1. Time evolution of scattering intensity I(q) at early stage in Al–9.5 at.% Li alloy aged at 423 K.

Fig.2. Time evolution of scattering intensity I(q) at early stage in Al–11.4at.% Li alloy aged at 423 K.

good. Since the atomic diffusion rate of Li in Al–11.4 at.% Li alloy is probably higher than in Al–9.5 at.% Li alloy and the phase separation in the former is initiated during quenching, a broad but week peak is observed in the as-quenched specimen of Al–11.4 at.% Li alloy. The scattering in Al–9.5 at.% Li alloy shows the typical behavior of phase separating sample; the peak position (q_{max}) shifts towards lower values of q as time elapses. In contrast, the Al–11.4 at.% Li alloy shows abnormal behavior at the early stage (t < 200 s); when it varies from the as-quenched state to the state aged at 423 K for 20 minutes, the peak position shifts slightly towards higher values of q.

Fig. 3 shows the typical scattering intensity in log–log plot at late stage of ageing. The peak intensity (I_{max}) is more than 2 orders of magnitude larger than the earlier one, and q_{max} is about 4 times smaller than the earlier one. Fig. 4 shows the time evolution of q_{max} amd I_{max} in Al–9.5 at.% Li alloy. After ageing for longer times than 7 ks, the following power laws hold: $q_{max} \sim t^{-1/3}$ and $I_{max} \sim t^{1}$. The results are well explained by the scaling theory.[12] We define this region as the late stage for the case of this sample. This definition is different from the previous one, because the q dependence of the scattering intensity is very much different from other alloys such as Al–Zn and Fe–Cr alloys as described hereafter. Before the time at which the power laws hold, there is another region where different power laws, i.e. $q_{max} \sim t^{-1/6}$ and $I_{max} \sim t^{5/8}$ are observed. The results are not explained by the scaling theories. We define this stage as the early stage. The similar behavior is observed in Al–11.4 at.% Li alloy, although it is complicated by the anomaly in the time dependence of q_{max} at the very early stage.

Fig. 5 and Fig. 6 show the logarithmic plots of the scattering intensity at the very early stage in Al–9.5 and 11.4 at.% Li alloys,

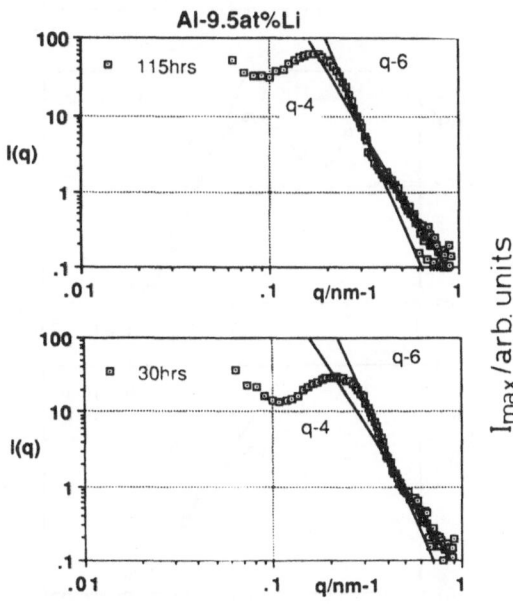

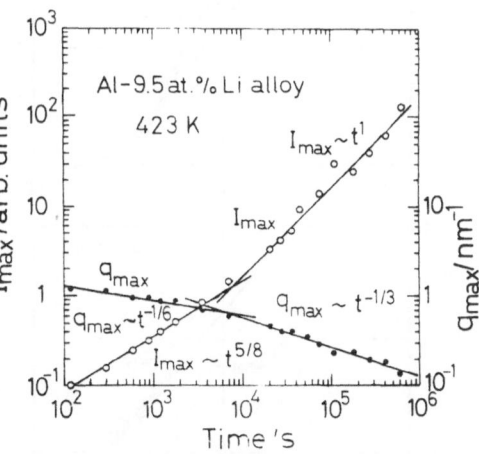

Fig.3. Logarithmic plot of time evolution of scattering intensity I(q) at the late stage in Al–9.5 at.% Li alloy.

Fig.4. Time dependence of peak intensity I_{max} and peak position q_{max} of scattering intensity in Al–9.5 at.% Li alloy.

respectively. The straight line is drawn to the intermediate q range
between the peak position and the tail part. The most significant
features of these results are as follows; the scattering intensity I(q) in
the intermediate q range is not proportional to q^{-2} nor q^{-4}, but it obeys
$q^{-a(t)}$ where a(t) is not necessarily integer. The values of a(t) varies
1.5 to 6 in Al-9.5 at.% Li alloy and from 2 to 6 in Al-11.4 at.% Li alloy.
The time variations of a(t) is shown in Fig. 7. The similar change has
been recently reported by Windsor[11] in this conference. The variation of
a(t) in the present work characteristically occurs in very early stage; it
takes only less than 60 minutes to attain to a(t) = 6. The peak intensity
increased more than 2 orders of magnitude after a(t) reaches 6. This
behavior at short time is very different from that observed in Fe-Cr and
Al-Zn alloys. In these alloys, the values of a(t) stay 2 or 4 for
sufficiently long period and the transition is not rapid.

One possible of this peculiar scattering function is the competition
between the ordering process and the phase separation process. Therefore,
the phase separation in Al-Li alloys can not take the normal path which is
observed in Al-Zn, Fe-Cr and some other alloys. The competition is
expected to exist only in the very early stage. Since at the later stage
a coarsening process mainly occurs, there might be no competition between
both processes. Therefore, it is not so surprising that the scaling
precicted by Furukawa holdsat late stage. More detailed studies must be

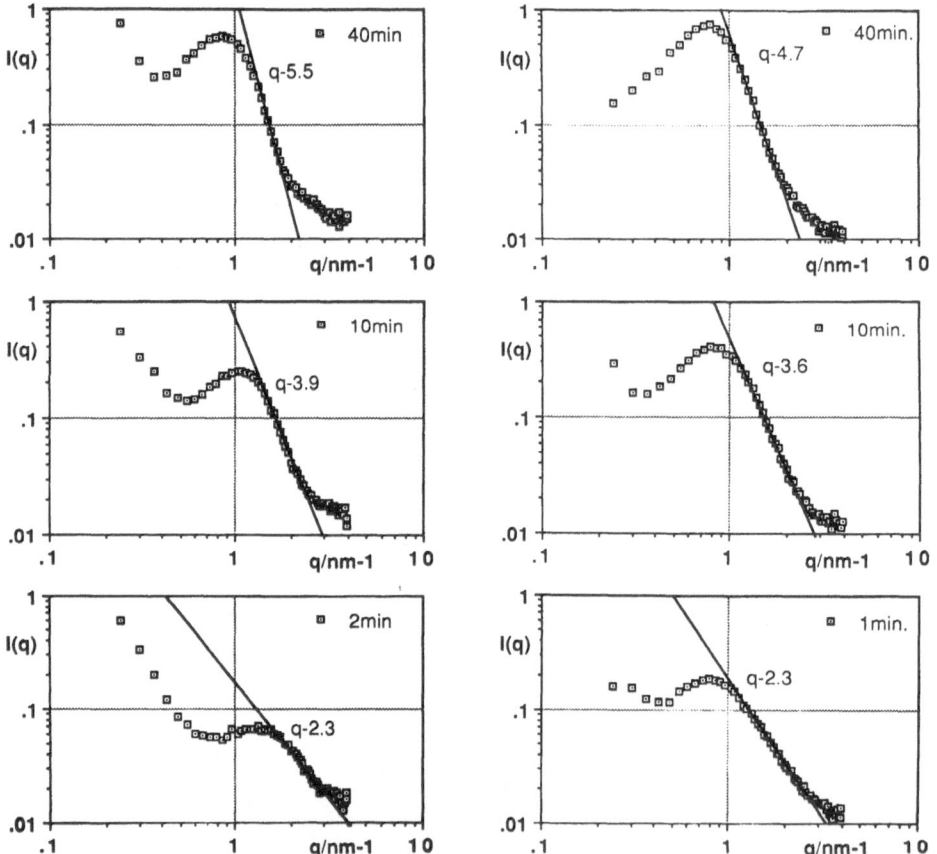

Fig.5. Logarithmic plot of time
evolution of scattering intensity
I(q) at early stage in Al-9.5 at.%
Li alloy.

Fig.6. Logarithmic plot of time
evolution of scattering intensity
I(q) at early stage in Al-11.4 at.%
Li alloy.

still carried out about the peculiar scattering behavior at early stage in this alloy.

To check the validity of Cahn's theory in the phase separation in Al-Li alloys, the logarithm of the scattering intensity was plotted[13] linearly aginst some fixed values of q, using the same results in the present work.

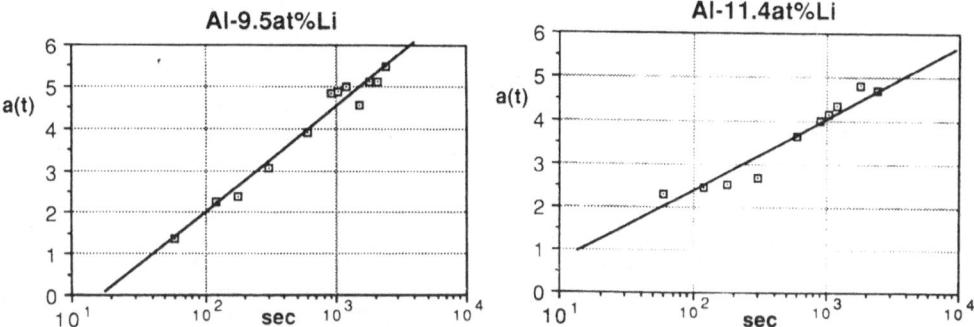

Fig.7. Time dependence of power a(t) in equation $I(q) \sim q^{-a(t)}$ in intermediate q range between peak position and tail part in Al-9.5 and 11.4 at.% Li alloys.

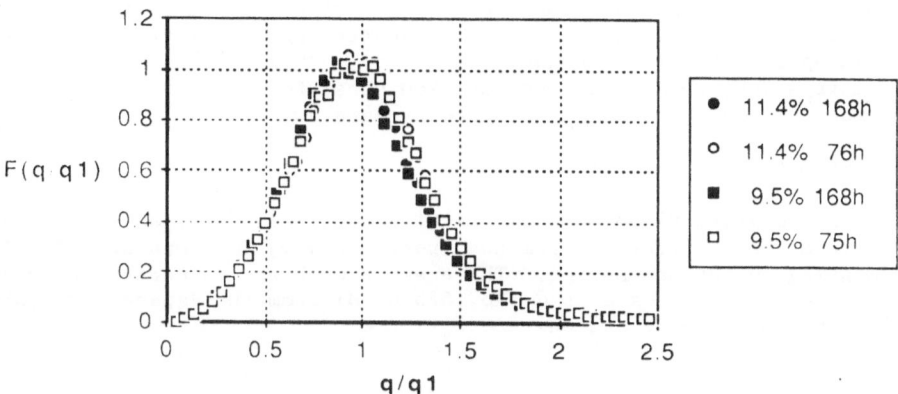

Fig.8. Scaling function $F(q/q_1)$ plotted against scaled momentum transfer q/q_1 in Al-9.5 and 11.4 at.% Li alloys at later stage.

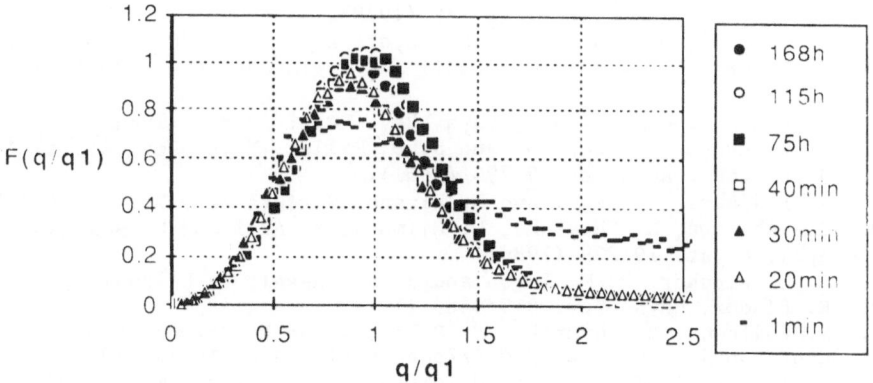

Fig.9. Scaling function $F(q/q_1)$ plotted against scaled momentum transfer q/q_1 in Al-9.5 at.% Li alloy at early and later stages.

It shows exponential growth at the very early stage at rather low q range. This is an almost unique fact that seems to support the theory. The peak position shifts as a function of ageing time, as shown in Fig. 1 and Fig. 2. The cross-over behavior is not detected in the very early stage. The ratio of the peak position of amplification factor q_m to the cross-over point q_c where the amplification factor is equal to zero, is almost 2 even in the very early stage.[13] It is much bigger than $\sqrt{2}$ predicted by the theory. Therefore, we conclude that it is difficult to explain rigidly by Cahn'theory even the earliest stage in Al-Li alloys that we have measured.

The results of scattering at late stage of ageing are shown in Fig. 3. The scattering function consists of q^{-4} and q^{-6} dependences. The contribution from q^{-4} dependece increases with ageing time. The scattering intensity I(q) indicates a q^{-6} bahavior, caused by scattering from intricate interfaces between the precipitates and matrix, which is explained by the Furukawa'theory.[3] This behavior is expected when the volume fraction (f) of precipitates is high. It is applicable to this case, since the value of f for the $\delta'(Al_3Li)$ in Al-9.5 at.% Li alloy is shown to be sufficently high by X-ray diffraction.[14] As proposed by the theory, the scaling holds at late stage, as shown in in Fig. 8. On the other hand, at the early stage, scaling does not hold, as shown in Fig. 9. It is natural, because at the early stage the S(q,t) itself changes with ageing time.

In conclusion, we have studied the very early stage as well as the late stage of the phase separation process in Al-9.5 and 11.4 at.% Li alloys by small-angle neutron scattering instrument. It shows anomalous behavior at the very early stage; the scattering intensity I(q) shows a non-integer power of q, in addition to q^{-2}, q^{-4} and q^{-6} dependences. At the late stage, the scattering function consists of q^{-4} and q^{-6} dependences and the scaling proposed by Furukawa holds.

ACKNOWLEDGEMENTS

We are grateful to Sumitomo Light Metals Ind. for the preparation of specimens, and to Prof. Fratzl and Prof. Furukawa for reading and commenting on the manuscript. The present work was supported by Grant-in Aid for General Research C (No. 62550458) from the Japanese Ministry of Education, Science and Culture.

REFERENCES

1. M. Furusaka, Y. Ishikawa and M. Mera, Phys. Rev. Lett. 54:2611 (1985).
2. M. Furusaka, Y. Ishikawa, S. Yamaguchi and Y. Fujino, J. Phys. Soc. Jpn. 55:2253 (1986).
3. H. Furukawa, Phys. Rev. 43:136 (1979).
4. M. Hennion, D. Ronzaud and P. Guyot, Acta Met., 30:599 (1982).
5. S. Komura, K. Osamura, H. Fujii and T. Takeda, Phys. Rev. B31:1278 (1985).
6. S. Katano and M. Iizumi, Phys. Rev. Lett. 52:835 (1984).
7. J. S. Langer, M. Bar-on and H. D. Miller, Phys. Rev. A11:1417 (1975).
8. J. W. Cahn, Acta Met. 9:795 (1961).
9. S. Fujikawa, Y. Izeki and K. Hirano, Scripta Met. 20:1275 (1986).
10. Y. Ishikawa, M. Furusaka, N. Niimura, M. Arai and K. Hasegawa, J. Appl. Cryst. 19:229 (1986).
11. C. G. Windsor, R. M. Baron and J. R. Russel, to be published.
12. K. Binder, Phys. Rev. B15:4425 (1977).
13. S. Fujikawa, M. Furusaka, M. Sakauchi and K. Hirano, Proc. 4th Int. Conf. on *Aluminum-Lithium Alloys*, Paris, June 10-12, 1987.
14. M. Tamura, T. Mori and T. Nakamura, Jpn. J. Inst. Met. 34:919 (1970).

PRECIPITATE COARSENING UNDER THE INFLUENCE OF ELASTIC ENERGY

IN Ni-BASE ALLOYS

Minoru Doi, Toru Miyazaki, Shigeru Inoue* and Takao Kozakai

Department of Materials Science and Engineering
Metals Section, Nagoya Institute of Technology
Gokiso-cho, Showa-ku, Nagoya 466, Japan
*Present address: IBM Japan Co. Ltd., Tokyo, Japan

INTRODUCTION

Man has been making every effort to produce a particular microstructure with desirable properties by utilizing the phase transformation. The desired structure is usually obtained by interrupting the transformation in the course of heat treatment and hence it is almost always in a thermodynamically metastable state. Therefore, during further heat treatment, such a desirable structure potentially transforms into a less desirable structure which is thermodynamically stabler.

The process of the precipitate coarsening due to the surface energy of the precipitate particle is well-known as 'Ostwald ripening'. In this process, the larger particles can coarsen by absorbing the smaller particles and the microstructure releases its excess surface energy. The mean particle radius $\bar{r}(t)$ at an ageing time t is expressed by the equation $\bar{r}(t)^m - \bar{r}(0)^m = kt$ where $\bar{r}(0)$ is the mean particle radius in the as-quenched state, and m and k are the constants. Since $\bar{r}(0)$ is virtually 0 for the present experiments, the above equation reduces to $\bar{r}(t) = kt^{1/m}$. Lifschitz and Slyozov,[1] and Wagner[2] have independently reported the theoretical treatment of Ostwald ripening which is well-known as 'LSW-theory'. Some of the predictions given by the LSW-theory are as follows:
(1) the precipitate particles coarsen as the cube root of the time, i.e.,
$\bar{r}(t)^3 - \bar{r}(0)^3 = kt$ or $\bar{r}(t) = kt^{1/3}$,
(2) the particle size distribution is scaled by the mean particle size.

Some experiments support the LSW-theory, e.g., sometimes the $t^{1/3}$ law holds. But there are a large number of experiments especially on the particle size distribution which disagree with those predicted by the LSW-theory: i.e., the experimentally observed distributions are almost always wider than the predicted one. Such a discrepancy is considered to be due to the fact that the LSW-theory neglects the effect of volume fraction of precipitate during coarsening. Many attempts have been made so far to modify the LSW-theory with respect to the volume fraction,[3-7] but their results leave something to be desired.

Another cause of such a discrepancy is likely to be due to the effect of elastic energy. The precipitate coarsening actually takes place in the elastically constrained metallic system. Therefore, the elastic energy due to the lattice misfit between the precipitate particle and the matrix is likely to affect the coherent precipitation. In fact, a number of phenomena which cannot be understood without considering the effects of elastic energy have been reported so far; e.g., the splitting of a single particle into several small particles,[8-10] the structural bifurcations,[11,12] etc. We can easily imagine that the elastic energy should actually affect the coarsening process also.

In the present studies, (i) the changes in the size and the size distribution of precipitates during ageing of some Ni-base alloys were investigated by means of transmission electron microscopy (TEM), and (ii) the effect of elastic energy on the precipitate coarsening was examined.

EXPERIMENTAL PROCEDURES

Ni-18.2%Cr-6.2%Al, Ni-7.0%Si-6.0%Al, Ni-36.1%Cu-9.8%Si, Ni-47.4%Cu-5.0%Si and Ni-16.3%Mo alloys were used in the present studies. All the compositions are given in atomic %. Disc-shaped samples (11 mm in diameter and 0.5 mm in thickness) of each alloy were at first solution-treated (homogenized) for 3600 sec at 1273 K (Ni-Mo alloy) or 1523 K (the other alloys) followed by quenching into iced water, and then aged at a temperature lower than the precipitation line. Thin foil specimens for TEM observations were prepared by electropolishing the aged samples. The changes in the size and the size distribution of precipitates during ageing were measured from TEM photographs.

EXPERIMENTAL RESULTS

Alloy Systems with Small Lattice Misfit

Figures 1-a and -b show the TEM images of the γ' precipitate particles in the Ni-Cr-Al and the Ni-Si-Al alloys. The lattice misfits are 0.008 % for the former and 0.10 % for the latter, which are relatively small. The shape of the particles remains spherical throughout the ageing except for the prolonged ageing ($t>10^6$ sec) for the Ni-Si-Al. The volume fractions f of the γ' particles are 0.11 and 0.16 for the Ni-Cr-Al and the Ni-Si-Al alloys aged at 1073 K, respectively.

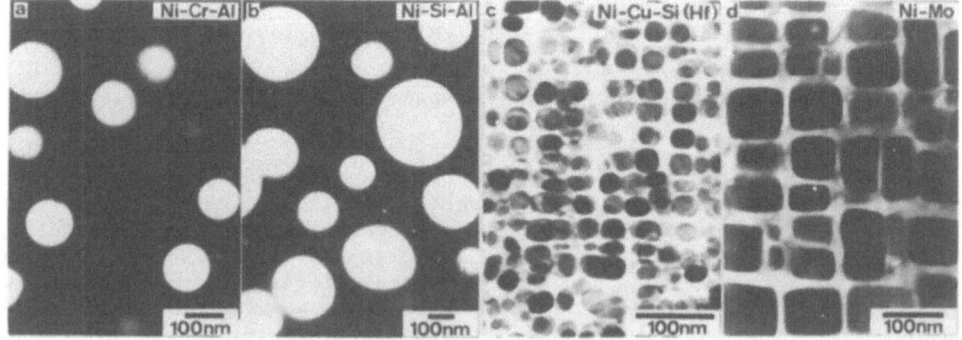

Fig. 1. TEM images of precipitate particles: a) γ' in Ni-7.0%Si-6.0%Al (1073 K, 691200 sec); b) γ' in Ni-7.0%Si-6.0%Al (1073 K, 345600 sec); c) γ' in Ni-36.1%Cu-9.8%Si (823 K, 1800000 sec); d) Ni$_4$Mo in Ni-16.3%Mo (973 K, 518400 sec).

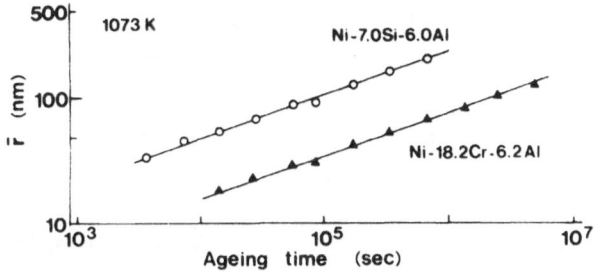

Fig. 2. Coarsening kinetics of γ' precipitates in Ni-base alloy
systems with small lattice misfit.

Figures 2, 3-a and 3-b illustrate the coarsening kinetics and the
size distribution of γ' particles in the Ni-Si-Al and the Ni-Cr-Al alloys
aged at 1073 K. In either alloy, there exists a linear relation between
the logarithms of $\bar{r}(t)$ and of t, and the slope 1/m is 0.32 for the former
and 0.33 for the latter, which agree with the value of 1/3 predicted by
the LSW-theory. Regarding the size distribution, the standard deviation σ
and the skewness K_S obtained here are as follows: σ=0.2425 and K_S=
-0.6433 for Ni-Cr-Al; σ=0.2755 and K_S=-0.2260 for Ni-Si-Al. The values of

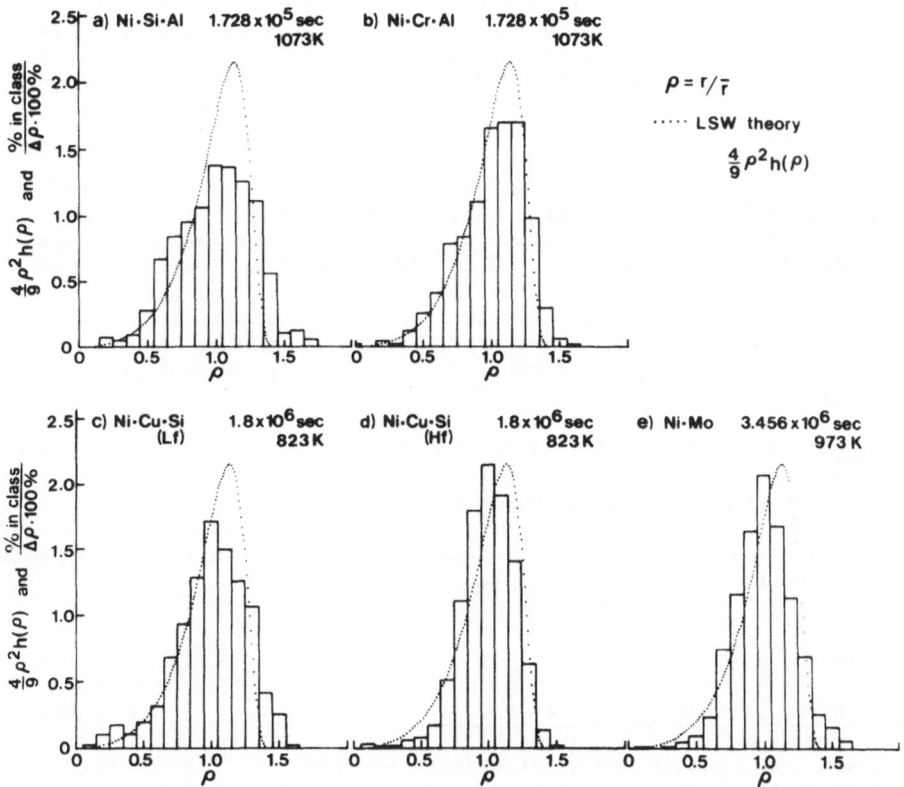

Fig. 3. Size distributions of precipitates in Ni-base alloys: a,b) γ'
(small misfit); c,d) γ' (large misfit); e) Ni_4Mo (large misfit).

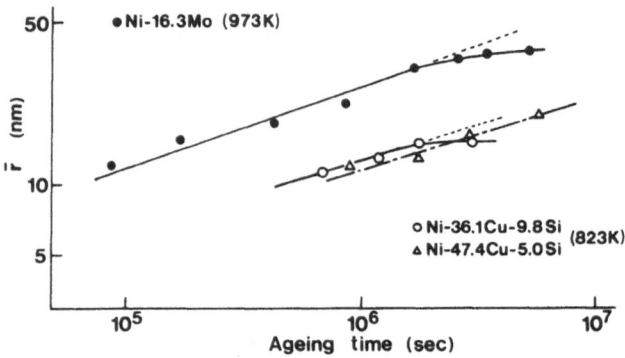

Fig. 4. Coarsening kinetics of precipitates in Ni-base alloy systems with large lattice misfit: γ' in the Ni-Cu-Si's and Ni_4Mo in the Ni-Mo.

σ and K_S predicted by the LSW-theory are 0.215 and -0.92, respectively.

Alloy Systems with Large Lattice Misfit

Figure 1-c and -d show the TEM images of the γ' precipitate particles in the Ni-Cu-Si and the Ni_4Mo precipitate particles ($D1_a$ bct structure) in the Ni-Mo alloys. The lattice misfits are -1.292 % for the former and 1.335 % for the latter, which are relatively large. The shape of the particles remains in substance cuboidal throughout the ageing. In this case, the half of the edge of the cuboid is regarded as the r(t) value. The f values are 0.18 and 0.50 for Ni-47.4%Cu-5.0%Si and Ni-36.1%Cu-9.8%Si alloys aged at 823 K respectively (this is called Ni-Cu-Si(Hf) and that is called Ni-Cu-Si(Lf)), and 0.48 for the Ni-Mo alloy aged at 973 K.

Ni-Cu-Si alloys. The open symbols (O,Δ) in Fig. 4 indicate the coarsening kinetics of γ' particles in the Ni-Cu-Si alloys. During the coarsening of γ' particles in either alloy, a linear relation between the logarithms of r(t) and of t (its slope 1/m is 0.3) can be seen, but the prolonged ageing causes a remarkably decelerated coarsening in the Ni-Cu-Si(Hf) alloy with higher f. Figures 3-c and -d illustrate the size distribution of γ' particles in the Ni-Cu-Si alloys aged at 823 K. The σ values calculated from the figures are 0.19 for the Ni-Cu-Si(Hf) and 0.26 for the Ni-Cu-Si(Lf) alloys.

Ni-Mo alloy. It is clear from Fig. 4 that a linear relation between the logarithms of r(t) and of t (1/m=0.34) is observed during the coarsening of Ni_4Mo particles in the Ni-Mo alloy aged at 973 K, but the coarsening is remarkably decelerated by the prolonged ageing. Furthermore, Fig. 5 indicates that the σ value gradually decreases while the coarseing is slowing down.

DISCUSSIONS

Alloy Systems with Small Lattice Misfit

Since the elastic energy is originated from the lattice misfit, the effect of elastic energy is not considered to be so dominant in the alloy systems with small misfit like γ' particles in Ni-Cr-Si and Ni-Si-Al alloys. In fact, it can be seen from Fig. 2 that the particle coarsening

290

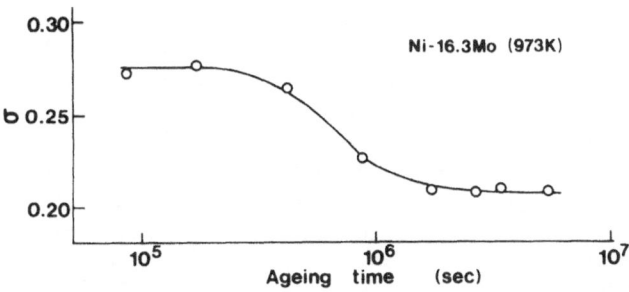

Fig. 5. Change in the standard deviation σ of Ni_4Mo particles with ageing.

in such systems obeys the $t^{1/m}$ law and the $1/m$ values are 0.32 and 0.33, which are practically not different from the value 1/3 predicted by the LSW-theory. Regarding the size distribution of particles, however, there are significant differences between the actual observations and the prediction by the LSW-theory, as seen in Figs. 3-a and -b: that is, the observed distribution is wider, more symmetric and less peaked than the predicted distribution. Furthermore, Figs. 3-a and -b also suggest that the distribution tends to become wider and more symmetric as the volume fraction increases. These facts fall in line with other experimental and theoretical results obtained by many investigators: that is, in the system with small lattice misfit, the driving force for coarsening is the excess surface energy and the general idea about the reason for the above discrepancies is based on the consideration of the effect of volume fraction.

Alloy Systems with Large Lattice Misfit

A notable point of the precipitate coarsening in the alloy systems with large lattice misfit is that the coarsening sometimes slows down at the later stage of ageing (see Fig. 4). This tendency to slowdown is obvious when the volume fraction f of precipitates is high. Another remarkable point of the precipitate coarsening in the systems with large misfit is that the standard deviation σ of the size distribution of precipitates gradually decreases during ageing (see Fig. 5). This is also obvious when the f is high, which is the opposite of the cases with small misfit where the σ is large when the f is high.

The above two points imply the important thing that the precipitate sizes become less scattered (i.e., uniform) and converge to a particular state when the precipitates are distributed in the matrix with large lattice misfit and high volume fraction. This hardly coincides with the already well-known theories of coarsening like the LSW-theory and its modified theories (in these theories, large particles continue to grow by absorbing small particles) and is really due to the effect of elastic energy. The elastic energy consists of the following two energies: the elastic strain energy due to lattice misfit, and the elastic interaction energy due to overlapping the elastic strain fields around particles. Especially, the latter interaction energy plays an important role here. Recently, we[12] have proposed the so-called 'stability bifurcation theory' based on the effect of elastic energy. This new idea predicts that small particles can grow at the expence of large particles to form uniform size distribution when the alloy system has large lattice misfit, high volume fraction and low surface energy. The unbelievable phenomena observed in the present studies are well explained by the new idea.

As the particles coarsen, the absolute value of the elastic energy (elastic interaction energy) increases contrary to the decrease in surface energy. Therefore, the elastic energy is dominant at the later stage of ageing. The deviation from the $t^{1/3}$ law and the remarkable slowdown are results of the change in coarsening mechanism, i.e., the change from the mechanism controlled by surface energy to that by elastic energy. This makes an important suggestion that not the exponent 1/m but the k of the equation $\bar{r}(t)=kt^{1/m}$ should change during ageing, which results in the slowdown of the coarsening, although we have usually directed our attention to the change in 1/m so far. Moreover, the size distribution of particles changes with ageing under the influence of elastic energy. This also makes another important suggestion that the size distribution can be no longer scaled by the mean particle size.

SUMMARY AND CONCLUSIONS

The important results obtained by the present studies are as follows:
(a) the coarsening is remarkably decelerated at the later stage of ageing,
(b) the size distribution of particles becomes less scattered and more symmetric with ageing.

These phenomena cannot be explained by the LSW-theory or its modified theories because such theories take notice only of the surface energy of the particle. The actual metallic systems are elastically constrained more or less. Therefore, when understanding the precipitate coarsening in such systems, we should not neglect the important role of elastic energy.

REFERENCES

1. I. M. Lifshitz and V. V. Slyozov, The kinetics of precipitation from supersaturated solid solutions, J. Phys. Chem. Sol., 19:35 (1961).
2. C. Wagner, Theorie der Alterung von Niederschlagen durch Umlosen (Ostwald-Reifung), Z. Elektrochem., 65:581 (1961).
3. A. J. Ardell, The effect of volume fraction on particle coarsening: theoretical considerations, Acta Metall., 20:61 (1972).
4. A. D. Brailsford and P. Winblatt, The dependence of Ostwald ripening kinetics on particle volume fraction, Acta Metall., 27:489 (1979).
5. J. A. Marqusee and J. Ross, Theory of Ostwald ripening: Competitive growth and its dependence on volume fraction, J. Chem. Phys., 80:536 (1984).
6. P. W. Voorhees and M. E. Glicksman, Solution to the multi-particle diffusion problem with applications to Ostwald ripening—I. theory, Acta Metall., 32:2001 (1984).
7. Y. Enomoto, M. Tokuyama and K. Kawasaki, Finite volume fraction effects on Otwald ripening, Acta Metall., 34:2119 (1986).
8. M. Doi and T. Miyazaki, The effect of elastic interaction energy on the shape of γ'-precipitate in Ni-based alloys, in: "Superalloys 1984," M. Gell et al. ed., The Metallurgical Society of AIME, Warrendale, Pa. (1984).
9. M. Doi, T. Miyazaki and T. Wakatsuki, The effect of elastic interaction energy on the morphology of γ' precipitates in nickel-based alloys, Mater. Sci. & Eng., 67:247 (1984).
10. M. Doi, T. Miyazaki and T. Wakatsuki, The effects of elastic interaction energy on the γ' precipitate morphology of continuously cooled nickel-base alloys, Mater. Sci. & Eng., 74:139 (1985).
11. W. C. Johnson and J. W. Cahn, Elastically induced shape bifurcations of inclusions, Acta Metall., 32:1925 (1984).
12. T. Miyazaki, K. Seki, M. Doi and T. Kozakai, Stability-bifurcations in the coarsening of precipitates in elastically constrained systems, Mater. Sci. & Eng., 77:125 (1986).

OSCILLATORY BEHAVIOR OF SOLUTE CONCENTRATION DURING PRECIPITATION

IN AN Al-1.14 at% Si ALLOY UNDER THE INFLUENCE OF ELECTRIC CURRENT

Yukio Onodera and Ken-ichi Hirano

Department of Materials Science
Faculty of Engineering
Tohoku University, Sendai 980, Japan

INTRODUCTION

We have investigated the effect of electric current on precipitation in various aluminum alloys over a wide temperature range[1-3]. In the course of these experimental studies, it was found that in an Al-1.14 at% Si alloy, the concentration of solute Si in the matrix oscillated at random with a large amplitude when an alternating current(a.c.) of 4×10^3 A/cm^2 was passed through the specimen, as reported in our previous note[ ].

Recently intensive studies of nonlinear dynamics have provided a very useful method to distinguish signals responsible for "determistic chaos" from that caused by mere external noise. Thus, it is very interesting to investigate whether the oscillatory behavior observed originates from deter- mistic chaos or merely from external noise.

The purpose of the present work is to examine the various experimental conditions which may affect the oscillatory behavior observed. Furthermore, the same kind of experiment was also carried out using a specimen of Al-1.8 at% Cu to compare the effect of electric current on precipitation with that in Al-1.14 at% Si. Here we used the same apparatus as in the ex- periment on the Al-Si alloy.

In the present work, direct electric current(d.c.) was used instead of a.c. for the following reasons: (i) in a preliminary test, it was con- firmed that the effect of a.c. stress was not so different from that of d.c. stress. (ii) a more stable d.c. current source was available. (iii) the elecrical effect on precipitation may be attributed to the electric field itself[4] and not to the passage of the electric current; for example, var- ious phenomena related to electromigration may intervene.

EXPERIMENTAL METHOD

An Al-1.14 at% Si alloy was obtained in the form of ingot bars from Sumitomo Light Metal Corporation. The main impurity in the bars was 0.002 at% Fe. A portion of the ingot bar was cold-swaged and finally drawn into wires of 0.44 mm in diameter. The wire was annealed for about 10 days at 580°C in a vacuum of about 10^{-4} mmHg. A piece of wire was fixed as speci- men to a quartz holder. Two pieces of the same material were spot-welded or brazed to the specimen wire as potential leads. The specimen thus prepared was solution heat-treated in air at 580°C for about 30 minutes in a horizontal furnace. After the specimen was extracted from the furnace for quenching, it was kept at room temperature for 20 seconds to reduce the influence of excess vacancies on the nucleation kinetics of solute Si. Then, it was immediately transferred into liquid nitrogen, where the electrical resistance of the specimen, R_q, was measured by the standard d.c. potentiometric method. Next the aging was done in air at various temperatures below the solvus(537°C) without and with the current of 3.86×10^3 A/cm^2 d.c. The current source was stabilized within 0.1% of the output. Here the specimen temperature during power aging was indirectly estimated by the same method described in Ref.4. After interrupting the precipitation reaction every 20 minutes, the isothermal change in the solute concentration in the matrix was followed by measuring the specimen resistance, R, at liquid nitrogen temperature.

EXPERIMENTAL RESULTS

A main part of the isothermal curves of the Al-1.14 at% Si alloy without and with current is shown in Fig.1a and Fig.1b, respectively, where

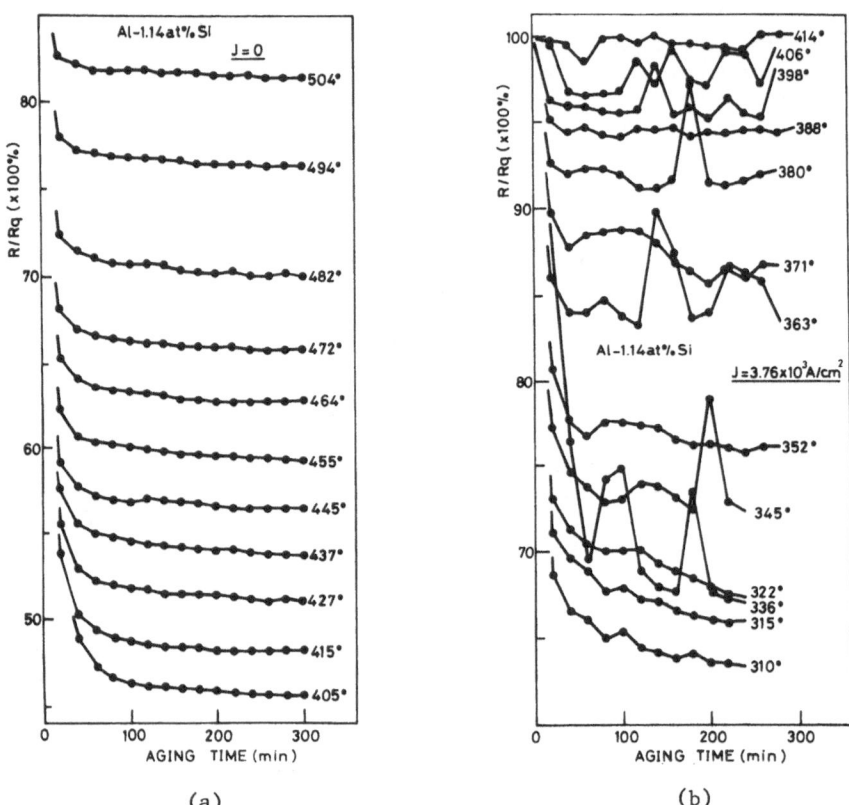

(a)   (b)

Fig.1. The specimen with spot-welded potential lead wires was
used in both (a) and (b).
(a) Isothermal aging curves without current stress.
Numbers near the curves represent aging temperatures in °C.
(b) Isothermal aging curves with 3.76×10^3 A/cm^2 d.c.
Numbers near the curves represent the furnace temperatures
during power aging, so that we must add 140°C in order to
obtain the true specimen temperature.

the ordinate indicates the percentage chage in the resistance R normalized
by the as-quenched value Rq and the abscissa, the aging time. The furnace
temperature was measured by the thermocouple placed at a distance of 1 cm
in the radial direction of the specimen wire. The real temperature of the
specimen was estimated from another independent measurement as described in
Ref.4. In the present case, the specimen temperature was 140°C higher than
furnace temperature owing to Joule heating.

The ordinary isothermal aging curves(Fig.1a) decrease monotonically
with aging time. On the other hand, in the case of power aging(Fig.1b), many
curves show random oscillations with sharp peaks. This tendency becomes appreciable above 336°C. However, no sharp peaks are observed at 352,371, and
388°C. This may be due to the fact that unfortunately we could not observe
these peaks since the time interval(every 20 minutes) of our measurement was
too coarse. The isothermal curve at 336°C is particularly anomalous if compared with the other curves. It decreased slowly until 40 minutes and rapidly decreases after that time. The region of the specimen temperature where
the oscillatory behavior is often observed is from 476 to 537°C. To clarify
this anomalous behavior, the isothermal curves were converted to isochronal
curves of 20 minutes. As shown in Fig.2, the ordinary isochronal curve is a
smoothly varying function of temperature over the entire range investigated.
On the other hand, the isochronal curve under current stress has an anomalous kink point at 336°C, which corresponds to 476°C of the specimen temperature.

Furthermore, to examine the existence of this anomalous behavior near
336°C, isothermal aging curves were again measured using specimens with
brazed potential wires in the furnace temperature region from 305 to 386°C.

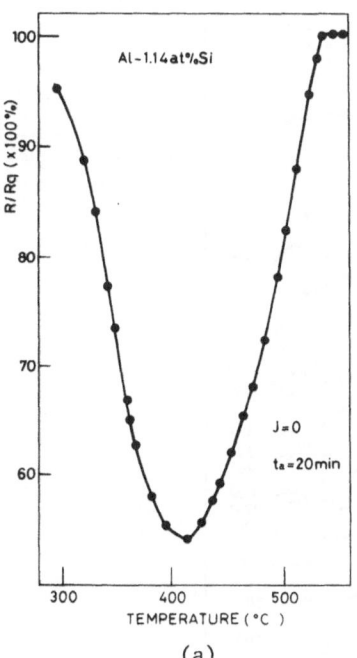 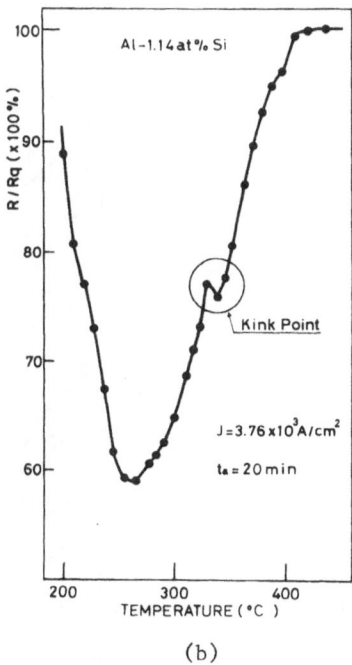

          (a)   (b)

Fig.2. (a) Isochronal aging curve of 20 minutes obtained
from the result of Fig.1a.
(b) Isochronal aging curve of 20 minutes obtained
from the result of Fig.1b.

The specimen thus prepared can be more homogeneously heated during power aging than the specimen with spot-welded potential lead wires because brazing gives rise to less damage to the specimen in the vicinity of the connected portion. The result is shown in Fig.3. It is clearly observed that the oscillatory behavior begins around 320°C or above although there is a little difference in the onset temperature of the oscillation between that observed in Fig.2b and Fig.3. This is attributable to the subtle difference of the specimen arrangement in the furnace.

In Fig.4, the isochronal aging curve of an Al-1.8 at% Cu alloy without and with the current of 3.84×10^3 A/cm^2 d.c. are presented, respectively. The curve given in Fig.4b shows no kink in contrast to the curve in Fig.2b.

DISCUSSION

It is required to obtain a vast amount of experimental data of time evolution(change in solute concentration in the matrix) in order to characterize that the random oscillation of solute concentration in the matrix observed in our experiment is due to the determistic chaos governed by a few numbers of degree of freedom, or to the noise caused by un-controllable experimental conditions. In the present work, however, a data set of time evolution(an isothermal curve) consists of at most several tens of points. Therefore, we could not discriminate whether the random oscilla-

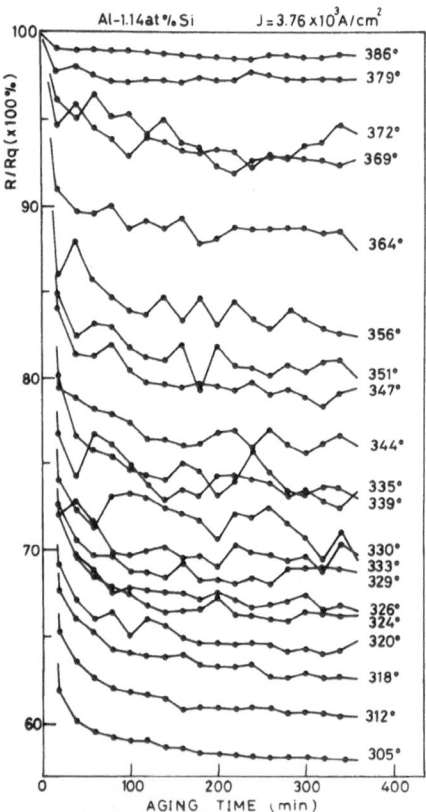

Fig.3. Isothermal aging curves at various temperatures under a d.c. current of 3.76×10^3 A/cm^2 for the specimen with brazed potential lead wires.

tion is due to chaos or not, *e.g.*, by such a method[5] as to estimate a fractal dimension of attractor from the correlation integral plot using a single-variable time series. Nevertheless, we can consider that there are several pieces of evidence which may suggest the existence of chaos resulting from the system itself: (i) both the isothermal and the isochronal curves under current stress are distinctly different from the ordinary curves, (ii) the oscillatory behavior with large amplitude is observed only in a particular temperature region above 320–340°C which corresponds to 460–480°C for the specimen temperature, (iii) in an Al-1.8 at% Cu alloy the isochronal curve under current stress does not differ appreciably from that of the ordinary aging as shown in Fig.4.

It should be noted that the amplitude of the oscillation shown in Fig.3 is approximately 1/3 as small as that in Fig.2b. This is possibly due to the fact that the potential lead wires attached by brazing play a roll of so good thermal path that the specimen temperature may decrease at both ends of the specimen. As a result, contrary to our first expectation, the specimen temperature of the central portion must become higher. Then, the peak amplitude of the oscillation is apparently rounded because of the ambiguity brought about by this inhomogeneous temperature distribution in the longitudinal direction of the specimen wire. On the other hand, the specimen with spot-welded potential leads is often subjected to damage near the connected portions as mentioned before. Accordingly, the electrical resistance near the potential leads becomes so large that the local temperature

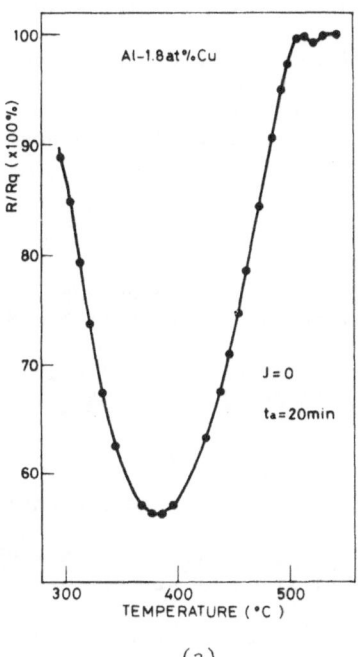
(a)

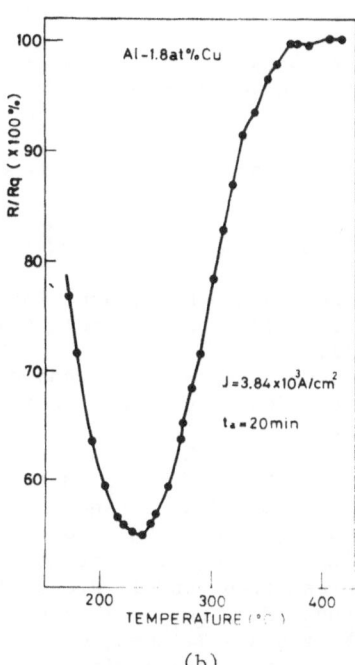
(b)

Fig.4. (a) Isochronal aging curve of 20 minutes without current in Al-1.8 at% Cu.
(b) Isochronal aging curve of 20 minutes with a d.c. current of 3.74×10^3 A/cm^2 in Al-1.8 at% Cu. We must add 140°C to the abcissa values in order to obtain the true specimen temperatures.

arises owing to Joule heating at these connected portions. Since this additional heat evolution can compensate the heat loss through the potential lead wires, the specimen temperature can be kept more homogeneous than that with the brazed potential lead wires.

On the basis of the present result, we can again propose the following two possible mechanisms for the oscillatory behavior of solute concentration in the matrix under current stress, which were already suggested in Ref.4: (i) Inhomogeneous temperature distribution caused by the Joule heating accompanying heat flow from the specimen to an outer heat bath. This situation is quite analogous to Nitzan-Ross model for photochemical reactions[6]. After a reactant A is excited by light absorption, it converts to a product species B. They analysed this simple chemical reaction by using the rate equation for the species A which is coupled nonlinearly to the equation for the rate of change of temperature. In the present case, the reactant A corresponds to the Si atom in the matrix and the product B, to the Si atom in a precipitate. The Si atom in the matrix must be "excited" by the Joule heating which will result from the difference of electrical resistivity between the matrix and the precipitate. (ii) Free energy change caused by the inhomogeneous distribution of the electric field in the specimen. Although the initial free energy of a supersaturated solid solution can be considered to be distributed homogeneously in the specimen, the strength of the electric field in a precipitate becomes stronger than that in the matrix with progressing precipitation because of the large difference of electrical resistivity between the matrix and the precipitate. Then, this additional field energy contributes inhomogeneously to the total free energy of the system. Thus it is probable, under specific conditions, to apply the formation theory of water droplets in the presence of electric field[7,8] to the present problem. More quantitative discussion is given in Ref.4.

At present, we have no conclusive evidence to determine which model described above can give a more suitable interpretation of the experimental result. Finally it should be noted that the onset temperature of the oscillatory behavior(460-480°C) might suggest a certain relation to a possible instability appearing at a spinodal temperature.

REFERENCES

1. Y. Onodera and K. Hirano, The effect of direct electric current on precipitation in a bulk Al-4 wt% Cu alloy, J. Mater. Sci. 11: 809 (1976).
2. Y. Onodera, J. Maruyama and K. Hirano, Retardation of the precipitation reaction by d.c. stress in an Al-12.5 wt% Zn alloy, J. Mater. Sci. 12: 1109 (1977).
3. Y. Onodera and K. Hirano, The effect of a.c. frequency on precipitation in Al-5.6 at% Zn, J. Mater. Sci. 19: 3935 (1984).
4. Y. Onodera and K. Hirano, Anomalous fluctuation of solute concentration during decomposition of supersaturated Al-1.14 at% Si under the influence of electric current, J. Mater. Sci. Lett. 5: 1048 (1986).
5. P. Grassberger and I. Procaccia, Characterization of strange attractors, Phys. Rev. Lett. 31: 346 (1983).
6. A. Nitzan and J. Ross, Oscillations, multiple steady states, and instability in illuminated systems, J. Chem. Phys. 59: 241 (1973).
7. D. Kashchiev, On the influence of the electric field on nucleation kinetics, Phil. Mag. 25: 459 (1972).
8. J. O. Isard, Calculation of the influence of an electric field on the free energy of formation of a nucleus, Phil. Mag. 35: 817 (1977).

ANTIPHASE BOUNDARY MIGRATION AND DOMAIN COARSENING

IN BULK AND THIN-FOIL Fe-Al SPECIMENS

Samuel M. Allen and Woonsup Park

Department of Materials Science and Engineering
Massachusetts Institute of Technology
Cambridge, Massachusetts 02139 USA

INTRODUCTION

The dynamics of microstructural coarsening processes has interested the materials science and condensed matter physics communities for a number of years now. There are many fascinating aspects of these problems, and much theoretical work on such problems has been done. One of the simplest processes to model is the coarsening of antiphase domains in ordered crystal structures like B2 (CsCl) having only two sublattices. The domain structure coarsens through capillarity-driven migration of antiphase boundaries (APBs), with the local velocity V of an APB given by [1]:

$$V = M \ (K_1 + K_2),\tag{1}$$

where $K_1 + K_2$ is the sum of the principal curvatures of the APB, and M is a "mobility" equal to $2\alpha\kappa$, where α is a coefficient related to the kinetics of ordering, and κ is a gradient energy coefficient.

This velocity expression can be incorporated [1] into a law for domain coarsening with only one assumption, namely that during domain growth, the domain structure remains essentially self-similar, giving a relation between K_m^2, the surface averaged square of $K_1 + K_2$, and S_v, the surface area per unit volume of the APBs:

$$K_m^2 = S_v^2 \ \varphi,\tag{2}$$

where φ is a topological constant of unknown magnitude. The resulting expression for the domain coarsening kinetics is [1]:

$$[S_v(t)]^{-2} - [S_v(t_0)]^{-2} = 2M\varphi(t-t_0),\tag{3}$$

where $S_v(t)$ and $S_v(t_0)$ are the values of S_v at times t and t_0, respectively.

Numerous experimental studies of antiphase domain coarsening kinetics in bulk alloy specimens have provided evidence for the domain growth law of eqn.(3). The study of Krzanowski and Allen [2] is of special interest because it presented evidence for a change of the mobility M with domain size in domain coarsening experiments on an Fe-24%Al alloy. This change was attributed to a solute drag effect.

The major disadvantage of experimental studies on bulk alloys is that they do not allow for a direct measurement of M, because K_m^2 is not known (and would be difficult to measure as it would require analysis of the topology of the APBs in three dimensions). In this paper, we present and discuss recent results of an in-situ study of APB migration carried out in a heating stage and observed with a 200 kV transmission electron microscope (TEM). In this type of experiment, a thin-foil specimen of approximately 100 nm thickness is used, and the APBs rotate into orientations perpendicular to the specimen surface early in the course of the experiment [4]. Thus, the sample contains a domain structure that is essentially two dimensional (one of the principal curvatures is near zero everywhere). The experiments allow direct determination of the mobility M from eqn. (1).

The experiments to be described were carried out on an Fe-Al alloy with a composition near Fe$_3$Al. These alloys are disordered bcc solid solutions above about 1200 K, and at somewhat lower temperatures, order to the B2 (CsCl) structure by a higher-order transition. In this structure, there is a single type of APB, characterized by a fault vector of the type (a/2)<111>, where a is the lattice constant of the crystal. On cooling to approximately 830 K, there is another higher-order transition to the DO$_3$ (BiF$_3$) structure. In this structure, there is a second type of APB, with a fault vector of the type a<100>. Other details about the alloy system and phase equilibria have been given by Swann et al. [5].

EXPERIMENTAL PROCEDURES

All experiments were carried out on an Fe-26%Al (atomic percent) alloy that was prepared by arc melting three times, followed by an isothermal forging treatment at 1173 K. The material was then homogenized 2 h at 1373 K and air cooled. In this condition, a relatively coarse structure of <111> domain boundaries with $S_v \sim 10^{-6}$ m^{-1} was produced, inside of which was a very fine structure of <100> domain boundaries with $S_v \sim 10^{-8}$ m^{-1}, where S_v is the surface area per unit volume of domain boundaries. (Note: The mean linear intercept domain size D is given by the relation D = 2/S_v).

For bulk domain coarsening kinetics, studies were carried out at 745 and 778 K (within the DO$_3$ phase field) by annealing small pieces of material for various times, quenching, then preparing thin-foil specimens for TEM examination. Antiphase domain sizes were measured using standard techniques of quantitative metallography [2].

For the in-situ studies, thin-foil specimens of the homogenized and quenched material were prepared and observed in a double-tilt heating holder in a JEOL 200-CX TEM. Observations of APB migration were made in separate experiments at 745, 778 and 800 K in the DO$_3$ phase, and at 838, 853 and 910 K in the B2 phase. Sequential series of images of the domain structure were recorded on photographic plates. Temperature calibration of the experiments was done by determining the indicated temperature of the critical temperature of the B2 -> DO$_3$ transition, as observed in <110> selected-area diffraction patterns. In this way, the temperature was known to within 5 K.

Analysis of the APB migration kinetics was done with the aid of a Magiscan II image analysis system. Local curvatures of APBs were measured from digitized images of the antiphase domain structure by marking three points on the digitized image with a light pen, then computing the radius of the resulting osculating circle. Migration distances were measured by reference to small high-contrast contamination spots on the specimens. Additional details are given elsewhere [4].

RESULTS

Domain Coarsening in Bulk Specimens

Results of the coarsening experiments done with bulk alloy specimens are presented in Figs. 1(a) and 1(b), which plot S_v^{-2} vs. t for the experiments carried out at 745 and 778 K, respectively. Eqn.(3) predicts that the plots should be linear if M is constant, but the plots suggest that there is a transition from rapid domain coarsening at small domain sizes (high driving-force regime) to slower kinetics at larger domain sizes (low driving-force regime). These results are similar to Krzanowski's results for coarsening of <111> domains in a B2 ordered Fe-24%Al alloy [2]. The slopes of the linear portions of the plots in Figs. 1(a) and 1(b) give values for $2M\varphi$, as seen by eqn.(3). These values are presented in Table 1.

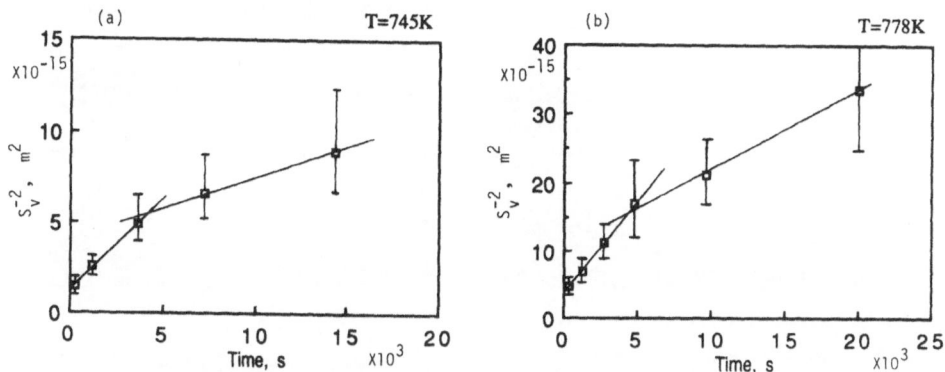

Fig. 1. Bulk domain coarsening kinetics observed in DO$_3$ ordered specimens at (a) 745 K and (b) 778 K.

Table 1. Observed Values of $2M\varphi$ in High and Low Driving-Force Regime of Migration

Temperature, K	$2M\varphi$, m^2/s	
	High Driving Force	Low Driving Force
745	1.0×10^{-18}	3.6×10^{-19}
778	2.7×10^{-18}	1.1×10^{-18}

Domain Coarsening in Thin-Foil Specimens

A typical sequence of electron micrographs showing APB migration is presented in Fig. 2, which shows results from an experiment performed at 778 K where the DO_3 structure is stable. The pictures show a number of small circular domains that shrink and disappear during the course of the experiment. These pictures, and others from similar experiments at other temperatures, were analyzed at numerous locations to obtain local values of curvature and velocities, thus enabling the mobility M to be determined from eqn.(1). An attempt was made in all cases to determine M over as wide a range of curvatures as possible, so as to pick up any transitional behavior of the migration kinetics. Results from

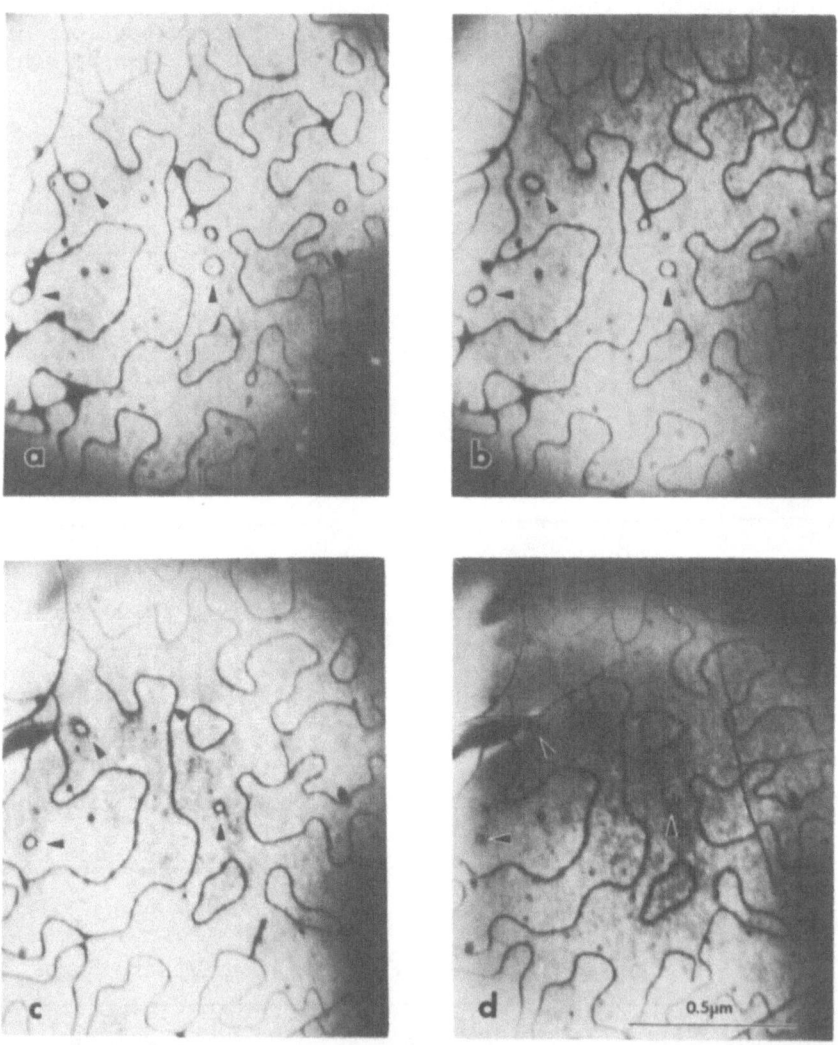

Fig. 2. In-situ observation of coarsening of a<100> APBs at 778 K. Time scale of the micrographs are: (a) 0 s; (b) 69 s; (c) 164 s; (d) 224 s.

experiments at all six temperatures studied are presented in Figs. 3(a)-3(f). These plots show observed values of M vs. curvature at each temperature. Note that the lowest three temperatures are for <100> APBs in the DO$_3$ phase, and the higher temperatures are for <111> APBs in the B2 phase. None of the plots shows a dramatic systematic change of M with curvature, but the data do seem to show a slight drop in mobility at the lowest values of curvature in most cases.

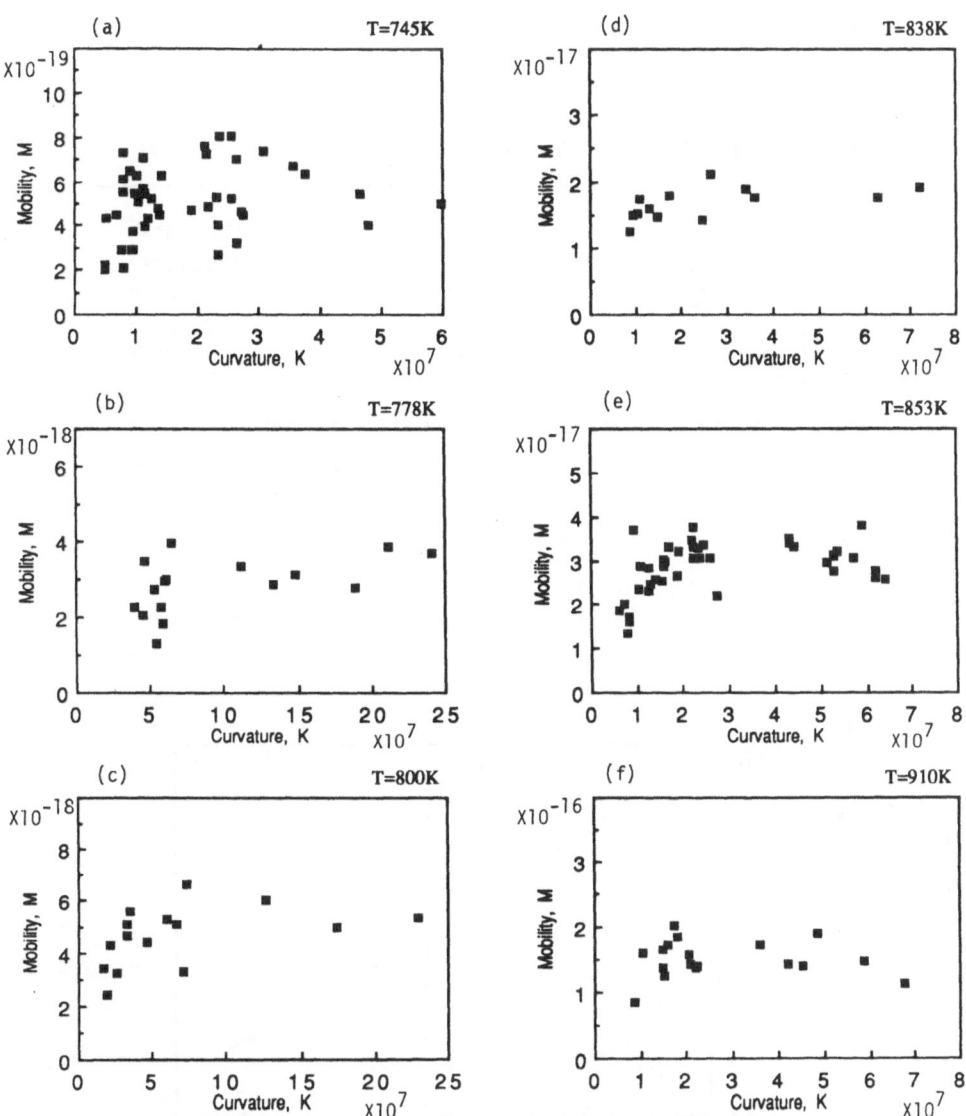

Fig. 3. Mobility M (m^2/s) vs. curvature K (m^{-1}) of APBs as measured from in-situ experiments. Curves on the left are for <100> APBs in the DO$_3$ phase at temperatures indicated; those on the right are for <111> APBs in the B2 phase.

A least-squares fit of the data for M at each temperature was performed, and the results are presented in Table 2 and as an Arrhenius
plot in Fig. 4. The error bars in Fig. 4 give the standard deviations
of the measurements. Also indicated on Fig. 4 is the critical temperature T_c for the B2 -> DO_3 transition, 826 K [5].

Table 2. Observed Values of APB Mobility M in In–Situ
Migration Experiments

APB Type and Phase	Temperature, K	M, m^2/s
<100>, DO_3	745	4.8×10^{-19}
<100>, DO_3	778	2.9×10^{-18}
<100>, DO_3	800	4.7×10^{-18}
<111>, B2	838	1.7×10^{-17}
<111>, B2	853	2.8×10^{-17}
<111>, B2	910	1.5×10^{-16}

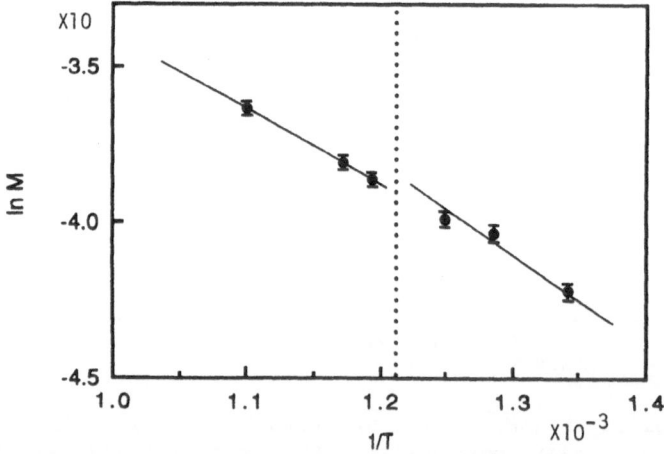

Fig. 4. Arrhenius plot of mobilities M determined from in–situ
migration experiments.

DISCUSSION

Bulk Domain Coarsening Experiments

Domain coarsening experiments on bulk specimens give results similar to those obtained by Krzanowski and Allen [2], namely that over a limited range of domain sizes, eqn.(3) was obeyed. The plots presented in Fig. 1 indicate a change in growth kinetics that occurs at approximately $S_v \sim 1.4 \times 10^7$ m^{-1} at 745 K and at $S_v \sim 7 \times 10^6$ m^{-1} at 778 K. These values correspond to domain sizes (mean linear intercept) of 140 and 260 nm, respectively. Although significant statistical errors are present in the measurements at the largest domain sizes, it is nonetheless clear that the domains coarsen more rapidly at small domain sizes, particularly at 745 K.

Such behavior would be expected if solute drag processes influenced the migration of these boundaries [6]. Small domain sizes correspond to high driving forces for APB migration, since the driving force for capillarity-driven migration is proportional to interface curvature. Conversely, the behavior at large domain sizes is in response to low driving forces for migration. If solute drag processes are important, the migration kinetics in the low driving-force regime would be controlled by diffusion of the solute "atmosphere" with the migrating boundary, resulting in what is termed "extrinsic" behavior. In the high driving-force regime of solute drag behavior, the boundary velocities are too high for solute atoms to keep up, and the kinetics are governed by the kinetics of reordering that occurs in the vicinity of the APB. The resulting migration kinetics are said to be "intrinsic."

Krzanowski and Allen [3] presented a calculation of the expected solute drag force to support their conclusion that the change of domain coarsening kinetics observed in the B2 ordered Fe-24%Al alloy was due to a solute drag effect. Although we have not completed a similar calculation for the DO_3 structure in the Fe-26%Al alloy, the work is in progress. Lacking this result, we make the tentative conclusion that the transitional behavior observed in the data presented in Fig. 1 are due to a solute drag process.

In-Situ Antiphase Boundary Migration Experiments

The in-situ boundary migration experiments allow for the first time measurement of the mobilities of APBs, over a wide range of curvatures (and hence driving forces). It is a versatile technique, in that the mobility data obtained span more than three orders of magnitude (see Table 2). There is also relatively little scatter in the data at a given temperature, because of the large number of data points that can be analyzed from a series of micrographs (see Figs. 3 and 4). Another virtue of the technique is that the inevitable shrinking and disappearance of single domains will always result in an increase of driving force to extremely high values as the domain collapses on itself and disappears. This contrasts with more conventional experiments in which the driving force steadily decreases [7]. The present technique thus guarantees access to the intrinsic regime of solute drag behavior.

The data presented in Fig. 3 show no evidence for a dramatic change of mobility over the range of curvatures accessible in the experiment. Since the data extend to domain radii as low as 5 nm, we believe that the bulk of the observations yield mobilities for the intrinsic regime of migration. The only possible indication of extrinsic behavior is a slight, but systematic, tendency for some data at the lowest curvatures

in the plots in Fig. 3 to show reduced values of mobility. This is
discussed more in the following section.

The observed temperature dependence of the mobilities, shown in
Fig. 4, is quite surprising. As mentioned earlier, the mobility M is
proportional to the product of a gradient energy coefficient κ, and a
coefficient α that is related to the kinetics of ordering. Values for
the gradient energy coefficient κ have been calculated for a given
phase and type of APB, using the Bragg-Williams model and considering
interactions out to third neighbor distances [8,9]. These quantities
are independent of temperature. Values for the Fe-26%Al alloy are presented in Table 3. Values for the <100> APBs in the DO_3 phase are seen
to be lower by a factor of 2.3 compared with those for <111> APBs in
the B2 phase. We have no detailed prediction for the expected temperature variation of the kinetic coefficient α. To a first approximation,
we expect it to have a simple Arrhenius dependence within a given
phase, similar to an interdiffusion coefficient. But if α does behave
like the interdiffusivity, a change to a higher activation energy in
the DO_3 phase would be expected [10,11]. Thus, the generally expected
form of Fig. 4 would be for the right-hand (DO_3) portion of the curve
to have a steeper slope, and for a discontinuity to exist at the critical temperature, the magnitude of which depends on the difference
between the gradient energy coefficients.

Table 3. Calculated Values of Gradient Energy Coefficients
for APBs in Fe-26%Al

APB Type and Phase	Gradient Energy Coefficient κ, J/m
<111>, B2 and DO_3	4.7×10^{-11}
<100>, DO_3	2.0×10^{-11}

Comparison of Bulk and In-Situ Studies

A comparison of the data obtained in the bulk and the in-situ
migration experiments carried out at 745 and 778 K allows the topological coefficient φ defined in eqn.(2) to be determined. In making this
comparison, we assume that the in-situ experiments give data for
intrinsic migration kinetics, and we use the data from the high
driving-force regime of Table 1. The calculations give $\varphi = 1$ at 745 K
and $\varphi = 0.5$ at 778 K. These results seem surprisingly large. For an
idealized structure of right circular cylindrical domain boundaries
occupying half of the volume of the material, it is easily shown that
$\varphi = 1$. Thus, the analysis of the experimental data implies that the
domain boundary shapes in bulk specimens are on average cylindrical.
This is at variance with the long-recognized concept of a multiply-con-
nected morphology for the domain structure [12]. The values of φ
implied by this study could be too large if some unanticipated drag
force were operating in the thin-foil experiment. The most obvious
possibility, grooving at the intersections of the APBs with the surfaces of the thin-foil specimens, should be negligible because the APB
energies [3] are so small in comparison to the free surface energies.
Additional analysis is required to fully interpret the values of φ
reported here.

Given the value of S_v at the point at which the bulk domain coarsening kinetics change, as seen in the plots of Fig. 1, plus knowledge of φ, it is possible to estimate the curvature K* at which the transition to high driving-force behavior would be expected in the in-situ experiments. The calculation gives estimates of the conditions for high driving-force behavior when $K^* > 1.4 \times 10^7$ m^{-1} at 745 K, and when $K^* > 0.5 \times 10^7$ m^{-1} at 778 K. From Fig. 3, it is seen that a portion of the data points at 745 K falls below this curvature value, while all of the data points at 778 K lie well above this curvature value. The most basic conclusion from this analysis is that the bulk of the in-situ data were indeed taken in the high driving-force "intrinsic" regime. The data at 745 K shown in Fig. 3 are not consistent with a change of M when $K^* = 1.4 \times 10^7$ m^{-1}. One possible reason for this is a rather large uncertainty in the value of S_v where M changes in the bulk domain coarsening experiments (i.e. the point where the slope changes in Fig. 1). There is a tendency in the in-situ data at 745 K for reduced mobilities when $K < 0.7 \times 10^{-7}$ m^{-1}, but for such gentle curvatures migration distances are small and hence errors in the measurements are larger than for the high-curvature data points. A second possibility is that the actual value of φ is less than that reported above. This would make the true value of K* smaller.

ACKNOWLEDGEMENTS

This research was supported by the National Science Foundation, Grant DMR-8606706. The alloy used in this study was graciously provided by William Kerr of the U.S. Air Force Materials Laboratory.

REFERENCES

1. S.M. Allen and J.W. Cahn, Acta metall. 27, 1085 (1979).
2. J.E. Krzanowski and S.M. Allen, Acta metall. 34, 1045 (1986).
3. J.E. Krzanowski and S.M. Allen, Acta metall. 34, 1035 (1986).
4. W. Park and S.M. Allen, in: Materials Problem Solving with the Electron Microscope, L.W. Hobbs, K.H. Westmacott and D.B. Williams, eds., Materials Research Society Symposium Vol. 62, p. 303, Materials Research Society, Pittsburgh, PA (1986).
5. P.R. Swann, W.R. Duff and R.M. Fisher, Metall. Trans 3, 409 (1972).
6. J.W. Cahn, Acta metall. 10, 789 (1962).
7. R.C. Sun and C.L. Bauer, Acta metall. 18, 635 (1970).
8. W. Park and S.M. Allen, unpublished research (1987).
9. J.W. Cahn and J.E. Hilliard, J. Chem. Phys. 28, 258 (1958).
10. A.B. Kuper, D. Lazarus, J.R. Manning and C.T. Tomizuka, Phys. Rev. 104, 1536 (1956).
11. K. Nishida, T. Yamamoto, and T. Nagata, Trans. Jap. Inst. Metals 12, 310 (1971).
12. A. English, Trans. Met. Soc. AIME 236, 14 (1966).

THE REMARKABLE INCUBATION TIME

ON THE ORDER-DISORDER PHASE TRANSITION OF Mg_3In

Hiroyuki Konishi and Yukio Noda

Faculty of Engineering Science
Osaka University
Toyonaka, Osaka 560, Japan

INTRODUCTION

In a first order phase transition, the system transforms from a metastable state to a stable state, where the probability distribution function of the system in the phase space must ride over the barrier of the local free energy potential between the stable state and the metastable state associated with thermodynamical fluctuations. This stochastic process is often expressed by a nucleation-growth picture in a real space. Avrami,[1,2,3] Johnson and Mehl[4] found that the time development of the fractional volume of a stable state phase was given by the equation

$$V(t)=1-\exp(-kt^n) , \qquad (1)$$

where k and n are constants and related to the mechanism of the nucleation-growth process. This equation suggests the existence of scaling properties associated with the kinetics of the first order phase transition. It is almost established that the characteristic time τ_c, which depends on the metastability of the system showing divergent behavior near the transition point, is defined to describe the ordering process. An universal function expressing the time development of the phase transition should be constructed in terms of the scaled time $\tau=t/\tau_c$.

We have investigated the kinetic process of the order-disorder phase transformation in a Mg_3In alloy by a time resolved X-ray diffractometry. Mg_3In is one of the Cu_3Au-type ordered alloy. It is well known that an A_3B type binary alloy, such as Cu_3Au, Ni_3Mn etc., takes an ordered structure at the low temperature phase and undergoes an order-disorder phase transformation concerned with the atomic rearrengement at the appropriate temperature.[5,6] Many studies have been made extensively on the order-disorder phase transformation by diffraction methods.[7,8] The change of the crystal symmetry on the order-disorder phase transformation is found as the appearance of X-ray, neutron and electron new diffraction spots, that is, a superlattice reflection is given as the order parameter of the phase transition. If the ordering or the disordering process is controlled by the nucleation-growth process, the intensity of the superlattice reflection is proportional to the entire volume of the ordered region in a nonequilibrium state. The width of a superlattice reflection profile along the direction of the scattering angle is concerned with the average radius of ordered droplets. Therefore, the kinetics of the ordering or the disordering process can be directly observed by a diffraction method when the 'pattern formation' of the system down to the nucleation-growth level is acceptable.

We can categorize nucleation processes into two classes. One is a homogeneous nucleation process and the other is a heterogeneous nucleation process. In the latter case, crystal imperfection (impurity, defect, dislocation etc.) will decrease the local potential barrier, so that a nucleation occurs at such a specific site. The crystal perfectibility may be different between a single crystal and a polycrystal. The kinetics of the ordering and the disordering processes of the Mg_3In alloy has not been studied previously. The purpose of the present study is to ascertain universality of equation (1) and the existence of the divergence property of the characteristic time in the vicinity of the phase transition point by using a single crystal and a polycrystalline sample of the Mg_3In alloy.

EXPERIMENTAL

Magnesium(99.99%) and indium(99.999%) pure metals were melted under flux in an alumina crucible with atmosphere at about 800°C. We used the mixture of LiCl and KCl as flux. The crucible containing these materials was sealed in a quartz glass tube in vacuum and settled in a Bridgman type furnace at 800°C for 20 hours. After this heat treatment, the melted alloy was cooled down to the room temperature within 15 hours to obtain a poly-crystalline sample. We adopted the Bridgman method in order to obtain a single crystal. The melting alloy sealed in a quartz tube was crystallized by bringing slowly to the end of the furnace with the temperature gradient. Two kinds of ingots were annealed for 2 days at 400°C in a evacuated quartz tube.

The ingots were cut to get plate-like samples having a (110) surface. These plates were polished by an emery paper and etched with HNO_3 aqueous solution. A typical size of samples used for the X-ray experiment was $7x7x1mm^3$. The sample was mounted in a small vacuum furnace. An alumel-chromel thermocouple was attached on the sample. The sample temperature was controlled within 0.05K by a microcomputer system. In the present work, we needed to change the sample temperature stepwise across the phase transition point. On the stepwise change of the temperature, less than 10 seconds were required to get the aimed temperature after a small damping oscillation with a maximum deviation of 0.15K.

The X-ray source was a rotating anode generator operated at 50kVx60mA with a fine focused filament and a copper target during the experiments. CuKα radiation was monochromatized with a pyrolitic graphite (002) plane. A position sensitive proportional counter (abbreviated as PSPC) was used. In our X-ray optical system, one channel of PSPC is chosen to be 0.02 degree of 2θ. The typical mosaicness of the single crystal was 0.3 degree (FWHM). When the single crystal was annealed more than one month at 370°C in the vacuum furnace, very fine crystal grains developed in the sample, which size was found to be about $0.1x0.1mm^2$ by measuring the diffracted beam size. The mosaic spread within each grain was less than the instrumental resolution of our X-ray optic system. In the present study on single crystals, a tightly collimated incident beam was used to perform 'the X-ray diffraction experiment from the selected area' by picking up one of grains. Alternative experiments were performed on the polycrystalline sample by using the X-ray beam radiated on the almost entire surface of the sample.

Before performing experiments of kinetics, we investigated the proper-ties of each sample in the thermal equilibrium state. The observed temperature dependence of the integrated intensity of the (110) superlattice reflection from a grain within the single crystal was shown in Fig.1. We

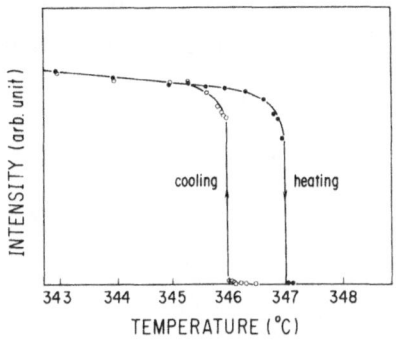

Fig. 1. The temperature dependence of the integrated intensity of
the (110) superlattice reflection of the single crystal.

waited for 30 minutes at each temperature in order to accomplish the
equilibrium state after the sample temperature was changed successively.
The stepwise change and hysteresis phenomena of order parameters shown in
Fig.1 clearly indicate that the order of the phase transformation in Mg_3In
is of first order. We also waited for several hours near the phase
transition point, but the temperature dependence of the order parameter was
invariable. There was 0.13% change of the lattice parameter through the
phase transition between the ordered simple cubic phase and the disordered
f.c.c phase. The lattice parameter also showed a stepwise temperature
dependence and hysteresis phenomena. Though the transition point in each
grain was determined exactly, there was the difference of the transition
temperature among each grain within about 1K. On the experiment of polycrystalline sample, the order parameter and the lattice parameter gradually
changed near the transition point over the temperature range of 10K. But
the hysteresis phenomena could be observed.

The kinetics of the ordering and the disordering processes of Mg_3In in
the vicinity of the phase transition point Tc are measured by the following
procedure: First, the sample was kept in the equilibrium state of the
disordered phase (Ts=Tc+1K) for the ordering process and in the ordered
phase (Ts=Tc-1K) for the disordering process. Then the temperature was
suddenly changed from Ts to the final temperature Tf across Tc. Tf was
below Tc (Tf=Tc-ΔT) for the ordering process and above Tc (Tf=Tc+ΔT) for the
disordering process. The time development of the X-ray diffraction profile
of the (110) superlattice reflection was observed successively in situ by
using the PSPC after the sample was brought to the nonequilibrium state.
Each result was labeled by the value of the parameter ΔT which corresponds
to the metastability of the system in the nonequilibrium state.

RESULTS

There was no detectable width change in FWHM of scattering profiles of
the (110) superlattice reflection during the ordering process, that is, the
intrinsic width of the profile is within the instrumental resolution
function of our X-ray optical system in both cases, the single crystal and
the polycrystalline sample. Therefore, the critical radius of the ordered
domain is more than 800Å. Similar results were obtained during the
disordering process of the single crystal and the polycrystalline sample,
respectively. On the present experiment, we only investigated the time
evolution of the integrated intensity.

The time evolution of the integrated intensity of the (110) superlattice reflection I(t) of the polycrystalline sample is shown in Fig.2

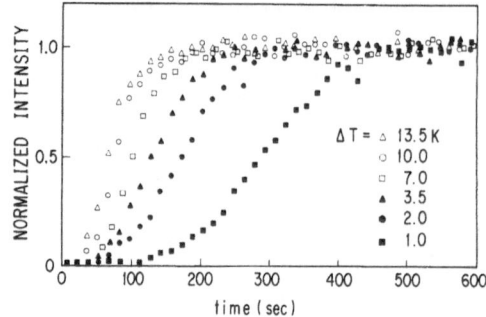

Fig. 2. The time evolution of the integrated intensity of the (110)
       superlattice reflection during the ordering process of the
       polycrystalline sample.

(time resolution Δt=15 seconds). All curves in Fig.2 are those of ordinary
Avrami type expressed by equation (1). Let us define the characteristic
time τ_c by $I(\tau_c)=0.5$. It is easily seen from the figure that τ_c behaves
divergently near Tc as a function of ΔT. It is also easy to find that a
scaling property of this kinetic process is given in terms of the scaled
time $\tau=t/\tau_c$. On the other hand, in the ordering process of the single
crystal, the behavior of I(t) from selected area is extraordinary. An
example is given in Fig.3. As shown in the upper side figure (Δt=3sec), the
intensity of the (110) superlattice reflection was not detected at all
during comparatively long time after the sample temperature was changed

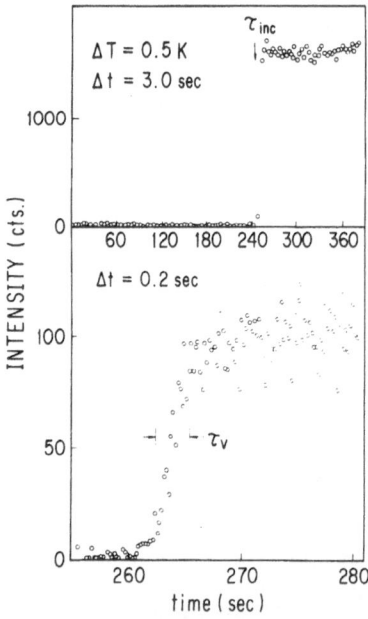

Fig. 3. The time evolution of the integrated intensity of the (110)
       superlattice reflection during the ordering process of the
       single crystal. Upper figure is taken with the time resolu-
       tion Δt=3sec and shows the data from the origin of time. The
       lower shows only the time region where the intensity changes
       abruptly, by taking with better time resolution.

stepwisely (within 10 seconds). After some finite characteristic time, the intensity abruptly increases like as a step function. Such a characteristic time is often observed on various phase transition phenomena and called as an incubation time. However, there is no example of such remarkable incubation time as observed in the present work. The incubation time τ_{inc} behaves divergently in the vicinity of Tc as a function of ΔT;

$$\tau_{inc} = A\Delta T^{-n},$$

where n is about 0.7 and A strongly depends on grains. We investigated the reproducibility of τ_{inc} at the same ΔT and with the same grain by repeating the same experiment many times. It was found that τ_{inc} distributes for the each experiment slightly beyond the instrumental error, but within 10% of the mean value of τ_{inc} for the given ΔT.

Using shorter time resolution Δt, we measured the growth time of the ordered phase more precisely. The result is shown in the lower side in Fig.3 ($\Delta t=0.2sec$). It takes a finite time τ_v (3 seconds in the example) until to be occupied by the ordered phase the almost entire volume of the X-ray irradiated region. This growth time τ_v must be another characteristic time. In fact, the behavior of τ_v, that is, universality, the divergence property and the grain dependence, are similar to those of τ_{inc}.

We can construct an universal function $I(\tau)$ in the ordering process of the single crystal in term of $\tau=t/\tau_{inc}$. However, if we try to fit the equation (1) to the observed universal curve shown in Fig.3, the value of index n must be larger than several hundreds. Such a large n loses physical meanings in the nucleation-growth picture.

The disordering process of the selected area is also extraordinary. As shown in Fig.4, the intensity of the superlattice reflection gradually decreases at the early stage of the transition process, and at some finite time τ_{inc}, the intensity suddenly disappears. The observed τ_{inc} again shows the critical divergence character as a function of ΔT.

CONCLUSION AND DISCUSSION

We performed the time resolved X-ray experiment during the ordering process and the disordering process of the Mg_3In alloy, both of the polycrystalline sample and the single one. The remarkable result is that

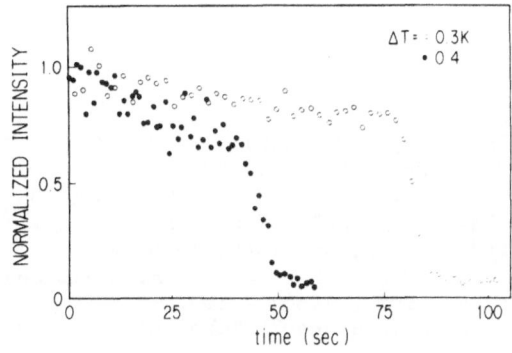

Fig. 4. The time evolution of the integrated intensity of the (110) superlattice reflection during the disordering process of the single crystal.

the incubation time is detected as shown in Fig.3 and Fig.4 when the very good quality grain was picked up by the tightly collimated beam. Furthermore, the incubation time depends on grain to grain. One will suspect that the growth process comes from the outside of the grain, just passing the selected area for the period τ_v. In this idea, τ_{inc} is nothing but the nucleation point is at some finite distance from the present area. However, this idea has following difficulties. First, the distribution of the observed τ_{inc} on the same ΔT is considerably small so that the starting time of the nuclei at the outside position must have again their own incubation time. Moreover, we can estimate the position where the nucleation starts, with the velocity of the grain front and the incubation time. The growth velocity is estimated by dividing the grain size by τ_v. The answer we obtained is that the distance between the present grain and the suspected nucleation site is larger than the sample size. Consequently, the incubation time is really intrinsic on the transition process in the Mg_3In alloy.

When we accept the idea of the incubation time within the grain, it is easy to introduce the conclusion that only one of the nucleation site will contribute for the growth process, by comparing τ_{inc} and τ_v. That is, one of the grains started from the particular nucleation site covers the whole area of the present qualified crystal and other embryos must be taken into the growing domain. This consideration is favorable to compare with the fact that we could not observe the width of the superlattice reflection at all even at the early stage of the growing process. If there are many stable nuclei in a given area, one would observe the width corresponding to the averaged size of the nuclei. One of the possibility is that the observed transition process in the Mg_3In alloy is governed by the heterogeneous nucleation as an elemental process. The behavior of the transition process frequently seen in a polycrystalline sample described by equation (1) is the result of the statistical average of such an elemental process distributed in a sample. Other possibility is that the system is really homogeneous in the grain and the homogeneous nucleation process under isothermal conditions are realized. According such a model,[9] the incubation time is interpreted to be the formation time of the steady state distribution of clusters. The origin of the incubation time is now under the consideration.

ACKNOWLEDGEMENTS

We are grateful to Professor H. Iwasaki for providing the sample used at the very early stage of the experiment and to Dr. S. Nasu for his helpful advice to grow the single crystal. We also wish to thank Mr. H. Satake for his technical support. Grateful acknowledgement is made to Professor Y. Yamada for several helpful discussions.

REFERENCES

1. M. Avrami: J. Chem. Phys. **7** (1939) 1103.
2. M. Avrami: J. Chem. Phys. **8** (1940) 212.
3. M. Avrami: J. Chem. Phys. **9** (1941) 177.
4. W. A. Johnson and R. F. Mehl: Trans. AIME **135** (1939) 416.
5. T. Muto and Y. Takagi: in Solid State Physics, ed. F. Seitz and D. Turnbull (Acamedic Press, New York, 1955) Vol.1.
6. L. Guttman; in Solid State Physics, ed. F. Seitz and D. Turnbull (Acamedic Press, New York, 1956) Vol.3.
7. T. Hashimoto, K. Nishimura and Y. Takeuchi: J. Phys. Soc. Jpn. **45** (1978) 1127.
8. Y. Noda, S. Nishihara and Y. Yamada: J. Phys. Soc. Japan **53** (1984) 4241.
9. K. F. Kelton, A. L. Greer and C. V. Thompson: J. Chem. Phys. **79** (1983) 6261.

DYNAMICS OF ORDERING WITH PHASE SEPARATION IN IRON-SILICON ALLOYS

Syo Matsumura, Hitoshi Oyama and Kensuke Oki

Department of Materials Science and Technology, Graduate
School of Engineering Sciences, Kyushu University 39
Kasuga-shi, Fukuoka 816, Japan

INTRODUCTION

In the phase diagrams of some alloys which undergo order-disorder transition of higher-order, there are sometimes two-phase fields where an ordered phase with lower-symmetry coexists with a higher-symmetry phase. In such a mixed-phase field, the ordering is expected to proceed at the same time as phase separation. Iron-aluminum alloys with Al content near 25 at% are a typical alloy-system in which we can observe such type of ordering reaction [1-4]. The corresponding phase diagram [5] has a triangular mixed-phase field of disordered A2(bcc) and B2(Pm3m), and also a phase field of (A2+DO$_3$(Fm3m)) in the lower temperature region. The transitions from A2 to B2 and from B2 to DO$_3$ are of higher-order at their critical temperatures, and eliminate the symmetry elements of translation along <111>a/4 and <100>a/2, respectively, where a is twice the lattice dimension of basic bcc lattice. The ordering reactions occuring in the alloys quenched into the two-phase fields were extensively observed using an electron microscope by Allen et al. [2,3] and the present authors [1,4]. The observation revealed complicated processes strongly depending on alloy-composition, temperature and initial state of order. Prior to the experiment, Allen and Cahn [2] predicted sequence of mechanisms during the processes, through discussion on the form of the equilibrium phase diagram of Fe-Al using Landau's mean field model for higher-order transitions. Eguchi and the present authors [6] have developed a simple kinetic theory based on a time-dependent Ginzburg-Landau model in order to understand the interplay of modulations in composition and degree of order for binary alloys. Using the kinetic equations derived, Eguchi and Ninomiya [7] and Ohta, Kawasaki and Sato [8] made computer-simulation on the pattern evolution within the miscibility gap between A2 and B2, as described in this proceedings. These thermodynamic considerations are in good agreement with the experimental results, which demonstrates that the models will provide useful insight into the dynamics of ordering with phase separation in binary alloys.

Both B2 and DO$_3$ types of ordered phase also appear in Fe-Si alloys less than 20at%Si. The phase diagram of the system contains a mixed-phase field of (B2+DO$_3$) in addition to their single phase regions [9], as shown in **Fig. 1**. The miscibility gap between B2 and DO$_3$ is absent from the equilibrium phase diagram of Fe-Al. It was previously shown in Fe-Si alloys [9] that the modulated structure of alternating B2 and DO$_3$ phases is formed along <100> directions after the phase separation. On the contrary, no modulated structure has been found in Fe-Al alloys. We are therefore interested in the dynamics of ordering processes occuring inside the miscibility gap of (B2+DO$_3$) in Fe-Si alloys, and have investigated transformations of B2→(B2+ DO$_3$) and DO$_3$→·(B2+DO$_3$) by electron microscopy and X-ray powder diffraction. In this paper, we briefly describe our results concerning only the transformation of B2→(B2+DO$_3$).

EXPERIMENTAL PROCEDURE

For X-ray diffraction and electron microscope experiments, specimens of powder finer than 200 mesh and discs with 3 mm diameter were obtained from bulks of Fe-13.8 and 14.6 at%Si alloys homogenized at 1173 K for three days. The specimens were kept at 1073 K for 22 ks to be in a single phase state of B2 type-order, and then decomposed into (B2+DO$_3$) by isothermal annealing within the miscibility gap. X-ray diffraction was carried out using a Philips PW-1730 diffractometer with filtered CoKα radiation. An electron microscope JEOL JEM-200B (HVEM Lab., Kyushu Univ.) was operated at 200 kV. Microstructures in the specimens were imaged with each of 111 and 222 superlattice reflections: the former reflection is characteristic of DO$_3$ phase, while the latter arises from both B2 and DO$_3$ phases.

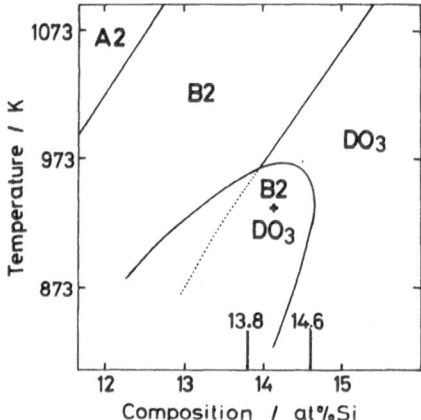

Fig. 1: A part of the phase diagram for Fe-Si system. The dotted line refers to metastable extension of the transition line of B2-DO$_3$ into miscibilty gap.

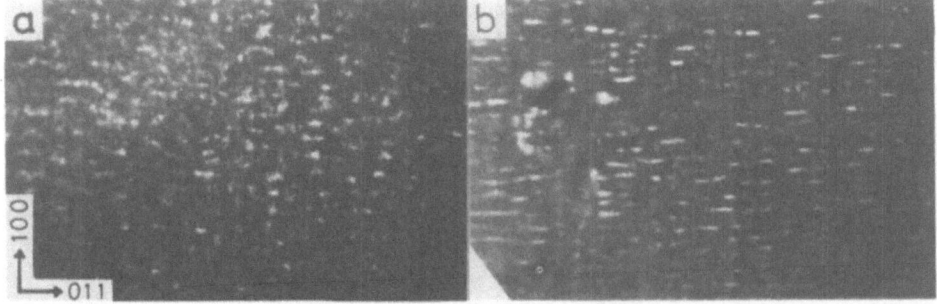

Fig. 2: 111 dark field images of Fe-13.8 at%Si annealed at 955 K for 0.3 (a) and 0.6 ks (b).

316

RESULTS AND DISCUSSION

Figure 2 shows 111 dark field images of an Fe-13.8 at%Si alloy annealed at 955 K for 0.3 and 0.6 ks. Regions with DO_3-type order are imaged
with white contrast in the micrographs. Cloud-like regions of DO_3 are
formed in the B2 matrix by annealing for 0.3 ks (a), and then the faces of
DO_3 regions become sharp and parallel to {100} with the annealing (b). The
grown plates of DO_3 were observed in images of specimens after prolonged
annealing.

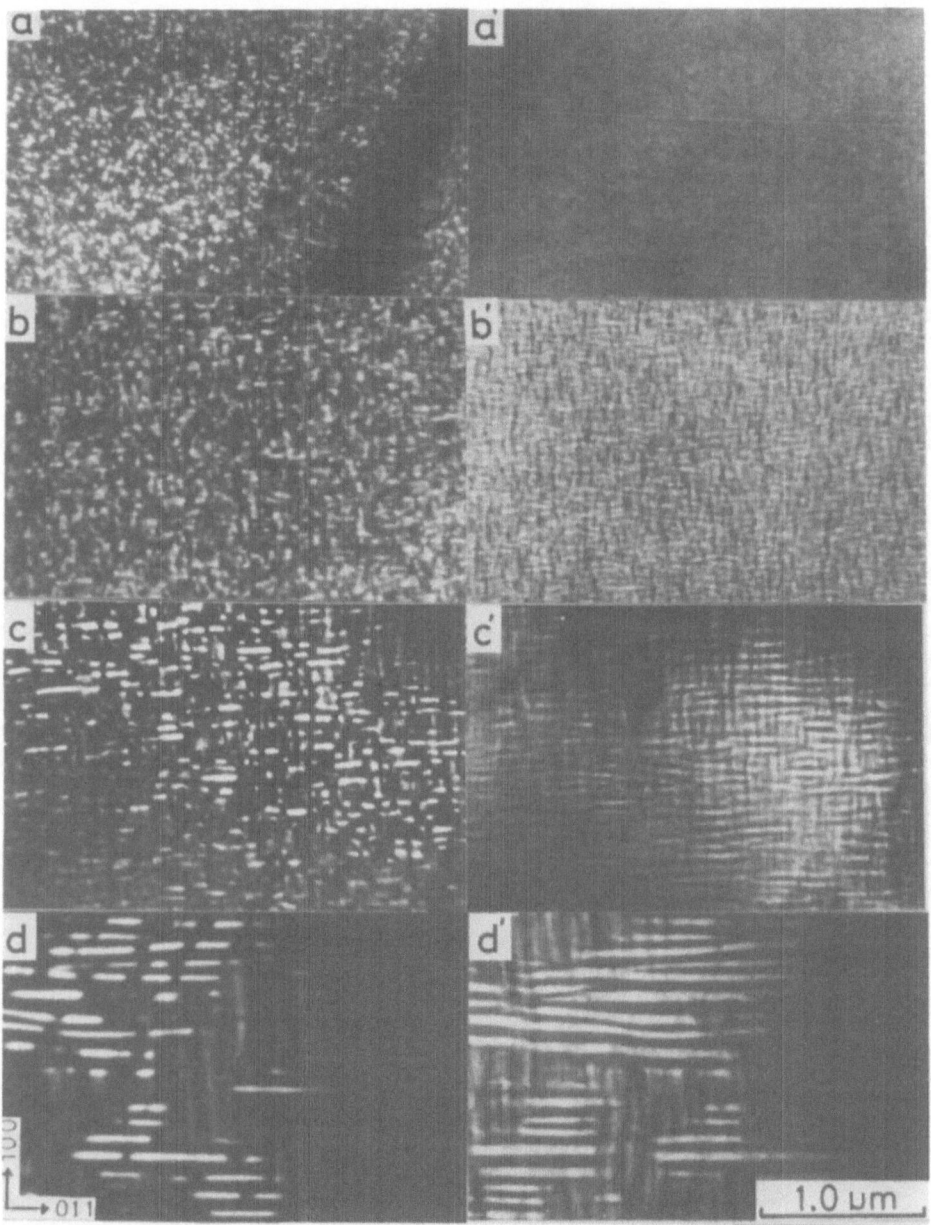

Fig. 3: 111 dark field images (left column) of Fe-13.8 at%Si annealed
at 923 K for 0.3 (a), 0.6 (b), 3.0 (c) and 100 ks (d), and their
corresponding 222 dark field images (right column: a',b',c'and d').

The change in microstructure during annealing at 923 K is demonstrated in **Fig. 3.** We can see fine black contrast corresponding to antiphase boundaries (APBs) with translation vector of $<100>a/2$ in Fig. 3(a), which reveals that DO_3-type ordering first occurs throughout the alloy, resulting in a fine-domain structure of DO_3. The APBs become flat parallel to {100} by annealing for 0.6 ks, as shown in Fig. 3(b). At the same time, tweed-like contrast appears along <100> directions in the 222 dark field image of Fig. 3(b'). The period of the tweed is nearly equal to the size of DO_3 domains. It seems that the B2 phase is formed mainly on the APBs of DO_3 domains. Thus the modulated structure of (B2+DO_3) develops along all three <100> directions: Fig. 3(c,c'). The microstructure after aging for a sufficient time consists of colonies in which B2 and DO_3 plates alternatively align along only one of the three <100> directions, as seen in Fig. 3(d,d'). We also investigated the process of ordering with phase separation at 873 K, and the process has been found to be similar to the one at 923 K.

The sequences of transformations from B2 to (B2+DO_3) in the Fe-13.8 at%Si alloy can be summarized into two types: (1) The coexisting phase state of (B2+DO_3) is directly established by the precipitation of DO_3 domains in the B2 matrix (at 955 K); (2) The DO_3-type order once prevails with numerous APBs in the alloy, and then the B2 phase is formed preferentially on the APBs (at 923, 873 K). The similar sequence of mechanisms has been found in Fe-Al alloys inside the miscibility gap, which has been explained by metastable extension of the higher-order transition lines into the miscibility gap retaining their meaning for single phase states [1,2,6]. Therefore we can consider that the transition line of B2-DO_3 also exists within the mixed-phase field of (B2+DO_3) for the Fe-Si system, as illustrated by the dotted line in Fig. 1. At temperatures above the transition line within the miscibility gap, the uniformly ordered state of B2 is metastable and the ordering of DO_3-type proceeds by nucleation and growth mechanism. While below the transition line, the single phase state with DO_3-type order is once set up and then decomposed into (B2+DO_3), because the B2 state becomes unstable with respect to the DO_3-type ordering. Microstructure in the alloy annealed at 953 K for 1 ks is shown in **Fig. 4**, in which fine-contrast is imaged in addition to the contrast from a coarser phase-separated structure. Such fine-contrast is absent from both Figs. 2 and 3. The characteristic fine-contrast is most likely caused by critical point fluctuation in degree of order appearing just below the B2-DO_3 transition temperature. Figure 4 is therefore a direct evidence for the existence of higher-order transition line of B2-DO_3 within the miscibility gap. However we suspect that the actual transition between metastable and unstable states is not so sharp as characterized by the transition line. It should be noted that the nucleation just above the transition line occurs in a considerably different way from the classical droplet formation, as seen in Fig. 2(a).

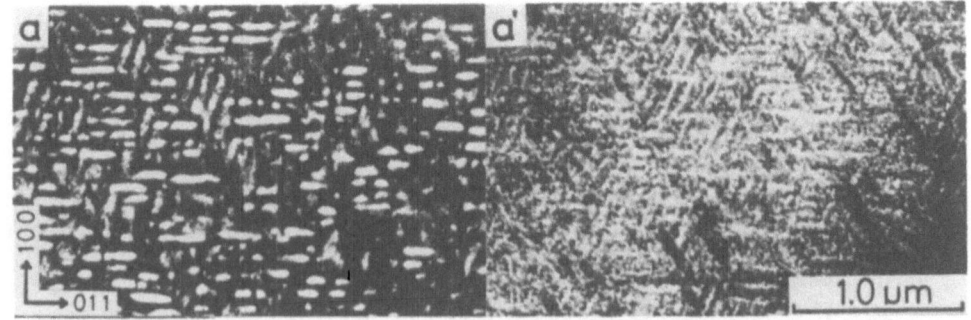

Fig. 4: Microstructure of Fe-13.8 at%Si annealed at 953 K for 1 ks. 111 (a) and 222 (a') dark field images.

Figure 5 gives microstructures in an Fe-14.6 at%Si alloy annealed at 953 K. The ordering is expected to proceed in a similar way to the second type observed in the 13.8 at%Si alloy, because the annealing temperature is lower than the B2-DO₃ transition temperature. Smoothly curved APBs with the translation vector of <100>$a/2$ are imaged in Fig. 5(a), which indicates that the specimen is in an ordered state of DO₃-type. Then the B2 phase is formed on the APBs by further annealing as expected (Fig. 5(b,b') and (c,c')). This observation therefore supports our consideration concerning the sequence of mechanisms. **Figure 6** shows the isothermal change in the mean values of degrees of order X and Y and in the line width of 440 fundamental reflection in the X-ray diffraction patterns. Here X and Y represent degrees of pair correlations of atoms at distances of <100> $a/2$ and <111> $a/4$ [10], respectively. In the B2 phase only the degree of order X should vanish. The initial state, however, has a non-zero value of X, as it is hard to suppress ordering in powder specimen during the quench from 1073 K. The value of X increases with annealing time as the ordering of DO₃-type proceeds, while that of Y is almost constant. The broadening of the fundamental line becomes remarkable at 6 ks. We regard this broadening is due to the coexistence of two phases whose lattice parameters are significantly different. The X-ray measurement also reveals that the DO₃-type ordering precedes the phase separation. However it should be emphasized that the phase separation starts before the completion of the ordering, as seen from Fig. 6.

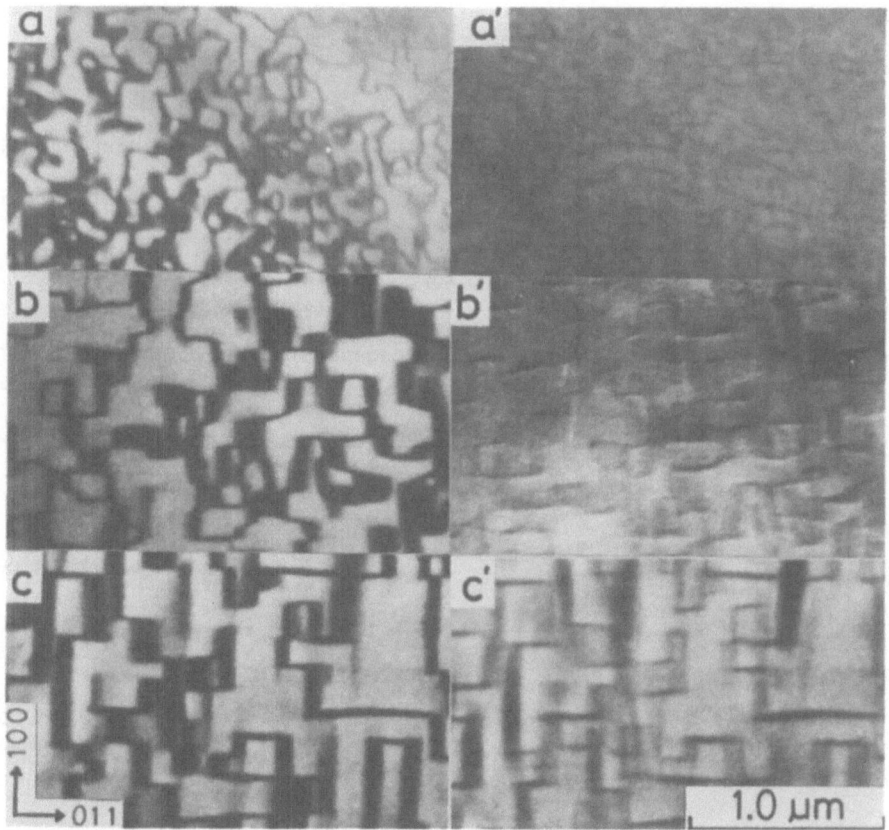

Fig. 5: 111 dark field images (left column) of Fe-14.6 at%Si annealed at 953 K for 0.3 (a), 2.0 (b) and 10 ks (c), and their corresponding 222 dark field images (right column: a',b',c').

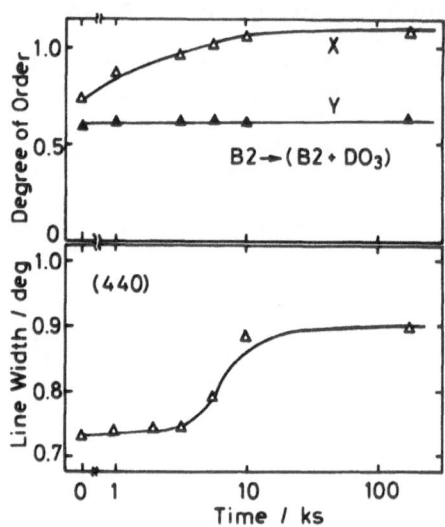

Fig. 6: Isothermal change in degrees of order X and Y, and in the 440 line-width of Fe-13.8 at%Si at 953 K.

REFERENCES

1. K. Oki, S. Matsumura and T. Eguchi,
    Phase separation and domain structure of iron-based ordering alloys,
    to be published in Phase Transitions, 9:35, (1987).
2. S.M. Allen and J.W. Cahn,   Mechanism of phase transformations within
    the miscibility gap of Fe-rich Fe-Al alloys, Acta Met., 24:425 (1976).
3. S.M. Allen,   Phase separation of Fe-Al alloys with Fe_3Al order,
    Philos. Mag., 36:181 (1977).
4. K. Oki, H. Sagane and T. Eguchi,   Separation and domain structure of
    α + B2 phase in Fe-Al alloys, · J. de Phys., 38:C7-414 (1977).
5. K. Oki, M. Hasaka and T. Eguchi,   Process of order-disorder transforma-
    tion in iron-aluminum alloys, Jpn. J. Appl. Phys., 12:1522 (1973).
6. T. Eguchi, K. Oki and S. Matsumura,   Kinetics of ordering with
    phase separation,   Mat. Res. Soc. Symp. Proc., 21:589 (1984).
7. T. Eguchi and H. Ninomiya, PC-visualization of phase transition
    in alloys,   in this proceedings.
8. T. Ohta, K. Kawasaki and A. Sato,   Ordering process in a quenched
    tricritical system,   in this proceedings.
9. P.R. Swann, L. Granäs and B. Lehtinen,   The B2 and DO_3 ordering reac-
    tions in iron-silicon alloys in the vicinity of the Curie
    temperature,   Metal Sci., 9:90 (1975).
10. T. Eguchi, H. Matsuda, K. Oki, S. Kiyoto and K. Yasutake,
    Order-disorder transformation in Fe-Al alloys,
    Trans. Jpn. Inst. Metals, 8:174 (1967).

DYNAMICAL SCALING IN THE KINETICS OF

PHASE SEPARATION AND ORDER-DISORDER TRANSITION

S. Katano and M. Iizumi

Department of Physics
Japan Atomic Energy Research Institute
Tokai, Ibaraki 319-11, Japan

INTRODUCTION

Recently there has been great interest in the dynamics of ordering as a system approaches equilibrium from an initial non-equilibrium state. Such studies are classified into two categories: the phase transitions with conserved order parameters and those with non-conserved order parameters. A typical example of the former cases is phase separation, and that of the latter is order-disorder transition. For both cases, an idea of self-similar growth have been ascertained; that is, the characteristic length $R(t)$ follows a power law as

$$R(t)=A(T)t^a, \tag{1}$$

where a is a universal exponent and $A(T)$ is a rate coefficient which depends on temperature T. Moreover, the non-equilibrium structure factor behaves according to

$$S(q,t)=R(t)^d F(qR(t)). \tag{2}$$

Here d is the dimensionality, q is the wave vector and F is a scaling function.

In order to verify this idea experimentally, we have done a neutron small-angle scattering study of phase separation in Fe-Cr alloy and a neutron diffraction study of order-disorder transition in Ni_3Mn alloy. For the following reasons these alloys were chosen. 1) The time constant is fairly long. This is advantageous to study the initial stage of the process. 2) Constituent atoms of these alloys have similar size, so the elastic strain involved in the transition should be small. This is good for a comparison with theories or computer simulations. 3) The neutron cross sections are favorable for measurements.

In this paper we will mainly discuss crossover phenomenon which is considered to be universal in dynamical scaling. A part of this paper was already published by the present authors (1984).

DYNAMICS OF PHASE SEPARATION

Figure 1 shows the time evolution of the structure factor $S(q,t)$ of Fe-Cr alloy annealed at $500°C$ for times up to 500 h. Measurements were

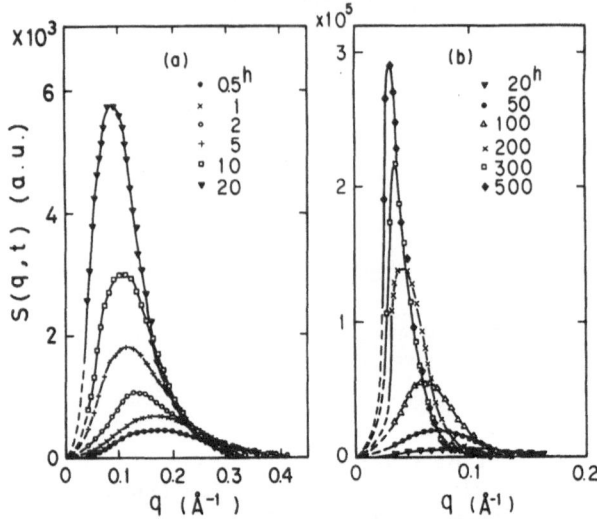

Fig. 1. Time evolution of the structure factor of Fe-Cr
alloy annealed at 500°C for aging times up to
500 h.

done at room temperature after the quench from the annealing temperature.
Neutron small-angle scattering was performed by a triple-axis spectrometer
in the diffraction arrangement. As shown in the figure, the structure
factor becomes sharp with increasing time. Moreover, the maximum position
shifts towards small vector and the peak intensity increases rapidly,
showing that the characteristic length $R(t)$ grows with time.

In order to examine the scaling property, we investigated the time
dependence of the wave vector $q_1(t)$ which is defined as the first moment of
the structure factor:

$$q_1 = \Sigma_q \, qS(q,t) / \Sigma_q \, S(q,t). \qquad (3)$$

In Fig. 2, q_1 is plotted as a function of time on the double logarithmic
scale. As is clearly seen, the entire time evolution cannot be expressed

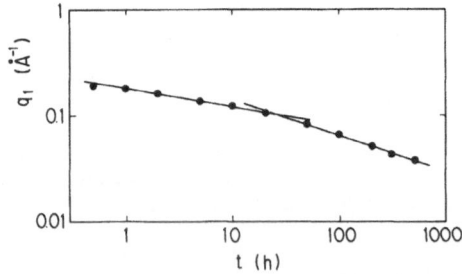

Fig. 2. Time dependence of the characteristic wave vector
$q_1(t)$. The full lines correspond to the exponent
$-1/6$ at the early stage and $-1/3$ at the late one.

by a single power law. The result indicates that the growth is slow at the initial stage, then it becomes faster. However, by choosing appropriate data points, we can fit the respective range by straight lines. The slope shown in the figure is -1/6 at the early stage, and -1/3 at the late one. Thus the experimental result indicates crossover of the exponent. The former value coincides with the exponent predicted by Binder and Stauffer (1974) for the coagulation process by the surface diffusion. The latter was obtained by Lifshitz and Slyozov (1961) for the condensation process. The result suggests that the mechanism of phase separation crossovers when the system approaches the equilibrium state.

In order to extend the comparison with the scaling theory, the normalized scaling function $\widetilde{F}$ was calculated. $\widetilde{F}$ is given as

$$\widehat{F}[q/q_1(t)]=[q_1(t)]^3\widetilde{S}(q,t),\tag{4}$$

where the normalized structure factor $\widetilde{S}(q,t)$ is defined as

$$\widetilde{S}(q,t)=S(q,t)/\sum_q q^2S(q,t)\,\delta q.\tag{5}$$

Here δq is the increment of the wave vector in the experiment. The plot of $q_1^3S(q,t)$ for different times shows also the change of the scaling function $F(q,t)$. This change seems to correspond with the crossover of q_1.

A similar crossover phenomenon was recently reported by Forouhi and de Fontaine (1987) in the study of phase separation in Al-Zn alloy by X-ray small-angle scattering and transmission electron microscopy.

DYNAMICS OF ORDER-DISORDER TRANSITION

Figure 3 shows the time evolution of the 211 superlattice reflection. The sample was heated to 600°C and kept at this temperature over 30 min to achieve the disordered state. The temperature was, then, abruptly changed to 470°C across the transition temperature around 510°C. The temperature of the sample was reached the final temperature in about 15 s by a furnace which can change the sample temperature very rapidly. The diffraction patterns were obtained by the Wide-angle Neutron Diffractometer (WAND) installed at the High Flux Isotope Reactor at the Oak Ridge National Laboratory. Since this diffractometer consists of a curved one-dimensional position-sensitive detector which covers a 130° angle, we are able to do time-resolved neutron diffraction experiments. In order to attain sufficient counting statistics, identical measurements were repeated several times and each of data was accumulated on the memory of the data acquisition system. As seen in the figure, a peak over a wide wave vector is observed at the initial stage, which suggests that fairly small ordered regions are formed in the sample. With increasing time the width of the peak becomes narrow, i.e. the size of the ordered region R(t) grows, and the background level decreases.

In order to compare the result with the power law (2), we calculated the second moment q_2

$$q_2(t)=\sum_q q^2S(q,t)/\sum_q S(q,t),\tag{6}$$

which corresponds to the mean square of the peak width. In this calculation the pure profile, which was obtained by the deconvolution of the diffraction pattern using the 220 fundamental reflection as a measure

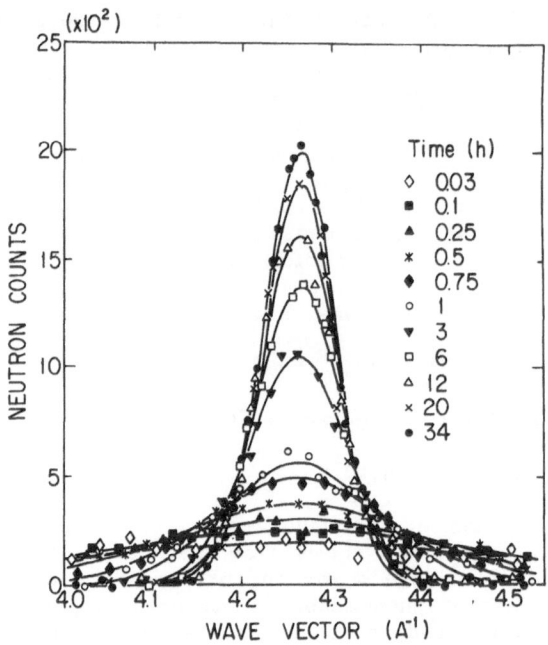

Fig. 3. Time evolution of the 211 superlattice peak of Ni$_3$Mn annealed at 470°C for times up to 34 h.

of the instrumental resolution, was used as the structure factor $S(q,t)$. The result is shown in Fig. 4 on the double logarithmic scale. As is clearly seen, $q_2^{1/2}$ decreases gradually but its entire time evolution cannot be expressed by a single power law. The result indicates that the growth of the ordered regions is slow at the initial stage, then becomes faster.

As mentioned above the entire data cannot be expressed by a single line. However, the data at the late stage can be fitted by a straight line

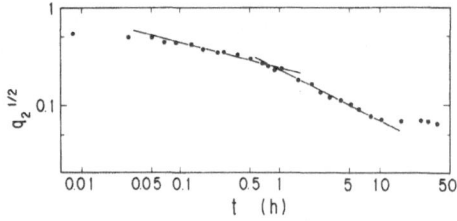

Fig. 4. Time dependence of $q_2(t)^{1/2}$. The solid lines show the exponent -1/4 at the early stage and -1/2 at the late one.

with the slope of -1/2. This exponent coincides with the prediction of Allen and Cahn (1979) for the process by the migration of domain walls, which was verified by many works of simulations or experiments. For the earlier stage we can fit the data to the straight line with the slope of -1/4. Thus the exponent in this case also shows crossover which was observed in phase separation of alloy systems. Therefore this kind of crossover is considered to be universal in the growth of the order. The exponent of -1/4 is explained by the theory of Furukawa (1983). For the non-conserved case, he predicted that the exponent for the coalescence process is -1/3 for the bulk diffusion and -1/4 for the surface diffusion. Therefore it is suggested that the surface diffusion takes place at this stage. The obtained result indicates that the ordering process changes from the stage of the slow growth by coalescence of clusters to that of the fast growth by the migration of the domain walls.

According to the equation (2), a calculation of the scaling function F is in progress. The detailed result of this work will be published elsewhere.

The latter part of this work was carried out at the Oak Ridge National Laboratory under the US-Japan Cooperative Program on Neutron Scattering. The authors would like to express our appreciation to the members of the neutron scattering group of the Oak Ridge National Laboratory for their hospitality.

REFERENCES

Allen, S. M., and Cahn, J. W., 1979, Acta. Metall. 27:1085.
Binder, K., and Stauffer, D., 1974, Phys. Rev. Lett. 33:1006.
Forouhi, A. R., and de Fontaine, D., 1987, Acta. Metall. 35:1863.
Furukawa, H., 1983, Phys. Lett. 97A:346.
Katano, S., and Iizumi, M., 1984, Phys. Rev. Lett. 52:835.
Lifshitz, I. M., and Slyozov, V. V., 1961, J. Phys. Chem. Solids 19:35.

TIME RESOLVED X-RAY SCATTERING STUDY OF THE ORDERING KINETICS IN THIN

FILMS OF Cu_3Au

S. E. Nagler, R. F. Shannon, Jr. and C. R. Harkless

Department of Physics
University of Florida
Gainesville, Florida 32611 U.S.A.

INTRODUCTION

A general understanding of the non-equilibrium ordering kinetics of first order phase transitions is an important goal in statistical physics and materials science. Many fundamental questions are unresolved,[1-3] including the existence and precise nature of "universality classes" that describe growth laws, the role of the ordering degeneracy p, and the effect of impurities.[4,5] Metallic alloys are convenient experimental systems for examining these phenomena. In this paper we report the preliminary results of time resolved x-ray scattering measurements of the kinetics associated with the order-disorder transition in thin films of Cu_3Au.

The equilibrium properties of Cu_3Au have been widely studied and there are many thorough reviews on the subject.[6] The alloy forms as FCC and orders in the Ll_2 structure (ie. Au on the conventional unit cell corners, Cu on the face centers) below $T_c \approx 663$ K. The degeneracy of the ordered state is therefore p = 4. Studies of the ordering kinetics in bulk Cu_3Au[7-9] found that at long times after quenching the typical domain size L(t) (as determined from the inverse radial width of the (1,1,0) superlattice peak) grows roughly as $L(t) \sim t^{1/2}$, consistent with a domain boundary curvature driven growth law.[10]

Dimensional effects in non-equilibrium ordering phenomena can be probed by systematic experiments on films of different thickness. Previous studies of ordering kinetics in thin films[11,12] of Cu_3Au have examined the effect of annealing on the "as deposited" films at $T < T_c$. In the work reported here measurements were made after quenching the sample from an initial temperature $T_i > T_c$ to a final temperature $T_f < T_c$. Time resolved x-ray scattering experiments have been performed in a transmission geometry allowing for different superlattice peaks to be characterized in the same sample. As discussed below, this probes anisotropy as well as different types of domain boundaries involved in the ordering process.

EXPERIMENTAL DETAILS

The films were deposited by sputtering onto a 0.13 mm thick HN kapton substrate. The final thickness was verified using a stylus profilometer. The films thicknesses used in the measurements reported here were 7500 ± 300 Å (film A) and 950 ± 50 Å (film B). Analysis of the films using Rutherford backscattering and electron microprobe techniques found the composition to be homogeneous with relative atomic concentrations Cu (0.75 ± .01) Au (0.25 ± .01). X-ray diffraction showed the "as deposited" films to be disordered, with a (1,1,1) orientation perpendicular to the substrate to within a mosaic spread of 3.5° HWHM (7500 Å film). The in-plane orientations were uniformly averaged. The lattice constant at room temperature was 3.74 Å. A preliminary annealing of the film for 12 hours at a temperature of roughly 350° C ordered the sample, but also caused the (1,1,1) orientation to sharpen to 3.0° HWHM. The enhanced orientational order explains some apparently anomalous results reported in previous studies on films.[11] It is important to anneal the specimens well to avoid confusing this effect with phenomena related to ordering.

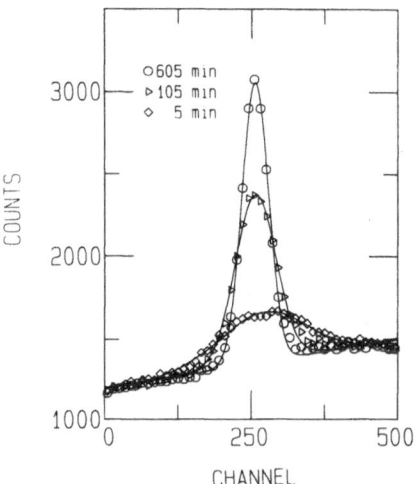

Fig. 1.   Typical (1,-1,0) profiles in a 7500 Å thick Cu_3Au film quenched to 638 K. For clarity of presentation, the data points have been averaged in groups of 10. Solid lines are fits as described in the text. Counting times are five minutes for each 1024 channel spectrum. The central 500 channels are shown.

For the time resolved x-ray scattering measurements the films were mounted in a furnace with kapton windows. The temperature was measured using a cold junction compensated K thermocouple mounted in contact with the film and control was achieved using a Micristar 828-D controller. The temperature stability realized was ± 0.3 K. After a typical temperature quench ($T_i - T_f \approx 30$ K) control was re-established at T_f in one minute.

The furnace was mounted in a large Huber eulerian cradle so that the sample lay at the center of rotation. X-rays were produced by a Rigaku RU-300 rotating anode source operating at 15-18 kw with a copper target in the line focus geometry. The x-rays were reflected from the (0,0,2) planes of a flat pyrolytic graphite monochromator and the resulting beam passed through a monitor counter and a slit system to the specimen position. Only the $CuK\alpha_1$ component was allowed to reach the specimen. X-rays passing through the film were scattered into a linear platinum wire based position sensitive detector (PSD) (Braun model OEO-50M) mounted on the 2θ scattering arm. The detector signal was binned into a 1024 channel multichannel analyser where each channel corresponds to $0.0125°$ 2θ. The resultant instrumental resolution was $0.06°$ HWHM in 2θ.

For each specimen T_c was determined by monitoring the temperature at which superlattice peaks disappeared. The time resolved measurements were typically made by heating the film well above T_c to ensure complete disorder, then, in some cases, cooling to a few degrees above T_c to achieve an equilibrium value for T_i. The film was then quenched rapidly to T_f. Spectra from the PSD were stored sequentially in memory. Detailed measurements were made at the (1,0,0) superlattice peak, as well as the (1,-1,0) peak. (The notation (1,-1,0) is used to indicate that the peak is perpendicular to the (1,1,1) direction.) Some typical data for the film A (1,-1,0) superlattice peak are shown in figure 1.

DISCUSSION

The observed peaks were fitted to gaussian lineshapes

$$I(x) = BG(x) + A \exp-\left((x-x_o)^2/2\sigma^2\right) \tag{1}$$

where I is the intensity in counts observed at channel x, BG(x) is a background arising primarily from the substrate, and the remaining parameters are the amplitude A, peak position x_o, and width σ. The gaussian function was found to give a better fit than either Lorentzian or Lorentzian squared functions for the (1,0,0) and (1,-1,0) superlattice peaks. Typical fits are shown in figure 1. The fundamental peaks such as (2,0,0) and (2,-2,0) also fit very well to gaussian lineshapes.

The resulting σ values for the superlattice peaks were corrected for finite size effects and instrumental resolution:

$$\sigma_c = \left(\sigma_o^2 - \sigma_s^2 - \sigma_R^2\right)^{1/2} \tag{2}$$

where σ_c is the corrected value, σ_o the fitted parameter, σ_s the contribution arising from finite size effects and σ_R the contribution from instrumental resolution. The dominant correction arises from the finite grain size. To account for anisotropy the superlattice peaks were corrected using the results obtained from fundamental peaks in the same direction in reciprocal space. The corrected values σ_c should be proportional to L^{-1}, and were fitted to a power law as a function of time

$$\sigma_c(t) = at^{-b} \tag{3}$$

Figure 2 shows the resulting fits for film A for a quench with $\Delta = T_c - T_f = 20$ K. The first important feature is the marked difference in the $(1,0,0)$ and (1,-1,0) widths. This arises from the type 1 domain boundaries that form easily in the ordered Cu_3Au structure.[6,13] Type 1 boundaries are formed by a half diagonal glide in a mixed Cu-Au layer. The corresponding scattering in reciprocal space is disk shaped. At the (1,0,0) peak the radial direction is along the disk axis while at the

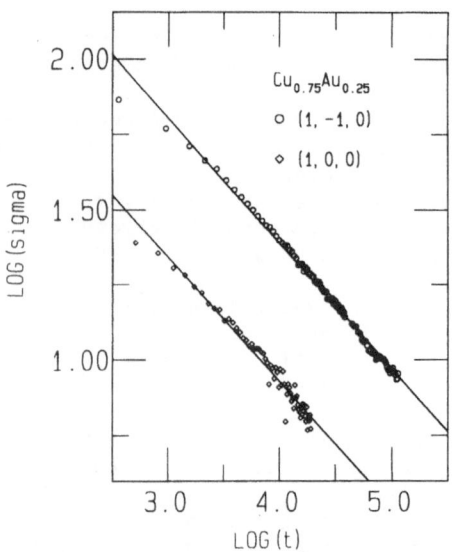

Fig. 2. log(σ_c) (channels) vs. log(t) (seconds) for a 7500 Å film with quench depth Δ = 20 K. The solid lines are power law fits as discussed in the text.

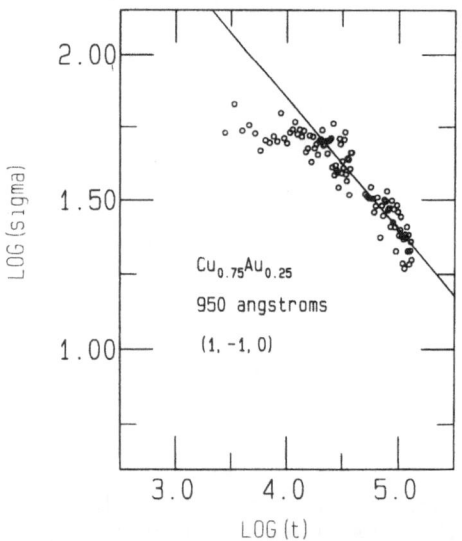

Fig. 3. log(σ_c) (channels) vs. log(t) (seconds) for a 950 Å film with quench depth Δ = 25 K. The solid line is a power law fit to the late time data.

Table 1. Results of Power Law Fits

Film	Peak	$T_f(K)$	$T_i(K)$	$T_c(K)$	Correction	b
A	(1,−1,0)	638	666	658	17.1	0.42 + .12 − .05
A	(1,0,0)	637	669	658	14.8	0.41 + .06 − .12
B	(1,−1,0)	638	675	663	21.3	0.45 ± .06

Note: correction = $\left(\sigma_s^2 + \sigma_R^2\right)^{1/2}$ is given in channels.
1 channel = .0125° 2θ.

(1,−1,0) peak the radial direction lies along the diameter.

Examination of the fits in figure 2 shows a systematic deviation from linearity at early times that is attributable to incomplete nucleation of the ordered phase. Small deviations at the latest times measured may arise from uncertainties in the finite size corrections. The corrections are the largest source of uncertainty in estimating the power law describing domain growth. In table 1 the error estimates for b include the largest uncertainties that we feel may arise from the resultant systematic errors. The fitted slopes in figure 2 yield the result b = 0.4 for both the (1,0,0) and (1,−1,0) data. The exponent b = 0.5 expected for an isotropic system with a non-conserved order parameter lies within the estimated error interval for the (1,−1,0) data. If the value of b is lower than that measured in the bulk[7] the difference may be due to finite size effects. In one sense it may be surprising that the exponents are equal.[14] The width of the (1,−1,0) peak provides a measure of the distance between type 1 domain boundaries; these may be expected to follow a curvature driven growth law leading to a $t^{1/2}$ time dependence. On the other hand, the (1,0,0) peak width indicates the presence of type 2 boundaries. The type 2 boundaries can have Au enriched or depleted layers across the entire boundary,[13] so diffusion of many atoms is necessary to conserve the correct local concentrations as the boundary moves. This might be expected to give a slower growth law. The data reported here cannot rule out an anisotropic growth law, and the uncertainties in the systematic corrections may be consistent with b ~ 0.3 for the (1,0,0) width.

Figure 3 shows a plot of the (1,−1,0) peak of film B. Although the peak is broader, the fitted growth law exponent at late times is equal within errors to that of the 7500 Å film. However, the onset of late stage behavior is significantly delayed. This may indicate that nucleation in the 950 Å film is significantly slower, even though the quench depth is slightly deeper than that of the data shown in figure 2. The difference probably does not arise from the presence of impurities, since the expected result would then be an early time $t^{1/2}$ law crossing over to a slower growth law at late times.[4,5] We do not yet have an explanation for this observation, and a more systematic experiment will be necessary to determine its significance.

CONCLUSIONS

Time resolved x-ray scattering studies of ordering in thin films of Cu_3Au allow many different aspects of domain growth processes to be investigated. The results to date find that late time ordering in Cu_3Au is probably isotropic, and growth of domain sizes in films down to 950 Å is consistent with a Cahn-Allen growth law as observed in bulk crystals. Early time ordering may be strongly affected by film thickness. We hope to obtain much improved data using newly available alloy films prepared by molecular beam epitaxy.[15]

ACKNOWLEDGEMENTS

We thank P.H. Holloway and J.K. Truman for their valuable help with the sample preparation, W. Ruby for his technical expertise, N. S. Sullivan for a critical reading of the manuscript, and G. Smith for typing. We have benefitted greatly by discussions with many colleagues. This work has been supported by the U.S. Department of Energy under award DE-FG05-86ER45280. One of us (SEN) is a NSF Presidential Young Investigator DMR-8553282.

REFERENCES

1. J. D. Gunton, M. San Miguel, and P. S. Sahni in Phase Transitions and Critical Phenomena, Vol. 8, ed. C. Domb and J. Lebowitz, Academic, New York, NY (1983).
2. A. Milchev, K. Binder, and D. W. Heerman, Z. Phys. B63:521 (1986).
3. G. Mazenko, O. T. Valls, and F. C. Zhang, Phys. Rev. B31:4453 (1985).
4. D. J. Srolovitz and G. N. Hassold, Phys. Rev. B35:6902 (1987).
5. D. Chowdhury, M. Grant, and J. D. Gunton, Phys. Rev. B35:6792 (1987).
6. For example, see B. Warren, X-ray Diffraction, Addison-Wesley, Reading, MA (1969).
7. Y. Noda, S. Nishihara, and Y. Yamada, J. Phys. Soc. Jpn. 53:4241 (1984).
8. S. Nishihara, Y. Noda, and Y. Yamada, Sol. St. Comm. 44:1487 (1982).
9. T. Hashimoto, K. Nishimura, and Y. Takeuchi, J. Phys. Soc. Jpn. 45:1127 (1978).
10. S. M. Allen and J. W. Cahn, Acta Met 27:1085 (1979).
11. K. N. Tu, Scripta Met 14:663 (1980).
12. V. ZH. Yelizarov, A. I. Shkurko, and V. G. Kazakov, Phys. Met. Metall. 52:(5)181 (1981).
13. R. Kikuchi and J. W. Cahn, Acta Met. 27:1337 (1979).
14. J. W. Cahn, Scripta Met. 14:93 (1980).
15. C. P. Flynn, private communication.

QUENCHING AND THE ORDERING PROCESS

OF DIPOLE MOMENTS IN $K_2Ba(NO_2)_4$

Yukio Noda

Faculty of Engineering Science
Osaka University
Toyonaka,Osaka 560,Japan

INTRODUCTION

The dynamical properties of phase transitions of various kinds are subjected to theoretical and experimental studies during the past decades. The typical situation widely studied is that a system is rapidly quenched from a disordered phase to an ordered phase through over the first-order phase transition point[1]. In such a quenched system of the first-order transition, the system transforms from a metastable state to a stable state by riding over the barrier of thermodynamical potential between two local minima. In addition to the above non-linearity of the potential in the system, the slow diffusion of atoms sometimes plays an important role to really observe the transition process by experiments, especially seen in the metal alloy system such as Cu_3Au [2-4], Ni_3Mn [5] *etc.*

Another situation is a sort of recrystallization process of glass or amorphous material. Extensive works were performed in this field by using a synthesized glass such as a metallic glass [6]. If one can obtain a frozen disordered phase, by some reason, it is easy to conjecture that the system will recover the long range order by annealing the system near the order-disorder phase transition point, just like as the above recrystallization process. In this case, the transition is no matter of first-order or of second-order, and also including the diffusion process or not.

As a candidate of the frozen disordered system, we investigated the frustrated triangle lattice of an electrical dipole moment. The material we studied was $K_2Ba(NO_2)_4$, which shows a peculiar phase transition scheme as is previously reported [7,8]. It contains NO_2 molecule groups accompanied by a permanent electric dipole moment. The highest temperature phase (Phase I) is characterized by the orientational disordering of the NO_2 groups sitting on the Kagome lattice in the hexagonal symmetry (P6/mmm). In Phase III below T_c=201K, those NO_2 groups order into a full ordered phase (FOD) by the second-order phase transition. The interesting point of this system is that there is an intermediate phase (Phase II) characterized by a partial ordering of the NO_2 groups (POD), which implies the frustration of the interaction between the NO_2 groups. In general, the transition from an ordered phase to another ordered phase is of first-order. However, the transition from the POD Phase to the FOD phase in $K_2Ba(NO_2)_4$ is of second-order and its behavior on the transition is quite similar to that of two-dimensional systems such as Ising anti-ferromagnetic K_2CoF_4 [8-10].

In Fig. 1, the schematic structure of $K_2Ba(NO_2)_4$ in Phase II is given, where the hexagonal lattice is also shown in addition to the orthorhombic lattice (Pbam). There are three types of NO_2 groups in the orthorhombic unit cell, ordered NO_2 groups on 4g site (m_z symmetry), disordered ones on 2c site (2/m symmetry) lying on the z=0 plane, and the disordered NO_2 groups pointing up and down on 4e site (2_z symmetry) of the z=1/2 plane. From the figure, one can see that the remained disordered NO_2 groups in the POD phase form a plane separated by ordered NO_2 groups and potassium atoms. It should also be noticed that the hexagonality is still there, approximately, because the b/a ratio of the orthorhombic lattice is 1.7303 and very close to $\sqrt{3}$.

From the view point of the diffraction experiment, the phase transition is characterized by the appearance of a new Bragg reflection or a

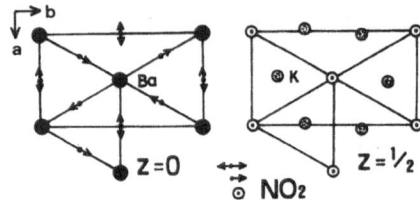

Fig.1.   Schematic structure of $K_2Ba(NO_2)_4$ in Phase II given by Harada. Arrows on z=0 plane represent the ordered and disordered NO_2 groups lying on the plane. Other disordered NO_2 groups are on z=1/2 and pointing up and down.

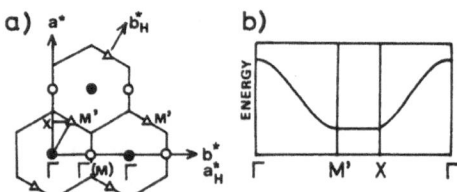

Fig.2.   a) Reciprocal space of Phase II. Closed circles are the lattice points of the hexagonal phase, open circles (M-point) a superlattice position on the transition from the disordered phase to the POD phase. M'-point represented by an opened triangle is the unstable point on the transition to the FOD phase (c*-direction is ignored).
b) Interaction energy between the remained disordered NO_2 groups caluculated by Yamada and Harada.

superlattice reflection at the M-point of the hexagonal reciprocal lattice depicted by open circles in Fig.2a), which turns to be a Γ- point in the POD phase. On the other hand, it is reported that new Bragg reflections appear at the (1/2 1/2 1/2) position in the orthorhombic reciprocal lattice of the POD phase on the transition to the FOD phase. In Fig.2a), the reciprocal point of the new Bragg reflection is shown by an opened triangle (M'-point), where the c^*-direction is ignored for the sake of simplicities.

Recently, Yamada and Harada discussed the mechanism of the phase transition of $K_2Ba(NO_2)_4$ by assuming pair interactions between NO_2 groups up to the third nearest neighbors [11]. The key point of the appearance of the POD phase is the second nearest neighbor interaction J_2 in their model. It is clear that there is a frustration between the disordered NO_2 groups on the triangle Kagome lattice if J_2 is negative, and indeed they showed that the interaction energy at the M-point is the lowest and the realized orientational pattern of the NO_2 group is that of the POD phase as shown on the z=0 plane of Fig.1), under the condition of $|J_2| > \frac{1}{3} \cdot J_1$. Moreover, they investigated the interaction energy between the remained disordered NO_2 groups by introducing the third nearest neighbor interactions J_3 and J_3' within the hexagonal symmetry. It is again obvious that J_3 and J_3' give the frustration character to line up the NO_2 groups in the b-plane and between the b-planes of the orthorhombic lattice. In Fig.2b), the interaction energy of the remained disordered system calculated by Yamada and Harada is schematically shown. It is remarkable that the interaction energy is not only the lowest at the M'-point but also dispersionless along the M'-X line. In other words, any kind of phase relations between the b-planes is a ground state within the above model. Obviously, this degeneracy of the ground state comes from the frustration character of J_3 and J_3' as stated above.

Now, we have a strategy to obtain the frozen disordered system. It is natural to consider that the ordering process of the NO_2 groups from the POD phase to the FOD phase is relatively fast within the b-plane but is slow between the b-planes because of the degeneracy of the ground state. If such a system is rapidly quenched, an enhancement of the slow transition process is expected because the rearrangement of the macroscopic order of the b-plane requires a long relaxation time.

In this paper, we investigate the possibility of quenching disordered NO_2 groups of the POD phase in $K_2Ba(NO_2)_4$ and its ordering process by annealing the system near the transition point.

EXPERIMENTAL

Time resolved X-ray diffraction experiments were carried out on single crystals of $K_2Ba(NO_2)_4$. Single crystals grown from the aquaous solutions, and also crystals kindly offered by Dr. M. Harada of the University of Tokyo were used. A single domain was carefully taken from the twined crystals under the polarising microscope. Typical size of a single domain used on the present X-ray experiments is about $1 \times 1 \times 1.5$ mm^3. A rotating anode X-ray generator (Rigaku RU-200) operated at 50KV and 60mA with a fine focused filament was used for the experiment. A copper target was used and a pyrolitic graphite (002) reflection was served as a monochromator. A position sensitive proportional counter (PSPC) which was settled along the vertical direction was effectively utilized for the present *in situ* experiments to measure the ordering process in $K_2Ba(NO_2)_4$. The crystal was mounted on the goniometer head so that the horizontal scattering plane was the (HOL)-plane and the vertical axis was

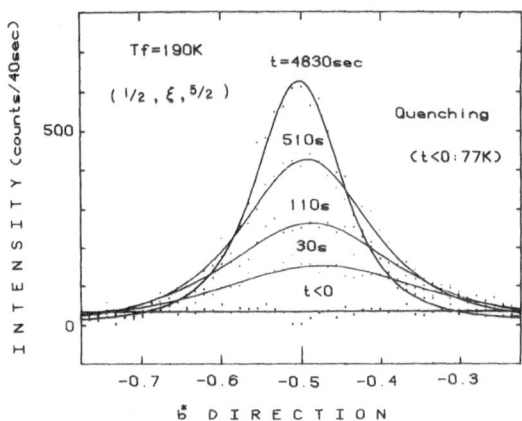

Fig.3.   X-ray Diffraction profiles around (1/2 -1/2 5/2) at
          77K and its time evolution at T_f=190K. Time resolu-
          tion of this particular measurement is 40 seconds.

along the b*-direction. Therefore, the PSPC can measure the b*-direction
simultaneously. We measured the (1/2 ξ 5/2) line of the reciprocal
space including the (1/2 ±1/2 5/2) new Bragg positions.

The cold nitrogen gas flow was used to cool down and to controle the
temperature of the sample. For the rapid quenching, the method to drop
the liquid nitrogen onto the crystal was also used. The temperature was
stabilized within 0.04-0.2K, depending on the quenching conditions, by a
computer-controled system.

Let us define the expression 'quenching' and 'rapid-cooling' for the
method to cool down the temperature on the present experiment. First, we
hold the temperature of the sample at 240K and then cool it down to the
final target temperature T_f very rapidly through T_c by using the nitrogen
gas flow; typically the period to get T_f is around two minutes. This is
the method of 'rapid-cooling'. Alternative method is 'quenching' where
the sample is rapidly quenched by dropping the liquid nitrogen, within
two seconds, from 240K to 77K. After the sample is kept at 77K for a
few minutes, the temperature is raised up to T_f for around 30 seconds.
During the temperature change of the sample and also after the time when
the sample temperature is held at T_f, that is, during the annealing the
sample at T_f, the diffraction pattern along the b*-direction is measured
by the time resolved X-ray diffractmetry with the PSPC.

In Fig.3, the example of the diffraction pattern around (1/2 -1/2
5/2) is shown, where the background observed at 240K was already sub-
tracted. Here, the origin of the time development is chosen to be the
time when the sample temperature is reached at T_f from 77K. Time resolution of this particular measurements was 40 seconds, or other words, the
intensity was accumlated for 40 seconds for each. Each point in the
figure is equivalent to the channel position of the PSPC lying along the
b*-direction. Reader will easily see from the figure that there is no
remarkable intensity of the new Bragg reflection at 77K when the sample
is quenched very rapidly. That means we can safely say that the orientational disordering of the NO_2 groups in the POD phase is successfully
frozen at 77K. Small amount of the intensity in the figure at 77K might
be the possibility of the small amount of ordering in the b-plane, not
along the b-direction. After the measurements at 77K, we raised up the
temperature to T_f=190K within a minute. Time evolution of the satellite
reflection is clearly seen in the figure during the annealing of the

sample at T_f. We systematically investigated the time evolution of the (1/2 ±1/2 5/2) Bragg reflection for the various T_f by two alternative methods. Essentially there was no difference between these two methods, especially at the later stage of the dynamics.

We studied not only the b*-direction but also the a*-direction by mapping out the intensity contour of (1/2+ζ 1/2+ξ 5/2) at T_f=190K. The results tell us that anomalous broadening of the Bragg reflection is limited along the b*-direction and there is no remarkable brodening along the a*-direction, that is, the width along the a*-direction is the same with the resolution function, about 0.02a*, measured by using the (003) Bragg reflection. The measurements along the b*-direction are, therefore, essentially important on the present experiments.

ANALYSIS

The observed diffraction profiles of (1/2 1/2 5/2) and (1/2 -1/2 5/2) superlattice reflections along the b*-direction were well fitted by a squared Lorentzian function. Solid lines in Fig.3 are the example of the fitted curves. As fitting parameters, the time development of the width and the peak intensity of the profile are given. The resolution correction was performed to obtain the intrinsic width; typically the instrumental resolution was 0.033b* measured at the (003) fundamental Bragg reflection.

Shown in Fig.4 are the time dependences of intrinsic widths on the rapid-cooling experiments, summarized as a logarithm form. It is seen from the figure that the time dependence of the widths behaves the straight line at the later stage, that is,

$$\Gamma(t;\Delta T) = \Lambda_o(\Delta T) \cdot t^{-n}$$

where $\Delta T = T_c - T_f$. Deviations from the above equation at the early stage strongly depend on the method how to quench the system. In the figure, the slope n at the early stage is definitely smaller than 1/4 and then approaches to around 1/4. On the other hand, the method of 'quenching' gives oposite behavior, that is, the larger value of n tends to 1/4 when

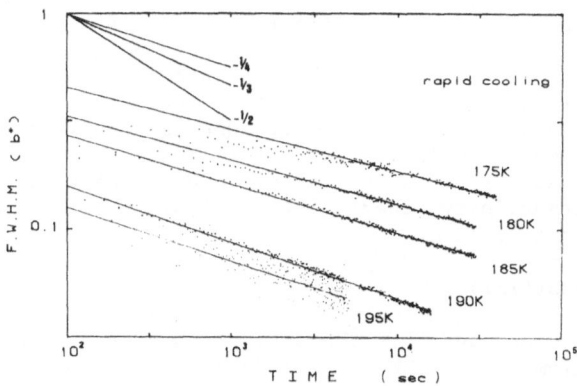

Fig.4.   Time dependences of the width of the (1/2 ±1/2 5/2) superlattice reflection on various T_f. Log Γ vs Log t is plotted. Slopes of -1/4, -1/3 and -1/2 are shown at the left-top to compare with the observed slope.

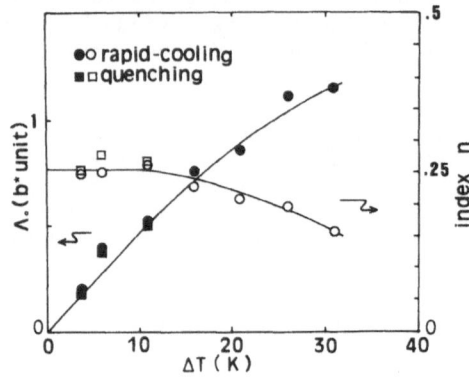

Fig.5.   Observed parameters Λ_0 and n for various T_f, of the
          asymptotic form of the width; $\Gamma(t;\Delta T)=\Lambda_0(\Delta T)t^{-n}$.
          Solid lines are drawn for the guide of eyes.

the ordering process goes on. In both methods, the absolute value of the
widths and the slope of the line at the later stage are consistent with
each other, and also the reproducibility of the data are satisfactory for
several runs at the same T_f. Obtained parameters $\Lambda_0(\Delta T)$ and n are
depicted in Fig.5. The value of n is about 1/4 when ΔT is relatively
small and then deviates to the smaller value on larger ΔT. Λ_0, whcih
gives the initial correlation length at the bigining of the time, is
proportional to ΔT on the smaller ΔT and again deviates from the linear
relation at the larger ΔT. Surprising points are that the correlation
length of the planes along the b*-direction are only two layers of the
plane around $\Delta T=20K$ and ten layers at $\Delta T=5K$. It should be remineded
that the transition of $K_2Ba(NO_2)_4$ at $T_c=201K$ is of second-order and one
can expect the long correlation just above the transition point due to
the fluctuation of the order parameter.

    The integrated intensity of the profile $I(t;\Delta T)$ is calculated
straight fowardly by using the obtained width and the peak intensity.
Let us introduce the characteristic time $\tau_{\frac{1}{2}}$ as the ordinary way;

    $I(\tau_{\frac{1}{2}};\Delta T)=\frac{1}{2}\cdot I(\infty;\Delta T)$.

It is interesting to see the scaling form of the fractional volume of the
ordered region. In Fig.6, the scaled integrated intensities are plotted
against the scaled time $\tau=t/\tau_{\frac{1}{2}}(\Delta T)$. The data points fall on to a
single universal curve, irrespective of the value of ΔT. In Fig.7, we
summarized the data $\tau_{\frac{1}{2}}$ for various ΔT. One can easily see from the
figure that the characteristic time $\tau_{\frac{1}{2}}(\Delta T)$ becomes systematically longer
as T_f is decreased, or ΔT is increased. One of the trial function to
explain the extraordinary behavior of $\tau_{\frac{1}{2}}(\Delta T)$ is given by the following
activation process [6],

    $\tau_{\frac{1}{2}}(\Delta T)=A\cdot\exp(E/kT)$,

where A is a constant and E is the activation energy. Let us rewrite the
equation as follow;

    $\tau_{\frac{1}{2}}(\Delta T)=\tau_0\cdot\exp(a\cdot\Delta T/T_c)$.

Here, we assumed that the activation energy is proportional to T_c and the
temperature factor is expanded around T_c by using the quench depth ΔT.

338

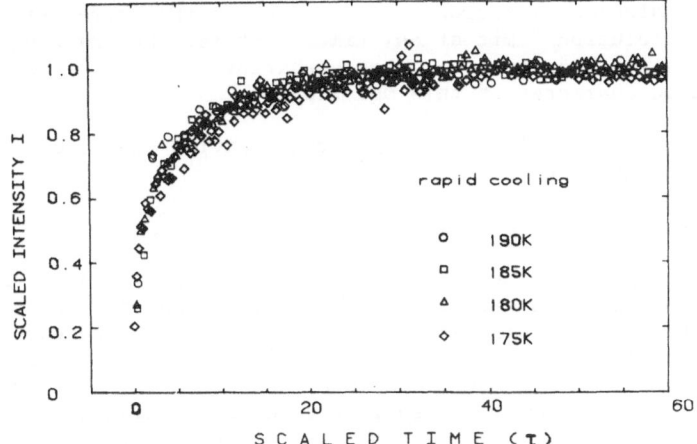

Fig.6.   Universal behavior of the scaled integrated intensity
          of the (1/2 -1/2 5/2) superlattice reflection against
          the scaled time.

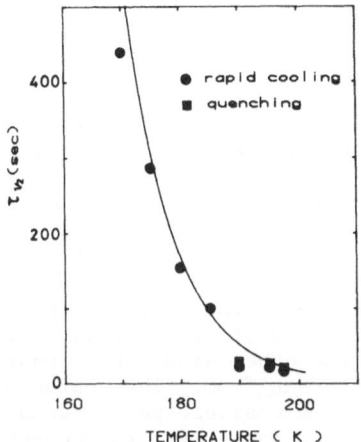

Fig.7.   The characteristic time $\tau^{\frac{1}{2}}$ for various T_f. Solid
          line is a calculated one by assuming the activation
          process.

τ_0 is interpreted as an elemental characteristic time at T_c. The solid
line in Fig.7 is the calculated curve by assuming $a = 0.12 \cdot T_c$ and $\tau_0 = 15$
seconds.

CONCLUSION AND DISCUSSION

We have seen that the orientational disorder of NO_2 groups in
$K_2Ba(NO_2)_4$ could be successfully frozen by rapid quenching. We have also
revealed that the quenched disordered system recovers its order during the
annealing around T_c. The characteristic of the ordering process of
$K_2Ba(NO_2)_4$ is that the system behaves as if it is two dimentional system.

The ordering along a-direction is very long and the correlation length is beyond the resolution limit at any time, that is, the ordering process is limited along the b-direction. These behavior are ones we expected from the frustrated character of this material.

We obtained the following asymptotic behavior of the correlation length at the later stage;

$$\xi = \Lambda_o^{-1} t^{\frac{1}{4}}.$$

Here, Λ_o is proportional to the quench depth ΔT as $\Lambda_o \sim 4.3 \times 10^{-3} \Delta T$ or $\Lambda_o^{-1} \sim 230/\Delta T$, when ΔT is small. It is not clear that the deviation from the linear relation of the data to ΔT is intrinsic or the experimental error due to the difficulties that the real late stage at the larger ΔT is too long and it is almost impossible to measure it. The scaling behavior of the ordered volume is also investigated by using the scaled integrated intensity and the scaled time, as depicted in Fig.6. The characteristic time τ_1 does not show any critical behavior at T_c but is expressed by a simple activation process as is shwon in Fig.7.

It is worthwhile to compare the above hehavior of the phase transition process observed in the present system with ones obtained in the first-order phase transition of Cu_3Au. In ref.2, the correlation length is given by

$$\xi = \Lambda_o^{-1} t^{\frac{1}{2}},$$

where Λ_o is about 1.4, irrespective of the value of ΔT. There are two remarkable differences compared with the present results. First of all, the index to the time in the present system is 1/4 and might be extraordinary. At the present stage, we have no explanation of the value, but we believe that the quantity 1/4 is the key value to consider the microscopic mechanism of the ordering precess. The difference of Λ_o between two cases are rather easy to consider. In the first-order phase transition, the system will ride over the thermodynamical potential where the energy difference is almost constant and relatively large, irrespective of the value of ΔT. On the other hand, the situation of the present system is completely different from. The representative point of the system in the phase space is unstable point because of the second-order phase transition character. Therefore, the energy difference between the starting point and the final stable point must be proportional to the quenched depth ΔT. When ΔT is small, the initial correlation length becomes long enough and, even at $\Delta T \sim 20K$, the initial correlation length is still ten times longer than that in the first-order phase transition. However it should be enphasized here that the correlation length at $\Delta T= 20K$ of $K_2Ba(NO_2)_4$ is only few layers distance along b-direction and the order of 10 Å. We have to rather say that the correlation length in the first-order phase transition of Cu_3Au is extremely short.

The difference of the critical behavior around T_c is also interesting. In the first-order phase transition, critical divergences of the characteristic time and the critical size of the droplet are reported just *below* the transition point T_c. In the second-order phase transition, on the other hand, the critical divergence of the correlation length is commonly given just *above* the transition point and there is no concept of the critical size. The temperature dependence of the observed characteristic time of $K_2Ba(NO_2)_4$ was explained by the ordinary activation process. Recently, Unruh and Bruckner [12] reported the dielectric measurement, in which the activation energy is given as 0.5eV and 0.2eV for two type of NO_2 groups, above T_c. It is interesting to compare their activation energy with one on the phase transition process below T_c;

about 0.3eV when the data at 201K and 170K in Fig.7 are used.

The other system we have to compare with our experiment is the two-dimentional magnetic system investigated by Ikeda [13,14], where the spin ordering process of the typical Ising antiferromagnets $Rb_2Co_xMg_{1-x}F_4$ was measured by neutron scattering technique. When the sample is rapidly cooled down below T_N, two dimentional ordering in the c-plane, in this case, takes place almost perfectly *and then* the interplanar ordering starts. This is really the example of the two-dimentional system and one should consider the one-dimentional rearrangement process. Theoretically, Nagai and Kawasaki [15] treated the system by a one-dimentional kink-antikink system to explane the observed time evolution of the peak intensity of the magnetic superlattice reflection. In their model, the averaged thickness of the kink develops by $t^{\frac{1}{2}}$ rule at the early stage and then tends to Log t behavior, partially because the interaction energy between kinks or antikinks is described by an exponential form.

When the above two systems, the two-dimensional magnetic system and the frustrated electric dipole moment system, are compared with each other, one will notice the similarity and disparity. The diffraction pattern is limited to one direction forming the Bragg rod for both systems; the former is on the strong diffuse scattering coming from the two dimensionalities. The remarkable difference is again the exponent of scattering. This is probably due to the difference of the degree of the two dimensionality. The remarkable difference is again the exponent of the time development at the late stage; the former gives Log t behavior and the latter gives $t^{\frac{1}{4}}$ behavior. This extraordinary value 1/4 suggests the existing of the different mechanism from the kink model how to line up the disordered system during the transition process. This point is the most important study we have to make clear in the future.

We are still standing at the starting point to understand the transition process of this system and experimentally we have to discover much more systems with a wide variety of new physical concepts.

Acknowledgments

The author would like to thank Mr. T. Yonekawa for his great technical help to perform the present experiments. He also wishes to express his sincerely thanks to Dr. M. Harada of the University of Tokyo for offering the excellent single crystal, and also to Professor Y. Yamada of the University of Tokyo for his fruitful discussions.

References

1) J.D.Gunton, M.S.Miguel and P.S.Sahni: *Phase Transition and Critical Phenomena*, edited by C.Domb and J.L.Lebowitz (academic Press, 1983, Vol.8,pp267).
2) Y.Noda, S.Nishihara and Y.Yamada: J. Phys. Soc. Jpn. 53(1984)4241.
3) D.G.Morris, F.M.Besag and R.E.Smallman: Philos. Mag.29(1974)43.
4) T.Hashimoto, K.Nishihara and Y.Takeuchi: J. Phys. Soc. Jpn. 45(1978)1127.
5) M.R.Collins and H.C.Teh: Phys. Rev. Lett. 30(1973)781.
6) M.G.Scott: J. Mater. Sci. 13(1978)291.
7) M.Harada: J. Phys. Soc. Jpn. 52(1983)3448.
8) M.Harada: J. Phys. Soc. Jpn. 55(1986)1051.

9) H.G.Unruh: Phys. Status Solidi (b) **126**(1984)115.

10) D.J.Breed, K.Gilijamse and A.R.Miedema: Physica **45**(1969) 205..

11) Y.Yamada and M.Harada: J. Phys. Soc. Jpn. **55**(1986)4315.

12) H.G.Unruh and H.J.Bruckner: Jpn. J. Appl. Phys. **24**(1985)suppl. 370.

13) H.Ikeda: J. Phys. Soc. Jpn. **52**(1983) suppl. 33.

14) H.Ikeda: J. Phys. C **16**(1983)3563.

15) T.Nagai and K.Kawasaki: Physica **120A**(1983)587.

SOFT ACOUSTIC MODES AND THE FORMING PROCESS OF

THE FERROELASTIC DOMAINS IN $KD_3(SeO_3)_2$

Toshirou Yagi and Zhi-li Lu

Department of Physics
Kyushu University 33
Hakozaki 6-10-1, Fukuoka 812, Japan

INTRODUCTION

Many studies of an unstable lattice vibration mode (soft mode) have been reported in the structural phase transition of crystals, for instance, ferroelectric or ferroelastic crystals. The ordered phase of these crystals usually shows, below the phase transition point Tc, a domain pattern in order to lower the bulk free energy of the crystal. The domain pattern of the ferroelectric crystal is composed of the spontaneous electric polarization Ps minimizing the electrostatic energy of the crystal. The ferroelastic domains are composed of the spontaneous strains x_s. The relation between the unstable (soft) lattice vibration and the domain has not been clarified enough though a close relation might be expected. The relation would give us a microscopic understanding of the macroscopic domain pattern being consistent with a phenomenological consideration.

An example which indicates a close relation between the soft acoustic mode and the domain has been reported recently in the structural phase transition of the scheelite type ferroelastic crystals, $BiVO_4$ and $LaNbO_4$, where the domain wall occurs along an "incommensurate" direction accompanied by freezing of the soft acoustic mode at Tc.[1-3] The direction of the domain wall is parallel to the direction of the propagation of the soft acoustic mode. Except for this example, though the relation is important for understanding the mechanism of the phase transition of crystals from both of the macroscopic and microscopic viewpoints, few examples of studies have been reported previously. The purpose of the present study is to investigate the relation between the softening process of the acoustic modes and forming process of the domain pattern in the ferroelastics.

FERROELASTIC PHASE TRANSITION OF $KD_3(SeO_3)_2$

The deuterated ferroelastic $KD_3(SeO_3)_2$ (hereafter abbreviated as DKTS) crystal is very adequate for the present purpose, because it is colorless and transparent with a good optical quality in addition to an appropriate value of Tc ($\sim$ 302 K).[4,5] DKTS undergoes a ferroelastic structural phase transition, as the nondeuterated $KH_3(SeO_3)_2$ crystal (abbreviated as KTS), changing its crystal symmetry from the orthorhombic D_{2h}^{14}-Pbcn to the monoclinic C_{2h}^5-$P2_1/b$ with decreasing temperature.[5,6] The lower C_{2h}^5 phase is ferroelastic with a spontaneous shear strain x_{4s}, which forms a 90 degree twined domain structure as shown in Fig.1. The spontaneous

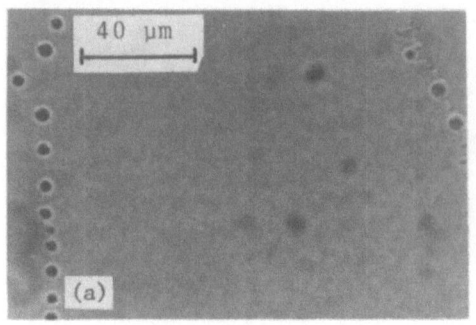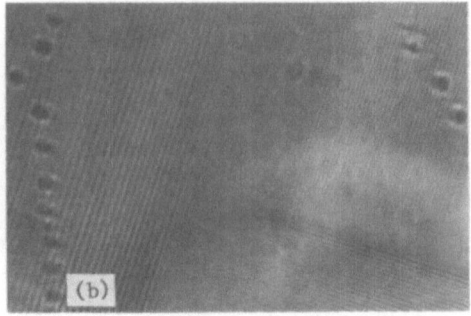

Fig. 1. The (100) plate of DKTS observed under a crossed nicol. (a)
No domain pattern is seen above Tc. (b) Twined domain pattern appears below Tc.

strains of DKTS form a domain structure the wall of which runs along the
[010] and [001] directions, similarly to the case the nondeuterated KTS
crystal.[5-7]

On the other hand a typical softening of the pure transverse acoustic
wave has been reported by the Brillouin scattering method[9-11] and also by
the ultrasonic method.[12-13] Figure 2 shows the temperature dependence of
the Brillouin frequency shift of the transverse acoustic mode of DKTS
reported by Tanaka et al.[9] The transverse acoustic mode propagating along
the [010] direction with the polarization along the [001] direction shows
a complete softening near the second order transition point Tc. The
freezing of the amplitude of the transverse acoustic mode might be closely
connected to the occurrence of the shear spontaneous strain below Tc.

EXPERIMENTS

The DKTS crystal was grown by a slow evaporation method from a D_2O
solution of the stoichiometric compounds at room temperature. The directions of the crystal axes were determined by reference to the crystal
habit. The (100) plane of the crystal is a strong cleavage plane. Using
the natural cleavage of the (100) plane, several (100) plates were prepared for the present experiment. The domain pattern in the (100) plate
was observed under a crossed nicol in a polarizing microscope with an oven
on the stage for the temperature control of the DKTS crystals.

In order to detect a dynamical property of the domain formation, we
observed the domain pattern under a several conditions.

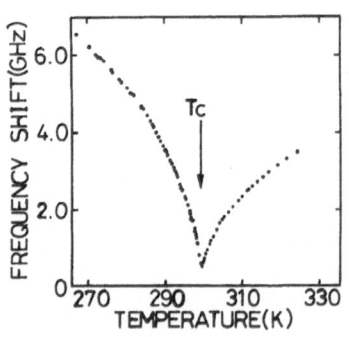

Fig. 2. Temperatrue dependence of
the soft acoustic mode frequency observed by the
Brillouin scattering.[9]

EXPERIMENTAL RESULTS

Time Dependence of the Growth of Domains

A rapid cooling (quenching) of the sample crystal through Tc, from the paraelastic phase to ferroelastic phase, induces a growth of the ferroelastic domain. The growing patterns with time are indicated by a sequence of the photos shown in Fig.3a-f. A narrow, spindle-shaped ferroelastic region (white area) grows quickly in the area remained still in the phase above Tc (black area). The direction of the growth seems to be normal to the direction of the soft acoustic mode propagation. A spindle-shaped island of spontaneously strained region is formed in the paraelastic region. One island induces another island nearby it in parallel to the direction. Connecting the islands each other, the ferroelastic region grows into the size of the whole crystal. Finally a stripe of the ferroelastic domain structure covers the whole crystal. These figures suggests a close relation between the soft acoustic mode and ferroelastic domains.

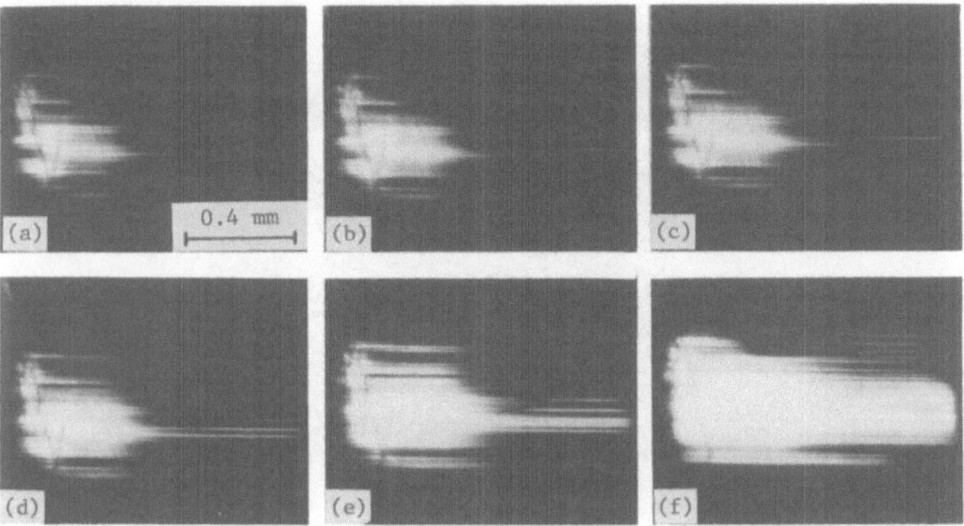

Fig. 3. Time dependence of the growing ferroelastic domains. The white area is a ferroelastic region in the paraelastic region (black area). The photos were taken in sequence at (a) 12 sec., (b) 4 min., (c) 5 min., (d) 6 min., (e) 7 min., and (f) 17 min. after the quenching.

Stress Induced Pattern

Just after polishing of the surface of the (100) plate, a domain was induced in the other orthogonal domain region by the stress applied in the polishing process. The induced area changes rapidly its shape with time as shown in Fig.4a-e. Soon the whole area becomes a monodomain state as before the polishing. This result indicates a small difference of the elastic energy between the two orthogonal domains. In these figures the cross shaded (checkered) area is the region where two orthogonal domains overlapped each other as a layer structure along the [100] direction. This area does not change with time as if both domain boundaries were coupled tightly by a friction force.

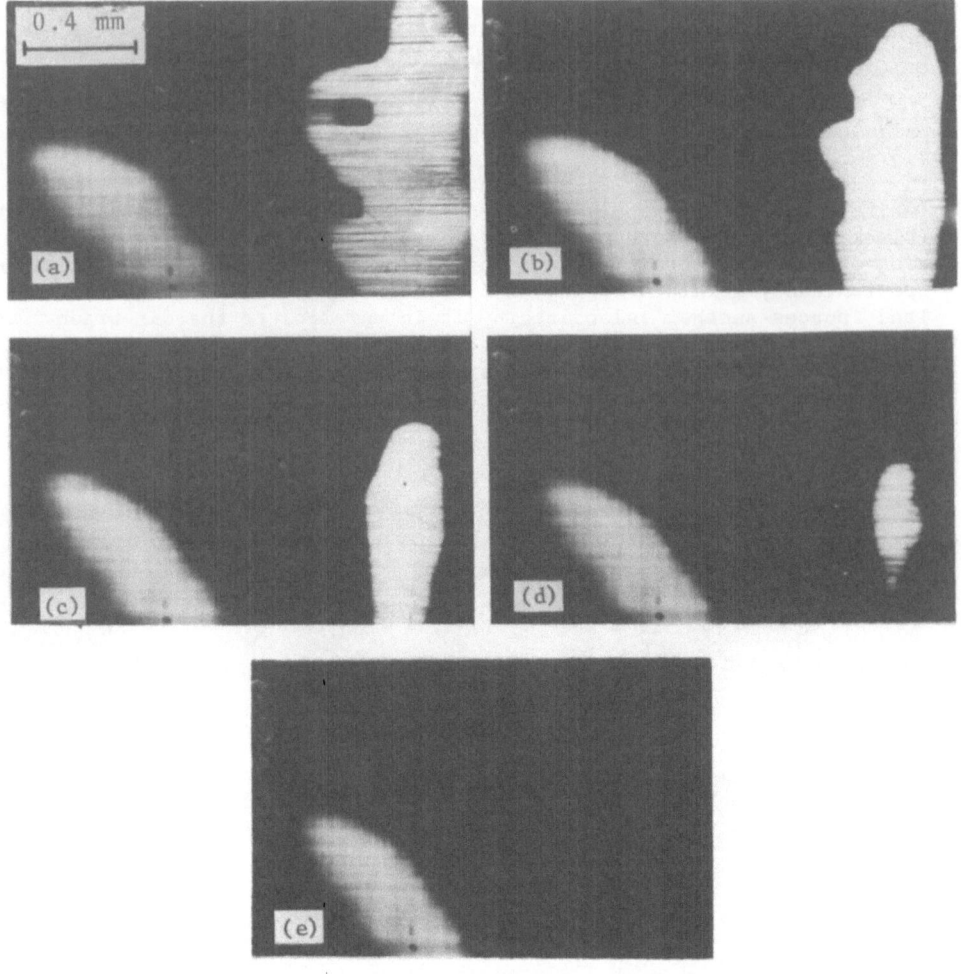

Fig. 4. Growing domain (black area) diminishes the crossed domain
(horizontally shaded white area) induced by the stress applied
by polishing. Cross shaded (checkered) area does not show any
change with time. Photos were taken in sequence (a) 1 min.,
(b) 2 min., (c) 3 min., (d) 3.5 min., and (e) 4 min., just
after polishing.

Effect of Shear Stress

With increasing shear stress conjugate to the spontaneous strain x_{4s},
the whole crystal becomes a single domain crystal. The change of the domain
pattern is very similar to the growing process as shown in Fig.3. Figure
5a-g shows the stress dependence of the forroelastic domain pattern below
Tc. The single domain region (black area) increases gradually with the
applied stress. The island also appears and then disappears finally.

Comb-Teeth Shaped Pattern

An interesting pattern appears just after cleaving the (100) plane at
a temperature below Tc as shown in Fig.6a-d. By an instant stress applied
for the cleavage, the crystal occasionally became a monodomain state, then
just after cleavage the multidomain state began to grow in the monodomain

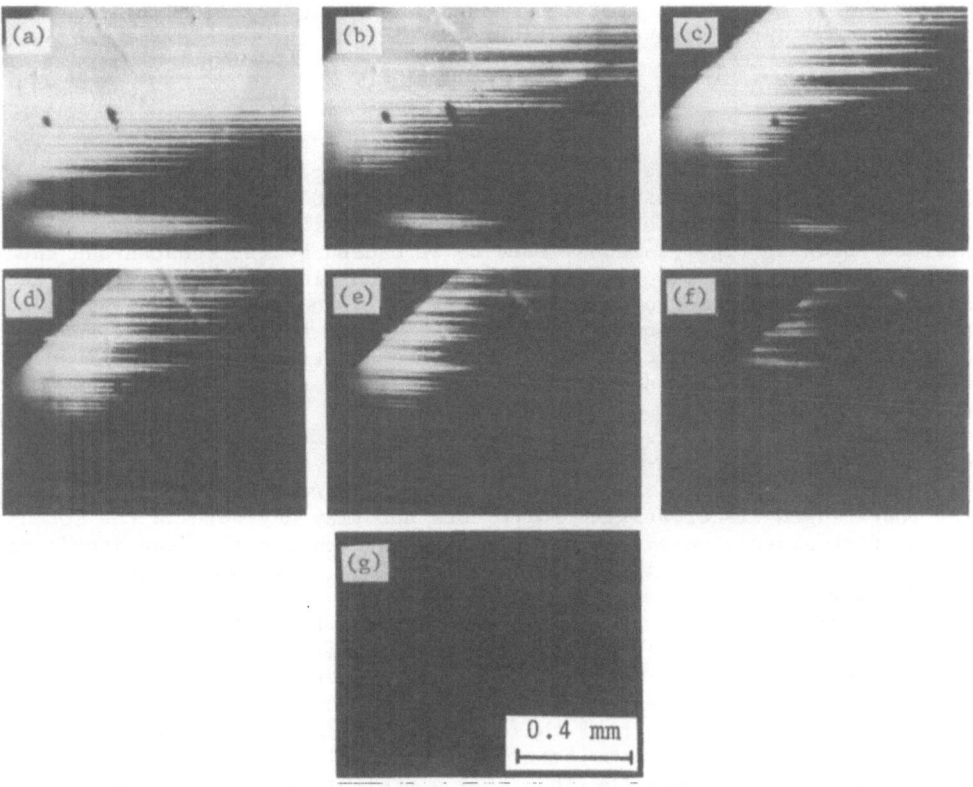

Fig. 5. With increase of the applied shear stress, the multidomain
region (white shaded area) becomes monodomain region (black
area). (a) Multidomain exists under zero stress, (b)-(f) the
multidomain decreases with increasing stress, (g) the whole
crystal becomes finally a monodomain state.

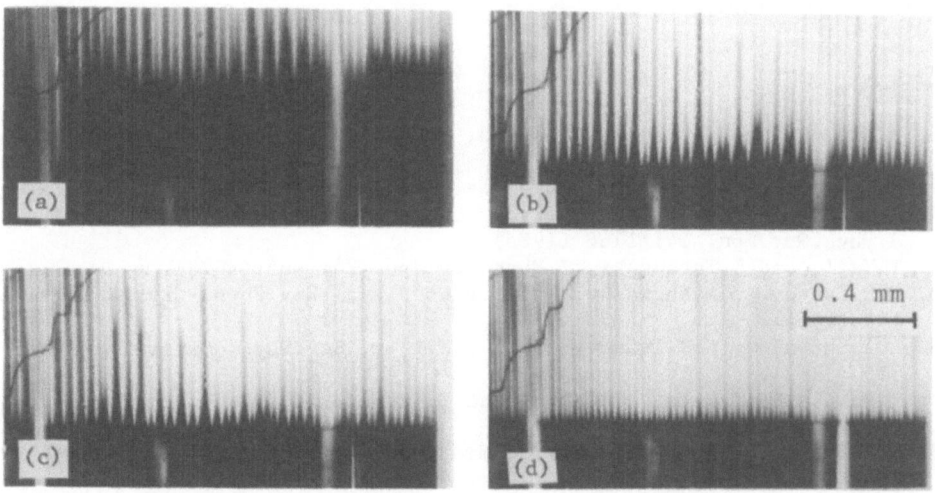

Fig. 6. Triggered by the cleavage, a comb-teeth shaped domain begins
to grow in the single domain region. Photos were taken in
sequence as a function of time, (a) 1 min., (b) 3 min.,
(c) 5 min., and (d) 7 min., after the cleavage.

region. They formed a comb-teeth shaped pattern in the (100) plate as shown in Fig.6a-d. The growth stops at the boundary of the monodomain region.

DISCUSSION

The monoclinic angle α of KTS has been reported to be 91.2 degrees at 144 K.[7] The two kinds of domains exist in the (100) plane according to the sign of the spontaneous shear strain, $\pm x_{4s}$. The fine band shaped domain pattern shown in Fig.1 is considered to be caused by the spontaneous shear strain changing its sign alternatively. In addition to this, because the lattice constants b and c of KTS are very near each other, b = 6.206 A and c = 6.259 A, the direction of the domain band can take one of the two directions with an almost equal weight. The difference of the free energy between the two states seems to be small as suggested in Fig.4. Thus we observe two domain regions orthogonal to each other. The static feature of the domain pattern of DKTS is similar to that of the ferroelectric Rochelle salt and KDP.[14] Because DKTS has a center of symmetry below Tc, the shear strain of DKTS does not produce any surface/boundary charges. Then the long-ranged electrostatic force does not play any role in the formation of the ferroelastic domain pattern. It is interesting that the static patterns of both types of crystal are very similar to each other in spite of the difference in the electrostatic energy.

The origin of the nucleation process of the ferroelastic "islands" in the paraelectric region is not certain at the present stage. The nucleation process associated by the strongly anharmonic interaction between soft= overdamped modes is suggested from the result shown in Fig.3. A quantitative analysis of the growth of these islands is expected for the next step.

The relation between the soft mode and domain pattern is elucidated in the present study, though the result is a preliminary one; The growth rate of the "comb-teeth" region corresponds to the freezing rate of the amplitude of the acoustic soft mode as in the scheelite type crystals.[3] Because, in the present case, the soft mode is a pure transverse mode, the freezing of its amplitude causes directly a shear-strained region. The direction of the mode propagation is perpendicular to the ferroelastic domain band (comb-teeth). Then the interval of the comb-teeth shown in Fig. 6 corresponds to a characteristic value of the wavelength of the freezing acoustic soft mode.

REFERENCES

1. M.Cho, T.Yagi, T.Fujii, A.Sawada, and Y.Ishibashi, J.Phys.Soc.Jpn. 51: 2914 (1982).
2. H.Tokumoto, and H.Unoki, Phys.Rev. B27: 3748 (1982).
3. K.Hara, A.Sakai, S.Tsunekawa, A.Sawada, Y.Ishibashi, and T.Yagi, J.Phys.Soc.Jpn. 54: 1168 (1985).
4. T.Yagi, and I.Tatsuzaki, J.Phys.Soc.Jpn. 26: 865 (1978).
5. N.R.Ivanov, L.A.Shuvalov, and N.V.Gordeeva, Sov.Phys.-Crystallogr. 13: 145 (1968).
6. L.A.Shuvalov, N.R.Ivanov, and T.K.Sitnik, Sov.Phys.-Crystallogr. 12: 315 (1967).
7. Y.Makita, F.Sakurai, T.Osaka, and I.Tatsuzaki, J.Phys.Soc.Jpn. 42: 518 (1977).
8. N.R.Ivanov, L.A.Shuvalov, H.Schmidt, and E.Stolp, Izv.Akad.Nauk SSSR Ser.Fiz. 39: 933 (1975).
9. H.Tanaka, T.Yagi, and I.Tatsuzaki, J.Phys.Soc.Jpn. 44: 1257 (1978).
10. T.Yagi, H.Tanaka, and I.Tatsuzaki, Phys.Rev.Lett. 38: 609 (1977).
11. M.Copic, M.Zgonik, D.L.Fox, and B.B.Lavrencic, Phys.Rev. B23: 3469 (1981).
12. Y.Makita, T.Osaka, and A.Miyazaki, J.Phys.Soc.Jpn. 44: 225 (1978).
13. C.W.Garland, G.Park, and I.Tatsuzaki, Phys.Rev. B29: 221 (1984).
14. T.Mitsui, and J.Furuichi, Phys.Rev. 90: 193 (1953).

GLASS LIKE NATURE OF MAGNETIC ORDERING IN MCl_2-GICs

M. Matsuura

Department of Material Physics, Faculty of
Engineering Science, Osaka University
Toyonaka 560, Japan

INTRODUCTION

Graphite intercalation compounds (GIC) form an interesting family of sandwich type layer structure compounds of graphite with intercalated atoms or molecules (intercalant, I), which are regularly separated by a certain number of carbon layers from the adjacent I-layers. Owing to the so called staged structure, each I-layer, especially of high stage GIC is well approximated a two-dimensional (2D) lattice system[1].

Among these, transition metal chlorides (MCl_2) -GICs are interesting from statistical mechanical viewpoint. The intraplane interaction of $CoCl_2$- and $NiCl_2$-GICs is ferromagnetic and of easy plane anisotropy[2]. Therefore, a topological order of Kosterlitz and Thouless (KT) type[3] has been expected to be realized in these coppounds. Actually, a successive two-step phase transition was observed with a purely 2D ordered state[4] of long range order (LRO) character[5] in the intermediate temperature range (See Fig.1). The intermediate state has once been suspected to be a KT state predicted by Jose et al.[6], refering to the hexagonal symmetry of the intercalant plane. It is now understood thermodynamically to be an intermediate state of a hierarchical two-step ordering[7] on the basis of the intraplane island structure of each I-layer which is not extended infinitely but divided into a number of finite size 2D clusters.

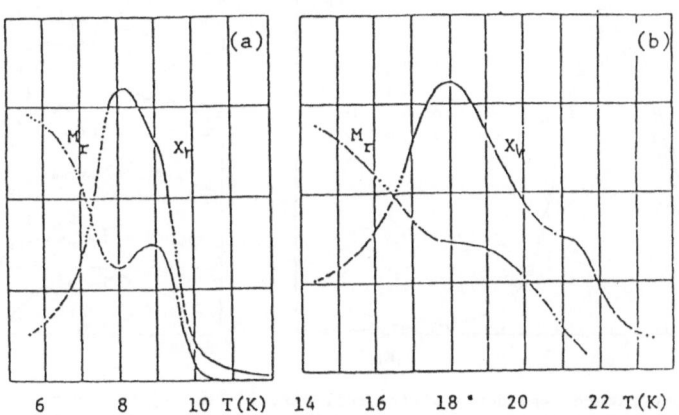

Fig.1 Temperature dependence of remanent magnetization M_r and simultanneously measured susceptibility X_r of (a) $CoCl_2$- and (b) $NiCl_2$-GICs (from Ref.7)

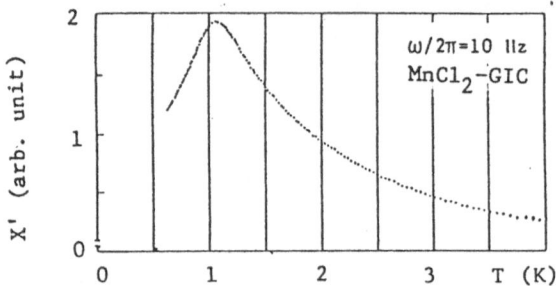

Fig.2 Temperature dependence of AC susceptibility (from Ref.15).

The state above the upper critical point T_{cu} is the disordered one. Then a LRO is established within each island but not among the ordered islands. The transition at T_{cu} is thus of a purely 2D system and the behaviour above T_{cu} could be well described by that of a 2D XY-type ferromagnet[8]. Below the lower critical point T_{cl}, the system goes into a 3D ordered state by interisland interactions both within each plane and between different planes. The nature of the transition at T_{cl} is, however, not well understood at present. Recent experimental investigations on the dynamical and non-equilibrium aspects[9,10] have revealed an interesting feature or a spin glass like nature of magnetic ordering as mentioned below.

As for $MnCl_2$-GIC, the intraplane interaction is antiferromagnetic and of nearly Heisenberg-type[11]. On the triangular lattice, Mn ions form a frustrated system on which a topological order has been predicted to appear[12]. A single transition into the order is expected for a purely Heisenberg system and a successive two-step one for an anisotropic one with distinguishable heat capacity anomalies in both cases. With some interplane interaction, a phase transition into a 3D LRO is predicted to appear but with a novel critical indices[13].

Experimentally, a single phase transition was observed by magnetic measurement. Below the critical temperature T_c, a 3D correlation was distinguished by neutron diffraction[14]. However, any heat capacity anomaly has not been found so far around T_c[15] in spite of repeated precise measurements. Any further transition has not been noticed above T_c. The behaviours of so called field cooled and zero-field cooled magnetization[15] as well as the cusp like susceptibility anomaly as shown in Fig.2 suggest that the ordering around T_c is of a spin glass type.

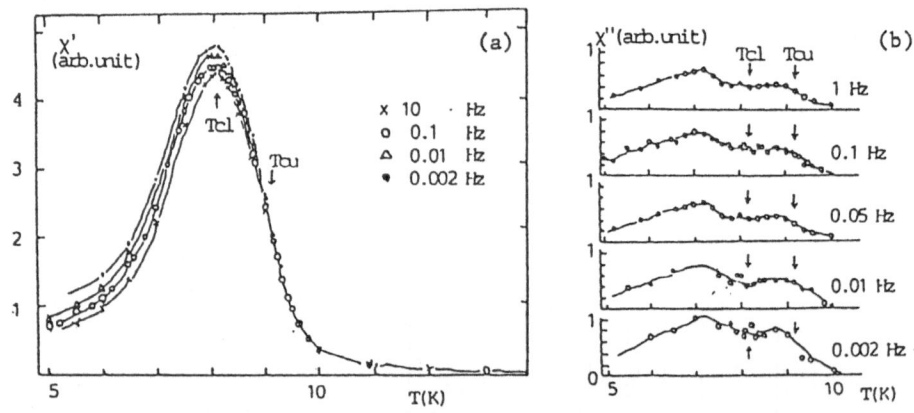

Fig.3 Temperature dependence of (a) real part X' and (b) imaginary part X" of complex susceptibility X*(ω) at very low frequencies (from Ref.5).

$CoCl_2$- and $NiCl_2$-GICs

Figure 3 shows the temperature dependences of AC susceptibility $X^*(\omega)$ for $CoCl_2$-GIC at various measuring frequencies ω[9]. A weak ω dependence is noticed for the real part X' below about T_{cu}. Correspondingly, the imaginary part X'' appears in the temperature range although the magnitude of X'' is only several % of X'. A remarkable is the nearly independent feature of X'' on ω in the measured wide frequency range. It is quite different from the Debye type absoption. Essentially similar result is found also for $NiCl_2$-GIC[16].

According to the fluctuation-dissipation theorem, the power spectrum of magnetic fluctuation is proportional to X''/ω for negligibly weak AC field. The result for $CoCl_2$- or $NiCl_2$-GIC tells us therefore a $1/\omega$ type magnetic fluctuation. Accordingly, we may expect a non-exponential relaxation of magnetization. Actually the decay of thermoremanent magnetization M_r for $CoCl_2$- and $NiCl_2$-GICs is found to be logarithmic[17], although a slight deviation is found below T_{cl}.

Figure 4a and b are the temperature dependence of M_r measured in a

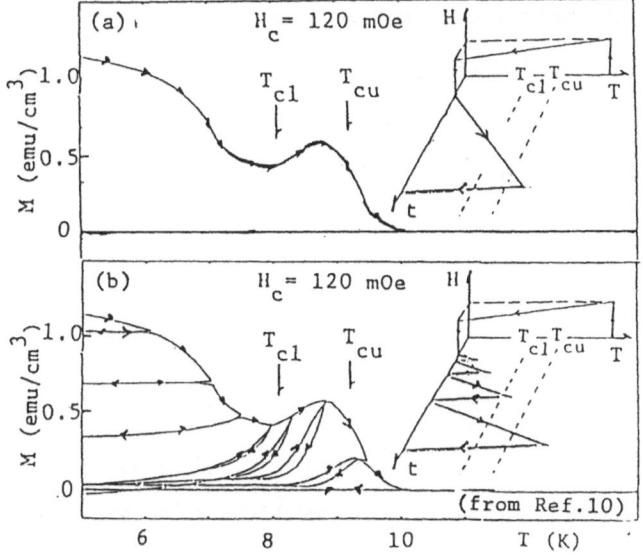

Fig.4 Temperature dependence of remanent magnetization in the process shown in the insert (Ref.10).

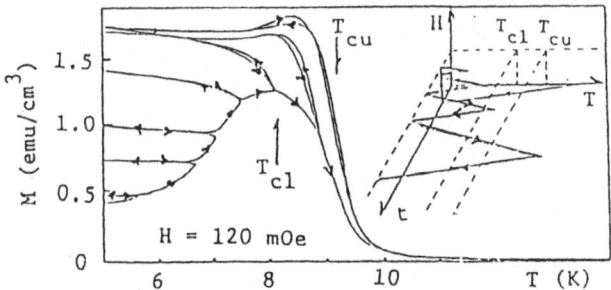

Fig.5 Temperature dependence of field induced magnetization in the process shown in the insert (from Ref.10).

simple heating process and in a series of heating and cooling processes (see the insert), respectively[5,10]. The whole measurement is undertaken within a time interval much shorter than the total relaxation time and the system is far from thermal equilibrium. As seen in Fig.4b, if the system is cooled down from a temperature T_r below T_{c1}, M_r is almost constant. When the system is heated up again from the lowest temperature M_r changes along the same pathway on the M_r-T curve up to T_r. The process is thus reversible. Then M_r changes along the M_r-T curve in Fig.4a. If T_r is higher than T_{c1}, M_r decreases to zero as temperature decreases across T_{c1}. The process is also reversible and M_r comes back to the initial value by heating up to T_r.

The M_r-T curve as shown in Fig.4a depends on the cooling field H_c, but the memory phenomena of M_r are qualitatively unaltered. Essentially similar phenomena are also observed for the field induced magnetization M_i after zero-field cooling process as shown in Fig.5.

MnCl$_2$-GIC

Figure 6 shows the temperature dependence of $X^*(\omega)$ at various value of ω[15]. A ω dependence is found for X' below about T_c and X'' appears in this temperature range. A remarkable feature is that X'' is roughly independent of ω in the measured wide frequency range. Therefore, the magnetic fluctuation in MnCl$_2$-GIC is also characterized by a $1/\omega$ like spectrum.

Figure 7 shows the temperature dependence of M_r in a series of heating

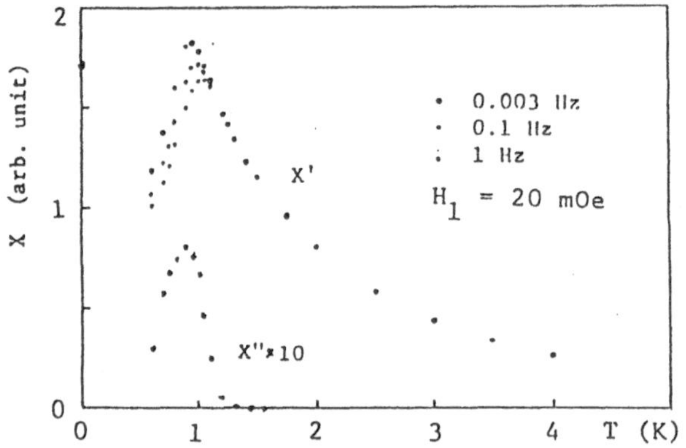

Fig.6 Temperature dependence of AC susceptibility at various frequencies (from Ref. 15).

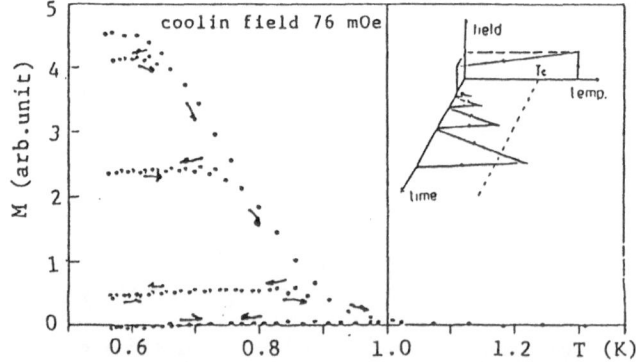

Fig.7 Temperature dependence of remanent magnetization in a series of heating and cooling processes shown in the insert (from Ref.15).

and cooling processes as in the case for $CoCl_2$-GIC[15]. The chracteristic
memory phenomenon is similar to that for $CoCl_2$-GIC in Fig.4b below T_{cl}.

GLASS LIKE NATURE OF MAGNETIC ORDERING

As mentioned in the previous paragraph, AC susceptibility at very low
frequencies reveals that the magnetic fluctuation in the ordered state is
characterized by a $1/\omega$ like spectrum in a wide frequency range for all
$CoCl_2$-, $NiCl_2$- and $MnCl_2$-GICs. It is largely different from the Lorenz-
ian spectrum which is characterized by a single relaxation time τ and
suggests a rather complicated relaxation mechanism. Consistently, a
logarithmic decay of M_r has been observed for these GICs. Such nonlinear slow fluctuation and relaxation have not been expected for any
simple systems like regular ferro- or antiferromagnets. In a frustrated
antiferromagnet $MnCl_2$-GIC, it may be expected but never in ferromagnetic
$CoCl_2$- and $NiCl_2$-GICs.
Meanwhile, a similar nonlinear relaxation e.g. a stretched exponential
decay of M_r has recently been observed in some spin glass like alloys[18].
A $1/\omega$ type spectrum of magnetic fluctuation has also been found in some
spin glass like mixed compounds[19]. Thus one may think that the above
nonlinear dynamical phenomena are attributable to a more or less random
or frustrated situation of ordering process in the present GICs. As for
$CoCl_2$- and $NiCl_2$ GICs, however, no frustration exists microscopicaly and
thus the island structure in the intermediate scale may be responsible.
Under such a circumstance, it is interesting to compare the memory phenomena of M_r in the present GICs and spin glasses.
Recent theoretical inspection on ordering in spin glass revealed that
the memory phenomena as shown in Fig.8a is well described if a successive
branching of energy level could occur in the system below the critical
temperature T_g as schematically shown in Fig.8b[20]. A qualitative similarity to spin glass is seen in the part below T_{cl} of Fig 5. It is easy
to speculate the memory phenomenon below T_{cl} in Fig.4b and Fig.7, on
the basis of the energy branching model. Figure 9 is the experimental

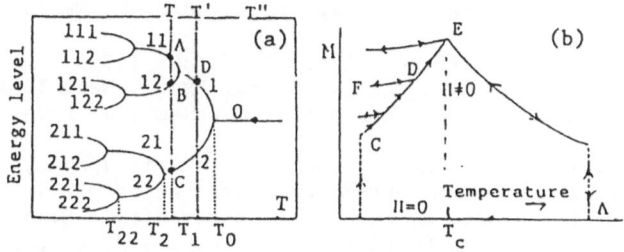

Fig.8 Schematic view of (a) temperature dependence of field induced magnetization of spin glass, speculated from the successive branching of energy
level like (b) (from Ref.20).

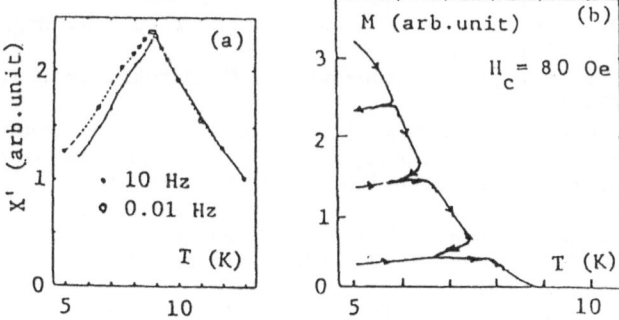

Fig.9 (a) X'-T curves and (b) memory phenomenon of CuMn(1at%) alloy.

results for a well known spin glass like alloy CuMn(1%)[17]. Fig.9a is the $X'-T$ curve and b is the memory phenomenon of M_r for the same sample. The latter is quite similar to those below T_{cl} for CoCl$_2$-GIC and for MnCl$_2$-GIC as seen in Fig.4b and Fig.7.

CONCLUDING REMARKS

As seen in the previous paragraphs, the ordering behaviours of CoCl$_2$-, NiCl$_2$- and MnCl$_2$-GICs are qualitatively analogous to a spin glass in the dynamical and non- quilibrium aspects. It has never been expected so far refering to the regular lattice structure in the microscopic scale, except MnCl$_2$-GIC. Probably, it should happen that the characteristic lattice structure of GIC family or the intralayer island structure in the intermediate scale of several hundred angstrom, is responsible for such a glass like nature of magnetic ordering. However, the origin of the present similarity still remains to be solved.

In concluding this article, the author expresses his sincere thanks to Dr. Y. Murakami, Dr. Y. Karaki, Mr. T. Yonezawa and Mr. N. Tanaka for their experimental works, to Prof. H. Ikeda of Ochanomizu Univ., Prof. M. Suzuki of the State Univ. of New York for their sample preparation and stimulating discussions and to Prof. H. Takayama of Kyoto Univ. for his helpful discussion on spin glass dynamics.

REFERENCES

1. See e.g. Proc. Int. Symp. GICs, Tsukuba 1985.
2. M. Matsuura. Y. Murakami, K. Takeda, H. Ikeda and M. Suzuki: Synth. Met. 12 (1985) 427.
3. J. M. Kosterlitz and D. J. Thouless: J. Phys. C6 (1973) 1181.
4. H. Ikeda, Y. Endoh and S. Mitsuda: J. Phys. Soc. Jpn. 54 (1985) 3232. D. G. Wiesler, M. Suzuki, H. Zabel, S. M. Shapiro and R. M. Nicklow: Physica 136B (1986) 22.
5. Y. Murakami, M. Matsuura and T. Kataoka: Synth. Met. 12 (1985) 443.
6. J. V. Jose, L. P. Kadanoff, S. Kirkpatrick and D. R. Nelson: Phys. Rev. B16 (1977) 1217.
7. M. Matsuura: Ann. Phys. (France) 11 No.2S (1986) 117.
8. G. Dresselhaus, S. T. Chen and K. Y. Szeto: Synth. Met. 12 (1985) 433.
9. M. Matsuura, Y. Endoh, T. Kataoka and Y. Murakami: J. Phys. Soc. Jpn. 56 (1987) 2233.
10. M. Matsuura, N. Tanaka, Y. Karaki and Y. Murakami: Jpn. J. Appl. Phys. 26S (1987) 797. Y. Murakami and M. Matsuura: to be published.
11. O. Gonzalez, S. Frandrois, A. Maaroufi and J. Amiell: Solid State Commun. 51 (1984) 499. D. G. Wiesler, M. Suzuki, P. C. Chow and H. Zabel: Phys. Rev. 34B (1986) 7951.
12. H. Kawamura and S. Miyashita: J. Phys. Soc. Jpn. 53 (1984) 9. S. Miyashita and H. Shiba: J. Phys. Soc. Jpn. 53 (1984) 1145.
13. H. Kawamura: J. Phys. Soc. Jpn. 54 (1985) 3220.
14. M. Suzuki, D. G. Wiesler, P. C. Chow and H. Zabel: J. Magn. Magn. Mater. 54-57 (1986) 1275.
15. M. Matsuura, Y. Karaki, T. Yonezawa and M. Suzuki: Jpn. J. Appl. Phys. 26S (1987) 773.
16. T. Yonezawa et al: Autumn Meeting of Jpn. Phys. Soc. Nishinomiya 1986.
17. N. Tanaka et al: Autumn Meeting of Jpn. Phys. Soc. Nishinomiya 1986.
18. R. V. Chamberlin, G. Mozurkewich and R. Orbach: Phys. Rev. Letters 52 (1984) 867.
19. M. Ocio, H. Bouchiat and P. Monod: J. Magn. Magn. Mater. 54-57 (1986) 11. W. Reim, R. H. Koch, A. P. Malozemoff and M. B. Kechen: Phys. Rev. Letters 57 (1986) 905.
20. R. G. Palmer: Heiderberg Colloquim on Spin Glasses, Lecture Notes in Physcs 192 (1983, Springer-Verlag).

TIME EVOLUTION OF THE MISFIT

PARAMETER IN K_2ZnCl_4

Hiroyuki Mashiyama

Faculty of Science
Yamaguchi University
Yamaguchi 753, Japan

ABSTRACT

The incommensurate-commensurate phase transition in ferroelectric K_2ZnCl_4 has been investigated by means of X-ray scatterings. The misfit parameter δ has been determined from the peak position of the satellite reflections. After the commensurate-incommensurate transition, the parameter δ changes very slowly with a logarithmic law, which is theoretically predicted by Kawasaki with taking account of the activation energy for creating stripples or antistripples.

INTRODUCTION

The incommensurate(IC)-commensurate(C) phase transition in dielectric materials have been widely interested in the last decade.[1] Near the IC-C transition point T_C, the IC structure can be characterized by well-defined discommensurations(DCs) which separate almost C regions.[2] DCs are arrayed in stripe with the spacing L in most materials. If L increases indefinitely at T_C, the C phase is realized. But this process would require a rearrangement of atoms all over the crystal. It has been demonstrated theoretically that n DCs can be incorporated at a dislocation and swept away as antistripples.[3] Here n is the order of the commensurability. On the other hand, the C-to-IC transition takes place with nucleating and growing of stripples. Thus the kinetic process of the IC-C transition is expected to be quite different from that of the usual domain wall formation in proper ferroelectrics.

Potassium tetrachlorozincate K_2ZnCl_4 is one of the K_2SeO_4-type materials and its superstructure with threehold cell dimension(along the c-axis) transforms to the IC structure above $T_C \simeq 402K$.[4] It has been observed that the dielectric constant ε of K_2ZnCl_4 has pronounced thermal hysteresis both in the C and IC phases as in the case of Rb_2ZnCl_4, in which the hysteresis is explained that imperfections in the crystal pin the DCs and prevent from reaching thermal equilibrium.[5] Moreover, ε of K_2ZnCl_4 changes with time t after the temperature is held constant near T_C.[6,7]

X-ray scattering experiments revealed that the change of the DC spacing L is accompanied with the variation of δ.[8,9] The reason of such a slow relaxation in K_2ZnCl_4 is not fully understood, however, it is considered that the time evolution reflects the kinetic process of creation or annihilation of

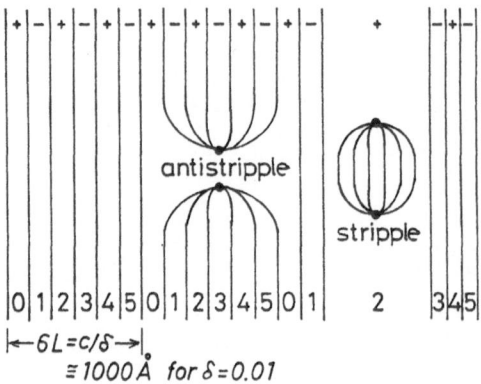

Fig. 1. Schematic picture of regular stripe discommensurations, antistripples and stripples for the commensurability n=6. The polarization in almost commensurate regions changes alternatively as shown by (+,-). The phase of the modulated structure changes by 2π/6 in the adjacent domains.

DCs under the influence of crystal imperfections.

By taking account of the activation energy for creating stripples or antistripples (see Fig. 1), Kawasaki calculated the time evolution of the misfit parameter as

$$\delta - \delta_\infty \simeq g \, / \, \ln(t/t_0) \, , \tag{1}$$

if δ is initially deviated from its final equilibrium value δ_∞.[10] Here we present the experimental results of the misfit parameter by means of X-ray scatterings, which is shown to be consistent with the theoretical eq. (1).

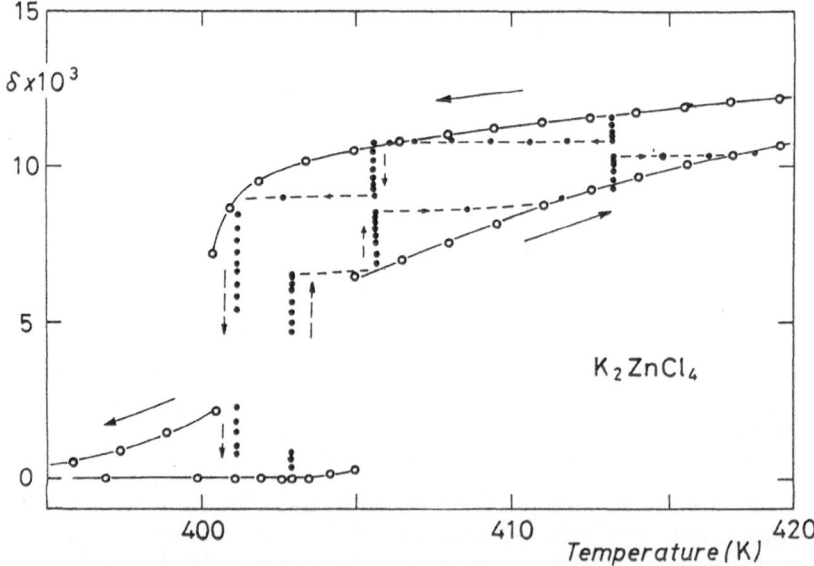

Fig. 2. The misfit parameter δ. The arrows show the direction of changes.

A speciman was mounted on a standard two-circle diffractometer so as to get (ξ 0 ζ) reflections and the major scans were performed around (2 0 2/3+δ) by the step scan method with the step width 0.01° in ω. Crystal directions and reciprocal space coordinates are referred to the unit cell of the normal phase (space group Pmcn) with lattice parameter a<c<b.

The misfit parameter δ has been determined from the peak position of the satellite reflection (2 0 2/3+δ) by fitting the scattering profile to a Gaussian form. The temperature dependence of δ is shown in Fig. 2 by open circles when the temperature is changed stepwise with a mean rate of 0.2K/min. With cooling the temperature from the IC phase, δ decreases and the peak width increases about 80% around T_c. Just below T_c, the C reflection is broad with accompanying the diffuse scatterings at the wings and its peak position is a little deviated from the C value. These facts suggest that many DCs remain in the crystal as ferroelectric domain walls (the mean spacing of walls is estimated as about 1000A at 400K). When the crystal is heated from the room temperature, the peak width of the satellite reflection remains constant in the C phase and is the same as that of the normal Bragg reflections. Just after the C-to-IC transition, the width increases about 20% and then becomes as sharp as the normal reflections.

The IC-C transition takes place gradually in the temperature interval between 400 and 405K. If the temperature is held constant at 401.2K on a cooling run, then δ decreases at first with increasing the width of the reflection. Another C peak emerges after 10 hours, and its intensity increases. The peak position shifts gradually as shown in Fig. 3 (a). On the other hand, Fig. 3 (b) shows the change at 402.9K on a heating run. The peak profile at t=0 in Fig. 3 (b) indicates the scattering resolution. The total integrated intensity of both the C and the IC reflections remains constant through the transition. The broad double peaks suggest the complicated distribution of DCs at the IC-to-C transition.

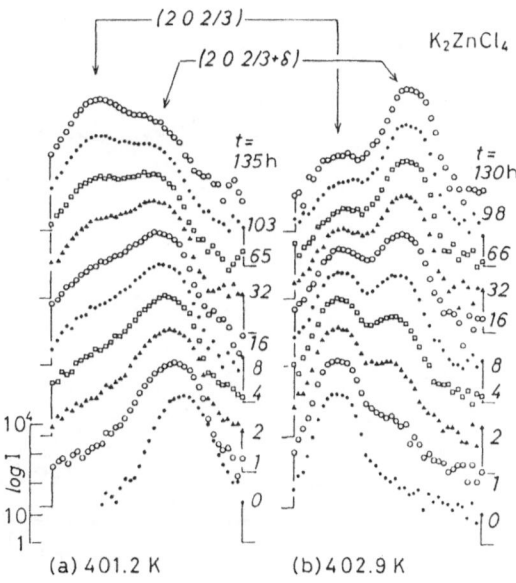

Fig. 3. Scattering along c* about (2 0 2/3+δ) of K_2ZnCl_4, showing time evolution with holding the temperature at (a) 401.2K on cooling and (b) 402.9K on heating.

When the IC phase coexists with the C phase, the time variation is very slow and obeys logarithmic relations as shown in Fig. 4, except for the initial hours where it can be fitted to a usual relaxation function of $\exp(t/\tau)$.[9] Although δ for the stable phase should tend to its final equilibrium value, the saturation is not clearly detected in the observed time until 135 hours.

If the temperature is held constant in the IC phase, the width of the satellite reflection changes little, and the misfit parameter deviates from the initial value rather steeply as $\exp(-\sqrt{t/\tau})$ for a few hours. For a longer time scale, however, δ evolves in accordance with eq. (1).[9] Figure 5 shows the logarithmic dependence. On a heating run from the C phase, the temperature is held at 407.1K after the C-to-IC transition, then δ increases and approaches its final equilibrium value of about 0.0094. If the temperature is held at the same value on a cooling run, then δ decreases and tends to the same final value. The fitted values of parameters in eq. (1) crucially depend on the estimation of δ_∞, however, they take the values of $t_0 = 0.1 \sim 0.2$h, $g = 0.001 \sim 0.002$ for 405~413K. At higher temperature, g is larger and the initial change of δ is faster.

DISCUSSIONS

In charge density wave structure, the growing process of DCs from stripples have been directly observed by transmission electron microscopy (TEM).[11,12] The regular stripe DCs as well as the ferroelectric domain walls in K_2SeO_4-type materials have been observed in Rb_2ZnCl_4 by TEM.[13] The observed DC spacing L is the same order as our X-ray results. A stripple pinned by defects has only been found in the C phase at T_C-80K, far below T_C.[14]

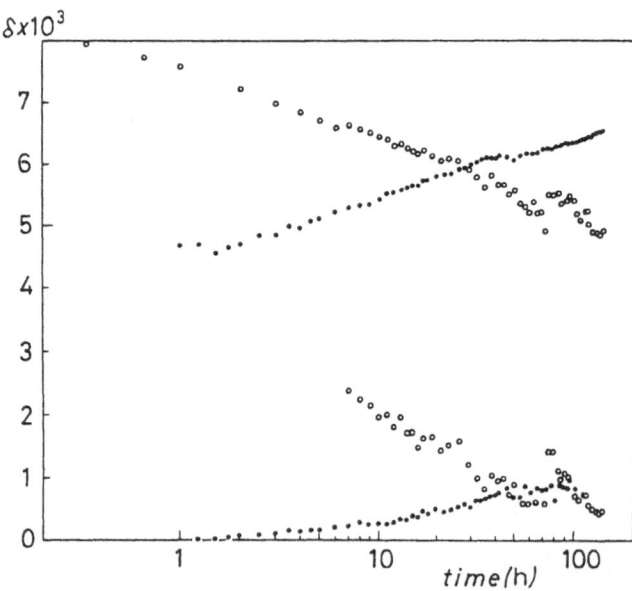

Fig. 4. Time evolution of the misfit parameter δ with holding temperature at 401.2K (open circles) on cooling and at 402.9K (closed circles) on heating.

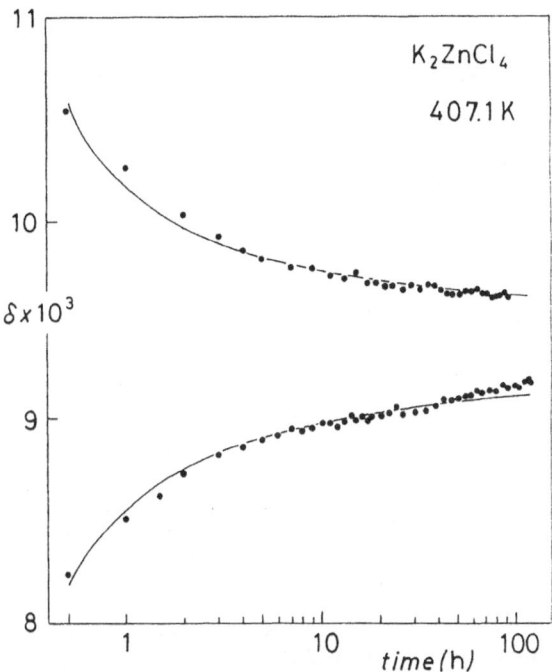

Fig. 5. Time evolution of the misfit parameter δ with holding temperature in the incommensurate phase at 407.1K. The solid curves show a relation $\delta = \delta_\infty + g/\ln(t/t_0)$.

From the X-ray diffraction study, we can usually get the modulation wave-number $k=1/3-\delta$. The existance of higher order harmonics is one of the indirect evidence of DCs.[15] Since the temperature dependence of δ in K_2ZnCl_4 is similar to that of Rb_2ZnCl_4, we consider that there exist well difined DCs and the misfit parameter δ directly relates to the mean spacing of DCs as shown in Fig. 1.

The time variation shown in Fig. 3 is very similar to the temperature dependence. Thus the thermal hysteresis in K_2ZnCl_4 is closely related to the slow relaxation. High activation energy to depin and/or nucleate stripples or antistripples would cause the slow time evolution.

The discontinuous jump of δ at T_C in Fig. 4 can be explained in terms of antistripples. If six adjacent DCs disappear through antistripples for every seven DCs, then δ decreases by 1/7. But there would remain odd DCs at random, which causes the diffuse scatterings. On the heating run, a small number of created DCs would diffuse in the C regions since the interaction between DCs is a repulsive one. If a stripple nucleates between DCs whose spacing is characterized by $\delta=0.001$, then the new spacing is $\delta=0.007$, in agreement with the experiments.

The IC-C transition has been described by creation or annihilation of DCs and the time evolution of the misfit parameter has been given by eq. (1) by the estimation of activation energy.[10] Our experimental results are consistent with the logarithmic dependence of the theoretical prediction in the IC phase. In real systems, the transition temperature crucially depends on impurities in K_2ZnCl_4. Therefore the observed slow time evolution of δ should also be affected by imperfections in the crystal. However, since the DCs are the topological defects, the imperfections may just alter the microscopic parameters quantitatively, and the qualitative character of the kinetics may be unchanged.

ACKNOWLEDGEMENT

This work was supported by a Grant-in-Aid for Education, Science and Culture.

REFERENCES

1. "Incommensurate phases in dielectrics", R. Blinc and A. P. Levanyuk eds., North-Holland, Amsterdam (1986).
2. W. L. McMillan, Phys. Rev. B 14 : 1496 (1976).
3. K. Kawasaki, J. Phys. C 16 : 6911 (1983).
4. K. Gesi and M. Iizumi, J. Phys. Soc. Japan 46 : 697 (1979).
5. K. Hamano, Y. Ikeda, T. Fujimoto, K. Ema and S. Hirotsu, J. Phys. Soc. Japan 49 : 2278 (1980) ; K. Hamano, in ref. 1.
6. H. -G. Unruh, Ferroelectrics 53 : 323 (1984).
7. G. Zhang, S. L. Qiu, M. Dutta and H. Z. Cummins, Solid State Commun. 55: 275 (1985).
8. H. Mashiyama and K. Kasatani, Jpn. J. Appl. Phys. Suppl. 24-2 : 802 (1985).
9. H. Mashiyama and H. Kasatani, J. Phys. Soc. Japan 56 : 3347 (1987).
10. K. Kawasaki, Physica 124B : 156 (1984).
11. K. K. Fung, S. McKernan, J. W. Steeds and J. A. Wilson, J. Phys. C 14 : 5417 (1981).
12. C. H. Chen, J. M. Gibson and R. M. Fleming, Phys. Rev. Lett. 47 : 723 (1981); Phys. Rev. B 26 : 184 (1982).
13. H. Bestgen, Solid State Commun. 58 : 197 (1986).
14. K. Tsuda, N. Yamamoto, K. Yagi and K. Hamano, Proc. XIth Int. Cong. on Electron Microscopy, Kyoto (1986) p.1233.
15. S. R. Andrews and H. Mashiyama, J. Phys. C 16 : 4985 (1983).

PHASE SEPARATION IN POROUS MEDIA

Walter I. Goldburg
Department of Physics and Astronomy
University of Pittsburgh
Pittsburgh, PA 15260 U.S.A.

INTRODUCTION

The equilibrium behavior of homogeneous solids and fluids near the critical point is a subject that is quite well in hand, but the dynamics of phase separation is a subject still alive with complexity and interest. Here we discuss an embellishment of the above two problems that has caught the interest of the phase transition community in recent years, namely phase transitions in the presence of externally imposed random fields[1]. In a magnetic system, this field is indeed an externally imposed magnetic field, its direction varying randomly from spin site. In binary mixtures, with which we are concerned here, the "field" is actually a chemical potential difference $\mu(r)$ that plays the role of the field. Instead of favoring one spin direction over another, it favors, at the point r, the presence of a molecule of species A, say, over that of B in an A–B mixture. In the present instance, the source of this random chemical potential $\mu(r)$ is porous glass or a gel, saturated with the liquid mixture. If the glass were, say, in the form of a flat sheet or a test tube instead of a porous block, its preference for one of the two components would be described by the theory of wetting[2]. One might reasonably expect that such systems should be described by random field theory, rather than wetting theory, when the correlation length ξ of the mixture is large compared to the pore size; in this case the attractive force for one of the components over the other is random on all relevant spatial scales. The experiments to be described do not fully bear out this expectation, in that some of the effects seen in large pore-systems also seem to fit well into the random field picture.

In the canonical version of the random field problem[1], one imagines that at each spin site, the direction of the magnetic field h(r) is determined by, say, the flip of a coin, and whose magnitude is established by, say, the roll of a die. If this field is much stronger than the internal field that one spin effectively imposes on its neighbor, each spin will line up with its own local field, and any tendency to form an ordered phase at low temperature, will surely be suppressed. The interesting case is when the random field is weak, in which case there will be a sharp (second order) phase transition in three dimensions, even though the temperature at which it occurs will be presumably be decreased. By now the literature on this subject is vast.

The randomness of the field or chemical potential requires, by definition, that its average value must be zero. In a mixture, this condition is equivalent the vanishing of the mean value of the local chemical potential $\mu(r)+\Delta\mu_{cx}$. Here the first term is from the random medium, and by definition its spatial average is zero. The second term is the equilibrium contribution from the mixture itself and is equal to zero at the critical composition. It is required, then, that the mixture within the pores of the random medium be of critical composition, where $\Delta\mu_{cx}$ vanishes.

Of necessity, a random field introduces metastability into the phase separation problem. Sometimes the lifetime of the metastable states is too long to be averaged over by the measurements, and yet too short to be ignored, as one can, say, studying sound propagation through a piece of glass or a rock. In some of the experiments reviewed here, particular attention will be paid to the measurement of lifetimes of these metastable states; in others the relaxation times appear fast enough to yield results that may be compared with equilibrium theories. At the present stage of this subject, there is rather little that can be understood quantitatively about the behavior of near-critical fluid mixtures contained in the interstices of porous materials.

PHASE SEPARATION IN GELS AND RANDOM FIELD THEORY

It requires only visual observation to establish that phase separation of a binary mixture in a gel is qualitatively different from that in the free mixture[3]. The starting point is the making of the gel itself, say an agarose+water gel, which has the physical consistency of desert gelatin. A sample is made merely by immersing roughly a cubic cm. of the gel (water+polymer) in a large reservoir of near-critical binary mixture and keeping it there for several diffusion times, which this might be a few days or even weeks. In the experiments to be discussed here, the second component was isobutyric acid. Many critical parameters of this mixture are known, and the critical temperature T_c of the free mixture is conveniently located at 26°C.

The entangled strands of agarose are wetted by water and not isobutyric acid (one cannot form a gel by mixing the polymer with isobutyric acid), hence a slight excess of acid might be required in the reservoir to drive the concentration of mixture inside the gel back to the critical value. Because a meniscus is never formed in any of the gel- mixtures, it is not possible to identify the phase separation temperature in the gel with precision. Rather, one makes up a series of samples of varying acid concentrations and identifies the critical sample as the one with the highest cloud point temperature[3, 4]. It turns out that the critical concentration, as measured in this way, differs little from that of the free mixture in the various gel- mixtures that have been studied to date. The critical concentration in the porous medium can also be determined analytically by Raman scattering or UV absorption[5, 6]. The critical temperature is now identified somewhat crudely with the cloud point temperature of the critical sample. This temperature can be strongly shifted from that of the free mixture for two reasons; The gel component in this three-component system may be thought of as an impurity, and impurities are known to shift the critical temperature of mixtures either up or down. Secondly, the random field from the gel network introduces disorder, and this externally imposed disorder delays the formation of the ordered two-phase state on cooling the sample. Quite generally, a random field will decrease the ordering temperature of systems with an upper consolute point and increase it in systems with an inverted coexistence curve, such as 2,6-lutidine+water (LW)

In the measurements to be described in further detail below, the cloudiness which is created by the cooling (or heating in LW) should ultimately disappear. A trivial reason for this decrease in scattering is to be found, if the sample is allowed to remain in contact with reservoir of the free mixture; the reservoir separates quickly into two phases, whose composition will differ greatly from the critical value c_c, and its contact with the sample will, in time, pull the mixture in the gel away from the critical value. where the scattering is weak. The characteristic time τ_D for diffusion is a^2/D, where a is the shortest sample dimension, and D(T) is the composition diffusivity. A few degrees from the critical point, τ_D will be hours to months, depending on sample size and temperature relative to the critical temperature. To avoid this effect, one need only isolate the sample from the reservoir before cooling it through T_c.

A more interesting reason for the decrease in scattering is to be found in random field ideas. In the absence of the random field, cooling the sample into the ordered state produces droplets of the opposing phases, and these domains grow, so as to minimize their surface energy. Their growth moves the scattering into the forward direction, and scattering at appreciable angles will ultimately vanish as the droplets become much larger than the wavelength of light.

The random field, however, will block this growth, as may be seen by focusing attention on a small "droplet" of up-spins in a sea of down-spins. For the system to achieve a state of uniform down-spin magnetization, the up-spin droplet must vanish. If however the up-spin droplet happens to be located in a region where the direction of the local random fields is predominantly up (and such statistical fluctuations will surely occur, even though h(r)=0 on the average), the dissolution of this small drop will be retarded. One has here a competition between surface energy and Zeeman energy. The latter favors the presence of this small the small domain, (and all others scattered througout the sample), while the surface energy is decreased by their disappearance. Fluctuations of order k_BT will overcome these local energy barriers, and in time the system will be one of uniform magnetization. As with all metastability effects the achievement of the equilibrium state can take a very long time if the barriers are high. In the equilibrium state, the Zeeman energy averages to zero, since the mean value of the random field, averaged over the sample volume, is zero. As equilibrium is approached through the conversion of many small droplets to larger and larger ones, the surface energy decreases. As a result, and the random field fluctuations become increasingly effective at slowing down further droplet growth[1]. This impediment to the reaching of a uniformly ordered state is probably why nucleated regions never grow to visible size in either quenched gel-mixtures or in mixtures which saturate the pores of Vycor.

Quantitatively, the above discussion of metastability in magnetic random-field systems is really not applicable to the present experiments, where the order parameter is conserved. In this case equilibration on long spatial scales can be also retarded by diffusion, which can be a slow process at small wavenumbers[7]. This issue will be returned to in the discussion of an important experiment on phase separation of LW in Vycor[5].

Using a mean-field model, DeGennes[8] and others[9] have calculated the structure factor S(q,t) for a binary mixture in a porous medium. The theory is a modification of random field theory that takes into account that the source of the random field, namely, the randomly oriented gel strands, has a non-zero pore size and a slight preference to be wetted by one component of the mixture more than the other. It is instructive to compare their results with light scattering experiments on isobutyric acid contained in a gel with very small pores[3] (IBW in polyacrylamide) and in another gel (gellan gum), in which the mean pore size L is much larger than than ξ. Though random field theory would seemingly apply only in the former situation, the measured scattering intensity I(q) is similar in two systems.

In the mean field approximation, the static structure factor S(q) (which is proportional to I (q) in the absence of multiple scattering) was calculated by DeGennes to be[8]

(1) $S(q,t)=<c(q,0)c^*(q,t)> \propto e^{-Dq^2t}/[q^2+\kappa^2]+ \ell^{-2}/[q^2+\kappa^2]^2$.

Here c(q,t) is a Fourier component of a composition fluctuation, c(r,t), and q is the photon momentum transfer, $q=(4\pi/\lambda)\sin(\theta/2)$, with λ being the wavelength of light in the mixture. The first term is the standard Ginzburg-Landau result, and shows the usual diffusive relaxation of the fluctuations (When $q\xi>>1$, D is itself q-dependent). When $q\xi$ is small,

(2) $D=k_BT/6\pi\eta\xi$,

where η is the viscosity of the mixture. In a porous medium, one would expect the viscosity of the free mixture should be replaced by a larger effective viscosity, representing the increased difficulty of fluid flow through small pores of the medium. This enhancement effect has been seen in LW in Vycor[5]. In a free mixture, the diffusivity vanishes as the critical temperature is approached, since $\xi=\xi_0\varepsilon^{-\nu}$. The constant ξ_0 is an atomic length (~.2 nm) and $\nu=0.63$[10]. In a free mixture of LW or IBW, the fluctuation rate $\Gamma=Dq^2$ is of the order of Khz one degree from T_c, and at a scattering angle of 90^0.

The second term in (1) is the random field Zeeman energy contribution, ℓ^{-1} being the rms random field strength in length units. Note the absence of a time-dependence here; these fluctuations in the order parameter are frozen. In magnetic language this term corresponds to the mean magnetization at each spin site, resulting from the precession of each spin about the direction of the local random field there. Because its average value is non-zero at each lattice point, it gives a finite contribution to the structure factor S(q,t).

The parameter κ^{-1} in Eq. (1) is a temperature-dependent length that is equal to the correlation length ξ far from the critical point, and to ℓ near it. In the first of these temperature ranges, the κ^{-1} is so short that the mixture is oblivious to the presence of the gel network, so the parameter ℓ^{-1} is irrelevant. On the other hand, near the critical temperature, the correlation length is cut off at the value ℓ [which is greater than the pore size L] by the presence of the gel strands. An interpolation formula for $\kappa(T,L)$ is given by DeGennes[8].

Recently Huse[7] has calculated S(q,t), taking into account the energy barriers that must be crossed to relax a composition fluctuation. The new aspect of this work is that in a system with conserved order parameter, diffusion must be taken into account, as well as the above mentioned metastability effects produced by the random field. Under certain approximations his result is[5]

(3) $\qquad S(q,t)=S_1(q)[1-A(q,T)]e^{-\Gamma t} + A(q,T)e^{-[\ln(t/t_0)/\ln(\tau_A/t_0)]^3}$.

The first term is the usual diffusive term discussed above, with $\Gamma=Dq^2$. The factor A(q,T) is the ratio of the amplitude of the first and second terms, and the product $S_1(q)A(q,T)$ corresponds to the second (random field) term in Eq (1). In the mean field calculation of deGennes, there is no analog to the exponential factor. This exponential contribution is associated with the crossing of local barriers, as discussed above. The relaxation of spontaneous composition fluctuations of small q is limited by the diffusion rate of composition through large distances, q^{-1}, which is slow. At large q, diffusion is no longer the bottleneck, and the relaxation rate will be given by the second term in (3). If one then measures the relaxation time at various q, there will appear a crossover from exponential relaxation (first term in (3)) to "activated" relaxation (second term) at some $q=q_x$. This cross-over has been seen in binary mixture-Vycor experiments of Dierker and Wiltzius to be discussed below. Because τ_A is a relaxation time arising from the surmounting of local energy barriers, rather than from diffusion, it will be q-independent in this simplied version of Huse's theory. In actuality, when the above equation was fitted to the experiments of Dierker and Wiltzius, τ_A^{-1} was also found to vary as q^2, contrary to Eq. (3)

Next we turn to the experiments on binary mixtures contained in the pores of a gel. The first experiments were in a 7 wt.% polyacrylamide-IBW gel[3]. Polyacrylamide (PAA) is a chemically cross-linked gel, with a pore size L of roughly 1 nm. This length is much less than $\xi(T)$, over the full temperature range of the scattering measurements. The strongest evidence that the gel strands are imposing a random field on the mixture is the absence of temporal fluctuations in the scattered intensity. The q-dependence of the scattering in this system is consistent with the Lorentzian-square (L-S) form if the second term in Eq.(1) is dominant, and with such a fit one obtains the parameter κ and its temperature dependence. One finds that κ varies in a manner consistent with DeGennes' theory discussed above, with κ^{-1} increasing from 20 nm at 23.5°C to 40 nm at 21.5°C. As expected, this length should increase toward ξ as the critical temperature inside the gel is approached. Because ξ is much smaller than the wavelength of light, it was not possible to make measurements in the range $q>>\kappa$, where a Lorentzian and square-Lorentzian forms of I(q) could be unambiguously distinguished. The scattering data could, in fact, also be fitted to a Lorentzian form.

Recently, Xia and Maher[4] have made similar measurements on IBW in a large-pore, entanglement gel (gellan gum), whose pore size L is of the order of 0.1 micron. In this system random field theory should apply only at temperatures within 0.1K of the critical point in the gel. Nevertheless the scattering is quite similar to PAA-IBW; the gel becomes cloudy on cooling, the temporal fluctuations are quenched, and the measurements can be slightly better fitted to a L-S form, with κ^{-1} of the order of 50 nm. There is, however a striking difference in the temperature dependence of κ. In gellan gum-IBW, κ is independent of temperature over the full temperature range in which the measurements could be made, $26^{\circ}C<T<28^{\circ}C$. This finding was in a sample of near-critical critical concentration, $c_c= 41$ wt.%.

From a measurement of the relative scattered intensity at small angles, where, $I(q)\sim\ell^{-2}/\kappa^4$ (It is assumed that the second term in Eq. (1) dominates. Measurements on PAA-IBW are consistent with the random field strength ℓ^{-1} being independent of temperature, as expected from random field theory. On the other hand, in gellan gum-IBW, where the pore size is large, the strong temperature dependence of the scattering intensity and the temperature independence of κ imply that ℓ is a strong function of temperature, according to the observations of Xia and Maher. They also found that critical and off-critical samples behaved similarly, though they differed in the following respect: In the critical sample, ℓ^{-2} increased as the gel-mixture was cooled by 3K below the temperature at which the scattering first became measurably large. In an off-critical sample (with isobutyric acid concentration 15 wt% larger than c_c), the temperature dependence of this field strength was weaker. These observations may provide a valuable clue to the difference between porous media of large and small pore sizes, but it has yet to be deciphered.

In the two systems, gellan-gum and PAA, the useful temperature range of light scattering measurements is only a few K. Neutron scattering measurements in this type of system, on the other hand, probe a very wide temperature range at smaller q-values[11]. This promising technique has not yet been fully exploited in porous media saturated with a critical mixture.

The observed behavior in gellan gum-IBW would seem consistent with random field theory ideas only if one assumes that the cloudiness in these samples is produced by nucleated droplets, whose growth is increasingly slowed down as their size increases. Presumably these domains are larger than the 1-micron pore size, assuming that the pore size is the smallest length scale in the gel; if not, there would be no small-scale random field to impede droplet growth. On the other hand, droplet growth must be halted at the smallest observable size (~10 μ), since they were never seen with a microscope. It is possible, however, that the domains have a very irregular surface, resulting from their energy preference to escape local energy barriers. This surface irregularity might make even rather large domains optically unresolvable.

Even though the gel-mixtures may not be in a true equilibrium state on large spatial scales, it is possible that on the scales available with light scattering, (≈500 nm), the equilibration time may be adequately short. If so, a comparison with equilibrium theories such as that of DeGennes would still be appropriate.

To understand the experimental results, it is therefore important to have in mind the approximate time required for the supernatant mixture and the mixture in the gel pores to come to a common chemical potential. Presumably the equilibration time τ

will be of the order of the square of the sample size divided by the composition diffusivity, and this time is of the order of hundreds of hours. The experiments themselves were usually completed in a time much shorter than this, in which case the gel samples may be thought of as effectively being in equilibrium. (In the measurements discussed above, the samples were held in the initial one–phase state for many days prior to the commencement of a run). Cooled and cloudy samples of gellan gum and PAA would clear up in months, starting from the outer surface of the gel in contact with the surrounding reservoir. Moreover smaller gel samples cleared up faster, providing further evidence that the samples were effectively isolated during in the experiments discussed above.

In summary, two sets of experiments have been reviewed in this section; in one (PAA–IBW), random field ideas would seem to be applicable, and the results are reasonably consistent with a mean field form of random field theory. In the other (gellan gum IBW), the pore size would seem to be too large for random field theory to be applicable. Yet the two systems exhibit similar qualitative behavior, though quantitative differences are clearly seen.

BINARY MIXTURES IN VYCOR

If the goal of the scattering experiments discussed in this review were to study random field effects, a rigid porous medium would be preferable to a gel. In a flexible gel the polymer strands can possibly shift their positions in the mixture in order to lie in the phase which is richer in the component to which they have a higher affinity (water in the present case). The local composition of the mixture, c(r), could then alter the local random field. This complication introduces interesting physics[12], but the simple assumptions of random field theory would no longer be applicable.

A closer approximation to the simple random field model would presumably be achieved if the gel were replaced by a rigid random medium, such as porous glass. Vycor is made by quenching a borosilicate glass into the spinodal region, and etching out the softer phase with acid. The product, in its commercially available form (Vycor 7930, Dow Corning), has a pore radius of 3 nm. and a porosity of 28%[13]. The experiments described below were in 7930 Vycor and also in a commercially unavailable type[14], with a pore diameter roughly equal to 50 nm[15], which has a similar porosity. Of course, neither of these systems truly mimics a random field system, because the glass takes up space, whereas in idealized theoretical models, the random field does not.

Recently Goh et al.[15] have measured the turbidity and 90^{0}–scattering from a small 50 nm Vycor rod immersed in a mixture of n–hexane and n–pei fluorohexane [HF]. This mixture has a normal coexistence curve with $T_{c}=21.5^{0}C$. The Vycor rod was a long, thin cylinder, with diameter d=2 mm, and a length of 1 cm. The rod was oriented so that the laser beam passed down its symmetry axis.

In a typical run the sample was cooled or heated at various rates (R), through the (bulk) critical temperature, T_{c}. The scattered intensity at 90^{0}, here designated as S, was a strong function of the magnitude and sign of R and on the history of the sample prior to the start of a run. The sample, was usually placed in the lower portion of the sealed sample tube, was held at a one-phase temperature many hours before a scattering measurement was started. Typically the cooling rate R was of the order of 10 to 50 mK/min. At, say R=−22 mK/min, as in Fig. 1, the scattering remains constant until a certain temperature was reached, and then peak appeared in S(see Fig. 1). The temperature T_{pk} at which S went through a maximum was roughly $2^{O}C$ below T_{c}. It was easy to identify the phase separation temperature T_{c} of the surrounding mixture, because the large density difference of the two components, produces a sharp meniscus between the two components develops in a few minutes, and T_{pk} is not reached for an hour at the above cooling rate. Once T_{c} has been crossed in a cooling run, the sample will be sitting in one of the two phases.

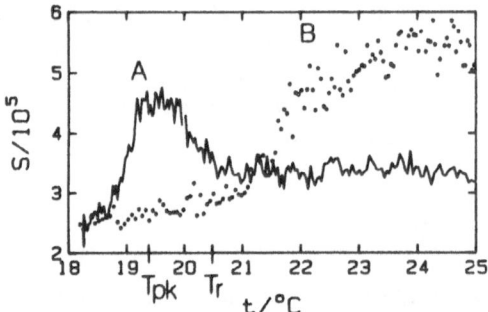

Fig. 1.   Temperature dependence of scattered intensity (in arbitrary units) at $\theta=90^{O}$. Curve A, cooling at −22mK/min; curve B, heating at 37 mK/min.

The experiment now to be described, leaves little doubt that the mixture in the pores of the Vycor undergoes a sharp transition from one to two phases on cooling, and that the transition temperature is less than T_{c}. The system was again taken far above T_{c}, so that the mixture is in one phase, and cooled in discrete steps. In one of the first steps in the series, the temperature is dropped from above T_{c} to one degree below, and the scattering shows little change. Only in a subsequent quench, when the temperature is suddenly dropped through T_{pk}, does one observe a burst of scattering intensity, presumably produced by the creation of a shower of droplets. Subsequent temperature steps also produce an intensity burst, as in a pure binary mixture. This behavior is just what one would expect of the phase separation temperature is indeed T_{pk}. The finite lifetime of these intensity peaks suggests that droplets are growing, as was argued above. Presumably the droplets which are generated in the process cannot grow to macroscopic size, for they never become visible by a 50-power microscope. One explanation that is consistent with this result is the Vycor pores, though too large to impose a random field on composition fluctuations of size ξ, do act as a random field on the nucleated droplets, once they have grown larger than the pore size.

Unexpected behavior is encountered when the rod is slowly heated from below T_{c} to above it. The dots in Fig. 1 correspond to a heating run at 37 mK/min, the sample being in the lower phase region of the external reservoir below T_{c}. Here one encounters very long metastable states whose origin is not understood. The large scattered intensity at high temperature $T \gg T_{c}$ persists for the order of 15 hours, which is orders of magnitude larger than the composition diffusion time through the rod diameter. With the rod in the upper phase, or the vapor phase, S remains constant on heating.

The above observed asymmetry suggests that wetting is playing a role here. The upper (hydrocarbon-rich) phase slightly prefers to wet glass and quartz. We do not have an adequate explanation for this rise in scattering on heating. Perhaps, there are present in the lower phase, droplets of the upper phase, which break up as the mixture is heated toward the critical point, where surface energy barriers are decreasing. In this case, large domains may break up into smaller ones, increasing the interfacial surface area in the sample, and hence the large-angle scattering. This explanation hinges on the assumption that wetting or random field effects favor small domains over larger ones.

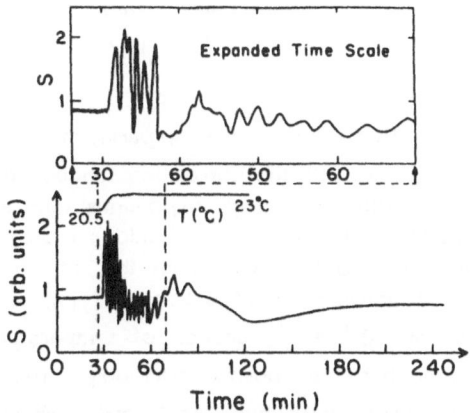

Fig. 2. Transient evolution of the 90^0-scattering on heating the 50-nm Vycor rod by 0.2K. The sample is immersed in a critical mixture which is in one phase at beginning (and end) of quench.

There is, however, evidence that adsorption rather than wetting (in the Cahn sense) is responsible for some of the above effects. When the rod is held for several days at a temperature well above T_c, appreciable scattering persists, but fluctuations in S die out (short, straight line segments at left in Fig. 2, $T=20.5^0C$). If now the temperature is abruptly raised even further (to 23^0C in Fig. 2), the mean value of S changes only slightly, but temporal oscillations develop. Invariably these fluctuations initially are relatively rapid, the correlation time (Γ^{-1}) being of the order of seconds. This characteristic time increases to hours, as may be seen in the figure. Ultimately this low-frequency composition noise damps out. The diffusive equilibration time, d^2/D, between the mixture and the reservoir is of the order of an hour, which is consistent with the data in Fig. 2. However, we do not understand the origin of the chemical difference that drives the diffusion (if that is what we are seeing), nor have we identified the feedback mechanism that is responsible for the oscillations. The pore size would seem to be too large to call on random field effects, and the Cahn theory of wetting would seem irrelevant as well.

The dynamics of random field behavior in a 3nm-Vycor saturated with a critical mixture of 2,6-lutidine and water (LW), has recently been studied by Dierker and

Wiltzius[13]. The pore size of this commercial Vycor is small enough to justify comparing the experimental results with random field theory. By measuring the correlation time as a function of temperature, it was clearly seen that the relaxation of composition fluctuations is impeded by local energy barriers imposed by the Vycor.

The mixture LW has an inverted coexistence curve with $T_c=33.3^{\circ}$C. In this small-pore Vycor the absence of a sharp meniscus or visible droplets makes the phase separation temperature difficult to identify, as in all the large-pore Vycor and the gels. Strong scattering does not appear until $T \approx 60^{\circ}$C, a temperature shift that is very large, and in the expected direction for a system with inverted coexistence curve. Above 60°C, I(q) reaches a plateau and does not relax away for weeks. This persistence of the scattering is presumably a random-field effect; this field pins the droplets, preventing them from growing to visible size (~40 μ, as in the gels and in the 50 nm-Vycor).

The most detailed information in this experiment comes from the intensity correlation measurements. Below 58° the correlation function g(t) (which is proportional to S(q,t) at fixed q), can be fitted to a single exponential, and a single relaxation rate Γ can be extracted. Above this temperature, g(t) develops a slowly decaying contribution, and at all T, Γ can be fitted to Eq. (3). From this fit one extracts the fractional contribution, A(q,T), of the activation contribution. In Fig. 3, this relaxation rate is plotted in dimensionless units (Γ^*) as a function of temperature. Plotted here is the relaxation rate associated with the exponential part only. The scattering angle is 30°, where $q^{-1}=112$ nm. The inset shows how the inverse relaxation rates, τ and τ_A of the exponential and activated terms vary with temperature. Here A(T) is identical to A(q,T) in Eq. (3)

The monotonically rising curve at the right side of Fig. (3) is the fractional contribution of the non-exponential, activation term to g(t) in Eq. (3). Below 53°C the relaxation is purely exponential (A=0) as in a free binary mixture. However the temperature dependence of Γ^* is very complex. We discuss the measurements for $T<53^{\circ}$C first.

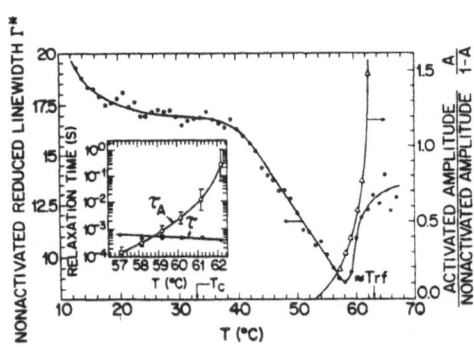

Fig. 3. Temperature dependence of the nonactivated reduced relaxation time, Γ^*, and the relative weight of activated to nonactivated relaxations A/((1−A). Inset: Temperature dependence of the nonactivated and activated relaxations times, τ and τ_A. The solid lines are a guide to the eye. The bulk mixture's T_c is indicated as well as the approximate vicinity of the random field transition, T_{rf}. $q=8.85\times10^4 cm^{-1}$

Below 20°C Γ^* is of the same order of magnitude as in the pure mixture at the same temperature, though measurably smaller. The authors credit the decrease in Γ^* to the small size of the Vycor pores; they restrict fluid flow and thus increase the effective viscosity of the mixture.

Between 20°C and 35°C, Γ^* is almost temperature independent. This suggests that ξ, on which Γ^* depends, has ceased varying with temperature; presumably the composition fluctuations are pinned at the pore size, L. Dierker and Wiltzius argue that as the temperature is further raised toward T_{rf}, large spatial fluctuations are increasingly favored energetically, and they now grow to encompass many pores of the Vycor. To explain the subsequent decrease in Γ^*, as T is increased to 58°C., we must assume that the growth of the fluctuations is not appreciably impeded by random energy barriers, since g(t) continues be a single exponential. This pure exponential relaxation persists out to ~55°C. Near this temperature, A commences to increase, i.e. g(t) develops a slowly decaying tail. Wiltzius and Dierker suggest that the system crosses from one to two-phases at $T_{rf}=63^\circ$, indicated by the arrow in Fig. 3. It is also possible that the minimum in Γ^* identifies T_{rf}; above this temperature, the diffusive relaxation rate Γ^* increases, as in a free binary mixture.

Dierker and Wiltzius also studied how the relaxation rate depends on q. As already noted, a crossover from diffusive to activated relaxation is expected as one increases q, with $\xi(T)$ held constant; at large enough q, diffusion through large distances, q^{-1}, ceases to be a bottleneck. This crossover was indeed observed by Dierker and Wiltzius. Contrary to Eq. (3), they found that τ_A^{-1} varied as q^2, just as Γ does. In actuality Huse's detailed calculation[7] allows for a term in S(q,t) which contains a relaxation rate proportional to q^{*2} but which also has an activation contribution.

SUMMARY

There are striking qualitative differences between the scattering behavior of pure critical mixtures and mixtures contained within the interstices of a porous medium. History T- dependent effects and slow relaxation are seen in systems with both large and small pore sizes. Both the large and small pore-systems exhibit the metastability that one associates with random field behavior, but invoking this theory is hard to justify for the 50 nm-Vycor-LW and gellan gum-IBW. In the experiments of Dierker and Wiltzius, random field theory provides a consistent explanation for the observations, though a quantitative understanding of the results is not yet within reach. In the large-pore Vycor, an explanation is lacking for the extremely long relaxation lifetimes that are observed on heating (Fig. 1), and the slowing compositions fluctuations in the one-phase state. The questions raised by some of these experiments will probably not be resolved until the light scattering measurements are supplemented by further neutron scattering experiments, which are ideal for probing the relatively small spatial scales which are encountered in the experiments we have described here.

ACKNOWLEDGMENTS

I have profited from discussions with my colleague J. V. Maher and from conversations with a good fraction of those whose names appear in the references.

An illuminating discussion with P. Wiltzius was particularly helpful. Thanks are also due to E. To for discussions of his recent experiments. Some of this work is the result of a very enjoyable collaboration with N. C. Goh and C. M. Knobler. This paper has greatly benefited from careful reading of the manuscript by P. Tong and K.-Q. Xia. The essential technical contributions of R. Tobin and J. Zagorac are gratefully acknowledged. This work was supported by The National Science Foundation under Grant No. DMR 12374.

References

[1] J. Villain, in 'Scaling Phenomena in Disordered Systems', Ed. by R. Pynn and A. Skjeltorp (Plenum,N.Y., 1985) p. 423.

[2] J. W. Cahn, J. Chem. Phys. 66, 3667 (1977); D. Sullivan and M. M Telo da Gamma, in 'Fluid Interfacial Phenomena' Ed. by C. S. Croxton(John Wiley, New York, 1985).

[3] J. V. Maher, W. I. Goldburg, D. W. Pohl, and ·M. Lanz, Phys. Rev. Lett. 53,60 (1984), S. K. Sinha, J. Huang, and S. K. Satija, in 'Scaling Phenomena in Disordered Systems', Ed. by R. Pynn and A. Skjeltorp (Plenum, N. Y., 1985) p.157; W. I. Goldburg, p151 of same volume; J. V. Maher in 'Physics of Finely Divided Matter', Ed. by N. Boccara and M. Daoud(Springer-Verlag, Berlin, 1985) p.252; J. S. Huang et al.,Physica, 136B, 291 (1986).

[4] K.-Q. Xia and J. V. Maher, Phys. Rev. A, in press (1987).

[5] S. B. Dierker and P. Wiltzius, Phys. Rev. Lett. 58, 1865 (1987).

[6] W. I. Goldburg and D. Bideau, preprint (1987).

[7] D. Huse, preprint(1987).

[8] P. G .deGennes, J. Phys. Chem. 88, 6469 (1984).

[9] D. Andelman and J.-F. Joanny in 'Scaling Phenomena in Disordered Systems', Ed. by R. Pynn and A. Skjeltorp(Plenum,N. Y., 1985)p. 163; Andelman and J. F. Joanny, in 'The Physcs of Finely Divided Matter' (1985), Ed. by N. Boccara and M. Daoud,(Springer -Verlag, Berlin, 1985); D. Andelman and J.-F. Joanny, Phys. Rev. B32, 4818 (1985).

[10] H. E. Stanley, 'Introduction to Phase Transitions and Critical Phenomena' (Oxford University Press, New York, 1971).

[11] J. S. Huang et al., Physica 136B, 291 (1986).

[12] F. Brochard and P. G. DeGennes, Ferroelectrics, 30, 33 (1980).

[13] P. Wiltzius et al, preprint (1987).

[14] K. Kodakura, Ph.D thesis, University of California, Los Angeles, 1983 (unpublished).

[15] M. C. Goh et al., Phys. Rev. Lett. 58, 1008 (1987).

PHASE SEPARATION IN FLUIDS IN THE ABSENCE OF GRAVITY EFFECTS

D. Beysens, P. Guenoun and F. Perrot

Service de Physique du Solide et de Résonance Magnétique
CEN-Saclay, 91191 Gif-sur-Yvette Cedex, France

ABSTRACT

Gravity effects during the phase separation of binary fluids in the critical region have been suppressed by two means: annuling the gravity in a space experiment and/or using a strictly density-matched mixture on earth. Such an isodensity system can be produced by partially deuterating one component in a binary mixture of cyclohexane and methanol. It has been demonstrated that this does not affect the phase transition.

Periodic-like patterns can thus be observed. They grow from a microscopic scale up to the final equilibrium stage, determined by the competition between wetting forces, finite volume effects and the remaining gravity influence. These structures are measured as a 2-D section of the 3-D pattern of interfaces between the phase domains. However only interfaces orientated perpendicular to the plane of observation can be detected. This combines with the interconnectivity of the domains to make the visible interface periodicity (L_m) the same as that of domains. After a numerical analysis, the structure factor $\hat{S}$ of these interfaces can be obtained, and its scaling properties can be checked.

Light-scattering experiments are also reported, so that the scaling properties of the corresponding 3-D structure factor S can be compared to those of $\hat{S}$; especially the equivalence between L_m, as measured from the direct observation or from light-scattering, can be tested, and the difference and similarities between S and $\hat{S}$ can be analyzed.

Finally emphasis is placed on the possibility of studying quantitatively the phase separation of fluids through this direct observation, and so, even at a microscopic level.

When fluids or mixtures of fluids phase separate on earth, sedimentation and flow convection induced by gravity (g) sooner or later drive the phase separation so that the denser phase goes to the bottom of the container, and the lighter to the top. The advent of space experiments has made possible recently the performance of experiments under very low gravity fields (microgravity μg). The above gravity-induced phenomena can thus be eliminated or at least very much attenuated, allowing the investigation of the late stages and of the ultimate equilibrium state of the phase separation process to be made. Note that this phenomenon is one of the rare process in condensed matter physics where the earth gravity does not act as a weak field.

Gravity, as we will see below, acts in principle only through the product (gδρ), with δρ the difference in density of the two phases at coexistence. It was therefore tempting to try to suppress -or at least to greatly reduce- the gravity influence by working with systems where the density of both phases can be adjusted to zero. This can be performed by adding a third component to a binary mixture (BM). Although such ternary mixtures do not generally behave as binary systems with respect to the phase transition properties, the 3rd component perturbing the coexistence of phases, in some conditions they can be considered as real binary systems, as it will be demonstrated below.

In order to show that in such density-matched fluids the influence of the earth gravity is effectively suppressed, it is necessary to perform at least one experiment under μg. Such experiments are reported in the following.

On the other hand, the late stages of phase separation are naturally associated with domains of each phase which are no longer microscopic, and cannot be studied by X-ray or Light Scattering (LS). Direct observation (DO) is absolutely necessary, and many questions which have not yet been raised about the formation of the corresponding images, their significance, and the way they can be interpreted (correlation functions, structure factors...) have to be solved.

In this lecture we will limit ourselves to the study of the phase separation of BF in the critical region, where the separation is known to occur through Spinodal Decomposition (SD). The following questions will be investigated:

What is expected for the phase separation in the critical region (Part II)? What is the influence of gravity and how to supress it (part III)? What are the relations between the direct observation study and the light scattering investigation? Especially, what are the respective roles of the domains and of their interfaces (Part IV)?

Some remarks will be made concerning the difference and similarities of the statistical functions obtained by LS and DO, how they compare to experiments, and what new type of investigations can be undertaken with such isodensity systems when associated with DO techniques.

II. PHASE SEPARATION IN THE CRITICAL REGION
1. Phase coexistence at equilibrium

In the domain of phase transitions, binary fluids, as liquid-gas systems, belong to the same universality class as the 3-D Ising model /1/, with respectively the concentration (c), the density (ρ) and the magnetization (M) as order parameter. In the following we will be concerned only with BF, and use the magnetic notation $M = c-c_c$, with c_c the critical concentration. The coexistence of phases at equilibrium is determined by the coexistence curve (CC) -the "miscibility gap" of metallurgists-, as shown in Fig.1. This curve delimits two regions: outside the CC, the fluid is monophasic at equilibrium, and inside the CC, the fluid cannot exist in a stable equilibrium. Two phases can therefore coexist only if the temperature (T) is lower than the critical temperature T_c (top of CC in Fig.1), and if their concentration is such that they fit the CC. The aspect of phases at equilibrium on earth is mostly governed by gravity (Fig.1a), the denser phase at the bottom and the lighter at the top. The location of the interface is determined by a competition between gravity, which tends to make the interface horizontal, volume fraction of the phases, which determines the vertical position, and wetting and capillary properties, which bend the interface. A measure of this last contribution with respect to gravity is given by the capillary length:

$$\ell_c = \sqrt{\frac{\sigma}{g\delta\rho}} \tag{1}$$

where σ is the interfacial tension and $\delta\rho$ the density difference of the two phases.

The capillary length exhibits a universal behavior near the critical point, in relation to the CC through $\delta\rho$ and the surface tension behavior. Indeed,

$$\delta\rho \simeq \left(\frac{d\rho}{dc}\right) \delta c \tag{2}$$

where δc follows the universal behavior /1/

$$\delta c = c^+ - c^- \simeq B(-\varepsilon)^\beta . \tag{3}$$

Here c^+ and c^- are the concentrations of the phases at coexistence at temperature $T = T_c - \varepsilon T_c$, with ε the reduced temperature. The exponent β is universal ($\beta = 0.325$), and B is a non-universal amplitude.

According to 2-scale factor universality /2/, the surface tension behavior can be related to that of the correlation length (ξ):

$$\frac{k_B T_c}{\sigma} = R(\xi^+)^2 , \tag{4}$$

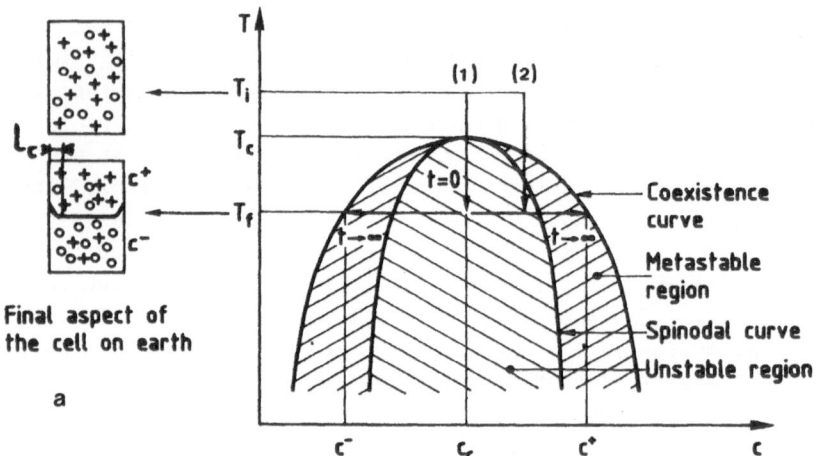

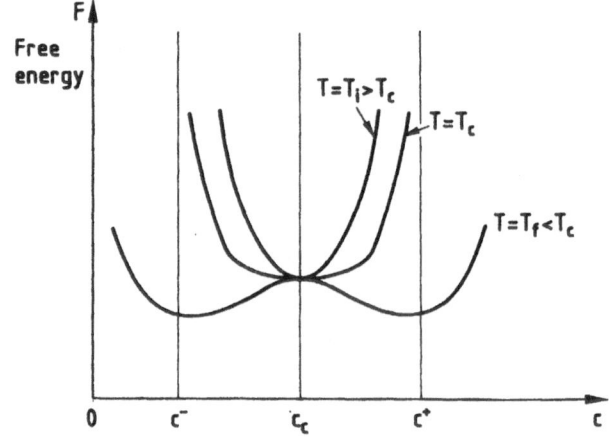

Fig.1. (a) Phase coexistence and phase separation region of a binary
system; ℓ_c is the capillary length. The quench (1) is at
criticality and is concerned with the unstable region. Spinoda
decomposition growth is the expected phase separation process.
The quench (2) is off-critical.
(b) Mean-field free energy corresponding to (a).

with R a universal ratio (R $\simeq$ 2.6 from experiments, but there is a disagreement by a factor of two with the theory). The correlation length behavior obeys the following:

$$\xi^{+,-} = \xi_0^{+,-} \ (\pm\varepsilon)^{-\nu} \ , \tag{5}$$

with the superscript +,- denoting the 1-phase or the 2-phase region respectively, $\nu = 0.630$ a universal exponent and $\xi_0^{+,-}$ a non-universal amplitude. These are related by a universal ratio /1/:

$$\xi_0^+/\xi_0^- \simeq 2 \tag{6}$$

From above, it follows that the capillary length behaves as:

$$\ell_c = \ell_c^0 \ (-\varepsilon)^\varpi \tag{7}$$

with

$$\ell_c^0 = \frac{1}{\xi_0^+} \sqrt{\frac{k_B T_c}{R \ g\left(\frac{d\rho}{dc}\right) B}}$$

and

$$\varpi = \nu - \beta/2 \simeq 1/2$$

Therefore ℓ_c goes to zero as T $\longrightarrow$ T$_c$, and the effect of gravity becomes important on a more and more microscopic level, in proportion to the distance from the critical point.

Phase separation

The state of a fluid mixture is well understood at equilibrium. But what is its behavior during a quench from the 1-phase region (temperature T_i) to the 2-phases region (temperature $T_f < T_c$)?

Two problems have to be distinguished: germination of the new phases, and growth of the corresponding domains which have reached a local equilibrium.

i) GERMINATION is in principle a static problem in the sense that it depends only on stability criteria. In the framework of a mean-field theory[3], a concentration fluctuation $\delta c(\vec{r},t)$ (here $\vec{r}$ is the space variable, and t is time) will decay or grow in time according to the sign of the associated susceptibility χ (osmotic susceptibility for BF): Below the CC the sign of χ determines two regions, delimited by the spinodal curve $\chi = 0$. Since χ is merely the 2nd derivative of the free energy of the

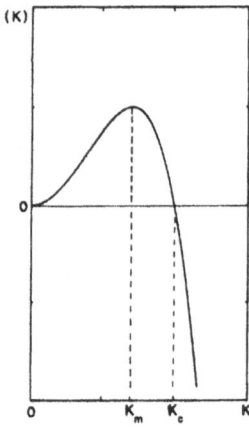

Fig.2. Amplification factor $\omega(K)$ for a fluctuation δc_K of wavenumber K, according to a linear theory : $\delta c_K \sim \exp[-\omega(K)t]$.

system with respect to concentration, it is the same as the curvature of the free energy (see Fig.1b). The curvature χ characterizes a region of metastability ($\chi > 0$, between the spinodal and the coexistence curves) and a region of instability ($\chi < 0$, inside the spinodal curve).

In the region of _metastability_, germination occurs through nucleation, and the formation of a new phase is a competition between the gain in free energy and the loss by the appearance of a surface tension energy. Only nuclei of radius larger than a critical radius will be allowed to grow .

On the other hand, in the region of _instability_, all fluctuations, in principle, are allowed to grow. However, here also, very small fluctuations, involving high concentration gradients, are so costly in energy that they must vanish. Since very large fluctuations require long times to grow (due to the mass diffusion), a typical size emerges which exhibits the maximum growth rate (see Fig.2).

ii) GROWTH problems take place immediately after germination and during all the period where the system is already at a local equilibrium and tends to reach, through a succession of local energy minima, the final equilibrium state, ("minimum minimorum" of the energy levels). This growth starts at a microscopic level, and there should not be any distinction between the unstable and metastable regions. On the other hand, at least two parameters seem to be important: The volume faction of the 2 phases (denoted by u and ℓ)

$$\Phi = V^u/(V^u + V^\ell) \; ,$$

and the capillary length. Volume fraction determines how dense-packed are the domains, and what can be their interactions, and the capillary length is a measure of the scale of the gravity influence. Typically, as explained below, gravity-induced effects occur when

$$L_m/\ell_c \gtrsim 1 \; , \tag{8}$$

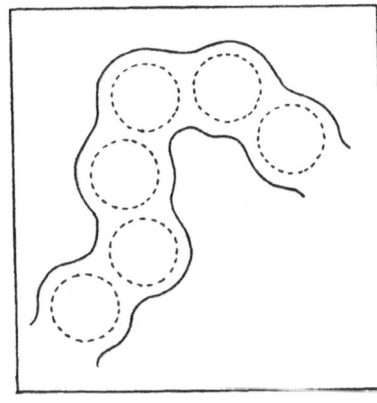

Fig.3. How a tube-like domain (full line) might be formed from interconnected clusters (dotted lines).

with L_m the typical size of a domain. The fact (Eq.7) that $\ell_c \rightarrow 0$ when $T \rightarrow T_c$ makes the effect of gravity act very early during the phase separation process in the critical region, typically when the domain size $L_m \sim$ a few μm.

Even though such gravity effects can be removed, the volume fraction Φ alone would determine the growth rate. In the critical region, Φ is close to 1/2 ($V_u = V_\ell$), making the density of the domains very high. The percolation threshold ($\Phi \sim 1/6$ at dimension 3) is likely to have a prominent role, deciding whether the clusters can be interconnected or not (Fig.3). If it is so, new types of growth, through capillary instabilities, can form and accelerate the growth very much.

A typical phase separation process is illustrated in Fig.4, corresponding to a SD process. The time is expressed in the natural time scale which is the typical lifetime (τ) of a concentration fluctuation, i.e. the time that a fluctuation of size ξ diffuses a length ξ :

$$\tau^{+,-} = \frac{6\pi\eta}{k_B T} (\xi^{+,-})^3 \tag{9}$$

Here η is the shear viscosity and k_B the Boltzmann constant. The superscript $(+,-)$ denotes the 1-phase or the 2-phase region; $\eta^+ \sim \eta^- \sim \eta$ close to T_c.

a) At time $t = 0$ (Fig.4a), the system is quenched from temperature T_i to $T_f < T_c$; it seems natural (but there is no verification of this point) to assume that the fluctuations in the fluid are those which should exist in the 1-phase region, at the "mirror" temperature $[-(T_f - T_c)]$. With the susceptibility of the system being negative, such fluctuations are growing (Fig.4b), and those which exhibit the higher growth rate (see above, Fig.2), will impose a characteristic size (L_m) in the system. According to the Cahn theory /4/, the corresponding wavenumber (K_m) is:

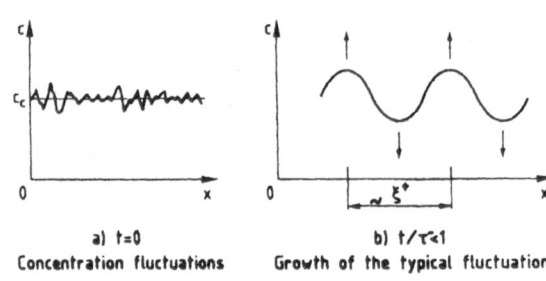

a) t=0
Concentration fluctuations

b) t/τ≪1
Growth of the typical fluctuation

Fig.4. Illustration of the growth of fluctuation in the first stages of the spinodal decomposition process.

c) t/τ≳1
Saturation of the typical fluctuation and growth of the typical length

d) t/τ~10

$$K_m = \frac{2\pi}{L_m} = \frac{1}{\sqrt{2}\,\xi}$$ (10)

and does not vary in time.

It is not clear, according to the above arguments, if it is ξ^+ or ξ^- which has to be considered. The fact that the domains are not yet at equilibrium makes us think that it is rather ξ^+ which matters.

b) This theory is however oversimplified in the sense that it is of a mean-field nature and that the free energy is linearized. The neglected non-linearities lead to a coupling of modes, and the dissymmetry in Fig.2 of the growth rate makes the mode K+K' more attenuated that the mode K-K'. Thus the maximum (K_m) decreases with time (Fig.4c). According to Langer, Bar-on and Miller /5-a/,

$$K_m \sim t^{-0.2}$$ (11)

However they neglected the influence of hydrodynamics. This effect has been considered by Kawasaki and Ohta /5-b/, and the growth (11) is accelerated.

c) At the end of this growth regime, the system has reached a local equilibrium, a surface tension between the domains can be identified, and other types of growth, related to the interactions between domains, can proceed (Fig.4d). As long as the Brownian motion of the domains is effective, one can imagine a diffusion-coalescence (diffusion-reaction)

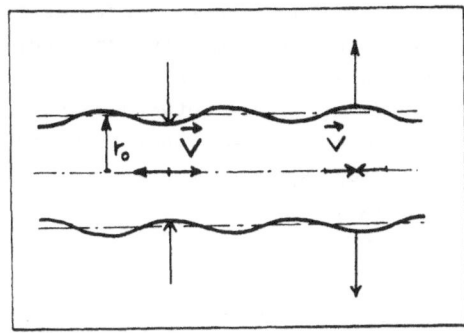

Fig.5. Deformation of a tube of average radius r_0 with an instability of wavelength Λ. $\vec{V}$ is the induced flow velocity.

process, according to Binder and Stauffer /6/, and Siggia /7/, where domains of size L_m (phase u) are separated by distance L_m (phase ℓ). The corresponding growth law should be:

$$K_m \sim t^{-1/3} \qquad (12)$$

One may question, however, the validity of such a mechanism when the density of clusters is so high that they percolate, as in Fig.3.

d) Concomitantly, in such tube-like domains, hydrodynamics plays a role (Siggia /7/) in the sense that capillary instabilities can form and induce capillary flows, as shown in Fig.5. The surface tension acts as an elastic membrane to accelerate the flow, which is however damped by viscosity, leading to:

$$K_m \sim \frac{\eta}{\sigma} \cdot t^{-1} \qquad (13)$$

e) The final stage of the growth occurs when the size (e) of the sample and/or the volume of the phases has been reached, i.e. when the finite size condition

$$K_m \, e \sim 1 \qquad (14)$$

has been fulfilled.

SCALING PROPERTIES OF THE SPINODAL DECOMPOSITION

In the above SD phase separation, two quantities seem to be the natural units: $\xi^{\cdot}$ (length) and $\tau^{\cdot}$ (time). The idea of using scaling (Binder et al./8/), originally used for phase transitions, was confirmed as a very useful tool.

One can therefore define two scaled variables,

and
$$L_m^* = L_m/\xi^{\cdot}, \quad K_m^* = 2\pi/L_m^*$$
$$t^* = t/\tau^{\cdot} \qquad (15)$$

with all quantities in the 2-phase region since most of the growth occurs with fluids at local equilibrium.

Other authors have tried to formulate the whole growth -at least in the domain where the fluid has reached a local equilibrium-, and we should note the interesting model of Kawasaki and Ohta /9/, which is based not on the growth of domains, but on the dynamics of *interfaces* between domains. This study will appear to be strongly connected to the one reported below.

A phenomenological approach by Furukawa /10/ leads to the growth law:

$$\frac{dL_m^*}{dt^*} = A^* L_m^{*-2} + B^* \; . \tag{16}$$

The first term of (16) represents the diffusion-reaction growth (12), and the last term the hydrodynamic growth (13). The solution of (16) leads to:

$$\left(L_m^* - 1\right) - \left[\left(\frac{A^*}{B^*}\right)^{1/2}\left\{\tan^{-1}\left[L_m^*\left(\frac{B^*}{A^*}\right)^{1/2}\right] - \tan^{-1}\left[\left(\frac{B^*}{A^*}\right)^{1/2}\right]\right\}\right] = B^* t^* \tag{17}$$

Here A^* and B^* are adjustable parameters, but an evaluation according to (12) is possible ,

$$A^* \simeq 2^{1/3} - 1 \simeq 0.26 \tag{18}$$

and, using the San-Miguel, Grant and Gunton /11/ estimation,

$$B^* \simeq 0.08 \tag{19}$$

The *structure factor* $S(K,t)$ of the domains can be written in a scaled form:

$$F(X,t) = \frac{K_m^3 \, S(K,t)}{\int_{K_A \ll K_m}^{K_B \gg K_m} S(K,t) \, K^2 \, dK} \tag{20}$$

where $X = K/K_m$, and K_A and K_B are two cut-offs. Practically, the experiments do not allow a K-range greater than $K_A \sim K_m/2$ and $K_B \sim 2K_m$ to be investigated.

The asymptotic behavior of F must follow:

- for $X \gg 1$, a Porod's law $F(x) \sim X^{-4}$; according to Furukawa /10/, the interconnectivity should modify this law, and this leads to:

$$F(X) \sim X^{-6}$$

- for $X \ll 1$, the local mass conservation $F(X) \sim X^2$.

The following form, which fits the above asymptotic behavior, has been proposed by Furukawa /10/:

$$F(X) = \frac{4X^2}{3+X^8} \tag{21}$$

Finally, another test of scaling is the time invariance of the 2nd reduced moment of $S(K,t)$ /11/:

$$r = \frac{k_2(t)}{k_1^2(t)} \tag{22}$$

where $k_n(t)$ is the intensity moment of order n:

$$k_n(t) = \frac{\int K^n S(K,t) \, dK}{\int S(K,t) \, dK} \tag{23}$$

When the form (21) of $F(X)$ is used, the following value can be calculated:

$$r_{FK} = 1.14 \tag{24}$$

GRAVITY INFLUENCE

1. Gravity and phase separation

How does the gravity affect the above process /12/ ?

Before and during the quench, one can reasonably ignore the sedimentation phenomenon which occurs in the 1-phase region. Though the osmotic compressibility diverges at T_c, the typical times to obtain a significant effect in BF are of order of one day or more , making the corresponding effects negligibly small in a typical experiment.

During the very early stages of the phase separation, sedimentation is still very slow, and can be neglected. However gravity can act on the interfaces between phases through capillary fluctuations; in the thermodynamic limit, it has been shown by Robert /13/ that the stability of interfaces in zero gravity is affected. However, the finite size of any real system generally prevents a divergence of the amplitude of capillary waves; less drastic modifications have thus to be expected .

Sedimentation can become efficient in a diffusion-reaction stage.

Stokes law once applied to a spherical droplet of diameter L_m leads one to expect an effect when

$$L_m^* > \sqrt{\ell_c^*} \; , \tag{25}$$

where $\ell_c^* = \ell_c/\xi$ is the scaled capillary length. Typically, for a mixture of deuterated Cyclohexane (C^*) and Methanol (M), this occurs for size $L_m \sim$ a few μm.

During the hydrodynamic growth, gravity-induced flows will compete with the capillary flows in the Navier-Stokes equation as long as :

$$L_m^* > \ell_c^* \tag{26}$$

In the same C^*M system as above, this occurs also for size $L_m \sim$ a few μm. Note that when (26) holds, (25) has been already verified; this means that gravity flows should have been preceeded by a former sedimentation of clusters.

However cooperative phenomena (instabilities) can also occur; a systematic study of these gravity phenomena has yet to be made. For an illustration, see below Fig.(7).

2. Removing the gravity influence

Since the domain size L_m is limited by the sample size (e), it seems sufficient to make the following inequality hold to remove all gravity influence:

$$\ell_c \gtrsim e \tag{27}$$

This is of course more difficult to fulfill close to T_c when ℓ_c goes to zero according to (7). However a very simple test of (27) can be performed by looking directly at the junction of the interface with the walls of the sample cell, as in Fig.(1) and Fig.(6) below.

To conclude, there seem to be two equivalent ways to remove the influence of g, either making the density difference of the phases $\delta\rho \sim 0$ or/and making gravity $g \sim 0$. Let us inspect in more detail these two possibilities:

i) Isodensity systems

There exist several BF where the density of the components does not differ very much. Among them, the mixture of Cyclohexane (C) and Methanol (M), where the densities of the components are

$$\rho_C \simeq 0.77$$
$$\rho_M \simeq 0.79$$

Natural cyclohexane is a mixture of isotopes, so a very natural idea /14/ was to change the isotopic proportion by adding partially deuterated cyclohexane (C^*), with density

$$\rho_{c^*} \simeq 0.89$$

so that a matching of density can be performed. However, matching the densities of the components does not mean that those of the phases at coexistence are matched. Indeed changes of volume occur during mixing and anomalies are expected close to T_c. In spite of an expected isodensity deuteration ratio ($C^*/C+C^*$ mass fraction) of about 12%, a large capillary length is found only for a deuteration ratio as small as 3-4%. With $\ell_c = 1$ cm at 30 mK from T_c, the density difference of the phases can be estimated as $\delta\rho \simeq 10^{-6}$ g.cm^{-3} at this temperature.

Nevertheless, a difficulty remains: a mixture of C, C^* and M is strictly speaking a ternary mixture. We have therefore performed (see Refs./14-15/) a systematic investigation of the mixtures CM, C^*M and C^*CM to determine whether such ternary mixtures could be considered as binary fluids in the field of phase transitions. The following properties have been investigated with the deuteration ratio of 0 , 100 and 11.6% respectively:

- <u>Coexistence curve,</u> which provides an estimate of the amplitude B and the exponent β, together with the critical concentration and critical temperature (see Fig.1 and Eq.3).
- <u>Susceptibility</u> and <u>correlation length</u> above T_c at criticality, allowing the amplitudes ξ_0^+ in Eq.5 to be determined, together with C^+ from the susceptibility behavior:

$$\chi^+ = C^+ \, \varepsilon^{-\gamma} \tag{28}$$

where $\gamma = 1.24$ is an universal exponent.

As shown in Table I, the exponent β exhibits the expected universal value, and the ratio of amplitudes:

$$\xi_0^+ (B^2/C^+)^{1/3}$$

has a value in agreement with theory (0.67). This would not have been the case if the mixture C^*CM were a ternary mixture, in particular the exponent β would have been renormalized .

Finally, the very small value of the deuteration ratio which makes the two phases density-matched in the critical region has the nice consequence of allowing the isodensity system to be considered as the C-M mixture itself for all the critical amplitudes and behavior.

Table I. Comparison of the universal exponent β and of the universal combination $\xi_0^+ \left(B_\phi^2/C^+\right)^{1/3}$ in the different mixtures.

System	β (0.325 theory)	B	ξ_0^+ $(10^{-8}\,\mathrm{cm})$	C^+ $(10^{-23}\,\mathrm{cm}^3)$	$\xi_0^+(B^2/C^+)^{1/3}$ (0.67 theory)
C-M	0.323 ± 0.003	0.755 ± 0.003	3.24 ± 0.23	6.19 ± 0.5	0.68 ± 0.07
C^*M	0.326 ± 0.002	0.713 ± 0.002	3.60 ± 0.30	5.55 ± 0.6	0.75 ± 0.09
C^*CM (11.6%)	0.322 ± 0.002	0.752 ± 0.002	3.30 ± 0.12	7.94 ± 0.35^c	0.64 ± 0.04^c

In such an isodensity system, no visible convection has been detected (see Fig.8 below and compared with Fig.7 where the non-isodensity C^*M system has been used). The phase separation pattern seems to be merely the extrapolation of what is detected by electron microscope in alloys (see Ref./16/). The origin of such images is however not straightforward, and we will analyze their formation below.

At the end, the final stage seems to be governed mostly by wetting (Fig.6); however the residual gravity competes to ultimately order the phases.

ii) Microgravity

Although the influence of gravity seems to have been qualitatively reduced by a very large amount, experiments in reduced gravity are necessary to try to answer the following questions:

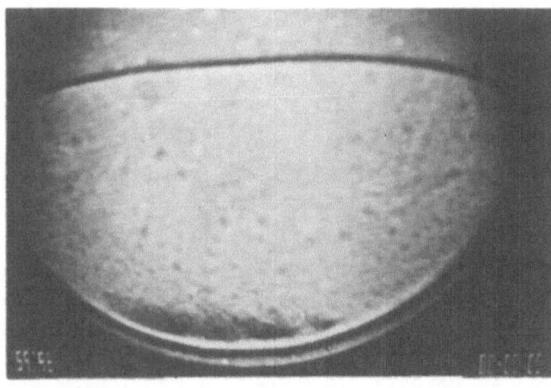

Fig.6. Final state after a quench of 5 mK depth ($t^* \sim 3\times10^6$) in an isodensity system of C^*CM. Note the large capillary length, and the total wetting. Small clusters are still included in each phase. They have been isolated during the growth, and they could not evolve as rapidly as the interconnected clusters.

- Are the remaining gravity flows still negligible in the late stages?
- Can the capillary fluctuations play a significant role ?
- Is there any other hidden influence of gravity?

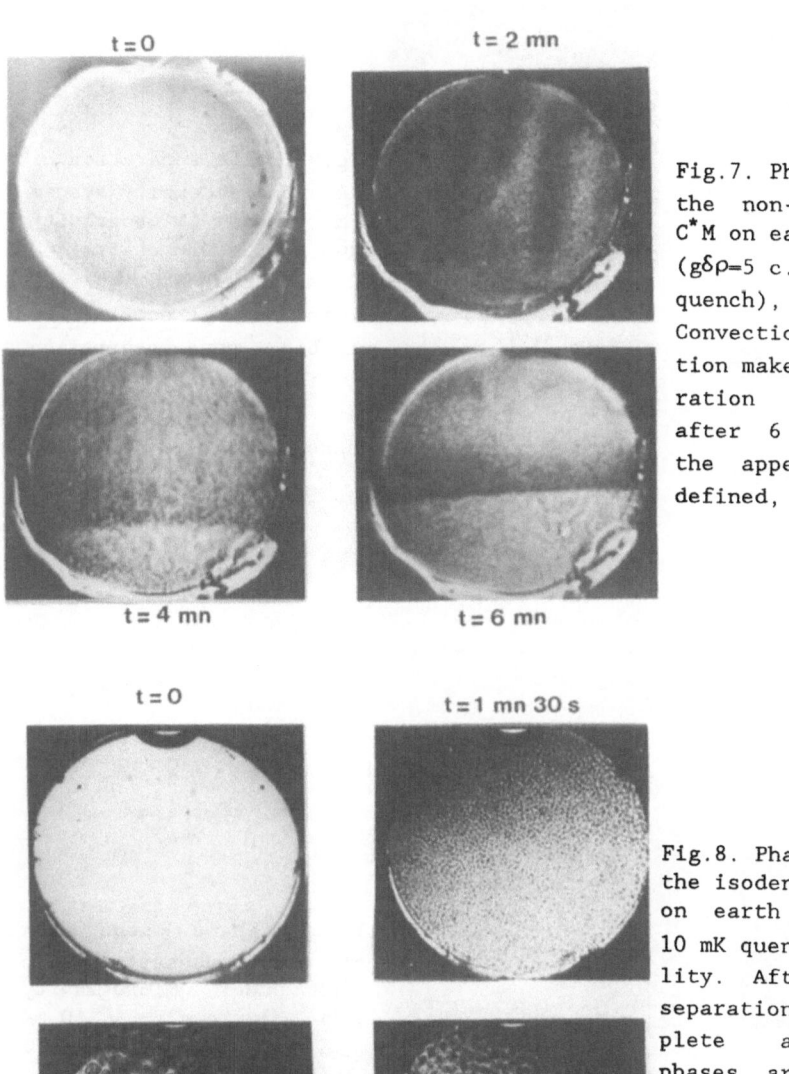

Fig.7. Phase separation of the non-isodensity system C^*M on earth ($g\delta\rho=5$ c.g.s., 10 mK quench), at criticality. Convection and sedimentation makes the phase separation nearly complete after 6 min, as shown by the appearance of a well defined, flat, meniscus.

Fig.8. Phase separation of the isodensity system C^*CM on earth ($g\delta\rho=10^{-3}$c.g.s., 10 mK quench), at criticality. After 6 min, phase separation is almost complete and well-defined phases are clearly visible. The interconnected morphology of the droplets is striking.

t = 0   t = 1 mn

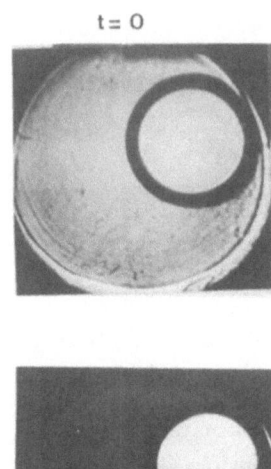 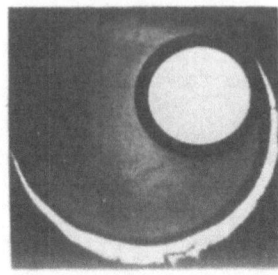 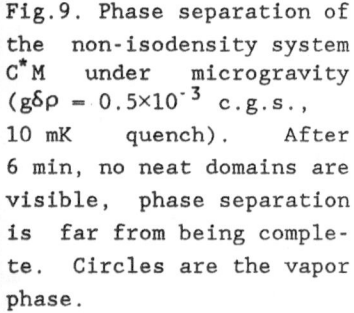

Fig.9. Phase separation of
the non-isodensity system
C*M under microgravity
($g\delta\rho = 0.5\times10^{-3}$ c.g.s.,
10 mK quench). After
6 min, no neat domains are
visible, phase separation
is far from being comple-
te. Circles are the vapor
phase.

t = 2 mn   t = 6 mn

t = 0   t = 1 mn 30 s

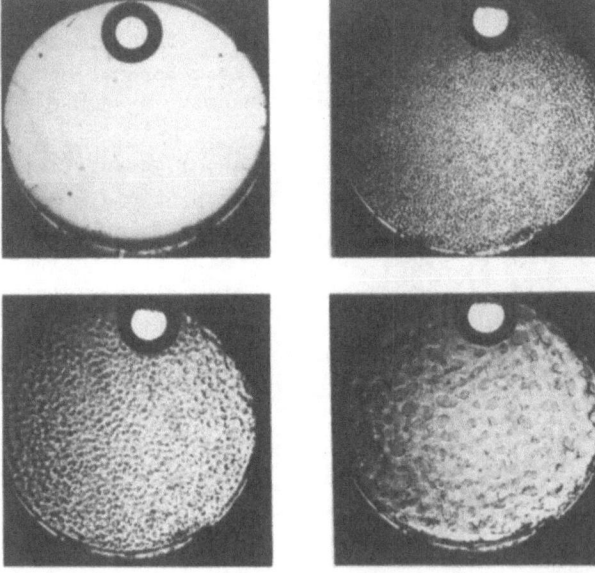

Fig.10. Phase separation
of the isodensity system
C*CM under microgravity
($g\delta\rho = 10^{-7}$ c.g.s., 10 mK
quench). The process is
qualitatively identical to
the corresponding earth
experiment. Circles are
the vapor phases.

t = 3 mn   t = 6 mn

Since the appearance of macroscopic ordered domains is the signature, at least qualitatively, of the absence of gravity effects, direct observation seems to be a good way to study phase separation. Since a few min is sufficient for a nearly full phase separation to be obtained on earth in a isodensity system with a quench of 10 mK below T_c, sounding rockets (TEXUS) from the Space European Agency program were used. They provide 6 min of a very good quality microgravity (residual accelerations lower than $10^{-4}g_0$, with g_0 the earth acceleration). We report in Figs.(7-10) the following experiments, performed in the same conditions (10 mKquench depth, same experimental module):

Earth experiments $(g_0=1)$

C^*M (Fig.7)
$\begin{cases} g\delta\rho \simeq 5 \text{ c.g.s} \\ \ell_c \simeq 25 \ \mu m \\ \text{Phase separation mostly driven by gravity} \end{cases}$

C^*CM (Fig.8)
$\begin{cases} g\delta\rho \simeq 10^{-3} \text{ c.g.s} \\ \ell_c \simeq 2 \text{ mm} \\ \text{No visible gravity influence} \end{cases}$

Space experiments $(10^{-4}g_0)$

C^*M (Fig.9) Texus 11
$\begin{cases} |g\delta\rho| \simeq 5\times10^{-4} \text{c.g.s} \\ \ell_c \simeq 3 \text{ mm} \\ \text{Very slow phase separation} \end{cases}$

C^*CM (Fig.10) Texus 13
$\begin{cases} |g\delta\rho| \simeq 10^{-7} \text{ c.g.s} \\ \ell_c \simeq 200 \text{ mm} \\ \text{Pattern nearly identical to that on earth} \end{cases}$

The results with the C^*M mixture under μg (Texus 11, Fig.9) were difficult to interpret, the phase separation being much slower than expected. A check of the criticality through susceptibility and correlation length measurements of the sample before and after the flight showed that it was indeed critical. It is why a *differential* experiment using the C^*CM mixture has been carried out in Texus 13 in order to try to observe any typical effect due to the absence of gravity.

This last experiment (Fig.10) demonstrated that phase separation on earth or under μg obeys the same laws, at least qualitatively. A quantitative comparison, as will be shown below (Fig.17), has confirmed this conclusion (see also /12/).

The striking result of Texus 11 can be understood as a very high sensitivity of the growth process to the volume fraction of the phases. A slight deviation from the critical concentration (1%, see Fig.11), invisible to the light scattering test (turbidity) that has been used, is able to shift the system far away from the percolated state and suppress the very rapid, linear in time, hydrodynamic growth process. This is in agreement with a study by Wong and Knobler /17/, where concentration was varied (however much more), and is qualitatively pictured in Fig.11: In a 1% off-critical cell, well defined clusters can be seen only after a few hours; this has to be compared with the few minutes corresponding to a sample at "criticality".

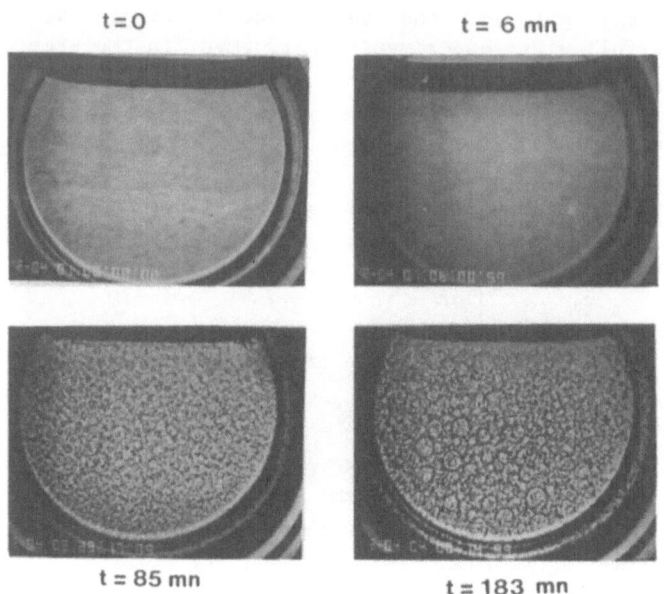

Fig.11. Phase separation of the isodensity system C^*CM on earth ($g\delta\rho = 10^{-3}$ c.g.s., 12 mK quench) at a concentration slightly non critical ($c-c_c = 1.0\times10^{-2}$). In 6 min, the domains are still microscopic and phase separation is not completed.

Now that μg experiments have validated the use of isodensity systems on earth to indeed remove the gravity effects, let us inspect in more details what are the new aspects which emerge when such an influence has been suppressed.

LIGHT SCATTERING AND DIRECT OBSERVATION STUDY

Only the early stages of the phase separation can be studied by light scattering, and the investigation of the late stages have to be performed by a direct imaging device. This raises questions about the meaning of patterns like those of Figs.8 and 10, and especially of their origin. For that purpose a set-up where one can perform both observation (with a resolution of a few μm and a field of 250 μm, or a resolution of a few tenth of mm and a field of 20 mm) and light scattering, has been used. It is described in Fig.12. It has the particularity of using an image analysis system, composed of a video camera and a computer, which digitizes the image (256×256 pixels and 6 bits) and allows various treatments to be made, among them the calculation of the structure factor of the image. The light scattering pattern can be analyzed also.

In order to understand what are the difference and similarities of the physical parameters that are obtained by direct observation and light scattering, let us discuss first what is effectively expected in both cases.

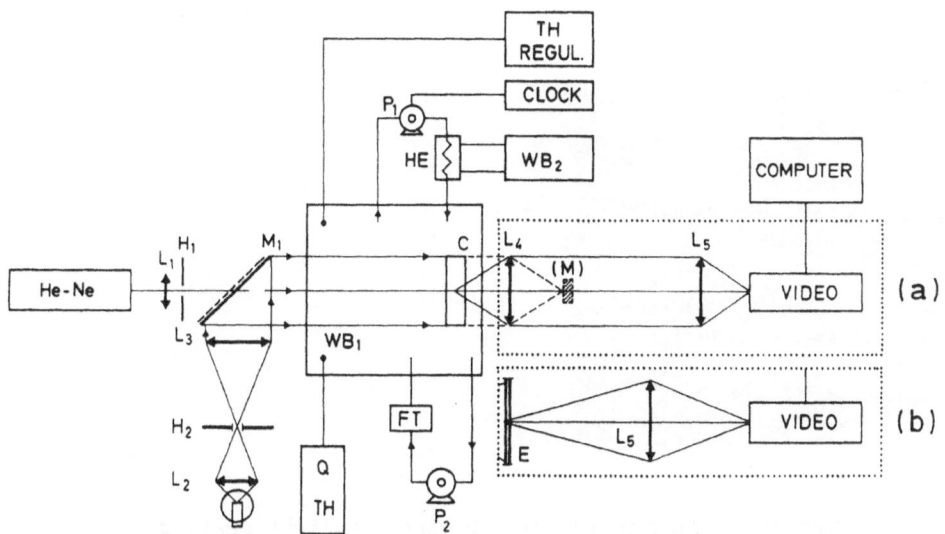

Fig.12. Experimental set-up used for both direct observation (a) and light scattering (b). L_1, L_2, L_3, L_4, L_5: lenses; He-Ne: helium-neon laser; H_1, H_2: pin-holes: M_1: semi-transparent mirror; Q.th.: quartz thermometer: WB_1, WB_2: temperature regulated water baths; P_1, P_2: pumps; FT: water filter; HE, heat exchanger; C: cell: M: mask; E: screen.

1. Light scattering /18,19/

Classically, the LS intensity $I(\vec{K}, t)$ (here $\vec{K}$ is the transfer wavevector between the scattered and incident radiation), as detected on a screen far from the sample (video camera here), represents the 3-D structure factor of the growing domains . This factor is symmetrical (Fig.13) and is a function only of $K = |\vec{K}|$:

$$I(K,t) \sim S(K,t)$$

In the SD process a peak (K_0) is seen (Fig.13), and K_0 is usually associated with K_m. This deserves some comment, since usually in classical theories of light scattered by particles /20/, such a peak is associated with the average distance between particles.

With this limitation in mind, the location of the peak determines K_m, and the scaling properties of S, through K_m, F and r, can be readily studied.

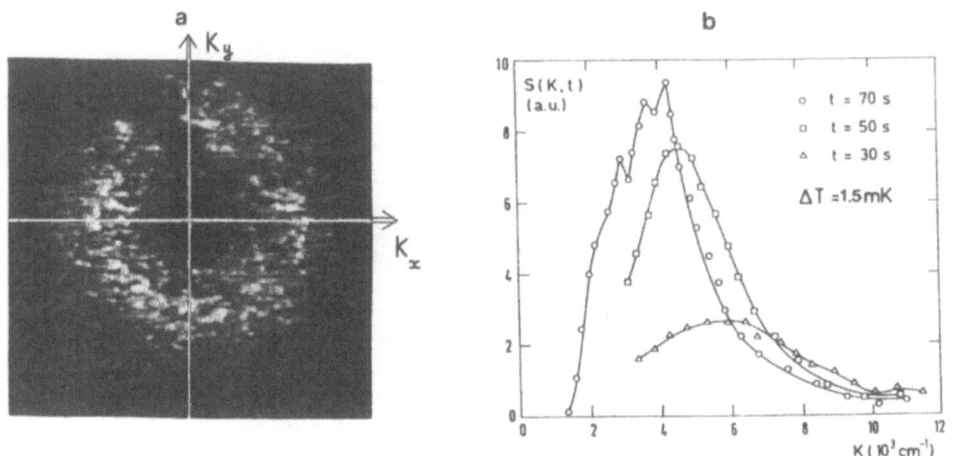

Fig.13. Typical structure factor obtained by light scattering. a) Aspect of the scattering pattern. b) Temporal evolution $\left(K = \sqrt{K_x^2 + K_y^2}\right)$.

2. Direct observation

What is the origin of images as reported in Figs.8-10-11?

Experimentally one finds that the contrast is maximum when the field of focus is located close to the exit window. This seems very natural, as the superposition of the structures makes the image very fuzzy if the focus is located too deep in the bulk.

Simple calculations, as performed in /21/, show that the modulation of the light intensity by the turbidity difference between both phases remains negligible; refractometry effects, due to the refractive index difference ("self-focusing") also give very small effects. A study with different magnifications shows that it is rather the refractive index gradient associated to the interfaces between domains which causes, by the scattering of light, the formation of these images. The situation is depicted in Fig.14. Here, moreover, the scale invariance of such patterns is striking: (c) is equivalent to (b) or (a) by a mere change of scale. For the latter case, the limited resolution of the optics has to be taken into account.

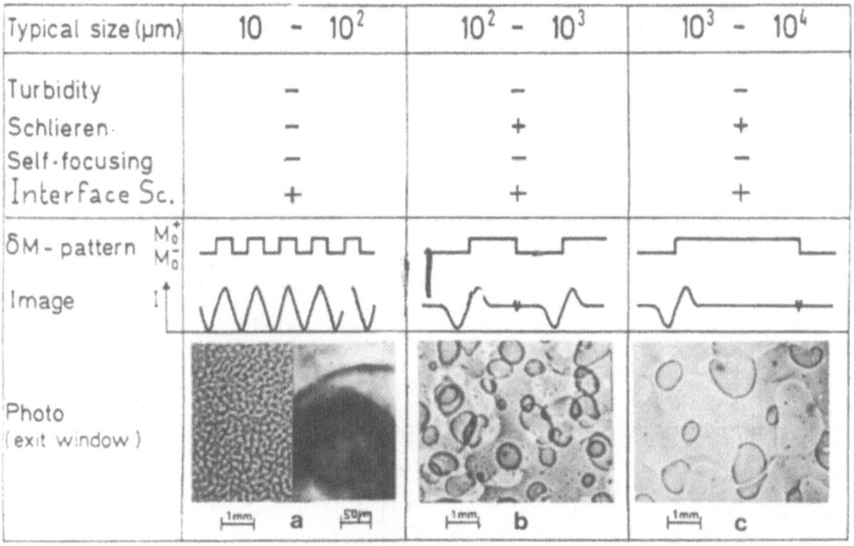

Typical size (µm)	$10 - 10^2$	$10^2 - 10^3$	$10^3 - 10^4$
Turbidity	−	−	−
Schlieren	−	+	+
Self·focusing	−	−	−
Interface Sc.	+	+	+
δM - pattern $\begin{smallmatrix} M_0^+ \\ M_0^- \end{smallmatrix}$			
Image $I\!\uparrow$			
Photo (exit window)	a	b	c

Fig.14. Various optical conditions encountered during the phase separation. For each size the respective influence of the mechanisms is listed, from − (low efficiency) to + (good efficiency).

What is obtained is therefore a 2-D section of the 3-D pattern of interfaces. However not all of these interfaces can be detected: scattering is more efficient at small angle, and only those interfaces which are nearly parallel to the direction of observation will be detected, as shown in Fig.15. The notion of interconnectivity becomes here very important; pictures as in Fig.14 cannot exist with an assembly of isolated droplets where all detected interfaces have to be closed (see Fig.11 and compare with Figs.10 and 14).

The digitization of such images gives an intensity $I(x,y)$ which is a function of the 2 space variables (x,y). A Fourier Transform (F-T) enables the corresponding structure factor $\hat{S}(K_x,K_y)$ to be determined:

$$\hat{S}(K_x,K_y) \propto |F\text{-}T[I(x,y)]|^2$$

a

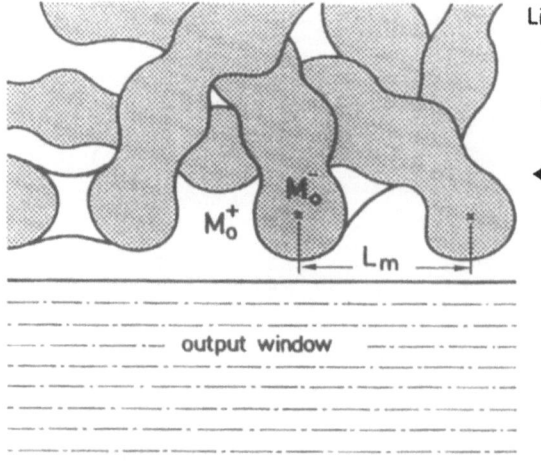

Fig.15. Sketch of the image formation. Small angle scattering by interfaces is the dominant process. a) Top view; b) Front view.

b

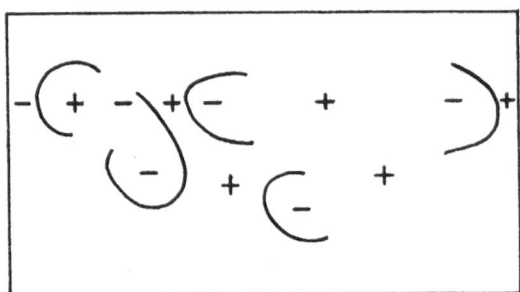

This structure factor, like the 3-D structure factor S, is symmetrical and exhibits a peak (K_m') (see Figs.13 and 16); it can be put into a reduced form, analogous to the Eq.(20):

$$\hat{F}(X) = \frac{(K_m')^2 \hat{S}(K,t)}{\displaystyle\int_{K_A \gg K_m'}^{K_B \ll K_m'} S(K,t) K \, dK} \tag{28}$$

where $K = \sqrt{K_x^2 + K_y^2}$; one notes the difference in normalization, the F-T being here 2-dimensional.

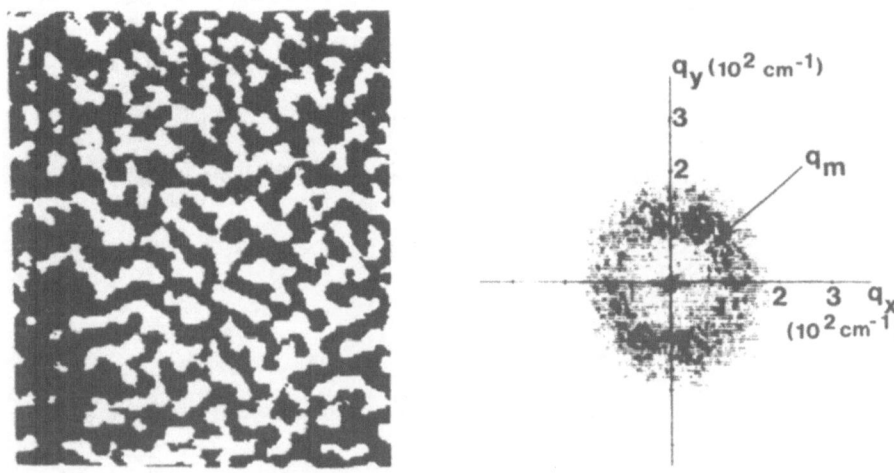

Fig.16. a) Digitized picture of a pattern like those found in Figs.8 or
10.
b) Structure factor obtained through the Fourier transform of a).

A reduced ratio $\hat{r}$, constructed from the n-th moments $(\hat{k}_n)$ of $\hat{S}$, defined as in (23), can be considered:

$$\hat{r} = \frac{\hat{k}_2(t)}{\hat{k}_1^2(t)} \qquad (29)$$

Scaling can then be checked, as for S, provided that the meaning of the peak wavevector, (K_m'), is well understood.

In fact, this wavevector K_m' corresponds to the average distance between the interfaces of the domains. Were these domains disconnected, as in Fig.11, the typical wavevector K_m' would have been twice the wavevector K_m corresponding to the distance between domains. The fact that in this study the domains are percolated, and that only interfaces parallel to the direction of observation can be detected, reduces the number of detected interfaces by roughly a factor of two, as far as statistics can apply (sufficient number of domains). This has therefore the nice consequence of giving the same spatial frequency:

$$K_m' \simeq K_0 \simeq K_m \qquad (30)$$

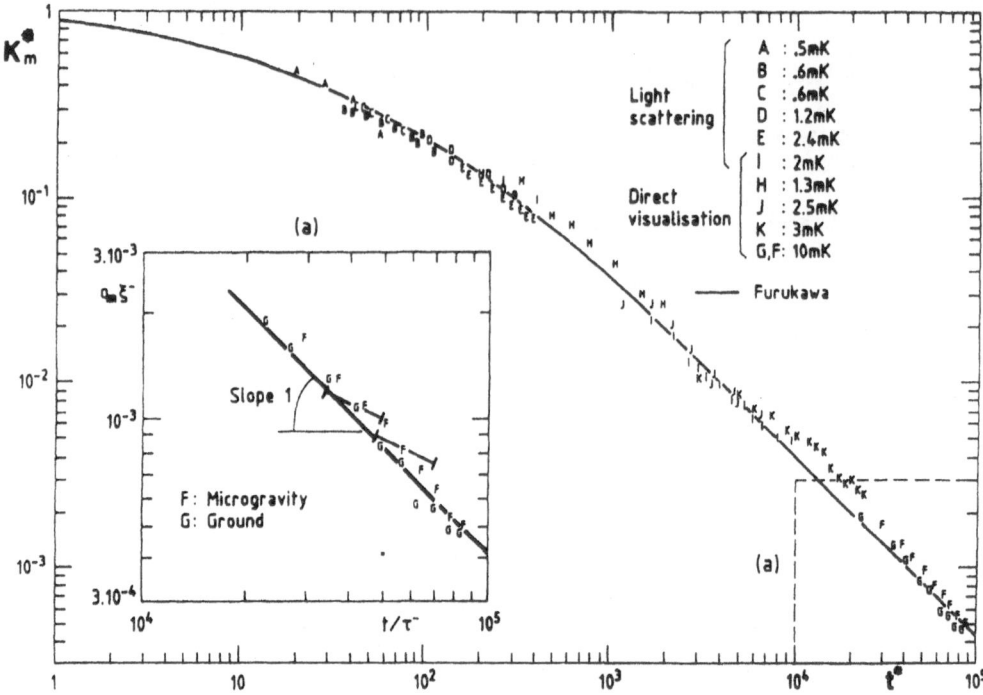

Fig.17. Behavior of the typical wavevector K_m^* in reduced units ($K_m^* = K_m \xi^-$), reduced unit ($t^* = t/\tau^-$). Scaling implies a universal behavior. In the insert a), we have reported the ground (G) and flight (F) experiments. The bar corresponds to a 1 mK uncertainty. The line with slope 1 indicates that the separation process is due to capillary flows. We have also reported data from light scattering and direct visualization techniques at various quench depths. They have been fitted to a universal function by Furukawa.

3. Comparison of light scattering and direct observation data

First, it must be noted that LS provides a 3-D average in all the illuminated volume, whereas DO is concerned with a thin slab close to the exit window. One can ask questions about the influence of the wetting properties of such a wall. Nevertheless, with complete wetting of one phase being generally seen near T_c /22/, if the corresponding film does not extend up to the focus plane, one can expect only very weak interactions, and neglect the wall influence.

Second, what is the domain of validity of the relation (30) between the periodicity of detected interfaces and that of domains? To answer this question, we have performed LS and DO experiments in the same reduced time region. Fig.17 shows that, within the experimental uncertainty, relation (30) does hold. Also reported are the μg experiments, nearly indistinguishable from the earth's ones (the apparent small difference is presumably due to the 1 mK quench thermal resolution of the space experiment).

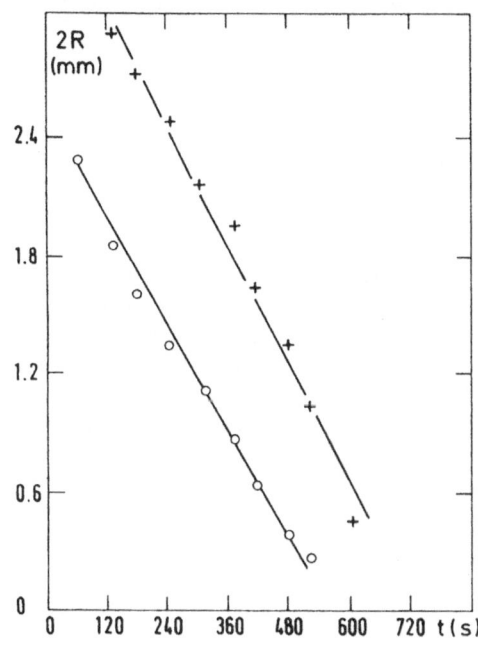

Fig.18. Experimental measurement of the interface velocity through the time evolution of the size (2R) of a tube-like domain.

This unique curve, which demonstrates scaling, can be fitted to the Furukawa behavior (17), and the following parameters can be determined:

$$A^*_{exp} = 0.14 \pm 0.01$$
$$B^*_{exp} = 0.022 \pm 0.001$$

They agree reasonably with those expected from a diffusion-reaction mechanism ($A^* \simeq 0.26$) and hydrodynamic growth ($B^* \simeq 0.08$).

An independent determination of B^* can be made by looking directly at the motion of interfaces, for instance by determining the rate at which the radius (R) of a tube as in Fig.3 can diminish. Fig.18 shows that R varies linearly in time; this would not have been the case for other systems where the order parameter is not conserved during the phase separation /3/. Another parameter B^*_{D0} is obtained through this linear variation:

$$B^*_{D0} \simeq 0.018$$

which compares well with the above determinations.

Tests of scaling can be also performed through the reduced factors F and $\hat{F}$. In Fig.19 it is clear that they are both time invariant, and that their shape is not very different -at least in this linear representation.

An attempt to demonstrate discrepancies for large X-values had led to the following X-behaviors (Fig.20):

$$F \sim X^{-[4.3\pm0.7]}$$

$$\hat{F} \sim X^{-[2.5\pm0.5]}$$

The large uncertainty in the exponents is connected to the small X-range available for the asymptotic behavior. The limiting behavior found in LS experiments suggests that the Porod's law (X^{-4}) is more likely to occur than the asymptotic behavior suggested by Furukawa (X^{-6}). The lowest exponent found with the DO experiments is in agreement with rigorous calculation /23/ that has been performed with a simple 1-D model, and where asymptotically $\dfrac{\hat{F}}{F} \sim K^{-2}$.

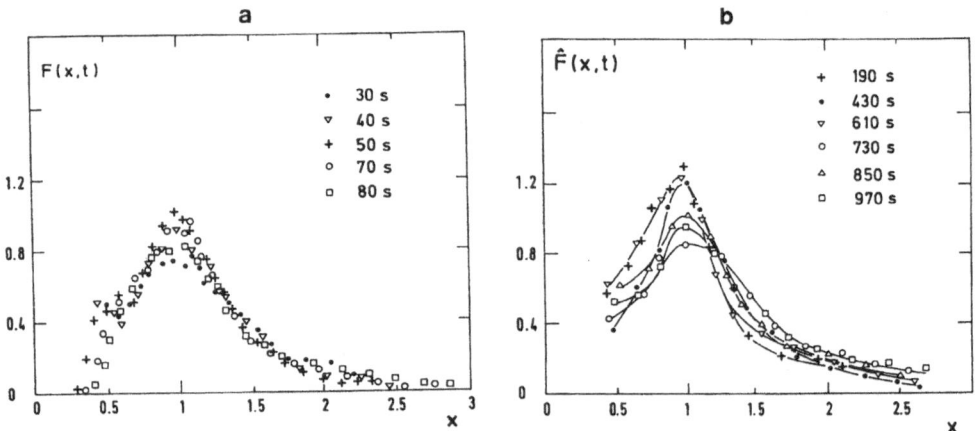

Fig.19. Reduced structure factors obtained from light scattering (F in a)) or through direct observation ($\hat{F}$ in b)).

Finally, the scaling of the reduced moments r and $\hat{r}$ is checked (Fig.21) to be time independent. It is striking that the values inferred are nearly the same, and remain very close to that calculated from the reduced profile (21):

$$r \simeq \hat{r} \simeq r_{FK} \simeq 1.1$$

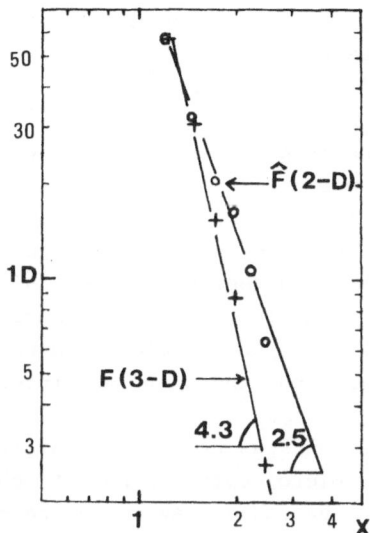

Fig.20. Two typical reduced structure factors in a log-log plot, obtained
by light scattering (F) or by direct observation ($\hat{F}$), showing the
limiting behaviors for large X.

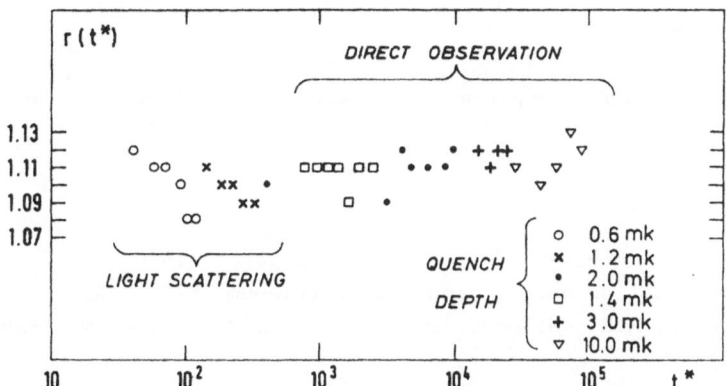

Fig.21. Evolution of the reduced second moments versus t^*. The results of
6 distinct quenches are shown and the analysis was done by light
scattering (r), by direct observation ($\hat{r}$) or by both techniques. An
average value is around 1.1.

V. CONCLUDING REMARKS

The use of the density-matched system C^*CM, after checks in a microgravity environment, has enabled practically all the gravity effects encountered during the phase separation in the critical region to be removed. "Practically" means that the final equilibrium stage is still partially governed by gravity, wetting forces competing with it. It seems difficult at this stage to suppress entirely this ultimate ordering role of earth gravity, and μg experiments seem to be the only remaining alternative to suppressing it.

Direct observations of phase separation have led to the quantitative study of the corresponding images. These images do not reproduce the domains, but the interfaces between them, in such a way that it is a 2-D section of the interfaces which are directed perpendicular to the observation plane which is detected. However this particular pattern was seen to obey the same scaling laws as the 3-D domain pattern, and it seems not unreasonable to look for precise relationships between the correlation functions of the domains and those of such interfaces -as has been done for the 1-D case.

Now direct observation was used here mainly in the late stages of phase separation, but microscopic means of observation are also very promising for studying the early stages and, perhaps, directly observing the growth of fluctuations during the very early stages. Clearly these optical means are very useful for relating the morphology of the separating domains to the growth, e.g. a percolated structure exhibits characteristics of growth much different from those of isolated clusters.

Finally, it should be noted that such phenomena where gravity effects have been suppressed provide much information on the microgravity environment, and should allow future space experiments to be rationalized and optimized.

ACKNOWLEDGMENTS

This work was supported in part by the Centre National d'Etudes Spatiales.

REFERENCES

/1/ See e.g. "Phase Transitions" Cargèse 1980 ed. by M. Levy, J.C. Le Guillou and J. Zinn-Justin (Plenum, N-Y, 1982).

/2/ H. Chaar, M.R. Moldover and J.W. Schmidt, J. Chem. Phys. 85, 418 (1986)

/3/ See e.g. "Phase Transition and Critical Phenomena" ed. by C. Domb, J.M. Lebowitz (Academic, 1983) Vol.8.

/4/ J.W. Cahn, J. Chem. Phys. 42, 93 (1965).

/5-a/ J.S. Langer, M. Bar-On and H.D. Miller, Phys. Rev. A11, 1417 (1975)

/5-b/ K. Kawasaki and T. Ohta, Progr. Theor. Phys. 59, 362 and 1406 (1978)

/6/ K. Binder and D. Stauffer, Phys. Rev. Lett. 33,1006 (1974).

/7/ E.D. Siggia, Phys. Rev. A20, 595 (1979).

/8/ K. Binder, C. Billotet and P. Mirold, Z. Phys. B30, 183 (1978).

/9/ K. Kawasaki and T. Ohta, Physica 118A, 175 (1983).

/10/ H. Furukawa, Adv. Phys. 34, 703 (1985) and references therein.

/11/ M. San Miguel, M. Grant and J.D. Gunton, Phys. Rev. A31, 1001 (1985).

/12/ D. Beysens, P. Guenoun and F. Perrot, Submitted to Phys. Rev.A (1987).

/13/ M. Robert, Phys. Rev. Lett. 54, 44 (1985).

/14/ D. Beysens, Acta Astron. 12, 525 (1985).

/15/ C. Houessou, P. Guenoun, R. Gastaud, F. Perrot and D. Beysens, Phys. Rev. A32, 1818 (1985).

/16/ J.W. Cahn, J. Chem. Phys. 66, 3667 (1977).

/17/ N.C. Wong and C.M. Knobler, Phys. Rev. A23, 858 (1981).

/18/ J.S. Huang, W.I. Goldburg and A.W. Bjerkaas, Phys. Rev. Lett. 32, 921 (1974).
    Y.C. Chou and W.I. Goldburg, Phys. Rev. A20, 2105 (1978);

/19/ N.C. Wong and C.M. Knobler, J. Chem. Phys. 69, 725 (1978);
    J. Phys. Chem. 85, 1972 (1981).

/20/ R. Hosemann, and J.N. Bagchi, Direct Analysis of Diffraction by Matter (North Holland, Amsterdam 1962).

/21/ P. Guenoun, R. Gastaud, F. Perrot and D. Beysens, Phys. Rev. A (1987, to appear).

/22/ J.W. Cahn, J. Chem. Phys. 66, 3667 (1976)
    For a recent review on "Fundamental Problems in Statistical Mechanics VI (E.G.D. Cohen ed., Elsevier, 1985).

/23/ P. Guenoun, Thesis (1987, Paris, unpublished).

GROWTH OF A DROPLET PATTERN (BREATH FIGURES)
ON A SURFACE

Daniel Beysens
Service de Physique de Solide et Résonance Magnétique
CEN Saclay, 91191 Gif-sur-Yvette Cedex (France)

Daniela Fritter, Didier Roux[*] and Charles M. Knobler
Department of Chemistry & Biochemistry
University of California
Los Angeles, CA 90024-1569 (USA)

Jean-Louis Viovy
ESPCI, rue Vauquelin
75235 Paris Cedex (France)

ABSTRACT

The nucleation and growth of "Breath Figures", the patterns formed
when vapor condenses onto a cold surface, are investigated for water. They
have been studied by simultaneous microscopic and light-scattering
observations. The parameters of the growth: contact angle θ of the drops
with the surface, incident gas flow, degree of supersaturation, have been
varied. For $\theta = 90°$, the growth of the pattern (after an initial period
during which the surface coverage and the droplet polydispersity reach a
constant value), is self-similar in time. The radius of an average droplet
grows with a power law with exponent $n_p \simeq 0.75$, whereas the growth of a
single droplet between two coalescences obeys a power law with a lower
exponent $n_s \simeq 0.23$. Comparisons are made between the experiments, theory
and numerical simulations.

[*]Permanent address: Centre P. Pascal, CNRS, Domaine Universitaire,
                      33404 Talence (France)

INTRODUCTION

The condensation of liquid droplets on a surface is not only a daily experience (breathing on a cold window), it is also a basic phenomenon that occurs in heat transfer technology [1]. Such patterns also involve new kinds of growth, mixing a 3-D condensation from the bulk and 2-D interactions, through the surface properties (heterogeneous nucleation).

We will report here a selection of recent experimental results that we have obtained when water was condensed on a silanized glass slide, and discuss the new ideas that this study has suggested. We have described elsewhere in greater detail the experimental set-up [2], the experiments [3] and the theory [4].

EXPERIMENT (Fig.1)

The experiment consists basically of sending, at (room) temperature T_R, a flow of water-saturated gas (nitrogen) onto a well characterized surface at lower temperature T_S. This paper will be concerned

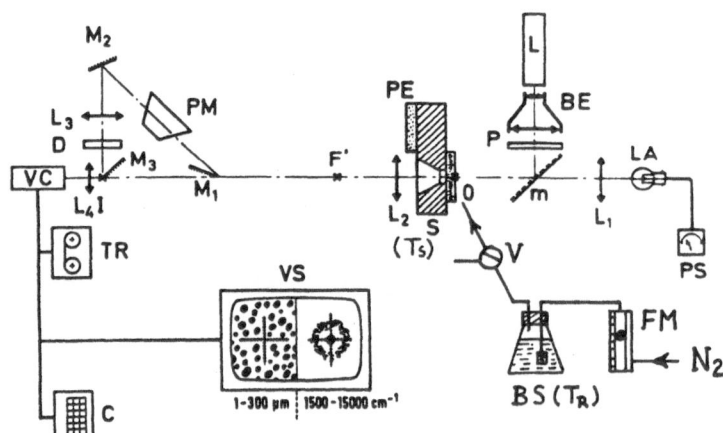

Fig.1- Set-up to obtain in a single image both the direct space and the momentum space as provided by the light scattering pattern. L_1, L_2, L_3, L_4: lenses; M_1, M_2, M_3: mirrors; m: semi-transparent mirror; PS: power supply; LA: white light source; L: He-Ne laser; BE: beam expander; P: polarizer; O: slide; S: supporting device (temperature T_S); PE: Peltier element; PM: prism; D: neutral density filter; VC: video camera; TR: tape recorder; VS: video screen; C: computer; N_2: nitrogen; FM: flow meter; BS: bubbling system in water (temperature T_R); V: by-pass valve.

only with silanized slides that, with liquid water and air or nitrogen, exhibit a contact angle of 90° [5], with a weak hysteresis. Both T_s and the gas flow can vary. A direct observation (DO) is performed with a typical field of view of 300μm and a resolution of a few μm. Light scattering (LS) can be performed simultaneously; data can usually be obtained in the range of wave-numbers K = 1500-15,000 cm^{-1}. From DO, the statistical properties of the droplet pattern can be calculated: mean distance a_m between droplets, mean radius R_m, polydispersity g, and surface coverage ϵ^2, where $g = \Delta R_m/R_m$ and $\epsilon^2 = 4R_m^2/a_m^2$. From LS, related quantities [6] can be deduced from the structure factor S(K), e.g., a_m from the peak at $K_m = 2\pi/a_m$ (see Fig.2).

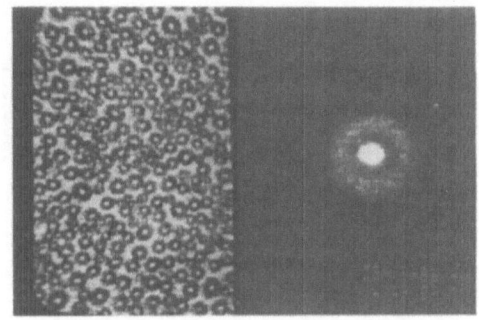

Fig.2- Photograph of television screen showing simultaneous display of images in direct and transform space. The average size of the droplets is 15μm.

OBSERVATIONS

Three growth domains can be distinguished.

(i) *Early Stage*. This period starts from the beginning of condensation, and typically lasts a few seconds. In this domain, DO does not generally allow the droplets to be well resolved; they remain smaller than a few μm. However, a furious activity results from motions caused by many coalescence events. From LS, one observes a ring whose intensity increases and whose diameter decreases with time. However, simple power laws are usually not sufficient to account for the full behavior. The pattern shows pulsations in which the intensity of the ring decreases, and its diameter shrinks very rapidly (see Fig.3 below).

Fig.3- (a) Typical K_m variation vs time (log-log plot) showing the initial state (i), where the droplet coverage increases, and the average stage (ii), where the pattern remains self-similar. The dotted letters indicate the corresponding stages for a "dirty" surface, where the nucleation sites are widely separated. At time t_f, the flow was stopped and the evaporating droplets left submicroscopic dust particles on the surface. The experiment was then repeated and the single-drop growth law was observed for $t < t_f$; average growth was seen for $t > t_f$.

(b) - Surface coverage under similar conditions, showing an increase up to a presumably universal value of 0.55.

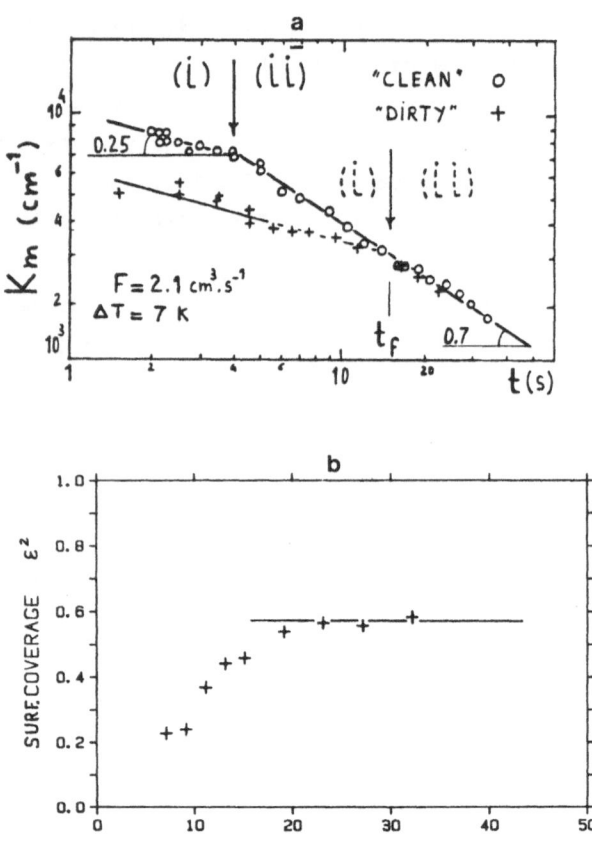

(ii) *Self-Similar Stage*. After the surface coverage has reached a large value, growth is seen to occur through an intermittent process (see Fig.4); growth by condensation of isolated droplets alternates with coalescence. Throughout this time domain, which lasts several minutes, the pattern remains remarkably self-similar. If K is replaced by K/K_m, all the droplet distributions are equivalent, as demonstrated by the constancy of the surface coverage ϵ^2 (see Fig.3b) and of the polydispersity of the droplet radii (see Fig.5a). The reduced structure factor $\tilde{S}(x) = [S(K/K_m)]/S(K_m)$ (see Fig.5b), which is a function of all the above parameters, also remains constant.

(iii) - *Late Stages: Steady State*. The distance between droplets increases with time so that, when droplets are separated by a distance of a few tens of μm, a new family of droplets nucleates and grows. This new pattern of droplets also goes through stages (i) and (ii). After another cycle, new families appear successively (see Fig.6) and the final stage is characterized by a broad dispersion of droplets belonging to all families.

When the effect of gravity becomes significant, the drops sweep off the slide (if vertical); this favors the appearance of new families. This is the steady state, which has been extensively studied by engineers [7].

Fig.4- Growth of a single drop. The numbers above the vertical segments indicate the number of droplets that have coalesced. The lower line of slope 0.23 has been obtained by displacing the segments between coalescences.

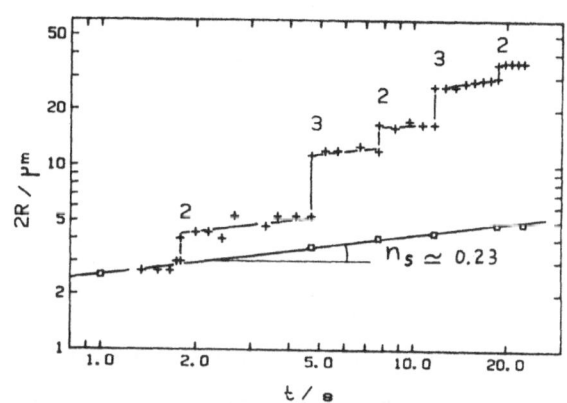

Fig.5- (a) Droplet radius polydispersity corresponding to Fig.3b, and analogous to the experiment shown in Fig.3a.

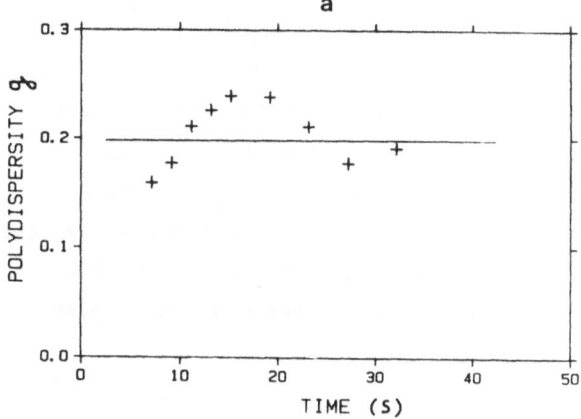

(b) Corresponding reduced structure factor $\tilde{S}(x) = [S(K/K_m)]/S(K_m)$ at different times of growth.

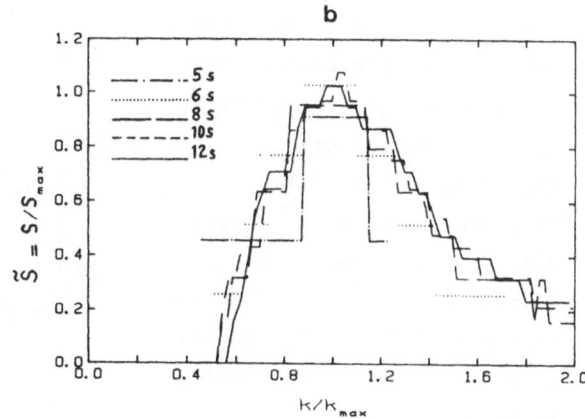

407

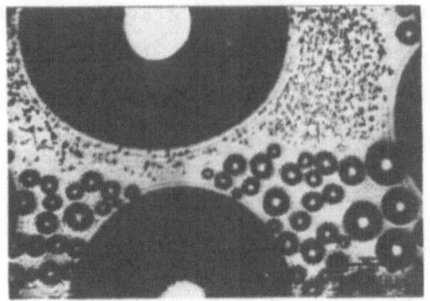

Fig.6- Three coexisting families of droplets whose mean radius is proportional to their age.

50 μm

We will now consider in detail the self-similar growth, stage (ii). Figure 7 shows the behavior of K_m with time (t) at various values of the flow rate F and $\Delta T = T_R - T_S$. Clearly a power law fits the data well:

$$K_m \sim t^{n_p}, \quad \text{with } n_p \simeq 0.75.$$

The constancy of this behavior under various temperature and flow conditions suggests that all the data can be placed on a single universal curve, provided that the time units are correctly scaled. Let us therefore define a typical time t_0 necessary to observe a ring of wave number K_m^0 (= 3750 cm^{-1}), and let us study its F- and ΔT-dependence.

When considering the flow rate variation (see Fig.8a), one must distinguish three regions. One expects a linear variation with time of the

Fig.7- Variation of K_m with the time t for nine runs. Values of F in cubic centimeters per second and ΔT in kelvins are, respectively, (plusses) 5.2, 5.0; (triangles) 12.4, 5.1; (circles) 5.2, 9.7; (squares) 2.1, 9.6; (crosses) 2.1, 13.5; (lozenges) 5.2, 13.3; (dotted circles) 5.2, 16.3; (dotted squares) 2.1, 16.3; (dotted triangles) 5.2, 19.7.

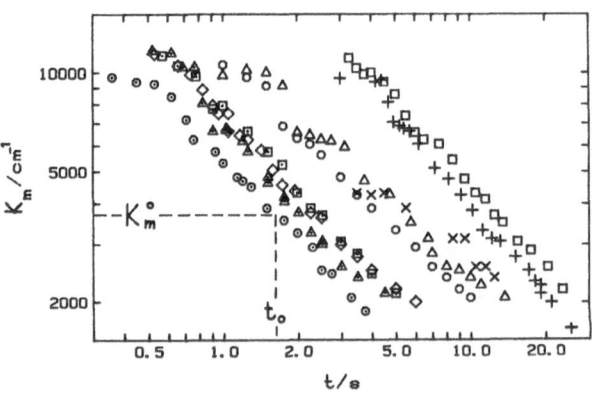

mass that condenses, $t_0 \propto F^{-1}$. At low flow rates, however, the velocity of the gas impinging on the slide is so small (~ 0.1 cm s^{-1}) that a non-negligible amount of water cannot reach the surface. This increases the time (t_0) to reach K_m^0. On the other hand, when F is large (velocity ~ 10 cm s^{-1}), a limitation also occurs because the growth is not sufficiently rapid to condense all of the vapor. A saturation thereby results. This limitation, of course, is a function of the supersaturation ratio, i.e., of ΔT, as shown in Fig.8a.

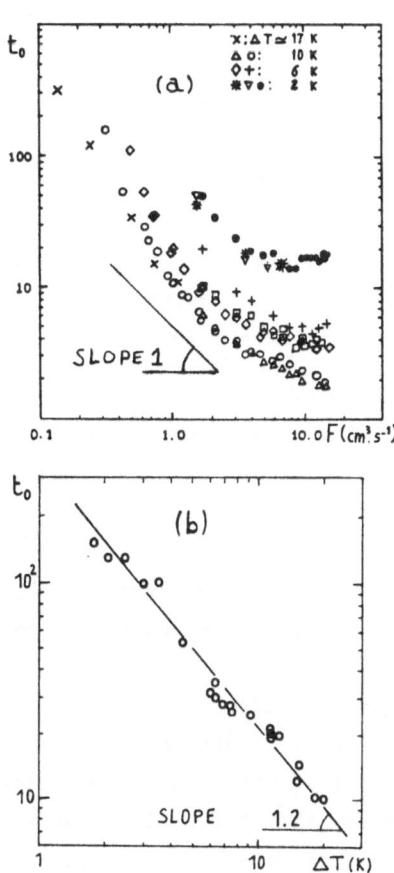

Fig.8- (a) Flow-rate variation of the time (t_0) necessary to obtain a wave number $K_m^0 = 3750$ cm^{-1} (mean droplet radius $\simeq 6$ μm). Saturation occurs at large F, retardation at small F; thus only an intermediate region exhibits a linear behavior $t_0^{-1} \propto F$.

(b) Temperature dependence of t_0 (ΔT is the temperature difference between the condensing surface and the gas).

By limiting the study to the region where t_0^{-1} is linear with F, one readily obtains the temperature variation of t_0^{-1} (see Fig.8b): $t_0 \sim \Delta T^{-1.2}$. It is in fact the supersaturation ratio Δp_s that is the important variable. It is related to ΔT by $\Delta p_s \sim \Delta T^{0.8}$, so that $t_0 \sim \Delta p_s^{-1.5}$.

Scaling of the growth is thus performed by using the reduced time unit:

$$t^* = (t/t_0) \sim t \ F \ \Delta p_s^{1.5}$$

DISCUSSION

Growth of a Droplet between Two Coalescences

Two mechanisms can account for the growth of a droplet (j) without considering coalescence: (i) condensation on the droplet surface (with the accommodation coefficient σ_{VL}) and/or (ii) nucleation on the substrate and diffusion of the embryos to the circumference of the droplets (accommodation coefficient σ_{vs}). One can easily show [3] that in case (i) the mean radius (R_j) of a droplet must grow according to:

$$R_J \sim \sigma_{VL} \, F \, \Delta p_s t,$$

whereas, in case (ii):

$$R_J \sim \sigma_{VS} \left(t \, F \, \Delta p_s^{3/2} \right)^{1/3}.$$

The last mechanism accounts for nucleation of embryos on a random distribution of nucleation sites with a given distribution of contact angles. The latter determines the nucleation barrier and thus the local nucleation rate. When compared to experiments (see above), the flow and supersaturation dependence is well described. However, some discrepancy still remains between the exponent given by theory (1/3) and that found in the experiments (0.23).

Growth of an Assembly of Droplets

The self-similarity of stage (ii) can be derived provided that the growth of an isolated droplet is scale-invariant, as it is for the two limiting cases above. Coalescence, in this case, simply rescales the pattern [4]. This theory allows the growth exponent (n_p), including the effect of coalescences, to be estimated for all dimensionalities; here, $n_p = 1$. Furthermore, the growth of a droplet where the coalescence effect has been suppressed, as in Fig.3, depends only on dimensionality, and one gets $n_s = 1/3$.

This result simply expresses the conservation of mass between the incoming vapor and the condensed liquid. That n_p is found lower than 1 means that the scaling has to break down, as shown by the formation of new generations of droplets in stage (iii). These arguments do not explain why, in the first growth stage (i), a pattern that starts growing with a low surface coverage will necessarily reach a coverage $\epsilon^2 \simeq 0.55$ and achieve self-similar growth.

Numerical Simulations

Numerical simulations based on a simple model have been performed in order to try to understand the experimental behavior. The initial configuration consists of N = 1800 nucleation sites randomly distributed on a unit square. A hemispherical droplet of critical radius is placed on each site and is allowed to grow according to a power law $R \propto t^{n_s}$.

Fig.9- Results of numerical simulations. Individual droplets grow with an exponent $n_s = 0.23$.

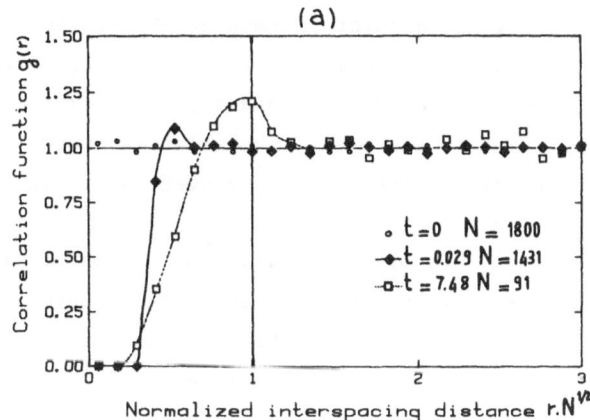

(a) Time dependence of the position of the peak, which corresponds to the distance $a_m = 2\pi K_m^{-1}$.

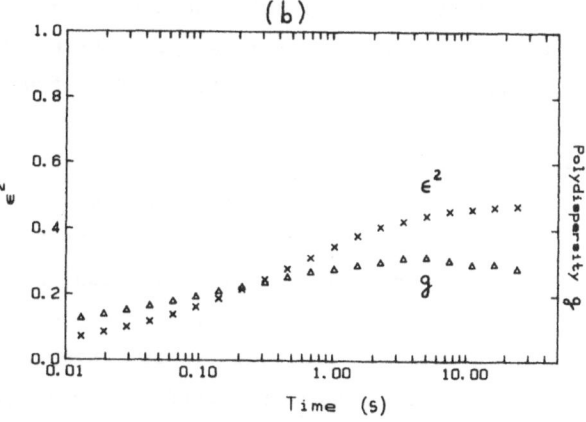

(b) Surface coverage ϵ^2 and polydispersity g as a function of time.

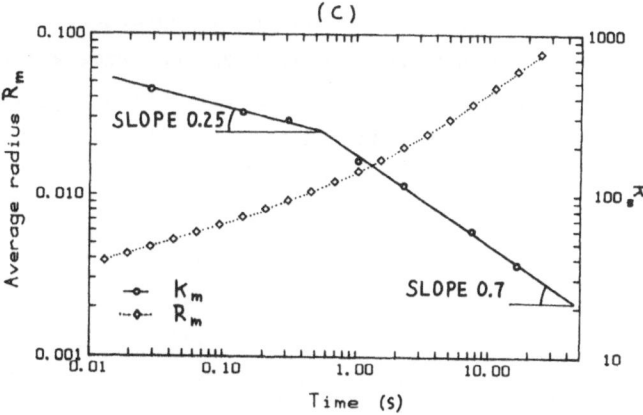

(c) Variation of average radius R_m with time. The behavior should be compared with the experimental results in Fig.3.

When two droplets come into contact, a coalescence occurs and the resulting droplet is located at the position of the larger droplet. The new radius is calculated from volume conservation. If the coalescence results in other droplet contacts, additional coalescences may occur. The average radius, the pair correlation function $g(r)$, the coverage and the polydispersity are calculated as a function of time, they are shown in Fig.9.

It is striking that the computer simulations, like the experiments, show both stages (i) and (ii), where the surface coverage increases and self-similarity begins. The limiting values of the polydispersity (0.30) and of the surface coverage (0.47) are in good agreement with values found in the experiments.

The pair correlation function, which is equal to 1 at $t = 0$ (random distribution of sites), begins to develop structure as a result of the coalescences, (Fig.9a). A peak arises whose amplitude increases with the coverage and whose position moves from a reduced distance $rN^{1/2} < 1$ to $rN^{1/2} = 1$.

It is also striking that both the average distance a_m and the average radius R_m initially exhibit slow growth (stage (i), limiting exponent 0.25), and reach a steady state where the growth is more rapid (stage (ii), limiting exponent 0.7). This can be understood qualitatively if one realizes that initially the coverage is low, so that the acceleration of the growth by coalescences is unimportant. A crossover to a more rapid growth occurs when the surface coverage has become large enough for the interaction of the droplets to be efficient.

From this simulation and the above experiments, it follows that stages (i) and (ii) can be well reproduced. It is intriguing that the limiting polydispersity and surface coverage values are nearly the same in both experiment and simulation, which suggests that some universality lies behind this phenomenon.

REFERENCES

1. See, e.g., C. Graham and P. Griffith, Int. J. Heat Mass Transfer, **16**, 337 (1973); and references therein.

2. F. Perrot and D. Beysens, Rev. Sci. Instrum. **58**, 183 (1987).

3. D. Beysens and C.M. Knobler, Phys. Rev. Lett. **57**, 1433 (1986).

4. J.L. Viovy, D. Beysens and C.M. Knobler, Phys. Rev. A, to appear (1988).

5. J. Sagiv, J. Am. Chem. Soc. **102**, 92 (1980);
    C. Allain, D. Aussere and F. Rondelez, J. Colloid Interface Sci. **107**, 5 (1985).

6. R. Hosemann and S.N. Bagchi, "Direct Analysis of Diffraction by Matter" (North-Holland, Amsterdam, 1962), p.403.

7. See, e.g., J.W. Rose and L.R. Glicksman, Int. J. Heat Mass Transfer **16**, 411 (1973) and references therein;
    H. Merte, C. Yamali and S. Son, Proc. 8th Int. Heat Transfer Conf. **4**, 1659 (1986).

ACKNOWLEDGEMENTS

This work was supported in part by the National Science Foundation and by NATO (grant # 86-0658).

DYNAMICS OF INTERFACES AND SELF-SIMILARITY

Hiroshi Orihara and Yoshihiro Ishibashi

Synthetic Crystal Research Laboratory
Faculty of Engineering, Nagoya University
Chikusa-ku, Nagoya 464, Japan

We discuss the dynamics of interfaces in a non-conserved system from a geometrical point of view and show that the temporal dependence of the interface area depends on the fractal dimension of the initial pattern. The dynamical behavior of disclinations in a quenched nematic liquid crystal is presented as an experimental example.

1. INTRODUCTION

The dynamics of interfaces separating two coexisting phases has been widely studied. For a non-conserved system Allen and Cahn predicted that area of the interfaces per unit volume decreases with time as $t^{-1/2}$.[1] Recently the $t^{-1/2}$-behavior has been extended by Toyoki and Honda(TH).[2] They showed that the exponent with respect to the time depends on the initial distribution of the interfaces and derived the following relation;

$$A(t) \propto t^{-b} \qquad (1)$$

with

$$b = (D+1-d)/2, \qquad (2)$$

where $A(t)$ is the total area density of the interface, d the Euclidean dimension of the space, and D the fractal dimension of the interfaces at $t=0$. Allen and Cahn's result appears only at $D=d$, i.e., when the initial interfaces cover the space completely. In deriving eqs. (1) and (2), however, TH assumed that 1) the initial interface is a nonrandom self-similar fractal(e.g the Peano curve) and 2) the time evolution of the interface should be self-similar. On the contrary, the present authors have obtained the same expression as eqs. (1) and (2) without assuming 2) in the case of a two-dimensional system, and furthermore derived the correlation function of tangential vectors for a random fractal interface(e.g. coastlines).[3]

In this paper we discuss the dynamics of the interfaces in a non-conserved system from a geometrical point of view. Equations (1) and (2) will be derived in a somewhat different form in §2 and a disclination system will be presented as an experimental example in §3.

2. DYNAMICS OF INTERFACES AND COARSE-GRAINING

The motion of an interface in a non-conserved system is described as

$$v = - \Gamma K, \tag{3}$$

where v is the velocity of the interface along its outward normal, Γ the kinetic coefficient and K the total curvature.[1] The above equation, however, is not suitable to treat an assembly of interfaces. Ohta, Jasnow and Kawasaki(OJK) solved the problem introducing an auxiliary field $u(r,t)$, the node of which represents the interfaces.[4] The time evolution of $u(r,t)$ is approximately given as a simple diffusion equation:

$$\frac{\partial u}{\partial t} = \Gamma \frac{d-1}{d} \nabla^2 u \tag{4}$$

and the solution is

$$u(r,t) = (2\pi\xi^2)^{-d/2} \int \exp(-(r-r')^2/2\xi^2)u(r',0)dr' , \tag{5}$$

where $\xi^2 = 2\Gamma t(d-1)/d$ and $u(r,0)$ is the initial ditribution of interfaces. As is seen from eq. (5), the time evolution is equivalent to the coarse-graining of the interfaces over the characteristic length ξ.

In the first place, we calculate the time dependence of the area of a single interface, which is a random fractal at the initial moment. Such an interface may be obtained by picking up the largest interface from random interfaces. Since the evolution is equivalent to the coarse-graining over ξ, it is enough to calculate the area of the initial interface coarse-grained. To do so, let us cover it with cubes of sides ξ. The number of cubes required is given as $N(\xi) \propto \xi^{-D_S}$. Therefore, the area $A_S(t)$ may be expressed as $A_S(t) \propto \xi^{D_T} N(\xi) \propto \xi^{D_T - D_S}$, where D_T is the topological dimension of the interface. With use of $\xi \propto t^{1/2}$, the exponent b in eq. (1) is obtained as

$$b = (D_S - D_T)/2 . \tag{6}$$

Equations (2) and (6) are equivalent because $D_T = d - 1$ in the case of interfaces. However, eq. (6) may be also applicable to the case $D_T < d-1$ (e.g. a string in a three-dimensional system, the driving force of which arises from the tension), if the geometrical representation mentioned above holds also in that case.

Next, we proceed to the case of an assembly of interfaces. Such a system may consist of many closed interfaces (of course, there exist infinitely extended interfaces of percolation clusters). According to Mandelbrot[5] in the case of a fractal pattern the following size-number relation holds:

$$N(\Lambda > \lambda) \propto \lambda^{-D_W} , \tag{7}$$

where λ is the average size for the region enclosed with an interface, $N(\Lambda > \lambda)$ the number of the closed interfaces per unit volume whose size is larger than λ and D_W the fractal dimension of the whole system. D_W does not coincide with D_S of a single interface and $D_S < D_W$ in general[5]. From eq. (7) the number density $n(\lambda)$ for λ is expressed as

$$n(\lambda) \propto \lambda^{-D_W - 1} . \tag{8}$$

Here, we apply the geometrical representation to the assembly of interfaces. Let us coarsen an interface of a size λ over ξ and express its area as $A_S(\lambda, \xi)$. Since the initial pattern considered is a fractal, we may accept the following scaling relation:

$$A_S(\lambda, \xi) = \xi^{D_T} f(\lambda/\xi).$$ (9)

We finally have the total area $A_W(t)$ as

$$A_W(t) = \int_0^\infty A_S(\lambda, \xi) n(\lambda) d\lambda$$

$$\propto \xi^{D_T - D_W}$$

$$\propto t^{-(D_W - D_T)/2},$$ (10)

where we have used eqs. (8) and (9). It should be noted that this relation corresponds formally to eq. (1) supplemented with eq. (6). From the above mentioned inequality $D_S < D_W$, we obtain

$$b_S < b_W$$ (11)

with

$$b_{S,W} = (D_{S,W} - D_T)/2$$ (12)

for the respective exponent to the single interface area and the total area.

As has been mentioned before, eq. (10) may be applicable to the case $D_T < d - 1$. For example, let us consider an assembly of strings with tensions ($D_T = 1$) covering a three-dimensional space completely ($D_W = 3$) at the initial time. In this case $A_W(t) \propto t^{-1.6}$.

It should be noted that the characteristic length ξ is always proportional to $t^{1/2}$ not depending on the fractal dimension of the initial pattern and the $t^{1/2}$-behavior arises directly from the equation of motion (3). Furthermore, we point out that while in the case of a single interface the dynamical scaling law holds irrespective of the fractal dimension as has been shown by the present authors,[3] in the case of an assembly of interfaces it does only at $D_W = d$. This arises from the fact that the size distribution $n(\lambda)$ at the initial time is invariant by a scale transformation only at $D_W = d$.

3. AN EXPERIMENTAL EXAMPLE

There appear many complicated disclinations in a TN(Twisted Nematic) cell after quenching from the isotropic phase to the nematic phase.[7] Such a system is suitable to study the ordering process, since it is readily observable in the real space and in the real time. In addition, the liquid crystal cell used is so thin that the system can be regarded as a two-dimensional one and the disclination as a one-dimensional interface. In the previous work[7] we have shown that the total disclination length $L(t)$ obeys the power law; $L(t) \propto (t-t_0)^{-0.44}$, where the origin of time (0 sec) is taken at the time when the disclinations begin to visually appear and $t_0 = -0.1$(sec). The exponent with respect to time is 0.44, resulting in $D_W = 1.88$ with the use of eq. (10). In the experiment, however, there is a possibility that the temperature in the cell gradually decreases after quenching, bringing about the effective temporal change of

417

the kinetic coefficient. Therefore, the exponent may be less than 2 even if D_W=2. To clarify this point further experiments are now in progress.

In the above mentioned experiment we analyzed the collective motion of an assembly of disclinations. In the following, on the contrary, we turn our attention to the motion of a single disclination. The disclination which appeared immediately after quenching is so complicated that it may be regarded as a fractal curve. In Fig. 1 we show the time evolution of the longest disclination line picked up from the photographs in the previous paper. The complicated form of the disclination line at 1 sec, reminding us of a coastline, becomes smoother with time. The time dependence of the disclination length $L_s(t)$ is shown in Fig. 2. The data fit to the power law:

$$L_s(t) \propto (t+0.1)^{-0.32}. \tag{13}$$

The exponent 0.32 is considerably smaller than 0.44 obtained for the total length, supporting the inequality $b_s < b_W$, which may be valid even if the kinetic coefficient Γ changes with time. In order to examine whether the initial disclination is really a fractal or not, we measured the disclination length by the divider method. We show in Fig. 3 a plot of the number N into which the disclination is divided vs. the opening ε of the divider.

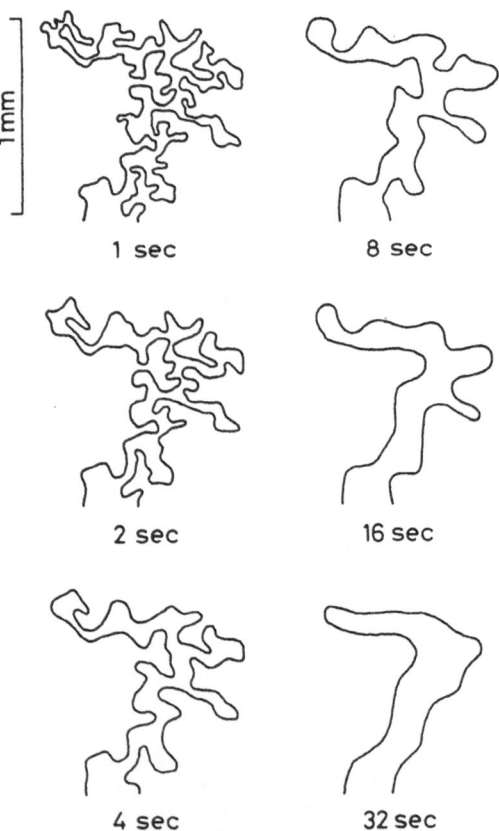

Fig. 1. The time evolution of a disclination line.

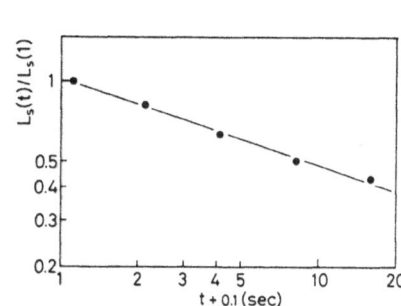

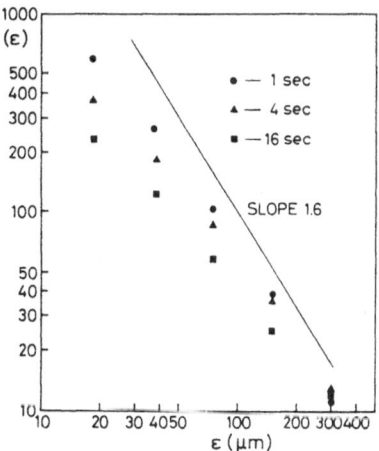

Fig. 2. Dependence of the disclination length on time.

Fig. 3. The plot of $N(\varepsilon)$ vs ε at various times.

The data taken at 1 sec do not fall on a straight line and the slope becomes steeper as ε increases. This can be ascribed to the coarse-graining of the disclination over the length ξ. This situation is more clearly shown in the fact that the slope becomes less steep as the time passes. Accordingly it is quite possible that the initial interface is similar to a fractal curve. Making use of the relation (12), we obtain D=1.64. The slope at 1 sec is smaller than this value. This discrepancy, however, may also be ascribed to the coarse-graining. It is worth noting that the fractal dimension 1.64 is considerably larger than those of actual coastlines and the self-avoiding random walk, which are almost 4/3.[5] Further experimental studies should be made to clarify whether this large fractal dimension is the characteristic of the disclination or not, i.e., if it is peculiar to the present experiment or not.

In this paper we have clarified the dynamical behavior of interfaces in a non-conserved system from a geometrical point of view. In the above discussion, however, we assumed that the initial pattern was a fractal. There remains a question as to whether the initial state of interfaces emerging after quenching is really a fractal or not in general. Studies on the geometrical nature of the quenched system will be required to clarify this point.

References

1. S. M. Allen and J. W. Cahn: Acta. Metall. 27 (1979) 1085.
2. H. Toyoki and K. Honda: Phys. Letters 111 (1985) 367.
3. H. Orihara and Y. Ishibashi: J. Phys. Soc. Jpn. 56 (1987) 2340.
4. T. Ohta, D. Jasnow and K. Kawasaki: Phys. Rev. Lett. 49 (1982) 1223.
5. B. Mandelbrot: The Fractal Geometry of Nature(Freeman San Fransisco, 1982).
6. H. Toyoki and K. Honda: Prog. Theor. Phys. 78 (1987) No. 2.
7. H. Orihara and Y. Ishibashi: J. Phys. Soc. Jpn. 55 (1986) 2151.

ORDERING DYNAMICS OF POLYMER MIXTURES AT PHASE TRANSITION

Takeji Hashimoto

Department of Polymer Chemistry, Kyoto University
Kyoto 606, Japan

ABSTRACT

Experimental studies on ordering dynamics of liquid polymer mixtures at spinodal decomposition by using time resolved light scattering are presented. Particular emphasis is laid on the kinetics of the early stage and late stage spinodal decomposition of polymer mixtures, and on the effects of shear deformation and shear flow on the concentration fluctuations of the mixtures. Universality and unique features in the dynamical behavior of the polymers were also discussed in comparison with the dynamical behavior of small-molecule systems.

I. INTRODUCTION

Many theoretical and experimental studies have been made on the kinetics of the phase separation of mixtures quenched into the miscibility gap[1,2]. The dynamics of the ordering processes must be studied to design and control the structure and properties of multicomponent polymeric materials, sometimes called "polymeric alloys" in contrast to metallic alloys, and for deeper understanding of nonequilibrium statistical physics in general. The most important objectives in such studies of polymers are (i) to explore "universality" in the polymer dynamics, i.e., common laws found for the dynamics of the ordering processes in other fields such as fluid mixtures, metallic alloys and inorganic glasses and (ii) to find some unique "characteristics" associated with the "connectivity" of elements such as atoms and molecules which are inherent to the long chain molecules. The connectivity of elements is a new physical factor which has to be incorporated for generalization of nonequilibrium statistical physics and a key factor to understand the polymer behavior. Here we review some of our recent experimental results on early-to-late stage spinodal decomposition of polymer systems and discuss the universality and the characteristics in the ordering dynamics of polymer systems.

II. EARLY STAGE SPINODAL DECOMPOSITION

II-1. Results

Early stage spinodal decomposition has been extensively studied for a

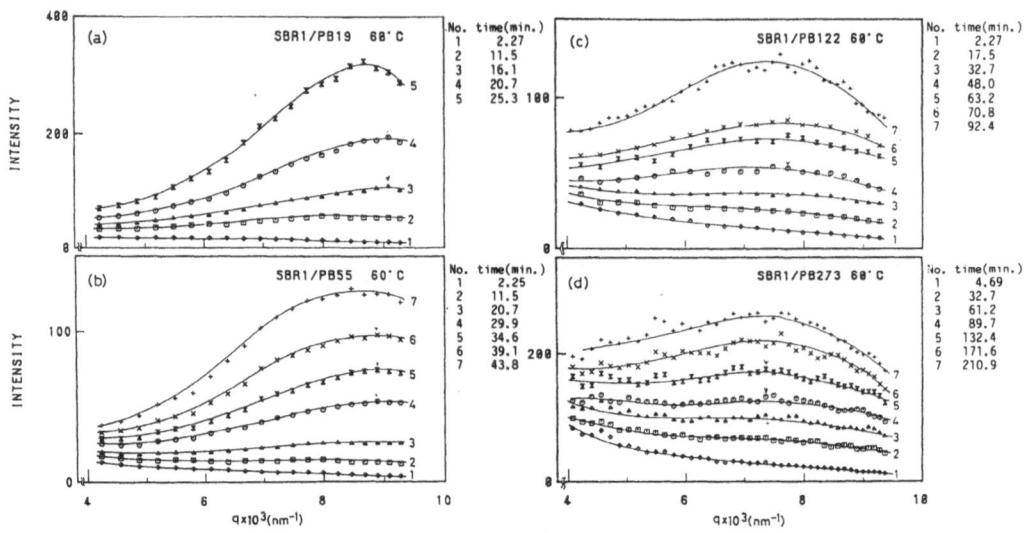

Fig.1. Time evolution of light scattering profiles during isothermal phase separation at 60°C for the binary mixtures of (a) SBR1/PB19,(b) SBR1/PB55, (c) SBR1/PB122, (d) SBR1/PB273. From ref.4.

number of polymer systems[3-5,23,35-38]. Here we present typical results obtained for a binary mixture of poly(butadiene) (PB) and poly(butadiene-ran-styrene) (SBR) with the ratio of 50/50 by weight. The SBR is a random copolymer in which the two monomers, butadiene and styrene, are expected to be distributed in a statistically random fashion in the polymer chain. The fraction of butadiene in SBR (ψ) determines the effective repulsive potential between PB and SBR. In the mean-field approximation, the repulsive potential is characterized by the thermodynamic interaction parameter χ. Then effective χ (χ_{eff}) can be given by

$$\chi_{eff} = \chi_o(1 - \psi)^2 \qquad (1)$$

where χ_o is the bare interaction parameter between PB and PS (polystyrene), and hence the effective interaction between the two polymers can be controlled by adjusting ψ, i.e., by the chemistry. The sample SBR used here (coded as SBR1) has ψ = 0.8, Mw = 1.18 x 10^5 and Mn = 1.00 x 10^5 where Mw and Mn are the weight and number average molecular weights. Four kinds of PB with different molecular weights were used; PB19 with Mw = 1.9 x 10^5 and Mn = 1.65 x 10^5, PB55 with Mw = 5.46 x 10^5 and Mn = 5.35 x 10^5, PB122 with Mw = 1.22 x 10^6 and Mn = 1.14 x 10^6 and PB273 with Mw = 2.73 x 10^6 and Mn = 2.19 x 10^6.

Thin film, about 100μ thick, was prepared by casting from a homogeneous solution with toluene, a neutrally good solvent. The as-cast films which contained the concentration fluctuations due to the phase separation during the solvent evaporation was brought into the single-phase state by the mechanical mixing as described later in sec. IV-1. This process is called "homogenization"[3b,6], and the homogenized films were then subjected to the isothermal phase separation.

Fig. 1 shows the typical time evolution of light scattering profiles during isothermal phase separation at 60°C from the homogenized mixtures of (a) SBR1/PB19, (b) SBR1/PB55, (c) SBR1/PB122, and (d) SBR1/PB273 where some of the time-sliced profiles only in the early stage unmixing process were shown and q designates the magnitude of the scattering vector as usually defined. Each mixture shows an increase of scattered intensity and appearance of the scattering maximum at the high q region. The q value at which the scattered intensity reaches maximum, $q_m(t)$, appears to be

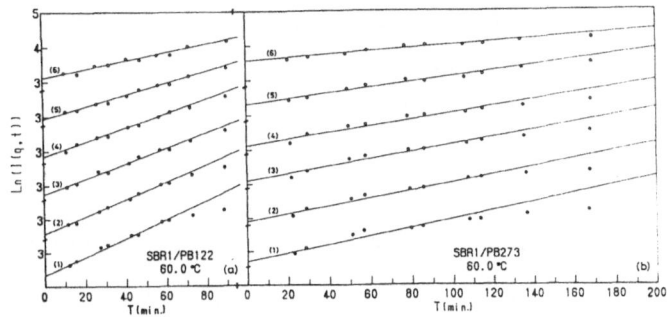

Fig.2. Time evolution of light scatttering intensity at various q's during the isothermal phase separation at 60 °C for the binary mixtures of (a) SBR1/PB122 and (b) SBR1/PB273. The logarithum of the intensity was plotted as a function of time t (min.). Curves 1-6 were obtained at q = 9.08 x 10^{-3}, 8.23 x 10^{-3}, 7.22 x 10^{-3}, 6.39 x 10^{-3}, 5.50 x 10^{-3}, and 4.24 x 10^{-3} nm^{-1}, respectively.

independent of time in the early stage unmixing process for each mixture,

$$q_m(t) = q_m(0)t^0 \tag{2}$$

The rate of the intensity increase at a given q may be more quantitatively investigated in Fig. 2 where the logarithms of the relative intensity, I(q,t), were plotted as a function of time, t, for some of the mixtures shown in Fig. 1. Each figure shows clearly that, in the early stage unmixing process, ln I(q,t) at a given q linearly increases with t, indicating the exponential growth of the concentration fluctuations of all-q Fourier modes, S(q,t), covered in this experiment,

$$I(q,t) = I(q,0) \exp[2R(q)t] \tag{3}$$

$$S(q,t) = S(q,0) \exp[R(q)t] \tag{4}$$

In the later stage of unmixing (e.g., in the time scale of t > 70 and 110 min. for SBR1/PB122 and SBR1/PB273 respectively), the intensity increase with t deviates from the exponential behavior at all q's, which is believed to be due to the onset of various coarsening mechanisms[1b,2,7-17,39-41]. In the later stage, the growth rates become much lower than that of the exponential behavior, the data points thus falling below the straight lines at all q's.

Fig. 3 shows the plots of R(q)/q^2 vs. q^2 for the four mixtures which provide a critical test of the linearized theory of spinodal decomposition(SD)[8]. All the experimental results seem to exhibit, with a good accuracy, the linearity. The arrows indicate q$_m$(0) estimated from the plots.

II-2. Discussion

The results presented in sec. II-1 indicate that the time-evolution of the concentration fluctuations in the isothermal unmixing process can be described remarkably well with the linearized theory of SD given by Cahn. The experimental observations for the particular systems described here as well as those for other systems[1b,1c,3,5,7] in the early stage SD are summarized below:

(1) The exponential growth of the scattered intensity and the concentration fluctuations as given by eqs. 3 and 4,

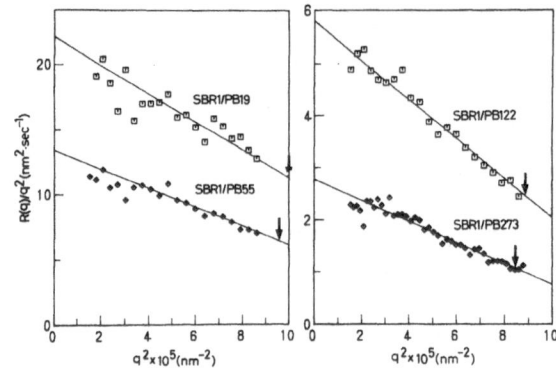

Fig.3. Plots of $R(q)/q^2$ vs. q^2 at 60°C for the binary mixtures of SBR1/PB19, SBR1/PB55, SBR1/PB 122, and SBR1/PB273. From ref.4.

(2) The constant q_m with time as given by eq. 2,
(3) The q-dependence of the growth rate, $R(q)$, is given by

$$R(q) = D_{app}q^2[1 - q^2/q_c^2] \tag{5}$$

as predicted by Cahn[18].

(4) The characteristic parameters D_{app} and q_c are given by the mean field model for the polymer mixtures[1c,19,21],

$$D_{app} = D_c\varepsilon \quad, \quad q_c^2 = 2q_m^2(0) = (18/R_0^2)\varepsilon , \tag{6}$$

$$\varepsilon = (\chi - \chi_s)/\chi_s \sim \Delta T \tag{7}$$

where D_c is the self-diffusivity, R_0 is the root mean squared end-to-end distance of polymer chain, χ is the thermodynamic interaction parameter between the constituent polymers, χ_s is the χ-value at spinodal point, and $\Delta T = |T - T_s|$ is the quench depth, and q_c is the crossover wavenumber.

(5) The upper limit of the reduced time , τ_c , below which the SD can be approximated by the linearized theory is universal[1b,7],

$$\tau_c \simeq 2 \tag{8}$$

The reduced time τ is defined here as

$$\tau \equiv t/t_c \quad , \quad t_c \equiv \xi^2/D_{app} \sim (R_0^2/D_c)\varepsilon^{-2} \tag{9a}$$

where t_c is the characteristic time of the systems, ξ is the correlation length which was defined here as $\xi \equiv q_m^{-1}(0)$, and D_{app} is the mutual diffusivity given by eq.6. The quantity $q_m(0)$ is also given by eq.6. The quantity ξ defined here is related to the correlation length in the single phase ξ_{single}, by eq.9b [10,21]. ξ_{single} rather than $\xi = q_m^{-1}(0)$ is used conventionally to define τ . Thus the reduced time τ defined here is related to the conventional reduced time, τ_{conv} , by

$$\xi \equiv q_m^{-1}(0) = \sqrt{2}\,\xi_{single} \quad , \quad \tau = \tau_{conv}/2 \tag{9b}$$

We do not intend to claim that the above observations are universal but rather propose that there are certain time, temperature, and q domains where the unmixing process is very weakly nonlinear in nature and can be approximated by the linearized theory. The conditions which lead to the unique observations are summarized as follows:

(A) small q range, $q \ll q_c$.

(B) relatively large quench depth $\mathcal{E}$ or ΔT.

(C) suitable time scale in the early stage;
    not too short after the isothermal phase separation where the thermal fluctuations[7,20,23] should contribute significantly to I(q,t) but not too long where the coarsening processes become important.

Under conditions A to C the effect of the thermal fluctuations is weak, which favors the above observations. Condition B is also important to satisfy the Ginzburg criterion[21],

$$|\mathcal{E}| \gg 1/N \qquad\qquad\qquad (10)$$

where the dynamics is described by the mean-field theory. Quantity N is the polymerization index of polymers. Observations (1) and (3) require conditions (A) to (C), while the observations (2), (4) and (5) require conditions B and C. For example eq. 5 is observed to fail at large q near q_c where, the effect of the thermal fluctuations is very important, even if condition (B) is satisfied[5,23]. Eqs. 3 and 4, and hence eq. 5 are found to fail even at a small q if the quench depth $\mathcal{E}$ or ΔT is very small, i.e., if condition B is not satisfied[23].

II-3. Some Effects Unique to Polymers

The effects unique to polymers in the early stage spinodal decomposition(SD) are:

(i) The long characteristic time t_c of the polymeric systems expands the time scale where the SD can be approximated by the linearized theory. The value t_c is very large because of very small $D_c \sim N^{-2}$ and very large molecular size $R_0 \sim N^{1/2}$.

(ii) The large polymerization index N of the polymer also acts to expand the mean-field regime (eq. 10).

(iii) The effect of the thermal fluctuations on I(q,t) is greater in polymers than in small molecules by a factor of N for a given $\mathcal{E}$[7].

(iv) The wavenumber $q_m(0)$ is very small due to the large R_0 and/or small $\mathcal{E}$ (small temperature dependence of χ).

In Fig. 2, the time scale where eq. 3 is observed is shown to extend up to t = t_{max} = 110 m for SBR1/PB273. The greater the molecular weight, the larger the value t_{max} and smaller the growth rate R(q). The values $q_m(0)$ estimated from Fig. 3 vary from 1.0×10^{-2} for SBR1/PB19 to 8.4×10^{-3} nm^{-1} for SBR1/PB273 which is predictable from the mean field theory, within experimental accuracy[16].

III. LATE STAGE SPINODAL DECOMPOSITION

III-1. Results for Critical Mixtures

We discuss here the late stage of the spinodal decomposition for the critical mixture of PS and poly(vinylmethylether) (PVME). Fig.4 shows molecular weight characterization and phase diagram. The phase diagram is skewed having ϕ_c = 0.8 (volume fraction of PVME at critical point) and shows LCST at T_c = 95.8°C. The early-to-late stage SD behavior at various quench depths ΔT = 1.0, 1.5, 2.4, 6.7 and 11.0°C, has already been reported in detail earlier [1b,7,10]. Here we present only some final results in Figs. 5 and 6.

Fig. 5 shows the scaled structure fractors $\overline{F}(x)$ to test the dynamical scaling hypothesis proposed by Binder[14] and the scaling functions predicted by Furukawa[24]. Here F(x) and x are defined as,

$$F(x) = I(q,t)q_m^3(t) \qquad\qquad\qquad (11)$$

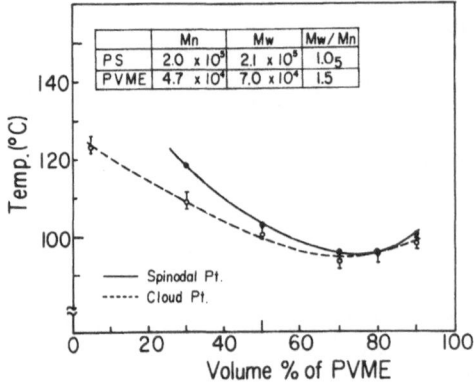

Fig.4. The cloud point curve and the spinodal curve for the PS/PVME.

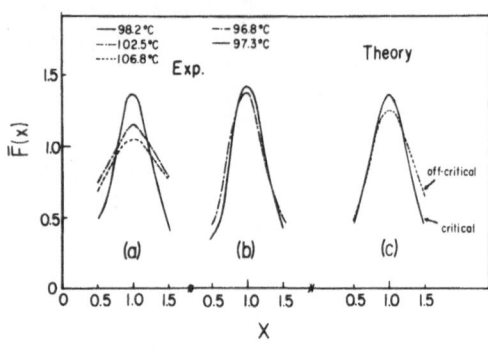

Fig.5. The normalized scaled structure factors $\overline{F}(x)$. (a) and (b) Experimental curves.(c) Theoretical curves.

$$F(x) = F(x)/\int_{0.5}^{1.5} x^2 F(x)dx \qquad (12)$$

$$x = q/q_m(t) \qquad (13)$$

Parts (a) and (b) show the experimental results for the five ΔT's and (c) shows the theoretical results obtained by Furukawa. Fig. 6 shows the scaling behavior of the reduced wavenumber Q_m of the dominant mode of the fluctuations with the reduced time τ where the experimental observations at the five ΔT's are shown by the data points together with the various theoretical predictions.

$$Q_m(\tau) = q_m(\tau)/q_m(0) \qquad (14)$$

Point (1) in the figure is the crossover point of $\tau_c \simeq 2$, and point (2) is the crossover point of $\tau \simeq 60$, above which the scaled structure factor $\overline{F}(x)$ can be constructed and the growth of the fluctuations become self-similar, satisfying the following relationship between the scaling exponents,

$$\beta = 3\alpha, \quad \text{with } q_m(\tau) \sim \tau^{-\alpha} \text{ and } \widehat{I}_m(\tau) \sim \tau^{\beta} \qquad (15)$$

where $\widetilde{I}_m(\tau)$ is the reduced maximum scattered intensity.

III-2. Discussions on the Critical Mixtures

The scaled structure factors can always be constructed experimentally

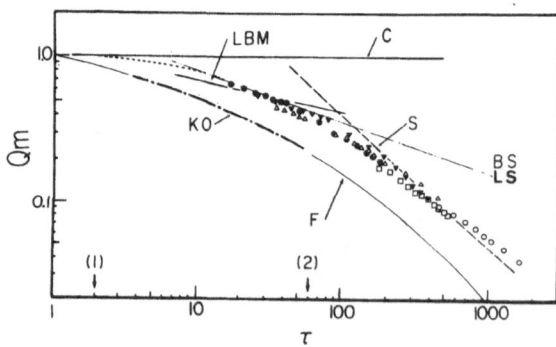

Fig.6. Comparison of the experimental growth behavior of Q_m with those predicted by some theoretical models. Note that the experimental data were plotted against the reduced time τ defined by eq.9a, while the theoretical curve, of LBM, KO and F were plotted against the conventional τ, i.e., τ_{conv} (eq.9b).

in the time scale $\tau > 60$ for this particular mixture. The scaled structure factors obtained for the three lower ΔT's are identical and are also in agreement with Furukawa's prediction for the critical mixture, i.e., $F(x) \sim x^2/(3 + x^8)$. Thus in certain temperature (T) and time (t) domains, the dynamical scaling hypothesis is valid even for polymers, and the structure factors are universal, independent of t and T. However, at the higher ΔT's (T = 102.5 and 106.8 °C) the scaled structure factors are much broader, and the self-similarity in the structure evolution occurs only at given T's and not at different T's. This might reflect a crossover of the structure evolution mechanism as will be presented in Fig.3 in the paper by Hasegawa et al.[25] in this volume. Takahashi et al.[41] also observed the broadering of the scaled structure factor.

The detailed coarsening mechanisms which are attributed to the intrinsically nonlinear nature of the ordering processes may be closely investigated by the reduced plot of Q_m shown in Fig. 6. The figure also includes the theoretical predictions by the linearized theory of Cahn (C)[18], Langer, Bar-on and Miller (LBM)[11], Kawasaki and Ohta (KO)[12], Furukawa (F)[15], Siggia (S)[16], Binder and Stauffer (BS)[14] for fluid mixtures, and Lifschitz and Slyozov (LS)[13]. As pointed out theoretically by Kawasaki-Ohta and Furukawa and experimentally by Snyder-Meakin[9] and Hashimoto et al.[7], there is no simple scaling relation over wide τ, and α depends on τ, which are believed to be due to crossover among various coarsening processes. The exponents α and β correspondingly increase:

$$0.2 < \alpha < 0.3, \quad 1.0 < \beta < 1.2 \quad \text{for } \tau_c < \tau < 60 \quad (16)$$

$$0.5 < \alpha < 0.8, \quad 1.5 < \beta < 2.4 \quad \text{for } 60 < \tau \quad (17)$$

Even in the long time limit of our experiments $\tau \simeq 2000$, the exponent α did not reach unity, contrary to the prediction by MacMaster[8] and Siggia[16] and to a number of experimental results[9,39,40], but rather the limiting value of 0.8 was attained.

III-3. <u>Polymer</u> <u>Effects</u> <u>on</u> <u>the</u> <u>Critical</u> <u>Mixtures</u>

LBM[11] gave a solution for the nonlinear time evolution equation,i.e, the time-dependent Ginzburg-Landau (TDGL) equation. Their results account, to some extent and over the limited τ, the coarsening of the real mixtures. The hydrodynamic interaction put forward in the TDGL equation by KO[12] gives a growth rate much faster than that predicted by LBM. Furukawa's theory predicts a coarsening law similar to that of KO but the prediction extends over a much wider time scale. Both theories predict well the relative change of Q_m with τ found experimentally, except for the shift of the absolute time scale τ. This discrepancy on the absolute time scale deserves further experimental studies. However it is of worth to note that the discrepancy may reflect fundamental problems, i.e., it is quite natural that the features unique to polymers (the polymer effects such as given below) were not intended to account for in the theories established for the small molecules :
(i) The effects of <u>molecular</u> <u>entanglements</u> on dynamics and viscosity of polymers in condensed state[26] , and (ii) the effects associated with the <u>interface</u> between the two coexisting liquid-phases of polymers not being necessarily identical to that of the small-molecule systems, especially in the strong segregation limit far from the criticality. The thickness of the polymer-polymer interface is much smaller than the size of the polymers, i.e., the gyration radii, while it is always larger than the molecule in the small-molecule systems. This would then imply that the cooperativity of the motions of the segments at the interface is essentially significant in polymers and is a unique physical factor not existing in small-molecule systems. This would also imply that the interfacial tension of the polymer-polymer interface is significantly affected by the loss of the

conformational entropy, the physical factor of which again does not exist in small-molecule systems.

The fact that the coarsening in real mixture occurs at a much slower rate than that predicted by KO and F is best interpreted as a consequence of extremely high viscosity in polymer systems. In Kawasaki's theory[12], the stochastic operator associated with the hydrodynamic interaction in the Fokker-Planck equation for the time-evolution of the probability functional of the order parameter contains viscosity η in the denominator. In polymers, η depends strongly on N, and further this N-dependence of η crossovers below and above Ne, the polymerization index between the entanglement couplings, i.e.,

$$\eta \sim N \quad \text{for} \quad N < Ne \tag{18}$$

$$\eta \sim Ne(N/Ne)^{3.4} \quad \text{for} \quad N > Ne \tag{19}$$

The value $(N/Ne) \simeq 10$ for the mixture studied here , and hence η is very high. Consequently the hydrodynamic interaction will be extremely suppressed and retarded. The retardation effect may appear even in the reduced time scale in the case of $N > Ne$. This is because the selfdiffusivity of a polymer in bulk, D_c, has different N-dependence for $N < Ne$ (Rouse mode) and for $N > Ne$ (reptation mode) , i.e.,

$$D_c \sim \begin{cases} D_1 N^{-1} & \text{for} \quad N < Ne \tag{20} \\ D_1 Ne^{-1}(N/Ne)^{-2} & \text{for} \quad N > Ne \tag{21} \end{cases}$$

From eqs. 9a, 20 and 21, one obtains

$$t_c^{-1} \sim \begin{cases} N^{-2}\varepsilon^2 & \text{for} \quad N < Ne \tag{22} \\ Ne^{-2}(Ne/N)^3\varepsilon^2 & \text{for} \quad N > Ne \tag{23} \end{cases}$$

For $N > Ne$, there appears to be a unique feature such that the two systems with identical N but different N/Ne may not have the same scaling behavior on the reduced plot. Thus the reduced time has to be rescaled to account for the difference of N/Ne in the case of $N > Ne$. If we define the reduced time for the high polymers as τ_p, $\tau_p = t/t_c \sim tN^{-2}\varepsilon^2(Ne/N)$, and hence

$$\tau_p = (Ne/N)^x \tag{24}$$

with x = 0 for $N < Ne$ and x = 1 for $N > Ne$.

Thus the scaling behavior of Q_m for the two systems with $N < Ne$ and $N > Ne$ would not come on to the same master curve but rather show a branching if Q_m is plotted with τ. This is called "N-branching" [7]. Since Ne depends on polymer concentration c if the solvent is incorporated into the system, the branching is expected to occur for the systems with different polymer concentration ("C-branching"[7]).

Onuki first explored theoretically the N-branching for the Siggia's coarsening processes. Hashimoto et al.[7] made a preliminary experimental analysis for the PS/PVME mixtures by comparing the scaling behavior of Q_m with τ for their own samples with that reported by Snyder and Meakin [9]. The results show a satisfactory agreement with the idea of N-branching. Fig. 7 shows the N-branching for a series of the SBR1/PB19 to SBR1/PB273 mixtures with different molecular weights at 60°C[28]. The results show a systematic slow down of the growth rate with increasing N even on the reduced plot. Fig.8 shows the C-branching for a given mixture SBR2/PB273 with 50/50 (w/w) composition having different total polymer concentration at 70°C[28]. The results show again a systematic slow down of the growth rate with increasing c even on the reduced plot. The SBR2 is an

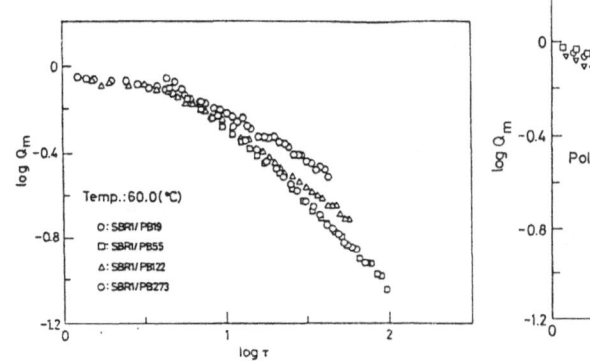

Fig.7. Reduced wavenumber Qm as a
function of the reduced
time τ, showing N-branching.

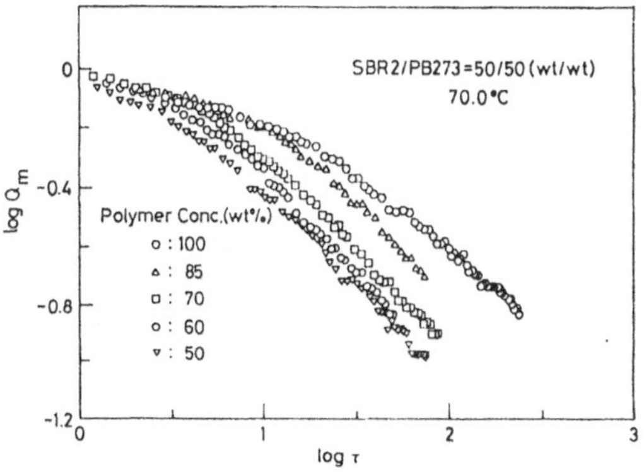

Fig.8. Reduced wavenumber Qm as a
function of the reduced
time τ, showing C-branching.

another SBR sample with $M_w = 1.90 \times 10^5$, $M_w/M_n = 1.21$ and $w_{ps} = 0.15$.

III-4. Results and Discussions on Off-Critical Mixtures

Fig. 9 shows the coarsening behavior at four different temperatures
for an off-critical mixture of SBR1/PB19 with 25/75 (w/w) composition as
observed by log q_m as a function log t[28]. The results show a crossover of
the coarsening behavior at t_{pn} such that

$$q_m \sim \begin{cases} t^{-0.4} & \text{for} \quad t < t_{pn} \\ t^{-0.02} & \text{for} \quad t > t_{pn} \end{cases} \tag{25}$$

where t_{pn} shortens with increasing temperature. The coarsening is virtually
pinned down at $t > t_{pn}$, and the level of the pinned-down wavenumber q_m
increases with increasing temperature. The crossover of the coarsening
behavior is expected to be due to a change in the phase-separated structure
from the percolated and bicontinuous network structure to the cluster
structure as reported by Hasegawa et al.(Fig. 3)[25]. The pinning of the
cluster growth may be attributed to an extremely slow cluster dynamics
resulting from the high viscosity, which represents again a feature unique
to high polymers[28] .

IV. SHEAR-INDUCED HOMOGENIZATION

Here we will report that the polymer mixtures are generally very
sensitive to shear deformation and shear flow, and the mixtures can be
brought into the single phase region under the applied external field.

IV-1. Shear Deformation

A mechanical mixing experiment was conducted for the as-cast films
prepared from the mixtures SBR1/PB19 with a composition of 70/30 (v/v). In
this experiment we folded the film in two and pressed it to the original
thickness repeatedly over n times, and we observed each time the light
scattering pattern with incident beam normal to the film surface. Fig. 10
shows the change of the light scattering patterns as a function of n. The
light scattering pattern which originates from the spinodal decomposition
and subsequent coarsening processes ("spinodal ring") shifts toward a small
angle with n and eventually disappears for n > 10. The mechanical mixing
process for the incompressible and rubbery mixtures involves shear

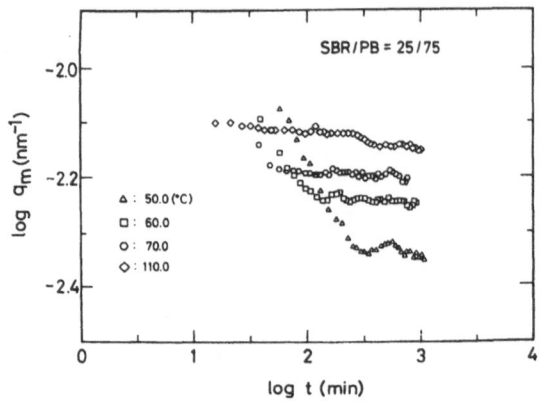

Fig.9. Pinning of the coarsening process for the off-critical mixture of SBR1/PB19.

deformation which expands and contracts the wavelengths $\Lambda_{\parallel}$ and $\Lambda_{\perp}$ of the dominant Fourier modes of the fluctuations in a direction parallel and perpendicular to the film surface, respectively. If $\Lambda_{\perp}$ is forced to be smaller than $\Lambda_c = 2\pi/q_c$, the crossover wavelength predicted by Cahn[18] by the mixing process, the fluctuations tend to be suppressed to the level of critical fluctuations. The dissolution of the fluctuations perpendicular to the film surfaces involves the dissolution of the fluctuations in an other direction, e.g., parallel to the film as observed experimentally. This implies a <u>nonlinear effect</u> involving a coupling of the Fourier modes. If the growth rate is sufficiently slow, as in our case, as discussed in sec. II, the homogenized state can be generated as a stationary state during the mixing process. In fact the homogenized state was found to give a scattering profile relevant to the critical scattering for mixtures in the single phase[6]. We believe that the "slow spinodal decomposition" is a key factor for easy achievement of the homogenization[6].

IV-2. Shear Flow

Experimental studies on the behavior of mixtures under shear flow were done by Silberberg and Kuhn[29], and Beysens et al.[30] and theoretically by Onuki et al.[31]. Fig.11 the shows small angle light scattering patterns from the 3 wt % solution of polymer mixtures PS/PB with a composition of 50/50 (w/w) with a neutral solvent (dioctyl-phthalate, DOP) in quiescent state and under a steady state Couette flow with shear rate S as indicated beneath each pattern. The experiments were conducted with an apparatus constructed in our laboratory[32] and at 43.2°C below the temperature $T_c = 54.0 \pm 0.7$°C on the coexistence curve. With increasing S, the solution is shown to change consecutively in state as follows; (i) the initial state of the two coexisting macroscopic phases, (ii) the state of

SBR/PB=70/30 at 25°C

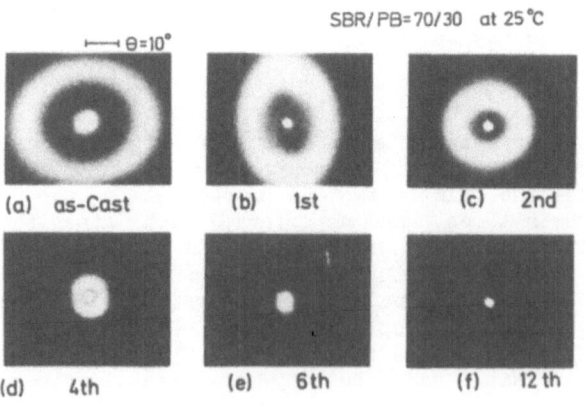

Fig.10.

Change of the light scattering pattern with the number of times over which the as-cast film is folded in two and pressed to the original thickness.

430

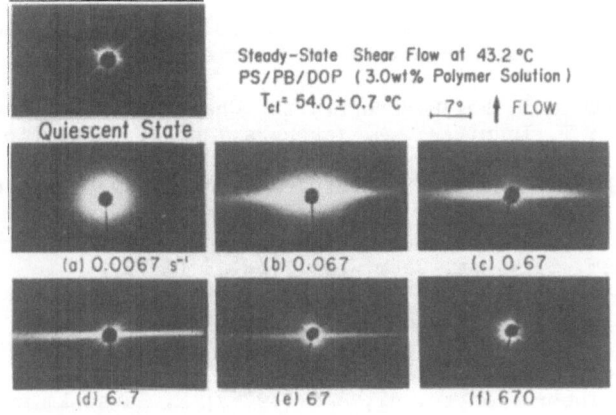

Fig.11. Light scattering patterns from dioctyl-phthalate solution of PS/PB mixture under quiescent state and steady-state Couette flow.

nearly spherical droplets of one solution dispersed in the matrix of the other solution (the pattern a), (iii) the state of the droplets being highly elongated in the direction of flow (b to d), (iv) the state of the scattering contrast of the droplets remarkably decreased due to an increasing degree of mixing of dissimilar segments in each phase (e) and finally (v) the state of shear-induced single phase with anisotropic critical fluctuations (f)[33]. At 43.2°C, the single phase was generated at $S > S_c = 165$ s^{-1}, the critical shear rate. In other words at $S = 165$ s^{-1}, T_c is equal to 43.2°C, a quite dramatic drop of T_c(54.0 ± 0.7 - 43.2 = 9.8 ± 0.7 °C) being observed. Quantitative analyses of $T_c(S)$ as a function of S indicated that

$$[T_c(S) - T_c(0)]/T_c(0) = -(2.0 \pm 0.15) \times 10^{-3} S^{0.55\pm0.02} \quad (26)$$

The prefactor of power S is about 3000 times that found by Beysens et al. for cyclohexane/anniline mixture, which is interpreted in terms of a long life time of the concentration fluctuations due to large correlation lengthξ and large viscosity η (the polymer effect)[33,34].

ACKNOWLEDGEMENTS

The author is grateful to his associates, especially H. Hasegawa, T. Izumitani, M. Itakura, M. Takenaka, and T. Takebe for their contributions. Thanks are also due to Professors A. Onuki, K. Kawasaki, and H. Furukawa for their useful and enlightening comments.

REFERENCES

1. See for example following reviews, a) T. Nose, Phase Transition, 8, 245 (1987); (b) T. Hashimoto, Phase Transition, in press (1987); (c) T. Hashimoto, in "Current Topics in Polymer Science Vol. II", R.M. Ottenbrite, L.A. Utracki, and S. Inoue, Eds., Hanser, N.Y., 1987, pp. 199-242.
2. See for example, J.D. Gunton, M. San Miguel, and P.S. Sahni, in "Phase Transitions and Critical Phenomena, vol.8", C. Domb and J.L. Lebowitz, eds., Academic Press, N.Y. 1983.
3. a) T. Hashimoto, J. Kumaki, and H. Kawai, Macromolecules, 16, 641 (1983); b) T.Izumitani and T.Hashimoto, J.Chem.Phys., 83, 3694 (1985).
4. M. Takenaka, T.Izumitani, and T. Hashimoto, Macromolecules, 20 (1987), in press.

5. T.Hashimoto, K.Sasaki, and H.Kawai, Macromolecules, 17, 2812 (1984); K.Sasaki and T.Hashimoto, Macromolecules, 17, 2818 (1984).

6. T.Hashimoto and T.Izumitani, Polym. Prepr., Am. Chem. Soc., Div. Polym. Chem., 26, 66 (1985); T. Izumitani, M. Takenaka, and T. Hashimoto, to be submitted to Macromolecules.

7. T.Hashimoto, M.Itakura, and N. Shimidzu, J. Chem. Phys., 85, 6773 (1986).

8. L.P. McMaster, in "Copolymers, Polyblends and Composites", N.A. Platzer Ed., Adv. Chem. Ser., Am. Chem. Soc.,142,43 (1975),

9. H.L. Snyder and P. Meakin, J. Chem. Phys., 79, 5588 (1983).

10. T.Hashimoto, M.Itakura, and H.Hasegawa, J. Chem. Phys. 85, 6118 (1986).

11. J.S. Langer, M. Bar-on, and H.D. Miller, Phys. Rev. A, 11,1417 (1975).

12. K.Kawasaki, Progr. Theor. Phys., 57, 826 (1977); K. Kawasaki and T. Ohta, Progr. Theor. Phys. 59, 362 (1978).

13. I.M. Lifshitz and V.V. Slyozov, J. Phys. Chem. Solids, 19,35 (1961).

14. K. Binder and D. Stauffer, Phys. Rev. Lett. 33, 1006 (1974); K. Binder, Phys. Rev. B. 15, 4425 (1977).

15. H. Furukawa, Phys. Lett. A, 98, 28 (1983); Adv. Phys. 34,703 (1985).

16. E.D. Siggia, Phys. Rev. A, 20, 595 (1979).

17. I.G.Voigt-Martin, K.H. Leister, R. Rosenau, and R. Koningsveld, J. Polym. Sci., Polym. Phys. Ed., 24, 723 (1986).

18. J.W. Cahn, J. Chem. Phys. 42, 93 (1965).

19. P.G. deGennes, J. Chem. Phys. 72, 4576 (1980).

20. H.E. Cook, Acta Metall. 18, 297 (1970).

21. K.Binder, J. Chem. Phys. 79, 6387 (1983).

22. G.R. Strobl, Macromolecules, 18, 558 (1985).

23. M. Okada, and C.C. Han, J. Chem. Phys. 85, 5317 (1986).

24. H. Furukawa, Progr. Theor. Phys. 59, 1072 (1978); Phys. Rev. Lett. 43, 136 (1979); Physica A, 123, 497.

25. H. Hasegawa, T. Shiwaku, A. Nakai, and T. Hashimoto, in "Dynamics of Ordering Processes in Condensed Matter", S. Komura Ed., Plenum Press. 1988.

26. See for example, J.D. Ferry, "Viscoelastic Properties of Polymers", Wiley, N.Y. (1980); M. Doi and S.F. Edwards, "The Theory of Polymer Dynamics", Clarendon Press, London (1986).

27. A. Onuki, J. Chem. Phys., 85, 6118 (1986).

28. T. Izumitani, M. Takenaka, and T. Hashimoto, in preparation.

29. A. Silberberg and W. Kuhn, J. Polym. Sci., 13, 21 (1954).

30. D. Beysens, M. Gbadamassi, J.Phys.Lett., 40, L565 (1979); D.Beysens, M. Gbadamassi, and L. Boyer, Phys. Rev. Lett., 43,1253 (1979); D.Beysens and F. Perrot, J. Phys. Lett., 45, L31(1984).

31. A. Onuki, Physica, 140A, 204 (1986), and the references quoted therein.

32. T. Hashimoto, T. Takebe, and S. Suehiro, Polym. J., 18, 123 (1986).

33. T. Takebe, S. Suehiro, and T.Hashimoto, J. Chem. Phys. to be submitted.

34. T. Takebe and T. Hashimoto, in preparation.

35. T. Nishi,T.T. Wang, and T. Kwei, Macromolecules,18,227(1975).

36. S. Nojima, K. Tsutsumi, and T. Nose, Polym. J. 14,225.

37. H.L. Snyder, P. Meakin, and S. Reich, Macromolecules,16,757(1983).

38. R.G. Hill, P.E. Tomlins, and J.S. Higgins, Macromolecules, 18, 2555(1985).

39. J. Gilmer, N. Goldstein and R.S. Stein, J.Polym.Sci, Polym.Phys.Ed., 20,2219(1982).

40. S. Nojima,Y. Ohyama, M. Yamaguchi, and T.Nose, Polym.J.,14,907. (1982).

41. M. Takahashi, H. Horiuchi, S. Kinoshita, Y. Ohyama, and T. Nose, J.Phys.Soc. Japan, 55,2687(1986).

DYNAMICS OF CONCENTRATION FLUCTUATION ON BOTH SIDES OF PHASE BOUNDARY

Charles C. Han, M. Okada[1] and T. Sato[2]

Polymers Division, National Bureau of Standards
Gaithersburg, MD 20899

INTRODUCTION

Phase separation kinetics of binary polymer mixtures has become an increasingly interested subject in recent years because of its fundamental and practical importance[1-9]. The Cahn-Hilliard[10] or Cahn-Hilliard-Cook[11] theory has generally been used in the analysis of experimental data, but the results are still not conclusive. For example: the importance of the thermal fluctuation during spinodal decomposition and a quantitative comparison of the interfacial free energy density between experiment and theory are still unsettled topics, let alone the more subtle question of the nature of the instability.

In an earlier report[6], we have studied a deuterated polystyrene/poly(vinylmethylether), (PSD/PVME), system. We have demonstrated the necessity of including the so called "virtual" structure factor in the shallow spinodal region, which accounts for the effect of the thermal fluctuations, introduced into the original Cahn-Hilliard[10] theory by Cook[11] in 1970. Due to the crude instrumentation, data correction procedure and sample handling process in our earlier study, a more quantitative comparison between experiment and theory was unobtainable. We have since constructed a new photometer with a one dimensional radicon detector and time resolved capability to collect scattering data from about 70 degree of angular range simultaneously. Also, we were able to combine results on statics of the same PSD/PVME system studied by Small Angle Neutron Scattering (SANS) and the results on kinetics from both demixing and remixing experiments to obtain a consistent examination of phase separation behavior.

---

[1]Current Address: Department of Polymer Chemistry, Tokyo Institute of Technology, Tokyo, Japan.

[2]Current Address: Department of Macromolecular Science, Osaka University, Osaka, Japan.

EXPERIMENTAL

1. Samples

     Deuterated polystyrene (PSD) and poly(vinylmethylether) (PVME)
samples used in this study are the same as those used by our group for
small angle neutron scattering[12] and temperature jump light scattering
experiments.[6] The PSD sample has a weight-average molecular weight, M_w,
equals 4.02 x 10⁵ and the molecular weight distribution index M_w/M_n =
1.42, while for the PVME sample we have M_w = 2.10 x 10⁵ and M_w/M_n = 1.32.
Solvent casted film samples are carefully dried. The thickness of sample
for the temperature jump experiment was 0.1 mm, while that for the reverse
quench experiment was about 0.6 mm.

     The temperature jump experiment was performed on blends whose volume
fraction, ϕ_{PSD}, of deuterated polystyrene (PSD) was given by ϕ_{PSD} = 0.10,
0.19, and 0.34, while the reverse quench experiment was made for
ϕ_{PSD}=0.19. The critical ϕ_{PSD} for hydrogenated or deuterated
polystyrene/PVME system was reported to be about 0.2 regardless of
molecular weight of each component.[1,9,12-14].

2.   Instrument

     In order to make rapid measurement of scattering intensity we
constructed a light scattering instrument using a radicon photodiode array
tube with 512 pixel elements (PAR Model 1452) as detector. Figure 1
depicts its schematic diagram. A vertically polarized He-Ne laser with
wavelength λ = 632.8 nm was used as the incident light beam. Since
scattered light for a given scattering angle from different parts of the
scattering sample is focused at the detector, the desmearing of scattering
data is avoided. Data at different times were collected by an optical
multichannel analyzer. The sensitivity and position linearity of the
detector were calibrated and the refraction and tarbidity of experimental
data were corrected.[15]

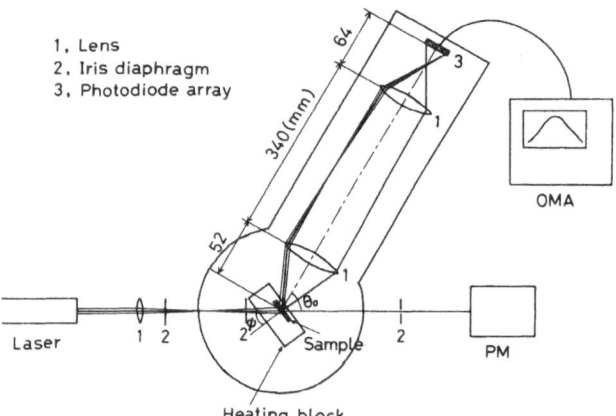

1, Lens
2, Iris diaphragm
3, Photodiode array

Figure 1. Schematic diagram of our light scattering instrument.

RESULTS AND DISCUSSION

1. Temperature Jump Experiment

Figure 2 shows an early time evolution of I(q,t) for a temperature jump experiment from 80.0 to 152.4°C (ϕ_{PSD} = 0.19). For viewing clarity, every curve except for θ=19.4° has been shifted vertically. The intensity growth is accelerative and almost exponential, with the higher angle intensity grows faster than at lower angle. The small insert in Figure 2 illustrates the change of I(q,t) over a wider time range. The intensity at larger angle stops increasing earlier than that at smaller angle and then it starts decreasing. For other temperature jump experiments, similar behavior in intensity changes have been observed.

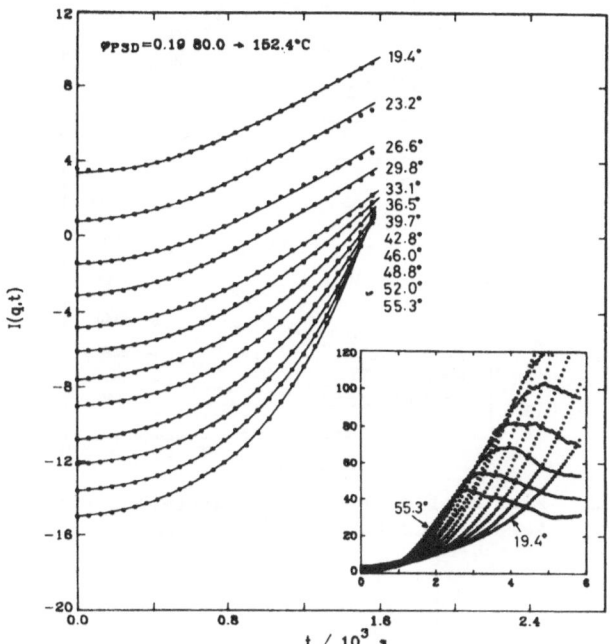

Figure 2.   Time evolution of scattering intensity I(q,t) for a temperature jump experiment from 80.0 to 152.4°C (ϕ_{PSD} = 0.19). Scattering angles shown in the figure are refraction-corrected values and each curve except for 19.4° is shifted vertically. The insert is the plot of I(q,t) vs. t over a wider time range.

Figure 3 displays the q dependence of I(q,t) at different t for the same temperature jump experiment as in Figure 2. At t < 2 x 10³ s (in the lower part of the figure), curves do not have any peak in the measured q range. An intensity maximum appears later and the peak position moves to lower q with increasing of time.

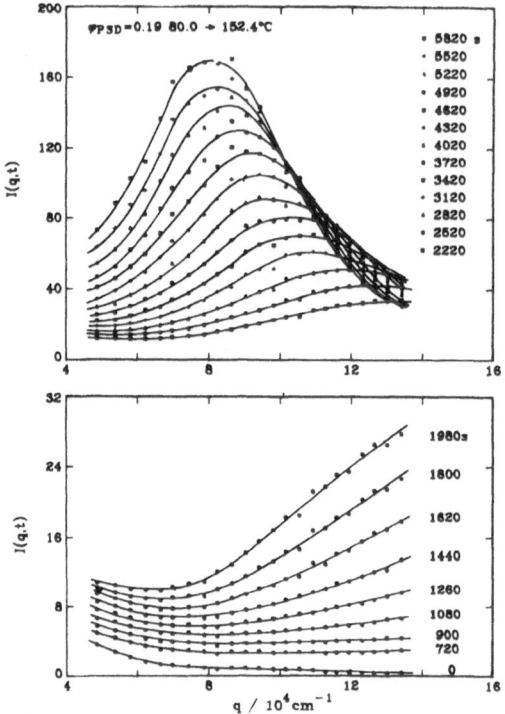

Figure 3.   Plot of I(q,t) vs. q at different time t for the same temperature jump experiment as in Figure 2.

According to Cahn-Hilliard-Cook's theory, the time evolution of the structure factor S(q,t) for a binary mixture when spinodal decomposition occurs is governed by the following differential equation:[16]

$$\frac{\partial S(q,t)}{\partial t} = -2Mq^2 \left[\left(\frac{\partial^2 f}{\partial \phi_0^2} + 2\kappa q^2 \right) S(q,t) + \frac{1}{2} \frac{\partial^3 f}{\partial \phi_0^3} S_3(q,t) \right.$$

$$\left. + \frac{1}{2} \frac{\partial^4 f}{\partial \phi_0^4} S_4(q,t) + - - - \right] + 2Mk_B Tq^2 \qquad (1)$$

where M is the mobility defined as the proportionality constant in the relation between interdiffusion current density and chemical potential gradient; q the magnitude of the scattering wave vector $[= (4\Pi n/\lambda)\sin(\theta/2)]$; f the free energy density for a uniform system; ϕ_0 the average composition; κ a interfacial free energy coefficient defined as the proportionality constant in the relation between interfacial free energy density and the square of concentration gradient; k_B the Boltzmann constant. $S_n(q,t)$ is the nth order density-density correlation function. By neglecting all S_n's for $n>2$, eq. 1 yields a solution:

$$S(q,t) = S_\infty + (S_0 - S_\infty)e^{2R(q)t} \tag{2}$$

where S_∞ is the virtual structure function in the unstable case and is the final equilibrium structure function in the stable case which comes from the thermal noise and S_0 is intensity at $t = 0$. The growth rate $R(q)$ is given as

$$R(q) = -Mq^2[\partial^2 f/\partial\phi_0^2 + 2\kappa q^2] \tag{3}$$

where the first term of the right-hand side arises from the bulk free energy and the second term from the interfacial free energy. It should be noticed that Eq. 2 is valid only in the early stage of spinodal decomposition when the higher order terms of S_n in eq. 1 can be neglected. Eqs. 2 and 3 were originally derived by Cahn, Hilliard, and Cook.[10,11]

As mentioned by de Gennes,[17] the mobility M for polymer blends generally depends on q. In the case of $qR_{gi} < 1$ (R_{gi} is the radius of gyration for component i), it can be taken as constant. Moreover, de Gennes has used the expression

$$\kappa = (k_B T/36)[b_A^2/(v_A\phi_0) + b_B^2/(v_B(1 -\phi_0))] \tag{4}$$

for a binary polymer mixture, where v_i is the monomer molar volume of component i and b_i is the effective bond length per monomer of component i.

Since the background stray light is still present in this experimental set up, the measured intensity, $I(q,t)$, is related to the time evolution of the structure factor, $S(q,t)$ by

$$I(q,t) = (KS_\infty + I_b)+K(S_0 - S_\infty)e^{2R(q)t}$$

$$= I_\infty + (I_0 - I_\infty)e^{2R(q)t} \tag{5}$$

where both I_∞ ($\equiv KS_\infty + I_b$) and I_0 ($\equiv KS_0 + I_b$) include a background intensity. Owing to this background intensity and the contribution from the virtual structure factor, S_∞[(6)], the quantity I_∞ is not necessarily negligible and the semi-logarithmic plot of $I(q,t)$ vs. t which was used by many authors[2-5,7,8,12] could include error in the determination of $R(q)$. We have attempted here to estimate $R(q)$ by a different method.

After expansion of the exponential term and some manipulations, eq.2 can be rewritten as follows:

$$\left[\frac{t}{I(q,t)-I_0}\right]^{1/3} = \frac{1}{[2(I_0-I_\infty)R(q)]^{1/3}} \left\{1 - \frac{1}{3}R(q)t + \frac{1}{81}[R(q)t]^3 + ---\right\} \tag{6}$$

The right-hand side of this equation does not have the second order term in t and can be well approximated by a linear equation in t at $R(q)t < 1$ (with an error of less than 2%). With this equation, $R(q)$ can be evaluated from the intercept and the initial slope of the plot of $[t/(I(q,t) - I_0)]^{1/3}$ against t. We will refer to this as the 1/3 power plot.

The 1/3 power plot made from the same data as in the Figure 2 is shown in Figure 4, where data from different scattering angles are shifted vertically except for data from $\theta = 19.4°$. The values of $R(q)$ for each q were calculated from the indicated solid lines. The linear region for each angle corresponds to the region where $R(q)t < 1$. In the top half of Figure 5, the calculated $R(q)$ are plotted against q^2 for different temperature jump experiments which all have composition $\phi_{PSD} = 0.19$. It can be seen in this figure that $R(q)$ is almost proportional to q^2 in our q range, i.e., the q^4 term (interfacial free energy term) in $R(q)$ can be neglected. Similar behavior has also been observed for other (off critical) compositions.

Several authors[2,3,5,6,8,12] have reported considerable contribution of q^4 term to $R(q)$. However, their $R(q)$ values have been obtained from the semilogarithmic plot of $I(q,t)$ vs. t, where the I_∞ terms have been neglected.

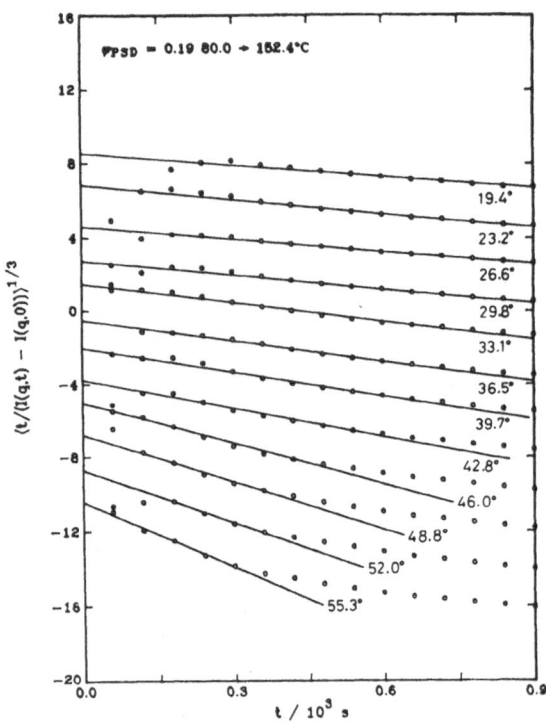

Figure 4.   1/3 power plot for temperature jump experiment from 80.0 to 152.4°C ($\phi_{PSD} = 0.19$). Each data sets except for 19.4° are shifted vertically.

We should point out that it is difficult to obtain the true initial
growth rate from a semilogarithmic analysis. A larger q^4 term in $R(q)$ has
often been obtained resulting from non-linear contributions at later time.
It should also be noted that the de Gennes theory for κ (see eq. 4)
predicts a very small q^4 term which will be discussed later.

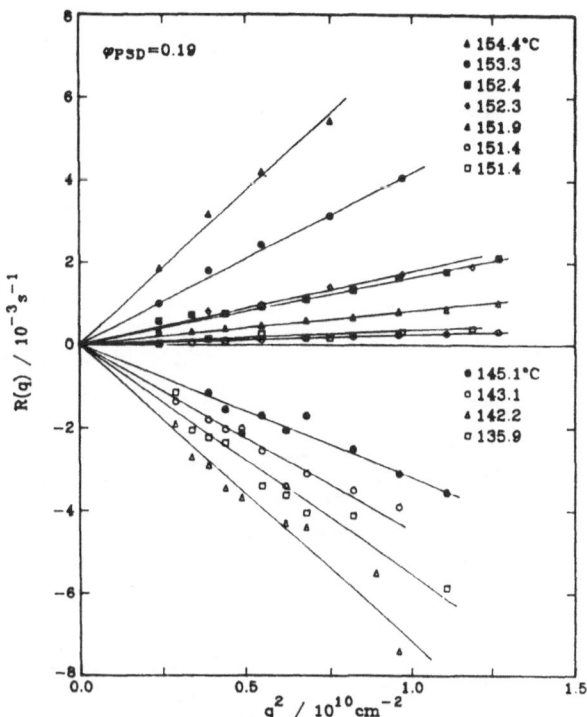

Figure 5.   Upper part: $R(q)$ vs. q^2 for seven temperature
jump experiments ($\varphi_{PSD} = 0.19$). Each data sets
except for 19.4° are shifted vertically.

2.   Reverse Quench Experiment

The change of structure function associated with concentration
fluctuation decay in a reverse quench experiment is also governed by eq.2,
if the concentration fluctuation is small enough. In this case, $R(q)$ is a
negative quantity referred to as a decay rate. The interdiffusion
coefficient,

$$D = M \; \partial^2 f / \partial \phi^2_o \qquad (7)$$

is positive which corresponds to a decay of concentration fluctuation;
while for temperature jump experiment, D is negative which corresponds to
a growth of concentration fluctuation.

Contrary to the temperature jump experiment, I_∞ in the reverse quench
experiment can be obtained from the final equilibrium intensity.
Therefore the decay rate $R(q)$ can be easily evaluated from a semi-
logarithmic plot of $I(q,t) - I_\infty$ against t.

Figure 6 displays intensity decay curves for a reverse quench 8
experiment from 149.9°C to 143.1°C ($\phi_{PSD} = 0.19$). Each curve except for
$\theta=21.6°$ is shifted vertically. After about 500s, intensity for each angle
equilibrates to some final value. Using this value as I_∞, the decay rates
$R(q)$ can be estimated from slopes of the $\ln[I(q,t)-I_\infty]$ vs t plots.

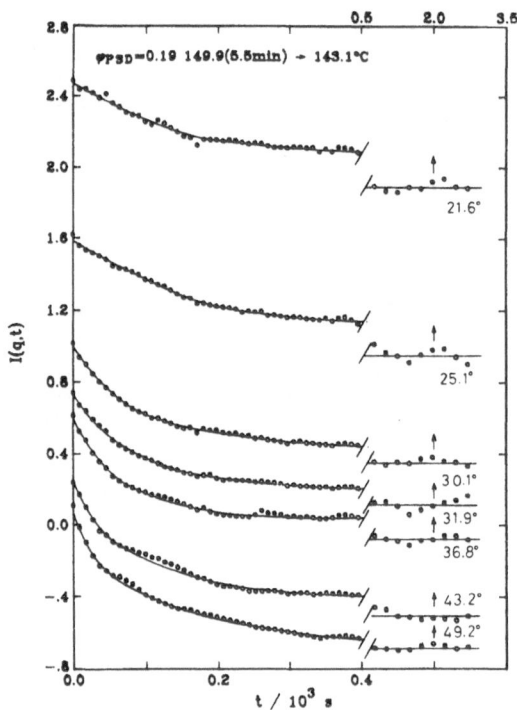

Figure 6.   Decay curve of $I(q,t)$ for a reverse quench
             experiment. The sample ($\phi_{PSD} = 0.19$) was brought
             to 149.9°C for 5.5 min and quenched to 143.1°C.
             Each curve except for 21.6° is shifted vertically.

The q dependence of the decay rate is depicted in the lower part of
Figure 5. The decay rate as well as the growth rate are shown to be
proportional to q^2 with perhaps a very small downward curvature arising
from the interfacial free energy contribution. This indicates that for
the PSD/PVME system the interfacial free energy hardly affects the
dynamics of concentration fluctuations in the early stage with wavenumbers
$< 1.4 \times 10^5$ cm^{-1}. q^2 dependence of the decay rate had already been found
by Kumaki and Hashimoto.[9]

3.   Diffusion Coefficient and Mobility

When the q^4 term in $R(q)$ is negligible, the interdiffusion
coefficient D can be evaluated from the slope of the $R(q)$ vs. q^2 plot (cf.
eq. 3 and 7).

The interdiffusion coefficients evaluated from the data in Figure 5
are shown in Figure 7, where squares represent D in the one phase region
obtained from the reverse quench experiment and circles represent those in

440

the two phase (spinodal) region obtained from the temperature jump experiment. Most of the data points can be represented by a single curve and the continuity of the interdiffusion coefficient at phase separation boundary is substantiated.

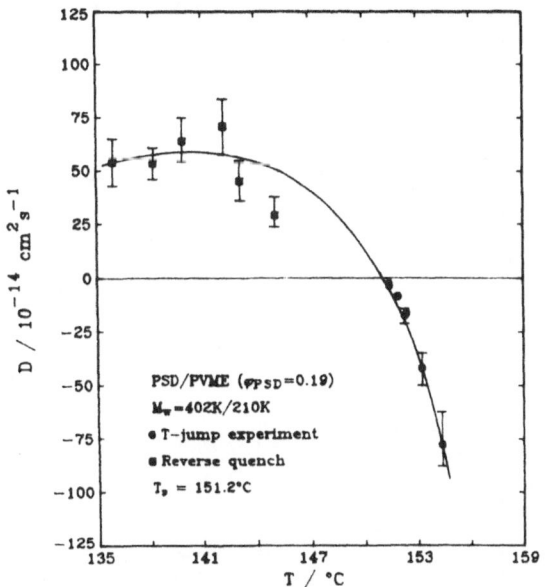

Figure 7.   Temperature dependence of D for ϕ_{PSD}=0.19 on both sides of phase boundary. The circles and the squares show D in two phase region and in one phase region, respectively.

The interdiffusion coefficient is a product of the mobility M and the second derivative of free energy density $\partial^2 f/\partial \phi^2$ as shown in eq. 7. The sign change of D is entirely due to that of $\partial^2 f/\partial \phi^2$. Data of $\partial^2 f/\partial \phi^2$ for PSD/PVME system have been obtained before from small angle neutron scattering experiment (SANS) by Muroga et al.[18] from this laboratory. Three different molecular weight pairs of PSD/PVME have been measured. Regardless of the sample molecular weight, data points can be fitted very nicely by the equation:

$$\partial^2 f/\partial \phi^2 = 2.18 \times 10^{10}(1/T - 1/T_s) \ \text{erg.cm}^{-3} \qquad (8)$$

We can calculate M from this equation and the interdiffusion data obtained earlier. A single Arrhenius type of equation has been obtained for M of the PSD/PVME system on both sides of phase boundary, with an activation energy of 37 kcal/mol.

According to the de Gennes theory[17], the coefficient for the q^4 term which is the interfacial free energy contribution to the total free energy density is small in our experimental q range. Actually, this coefficient can be easily obtained from the moment expansion of $S(q)$ because $R(q) = -Mk_B Tq^2/S(q)$.

In Figure 8, the experimental values of $R(q)/(q^2M)$ are plotted against q^2. Theoretical values of $R(q)/(q^2M)$ for the corresponding temperatures has also been calculated and shown by solid lines. The theoretical lines are almost horizontal which substantiate that q^4 term in $R(q)$ is negligible in the q range of our experiment.

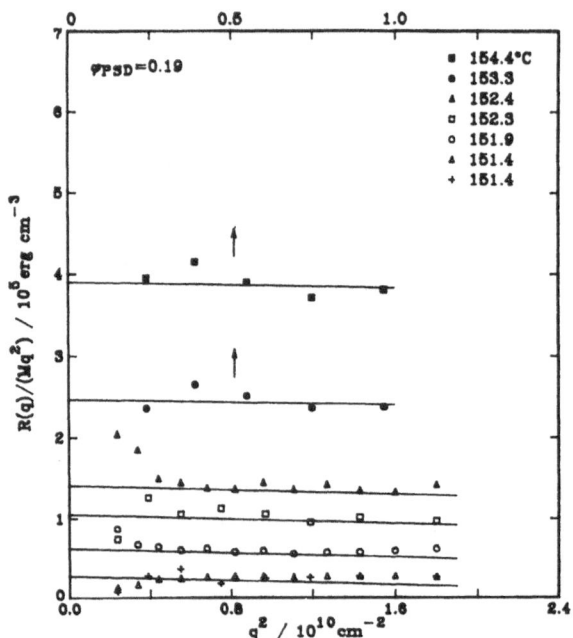

Figure 8. $R(q)/(q^2M)$ vs. q^2 for different temperature jump experiments ($\phi_{PSD} = 0.19$). The circles show the experimental results obtained from the data in Figure 5, while the solid lines show the theoretical values calculated from eqs. 3 and 4.

CONCLUSION

We have shown for PSD/PVME polymer blend system that interfacial free energy is small and consistent with theoretical prediction, which does not play an important role in the dynamics of concentration fluctuations in either one phase or two phase regions. The interdiffusion coefficient is continuous at the phase separation boundary. Furthermore, the mobility M has been extracted and an Arrhenius type of temperature dependence has been found. To the best of our knowledge, this is the first time that M has been obtained on both sides of the phase boundary, and the static results have been incorporated into the kinetics study for a consistent evaluation of the Cahn-Hilliard-Cook theory.

REFERENCES

1. T. Nishi, T. T. Wang, and T. K. Kwei, Macromolecules, $\underline{8}$, 227 (1975).
2. S. Nojima, K. Tsutsumi, and T. Nose, Polym. J., $\underline{14}$, 225 (1982).
3. T. Hashimoto, J. Kumaki, and H. Kawai, Macromolecules, $\underline{16}$, 641 (1983).
4. H. Snyder, P. Meaking, and S. Reich, Macromolecules, $\underline{16}$, 757, (1983).
5. R. G. Hill, P.E. Tomlins, and J. S. Higgins, Macromolecules, $\underline{18}$, 2555 (1985).
6. M. Okada and C. C. Han, J. Chem. Phys., $\underline{85}$, 5317 (1986).
7. T. Hashimoto, M. Itakura, and N. Shimidzu, J. Chem. Phys., $\underline{85}$, 6773 (1986).
8. H. Yang, M. Shibayama, R. S. Stein, N. Shimizu, and T. Hashimoto, Macromolecules, $\underline{19}$, 1667 (1986).
9. J. Kumaki and T. Hashimoto, Macromolecules, $\underline{19}$, 763 (1986).
10. J. W. Cahn and J. E. Hilliard, J. Chem. Phys., $\underline{28}$, 258 (1958); J. W. Cahn, J. Chem. Phys., $\underline{42}$, 93 (1965).
11. H. E. Cook, Acta Metall., $\underline{18}$, 297 (1970).
12. C. C. Han, M. Okada, Y. Muroga, F. L. McCrackin, B. J. Bauer, and Q. Tran-Cong, Polym. Eng. Sci., $\underline{26}$, 3 (1986).
13. M. Shibayama, H. Yang, R. S. Stein, and C. C. Han, Macromolecules, $\underline{18}$, 2179 (1985).
14. T. Hashimoto, M. Itakura, and H. Hasegawa, J. Chem. Phys., $\underline{85}$, 6118 (1986).
15. T. Sato and C. C. Han, (submitted to J. Chem. Phys.).
16. J. S. Langer, M. Bar-on, and H. Miller, Phys. Rev., $\underline{A11}$, 1417 (1975); J. D. Gunton, M. San Miguel, and P. S. Sahni, in "Phase Transition and Critical Phenomena," Chap. 3, Vol. 8, C. Domb and J. L. Lebowitz, Eds., Academic Press, New York (1983).
17. P. G. de Gennes, J. Chem. Phys., $\underline{72}$, 4756 (1980).
18. Y. Muroga, B. J. Bauer, M. Okada, Q. Tran-Cong, and C. C. Han, in preparation.

SMALL ANGLE SCATTERING EXPERIMENTS OF NEUTRONS

FOR THE POLYMER BLEND PVME/d-PS

D. Schwahn, T. Springer, K. Mortensen[a], and H. Yee-Madeira[b]

Institut für Festkörperforschung
Kernforschungsanlage Jülich GmbH, D-5170 Jülich, West-Germany
[a]Riso National Laboratory, DK-4000 Roskilde, Denmark
[b]on leave from ESFM-IPN Mexico

INTRODUCTION

This paper is a study by neutron small angle scattering of the polymer blend: polyvenylmethylether and deuterated polystyrene. The blend is a compatible system with a miscibility gap at high temperature, and a lower critical solution temperature (LCST). The aim of the investigations was primarily a study of the time dependent structure factor $S(Q,t)$ for neutron small angle scattering in the unstable region. This was achieved by means of a double crystal diffractometer (DKD). Its resolution of the scattering vector Q is of the order of $5 \cdot 10^{-5} \text{ Å}^{-1}$. The dynamics of precipitates – the early states and the regime of coarsening – are easily accessible because the growth rates are small and can be easily resolved. Furthermore, the early stages can be treated by theory in the meanfield approximation[1].

In order to obtain a reliable and accurate phase diagram of the system for these investigations, the spinodal and the coexistence curve (binodal) were studied by critical neutron scattering. This was carried out by a conventional pin-hole small angle scattering instrument (resolution 10^{-3} Å^{-1}). The investigations have lead to the observation of a meanfield behaviour of the temperature dependence of scattering for Q = 0. Furthermore, the measurements indicate a cross-over to an Ising-like curve very close to T_c[2]. The experiments yield the spinodal $T_s(\phi)$ as a function of composition ϕ, the Flory Huggins interaction parameter $\chi(T,\phi)$ as obtained from $T_s(\phi)$, and finally, the coexistence or binodal curve and the critical temperature T_c.

In the meanfield–random phase approximation[1], the structure factor is

$$S^{-1}(Q) = S_A^{-1}(Q) + S_B^{-1}(Q) - 2 \chi \tag{1}$$

$S_A(Q) = \phi \, N_A \, g_D \, (Q^2 R_A^2)$ where $g_D(x) = (2/x^2) \left[x - 1 - \exp(-x) \right]$ is the Debye function for an undisturbed chain, with a radius of gyration of R_A. The Flory–Huggins parameter χ describes the interaction term in the Gibbs potential of mixing in the cell model[3]

$$\Delta G_{mix}/kT = \phi(1-\phi) \, \chi(\phi,T) - \Delta S_{mix}/k \tag{2}$$

where $\Delta S_{mix}/k = (\phi/N_A) \, \ln\phi + \left[(1-\phi)/N_B \right] \ln(1-\phi)$. $N_{A,B}$ is the number of statistical chain segments. Expansion of eq. (1) up to Q^2 yields

$$S^{-1}(Q) = S^{-1}(0) \, (1 + \xi^2 \, Q^2) \tag{3}$$

where

$$\xi^2 = (1/3) \, S(0) \left[(R_A^2/\phi N_A) + R_B^2/(1-\phi)N_B \right]. \tag{4}$$

The spinodal curve $T_s(\phi)$ is obtained from $S(Q)^{-1}$ $(Q \to 0)$ extrapolated to zero. This corresponds to the thermodynamic stability condition

$$\delta^2 \, \Delta G_{mix}/\delta\phi^2 = 0. \tag{5}$$

The Flory–Huggins paramter χ is obtained from $T_s(\phi)$ which is determined by measuring $S(Q)$ for $Q \to 0$, with eqs. (2), (3), and (4). The radii R_A and R_B are determined experimentally[4].

THE STRUCTURE FACTOR $S(Q,t)$ AS A FUNCTION OF TIME
IN THE UNSTABLE REGION

Crossing the spinodal, the system starts to decompose into two phases with volume fractions ϕ' and $1-\phi'$. Precipitates or "domains" grow and coarsen for longer times, forming an interconnected three–dimensional structure. The resulting structure factor $S(Q,t)$ can be interpreted and characterized in terms of three different and independent lengths[5]:

(i) $S(Q,t)$ has a peak for a value of $Q = Q_{max}$. The characteristic length $R_m(t) = 1/Q_{max}$ is related to the mutual distance of neighbouring domains.

(ii)   At Q-values above Q_m, the structure factor should approach a Guinier function

$$S(Q) = V_0 \; \phi' \; (1-\phi') \; \exp(-R_g^2 \; Q^2)/3 \qquad (6)$$

assuming that the domains are separated. This defines the average radius of gyration R_g of the polymer molecules, and V_0, the volume of "domains" assuming they are separated.

(iii)   Independent of (i) and (ii), a length can be defined from eq. (6) by this volume, $V_0 = (4\pi/3) \; R_v^3$.

For large Q, Porod's law $S(Q) \propto Q^{-4}$ should hold, provided that the boundaries of the domains are sharp. Finally, the second moment of the structure factor yields the degree of decomposition, namely

$$\int S(Q) \; Q^2 dQ = 2\pi^2 \; \phi' (1-\phi'). \qquad (7)$$

For small decomposition times, the total scattering can be determined by a transmission experiment. Using the validity of the Cahn-Hilliard theory[6], one obtains approximately

$$\Sigma(t) = (1/k^2) \int S(Q) \; 2\pi \; Q \; d \; Q \simeq \exp(- D \; Q_m^2 \; t/2) \qquad (8)$$

where k is the neutron wavevector, D is the interdiffusion constant of the components; this is the kinetic parameter which drives the decomposition. The factor $D \; Q_m^2$ is the relaxation rate for the decomposition process.

EXPERIMENTS

    The blends under consideration in this work were
I.   72 % PVME (M_w = 89.000; M_w/M_n = 2.5) + 28 % d-PS
      (M_w = 473.000; M_w/M_n = 1.18), and
II.   79 % PVME (as before) + 21 % d-PS
      (M_w = 232.000; M_w/M_n = 1.08).

The small angle experiments to study the critical scattering were carried out at 10 Å by a pinhole instrument at the Risø research reactor, with a resolution of about 10^{-3} Å^{-1}. The experiments with the double crystal diffractometer DKD[7] at the Jülich research reactor were performed at λ = 1.8 and 2.52 Å with a resolution half-width of about 4.5 μ radian or $5 \cdot 10^{-5}$ Å^{-1}. This half-width is determined by the Darwin width of the reflection curves from two ideal Si(331) crystals in parallel position which serve as collimators. The DKD-data were de-smeared analytically. The calibration was performed by a water scatterer, by a Lupolen standard, and by the direct measurement of the primary beam in the case of DKD[8].

First of all, the <u>phase diagram</u> was investigated by critical scattering experiments. <u>Fig. 1</u> presents the extrapolated structure factor following eq. (3), namely $S(0)^{-1}$ as a function of temperature T, for a composition $\phi = 0.22$, which was <u>off</u> the critical composition $\phi_c = 0.21$. A meanfield behaviour was observed, with S(0) proportional to $(T_c - T)^{-1}$.

The sharp up-bending of the curve at 141.6 °C is caused by the beginning of nucleation and domain growth; it occurs at T_B where the coexistence curve (binodal) is crossed.

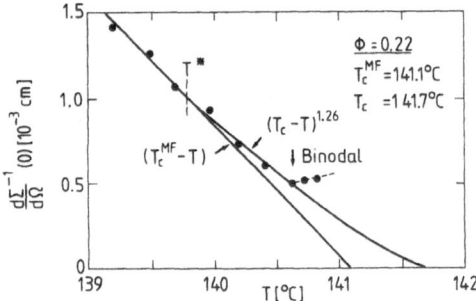

<u>Fig. 1.</u> Inverse neutron scattering intensity which is proportional to S(Q), for Q = 0, at $\phi = 0.22$. The straight line describes the meanfield behaviour. $T_s(MF)$ is the spinodal temperature extrapolated by the meanfield curve. The temperature is indicated, where the coexistence curve (binodal) is crossed and domain scattering appears.

<u>Fig. 2</u> shows $S(0)^{-1}$ <u>at</u> ϕ_c. One obtains the <u>extrapolated</u> critical temperature $T_c(MF) = 141.3$ °C. A deviation from the meanfield behaviour appears near T_c with a critical exponent as for an Ising system $\gamma = 1.26$. The corresponding "true" critical temperature $T_c = 141.9$ °C is also indicated.

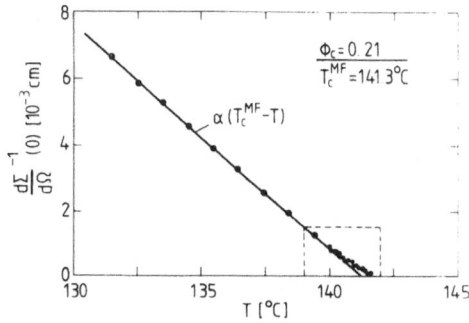

Fig. 2. The inverse scattering intensity for Q = 0, __at__ the critical concentration ϕ_c = 0.21. Close to T_c(MF) a deviation from the meanfield behaviour occurs.

Fig. 3 shows a small branch of the spinodal where $S(0)^{-1}$ = 0 or $\delta^2 \Delta G_{mix}/\delta\phi^2$ = 0. Also the __extrapolated__ meanfield spinodal is shown. The binodal or coexistence curve is obtained from the position of the kink as in Fig. 1. The observation, that the critical point is above the minimum of the spinodal, and the unusual shape of the coexistence curve can be explained by the polydispersity of one component in the blend[9]. The limits of the meanfield behaviour[2] are indicated by shaded lines. All our decomposition measurements were carried out in a region where the meanfield theory is expected to be valid.

DECOMPOSITION EXPERIMENTS

Fig. 4 combines two typical small angle patterns, i.e. the independently calibrated results for the structure factor from the conventional small angle scattering instrument, and, at small Q, from the double crystal diffractometer (for system II). The solid line is drawn as a guide line for the eye; presently, there is still a gap in the region between 0.8 and $2 \cdot 10^{-3}$ $Å^{-1}$.

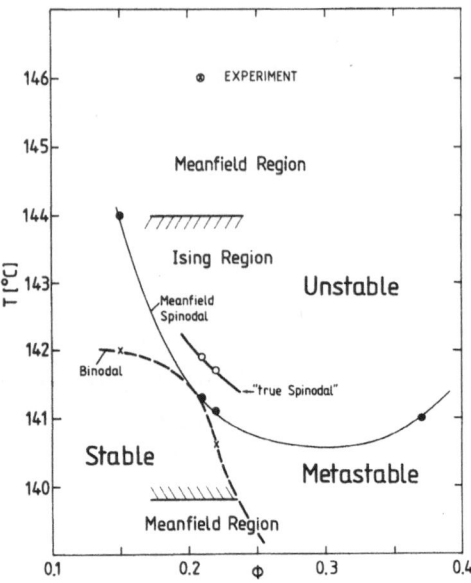

Fig. 3. Part of the phase diagram of PVME/d-PS for system II, in the region near T_c. Solid line = "true spinodal" obtained from $S(0)^{-1} = 0$. Thin line: Spinodal calculated by extrapolation of the meanfield behaviour. Dashed: coexistence curve or binodal. Crosses and circles are measured points (from small angle scattering). Shaded: approximate limits of the region where Ising theory applies.

The combination of both methods covers 4 orders of magnitude in Q, and nearly 8 orders in intensity! At small Q-values, one observes the peak caused by the <u>boundary</u> (<u>shape</u>) of the domains. On the other hand, at larger Q the critical scattering <u>in</u> the domains is clearly visible as well. At Q-values above the peak position, the Guinier region can be recognized, and the curve follows eq. (6).

<u>Fig. 5</u> summarizes measurements with the double crystal diffractometer for times between 4 and 120 minutes (for system I.). The coarsening (shift of the peak position and increase of its intensity), is clearly seen.

The different characteristic lengths are presented in <u>fig. 6</u> as volume $V_m = R_m{}^3$, volume $V_G = (4\pi/3) (1.30 R_g)^3$, and V_o from eq. (6). All three quantities are of similar magnitude. For $t > 25$ min, the volumes follow a t^3 law for the time t, whereas, at smaller times, they are very roughly linear in t.

A similar behaviour was observed for ordinary liquid mixtures[10], namely for smaller times

$$R(t) = \left[12 \, D^* \, R(t)\right]^{1/3} \, t^{1/3} \tag{9}$$

where D^*R does not depend on R. D^* is the diffusion constant of the domains. Furthermore, for larger times one finds[10]

$$R(t) = 0.10 \, (\sigma/\eta) \, t. \tag{10}$$

σ and η are the surface tension of the domains and the viscosity, respectively. The measured prefactor of eq. (9) yields $D^* = 3 \cdot 10^{-12}$ cm^2/sec for $R = 1$ μm. Values of $\eta = 5 \cdot 10^5$ pa, and $\sigma = 3 \cdot 10^{-5}$ N/cm^2 approximately lead to the experimental prefactors of eq. (10).

So far, at times below 7 minutes, measurements of S(Q,t) were not possible for intensity reasons. However, transmission experiments were carried out with the sample between the two crystals of the double crystal diffractometer. This yields the total scattering cross section Σ_s integrated over Q-values larger than 10^{-5} Å^{-1}. It was found that Σ_s shows an exponential growth as in eq. (8), with $D = 7 \cdot 10^{-12}$ cm^2/sec.

In a forthcoming paper, a more detailed report of part of our critical and domain scattering experiments will appear.

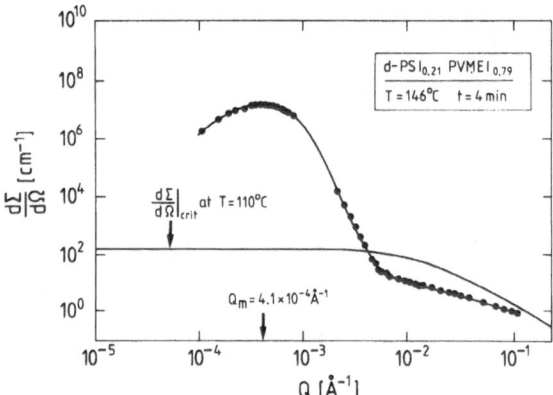

Fig. 4: S(Q) measured by a conventional small angle scattering and a double crystal instrument (DKD). For $Q < 10^{-2}$ $\overset{\circ}{A}^{-1}$: domain shape scattering. For $Q > 10^{-2}$ $\overset{\circ}{A}^{-1}$: critical fluctuation scattering in the domains. The data from both instruments are independently calibrated.

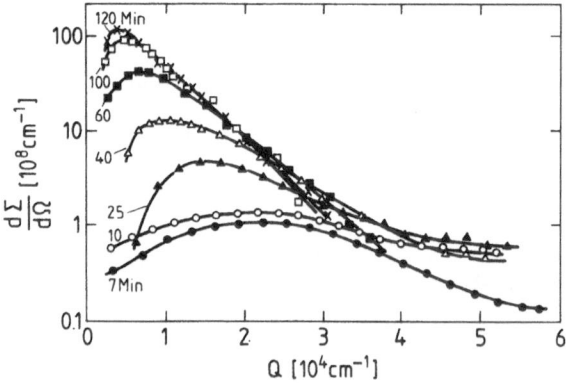

Fig. 5: Scattering intensities for the blend after various decomposition times t, as a function of Q.

CONCLUSIONS

(i)   Sufficiently far from T_c or from the spinodal, critical
scattering experiments on the PVME/d-PS blend reveal a
meanfield behaviour for the temperature dependence. However,
in the region $T_c - 2 K \leq T \leq T_c$ an Ising-like behaviour with
$\gamma \simeq 1.26$ was observed.

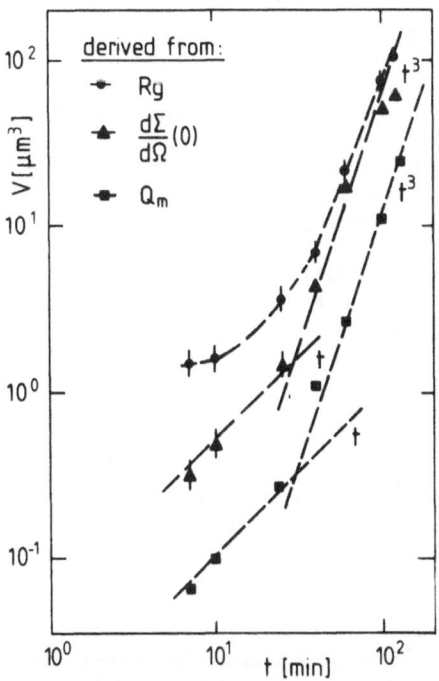

Fig. 6: The volumes as defined from the peak in S(Q) of fig. 5, by
$R_m = 1/Q_m$, from the radius of gyration, R_g, and from V_o in eq. (6).

The "true spinodal", where $S^{-1}(0)$ or $\delta^2 \Delta G_M / \delta \phi^2$ vanishes, is
about 0.6 K above the "meanfield spinodal", as found by an
extrapolation of $S^{-1}(0,T)$ in the meanfield region (see
Fig. 3).

The critical temperature T_c occurs above the minimum of the
spinodal, and the curvature of the coexistence curve is
opposite to that of the spinodal. This behaviour is expected
and typical for polydisperse systems.

(ii)   In the demixing region, experiments with a conventional
        small angle scattering instrument and with a double crystal
        diffractometer were carried out, covering four orders of
        magnitude in Q, for times between a few minutes up to hours.
        Critical fluctuations in the domains and, as well, shape
        scattering by the domain itself were observed.

(iii)   Domain shape scattering is understood by the coarsening
        process. The morphology of the domain structure is
        characterized by three independent characteristic lengths:
        R_m, R_g, and R_v. They all have nearly the same magnitude and
        time dependence, namely $R(t) \simeq$ const t for t > 25 min and
        roughly as $R(t) = $ const $t^{1/3}$ for 4 min < t < 25 min. This
        behaviour is established for ordinary liquid mixtures in
        terms of diffusive growth and growth governed by
        hydrodynamic effects. The proportionality of the different
        characteristic lengths indicates an affine growth of the
        precipitation morphology. At smaller times, transmission
        experiments indicate an exponential growth, as expected from
        Cahn-Hilliard theory.

The experiments in the miscibility region of blends will be continued
for shorter times and closer to T_c, to study the early states of
demixing and the region where Ising behaviour should determine the
kinetics.

ACKNOWLEDGMENT

We thank Mr. Beyer preparing the computer software, and Mr. Pohl for
his infatigable help during all our measurements.

REFERENCES

1.   P.G. de Gennes, "Scaling Concepts in Polymer Physics", Cornell
      U.P., Ithaca, N.Y. (1979); K. Binder, "Collective Diffusion,
      Nucleation, and Spinodal Decomposition in Polymer Mixtures",
      J. Chem. Phys. 79:6387 (1983); J.D. Gunton, M.S. Mignel and
      P.S. Sahni, "Phase Transitions and Critical Phenomena", Academic
      Press, London, (1983) Vol. 8; T. Hashimoto, "Structure Formation
      in Polymer Mixtures by Spinodal Decomposition" in: Current Topics
      in Polymer Science, Vol. 2, Hauser, München 1987.
2.   D. Schwahn, K. Mortensen, and H. Yee-Madeira, "Mean-Field and
      Ising Critical Behaviour of a Polymer Blend", Phys. Rev. Lett.
      58:1544 (1987).
3.   P.J. Flory, "Principles of Polymer Chemistry", Cornell U.P.,
      Ithaca, N.Y. (1953) Chap. 12.
4.   D. Schwahn, K. Mortensen, T. Springer, H. Yee-Madeira, and
      R. Thomas, "Investigation of the Phase Diagram and Critical
      Fluctuations of the System Polyvenylmethylether and d-Polystyrene
      with Neutron Small Angle Scattering", J. Chem. Phys., in press.
5.   A. Guinier and G. Fournet, "Small Angle Scattering of x-Rays",
      Wiley, N.Y. (1955).
6.   J.W. Cahn, "On Spinodal Decomposition", Acta Met. 9:795 (1961).

7. D. Schwahn, A. Miksovsky, H. Rauch, E. Seidl and G. Zugarek, "Test of Channel-Cut Perfect Crystals for Neutron Small Angle Scattering Experiments", Nucl. Instr. Meth. A 239:229 (1985).
8. D. Schwahn and H. Yee-Madeira, "Spinodal Decomposition of the Polymerblend Deuterious Polysterene (d-PS) and Polyvinylmethylether (PVME) studied with High Resolution Neutron Small Angle Scattering", Colloid & Polymer Sci. to be published.
9. R. Koningsveld, "Partial Miscibility of Mulitcomponent Polymer Solutions", Advan. Colloid Interface Sci., 2:151 (1968).
10. E.D. Siggia, "Late Stages of Spinodal Decomposition in Binary Mixtures", Phys. Rev. A 20:595 (1979).

MECHANISM AND KINETICS OF PHASE SEPARATION OF

A SEMI-RIGID AND FLEXIBLE POLYMER MIXTURE

Hirokazu Hasegawa, Toshio Shiwaku, Akemi Nakai
and Takeji Hashimoto

Department of Polymer Chemistry
Faculty of Engineering
Kyoto University
Kyoto 606, Japan

ABSTRACT

The mixture of a semi-rigid polymer and a flexible polymer undergoes phase separation via spinodal decomposition. The unmixing process of the polymer mixture can be divided into three characteristic regions depending on the difference in the growth behavior. In each region the growth of the wavelength of the concentration fluctuation obeys a scaling law.

I. INTRODUCTION

The ordering processes during phase separation of an incompatible polymer mixture consisting of a main-chain liquid-crystalline (semi-rigid) polymer and a flexible polymer were studied by means of polarizing optical microscopy and light scattering. The polymer mixture studied in this work has two great advantages for the phase separation experiment. One advantage is that a homogeneous mixture can be easily obtained in the form of thin film by quick casting from a dilute solution, and stays stable in the amorphous glassy state at room temperature. Unmixing between the two components is initiated by raising the temperature of the mixture above the glass transition temperature. The other advantage is that the two phases developed in the unmixing process can be clearly distinguished from each other. The semi-rigid polymer forms an optically anisotropic, liquid-crystalline phase, and therefore, appears bright under a polarizing optical microscope (POM) with crossed polarizers. On the other hand, the flexible polymer forms an optically isotropic phase above its melting point, and appears dark. In addition to these advantages, the unmixing process of this system is slow enough to detect and record the structure by conventional techniques. Therefore, both mechanisms and kinetics of phase separation can be clarified.

II. EXPERIMENTAL

The system used in this study was a binary mixture (50:50 by weight) of poly(ethylene terephthalate) (PET) and a copolyester composed of 60 mol% p-oxibenzoate (OBA) units and 40 mol% ethylene-terephthalate units (X-7G,

Tennessee Eastman Co.). PET is a popular semi-crystalline flexible polymer, and X-7G is one of the well-known main-chain thermotropic liquid-crystalline polymers[1-6]. Test film of the polymer mixture was prepared by casting from solution on microscope cover glasses. The details of the sample preparation were as reported elsewhere[7]. The as-cast films were clear, and did not exhibit optical anisotropy under a POM. PET forms an isotropic liquid comprising flexible coil chains above its melting point (ca. 275°C for pure PET and ca. 256°C for the 50:50 mixture), and X-7G is in the liquid crystalline state between the melting point (ca. 230°C for pure X-7G) and the nematic-isotropic transition temperature (ca. 270-360°C).[8] An annealing temperature of 270°C was chosen to make sure that both PET and X-7G are in the liquid state.

Phase separation process of the polymer mixture was observed under a POM with crossed polarizers equipped with a heating stage (TH-600 type, LINKAM Scientific Instruments). Temperature jump (T-jump) was made by placing a test film on the heating stage controlled at 270°C. Within a few seconds the test film reached the preset temperature. The POM images were recorded on photographic film or video cassette tape. Light scattering patterns were obtained from the test film annealed at 270°C for a particular time period and quenched to room temperature. The characteristic wavelengths of the concentration fluctuations (Λm) at various stages of phase separation were obtained by image analysis of the polarizing optical micrographs and by optical diffraction of the micrographs.[8]

III. RESULTS AND DISCUSSION

III-1. Three Characteristic Regions of Phase Separation Process

Figure 1 shows the selected pictures from a series of POM images obtained from the same area of the test film of the polymer mixture during the isothermal annealing at 270°C and reproduced from the video cassette using a video graphic printer. Immediately after the T-jump a bright area starts to develop suggesting the formation of optically anisotropic phase of the liquid-crystalline X-7G. Depending on the difference in the behavior of structural growth, the unmixing process can be divided into three characteristic regions.

Region I: A percolation-network structure of liquid crystal (rich in X-7G) in isotropic melt (rich in PET) grows in size with self-similarity[9,10], suggesting the late stage of spinodal decomposition.

Region II: Entire network of the liquid-crystalline phase breaks up into fragments, and the continuity of the network as a whole is lost. The fragments eventually change in shape to form spherical droplets.

Region III: The size and spacing of the droplets of X-7G liquid-crystalline phase dispersed in the matrix of PET isotropic liquid increase with time due to the diffusion-coalescence mechanism.

Figure 2 shows the V_v light scattering patterns and corresponding polarizing optical micrographs obtained from the test films annealed at 270°C for 1, 3, 6 and 10 sec and quenched to room temperature. The micrographs show that the percolation-network structures developed in region I are fixed in the film by quenching. The network structure of 10 sec seems to be just a magnified image of the network structure of 6 or 3 sec suggesting the self-similar growth[9,10]. The light scattering patterns in Figure 2 show a "spinodal ring"[11] which indicates the existence of periodic concentration fluctuation. The peak position shifts towards a smaller scattering angle with increasing annealing time indicating the increase in the wavelength of the concentration fluctuation. These observations strongly suggest that the unmixing process in region I is the late stage of spinodal decomposition[12]. The arc-like scattering patterns for the film for a longer annealing time are due to the unique feature of the phase separated structure with one phase being optically anisotropic, i.e., the OBA segments have a preferential orientation within the liquid-crystalline phase.[7,8]

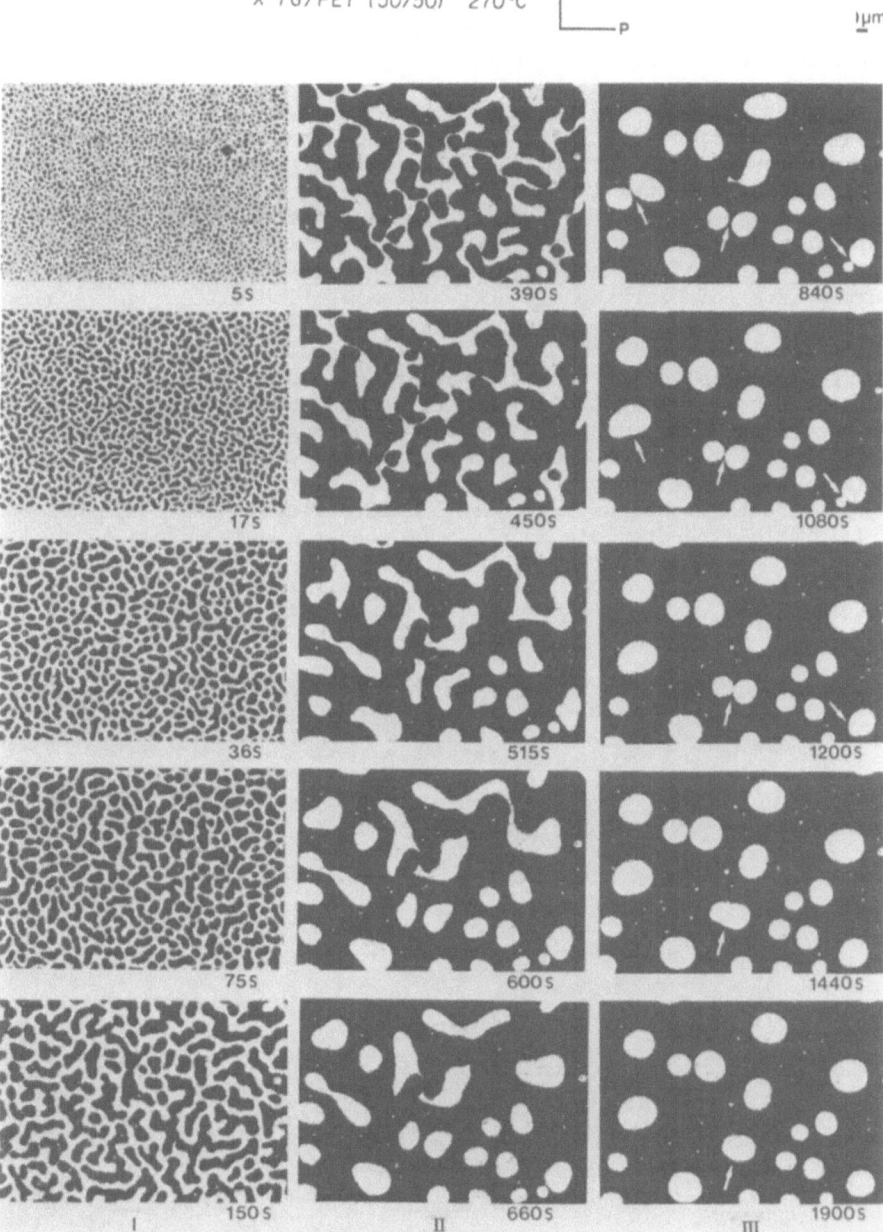

Figure 1 Polarizing optical microscope images obtained from the same area of a cast film of a binary mixture of X-7G and PET (50:50) during isothermal heat treatment at 270 °C. The phase separation process is divided into three characteristic regions.

III-2. Coarsening Mechanisms

Two mechanisms were observed for the coarsening process in region I.[8] One is local disruption of the liquid-crystalline network. A part of the liquid-crystalline network which happens to become thin due to the thermal motion of the polymer molecules at interface is torn by the interfacial tension. Each portion of the torn network is absorbed in the rest of the network to make it thicker. The other mechanism is the vaporization-condensation process of the isotropic melt. The PET molecules in small isotropic domains trapped in the liquid-crystalline network evaporate into and diffuse through the liquid-crystalline phase, and condense into the network of the PET isotropic melt.

The process of the disruption of the connectivity of the entire network and droplet formation in region II is shown in Figure 3a by the POM video images. A portion of the liquid-crystalline network (white area) in image no.1 transforms into four droplets in image no.6. However, the center of gravity of the mass in the four droplets seems to be unaffected by this transformation, i.e., the spacing between the centers essentially remains constant. Nevertheless, the characteristic wavelength of concentration fluctuation rapidly increases with time in region II because of the disappearance of intradomain concentration fluctuation as schematically illustrated in Figure 3b. It should be noted that the relative area occupied by the liquid-crystalline phase is almost constant (1/2) in region I, but rapidly decreases to 1/5 in region II.[8] This suggests that the structural growth in region I is two-dimensional, and the thickness of the film is unchanged, but the droplet formation in region II raises the liquid-crystalline phase from the film surface while the isotropic melt phase remains flat. This is confirmed by scanning electron microscopy.[8]

The growth mechanism in region III is diffusion and coalescence of the liquid-crystalline droplets as shown by the arrows in Figure 1.

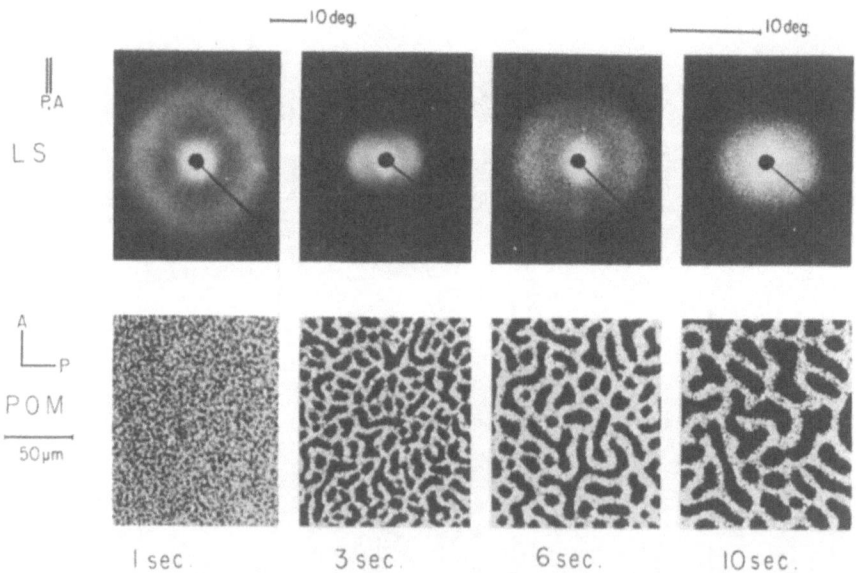

Figure 2 V_v light scattering patterns and polarizing optical micrographs obtained from the film of a binary mixture of X-7G and PET (50:50) annealed at 270 °C for 1, 3, 6 and 10 sec and quenched. Note that the camera length of the LS patterns for 6 and 10 sec is much longer than that for 1 and 3 sec. (10 deg bars are indicated.)

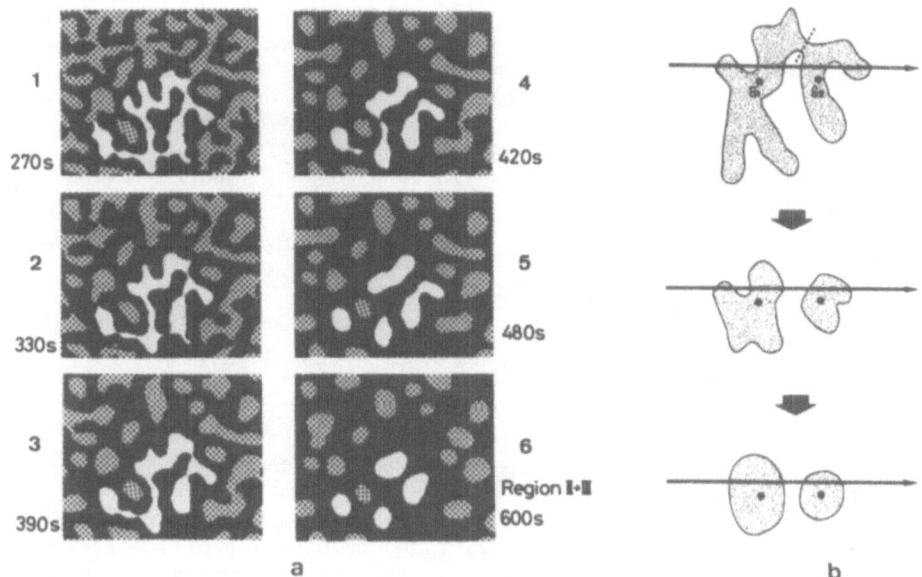

Figure 3 (a) Change in polarizing optical microscopic image of a binary mixture of X-7G and PET (50:50) in region II during isothermal heat treatment at 270°C. (b) An explanatory model of increase in the characteristic wavelength of concentration fluctuation in region II.

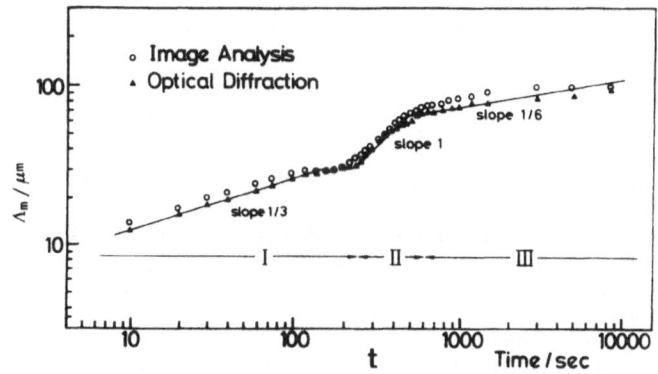

Figure 4 Time dependence of the characteristic wavelength of concentration fluctuation, Λm, obtained for isothermal annealing of a binary mixture of X-7G and PET (50:50) at 270°C by image analysis (designated by circles) and the optical diffraction (triangles).

III-3. Kinetics of Phase Separation

The characteristic wavelengths of the concentration fluctuations, Λm, obtained by image analysis and optical diffraction of the micrographs are plotted against annealing time, t, in Figure 4. Details of the analysis will be given elsewhere[8]. The value Λm evaluated by the optical diffraction tends to appear a little smaller than that obtained by the image analysis. Although the absolute value of Λm depends on the evaluation method, the relative change of Λm with t is independent of the method. The three linear portions of the curve with slope 1/3, 1 and 1/6 in Figure 4 exactly correspond to regions I, II and III, respectively, observed by the polarizing optical microscopy. Thus in each region, the following scaling law was obtained.

Region I: $\qquad \Lambda m \sim t^{1/3}$

Region II: $\qquad \Lambda m \sim t^{1}$

Region III: $\qquad \Lambda m \sim t^{1/6}$; $\quad \Lambda m \sim t^{0}$ (at later stage)

III-4. Comparison with the Theories

Among the theories dealing with the growth of the concentration fluctuation, those which give the exponent 1/3 are: (a) the theory by Kawasaki and Ohta based on Brownian coagulation in three dimensional space[13], (b) the theory by Kawasaki based on dynamics of interfaces in the conserved order-parameter system[14], and (c) vaporization-condensation theory by Lifshitz and Slyozov[15]. All of the mechanisms considered in these theories agree with our observations in region I. The coarsening in region II may be predicted by Siggia's model[16], but as shown in Figure 3a the morphological details in the ordering process of the real system contain information much richer than that treated by the model. Moreover, the partial instability of the network (i.e., the local disruption) which was dealt with in the Siggia's model occurs even in the region I, giving rise to the coarsening with the exponent 1/3. The theory by Binder and Stauffer[9] which gives the exponent 1/5 is based on diffusion and coalescence of droplets in solids in two-dimensional space, and is sufficient to explain our observation in region III. The pinning of the structural growth at a later stage of region III ($\Lambda m \sim t^{0}$) is due to the "polymer effect"[8].

REFERENCES

1. W.J. Jackson and H.F. Kuhfuss, J. Polym. Sci., Polym. Chem. Ed., 14, 2043 (1976).
2. M.R. Mackley, F. Pinaud and G. Siekmann, Polymer, 22, 437 (1982).
3. C. Viney and A.H. Windle, J. Mater. Sci., 17, 2661 (1982).
4. E.G. Joseph, G.L. Wilkes and D.G. Baird, Polymer, 26, 689 (1985); Polym. Eng. Sci., 25, 377 (1985).
5. C. Viney, A.M. Donald and A.H. Windle, Polymer, 26, 870 (1985).
6. T. Shiwaku, A. Nakai, H. Hasegawa and T. Hashimoto, Polym. Commun., 28, 174 (1987).
7. A. Nakai, T. Shiwaku, H. Hasegawa and T. Hashimoto, Macromolecules, 19, 3008 (1986).
8. A. Nakai, T. Shiwaku, H. Hasegawa and T. Hashimoto, to be submitted.
9. K. Binder and D. Stauffer, Phys. Rev. Lett., 33, 1006 (1974).
10. H. Furukawa, Prog. Theor. Phys., 59, 1072 (1978); Phys. Rev. Lett., 43, 136 (1979); Phys. Rev. A, 23, 1535 (1981); Physica A (Amsterdam), 123, 497 (1984).
11. T. Hashimoto, K. Sasaki and H. Kawai, Macromolecules, 17, 2812 (1984).
12. T. Hashimoto, M. Itakura and H. Hasegawa, J. Chem. Phys., 85, 6118 (1986).
13. K. Kawasaki and T. Ohta, Progr. Theor. Phys., 59, 362 (1978).
14. K. Kawasaki, Ann. Phys. (N.Y.) 154, 319 (1984).
15. I.M. Lifshitz and V.V. Slyozov, J. Phys. Chem. Solids, 19, 35 (1961).
16. E.D. Siggia, Phys. Rev., A 20, 595 (1979).

TRANSITION OF LINEAR POLYMER DIMENSION FROM THETA TO COLLAPSED REGIME

Benjamin Chu

Chemistry Department and Department of Materials Science
and Engineering, State University of New York
Stony Brook, Long Island, New York 11794-3400, U.S.A.

ABSTRACT

Coil-to-globule transitions of polystyrene (over a range of molecular weights with narrow molecular distributions) in cyclohexane and in methyl acetate showed that the collapsed state based on the static radius of gyration (R_g) could be achieved while those based on the hydrodynamic radius were yet to be realized by using the light-scattering technique. Master curves of $\alpha_s^3 |\tau| M^{1/2}$ versus $|\tau| M^{1/2}$ [with $\alpha_s = R_g(T)/R_g(\theta)$ and $\tau = (T-\theta)/\theta$] revealed solvent-dependent plateaus and the onset of phase-separation behavior in the metastable region of the coexistence curve at very dilute but finite concentrations. Our experimental results differ from most reported in the literature but compare favorably with the blob theory.

I. INTRODUCTION

Polystyrene in cyclohexane exhibits an upper critical solution temperature (UCST) while polystyrene in methyl acetate exhibits both an UCST and a lower critical solution temperature (LCST). The phase behavior of both polymer solution systems is shown schematically in Fig. 1 where no LCST is known to exist for the polystyrene/cyclohexane system under normal atmospheric pressures. The asymmetry of the coexistence (COEX) curve,[1] as schematically illustrated in Fig. 1, shows that $(\phi_{SD} - \phi_C) > (\phi_C - \phi_D)$ where ϕ_{SD}, ϕ_C and ϕ_D are the volume fraction of the polymer in the polymer-rich (semidilute; SD) phase, at the critical (C) solution concentration, and in the polymer-poor (dilute; D) phase, respectively. Thus, in studying the coil-to-globule transition of a linear polymer chain from the theta to the collapsed state, we are dealing with extremely dilute concentrations as the collapsed state (CS in Fig. 1) exists in a very narrow gap between infinite dilution and $\phi_D (\ll \phi_C)$ at temperatures far away from the theta (θ)

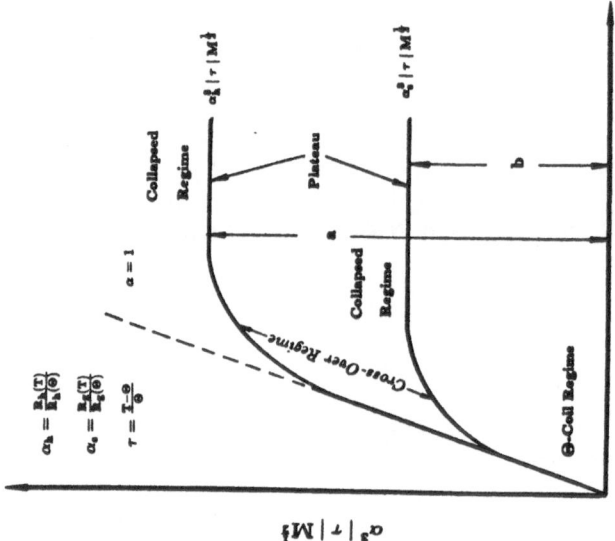

Fig. 2. Schematic representation of the master curves in plots of $\alpha^3 |\tau| M^{1/2}$ versus $|\tau| M^{1/2}$ with $\alpha_h = R_h(T)/R_h(\theta)$, $\alpha_s = R_g(T)/R_g(\theta)$, $\tau = (T-\theta)/\theta$. Based on the blob theory, $\alpha_s = 1.161 (N/N_c)^{-1/6}$ and $\alpha_h = 1.481 (N/N_c)^{-1/6}$. Thus, $a/b = \alpha_h^3/\alpha_s^3 = 2.08$.

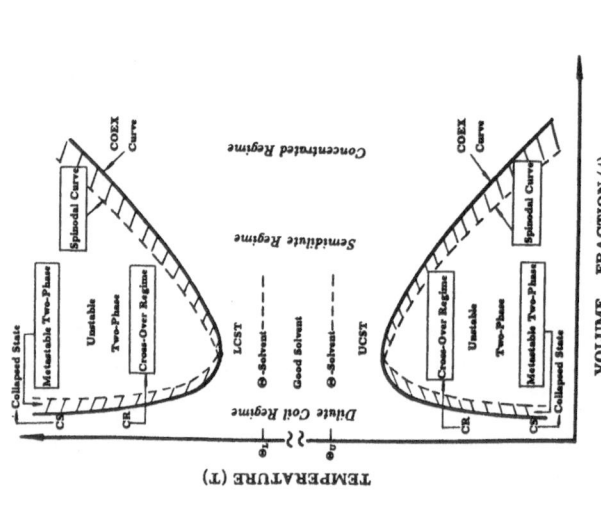

Fig. 1. Phase diagram (T, ϕ) of a generalized polymer solution exhibiting both an upper critical solution temperature (UCST) and a lower critical solution temperature (LCST), such as polystyrene in methyl acetate. For high molecular-weight polymer solutions, the coexistence (COEX) curves are extremely asymmetric. In this lecture, we are concerned mainly with light-scattering experiments on the coil-to-globule transition and effects of intra- and intermolecular attractive forces for polymer chains in a poor solvent from the theta temperature (θ_L or θ_U), through the cross-over (CR) regime, to the collapsed state (CS).

temperature or the critical solution temperature (T_C).

There have been many experiments on the transition from random-coil behavior in the theta state to a globule coil in the collapsed state.[2-17] Most of the results could not achieve a complete transition between the θ state and the asymptotic collapsed regime, especially if the collapsed state is based on hydrodynamic size. However, if we consider the collapsed state based on the radius of gyration R_g from static light scattering measurements, recent experiments on the polystyrene/cyclohexane (PS/CY) system below the θ_U temperature[16] and the polystyrene/methyl acetate (PS/MA) system both below the θ_U temperature and above the θ_L temperature[17] showed evidence for reaching the plateau region ($\alpha_s^3|\tau|M^{1/2}$ = constant) of the collapsed regime as presented schematically in Fig. 2.

II. THEORETICAL AND EXPERIMENTAL BACKGROUND

In the coil-to-globule transition, the single polymer coil size becomes smaller and the segmental density distribution of a single polymer chain changes from a Gaussian form to a more uniform globular form. In addition, as measurements have to be performed at finite concentrations, though so extremely dilute, the polymer solution may easily reach the vicinity of the coexistence curve whereby the interactions between polymer chains increase.[18]

In light scattering, physical quantities of importance are as follows.

1. Absolute scattered intensity extrapolated to zero scattering angle (θ). With concentration C approaching inifnite dilution, $HC/R_{vv}(0)$ = M_w^{-1} where H and $R_{vv}(0)$ are, respectively, an optical constant, and the excess Rayleigh ratio at scattering angle θ = 0 using vertically polarized incident and scattered light. A constant $R_{vv}(0)$ confirms that the polymer chains have neither degraded nor aggregated. It is a good criterion which we use to monitor the experiment during the course of our measurements.

2. Angular distribution of scattered intensity I(K) when extrapolated to infinite dilution yields information on the z-average radius of gyration R_g, where K is the scattering vector. The important point here is that we need to determine the initial slope which is related to R_g in a reciprocal excess scattered intensity versus K^2 plot. Unfortunately, in coil-to-globule transition studies, it is essential to

use very high molecular weight polymer chains. Thus, even in a theta solvent, R_g is fairly large resulting in a breakdown of the usual experimental condition requiring $KR_g < 1$. For example, with $\lambda_o = 488$ nm, n = 1.45 and $R_g \sim 150$ nm, we have $\theta < 20°$ for $KR_g < 1$, a fairly difficult experimental condition for most light-scattering intensity measurements using standard commercial light-scattering spectrometers. In the neighborhood of $KR_g \sim 1$, we must be quite careful in evaluating R_g.[16]

  3. Linewidth measurements to determine the translational diffusion coefficient. The hydrodynamic radius R_h is computed from the Stokes-Einstein relation ($D^o = k_B T/6\pi\eta_o R_h$) using the translational diffusion coefficient (D) which has been extrapolated to infinite dilution. For large polymer molecules, we must be concerned with internal motions in linewidth measurements at $KR_g > 1$. Here, we also use information on the characteristic linewidth distribution $G(\Gamma)$, often in terms of its variance $\mu_2/\bar{\Gamma}^2$ at $KR_g < 1$, to evaluate the polydispersity of the polymer; $\mu_2 = \int G(\Gamma)(\Gamma-\bar{\Gamma})^2 d\Gamma$ and $\bar{\Gamma} = \int \Gamma G(\Gamma) d\Gamma = \bar{D}K^2$.

  4. The presence of internal motions at higher $KR_g (>1)$ values signifies a deviation from the K^2 dependence of $\bar{\Gamma}$ yielding $\bar{\Gamma}/K^2 \approx \bar{D}^o (1 + fR_g^2 K^2)(1 + k_d C)$ with f being a dimensionless constant depending on chain structure, polydispersity and solvent quality[19] and k_d, a second virial coefficient for diffusion. The superscript zero denotes infinite dilution. The f-value can be used to monitor chain structure changes.

  Many discussions on the theoretical interpretation of coil-to-globule transition exist.[20-32] Table I lists the molecular weight and the temperature dependence of polymer dimension (R_g or R_h) in the theta as well as the asymptotic good and collapsed states. Our main interest is to provide experimental data for the schematic curves shown in Fig. 2; in particular, what are the magnitudes of the plateau regions for α_S and α_h, if they exist. How broad are the cross-over regimes? Furthermore, as our experimental results are somewhat different from most data reported in the literature, could we reconcile the differences? The molecular weights, polydispersities, and polymer dimensions at θ-temperature for the polystyrene solutions under discussion are listed in Table II.

III. RESULTS AND DISCUSSIONS

  Experiments on coil-to-globule transitions require a great deal of care. With the use of high molecular weight ($>10^6$ g/mole) polymers, it is important to be certain that the samples have narrow molecular weight distributions (MWD; $M_w/M_n < 1.05$). Yet in view of the high molecular weights especially when $M > 10^7$ g/mole, determinations of MWD are not trivial by using standard analytical techniques such as size exclusion chromatography. Thus, we have used measurements of the variance $(\mu_2/\bar{\Gamma}^2)$

Table I. Molecular Weight (M) and Temperature (T) Dependence
of Polymer Dimension

Temp.	Size	Expansion Factor	
$T > \theta$	$R \propto M^{3/5} \tau^{1/5}$	$\alpha \propto M^{1/10} \tau^{1/5}$	good solvent
$T = \theta$	$R \propto M^{1/2}$	$\alpha = 1$	theta condition
$T < \theta$	$R \propto M^{1/3} \tau^{-1/3}$	$\alpha \propto M^{-1/6} \tau^{-1/3}$	collapsed state

where $\tau = |T-\theta|/\theta$, $\alpha = R(T)/R(\theta)$

Table II. Molecular Weights, Polydispersities and Polymer
Dimensions at θ Temperature

Sample Designation	2N	4N	8N	17B	F-3	LF[a]
$M_w \times 10^{-6}$ (g/mole)	1.98	4.57	8.62	17.5	10.5	40.9
M_w/M_n	1.10	1.14	1.26	2.1	1.4	1.7
PS/CY[b]; $\theta = 35°C$						
R_g (nm)	44.3	67.0	95.3	167	112	230
R_h (nm)	33.0	48.7	69.5	110	79.9	164
PS/MA[b]; $\theta_U = 43°C$						
R_g (nm)	41.8	64.7				
R_h (nm)	29.7	45.5				
PS/MA (1 wt%)[c]; $\theta_U = 43°C$						
R_g (nm)	65.1	92.8				
R_h (nm)	48.3	66.2				
PS/MA (1 wt%)[c]; $\theta_L = 114°C$						
R_g (nm)	59.9	86.7				

(a) θ-temperature of PS/CY with 1 wt% antioxidant is lowered to 32.7°C.

(b) MA = methylacetate; CY = cyclohexane; PS = polystyrene.

(c) 1 wt% denotes 1 wt% of antioxidant: 2,6-di-tert-butyl-4-methylphenol.

from dynamic light scattering, the Fujita plot[33] from static light scattering, and behavior of R_g^2/M_w at θ-temperature[16] to estimate the polydispersity of polymer samples before we begin our coil-to-globule transition measurements by means of laser light scattering.

High molecular weight polymer samples are subject to shear degradation. Clarification of polymer solutions in preparation for light-scattering measurements becomes another possible source of error. A predetermined procedure has to be established, mostly by means of high-speed centrifugation instead of filtration, in order to insure that dust removal has not altered the polymer characteristics. Furthermore, we must be concerned with degradation produced by maintaining the polymer solution at high temperatures (above the θ_U temperature) during polymer dissolution and by laser irradiation during light-scattering measurements. Thus, the solvents were purified and degassed. The polymer solution was prepared under innert atmosphere and the final solution often flame-sealed in the cylindrical light-scattering cell.

All measurements were performed satisfying the requirements for initial slope determination in R_g and pure translational motion in characteristic linewidth (Γ). Although the polymer concentrations were often very low ($\lesssim 10^{-6}$ g/mL), we also considered the concentration dependence of R_g and R_h. It is difficult to be absolutely sure that intermolecular interactions[18] did not play an appreciable role in our final analysis of the expansion factors $\alpha_s [= R_g(T)/R_g(\theta)]$ and $\alpha_h [= R_h(T)/R_h(\theta)]$ because we can barely reach the globule state with the most dilute concentration we are able to manage. Finally, we monitored the absolute scattered intensity extrapolated to zero scattering angle [$\lim_{K \to 0} I(K)$] in order to insure a lack of phase separation. Values of R_g in the plateau region (see Fig. 2) could be reproduced by increasing or decreasing temperatures and remained stable for long periods of time.

For observation of phase separation in the metastable region (see Fig. 1), there will be slow intensity changes, with the rate of change dependent partially on polymer concentration. The details will be discussed in III.4.

III.1. Changes of static size with temperature

The variation of scaled expansion factor $\alpha_s^3 |\tau| M_w^{1/2}$ of static size as a function of scaled reduced temperature $|\tau| M_w^{1/2}$ at $T \leq \theta_U$, as shown in Fig. 3, reveals that (1) plateau regions (denoted by dashed lines) confirming the collapsed state of static size exist; (2) the asymptotic heights depend on the chemical nature of the solvent with values of 70.1, 50.3 and 20.1 $g^{1/2}$ mole$^{-1/2}$ for PS/MA (1 wt %), PS/MA and PS/CY, respectively; and (3) the cross-over (but possibly not θ-) regimes appear to increase with increasing asymptotic height. A universal curve covering the θ-, cross-over, and collapsed regimes of a polymer solution in a poor

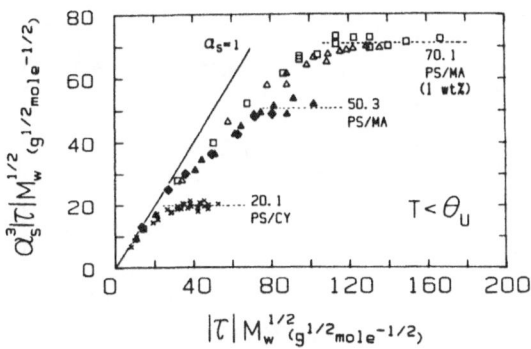

Fig. 3. Variation of scaled expansion factor $\alpha_s^3 |\tau| M_w^{1/2}$ of static size as a function of scaled reduced temperature $|\tau| M_w^{1/2}$ at $T < \theta_U$. Open and filled symbols indicate PS/MA (1 wt%) and PS/MA, respectively. (Diamonds) sample 2N ($M_w = 2.0 \times 10^6$ g/mole); (triangles) sample 4N ($M_w = 4.6 \times 10^6$ g/mole); (squares) sample 8N ($M_w = 8.6 \times 10^6$ g/mole); (X) PS/CY (Figure 10 of reference 16). PS, MA and CY denote polystyrene, methylacetate and cyclohexane, respectively. The dashed lines denote the asymptotic collapsed regimes and the numbers are the asymptotic heights in the collapsed regime in units of $g^{1/2}$ mole$^{-1/2}$. It should be noted that the scaled curves were obtained using PS with different molecular weights and at different concentrations. (Figure 2 of reference 17).

solvent can represent all polymer dimensions at different molecular weight and temperature below the θ_U-temperature. Figure 4 shows the collapsed regimes of the static size both above the θ_L temperature and below the θ_U temperature for the PS/MA (1 wt%) system. For $T > \theta_L$, the plateau value is 38.7 $g^{1/2}$ mole$^{-1/2}$, only slightly more than half of the plateau value (~ 70 $g^{1/2}$ mole$^{-1/2}$) for $T < \theta_U$. For example, at $|T-\theta| \sim 10°$, $\alpha_s = 0.883$ and 0.918 while $|\tau| M_w^{1/2} = 55.4$ and 67.9 at 10° above θ_L and 10° below θ_U, respectively. Thus, by a 10° change from the θ-temperature, R_g in the $\theta_L (= 114°C)$ region is reduced to 88% of the R_g ($= 60$ nm) value at θ_L while R_g in the $\theta_U (= 43°C)$ region is reduced to 92% of the R_g ($= 65$ nm) value at θ_U. $R_g(\theta_L)$ seems to be larger than $R_g(\theta_U)$ and the contraction appears to be faster in the θ_L region although the differences are almost within experimental error limits. On the other hand, $\alpha_s^3 |\tau| M^{1/2} = 38.1$ and 52.5 $g^{1/2}$ mole$^{-1/2}$ at 10° above θ_L and 10° below θ_U, respectively; yet 38.1 $g^{1/2}$ mole$^{-1/2}$ corresponds to the plateau region while a value of 52.5 $g^{1/2}$ mole$^{-1/2}$ for $T < \theta_U$ is still in the cross-over regime.

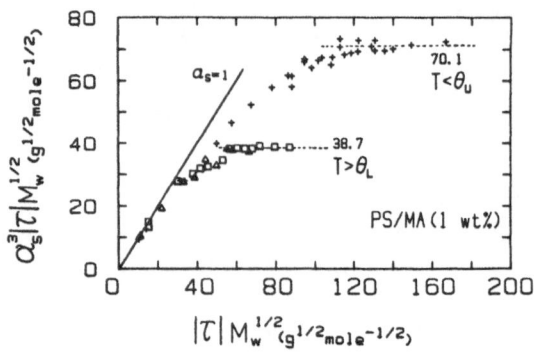

Fig. 4. Variation of scaled expansion factor $\alpha_s^3|\tau|M_w^{1/2}$ of static size as a function of scaled reduced temperature $|\tau|M_w^{1/2}$ in PS/MA with 1 wt% anti-oxidant. (Hollow triangles) sample 4N ($M_w = 4.6 \times 10^6$ g/mole); (hollow squares) sample 8N ($M_w = 8.6 \times 10^6$ g/mole). The static size data obtained at T $< \theta_U$ for the same system, as shown in Fig. 3, are denoted by (+) in this figure for comparison purposes. The dashed lines denote the asymptotic collapsed regime and the numbers are the asymptotic heights in the collapsed regime in units of $g^{1/2}$ mole$^{-1/2}$. (Figure 8 of reference 17).

III.2. Change of hydrodynamic size with temperature

The variation of scaled expansion factor $\alpha_h^3|\tau|M^{1/2}$ of hydrodynamic size as a function of scaled reduced temperature $|\tau|M_w^{1/2}$ at T $\leq \theta_U$, as shown in Fig. 5, suggests that the asymptotic collapsed regime based on hydrodynamic size is much more difficult to realize than that based on R_g. As $(\alpha_h^3|\tau|M^{1/2})_{plateau} > (\alpha_s^3|\tau|M^{1/2})_{plateau}$, it becomes necessary to perform measurements at higher values of $|\tau|M^{1/2}$. If we take a/b = 2.08 for the blob theory as a guideline, $(\alpha_h^3|\tau|M^{1/2})_{plateau}$ for the PS/MA system is expected to be 105 $g^{1/2}$ mole$^{-1/2}$ while we have reached a value of 72 $g^{1/2}$ mole$^{-1/2}$. Thus, Fig. 5 shows very broad cross-over regimes, suggesting that we still have some distances to go before reaching the

asymptotic collapsed state based on hydrodynamic size. In log-log plots of expansion factor α versus scaled reduced temperature $|\tau|M_w^{1/2}$ in PS/MA at T $< \theta_U$, we note that the asymptotic collapsed regime based on R_g (denoted by hollow inverted triangles) shows a limiting slope of 1/3 at $|\tau|M_w^{1/2} > 70$ $g^{1/2}$ mole$^{-1/2}$ while values of α_h (denoted by hollow squares) seem to approach the 1/3 limiting slope. At this time, although we have not yet reached the asymptotic collapsed regime based on hydrodynamic size, it appears that we are not far from achieving our final goal based on a more cautious approach. Similar tendency has been observed using sedimentation coefficient measurements where a claim on reaching the asymptotic region of the collapsed state has been reported.[14] Further discussions will be included in III.5. It is suffice to confirm that for polymer solutions with M_w up to 10^7 the cross-over regimes are smooth and continuous.

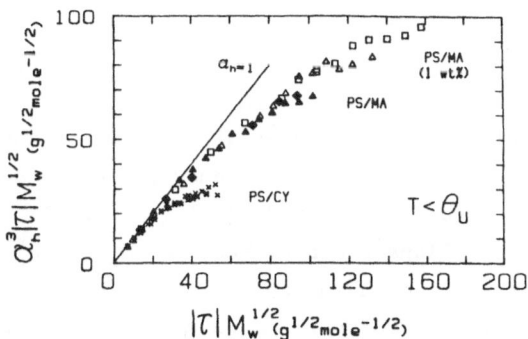

Fig. 5. Variation of scaled expansion factor $\alpha_h^3 |\tau| M_w^{1/2}$ of hydrodynamic size as a function of scaled reduced temperature $|\tau| M_w^{1/2}$ at $T < \Theta_U$. Open and filled symbols indicate PS/MA (1 wt%) and PS/MA, respectively. (Diamonds) sample 2N ($M_w = 2.0 \times 10^6$ g/mole); (triangles) sample 4N ($M_w = 4.6 \times 10^6$ g/mole); (squares) sample 8N ($M_w = 8.6 \times 10^6$ g/mole). (X) PS/CY (Figure 17 of reference 16). The collapsed regime based upon hydrodynamic size has not yet been reached in this experiment. It should be noted that the scaled curves were obtained using PS with different molecular weights and at different concentrations. With imagination, one may see a possible leveling in the scaled curves suggesting that we have almost reached the collapsed regime based upon the hydrodynamic size (Figure 3 of reference 17).

III.3. Comparison with literature results

Although many publications using a variety of techniques, such as small angle neutron scattering,[3] viscosity,[12,13] laser light scattering[2,4-11,15-17] and analytical ultracentrifugation[14] have reported successes in reaching the aymptotic collapsed state,[6,7,11-14] we shall illustrate the differences mainly on light scattering results and refer to possible explanations in III.4 and III.5.

Figure 7 shows a comparison of static size data for polystyrene in cyclohexane with literature values. Aside from the fact that most data did not reach a plateau region, three notable exceptions are data from the Akron group,[2] the MIT group[6,7] and the Prague group.[11] Their data exhibited maxima while ours[16,17] were barely able to reach the plateau. As the Prague group used a different solvent (dioctyl phthalate), we would expect its plateau value to be different from that of the PS/CY system (20.1 $g^{1/2}$ mole$^{-1/2}$ as shown in Fig. 3). The major difference is in the qualitative feature, i.e., the Prague group shows a maximum while we have difficulty achieving a plateau, not to mention the absolute magnitude of the plateau value. Figure 8 shows a comparison of hydrodynamic size data with literature. Our hydrodynamic data are in essential agreement with the viscosity results of Perzynski et al.[13] However, at higher values of $|\tau|M_w^{1/2}$ our results seem to show a tendency of not yet quite reaching the plateau value as also illustrated in Fig. 6, while the viscosity data could have reached a collapsed state because $\alpha_\eta^3 = \alpha_s^2\alpha_h$.[34] Without a combination of static and dyanmic data, it is difficult to evaluate the polystyrene/cyclopentane results.[14] A word of caution in considering the concentrations used for sedimentation coefficient measurements, i.e., they are of the order of 10^{-5} g/mL. We shall discuss this point in III.5. We cannot offer a definitive explanation on the marked difference between our results and those shown in Figs. 7 and 8. Possible reasons could be rationalized according to III.4 and III.5.

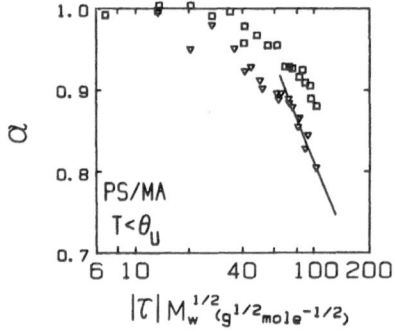

Fig. 6.   Log-log plots of expansion factor α versus scaled reduced temperature $|\tau|M_w^{1/2}$ in PS/MA at $T < \theta_U$. (∇) expansion factor of static size $\alpha = \alpha_s$; ($\square$) expansion factor of hydrodynamic size $\alpha = \alpha_h$. Solid line denotes an asymptotic slope of 0.29 ± 0.06 in the collapsed regime based upon the static R_g values. $\theta_U = 43°C$. Note the rate of change of α_h suggesting that we have almost reached the collapsed regime based on hydrodynamic size. (Figure 4 of reference 17).

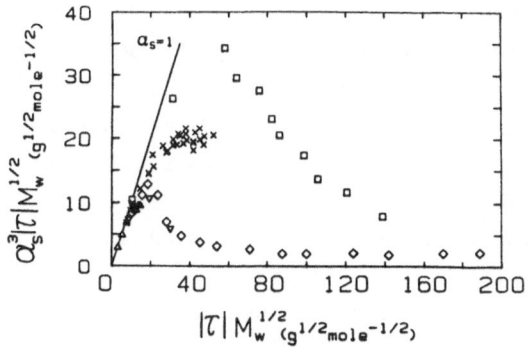

Fig. 7. Comparison of static size data (in Fig. 10 of reference 16 for PS/CY) with literature. (Inverted triangles) Akron group, ref. 2; (triangles) French I, ref. 3; (diamonds) MIT group, refs. 6 and 7; (squares) Prague group, ref. 11; (crosses) Stony Brook group, ref. 16. (Figure 16 of reference 16).

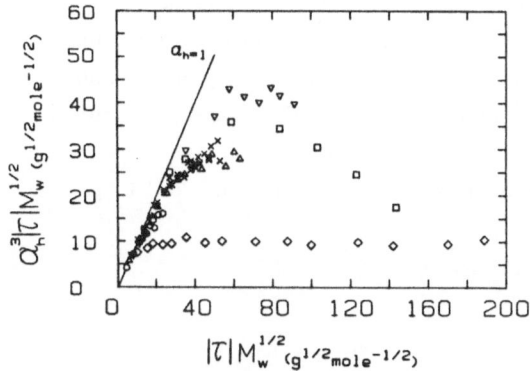

Fig. 8. Comparison of hydrodynamic size (Figure 17 of reference 16 for PS/CY) with literature. (Inverted triangles) French II, ref. 14; (triangles) French III, ref. 13; (diamonds) MIT group, ref. 7; (squares) Prague group, ref. 11; (hexagons) Ford group, ref. 4; (crosses) Stony Brook group, ref. 16. (Figure 18 of reference 16).

III.4. Effects of phase separation at finite but very dilute concentrations

Figure 1 illustrates the very narrow gap in concentration between infinite dilution and the polymer-poor side of the coexistence curve. At

$T \ll \theta_U$ (or $T \gg \theta_L$) phase separation could occur even at very small polymer concentrations. In examining a plot of the scaled expansion factor $\alpha_s^3 |\tau| M_w^{1/2}$ versus the scaled reduced temperature $|\tau| M_w^{1/2}$ for polystyrene in cyclohexane, as shown in Fig. 9, we have noted that at high molecular weights (17.5×10^6 and 24.7×10^6 g/mole) with fairly broad molecular weight distributions ($M_w/M_n \sim 2$), we could also create maxima as denoted by the filled symbols. The magnitude of the observed maximum depends on polymer concentration and temperature. The values, although reproducible on a short-term basis, change with time. Changes of structure on a long-term time sequence could show a relatively rapid (over a 10-hour period) shrink-down of size due possibly to strong intramolecular attractive forces (e.g., at 31°C, $C = 2.1 \times 10^{-6}$ g/g and $M_w = 24.7 \times 10^6$ g/mole, denoted by point B in Fig. 9); or for another high molecular weight polymer sample (at 30.5°C, $C = 1.8 \times 10^{-6}$ g/g, $M_w = 17.5 \times 10^6$ g/mole, denoted by A in Fig. 9) a longer time ($\sim$5 days) polymer aggregation due possibly to strong intermolecular attractive forces and finally resulting in precipitation over a very long time period (>10 days). Thus, we see that phase separation could occur even at very low concentrations. When it occurs due to a temperature drop (as for $T \ll \theta_U$), the kinetic behavior depends on polymer concentration. It took us continuous measurements over a period of $\sim$one month to realize the non-equilibrium behavior of very dilute polymer solutions in the metastable region as shown schematically in Fig. 1. At even lower concentrations as those reported by the MIT group or with high-viscosity solvent as used by the Prague group, non-equilibrium phase separation or aggregation behavior could take an even longer time to be realized.

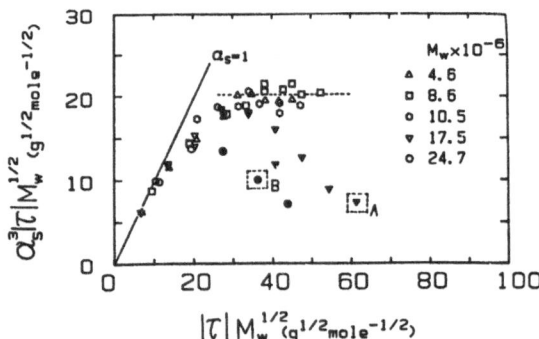

Fig. 9. A plot of scaled expansion factor $\alpha_s^3 |\tau| M_w^{1/2}$ of static size versus scaled reduced temperature $|\tau| M_w^{1/2}$ for polystyrene in cyclohexane. Filled symbols denote "metastable" collapsed regime. (Figure 10 in reference 16).

III.5. Coexistence curve of polystyrene in cyclohexane

In order to show that we have successfully achieved measurements in the collapsed state, we need to know the thermodynamic state of the system, i.e., the coexistence curve of the polymer solution under investigation. For polystyrene in cyclohexane, the coexistence curve of polystyrene with $M_w = 1.56 \times 10^6$ g/mole and $M_w/M_n < 1.03$ is known over a temperature range of $4 \times 10^{-5} < \epsilon < 4 \times 10^{-2}$.[35] At other concentrations, we can estimate the phase-separation concentration in the dilute solution regime of the coexistence curve[18] according to

$$C_D = 0.15(6.8 \ M^{-0.38}) \ \exp(-3.5\epsilon \ M^{0.31}) \qquad (1)$$

where $\epsilon = (T_c - T)/T_c$ with T_c being the critical solution temperature for polystyrene in cyclohexane. Although Eq. (1) could be considered as being quite approximate, it is interesting to examine the order of magnitude of C_D in order to reach $|\tau| M_w^{1/2} \sim 40$ for polystyrene with a molecular weight of 8.5×10^6 g/mole in cyclohexane: $C_D = 2.37 \times 10^{-3} \exp(-492\epsilon)$. With $C_D = 2.2 \times 10^{-5}$ g/mL, the phase separation temperature T_p is of the order of 29.5°C, while for $|\tau| M^{1/2} \sim 40$, $T \sim 30.8$°C. Thus, we can reach the collapsed state based on static size but not on hydrodynamic size where T has to be ~26.5°C $(< T_p)$ for $|\tau| M^{-1/2} \sim 80$. If we take the maximum of the curve for polystyrene with $M_w = 17.5 \times 10^6$ g/mole to be at $|\tau| M_w^{1/2} \sim 34$ $g^{1/2}$ mole$^{-1/2}$ in Fig. 9, $T_p \sim 32.5$°C at 1.4×10^{-6} g/mL.[16] However, according to Eq. (1), $T_p = 29.4$°C if $C_D = 1.4 \times 10^{-6}$ g/mL. The increase in the actually measured T_p (~32.5°C from 29.4°C) suggests a broadening of the coexistence curve due to polydispersity effects. Although the above discussions deal with estimates, we can see that we have to be very careful with the phase-separation phenomena in the coil-to-globule transition studies. The MIT experiments were correctly performed except for the polymer sample used, i.e., the polydispersity data from the manufacturer could likely be in error.

III.6. Change of structure with temperature

Ratio of R_h/R_g is a measure of polymer chain structure. The R_h/R_g ratio is 0.664 at θ-temperature and 0.864 and 1.29 at the collapsed state based on the blob theory and the hard sphere limit, respectively. Thus, the magnitude of R_h/R_g at the collapsed state normalized to R_h/R_g at θ-temperature is 1.27 and 1.94 based on the blob theory and the hard sphere limit, respectively. In practice, $R_h/R_g \sim 0.7$ at θ-temperature for PS/CY with polymer molecular weights ~10^6-10^7 g/mole and $(R_h/R_g)/(R_h/R_g)_\theta \sim 1.15$ at $|\tau| M_w^{1/2} \sim 45$. The relative change is only of the order of ~15%. If we take the blob theory as our basis for the collapsed state, we must try to reach a $|\tau| M^{1/2}$ value of ~80. From III.5, it is clear that we must try to perform the experiments using polystyrene samples of highest possible molecular weights ($>10^7$ g/mole) with very narrow molecular weight distributions so that we can reach an effectively large $|\tau| M^{1/2}$ value with a concentration still sufficiently high ($\sim 10^{-7}$ g/mL) for light scattering measurements but low enough for phase separation not yet to occur. All present-day commercial polystyrene samples with molecular weights $\sim 10^7$ g/mole that we have tried cannot satisfy the above requirments because of polydispersity problems. Samples from private sources (courtesy of Fetters and Fujita) also indicated a MWD too broad for our studies. Thus, further fractionations become necessary. However, from the estimates, we see that the collapsed state based on hydrodynamic size is within reach experimentally by means of photon correlation spectroscopy.

III.7. Universal curves in $\alpha(N/N_c)^{1/6}$ versus N/N_c

Figures 10 and 11 show plots of $\alpha(N/N_c)^{1/6}$ as a function of the reduced blob parameter N/N_c for all known systems.[16] In Figure 10, all data points are well superposed, not only in the collapsed regime

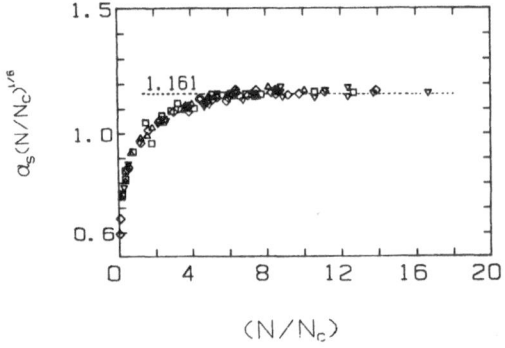

Fig. 10. A universal plot of static size versus reduced blob parameter (N/N_c). All data obtained from different systems could be superposed on one curve. (Squares) PS/MA (1 wt%) at $T > \Theta_L$, (diamonds) PS/MA (1 wt%) at $T < \Theta_U$, (triangles) PS/MA at $T < \Theta_U$, (inverted triangles) PS/CY at $T < \Theta_U$. Data points reached the asymptotic collapsed value (1.161) predicted by $\alpha_s = 1.16\,(N/N_c)^{1/6}$ above $(N/N_c) > 5$. (N/N_c) is the number of temperature blobs in a single polymer chain. $(N/N_c) = \tau^2 M/M_o (A^* \cdot N_1)$, with M_o and $A^* N_1$ being , respectively, the molecular weight of one monomer segment and a prefactor which we determined experimentally.[16] It should be noted that we have used the $(A^* N_1)$ value for each system based $h^2/(2.45\,M_o)$ with h being the asymptotic plateau value in Fig. 3. (Figure 10 in reference 17).

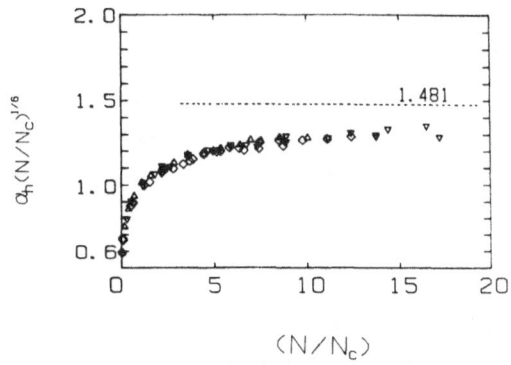

Fig. 11. A universal plot of hydrodynamic size versus reduced blob parameter (N/N_c). Although all data obtained from different systems (including the use of the (A^*N_1) value for each system as reported in Fig. 10, based on static size) could be superposed on one curve, they did not reach the asymptotic collapsed value (1.481) predicted by $\alpha_h = 1.481 \ (N/N_c)^{1/6}$. Symbols are the same as in Figure 10. (Figure 11 of reference 17.)

$(N/N_c>5)$, but also in the cross-over regime $(0 < N/N_c < 5)$, indicating that a reduced blob parameter can be considered as a universal system-independent variable with regard to the contraction of polymer sizes even if it needs an empirically adjustable parameter between theory and experiments in the asymptotic collapsed regime.

IV. CONCLUSIONS

The collapsed regime based on the static radius of gyration has been reached above the θ_L temperature for the PS/MA (1 wt%) system and below the θ_U temperature for the PS/CY, PS/MA and PS/MA (1 wt%) systems. The magnitude of the plateau region is solvent dependent and can change substantially by addition of even only 1 wt% antioxidant.

The collapsed regime based on hydrodynamic size remains elusive. However, from estimates of phase separation temperature and limits of detectability of present-day instrumentation, the experiment is feasible but requires a very high molecular weight polymer sample with a fairly narrow molecular weight distribution and a small-angle light scattering spectrometer.

The coil-to-globule transition has a broad and continuous cross-over regime which follows the universal curve based on a plot of $\alpha(N/N_c)^{1/6}$ versus (N/N_c). Differences in experimental results could possibly be attributed to reaching the matastable two-phase region of the coexistence curve. In view of the very steep cloud-point curve and extremely dilute polymer concentrations, phase separation in the metastable region reacts slowly with time, resulting in apparent size contractions which seem stable, but could eventually change after periods of days or weeks.

ACKNOWLEDGMENT

We gratefully acknowledge support of this work by the Department of Energy (DEFG0286ER45237A001) and the National Science Foundation (Polymers Program, DMR 8617820).

REFERENCES

1. I. C. Sanchez, _J. Appl. Phys._ 58:2871 (1985).
2. E. Slagowski, B. Tsai and D. McIntyre, _Macromolecules_ 9:687 (1976).
3. M. Nierlich, J. P. Cotton and B. Farnoux, _J. Chem. Phys._ 69:1379 (1978).
4. D. R. Bauer and R. Ullman, _Macromolecules_ 13:392 (1980).
5. M. J. Pritchard and D. Caroline, _Macromolecules_ 13:957 (1980).
6. G. Swislow, S. T. Sun, I. Nishio and T. Tanaka, _Phys. Rev. Lett._ 44:796 (1980).
7. S. T. Sun, I. Nishio, G. Swislow and T. Tanaka, _J.Chem. Phys._ 73:5971 (1980).
8. Y. Miyaki and H. Fujita, _Polymer J._ 13:749 (1981).
9. T. Oyama, K. Shiokawa and K. Baba, _Polymer J._ 13:167 (1981).
10. C. B. Post and B. H. Zimm, _Biopolymers_ 21:2139 (1982).
11. P. Stepanek, C. Konak and B. Sedlacek, _Macromolecules_ 15:1214 (1982).
12. R. Perzynski, M. Adam and M. Delsanti, _J. Physique_ 43:129 (1982).
13. R. Perzynski, M. Delsanit and M. Adam, _J. Physique_ 45:1765 (1984).
14. P. Vidakovic and F. Rondelez, _Macromolecules_ 17:418 (1984).
15. J. C. Selser, _Macromolecules_ 18:585 (1985).
16. I. H. Park, Q.-W. Wang and B. Chu, _Macromolecules_, in press.
17. B. Chu, I. H. Park, Q.-W. Wang and C. Wu, _Macromolecules_, in press.
18. P. Perzynski, M. Delsanti and M. Adam, _J. Physique_ 48:115 (1987).
19. W. H. Stockmayer and M. Schmidt, _Pure Appl. Chem._ 54:407 (1982).
20. O. B. Ptitsyn, A. K. Kron and Y. Y. Eizner, _J. Polym. Sci., Part C_ 16:3509 (1968).
21. Y. Y. Eizner, _Polymer Sci. USSR_ 14:1695 (1972).
22. C. Domb, _Polymer_ 15:259 (1974).
23. A. R. Massih and M. A. Moore, _J. Phys. A: Math Gen._ 8:237 (1975).
24. P. G. deGennes, _J. Physique Lett._ 36:L55 (1975); _ibid._ 39:L299 (1978).

25. M. A. Moore, _J. Phys. A: Math Gen._ 10:305 (1977).

26. C. B. Post and B. H. Zimm, _Biopolymers_ 18:1487 (1979); _ibid._ 21:2123.

27. I. C. Sanchez, _Macromolecules_ 12:980 (1979).

28. C. Williams, F. Brochard and H. L. Frisch, _Ann Rev. Phys. Chem._ 32:433 (1981).

29. G. Allegra and F. Ganazzoli, _Macromolecules_ 16:1311 (1983).

30. E. A. DiMarzio, _Macromolecules_ 17:969 (1984).

31. A. L. Kholodenko and K. F. Freed, _J. Chem. Phys._ 80:900 (1984).

32. G. Allegra and F. Ganazzoli, _J. Chem. Phys._ 83:397 (1985).

33. Y. Miyaki, Y. Einaga and H. Fujita, _Macromolecules_ 11:1180 (1978).

34. G. Weill and J. des Cloizeaux, _J. Physique_ 40:99 (1979).

35. M. Nakata, T. Dobashi, N. Kuwahara, M. Kaneko and B. Chu, _Phys. Rev._ 18A:2683 (1978).

DYNAMICS OF PHASE TRANSITION IN POLYMER GELS:

STUDIES OF SPINODAL DECOMPOSITION AND PATTERN FORMATION

Shunsuke Hirotsu and Akio Kaneki

Department of Physics, College of Science
Tokyo Institute of Technology, Ohokayama, Meguroku
Tokyo 152, Japan

INTRODUCTION

Gel is composed of network and liquid. It has been recognized[1] that a polymer gel is a kind of polymer solution, in which solute is a single molecule with an infinite molecular weight. Due to the cross-linked network structure of the solute, this "solution" exhibits several peculiar properties which are not common in ordinary solutions. For example, the concentration of gel depends on temperature. On changing temperature gel expels or absorbs solvent so as to attain the thermodynamic equilibrium. The macroscopic state of gel is characterized by the swelling ratio α defined as $\alpha=(\phi_0/\phi)^{1/3}$, where ϕ is the volume fraction of polymer and ϕ_0 is that when the network has a random walk configuration. Some gels undergo a discontinuous transition between two phases with different values of α, which we call a gel-gel transition or a volume transition of gel.[2-4]

It has been known[5-10] that N-isopropylacrylamide (NIPA) gel with water as a solvent undergoes a slightly discontinuous transition from the low-temperature swollen phase to the high-temperature shrunk phase around 33.6°C. At present NIPA/water gel is the only known example in which a discontinuous collapse occurs in a neutral (non-ionic) network. Although the phase transition of NIPA/water gel is discontinuous it is close to being continuous,[5,6] and some aspects of the critical phenomena of this transition have been investigated.[9,10] However, the spinodal decomposition of this system has not yet been studied. One interesting phenomenon related to the kinetics of swelling of gels is that a periodic pattern appears on the surface of gel and it evolves during swelling.[11] However, the correlation between the evolution of phase transition and the formation of patterns has not yet been understood.

In the present paper we first summarize briefly the phenomenological theory of gel-gel transition together with some recent experimental results on NIPA/water gel. Then we report some preliminary results of light scattering experiment and pattern observation which have been made to understand the kinetics of spinodal decomposition in NIPA/water gel.

PHENOMENOLOGICAL THEORY OF GEL-GEL TRANSITION

The free energy of gel can be written as[1,2,4]

$$F = F_0 + kTN_s\{\ln(1-\phi) + \chi\phi\} + (3kT/2)N_c(\alpha^2-1-\ln\alpha), \tag{1}$$

where F_0 is the sum of free energies of solvent and solid amorphous polymer, k the Boltzmann constant, T the absolute temperature, N_s and N_c are the numbers of solvent molecules and the polymer chains contained in a gel, respectively, and χ the polymer-solvent interaction parameter. The equilibrium condition of gel is that the osmotic pressure π exerted on the network is zero, which on the basis of Eq.(1) can be written as

$$\pi \simeq \phi + \ln(1-\phi) + \chi\phi^2 - (N_cv_1/NV_0)\{\tfrac{1}{2}(\phi/\phi_0)-(\phi/\phi_0)^{1/3}\} = 0. \tag{2}$$

Here v_1 is the molar volume of solvent, N the Avogadro's number, and V_0 the volume of gel when it was formed.

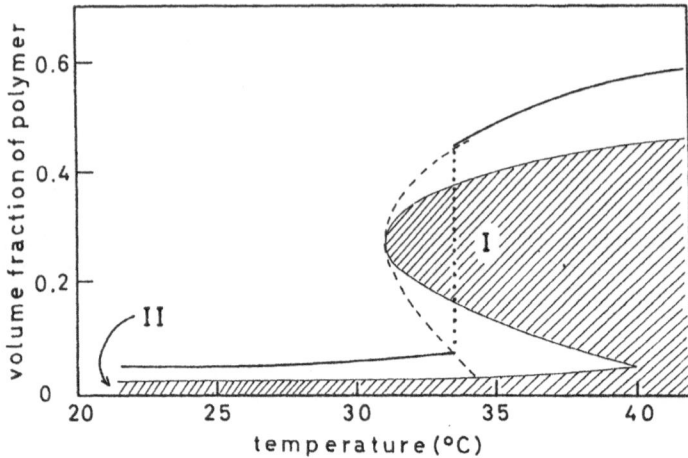

Fig.1 Swelling curve (thick line), coexistence curve (dashed line), and spinodal curve (thin line) of NIPA/water gel calculated from the phenomenological theory. Hatched regions represent spinodal regions and the dotted line a first-order transition.

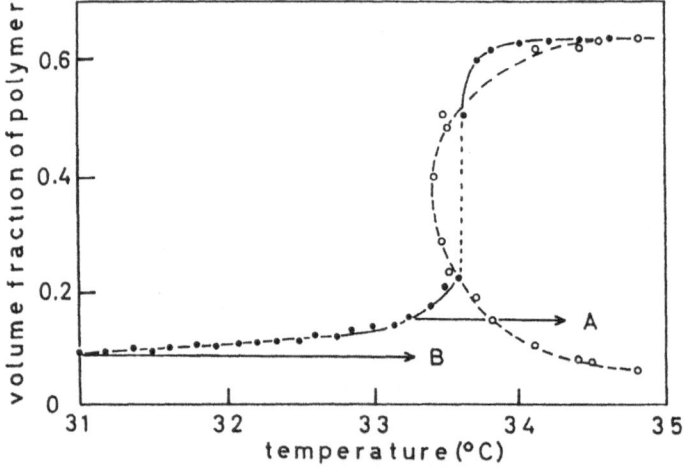

Fig.2 Swelling curve (thick line) and coexistence curve (dashed line) of NIPA/water gel determined experimentally. For the meaning of arrows see text.

The polymer-solvent interaction parameter χ depends on both temperature and concentration. It is these dependence which reflect the microscopic interactions among solvent and polymer molecules and are essential to the phase transition. Here we use the following form of χ as suggested by theoretical[12] and experimental[6] studies.

$$\chi = \chi_1 + \chi_2\phi = \frac{1}{kT}(\Delta h - T\Delta s) + 0.63\phi. \qquad (3)$$

The concentration dependence of χ is taken into account by expanding it in terms of the polymer concentration ϕ, and the terms higher than the second order are neglected. Δh and Δs are the changes of enthalpy and entropy, respectively, when polymer segment is brought into contact with solvent molecules.[1] On the basis of Eqs.(2) and (3) we can calculate the equilibrium swelling curve, the coexistence curve, and the spinodal curve for NIPA/water gel as shown in Fig.1. The parameters N_c and ϕ_0 were estimated from the preparation condition of the gel, and Δh and Δs were determined from the experimental swelling curve in the low-temperature region.[6] The experimental swelling curve[6] is shown in Fig.2 together with the coexistence curve, which has been deduced from the measurement of the first-order transition temperatures under negative osmotic pressure[10] and in ionic gels.[7,8] We see that the above phenomenological theory reproduces qualitative features of the gel-gel transition of NIPA/water gel.

EXPERIMENTS AND RESULTS

NIPA gels used in the present experiments were films of 0.2 ∿ 1.0mm in thickness and ∿ 1cm^2 in area. The method of preparation of gels was the same as that given in ref.6. The hatched regions in Fig.1 correspond to the negative compressibility, i.e. $K=\phi(\partial\pi/\partial\phi)_T < 0$. If a gel enters these regions it will undergo a spinodal decomposition. In both light scattering and pattern formation experiments, a gel which had been kept at the initial temperature T_i was inserted into the sample cell kept at the final temperature T_f and the measurements were made as a function of time after insertion. Details of the experimental method and result will be published elsewhere.[13]

Light Scattering

A far-field pattern was recorded with a video system. The relative intensity of the scattered light was measured later with a silicon photodiode on a static video image displayed on a color monitor TV. The linearity between the intensity of the scattered light and the brightness of the image was checked and was found to be satisfied within ±5% as long as the intensity was not too high.[13] The scattering pattern and its evolution with time differed markedly for different values of T_i and T_f.

(1) $T_i < T_0' < T_f$: Here T_0' is the temperature corresponding to the intersection of path A and the coexistence curve. The state of gel changes along path A in Fig.2. The change of the intensity profile with time is shown in Fig.3 for the case of $T_i=33.2°C$ and $T_f=34.2°C$. The wave number is defined as $q = (2\pi/\lambda)\sin(\theta/2)$, where θ is the scattering angle and λ is the wavelength of light in the medium. A circular ring appeared in the scattering pattern about 50 seconds after inserting the sample, indicating that the spinodal decomposition (SD) occurred. After about 90 seconds opacity began to develop in the gel so that the scattered light was soon masked completely. The higher T_f the sooner the gel became opaque. If T_f was in the range $T_0'<T_f<34.0°C$ no ring appeared at all, indicating that the phase separation proceeded via the nucleation and growth mechanism.

(2) $T_i < T_f < T_0'$: The state of gel changes along path B. In spite

of the fact that the gel is in the low-temperature phase throughout, a very clear circular ring appeared and it could be observed for a duration longer than twenty minutes after insertion of the sample. The gel never became opaque. The evolution of the scattering profile is shown in Fig.4(a) for the case of $T_i=30^\circ$C and $T_f=33.2^\circ$C. The scattered intensity in an early stage increased with time nearly exponentially as shown in Fig.4(b).

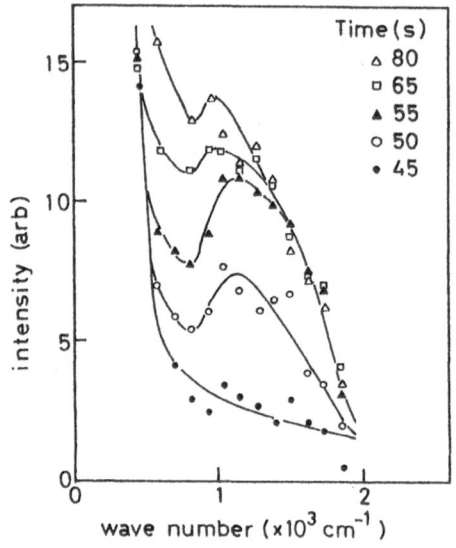

Fig.3 Evolution of light scattering profile obtained after the temperature jump along path A with $T_i=33.2^\circ$C and $T_f=34.2^\circ$C.

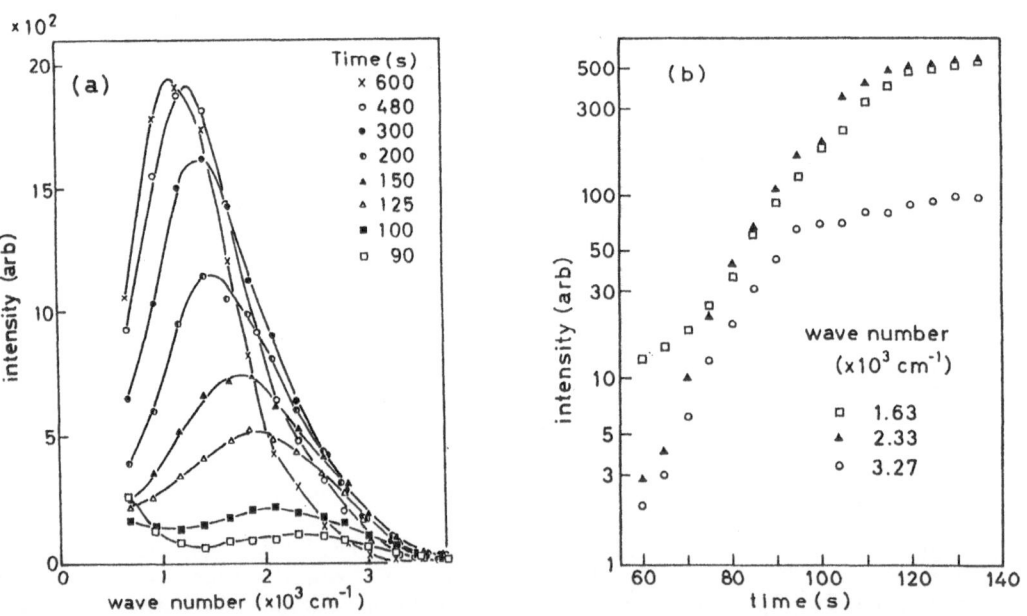

Fig.4 (a) Evolution of light scattering profile. (b) Time dependence of the scattered intensity in an early stage of the phase separation. The temperature jump was along path B with $T_i=30^\circ$C and $T_f=33.2^\circ$C.

Pattern Formation

The pattern observation was made with a microscope equipped with a camera and a temperature controlled sample cell. It was found that when the temperature of gel was changed along both of paths A and B characteristic patterns appeared on the surface as well as inside of a gel film and they evolved with time. Here we show typical examples of patterns appeared after the temperature change along path B. About one minute after the temperature jump fine thread-like patterns appeared and they covered whole of the gel sample. The "threads" were interconnected with each other constituting a structure like a maze (Fig.5(a)). As time passes a coalesceence and coarsening of "threads" occurred, after which many small droplets were left (Fig.5(b)). The droplets got smaller and smaller and finally disappeared. The gel became homogeneous again. On the other hand, patterns observed after the temperature jump along path A looked quite different from those shown in Fig.5. Although, the features such as interconnectivity and coarsening were also noted in them.

Due to limitation of space we cannot give details of these patterns here.[13] We want to stress that the size, shape, and evolution of these patterns are consistent with the results obtained by light scattering.

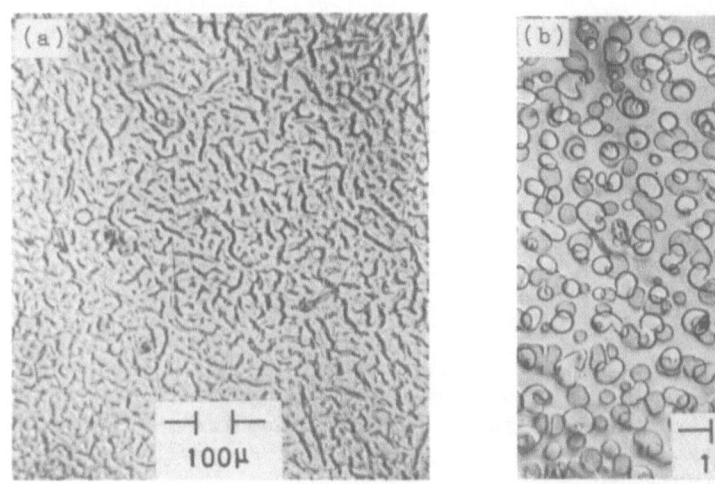

Fig.5 Photomicrographs showing patterns observed on a gel film ;
(a) about three minutes after, and (b) about 30 minutes after
the temperature jump along path B with $T_i=30^{\circ}C$ and $T_f=33.2^{\circ}C$.

DISCUSSION

Figure 1 shows two types of spinodal regions, I and II. Region I is associated with the phase separation between the two gel phases, whereas region II with that between the swollen gel and pure water phases. By the temperature jump along path A we could induce SD of the former type. On the other hand, the ring in the scattering profile appeared after changing temperature along path B may be due to SD of the latter type. While region II can exist in any gel,[14] region I exists only in a gel undergoing a gel-gel transition. It is clear from Fig.1 that for a gel to enter region II by a temperature jump the spinodal line must lie very close to the equilibrium swelling curve and must have an appreciable positive gradient. Our result indicates that these are realized in the vicinity of the gel-gel transition in NIPA/water gel. To the authors' knowledge neither types of

SD of gel have so far been observed directly, and thus the present results can be regarded as the first observation of SD of a gel.

According to the linear theory of SD[15] the wave number of the scattering maximum should be independent of time and the scattered intensity should increase exponentially with time. The present result shows that the ring shrinks rather rapidly and the intensity obeys the exponential law only in a very early stage (within a few tens of seconds) of the decomposition process. A more detailed analysis of the data is in progress.[13]

The pattern shown in Fig.5(a) and its evolution with time seem to visualize features of SD well. Transient patterns formed on NIPA/water gel near the gel-gel transition are rich in variety depending on T_i and T_f, as well as on the shape (possibly on the curvature of the surface) of a gel. Identification of the nature of elements constituting patterns (e.g., "thread" and "droplet" appearing in Fig.5) will be possible and it will be helpful for clarifying the mechanism of phase separation of a gel.

In summary experimental evidence showing that SD occurs in NIPA/water gel has been obtained for the first time. Interesting correlations between the evolution of patterns appearing on a gel and that of light scattering profiles were found. More detailed studies on these subjects will be of great value in elucidating the kinetics of gel-gel transition.

REFERENCES

1. P.J.Flory, "Principles of Polymer Chemistry," Cornel Univ. Press, Ithaca (1966).
2. K.Dusek and D.Patterson, Transition in Swollen Polymer Networks Induced by Intramolecular Condensation, J.Polymer.Sci C1:1209 (1968).
3. T.Tanaka, Gels, Scientific American, 244:124 (1981).
4. T.Tanaka, D.Fillmore, S.T.Sun, I.Nishio, G.Swislow and A.Shah, Phase Transitions in Ionic Gels, Phys.Rev.Lett., 45:1636 (1980).
5. Y.Hirokawa and T.Tanaka, Volume Phase Transition in a nonionic gel, J.Chem.Phys., 81:6379 (1984).
6. S.Hirotsu, Phase Transition of a Polymer Gel in Pure and Mixed Solvent Media, J.Phys.Soc.Japan, 56:233 (1987).
7. S.Hirotsu, Y.Hirokawa and T.Tanaka, Volume Phase Transition of Ionized N-Isopropylacrylamide Gels, J.Chem.Phys, 87:1392 (1987).
8. S.Hirotsu, unpublished.
9. T.Tanaka, E.Sato, Y.Hirokawa, S.Hirotsu, and J.Peetermans, Critical Kinetics of Volume Phase Transition of Gels, Phys.Rev.Lett., 55:2455 (1985).
10. S.Hirotsu, Critical Points of the Volume Phase Transition in N-isopropylacrylamide Gels, now submitted to J.Chem.Phys..
11. T.Tanaka, Kinetics of Phase Transition in Polymer Gels, Physica, 140A:261 (1986).
12. B.Erman and P.J.Flory, Critical Phenomena and transitions in Swollen Polymer Networks and in Linear Macromolecules, Macromolecules, 19:2342 (1986).
13. S.Hirotsu and A.Kaneki, in preparation.
14. P.G.deGennes, chapter V in :"Scaling Concepts in Polymer Physics," Cornell Univ. Press, Ithaca (1979).
15 J.W.Cahn, Phase Separation by Spinodal Decomposition in Isotropic Systems, J.Chem.Phys., 42:93 (1965).

STOCHASTIC APPROACH TO THE DYNAMICS OF PATTERNS

IN CONDENCED MATTER BY SCALING EXPANSION METHOD

Moyuru Ochiai

Department of Electronics
North Shore College
Atsugi 243, Japan

Yoshitake Yamazaki

Department of Applied Physics
Tohoku University
Sendai 980, Japan

Arno Holz

Fachrichtung Theoretische Physik
Universität des Saarlandes
D-6600 Saarbrücken, Fed. Rep. Germany

INTRODUCTION

Dynamics of patterns governed by such macroscopic laws as the Boltzmann equation or hydrodynamical equations near equilibrium and deterministic equations describing the system far from equilibrium, is discussed.

Such patterns are formed during the ordering processes for which coarse-graining or time smoothing is responsible.

In this paper, it is shown that macroscopic laws and properties of fluctuations can be easily determined by introducing a cumulant generating function with a scaling expansion method which carries out coarse-graining. By introducing scaling invariants which are obtained so that macrovariables are scaled by the characteristic sizes corresponding to the phenomena of interest, the cumulant BBGKY chain equation derived from the cumulant generating function, is automatically decoupled.

Thus, the macroscopic transport law and the equation describing the fluctuating deviations from the deterministic path and the variance of fluctuations, which have closed forms and obey inverted hierarchical relations, are immediately obtained.

In the case of a dilute neutral gas, it follows that the macroscopic transport law is reduced to a nonlinear Boltzmann equation and an inverted hierarchical representation relates the Boltzmann equation to the fluctuating deviation around it and the variance.

In the linear regime, we provide the law of regression of fluctuations, which signifies that the macroscopic relaxation processes are subject to the decay of spontaneous fluctuations on a deterministic path. This means the generalization of Onsager's famous assumption called regression of fluctuations. It follows that the self-similarity law in fractals are also confirmed explicitly.

In case of the system far from equilibrium, the Gaussian approximation around the deterministic path becomes irrelevant. Then, the usual perturbation method can not be carried out. The newly determined scaling exponents of fluctuations should be introduced.

As an illustration which gives a prescription in order to determine scaling exponents, the chemical reaction called Schlögl model is shown here.

FRAMEWORK OF THE THEORY

In this paper, we assume a Markov process for a system and take a master equation as a first principle. Many studies have been devoted to deriving the solution of the master equation.

By the use of a master equation one can easily derive moment equations for macroscopic variables. A set of the moment equations obeys a open hierarchy. Some literatures[1~3] have developed the formulations in order to decouple the chain of equations.

Here, we apply the scaling expansion method[3] to the hierarchy structure to obtain a closed set of macroscopic equations and furthermore, introduce the method of the cumulant generating function.[4]

The method we apply, leads directly to equations of motion for cumulants related to the macrovariables describing the system of interest. However, cumulant equations are also not closed. A systematic method should be available to decouple the chain of equations. The derivation is established in the second step of our approach, by means of a variant of the scaling expansion method which is essentially based on the following concept.

A system composed of a large number of constituents contains many kinds of space-time dependent modes from microscopic to macroscopic one's. The physical quantities we want to observe are characterized by some character-istic sizes.

Consider a macroscopic variable A at time t.
Then, we assume the power law of the characteristic sizes $\{\Omega^\nu\}$ corresponding to the phenomena in the form:

$$A = \sum_{\{\nu\}} \Omega^\nu a_{(\nu)} \quad , \tag{1}$$

where $a_{(\nu)}$ stands for a scaling invariant corresponding to the size and may depend on time t implicitly or explicitly. The parameter Ω refers to the system size and the exponents $\{\nu\}$ are not always integer. Rewriting the master equation in the scaling invariants and taking the dominant terms of the same order in the limit $\Omega \to \infty$, then we can attain the closed macro-scopic equations written by the invariants.

The theory works well, so far as the differences between characteristic sizes under consideration are clearly defined. For example, the size difference between the collective mode and the Gaussian fluctuation satisfies this condition. It is essential in our theory that the size of phenomena with which we deal must be separable.

When the fluctuations in the system grow up and the Gaussian approximation comes to fail, the usual perturbation method is not available. Several studies in this case have been successful.[5,6] Here, we consider the Schlögl model, of which macroscopic rate equation has several kinds of stationary states. In case that fluctuations grow up near unstable or critical one, we provide a new prescription to determine the scaling exponent for fluctuations so as to avoid the singularities in the limit $\Omega \to \infty$.

IN CASE OF A DILUTE GAS WITH THE GAUSSIAN FLUCTUATION

We consider a homogeneous dilute gas in the momentum space, which is divided into unit cells of the same size V . The i th unit cell contains n_i molecules at time t . Here, n_i stands for a macrovariable and follows the relation,

$$ n_i = V f_i \quad + \quad V^{\frac{1}{2}} u_i \quad , \tag{2} $$

where V and $V^{\frac{1}{2}}$ are the characteristic sizes of the deterministic and stochastic part of the process respectively. Let $P(\{n\};t)$ be a joint probability distribution and $W(n_i, n_j \to n_{p+1}, n_{q+1}) \equiv \alpha_{pqij} n_i n_j$ denote a transition probability per unit time which takes two molecules from cells i,j into cells p,q at time t . Thus, we can write the master equation in the form,

$$ \frac{\partial P(\{n\};t)}{\partial t} = -\frac{1}{2} \sum_{\substack{ijk\ell \\ (ij \neq k\ell)}} \alpha_{k\ell ij} n_i n_j P(\{n\};t) $$

$$ +\frac{1}{2} \sum_{\substack{ijk\ell \\ (ij \neq k\ell)}} \alpha_{ijk\ell} (n_k+1)(n_\ell+1) P(\{n'\}, n_i-1, n_j-1, n_k+1, n_\ell+1; t) \quad . \tag{3} $$

We now define a moment generating function $G(\{x\};t)$ as follows,

$$ G(\{x\};t) = \sum_{\{n\}} P(\{n\};t) \prod_i x^{n_i} \quad . \tag{4} $$

Putting $x_j \equiv e^{i\xi_j}$ and $G(\{e^{i\xi}\};t) = \langle \exp(i\sum_j \xi_j n_j) \rangle$, we can introduce the cumulant generating function as

$$ C(\{i\xi\};t) \equiv \ell n\, G(\{e^{i\xi}\};t) \quad , \tag{5} $$

and

$$ C(\{i\xi\};t) = \sum_{\{\ell\}} \prod_j \frac{(i\xi_j)^{\ell_j}}{\ell_j!} \lambda_{\{\ell\}}(t) \quad , \tag{6} $$

where the cumulants are given by

$$ \lambda_{\{\ell\}}(t) = \prod_j \left\{ \frac{\partial}{\partial(i\xi_j)} \right\}^{\ell_j} C(\{i\xi\};t) \Big|_{\{\xi\}=0} \quad . \tag{7} $$

Joint use of the master equation (3) with the relations (4)-(7) yields the equation of motion for the cumulant generating function

$$ \frac{\partial C(\{i\xi\};t)}{\partial t} = \frac{1}{2} \sum_{ijk\ell} \alpha_{k\ell ij}\, e^{i(\xi_k + \xi_\ell - \xi_i - \xi_j)} \quad \times $$

$$ \times \left\{ \frac{\partial C(\{i\xi\};t)}{\partial(i\xi)} \frac{\partial C(\{i\xi\};t)}{\partial(i\xi)} + \frac{\partial^2 C(\{i\xi\};t)}{\partial(i\xi)\,\partial(i\xi)} \right\} \quad . \tag{8} $$

The first few cumulant equations of motion are straightforwardly derived from Eq.(8). Introducing the scaling invariants,

$$ \lambda_{1p}(t) = V f_p(t) + V^{\frac{1}{2}} \langle u_p \rangle \equiv V f_p(t) + V^{\frac{1}{2}} u_p(t) \quad , $$

$$ \lambda_{1p1q}(t) = V \sigma_{pq}(t) \tag{9} $$

and

$$ \alpha_{ijk\ell} = V^{-1} A_{ijk\ell} \quad , $$

we can attain a set of the closed equations which gives the patterns corresponding to the stages of the process, as follows:
the Boltzmann equation for the deterministic path;

$$ \frac{d f_p(t)}{dt} = \sum_{\substack{\ell ij \\ (p\ell \neq ij)}} \left\{ A_{p\ell ij} f_i(t) f_j(t) - A_{ijp\ell} f_p(t) f_\ell(t) \right\} \quad , \tag{10} $$

it s deviation

$$\frac{du_p(t)}{dt} = \sum_{\substack{\ell\,ij\\(p\ell\,\neq\,ij)}} \Big[A_{p\ell\,ij}\{f_i(t)u_j(t) + f_j(t)u_i(t)\}$$
$$- A_{ij\,p\ell}\{f_p(t)u_\ell(t) + f_\ell(t)u_p(t)\}\Big] \quad, \tag{11}$$

and the time development of the covariance,

$$\frac{d\,\sigma_{pq}(t)}{dt} = \sum_{\substack{ij\\(p\ell\,\neq\,ij)}} A_{pq\,ij}\,f_i(t)f_j(t) + \sum_{\substack{k\ell\\(k\ell\neq pq)}} A_{k\ell\,pq}\,f_p(t)f_q(t)$$
$$- 2\sum_{\substack{kj\\(kp\neq qj)}} A_{kp\,qj}\,f_q(t)f_j(t) - 2\sum_{\substack{nj\\(nq\neq pj)}} A_{kq\,pj}\,f_p(t)f_j(t)$$
$$+ 2\sum_{\substack{\ell\,ij\\(p\ell\neq ij)}} A_{p\ell\,ij}\,\sigma_{qi}(t)f_j(t) + 2\sum_{\substack{\ell\,ij\\(q\ell\neq ij)}} A_{q\ell\,ij}\,\sigma_{pi}(t)f_j(t)$$
$$- \sum_{\substack{k\ell\,j\\(k\ell\neq pj)}} A_{k\ell\,pj}\{\sigma_{qp}(t)f_j(t) + \sigma_{qj}(t)f_p(t)\}$$
$$- \sum_{\substack{k\ell\,j\\(k\ell\neq qj)}} A_{k\ell\,qj}\{\sigma_{qp}(t)f_j(t) + \sigma_{pj}(t)f_q(t)\} \quad. \tag{12}$$

By expanding the nonlinear Boltzmann equation (10) around the equilibrium $f^{(e)}$ and taking the terms up to the linear regime in the deviation $\Delta f(t)$, Eq. (10) reduces to the linearized Boltzmann equation:

$$\frac{d}{dt}\Delta f_p(t) = \sum_{\substack{\ell\,ij\\(p\ell\neq ij)}} \Big[A_{p\ell\,ij}\{f_j(t)\Delta f_i(t) + f_i(t)\Delta f_j(t)\}$$
$$- A_{ij\,p\ell}\{f_\ell(t)\Delta f_p(t) + f_p(t)\Delta f_\ell(t)\}\Big] \quad. \tag{13}$$

Comparing Eq.(13) with Eq.(11), we can find that this shows the generalization of Onsager's famous assumption called regression of fluctuations.

Thus, it follows that , so long as the fluctuation has such a clearcut scaling exponent as in case of the Gaussian distribution, the joint use of the scaling expansion method with the cumulant generating function makes the derivation of the physically significant closed equations corresponding to the patterns of the process more straightforward.

IN CASE OF A CHEMICAL REACTION WITH SINGULARITIES

As an illustration, we take the chemical reaction called the Schögl model:

$$A \;\underset{\gamma_1}{\overset{k_o}{\rightleftharpoons}}\; N$$

$$B + 2N \;\underset{\gamma_3}{\overset{k_2}{\rightleftharpoons}}\; 3N \quad, \tag{14}$$

where the quantity of the chemical species A and B are cotroled from outside and N stands for the chemical intermediate. It is well known that the rate equation for the number of the molecules of intermediate N includes the third order term which predicts a cusp bifurcation.

The Schlögl model has the following master equation,

$$\frac{\partial P(N;t)}{\partial t} = k_o A \, P(N-1;t) + k_2 \frac{B}{V^2}(N-1)(N-2) P(N-1;t)$$

$$- k_o A P(N;t) - k_2 \frac{B}{V^2} N(N-1) P(N;t) - \gamma_1 N P(N;t)$$

$$- \gamma_3 \frac{1}{V^2} N(N-1)(N-2) P(N;t) + \gamma_1 (N+1) P(N+1;t)$$

$$+ \gamma_3 \frac{1}{V^2}(N+1) N(N-1) P(N+1;t) \qquad , \qquad (15)$$

where the volume V of the system plays a role of the system size Ω .

Using the moment generating function for single variable introduced in the previous section, the master equation can be written as follows,

$$\frac{\partial G(x;t)}{\partial t} = (x-1)\left\{ k_o V - \gamma_1 \frac{\partial}{\partial x} + k_2 V^{-1} x^2 \frac{\partial^2}{\partial x^2} - \gamma_3 V^{-2} x^2 \frac{\partial^3}{\partial x^3} \right\} G(x;t), (16)$$

where the scattering cross section can be scaled as

$$K_o = V^{-1} k_o A \qquad \text{and} \qquad K_2 = V^{-1} k_2 B \qquad . \qquad (17)$$

We now provide another formula to make it easy to extract the deterministic path and the behaviour of fluctuations.

Since the number of the molecules of the chemical intermediate N has the extensive property, the scaling invariant n for the intermediate is represented by

$$N = V n \qquad . \qquad (18)$$

Noting that

$$G(e^{i\xi};t)\Big|_{\xi=0} \sim O(v^o) \ , \ \frac{\partial}{\partial(i\xi)} G(e^{i\xi};t)\Big|_{\xi=0} \sim O(v) \ , \ \frac{\partial^2}{\partial(i\xi)^2}G(e^{i\xi};t)\Big|_{\xi=0} \sim O(v^2) \ , \cdots,$$

we can introduce a scaled moment generating function $G(e^{i\eta};t)$ written by such a parameter η as $i\xi \equiv iV^{-1}\eta$ in the moment generating function $G(e^{i\xi};t)$ and a newly defined generating function $C(i\eta;t)$ as

$$C(i\eta;t) \equiv \ln G(e^{i\eta};t) = \sum_{\ell=1}^{\infty} \frac{(i\eta)^{\ell}}{\ell!} \lambda_{s\ell} \qquad ,$$

which gives the relations

$$\lambda_{s_1} = \langle n \rangle \quad , \quad \lambda_{s_2} = \langle n^2 \rangle - \langle n \rangle^2 \quad , \cdots \qquad . \qquad (19)$$

Thus, Eq.(16) reads

$$\frac{\partial C(i\eta;t)}{\partial t} = (e^{i\eta V^{-1}} - 1)\left[K_o V + (-\gamma_1)e^{-iV^{-1}\eta} V \frac{\partial C}{\partial(i\eta)} \right.$$

$$+ K_2 V\left\{\left(\frac{\partial C}{\partial(i\eta)}\right)^2 + \frac{\partial^2 C}{\partial(i\eta)^2}\right\} + (-\gamma_3) V e^{-iV^{-1}\eta}\left[\frac{\partial C}{\partial(i\eta)}\left\{\left(\frac{\partial C}{\partial(i\eta)}\right)^2 + \frac{\partial^2 C}{\partial(i\eta)^2}\right\}\right.$$

$$\left.\left. + \left\{ 2\left(\frac{\partial C}{\partial(i\eta)}\right)\frac{\partial^2 C}{\partial(i\eta)^2} + \frac{\partial^3 C}{\partial(i\eta)^3}\right\}\right]\right] \qquad . \qquad (20)$$

Using the relations (19), the equation of motion which leads to the deterministic path is immediately obtained

$$\frac{\partial \langle n \rangle}{\partial t} = K_0 - \gamma_1 \langle n \rangle + K_2 \langle n^2 \rangle - \gamma_3 \langle n^3 \rangle \quad . \quad (21)$$

The equation above is not closed yet. In order to get the closed macroscopic equation, the scaling exponent of the fluctuation should be informed.

We now discuss the properties of the fluctuations around the deterministic path. Choosing a scaling invariant u for the fluctuation as

$$\Delta N \equiv N - \langle N \rangle = V^\nu u \quad , \quad (22)$$

we introduce a new generating function $G_\Delta(e^{i\xi}; t)$ for single variable, in which $\chi \equiv e^{i\xi}$ in

$$G_\Delta(\chi; t) = \sum_{N=0}^{\infty} \chi^{\Delta N} P(N; t) \quad . \quad (23)$$

If we put the parameter such as $\xi \equiv V^{-\nu}\zeta$ and introduce scaled generating functions

$$K(i\zeta; t) = \ln G_\Delta(e^{i\xi}; t) = \sum_{\ell=1}^{\infty} \frac{(i\zeta)^\ell}{\ell!} \delta_{S\ell}(t) \quad , \quad (24)$$

from which a few variables $\delta_{S\ell}(t)$ are given by

$$\delta_{S2} = \langle u^2 \rangle \quad , \quad \delta_{S3} = \langle u^3 \rangle \quad , \cdots , \quad (25)$$

the equation of motion of the generating function $G_\Delta(e^{i\xi}; t)$ is written by the scaling invariant. Taking a few dominant terms, we obtain

$$\frac{\partial G_\Delta(e^{i\xi}; t)}{\partial t} \simeq \left\{ \frac{1}{2}(i\xi)^2 V^{1-2\nu} A + (i\xi)B\frac{\partial}{\partial(i\xi)} + (i\xi)V^{\nu-1}C\frac{\partial^2}{\partial(i\xi)^2} + (i\xi)V^{2(\nu-1)}D\frac{\partial^3}{\partial(i\xi)^3} \right\}G_\Delta, \quad (26)$$

where A, B, C and D are the fuctions of the scaled reaction rates and $\langle n \rangle$. Thus, Eq.(26) suggests that the condition $1 - 2\nu \leq 0$ and $\nu - 1 \leq 0$ comes to make sense. One can find the scaling exponent ν as $1/2 \leq \nu \leq 1$. If we adopt the condition $1/2 \leq \nu < 1$, Eq.(21) comes to the closed deterministic equation:

$$\frac{\partial f(t)}{\partial t} = K_0 - \gamma_1 f + K_2 f^2 - \gamma_3 f^3 \quad . \quad (27)$$

From Eq.(26) , we also attain the Fokker–Planck equation

$$\frac{\partial P(u; t)}{\partial t} = -\frac{\partial}{\partial u} F(f, u; V^0, V^{\nu-1}, V^{2(\nu-1)})P + \frac{1}{2}V^{1-2\nu}A(f)\frac{\partial^2}{\partial u^2}P \quad (28)$$

Here, the authors have shown how to obtain the deterministic path and determine the scaling exponent describing the system of which fluctuation does not obey the Gaussian approximation. We, however, have not discussed the behavior of the system near the cusp bifurcation. It will be discussed later.

REFERENCES

1) N. G. van Kampen: Can. J. Phys., 39, 551 (1961)
2) R. Kubo, K. Matsuo and K. Kitahara: J. Stat. Phys., 9, 51 (1973)
3) M. Ochiai: Lett. Nuovo Cimento, 40, 433 (1984)
4) N. Hashitsume and M. Ochiai: Bussei Kenkyu (Kyoto), 25, 320 (1976)
5) M. Suzuki: Phys. Lett., 56A, 71 (1976)
6) G. Nicolis and J. W. Turner: Physica A89, 245 (1977)

PATTERN FORMATION IN MIGRATION PROCESSES

Yong Liu and Masuo Suzuki

Department of Physics, Faculty of Science
University of Tokyo
Hongo, Bunkyo-ku, Tokyo 113, Japan

Lots of studies on the migration processes of biological population and the spatial patterns thus induced can be found in the literature (references 1 through 4 give some examples), though not many of them explicitly took into account individuals' migrating behavior such as crowd-preferring habit. In this presentation, we propose two specific models for crowd-preferring behavior and study their stabilities.

We consider n sites located on a one-dimensional line and assume that individuals can migrate only between neighboring sites at one time. The equation of motion for the population in site i is as follows (except boundary sites 1 and n),

$$\dot{Y}(i) = G(Y(i)) + M(Y(i-1), Y(i)) - M(Y(i), Y(i+1)). \tag{1}$$

$Y(i)$ is the population in site i, $G(Y(i))$ is the self-growing term

$$G(Y(i)) = RY(i)(1 - Y(i)) \tag{2}$$

(capacity of each site is assumed to be unity) and $M(Y(i), Y(i+1))$ is the migration term from i to i+1

$$M(Y(i), Y(i+1)) = Y(i)p(Y(i), Y(i+1)) - Y(i+1)p(Y(i+1), Y(i)) \tag{3}$$

where $p(Y(i), Y(i+1))$ is the probability of one individual migrating from i to i+1 in unit time which will be defined below. To see the stability of the system, we assume a constant influx of population M0 from outside into site 1,

$$\dot{Y}(1) = G(Y(1)) + M0 - M(Y(1), Y(2)). \tag{4}$$

M0 makes the system diverge if $G(Y(i))$ is vanishing. To prevent this, we assume the population in the other boundary site to be constant

$$Y(n) = constant \tag{5}$$

so that it may act as an exit of population. Equation (5) can be realized if in site n the birth and death processes take place so fast that the population recovers to the constant instantly.

For the individual migrating probability, we consider two specific types,

$$p(Y(i),Y(i+1))=\exp(K(Y(i+1)-Y(i)))+D \qquad (6.1)$$

and

$$p(Y(i),Y(i+1))=1+\tanh(K(Y(i+1)-Y(i)))+D \qquad (6.2)$$

(the time constant of migration is assumed to be unity without loss of generality). K>0, K=0 and K<0 represent crowd-prefering, crowd-independent and crowd-avoiding behaviors, respectively (we focus on K>0 here since other cases give only uniform patterns). D is the random walk component. Type (6.1) has been studied extensively by Weidlich et al (with D≈0)[5].

The stationary pattern of our system (if any) is defined by

$$M(Y(i),Y(i+1))=MO \;, \quad \text{for all i's,} \qquad (7)$$

that is, if we plot Y(i+1) vs Y(i) for all i's we get a spatial return map (Fig. 1).

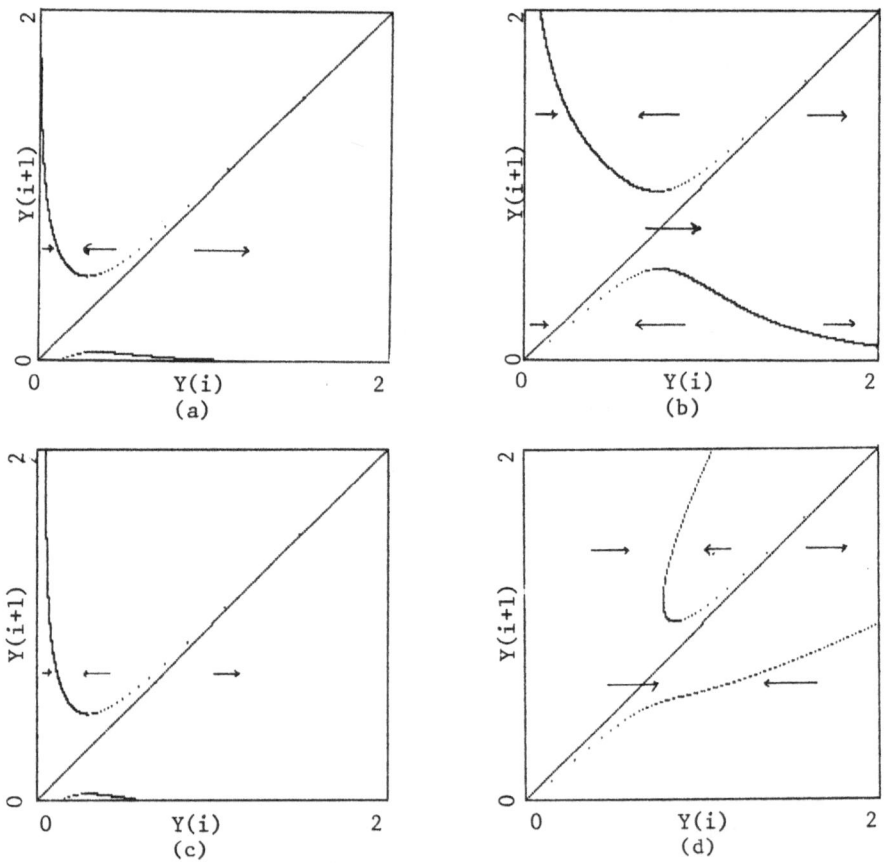

Fig. 1. Spatial return map defined in (7). (a) Type (6.1) with parameters
K=2, MO=0.1 and D=0; (b) Same as (a) except D=2; (c) Type (6.2)
with same parameters as (a); (d) Same as (c) except D=2.

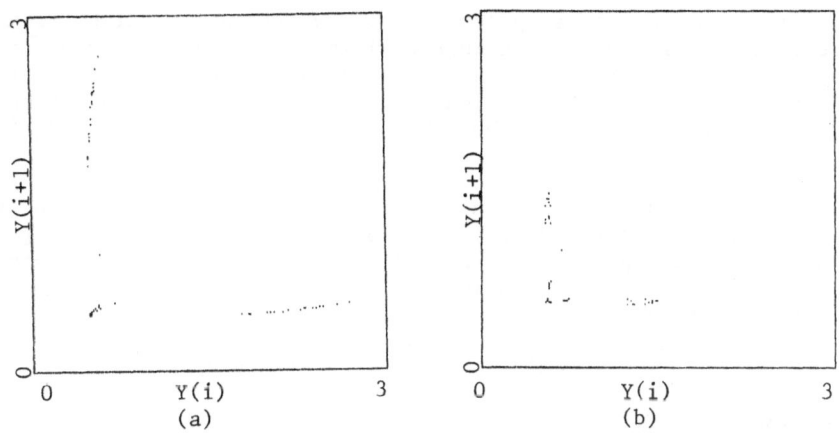

Fig. 2. Y(i+1) vs Y(i) plots for stationary pattern of type (6.2) with
101 sites. (a) K=1, M0=0.05, D=0.5 and R=0; (b) Same as (a)
except R=0.3.

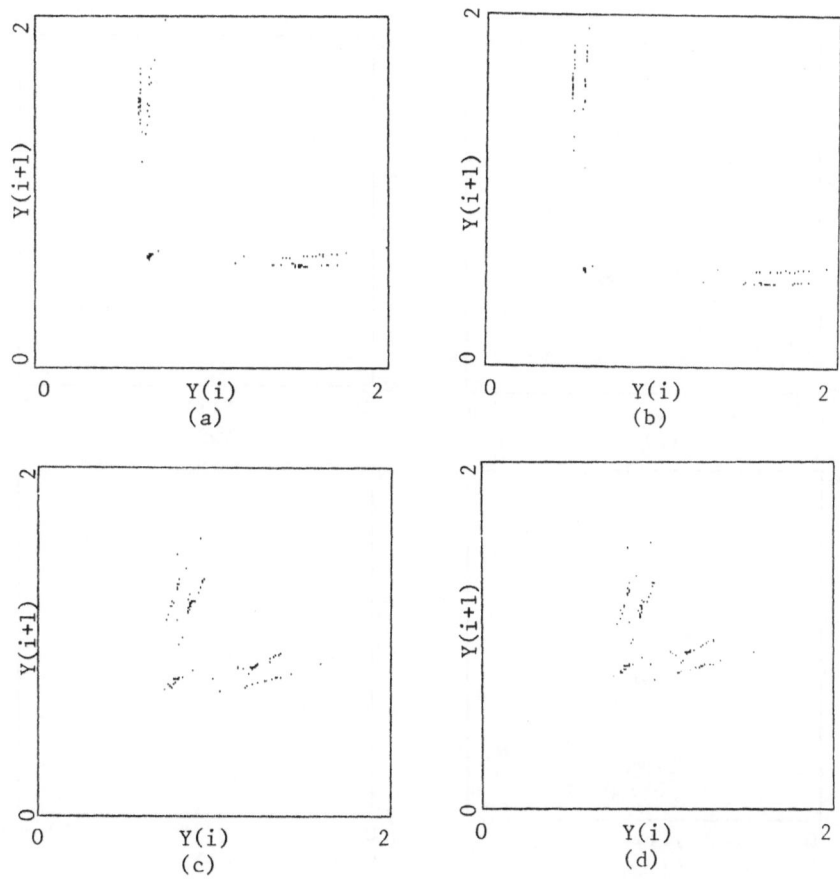

Fig. 3. Y(i+1) vs Y(i) plots for oscillating pattern of type (6.2) with
101 sites. System goes as (a)→(b)→(c)→(d) in one period.
K0=1, A=1, T=1, M0=0.05, D=0.5 and R=0.05.

The simplest case is a system of only two sites (one site fixed). The trajectory is a horizontal line in Fig. 1 (direction indicated by an arrow). For type (6.1), if Y(2) is fixed between the two extrema of upper and lower branches, there is no stationary state and the system diverges,whatever D is. For type (6.2), on the other hand, the system does not diverge for any value of Y(2), provided that D is not vanishing.

Fig. 2 is the numerical result of the stationary pattern of a system with 101 sites (Y(101) fixed) for type (6.2). Non-vanishing R brings the pattern out of the map (defined in (7)) and toward a uniform pattern, as expected. The numerical calculation for type (6.1) with the same parameters shows quick divergence.

Next, we observe system's response to oscillating K as

K=K0+Acos(2πt/T). (8)

Change of migrating behavior can be considered as the effect of environmental change such as the annual oscillation of weather. Fig. 3 shows an oscillating pattern for type (6.2). It requires much larger R to obtain an oscillating pattern for type (6.1) with similar amplitude in order to maintain stability. Fig. 4 is the case for type (6.1) with R 30 times as large as that of Fig. 3.

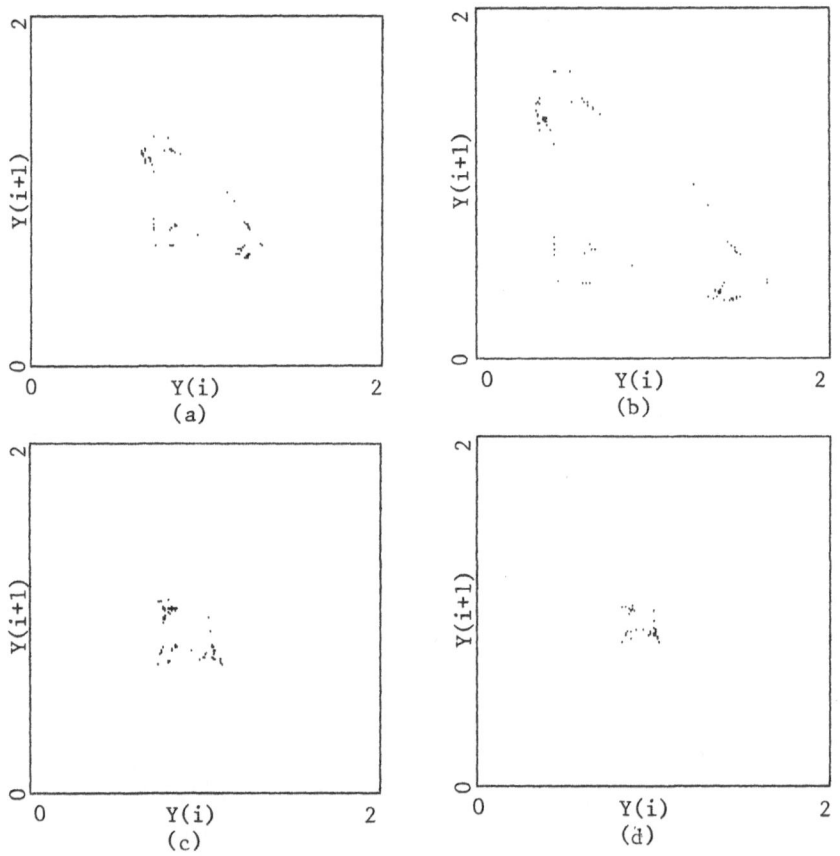

Fig. 4. Y(i+1) vs Y(i) plots for oscillating pattern of type (6.1) with 101 sites. System goes as (a)→(b)→(c)→(d) in one period. Parameters are the same as Fig. 3 except R=1.5.

496

Our conclusion is that, for models of crowd-preferring migration, individual migrating probability of a hypertangent form (type (6.2)) is much more stable than that of an exponential form (type (6.1)).

ACKNOWLEDGEMENT

One of the authors (Y. Liu) would like to thank Dr. T. Onogi, H. Ito, A. Terai and N. Ito for helpful discussions.

REFERENCES

1. N. T. J. Bailey, Biometrica 55 (1968) 189.
2. E. Renshaw, Biometrica 59 (1972) 49.
3. W. S. C. Gurney and R. M. Nisbet, J. Theor. Biol. 52 (1975) 441.
4. Y. Iwasa and E. Teramoto, J. Math. Biology 19 (1984) 109.
5. W. Weidlich and G. Haag, Concepts and Models of a Quantitative Sociology, Springer Ser. Synergetics, Vol. 14 (Springer 1983).

DYNAMICS OF INTERFACE DESCRIBED BY GENERALIZED RANDOM WALKS

Hiroaki Hara

Department of Engineering Science
Faculty of Engineering
Tohoku University, Sendai, 980, Japan

By modelling the dynamics of an interface, we show a procedure of obtaining an action which plays a role of the thermodynamic potential for an ecological system.   The interface is a boundary of a colony size of bacteria.   To specify the boundary, we propose a simple model process for growth rates of the colony size.

1. INTRODUCTION

An interface is the boundary between one phase and the other phase. The boundary, generally, moves and makes patterns due to the relevant dynamics in the system.   When the system is specified by the free energy, behaviors of interface have been studied by applying the TDGL equation /1/, or droplet models /2/,/3/, etc.

In some problems concerned with biological system, however, we can not specify states by the thermodynamic potential as the free energy. Usually, growth rates of population in the ecological system are modelled in discrete space-time descriptions, or in the diffusion equations /4/. In the colony of bacteria, the growth rates of boundary of the colony is evaluated by the growth rates of the cell mass per colony by assuming some growth patterns /4/-/6/.

In this paper, we show a procedure of obtaining an action specifying the dynamics of interface, where the interface is the boundary of a colony size of bacteria.   The action plays a role of a " thermodynamic potential " in the system composed of colonies.   In the modelling we use a recursion relation of generalized random walks(GRW) /7/-/9/ .   The first step is to specify the boundary of a colony size.   To this end, a simple model for growth rates of colony is proposed.   Next step is to set up an equation of a preserving profile of the interface, and then we convert the resultant to an expression which is derivable from a Lagrangian. Finally, we determine an action specifying the system, and yielding the equation of interface by minimizing the action.

2. GROWTH RATE OF COLONY SIZE

As a moving interface, we consider a boundary of colony size.   Let $W^{(\kappa)}(m,N)$ be the probability that the number of cells in a colony is m   at step N .   Suppose that growth rates of the number of cells are classified into three stages (initial(i) , exponential(e), and steady(s) stages), and formally they are described by a recursion relation of generalized random walks(GRW), see Fig.1.

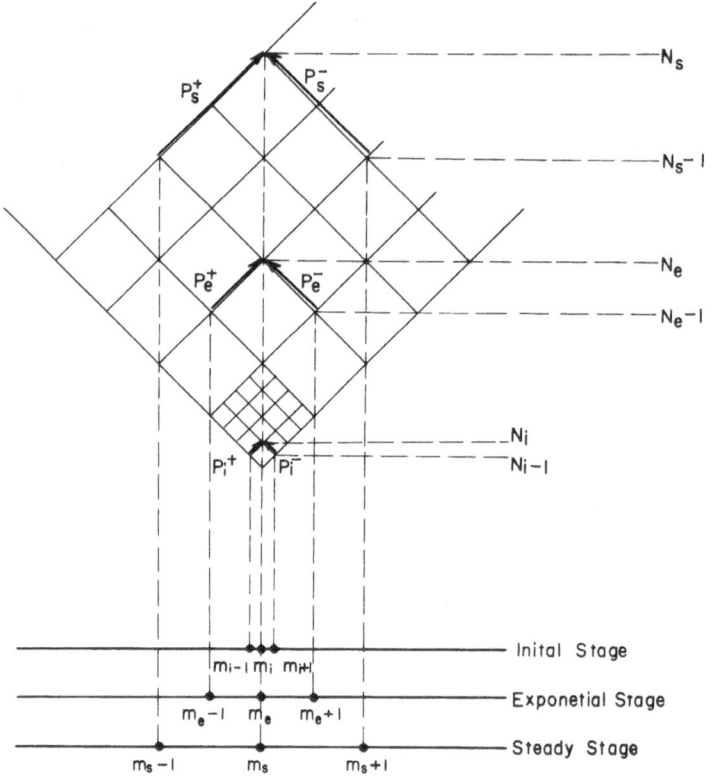

Fig.1. Number(m)–Step(N) space in generalized random walks (GRW), where P_ν^α means $P_{t_\nu}^\alpha(m_\nu | m_\nu - \alpha \cdot 1)$ for $\nu= i$, e, and s.

By taking continuum limits ($t_\nu = Nt_{o,\nu}$, $t_{\nu} \to 0$) in a recursion relation of the GRW and an additional relation for the jumping probabilities, we have a master equation expressing the three stages ($\nu= i,e,s$);

$$\frac{\partial W^{(c)}(m_\nu, t_\nu)}{\partial t_\nu} = \frac{1}{t_{o,\nu}} \left\{ P_{t,\nu}^+(m_\nu | m_\nu - 1) W^{(c)}(m_\nu - 1, t_\nu) + P_{t_\nu}^-(m_\nu | m_\nu + 1) W^{(c)}(m_\nu + 1, t_\nu) \right\}$$
$$- P_{t_\nu}^+(m_\nu + 1 | m_\nu) W^{(c)}(m_\nu, t_\nu) - P_{t_\nu}^-(m_\nu - 1 | m_\nu) W^{(c)}(m_\nu, t_\nu) \right\}, \quad (2.1)$$
$$(t_{o,\nu} : unit\ step)$$

with $\sum_{d=\pm,0} P_{t_\nu}^\alpha (m_\nu + d \cdot 1 | m_\nu) = 1$ and

$$P_{t_\nu}^\alpha (m_\nu | m_\nu + \alpha \cdot 1) = F \left(\{ P_{t_{\nu-1}}^{\alpha'}(m_\nu + \alpha \cdot 1 | m_\nu + d \cdot 1 + \beta \cdot 1) \} \right). \quad (2.2)$$

where F is a functional specifying t-dependence in $P_{t_\nu}^\alpha(m_\nu | m_\nu + \alpha \cdot 1)$, and a symbol $\alpha \cdot 1$ (or $\beta \cdot 1$) takes 1, 0, and −1 as α(or β)= +, 0, −, respectively.

We consider a case that $P_{t_{\nu-1}}^\alpha(m_\nu + \alpha \cdot 1 | m_\nu)$'s are divided into two parts,

$$P_{t_\nu}^\alpha (m_\nu + d \cdot 1 | m_\nu) / t_{o,\nu} = \Gamma_{o,\nu}^\alpha + m \Gamma_{1,\nu}^\alpha . \quad (2.3)$$

The first term is a constant part, and the second term is a m-dependent part. Symbols $\Gamma_{j,\nu}^\alpha$'s(j=0,1, ν =i,e,s) denote birth rates for α = + , death rates for α = − , and incubation rates for α = 0, respectively. Let $M(t_\nu)$ be an average number of cells at the respcetive stages, and define it by an expression

$$M(t_v) = \sum_{m_v=0}^{\infty} m_v \, w^{(c)}(m_v, t_v) .$$

(2.4)

With the aid of (2.1), we have

$$\frac{dM(t_v)}{dt_v} = (\Gamma_{0,v}^+ + \Gamma_{1,v}^+ - \Gamma_{1,v}^-) M(t_v) + \Gamma_{0,v}^+ .$$

(2.5)

By specifying the transition rates, we have behaviors: $M(t_i) \propto \exp(bt_i)$, ($b = \Gamma_{1,i}^+ - \Gamma_{0,i}^-$), $M(t_e) \propto \exp(ct_e)$, ($c = \Gamma_{0,e}^+ + \Gamma_{1,e}^+ - \Gamma_{1,e}^-$), and $M(t_s) \propto t_s \sim$ $1 - \exp(-c't_s)$, ($c' = \Gamma_{1,s}^- - \Gamma_{0,s}^+ - \Gamma_{1,s}^+$) . The specifications of transition rates mean to take an explicit form of functional F in (2.2).

By connecting the values of $M(t_v)$ piecewisely, we obtain a sigmoid curve for $M(t)$. To simplify the following analysis, we assume that before the inflection point, t_f , of the sigmoid curve, the colony grows as an active disk of constant height h , ($M(t) = \pi r^2(t) h \rho$, where ρ is number density), and after the inflection point, the colony behaves a quite disk specified by the parameter c' . Then we have behaviors of $r(t)$ as follows

$$r(t) \propto e^{\frac{b}{2}t} \sim e^{\frac{c}{2}t} \quad , \; (0 < t < t_f)$$

(2.6)

$$r(t) \propto \sqrt{t} \sim [1 - e^{-c't}]^{\frac{1}{2}}, \; (t_s < t) .$$

(2.7)

Note that in realistic models, there is an interesting period passing the inflection point; the active growing zone is restricted to the border of disk, and then one gets a behavior of $r(t) \sim t$.

3. SOLITARY WAVE MOVING WITH VELOCITY

An interface we consider is a boundary of colony size. The boundary is specified by the previous results (2.6) and (2.7). Let $\varphi(x,t)$ be a function representing a profile of interface. $\varphi(x,t)$ provides us a pattern similar to the distribution function in the probability theory. When the profile of interface is preserved in the course of time t , a quantity $\phi(x,t)$ ($= \varphi_x(x,t)$) obtained by differentiating $\varphi(x,t)$ with respect to x is regarded as a probability density function moving like a solitary wave. Its velocity is given by differentiating (2.6) and (2.7), that is $r_t(t)$ ($= dr(t)/dt$).
The behaviors are described by a recursion relation of the GRW and a suitable functional connecting jumping probabilities (cf. (2.1) and (2.2)). Namely, continuum limits of the recursion relation and the functional yield a non-linear differential equation after substituting the functional into the jumping probabilities in the recursion relation /9/;

$$\phi_t(x,t) = - [k_0 \phi - k_1 g(t)] \phi_x - k_2 \phi_{xxx}$$

(3.1)

where $g(t)$ is related to the velocity $r_t(t)$, see (2.6) and (2.7), by

$$g(t) = \frac{1}{k_1} \left[r_t(t) - \frac{k_0 A}{3} \right],$$

(3.2)

and A, k_0, k_1 , and k_2 are constant.
A solution of (3.1) is given by

$$\phi(x,t) = A \, \operatorname{sech}^2 \sqrt{\frac{k_0 A}{12 k_2}} \; (x - r(t)) .$$

(3.3)

After rewriting (3.1) by $z = x - r(t)$, we obtain an expression corresponding to the Euler-Lagrange(EL) equation for a Lagrangian $L(\phi, \phi_z)$;

$$L(\phi, \phi_z) = \frac{1}{2} \phi_z^2 - U(\phi) \tag{3.4}$$

where

$$U(\phi) = \frac{k_0}{6k_2} (\phi^3 + A\phi^2 - B\phi), \tag{3.5}$$

and B is an integration constant.
By transforming (3.3) by an integration over $[0,z]$, we get an expression for $\varphi(z)$.

4. GENERALIZED RANDOM WALKS FOR INFINITE DIMENSIONAL SPACE

Let us take a system composed of many colonies. Each colony is assumed to be specified by $\phi(z)$, or $\varphi(z)$. We express states of the system by a functional of $\phi(z)$. To get the functional, we devide a single coordinate z into d ($\gg 1$, many) small intervals, and associate values of $\phi(z)$, $m_1, m_2, \ldots, m_d$, with the respective intervals. Furthermore, to take into account fluctuations and correlations in the system, we regard the values as random variables. A probability density function $W(m_1, m_2, \ldots, m_d, N)$ is then expressed by a recursion realtion of the GRW in d-dimensional space /10/ ;

$$W(m_1, m_2, \ldots, m_d, N) = \sum_{i=1}^{d} \sideset{}{'}\sum_{\alpha=\pm} P_{N-1}^{\alpha}(m_1, m_2, \ldots, m_d \mid m_1, m_2, \ldots, m_i - \alpha \cdot 1, \ldots m_d)$$
$$\times W(m_1, m_2, \ldots, m_i - \alpha \cdot 1, \ldots, m_d, N-1) \tag{4.1}$$

Here note that a case $\alpha = 0$ is excluded, and its restriction is denoted by a prime on the symbol of summation. We have put no additional condition corresponding to (2.2).
After taking the continuum limit ($N \to (=N\tau_0)$, $m \to \phi_i(= m_i a)$, while a^2/τ_0 remains fixed), and another limit $d \to \infty$, we have an equation for a functional $W[\phi, \tau]$ instead of $W(\phi_1/a, \phi_2/a, \ldots, \phi_d/a, N/\tau_0)$,

$$\frac{\partial W[\phi, \tau]}{\partial \tau} = \int dz \frac{\delta}{\delta\phi(z,\tau)} \left\{ -K^{(1)}[\phi] + \frac{1}{2} \frac{\delta}{\delta\phi(z,\tau)} K^{(2)}[\phi] \right\} W[\phi, \tau] \tag{4.2}$$

where

$$K^{(1)}[\phi] = \frac{a}{\tau_0} (P^+[\phi] - P^-[\phi]), \quad (\tau_0 : \text{unit step}), \tag{4.3}$$

$$K^{(2)}[\phi] = \frac{a^2}{\tau_0} \equiv D. \tag{4.4}$$

In the derivation we have used the normalization condition; $p^+[\phi] + p^-[\phi] = 1$.
For steady state $\tau \to \infty$, we get a solution of (4.2), that is the functional of ϕ mentioned in the beginning, as follows

$$W[\phi] = e^{-S[\phi]/2D}, \tag{4.5}$$

$$S[\phi] = \int L(\phi, \phi_z) dz, \tag{4.6}$$

and $S[\phi]$ is an action such that $\delta S = 0$ yields the EL equation for ϕ, that is,

$$\frac{d}{dz}\left(\frac{\partial L}{\partial \phi_z}\right) - \frac{\partial L}{\partial \phi} = 0 \tag{4.7}$$

The action $S[\phi]$ plays a role of a " thermodynamic potential " for the system composed of colonies.

Based on (4.5), we can calculate an average of a quantity denoted by $G[\phi]$ by a formula,

$$\langle G[\phi] \rangle = \int G[\phi] e^{-S[\phi]} \delta\phi \Big/ \int e^{-S[\phi]} \delta\phi , \tag{4.8}$$

where

$$\int G[\phi] \delta\phi = \lim_{d \to \infty} \int G(\phi_1, \phi_2, \cdots, \phi_d) \prod_{i=1}^{d} d\phi_i (\Delta z_i)^{\frac{1}{2}}, \tag{4.9}$$

and Δz_i is a devided small interval of z.

From the expected value of $G[\phi] = \phi(z)\,\phi(z')$, we can study a correlation between the colonies in the system specified by (4.5). Actual calculations are performed on the WKB method so that the extremum condition yields (4.7) /10/.

5. CONCLUSION AND DISCUSSIONS

When the system is specified by the free energy, behaviors of interface have been studied by applying the TDGL equation. To solve the TD GL equation, however, we approximate the equation. Unfortunately, in some biological systems, we can not specify the states by the free energy with which the TDGL equation is expressed.

In this paper, we showed a procedure of obtaining an action for the ecological system. The action plays a role of the thermodynamic potential in the thermodynamic systems. The procedure started with a specification of the dynamics of interface. The interface is a boundary of the colony size. The first step is a modelling of a time dependence of the growing boundary, that is , the dynamics of interface. Next step is to set up an equation yielding a preserving profile of interface. Final step is to convert the resultant to a basic equation which is derivable from a Lagrangian. In a sense, the action (4.6) may be regarded as a generalization of the thermodynamic potential introduced by Kerner /11/ for an ecological system. The present procedure can be applied to more realistic dynamics by choosing suitable parameters in (3.5) or (4.6).

REFERENCES

1) K. Kawasaki and T. Ohta: Physica 118A (1983),175
    K. Kawasaki : Lecture Note in Busseiron Kenkyu 43 (1985), 181
2) H. Furukawa : Phys. Rev. A48 (1983), 1729
3) J. S. Langer : Fundamental Problems in Statistical Mechanics VI (ed.
    E. G. Cohen (1985), 313
4) A. Okubo : Diffusion and Ecological Problems Mathematical Models. Springer
    Berlin, New York. 1980
5) A. C. R. Dean and Sir C. Hishelwood : Growth Function and Regulation in
    Bacteria Cells. Oxford University Press. London. 1966
6) T. Hattori : Reports of Institute for Agricultural Research. Tohoku Univ.
    34(1985),1

7) H. Hara : Phys. Rev. B15(1979), 4062, B31 (1985), 4612

8) H. Hara : Z. Physik B43 (1981), 321, B45 (1981),159

9) H. Hara : Busseiron Kenkyu 44 (1985), 505

10) H. Hara, T. Obata and S.J. Lee : (to appear in Phys. Rev. A)

11) E.H. Kerner : Bull. Math. Biol. 19 (1957), 121, 21 (1959),159

PATTERNS IN THE PROTEIN MOTIONS

THROUGH THE BIOMEMBRANE

Takeo Izuyama

Institute of Physics
College of Arts and Sciences
University of Tokyo, Komaba
Meguro-ku, Tokyo 153, Japan

A biomembrane consists of the lipid bilayer and the proteins (and also some sugar chains, which are neglected here). The motion of the proteins are important subjects in the cell physiology. Only the lateral motion is considered here. In the physiological region of temperature, the lipid bilayer takes a kind of liquid crystal phase. Such layer is usually regarded as a two-dimensional viscous and incompressible fluid sandwiched by semi-infinite regions of water. This view was advocated by Saffman and Delbrueck[1] in their theory of the mobility of a membrane protein. The celebrated Saffman-Delbrueck theory predicts that the size dependence of the protein mobility is weak; the logarithmic dependence on the size of the protein. This result implies that the back flow in the two-dimensional fluid induced by the protein motion extends to a long-range region. Hence it is expected that there exist long-range interactions among membrane proteins. It is the purpose of this report to show the interprotein forces arising from the hydrodynamical origin. It will be seen that the proteins can not be regarded as independent Brownian particles.

According to Saffman and Delbrueck, the momentum dissipation from the two-dimensional fluid to the bulk water is rather weak. They ascribed, though implicitly, the momentum dissipation to a purely hydrodynamical mechanism; through the boundary condition at the interface separating the two fluids with different viscosities. Actually there exists a strong coupling between a water molecule and the alcohol head of a lipid molecule. Another problem is that the hydrocarbon chains of lipids may be mutually entangled. Hence there is another possibility that the lipid molecules themselves move through stochastic perturbations, by receiving (or emitting) momenta from (or into) the reservoir. The stochastic model of this type has been investigated by Y. Suzuki and myself[2]. The size dependence of the protein mobility can be weak even in our model, though not so remarkable as the Saffman-Delbrueck prediction. The weak size dependence appears for certain sizes of the proteins, which seems to be the actual dimensions of the proteins observed in the membranes. The weak dependence originates from an entropy reason; there is a characteristic pattern in the statistical distribution of the lipid molecules in the vicinity of the wall produced by the protein molecule. There is a tendency of surrounding the protein by an empty bag.

Consider first the hydrodynamical interprotein force.

Let us start with the hydrodynamical equation

$$(\vec{v} \cdot \text{grad})\vec{v} = - \frac{1}{\rho} \text{grad } p + \nu \nabla^2 \vec{v} - \frac{1}{\tau} \vec{v} , \qquad (1)$$

where $\vec{v}$ is the velocity field of the two-dimensional fluid, ρ is the two-dimensional density, ∇^2 and grad are the two-dimensional operators, p is the hydrostatic pressure, and τ is the relaxation time of the momentum of the lipid assembly. For the incompressible fluid, we have div $\vec{v} = 0$ and hence $\nabla^2 p = 0$. There is no retardation in the interprotein action, since the fluid is incompressible. A protein molecule is regarded as a hard disc with radius a placed in the two-dimensional fluid.

Consider a disc "A" moving with a constant velocity $\vec{U}_A$. We can take the coordinates (x,y) such that the origin $(0,0)$ is fixed at the center of the disc and the x-axis is taken along the direction of $\vec{U}_A$. The polar coordinates associated with this Cartesian coordinates are (r, θ). In this coordinate system, we write $\vec{v} = \vec{U}_A + \vec{u}$. Then the last term in Eq.(1) is to be replaced by $-\frac{1}{\tau} u$. The boundary conditions are $\vec{u} = 0$ for $r \to \infty$ and $\vec{v} = 0$ at $r = a$.

The important parameter in our problem is

$$\kappa \equiv \sqrt{(U_A/2\nu)^2 + (1/\nu \tau)} \underset{\sim}{=} (1/\sqrt{\nu \tau})$$

Consider the case; $\kappa r \gg 1$, $r \gg a$ and $\kappa a \ll 1$. Only the linear terms with respect to U_A are retained. Then the pressure at (x,y) given by the back flow is found to be

$$p = \frac{\rho}{\tau} \frac{U_A a^2}{\frac{1}{2} - \gamma - \log \frac{1}{2}\kappa a} \frac{x}{r^2}$$

where is the Euler's constant.

Consider another disc "B" centered at (x,y), which is moving with the velocity $\vec{U}_B$. We linearize the equations with respect to U_A and U_B. The field of the back flow is then expressed as a superposition of the two fields produced, respectively, by U_A and by U_B. The force acting on the disc "B" is hence obtained as

$$F_x^{(B)} = C \pi a \frac{\cos 2\theta}{r^2} ,$$

$$F_x^{(B)} = C \pi a \frac{\sin 2\theta}{r^2} ,$$

where

$$C = \frac{\rho}{\tau} \frac{U_A a^2}{\frac{1}{2} - \gamma - \log \frac{1}{2}\kappa a} .$$

The long range nature of the interacting forces leads to characteristic patterns in the motion of the many protein system. For example, there is a tendency towards the formation of lines (of proteins) or that of cycles in the form of rings. Such patterns are disturbed by the perturbation from the thermal reservoir. Details will be published elsewhere.

Let us turn our attention to the stochastic model(Ref.2). This is explained in Fig.1. The black disc represents a lipid molecule and the square is a protein molecule. They are placed on a squre lattice. The lipids are not allowed to overlap with each other and with the protein. A lipid sitting at a site can jump to the nearest neighbor site, if it is empty. Every jump is stochastic. The protein can shift its position through one lattice spacing, if such jump is possible without excluding or overlapping any lipid. For example, in the case of the protein shown in Fig.1, it can move to the right or to the downward direction. The protein motion in the other directions is prohibited in this example.

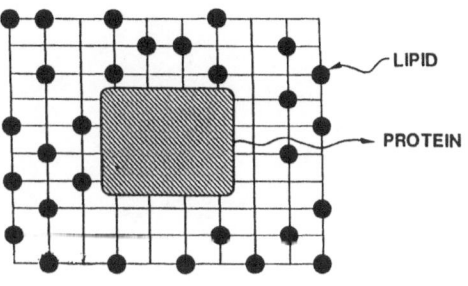

Fig. 1.

An extensive numerical study of this model has been carried out by Y. Suzuki. The diffusion coefficient D of the protein has been obtained. The dependence of D on the protein size is shown in Fig.2. The Saffman-Delbrueck prediction of the protein diffusion is also shown in Fig.2. The hydrodynamical result for D is shown for the two cases; [the viscosity of the two-dimensional lipid flow] = 1 poise and = 2 poise. The experimental data are also shown in Fig.2. Our result is not poor. The main reason for this is the entropy effect on the distribution of the lipids

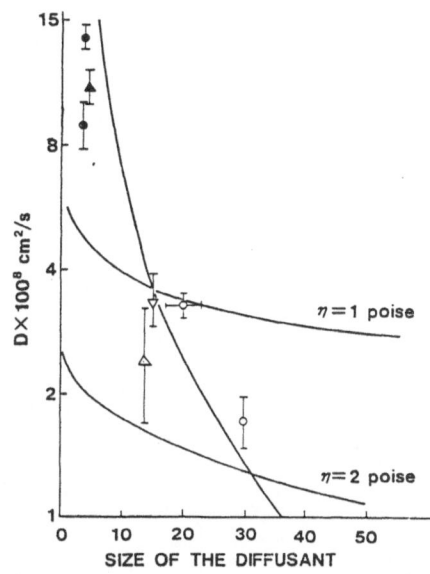

Fig. 2.

under the Brownian motion. As will be reported elsewhere, the entropy effect is seen in the pattern of the distribution of the lipids around the protein. The lipids tend to avoid the nearest neighbor positions from the protein wall.

References

1. P. G. Saffman and N. Delbrueck, Proc. Natl. Acad. Sci. USA, 72, 3111 (1975).   P. G. Saffman, J. Fluid Mech., 73, 593 (1976).

2. Y. Suzuki and T. Izuyama, to be published.

AGGREGATION SYSTEM WITH A SOURCE PREFERS A POWER LAW SIZE DISTRIBUTION

Hisao Hayakawa[†] and Hideki Takayasu[§]

Faculty of Science, Kobe University, Kobe 657, Japan
[†]Present address; Department of Physics, Kyushu University
Fukuoka 812, Japan
[§]Present address; Applied Physics, Yale University, New Haven
Connecticut 06520, USA

Solving the Smoluchowski equation we show that the cluster size
distribution tends to follow a power law in the aggregation system with a
permanent source and a negligible sink . This statement is confirmed by
simulations.

INTRODUCTION

Aggregaton is an irreversible physical process in which a number of
basic units (monomers) stick together to form clusters. This process is
one of the most characteristic processes in many fields of physical,
chemical and biological sciences.[1-6] We often observe statistical and
geometrical scaling properties in aggregation processes. It is important
to make clear the reason why aggregation has scaling properties.   In this
paper we show that the cluster size distribution tends to follow a power
law in the aggregation system with a permanent source and a negligible
sink.

In many cases, the aggregationis controlled by diffusion and reaction
processes with characteristic times, t_d and t_r. The t_d is the typical time
needed for clusters to come close together. The t_r is the time taken for
the formation of a chemical bond between contacted clusters. If one of the
time scales is much greater than the other, it is possible to find a
simplified kinetic description on the time scale of the slower process. We
refer to two extreme cases as the diffusion limited aggregation (DLA) if
$t_d \gg t_r$, and the reaction limited aggregation (RLA) if $t_d \ll t_r$.   In both
cases, the time evolution of clustering processes can be described by the
Smoluchowski equation without account of spatial fluctuations.

We first show that a universal power law of cluster-size distribution
results from the Smoluchowski equation.[6] When there occurs a sol-gel
transition (or gelation) , we obtain a self-consistent post-gel solution.
If there is no gelation (in a non-gelling system), we find a steady
solution obeying a power law of cluster size. We also perform numerical
calculations for two models of DLA and get asymptotic power laws. Finally,
we discuss the origin of power law of cluster-size distribution.

GENERAL THEORY

The aggregation processes with creation of monomers are often observed in aerosol systems[7]. Aerosols are created by various processes, such as smoke particles from fire, sand storm, condensation of water vapor in the atmosphere, and the nucleation through chemical reactions. Although the Smoluchowski equation has been extensively reviewed[8], there are only a few results for the aggregation equation with injection [9-10].

The Smoluchowski equation involving a source is represented by

$$\dot{c}_k(t) = \frac{1}{2} \sum_{i+j=k} K_{ij}c_i(t)c_j(t) - c_k(t) \sum_j K_{kj}c_j(t) + I_k . \qquad (1)$$

Here $c_k(t)$ denotes the concentration of clusters of size k (k-mer) , and K_{ij} , the aggregation kernel or the rate coefficient, denotes the probability of coalescence of i-mer and j-mer in unit time. We introduce I_k as the permanent source. Although the effect of sink often appears in real physical processes such as gravitational sedimentations, we restrict our interest to an idealized case that we can neglect the effect of sink. In the absence of sink, a steady solution obeying a power law has been obtained in the case of $K_{ij}=K(ij)^n$. [9-10]

Most of aggregation kernels used in the description of physical phenomena are homogeneous functions of i and j, at least for large cluster sizes [11]. Such a kernel has the following properties:

$$K_{ij} \simeq i^{\mu}j^{\nu} \qquad (j \gg i) \quad , \qquad (2a)$$

$$K_{ai,aj} \simeq a^{\lambda} K_{ij} \qquad (\lambda = \mu+\nu; \quad ,a:const.) \quad . \qquad (2b)$$

Since the average number of reactive sites on a cluster cannot increase faster than its size, we impose restrictions $\lambda \leq 2$ and $\nu \leq 1$ [11]. Under these restrictions three classes are distinguished according to the value of μ, $\mu>0$ (class I), $\mu=0$ (class II) and $\mu<0$ (class III).[11] A typical example of the class I is RLA, such as polymerization, and that of the class III is DLA, occurring in the aerosol aggregation.

To simplify our discussion, we restrict ourselves to the simplest case satisfying (2a) and (2b). We choose K_{ij} and I_k as

$$K_{ij} = K(i^{\mu}j^{\nu}+i^{\nu}j^{\mu}) \qquad (\mu \leq \nu) \quad , \qquad (3a)$$

$$I_k = I\delta_{k1} , \qquad (3b)$$

where K and I are constants.

STEADY SOLUTION AND POST-GEL SOLUTION

First, we consider the asymptotic form of cluster size for a long time limit without gelation. As time goes onto infinity, the time evolution of cluster-size distribution ceases. Therefore, the Smoluchowski equation is reduced to the following form by substituting Eqs.(3a) and (3b) into Eq.(1).

$$0 = K \sum_{i+j=k} i^{\mu}j^{\nu} c_i c_j - Kk^{\mu}c_k M_{\nu} - Kk^{\nu}c_k M_{\mu} + I\delta_{k1} , \qquad (4)$$

where M_n is the n-th moment of c_k

$$M_n = \sum_k k^n c_k \quad . \tag{5}$$

In order to solve Eq.(4), we introduce a generating function $f_n(x) = \sum k^n c_k e^{-kx}$ with the following behavior for small x.[12]

$$f_n(x) \sim M_n + a(n) x^{\alpha(n)} \quad , \tag{6}$$

where $\alpha(n)$ must be positive because of continuity of the generating function at x=0. We are interested in the case that $\alpha(n)$ is not equal to unity. In such cases M_{1+n} is divergent, and the second term of Eq.(6) expresses the leading sigularity of $f_n(x)$ at x=0. Multiplying e^{-kx} and 1 by Eq.(4), and summing them over all k and subtracting one from the other, we get from Eq.(6)

$$Ix \simeq K a(\mu) a(\nu) x^{\alpha(\mu)+\alpha(\nu)} \quad . \tag{7}$$

Then the generating functions are not regular and the following relations are derived;

$$\alpha(\mu)+\alpha(\nu) = 1, \quad K a(\mu) a(\nu) = I \quad . \tag{8}$$

As $\alpha(n)$ (n is equal to μ or ν) is not integer, Eq.(6) is identical to [12]

$$k^n c_k \simeq \frac{a(n) k^{-\alpha(n)-1}}{\Gamma(-\alpha(n))} \quad \text{(for } k \gg 1 \text{)}, \tag{9}$$

where $\Gamma(x)$ denotes the gamma function. From Eq.(9), we can see $\alpha(\mu)+\mu = \alpha(\nu)+\nu$. From the above arguments, the distribution of cluster size finally yields

$$c_k \simeq \left(\frac{I(1-4\alpha^2)\cos\pi\alpha}{4\pi K} \right)^{1/2} k^{-(3+\mu+\nu)/2} \quad , \tag{10}$$

where $\alpha=(\nu-\mu)/2$. The applicable region of Eq.(10) is $\mu \leq \nu$, $|\mu+\nu|<1$ and $|\mu-\nu|<1$ (see Fig.1) due to the divergency of $a(\mu)$ and $a(\nu)$, the finiteness of the total number of clusters and the regularity of the gamma function. When there is no injection, i.e. I=0, we obtain only the trivial solution from Eq.(10). Therefore, we conclude that the balance between injection and aggregation leads to an asymptotic power law distribution of cluster size. The Brownian aggregation in the continuum region with $K_{ij}=K(i^{1/3}+j^{1/3})(i^{-1/3}+j^{-1/3})$ is the most important in colloid and aerosol science.[13] We can apply our method directly to the Brownian aggregation and obtain an asymptotic power law of cluster size as $c_k \sim k^{-3/2}$. The power law tail can be observed in the intermediate range of cluster size, because Eq.(4) approximately holds for suitable k.

We also get a self-consistent post-gel solution at a finite time by the same method. We use two generating functions ; $f_n(x,t) = \sum k^n c_k(t) e^{-kx}$

and $g(x,t)=\Sigma c_k(t)e^{-kx}$. We get the solution

$$c_k(t) \simeq \left(\frac{(\dot{M}_1-1)(4a^2-1)\cos\pi a}{4\pi K}\right)^{1/2} k^{-(3+\mu+\nu)/2} . \tag{11}$$

The finite total mass of clusters at finite time leads to $\mu+\nu>1$. The conditions $\mu,\nu>0$ and $|\mu-\nu|<1$ are derived by the divergence of $a(\mu)$ and $a(\nu)$. Equation (11) expresses a post-gel solution, because nonzero $\dot{M}_1-1$ means a violation of mass conservation in the sol phase. This post-gel solution is essentially the same as that of noinjection obtained by Hendriks et al.[12]

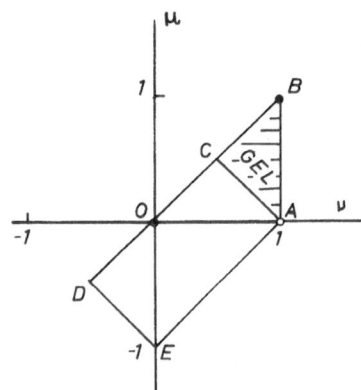

Fig. 1.   Solutions in Eqs.(10) and (11) in the parameter (μ,ν) space. Non-gelling steady solutions are inside of the ACDE and self-consitent post-gel solutions are inside of ABC. The border lines but AB and BD are excluded.

SIMULATION FOR SPECIAL MODELS

Two simple models of DLA are simulated to support the above results.[14],[15] One is a one-dimensional deterministic model of the aggregation system of inelastic particles (hereafter we call this model the ballistic model[14]). The other is a stochastic model of the aggregation system with uniform injection where particles walk at random in the d-dimensional Euclidean lattice (the STN model which is named after Scheidegger[16], Takayasu and Nishikawa[15]). When the number of clusters are kept constant, we find that the steady distribution of cluster sizes obeys a power law on each model.

The ballistic model consists of flying particles with linear trajectories. When a particle collides an adjacent particle, they form a cluster and a new particle is injected at a random position. The mass and momentum of each cluster are conserved in each collision. We obtain an asymptotic power law of cluster size as $c_k \sim k^{-4/3}$ (Fig.2). The power index of cluster-size distribution is insensitve to the velocity and mass distributions for initial and injected particles so far as we have examined. This model can be regarded as an idealized model of the forced Burgers turbulence or the growth of interfaces. When the mean free path is much larger than the linear size of partcles, the source induced aggregations like our model are also realized in aerosol systems.

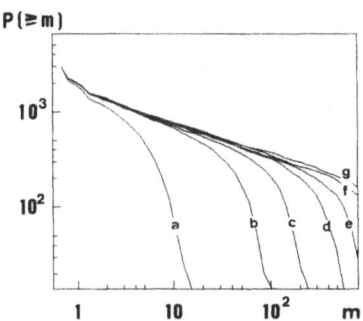

Fig. 2. Log-log plots of the time evolution of the cumulative distribution of cluster size from (a) to (g). We can see the asymptotic power law represented by $P(\geq m) \sim m^{-0.33}$.

On the other hand, the STN model can be regarded as a river model by Scheidegger[16] or a model of Brownian aggregation of colloidal particles. In this model, rain (injection) is falling steadily and uniformly on a slope. As rain drops drip down at random along the slope, they collide with each other to make a stream. We can consider this aggregation as the case K_{ij} is constant, since the mobility of each cluster is independent of its size (Stokes' frictonal force was not considered). In the one-dimensional case simulations give a power law $c_k \propto k^{-4/3}$, which is also analytically obtained. In the STN model, large dimensional simulations agree with Smoluchowski's prediction, $c_k \propto k^{-3/2}$. Our simulations suggest that there is an upper critical dimension, d_c, which gives a criterion for the validity of mean field theory or the Smoluchowski description .

The above simulations confirm the universality of power law size distribution in the aggregation system with source.

DISCUSSION AND CONCLUSION

Why does the presence of source lead to power law size distribution? We think that the power laws are connected with a kind of phase transition. In the aggregation system with source, the order parameter is M_0 which discriminates the presence or absence of the steady disitribution of clusters

Equation(1) has simple scaling properties with respect to the feed or nucleation rate I.[17] Indeed, introducing the time and concentration scaled as

$$\tilde{t} = I^{1/2}t \quad \text{and} \quad \tilde{c_k}(\tilde{t}) = c_k(t)/I^{1/2} \quad . \tag{12}$$

And substituting Eqs.(12) into Eq.(1), we can see that in the long time limit the solution is written in the scaling form

$$c_k(t,I) = I^{1/2}f_k(I^{1/2}t) \quad . \tag{13}$$

It follows that the steady state cluster distribution is described as $c_k \sim I^{1/2}$, which is consistent with Eq.(10). This scaling can be considered as a dynamical phase transition at I=0. Racz suggested the injection rate I corresponds to an external field.[17] In the critical phenomena, we are familiar with the scaling form of cluster size distribution . This scaling is related to the ordering by external field at the critical point.

We summarize our results. (I) the balance between injection and aggregation yields an asymptotic power law distribution of cluster size in the non-gelling system. This is confirmed by the analytic method and simulations. It seems that this property is connected with a kind of dynamical phase transition. (II) In the gelling system, the effect of injection is not essential. We emphasize the result that in mean field theory $c_k \propto k^{-a}$ with $a=(3+\mu+\nu)/2$ is a universal form in both gelling and non-gelling systems. Injection and negligible removal are common in nature. Our analysis suggests an origin of often observed power law distributions of cluster size. The power law distribution represents one of the characteristic aspects of fractals. We believe our results to be a first step towards the understanding why we often see the fractals in nature.

The authors are grateful to M.Yamamoto and J.J.Kim for stimulative discussions. One of the authors (H.T.) also thanks to I.Nishikawa for collaborating work on the STN model.

References

1. S.K.Friedlander, Smoke, Dust and Haze (New York: Willy-Int.Sc.1977).
2. R.M.Ziff, J.Stat.Phys 23, 241,(1980).
3. J.Silk and S.D.White, Astrophys.J. 223,L59,(1978).
4. R.W.Samsel and A.S.Perelson, Biophy.J. 37,493,(1982).
5. K.Binder and D.Stauffer, Adv.Phys. 25,343,(1976).
6. H.Hayakawa, J.Phys.A: Math.Gen. 20,L801,(1987).
7. R.L.Drake, in Topics in Current Aerosol Research, vol.3,ed. G.M.Hiddy and J.r.Brock (New York: Pergamon Press 1972).
8. M.H.Ernst, in Fractals in Physics, ed. L.Pietrorero and E.Tosatti, (Amsterdam: Elsevier Sci.Publ.B.V.1986).
9. J.D.Klett, J.Atmos.Sci.12,380,(1975).
10. W.H.White, J.Colloid Interface Sci 87,204,(1982).
11. P.G.J.van Dongen and M.H.Ernst, Phys.Rev.Lett. 54,1396,(1985).
12. E.M.Hendriks, R.M.Ziff and M.H.Ernst, J.Stat.Phys. 31,519,(1983).
13. S.Chandrasekhar, Rev.Mod.Phys. 15,1,(1943).
14. H.Hayakawa, M.Yamamoto and H.Takayasu, Prog.Theor.Phys. 78,1,(1987).
15. H.Takayasu and I.Nishikawa, in Proc.of 1st.Int.Symp.for"Science on Form" ,ed.S.Ishizuka (KTK Sci.Publ. 1986)
16. A.E.Scheidegger, Bull.I.A.12,(1),15,(1967).
17. Z.Racz, Phys.Rev.A 32,1129,(1985).

ONE-DIMENSIONAL AGGREGATION OF ANISOTROPIC PARTICLES IN AN EXTERNAL FIELD

Sasuke Miyazima
Department of Engineering Physics
Chubu University, Kasugai Aichi 487, Japan

Center for Polymer Studies and Department of Physics
Boston University, Boston, MA 02215

Paul Meakin
Central Research and Development Department, Experimental Station
E. I. duPont de Nemours and Company, Inc., Wilmington, DE 19898 USA

Fereydoon Family
Department of Physics, Emory University, Atlanta, GA 30322 USA

Introduction

The formation of clusters by diffusion-limited aggregation has become an important subject with a broad range of applications in physics, chemistry, biology and engineering.[1-3] The investigation of these phenomena has a long history, but it is only quite recently that interest has developed in simple non-equilibrium aggregation models which help us to describe and understand aggregation processes.

Interest in this area was initially stimulated by the discovery of Witten and Sander[4] that a diffusion-limited aggregation model in which particles are added, one at a time, to a growing cluster of aggregate of particles leads to the formation of scale invariant (fractal[5]) structures.

These structures closely resemble those formed in a variety of preocesses, such as dielectric breakdown, fluid-fluid displacement in Hele-Shaw cells and porous media, random dendritic growth, and the dissolution of porous materials. The diffusion-limited aggregation model may also provide a basis for understanding some types of biological growth processes and mechanical failure. A more realistic model for colloidal aggregation is provided by the diffusion-limited cluster-cluster aggregation model developed by Meakin[6] and by Kolb et al.[7] This model can be used to simulate both the kinetic (time-dependent) and structural aspects of diffusion-limited colloidal aggregation. For a wide variety of aggregation processes the time dependent cluster size distribution, $N_s(t)$, can be represented by the dynamic scaling form

$$N_s(t) \sim s^{-2} f\left(\frac{s}{S(t)}\right), \tag{1}$$

where $N_s(t)$ is the number of clusters of size s at time t and $S(t)$ is the mean cluster size[8]

$$S(t) = \frac{\Sigma s^2 N_s(t)}{\Sigma s N_s(t)}.$$

The realization that colloidal aggregates commonly exhibit a fractal geometry has stimulated renewed interest in aggregation kinetics, a subject that has been investigated throughout this century. A description of some of the earlier work in this field is provided in the works of Drake[9] and Friedlander.[3]

In this paper we are concerned with the aggregation of oriented particles to form linear structures. Here we assume that oriented particles aggregate to form oriented rods which can aggregate with other particles and rods only at their ends. This model has been investigated previously by de Gennes et al.[10] In this paper we will compare our results with experimental investigations of the aggregation of polystyrene spheres in a magnetic fluid[12] and rouleaux formation from red blood cells.

Simulations

Most of the simulations were carried out using 512×1024 (w $\times$ ℓ) square lattices and $64 \times 64 \times 512$ (w $\times$ w $\times$ ℓ) cubic lattices with periodic boundary conditions. Initially $N_o = \rho w^{d-1}\ell$ particles are placed randomly on an $w^{d-1}\ell$ site lattice. In all of our simulations the diffusion coefficient D_s for clusters of mass s is assumed to be given by $D_s = D_o s^\gamma$ where D_o is a constant. Clusters (rod or particles) are selected randomly and moved by one lattice unit in a randomly selected direction on the lattice if a random number x uniformly distributed over the range $0 < x < 1$ is smaller than $D_s/D_{\max}$. D_s is the diffusion coefficient of the randomly selected cluster of size s and $D_{\max}$ is the maximum diffusion coefficient for any cluster in the system. After each cluster has been selected the time is incremented by an amount δt given $\delta t = 1/(N(t)D_{\max})$ irrespective of whether or not the cluster is moved. If the randomly selected direction of motion would result in the overlap of one cluster with another the move is not made and another cluster is randomly selected (however the time in incremented by δt). After each cluster has been moved, its ends are examined; if the site at the end of the cluster is occupied by the end of another cluster, the two clusters irreversibly combined. In most cases the results from a number of simulations (10-100) were averaged to reduce the statistical uncertainty.

Results

In the two-dimensional case, 10,000 particles were distributed on square lattices of 512×1024 sites corresponding to a density of 0.019. Three kinds of diffusion coefficient, $D(s) \sim s^o$, $D(s) \sim s^{-0.5}$, and $D(s) \sim s^{-1.0}$ were investigated and the dependence of cluster-size distribution on the exponent γ was examined. Details concerning the total number of clusters, the mean size of clusters and the time-dependent cluster-size distribution were obtained, but only a snapshot of two-dimensional lattice for $\gamma = -1$ is shown in Fig. 1, where 8000 particles are initially distributed and the mean size of the cluster is 20. 2. The exponent (z) that describes the dependence of the mean cluster size $S(t)$ on t becomes smaller as γ increases. Above the critical dimension $(d_c = 2)$ the exponent z is given by[11] $z = 1/(1 - \gamma)$, and below the critical dimension,[12] $z = d/(2 - d\gamma)$. On a two dimensional square lattice we find that $z = 1$, 0.65 and 0.5 for $\gamma = 0$, 0.5 and 1.0 in accord with the theoretical values of 1, 2/3, and 1/2, respectively.

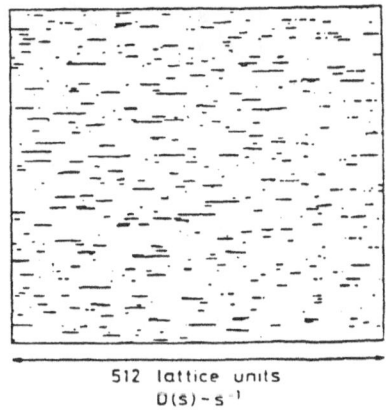

Fig.1. Pattern of formed linear roads in the 512x512 square lattice. Initially 8000 particles are distributed and the average cluster size is 21.16 and γ=0.

512 lattice units
$D(s) \sim s^{-1}$

Scaling Theory

Under conditions where there is only one cluster size in the system (which can be represented by the mean cluster size $S(t)$) it is natural to attempt to describe the cluster size distribution in terms of the scaling forms

$$N_s(t) \sim S^{-\theta} f\left(\frac{s}{S(t)}\right). \tag{2}$$

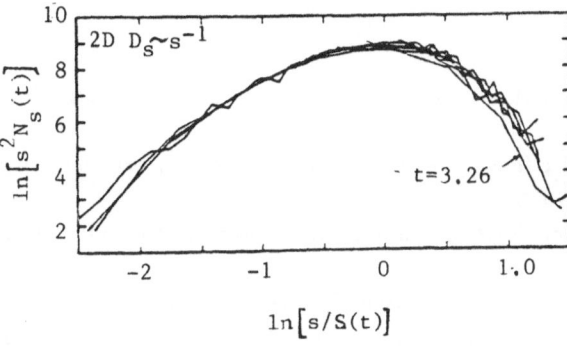

Fig.2. Test of the scaling equation using the results from two-dimensional simulation. The diffusion coefficient exponent γ is −1.

Since the total number of particles in the system is conserved the exponent θ in (2) must have a value of 2.0.

To test the validity of this scaling theory we have checked the data collapse by plotting $\ln[s^2 N_s(t)]$ versus $\ln[s/S(t)]$ for $\gamma = 1.0$ in the two-dimensional case. The results are shown in Fig. 2.

Comparison with Experiments

Many experimental and theoretical investigations of the structure[13] and kinetics of colloidal aggregation processes have been carried out. However very little of this has been concented with the quantitative aspects of the formation of oriented rod-like structures. Kernick et al[14] have measured the time dependence of the mean rouleaux length in the aggregations of red blood cells. Some of their results are shown in Fig. 3. In this case the mean rouleaux length reaches a constant value with increasing time as

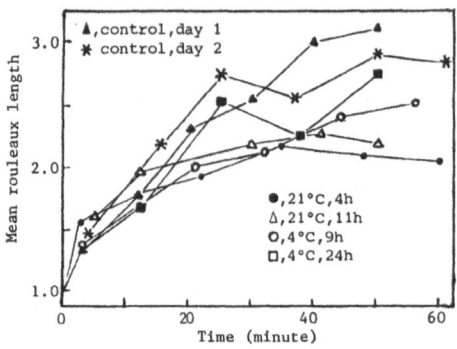

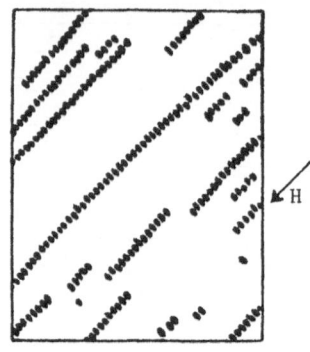

Fig.3. Mean rouleau length against t for fresh blood and blood stored for various times at room temperature and 4°C.

Fig.4. Polystyrene spheres in a magnetic-fluid film with an external field H =1200oe.

a result of dissociation processes which are not included in our model. A considerable amount of work has been carried out on aggregation processes in magnetic fluids which has been motivated by their practical importance. For example, Fig. 4 shows the formation of a two-dimensional oriented state formed by polystrene "holes" in a magnetic fluid oriented in a magnetic field.[15] However quantitative measurements of the aggregation kinetics which are suitable for comparison with our simulation results do not appear to be available. Irrespective of the outcome of any detailed comparisons between simulation results and experiments we expect the scaling form given in (2) to be valid for most simple systems.

References

1. F. Family and D. P. Landau (eds.), *Kinetics of Aggregation and Gelation* (North-Holland, Amsterdam, 1984).

2. H. E. Stanley and N. Ostrowsky (eds.), *On Growth and Forms* (Martinus Nijhoff, Dordrecht, The Netherlands, 1986).

3. K. Friedlander, *Smoke, Dust and Haze: Fundamentals of Aerosol Behavior* (Wiley, New York, 1977).

4. T. A. Witten, Jr. and L. M. Sander, *Phys. Rev. Lett.* **47**, 1400 (1981).

5. B. B. Mandelbrot, *The Fractal Geometry of Nature* (Freeman, San Francisco, 1982).

6. P. Meakin, *Phys. Rev. Lett.* **51**, 1119 (1983).

7. M. Kolb, R. Botet and R. Jullien, *Phys. Rev. Lett.* **51**, 1123 (1983).

8. T. Vicsek and F. Family, *Phys. Rev. Lett.* **52**, 1669 (1984).

9. R. L. Drake, in *Topics in Current Aerosol Research*, eds. G. M. Hidy and J. R. Brock (Pergamon, New York, 1972), Vol. III, p. 20.

10. P. G. de Gennes and P. A. Pincus, *Phys. Kondens. Matter* **11**, 189 (1970).

11. T. Viscek and F. Family in Ref. 1; M. H. Ernst and E. M. Hendriks, *Ibid.*

12. K. Kang, S. Redner, P. Meakin and F. Leyvraz, *Phys. Rev. A* **33**, 1171 (1986).

13. For a bibliography on ferrofluids, see M. Zahn and K. E. Shenton, *IEEE Trans. Magn.* **16**, 387 (1980); P. C. Jordan, *Mol. Phys.* **25**, 961 (1973).

14. D. Kernick, A. W. L. Jay, S. Rowlands and L. Skiko, *Can. J. Physiol. Pharmacol.* **51**, 680 (1973).

15. A. T. Skjeltorp, *Phys. Rev. Lett.* **51**, 2306 (1983)./end

INCOMMENSURATE-COMMENSURATE TRANSITION IN THE
LONG PERIOD SUPERLATTICE IN Ag-Mg ALLOYS

H. Sato[*], Y. Fujino[**], M. Hirabayashi[****], and Y. Koyama[***]

*   School of Materials Engineering, Purdue University
        W. Lafayette, IN 47907 USA
**   Department of Nuclear Engineering, Tohoku University
        Sendai, Japan
***   Department of Metallurgy, Tokyo Institute of Technology
        Tokyo, Japan
****   Institute for Materials Research, Tohoku, University
        Sendai, Japan

Long period superlattice (LPS) in alloys is a periodically modulated
structure of normal ordered structure found in noble metal alloys of
close-packed structure, and its modulation mode is basically that of
antiphase formation. Its period is specified by 2M, where M is the number of the unit cells of the normal superlattice included in one antiphase domain of the structure, and experimentally M is usually not an
integral number[1-3]. The period 2M which exists in one of the <100>
directions of the cubic fcc lattice (one dimensional (LPS) is determined
by the superposition of either two (AB type alloys) or four (A_3B type
alloys) Q vectors with the length of $2k_F$ in the <110> directions, in
which flat parts of the Fermi surface exist. At these parts, split
Brillouin zone boundaries contact.[1-3] By superposing charge density
waves (CDW) with these Q vectors and by comparing the resultant wave with
the atomic arrangement, it is found that the larger atoms of alloys
(such as Au atoms in the Cu-Au alloys) always occupy the maximum parts
of the charge density while the smaller atoms occupy the minimum parts.
Then antiphase boundaries exist where the deviation of the location of
atoms from this condition becomes maximum. Further, if atoms are assumed
to shift towards the charge density maximum,[4] the result agrees well
with experimental data on atomic shifts by Iwasaki, et al.[5-8] (Fig. 1).
In other words, the LPS is considered to be a modulated structure created
by the formation of CDW in the three dimensional crystal[1-3] and the antiphase formation is a more efficient way to compensate the charge distributions of CDW in ordered alloys than static phonons commonly observed
in one- and two-dimensional systems.

On the other hand, the period is also affected by the constraints
of the structure and can deviate from that derived from $2k_F$ when the constraints are strong. As the degree of order develops, the antiphase
modulation tends to be a square wave and to be a commensurate modulation.
In other words, in this system, the incommensurate-commensurate transition similar to that observed in transition metal dichalcogenides[9-12]
should occur under a proper experimental condition. However, in LPS,
because of the frustration problem in close-packed structures, the

disordered phase is favored down to low temperatures, and the transition temperature to the ordered state is generally lower than the normal-incommensurate transition point. In other words, LPS forms directly from the disordered state and is not expected to be observed.

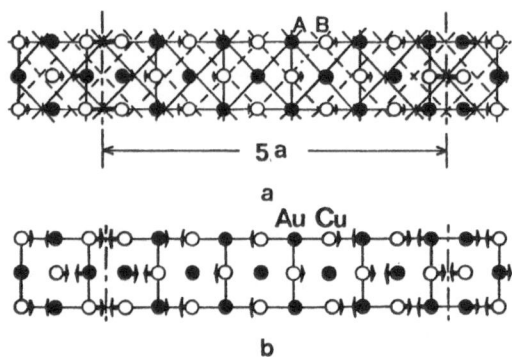

Fig. 1. Relation between the charge density distribution and atomic arrangement of CuAu II type LPS assuming that M=5. Under this condition, the wave length of the <010> component of the charge density (in the direction of the period of LPS) is given by 0.909a.

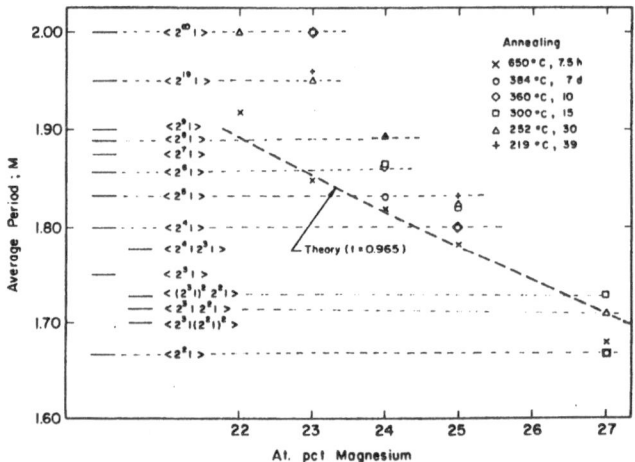

Fig. 2. Dependence of M of the Ag-Mg alloys upon heat treatment.

Non-integral (or incommensurate) values of M in LPS is generally due to the mixture of domains of different sizes and to the disordering (lowering of the degree of order) near the antiphase boundaries.[3] However, in LPS with smaller M, the disordering near the antiphase boundaries is energetically unfavorable compared to LPS with larger M, because in the former, a larger portion of the total volume has to be disordered, and commensurate structure tend to be favored. Indeed in alloys with short periods such as Ag-Mg alloys and Ag-Zn alloys, it has been found that the tendency for LPS to stick to the commensurate period such as M=2 is very strong.[13,14]

The LPS in the Ag-Mg alloys appears in the composition range near Ag_3Mg and the values of M ranges between ~1.6 and 2.0.[3,15] In the Ag-Mg alloys, in which the period M is smaller than two, the non-integral value of M is derived as the average value of the mixture of antiphase domains of M=1 in the matrix of the commensurate M=2 structure. Our experiment shows that, by quenching Ag-Mg alloys with compositions from 22% to 27% Mg from 650°C (far above the order-disorder temperature), the values of M of LPS thus formed are found to agree well with the theoretical values expected from the size of $2k_F$.[15] However, upon annealing, the values of M change towards the values of M close to the theoretical values but are represented by regular structures such as $<2^j1>$. In 22% and 23% Mg alloys in which theoretical M values are ~1.89 and ~1.85, they assumed the commensurate structure M=2 (Fig. 3). The $<2^j1>$ structures are those with an ordered arrangement of domains of M=1 at every jth domains of M=2. The tendency for the Ag-Mg alloys to take such regular structures was first pointed out by Guymont, et al.[16] The $<2^j1>$ type structures are those structures whose periods match with the period of the normal structure, but with M values different from the commensurate structure M=2. In this sense, such structures can be thought of as 'discommensurate' structures.

The incommensurate state of LPS in the Ag-Mg alloys is thus the modulation of the commensurate structure of M=2 by means of domains of the M=1 structure.[17] Generally, domains of the M=1 structure are isolated in the direction of the period but are connected in the direction perpendicular to it and thus form discommensuration lines.[15,18] A similar situation was observed in 2H-TaSe$_2$ in the lower temperature range of the incommensurate phase and was termed as the 'incommensurate modulation of the commensurate structure (by means of domains)'[12].

Observation of ordered discommensurate structures such as $<2^j1>$ [6] and others in this alloy system have been reported by Guymont, et al.[6], and de Fontaine, et al.[19,20] Also, the stability of this type of structure has been discussed in terms of the ANNNI (axial next-nearest neighbor Ising) model.[21,22] In Fig. 3, an example of $<2^5\overline{1}>$ structure which appeared in 24 at.% Mg alloy by annealing the alloy at 384°C for 7 days, is shown.[15] On the other hand, the introduction of the M=1 discommensuration lines in the matrix of the M=2 structure introduces the phase shift of $2\pi/4$ across each discommensuration line.[15] Therefore, the assembly of four discommensuration lines eliminates the phase shift. Thus, four discommensuration lines tend to meet at one point when there is a tendency for commensuration. We call this type of pattern a jellyfish pattern (JFP) from its shape.[12,13] In Fig. 3 we also show the observation of such jellyfish patterns (JFP) of four legs in the same alloy annealed at 252°C for 22 days and then annealed at 219.3°C for 39 days.[15] In other words, the alloy has a tendency for commensuration at lower temperatures than 384°C. The motion of JFP's leaves the commensurate structure M=2 in its wake and represents the commensuration process. We also observe JFP's with three legs in 2H-TaSe$_2$ at the temperature range where the commensuration process takes place.[9,10,12] An in-situ observation of the commensuration process of this alloy by TEM was thus planned by keeping this alloy annealed at 219.3°C at a temperature between 252°C and 384°C where the motion of these discommensuration lines was expected.[18]

The in-situ observation of the commensuration process was carried out in a JEM model 2000 EX electron microscope with a double-tilting, side entry heating stage at 325°C. A series of 17 photos was taken in a time span of around 40 minutes, and in Fig. 3, pictures Nos. 1, 5, 11 and 17 of this series are shown.[18] A receding motion of a number of JFP's of four legs is observed. However, special attention should be given to the changes of two different types of JFP's indicated by arrows. The

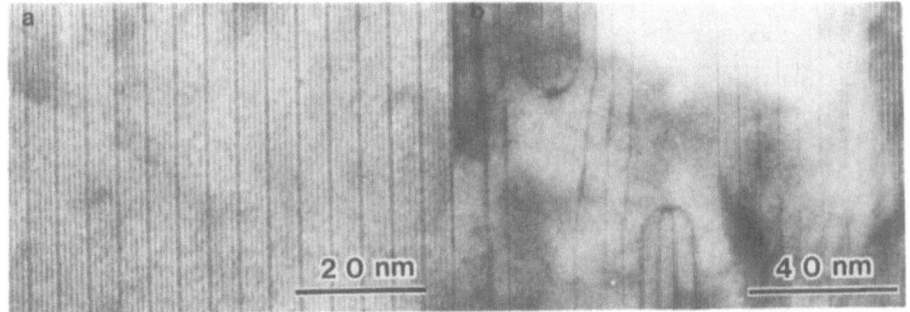

Fig. 3.   Dark field images of Ag-24 at.% Mg alloy using the 110
spot and surrounding satellites.
a) Annealed at 384°C for 7 days. The $\langle 2^5 1 \rangle$ structure
is shown.
b) Annealed first at 252°C for 22 days and then further
annealed at 219.3°C for 39 days. Jellyfish patterns
with four legs are seen.

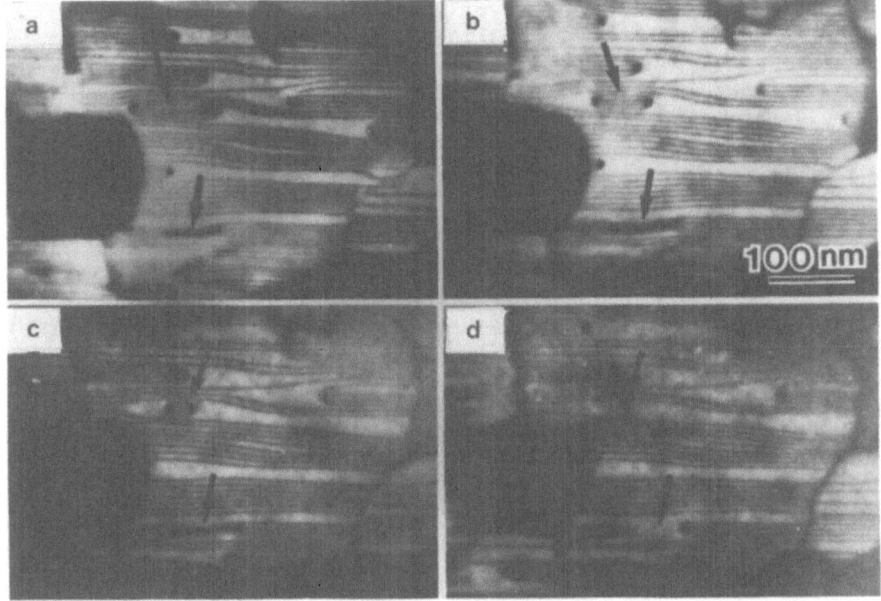

Fig. 4.   A series of in-situ micrographs taken at 325°C which show
the commensuration process of LPS in Ag-24 at.% Mg alloy
annealed at 219°C for 39 days. Only four out of seventeen
micrographs are shown, arranged in chronological order.

one type is that of an elliptical loop and the other is an incipient JFP with a joint part of four discommensuration lines. As the time elapses, it is clearly seen that the loop shrinks and then disappears, while the other type gradually separates into two opposing JFP's. More details are given in Ref. 18. It is, however, noted that the commensuration process is extremely similar to that of 2H-TaSe$_2$[9,10,12] and to that suggested by Kawasaki, et al.[24]

CONCLUSION

Long period superlattice is a periodically modulated structure of the normal ordered structure created by CDW formation. In the Ag-Mg alloys, the commensurate LPS is the M=2 structure, but a large number of periodic discommensurate structures exists which satisfies the matching with the background matrix (normal ordered structure). Because of this reason, the average period M has the main physical meaning and the periods of discommensurate structures are determined by the matching condition of the period and the background matrix. Irrespective of the existence of a large number of periodic structures, LPS in the Ag-Mg alloys thus has the same common characteristics as LPS with larger M, such as that in the Cu-Au alloys.

Commensuration process occurs in a very similar fashion to 2H-TaSe$_2$ and by means of the motion of JFP's with four legs. Here, the lines of the M=1 domains serve as the discommensuration lines. Fig. 3 represents the first in-situ observation of the commensuration process of any kind[18].

The work was supported by the National Science Foundation, Solid State Chemistry Program, grant number DMR 8304314, US-Japan Science Cooperation Program, grant number INT 84-12550, and by the grant of the Japan Society for the Promotion of Science.

REFERENCES

1. H. Sato and R. S. Toth, Phys. Rev. 124, 1833 (1961).
2. H. Sato and R. S. Toth, Phys. Rev. 127, 469 (1962).
3. H. Sato and R. S. Toth, in Alloying Behavior in Concentrated Solid Solutions, edited by T. B. Massalski (Gordon and Breach, New York, 1965, pp. 295-419.
4. Y. Koyama and H. Sato, to be published.
5. K. Iwasaki, M. Hirabayashi and S. Ogawa, J. Phys. Soc. Japan, 21, 1616 (1966).
6. K. Okamura, J. Phys. Soc. Japan, 28, 1005 (1970).
7. K. Okamura, H. Iwasaki and S. Ogawa, J. Phys. Soc. Japan, 24, 569 (1968).
8. H. Iwasaki, K. Okamura and S. Ogawa, J. Phys. Soc. Japan, 31, 497 (1971).
9. K. K. Fung, S. McKernan, J. W. Steeds and J. A. Wilson, J. Phys. C14, 5417 (1981).
10. C. H. Chen, J. M. Gibson and R. M. Flemming, Phys. Rev. B26, 184 (1982).
11. Previous paper.
12. Y. Koyama, Z. P. Zhang and H. Sato, Phys. Rev. V., Sept. 1987.
13. H. Sato and R. S. Toth, Phys. Rev. 139A, 1851 (1964).
14. M. Hirabayashi, S. Yamaguchi, K. Hiraga, N. Ino, H. Sato and R. S. Toth, in Proceedings of the Third Bolton Landing Conference on Ordered Alloys: Structural Applications and Physical Metallurgy, 1969, edited by B. H. Kear, C. T. Sims, N. S. Stoloff, and J. H. Westbrook (Claitor's, Baton Rouge, LA, 1970), pp. 137-148.

15. Y. Fujino, H. Sato and N. Otsuka, in Materials Problem Solving with the transmission electron microscope, edited by L.W. Hobbs, K.H. Westma-cott, and D.B. Williams, Materials Research Society Symposium Proceedings, vol. 62 (Materials Research Society, Pittsburgh, 1986), p. 349.
16. M. Guymont and D. Gratias, Acta Crystallogr. Sect., A35, 181 (1979).
17. K. Fujiwara, J. Phys. Soc. Japan, 12, 7 (1957).
18. Y. Fujino, H. Sato, M. Hirabayashi, E. Aoyagi and Y. Koyama, Phys. Rev. Lett., 58, 1012 (1987).
19. D. deFontaine and J. Kulik, Acta Metall. 33, 145 (1985).
20. J. Kulik, S. Takeda and D. de Fontaine, Acta Metall., to be pub-lished.
21. J. van Boem and P. Bak, Phys. Rev. Lett., 42, 122 (1979).
22. M. E. Fisher and W. Selke, Phys. Rev. Lett., 44, 1502 (1980).
23. T. Onozuka, N. Otsuka and H. Sato, Phys. Rev. B34, 3303 (1986).
24. K. Kawasaki and S. Iwamoto, Physica, 133B5, 76 (1985).

TEM OBSERVATION OF PATTERN FORMATION IN

THE INCOMMENSURATE PHASE OF QUARTZ

Naoki Yamamoto, Kazuya Tsuda* and Katsumichi Yagi*

Faculty of Engineering
·Technological University of Nagaoka
Nagaoka 940-21, Japan

* Department of Physics
  Tokyo Institute of Technology
  Oh-okayama, Meguro-ku, Tokyo 152, Japan

INTRODUCTION

Incommensurate (IC) phases have been found in many kinds of materials such as transition metal dichalcogenides, ferroelectrics, alloys and polymers. In the IC phase, a fundamental structure varies spatially by a modulation wave such as an atom displacement wave or an atom occupation wave. The modulation wave changes to an array of 'discommensurations (DC's)' at near the transition temperature (T_c) between an IC and commensurate (C) phases as predicted by McMillan[1]. They can be directly observed by transmission electron microscopy (TEM). The existence of the DC's in 2H-TaSe$_2$ was first confirmed by TEM [2,3], and characteristic defects of the DC lattice were also revealed in the TEM images. Recently the time evolution of DC patterns has been studied theoretically by several authors [4]. However, there are only a little experiment to be compared with the theories so far. Here we present the recent TEM observations for the pattern formation of the DC lattice appearing in the IC phase of quartz. The observation has been carried out using a 1MeV high voltage TEM in Tokyo Institute of Technology, which is equiped with a TV camera system. The pattern evolutions observed in the microscope were recorded on VTR and were analyzed after the observation.[5]

IC PHASE IN QUARTZ

The IC phase in quartz was recently found to exist in a narrow temperature range from 846 K(T_c) to 847.8 K(T_i). A modulated structure in the IC phase is a '3-q' state composed of three modulated waves whose wave vectors are slightly deviated from the < 10.0>* directions by an angle + ϕ and - ϕ. A period of the moduration changes from 45 nm to 14 nm with increasing temperature , and correspondingly the deviation angle ϕ changes from 10° to 1°. At near T_c the modulation consists of a 2 dimensional array of triangular Dauphiné twin domains (α_1 and α_2), where Dauphine twin boundaries are considered as DC's in this case. Due to the existence of the two deviation angles, there are two types of macrodomains (± ϕ domains) as schematically depicted in Fig. 1, where the lattice displacements at the boundaries (ΔR)[6] are also shown, which is very small (less than 0.01 nm) so will be ignored in the following diagrams. The permissible domain configurations in the IC phase were theoretically studied by Walker.[7]

α – IC PHASE TRANSITION

As for the phase transition from the α to IC phase, we noticed the following points; i) a nucleation shape and sites of the IC phase, ii) the process of successive nucleation and iii) the formation of regular DC lattices in $\pm\phi$ domains. Fig. 2 are VTR images taken in the (30.1) dark field on heating process, where the Dauphiné twin domains show strong contrasts. This process evolves through the characteristic four stages schematically shown in Fig. 3. The shapes of triangular and 3-point star-shaped domains appearing in these stages are restricted to those indicated in Fig. 4. All the shapes in Fig. 4 were observed in the present TEM study, and also are consistent with the domain configurations proposed by Walker. In the first stage (a) in Fig. 3), giant Dauphiné twin domains of irregular shapes appear suddenly at near T_c. They are about a few μm in size, and were frequently observed

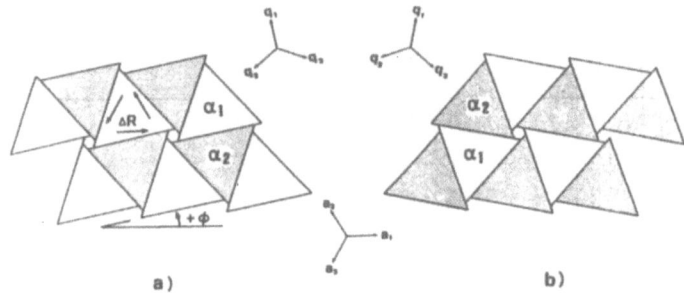

Fig.1 Dauphiné twin domain configurations in the modulated structure of the IC phase for a) $+\phi$ and b) $-\phi$ domain.

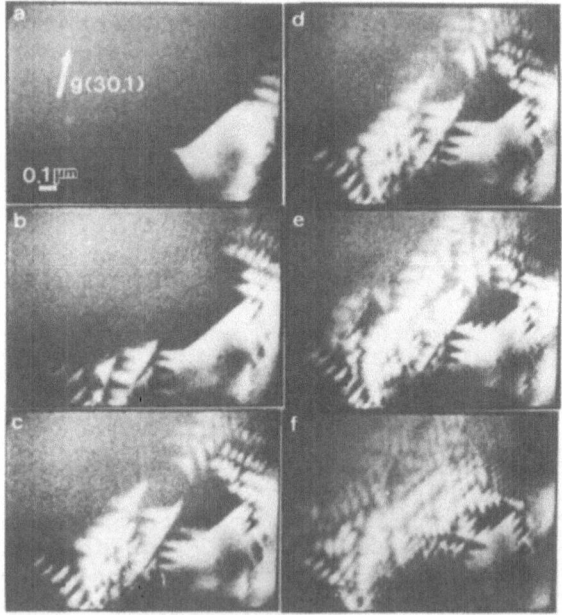

Fig.2 The α to IC phase transition on heating at near T_c.

to nucleate at thin edges of the sample crystals. This is plausible, since at the thin regions only a small volume changes to an opposite Dauphine twin domain surrounded by a narrow twin boundary during the nucleation. Small domains of less than 1 μm in size were also observed to nucleate inside a crystal near the large domains. As the process evolves, the corners of the giant domains tend to have an acute corner angle of 2ϕ ($\approx 20°$), so the domains are seen to expand into a matrix domain with sharp edges. In the second stage 3-point star-shaped domains of about 0.1 μm in size nucleate both in the matrix and nucleated giant domains. They preferencially appear at the sharp corners of the domains, so they often form 1-dimensional rows successively as seen in Figs. 2 b) and c) (Fig. 3 b)). The star-shaped domains in these rows change to acute triangular domains in the third stage (Fig. 3 c)), where $\pm\phi$ domain configurations are almost settled. Finally the rows of the acute triangular domains are reconstructed into a two-dimensional array of regular triangle domains in each $\pm\phi$ domains (Fig. 3 d)).

PATTERN FORMATION IN THE IC PHASE

Dislocations in the DC Lattice

A characteristic property in the pattern variation with temperature is that the DC lattice changes in spacing together with orientation. A mechanism of this process was theoretically treated by Koh and Yamada[8] very recently, in which they suggested that a 'dislocation' of the DC lattice (here abbreviated as DL) plays an impotant role in changing the pattern of the DC lattice. The DL is schematically depicted in Fig. 5 a) and was actually observed in the TEM image of b). According to Koh and Yamada, the pattern changes in the following way:
1) Three pairs of DL's nucleate at the same time inside a crystal to conserve the symmetry of the DC lattice.
2) The DL's move apart from each other in the selective directions, in order to produce a new DC lattice with proper spacing and orientation in the inside region among them.

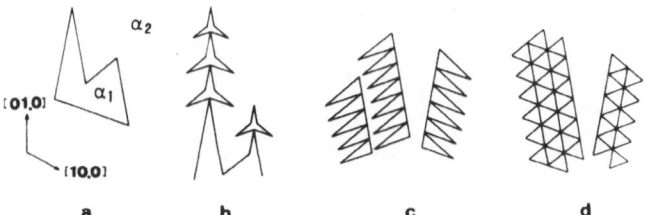

| a | b | c | d |

Fig.3 Four stages on the α to IC phase transition.

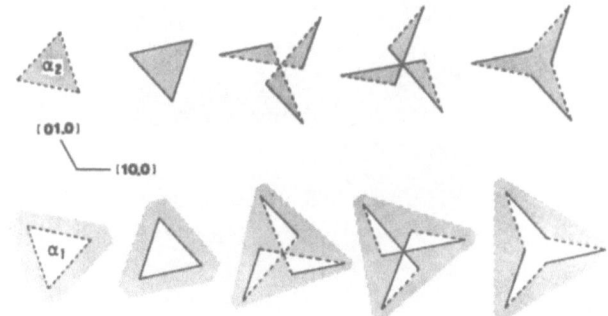

Fig.4 Permissible shapes of triangular and star-
shaped Dauphiné twin domains.

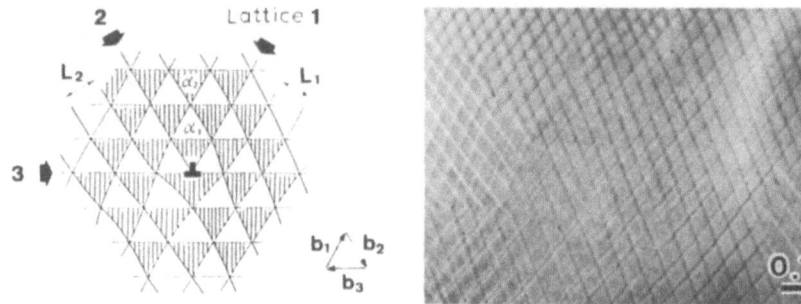

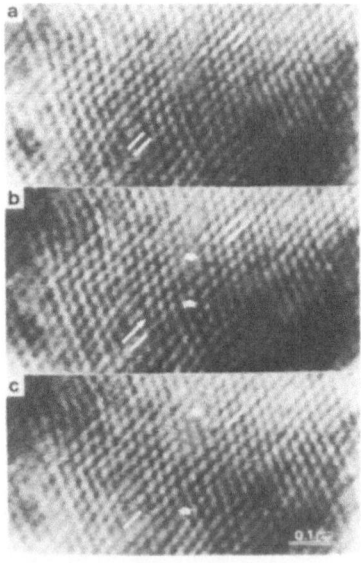

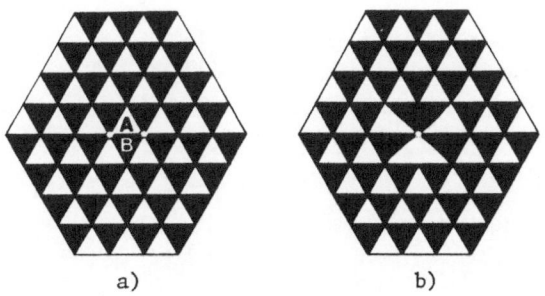

a)   b)

Fig.7 A mechanism of the pair nucleation of DL's.

Movement of DL's

After the nucleation each DL's are expected to move in a special direction so as to rotate a DC lattice into a proper orientation. Shaded areas in Fig. 8a) indicate the expected direction of the DL motion in $\pm\phi$ domains on cooling process; such motion can expand the spacing of the lattice 1 and 2 in Fig. 5a), and increase the deviation angle ϕ for the two lattices in the $\pm\phi$ domain. The observed results are shown in Fig. 8b), where solid circles indicate the directions of motion for the DL's in the $-\phi$ domains, and a open circle indicates for that in $+\phi$ domain. The data are only for the DL's which moved over sufficiently long distances. Though the number of the data is still small, the tendency of the preferencial movement of DL's is clearly indicated.

Interactions between DL's

There are six kinds of DL's with different Bergers vectors, $\pm b_1$, $\pm b_2$ and $\pm b_3$, where b_i's are indicated in Fig. 5a). The interactions between these DL's were frequently observed during their motions. Figs.9a) to d) show one of the examples on the cooling process, where the DL 1 collides with the DL 2 coming from below, and the two DL's react and change to another type of DL, 3. These interactions are classified into five casesas depicted in Fig. 10. The reaction occurs in the 1) and 2) cases, but there is no reaction in the 3) and 4) cases. The case 5) indicates a pair annihilation of DL's which were often observed on the cooling processes.

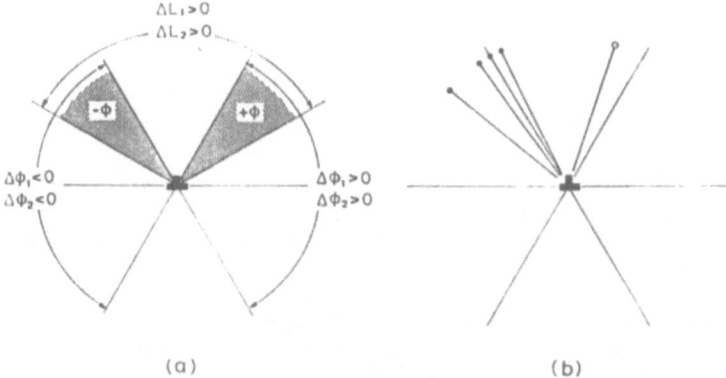

(a)   (b)

Fig.8 a) Preferential directions of motion for a DL in $\pm\phi$
domains, and b) observed ones. Open and solid circles indicate directions for the DL's in $+\phi$ and $-\phi$
domains, respectively.

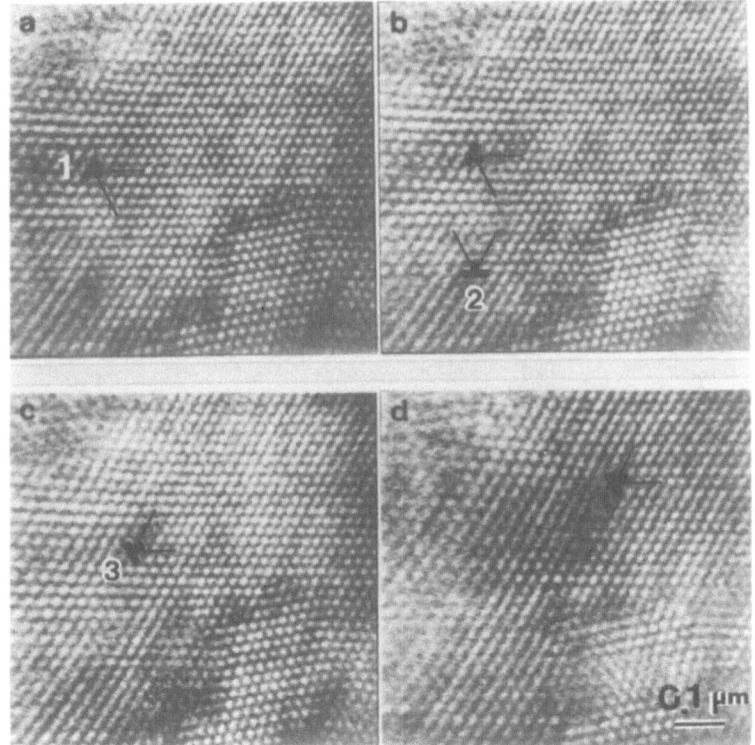

Fig.9 Pattern evolution on the cooling process.

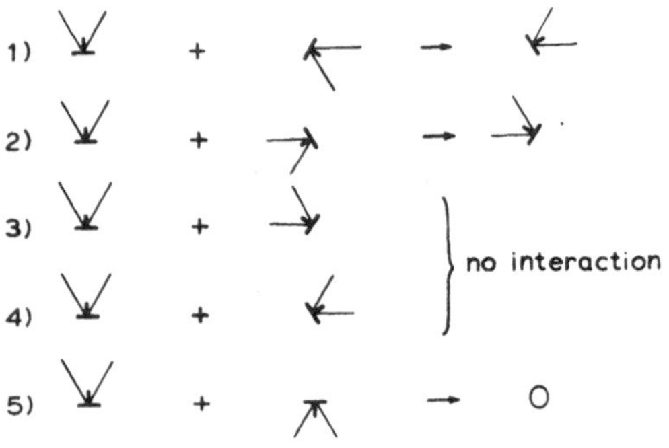

1) $\searmark$ + $\nwarrow$ → $\nwarrow$

2) $\searmark$ + $\nearrow$ → $\searrow$

3) $\searmark$ + $\nearrow$ ⎫

4) $\searmark$ + $\swarrow$ ⎬ no interaction

5) $\searmark$ + $\uparrow$ → O

Fig.10 Interactions between the 6 kinds of DL's.

REFERENCES

1. W. L. McMillan, Phys.Rev.B12, 1187 (1975).
2. K. K. Fung, S. McKernan, J. W. Steeds and J.A. Wilson, J.Phys.C14, 5417 (1981).
3. C. H. Chen, J.M. Gibson and R. M. Fleming, Phys.Rev.B26, 184 (1982).
4. K. Kawasaki and S. Yamanaka, Phys.Rev.B34, 7986 (1986).
5. N. Yamamoto, K. Tsuda and K. Yagi, Proc.XI-th Int.Cong. on Electron Microscopy, vol.2, 1231,Kyoto (1986).
6. N. Yamamoto, K. Tsuda and K. Yagi, Proc.6-th Int. Meeting on Ferro-electricity, Jpn.J.Appl.Phys. Suppl.24-2, 796 (1985).
7. M. B. Walker, Phys.Rev.B28, 6407 (1983).
8. S. Koh and Y. Yamada, J.Phys.Soc.Jpn.56, No.5, (1987).

INCOMMENSURATE-COMMENSURATE TRANSITION IN 2H-TaSe$_2$

Hiroshi Sato[*], Takashi Onozuka[**] and Yasumasa Koyama[***]

* School of Materials Engineering, Purdue University
  W. Lafayette, In 47907 USA
** Institute for Materials Research, Tohoku University
  Sendai, Japan
*** Department of Metallurgy, Tokyo Institute of Technology
  Tokyo, Japan

ABSTRACT

Incommensurate-commensurate transition which occurs in 2H-TaSe$_2$ is investigated in detail by means of high resolution transmission electron microscopy. The double honeycomb discommensuration model so far accepted for the incommensurate phase of 2H-TaSe$_2$ is not found to be appropriate. There is a change in the modulation mode in the incommensurate phase and the commensuration process is found to occur in the lower temperature range. The commensurate process takes place as the motion of the jellyfish patterns with three legs. The change in the modulation mode also explains the change in the incommensurability in two steps.

Among transition metal dichalcogenides, 2H-TaSe$_2$ exhibits a rich diversity of charge density wave (CDW) superlattice phenomena including transitions between fully commensurate, incommensurate and normal structures. The behavior of incommensurate-commensurate transition in this material was extensively investigated by TEM[1-3] and the results have been quoted in many theoretical treatments as one of most typical results on incommensurate-commensurate transitions. Unfortunately, some misinterpretations with respect to the incommensurate phase are involved in these results and, hence, some reinterpretations with respect to the behavior of the transition are inevitably necessary. Here, we present a more straightforward view on the transition based on our recent study of the incommensurate phase of this material.[4,5]

The model for the hexagonal (triple Q) incommensurate phase of 2H-TaSe$_2$ by earlier workers is called the double honeycomb discommensuration model.[1] Here, three kinds of domains of the commensurate structure of the orthorhombic symmetry are arranged in a hexagonal fashion so that the incommensurate phase is kept in the hexagonal symmetry, and the moiré-like fringes observed by dark field images utilizing one of superlattice spots in the upper temperature range of the incommensurate phase are interpreted as being the image of domains due to the dark field contrast.[1-3] In other words, based on their interpretation, domains arranged in the symmetry of a double honeycomb exist in the whole range of the incommensurate phase to a temperature at which the moiré-like fringes become invisible. First, we investigated the upper temperature range of the incommensurate phase, especially above ~95K, by means of a high resolution technique (with the resolution of ~3Å). By this technique, we could clearly observe the lattice images of the background normal structure with the spacing of ~3Å. Analysis of all data indicated that moiré-like fringes observed by dark field technique in this temperature range such as that shown in Fig. 1 were not due to the dark field contrast but were interference fringes due to the superposition of the first and the second order diffraction beams from the incommensurate state.[4] The superlattice spots used for dark field imaging are split, and the first and the second order diffraction spots are clearly detectable.[4] Especially convincing for the interpretation are the high resolution lattice images of the incommensurate state shown in Fig. 2. These high resolution images do not show the existence of any domain boundaries at the boundaries of moiré-like fringes. In addition, a close scrutiny of the images show that the spacing in the area of the weaker contrast (dark fringe) is somewhat larger (by about 3 Å in 100 A) than in the stronger contrast area. This result, along with the moiré-like fringes, is also interpreted to be due to the superposition of the first and the second diffraction beams.[4] That the variation of the spacing of the lattice fringes is due to the superposition of the first and the second diffraction beams (and possibly more weak higher harmonics) can be derived from the fact that this variation in the spacing occurs in phase with the moiré-like fringes (Fig. 14, Ref. 4).[4] This sinusoidal change in the spacing results in the phase shift of $\sim 2\pi/3$ across the region of weaker contrast. This type of phase shift in lattice images of $2\pi/3$ can be explained easily as a result of inclusion of the two split superlattice spots with the separation of δ (incommensurability) and 2δ around the position 1/3.(the position of the superlattice spot of the commensurate structure). Because of this phase shift, the existence of jellyfish patterns (JFP) with three legs (stripples or CDW dislocations[2] used by earlier workers) is sometimes found.

The appearance of domains in the incommensurate phase is, however, observed below around 92K.[5] Below this temperature, the contrast of the dark field images is found to be due mostly to the dark field contrast. The contrast between domains of different orientation due to the dark field contrast also supports that the domains have the commensurate structure[5] of the orthorhombic symmetry as discussed by earlier workers.[1-3] As the temperature is lowered, domains of parallel stripes develop in two directions (two types of domains) making 120° with each other, leaving rhombus-shape domains (the third type domains) as shown in Fig. 3. These stripe domains serve as the discommensuration lines in the commensurate structure.[5] There is, geometrically, a phase slip of $2\pi/3$ across each domain.[1,5] Therefore, the assembly of three discommensuration lines eliminates the phase shift. In other words, JFP's with three legs consisting of three discommensuration lines which meet at one point appear when commensuration process begins. The motion of JFP's leaves the commensurate structure in its wake and represents the commensuration process. The commensuration process in this temperature range is thus essentially the same as that proposed by earlier workers.[1-3] A similar process takes place in the commensuration stage of long period superlattice (LPS) in the Ag-Mg alloys, but by means of JFP's with four legs.[6,7]

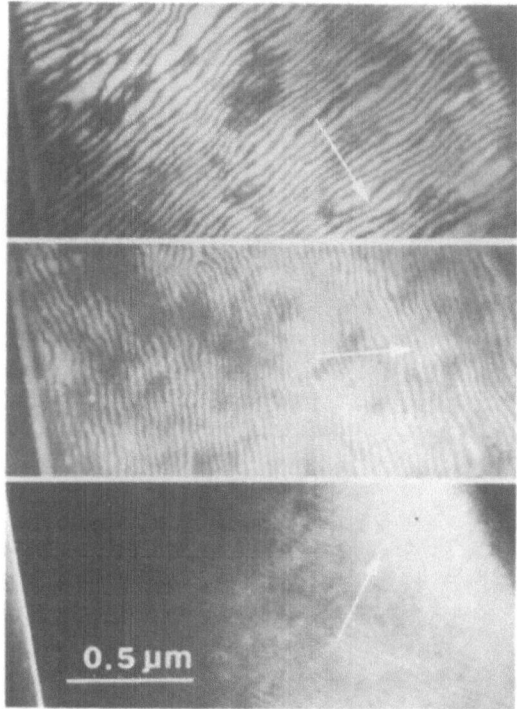

Fig. 1. Moiré-like fringes obtained by the dark-field technique from
the same area utilizing three different diffraction vectors
taken at 97K. Arrows indicate the directions of the diffraction
vector of each micrograph.

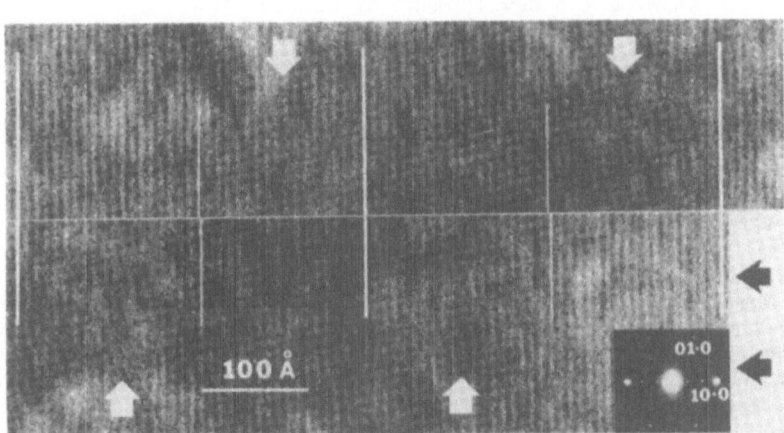

Fig. 2. Periodic change of CDW lattice spacings, showing the existence
of the phase slip region. The lower half of the picture is
shifted to the left by approximately 150Å along the solid
arrows in order to compare the spacing of lattice fringes in
the phase slip region and other regions. Areas of weaker contrast in the moiré-like fringes are indicated by open arrows.

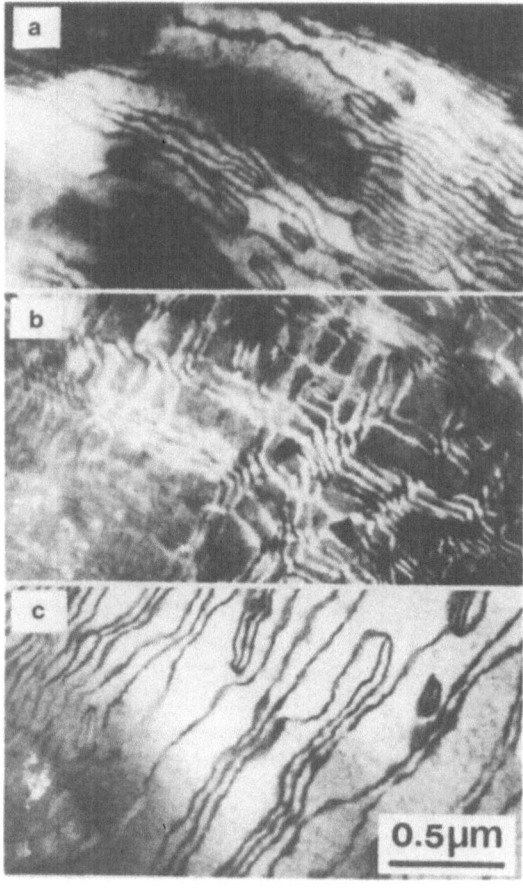

Fig. 3. Dark field images at 98K in the three imaging directions, showing the development of domains.

There are, however, several things to be noted. In the first place, stripe domains develop only in two directions, leaving rhombus shape domains in the matrix. Therefore, different from the double honeycomb model, the domain distribution does not have a hexagonal symmetry locally. Secondly, the heads of jellyfish patterns, based on Fig. 3, (and Fig. 7 of Ref. 5) seem to be stuck at the domain boundaries of stripe domains in the other direction. In other words, the modulation mode of the incommensurate phase changes at around 92K. Below this temperature, the commensurate structure is modulated by means of domains and the incommensurability δ in this temperature region is given by the average separation of stripe domains. The modulation mode of this sort is called the incommensurate modulation of the commensurate structure (by means of domains), as against the incommensurate modulation of the normal structure (by CDW) just below the normal-incommensurate transition point.[5] The incommensurate modulation of the commensurate structure is commonly found in LPS.

We observe that the incommensurability δ changes in two steps.[7,8] The change of δ with T is first concave and seems to become stationary at $\delta \sim 0.3$, and then the change occurs rapidly from around 92K to zero. This type of change in δ corresponds to the change in the modulation mode mentioned above. Most discommensuration theories[9,10] predict a second order transition for the incommensurate-commensurate transition, but, at the same time, a large number of harmonics is required in these theories in order to reach the commensurate state in one step.[9,10] On the other hand, although the contribution of the second order harmonic is large, the existence of higher harmonics has not been experimentally confirmed.[4,5] In other words, although the contribution of some higher harmonics can be rationalized in order to explain the contrast of moiré-like fringes, the amplitudes of higher harmonics seem to be extremely small, and, hence, the number of higher harmonics, N, to be included in the discommensuration theory should be limited. If N is small, it is possible to make the δ-T curve concave and to let δ approach a finite value.[5] The best fit to the experimental curve is obtained for N=4[5]. The relation is shown in Fig. 4.

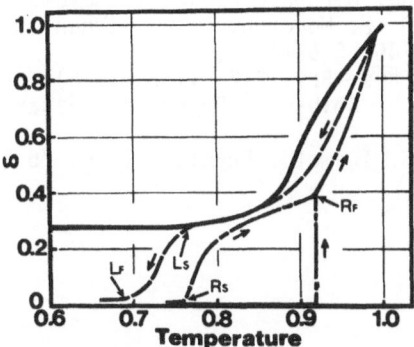

Fig. 4. Comparison of calculated δ-T curve (solid line) for N=4 with experimental one (broken line). A reduced scale of the incommensurability, δ/δ_{NI}, is used here, where δ and δ_{NI} are the incommensurability at T and at the normal-incommensurate transition temperature T_{NI}, respectively.

The commensuration process takes place in the temperature range where the modulation mode is 'the incommensurate modulation of the commensurate structure', and in the form of the motion of JFP's. This mechanism is very similar to that suggested by Kawasaki, et al.[11] and is common with that which occurs in the commensuration process observed in LPS in the Ag-Mg alloys.

CONCLUSION

The concept of the double honeycomb discommensuration model for the incommensurate phase of 2H-TaSe$_2$ has to be modified. The modulation mode changes from the 'incommensurate modulation of the normal structure' to the 'incommensurate modulation of the commensurate structure' at ~92K in the incommensurate phase and the commensuration process takes place in the latter temperature range as the motion of JFP's with three legs.

This commensuration process is essentially the same as that stated in earlier works.[1-3] The change of the modulation mode, however, explains the change of the incommensurability δ in two steps.

The work was supported by the National Science Foundation, Solid State Chemistry Program, grant number DMR 8304314 and US-Japan Science Cooperation Program, grant number INT 84-12550.

REFERENCES

1. K. K. Fung, S. McKernan, J. W. Steeds and J. A. Wilson, J. Phys. C14, 5417 (1981).
2. C. H. Chen, J. M. Gibson and R. M. Fleming, Phys. Rev. B26, 184 (1982).
3. S. McKernan, J. W. Steeds and J. A. Wilson, Physica Scripta, T1, 74 (1982).
4. T. Onozuka, N. Otsuka and H. Sato, Phys. Rev. B34, 3303 (1986).
5. Y. Koyama, Z. P. Zhang and H. Sato, Phys. Rev. B36, September (1987).
6. Y. Fujino, H. Sato, M. Hirabayashi, E. Aoyagi and Y. Koyama, Phys. Rev. Lett. 58, 1012 (1987).
7. H. Sato, Y. Fujino, Hirabayashi and Y. Koyama, in this volume.
8. R. M. Flemming, D. E. Moncton, D. B. McWhan and F. J. DiSalvo, Phys. Rev. Lett. 45, 576 (1980).
9. W. L. McMillan, Phys. Rev. B12, 1187 (1975); B14, 1496 (1976).
10. K. Nakanishi and H. Shiba, J. Phys. Soc. Japan, 44, 1465 (1978); 45, 1147 (1978).
11. K. Kawasaki and S. Iwamoto, Physica, 133B, 76 (1985).

ULTRASONIC STUDIES ON PSEUDO-CRITICAL DYNAMICS

AND PSEUDO-DYNAMICAL SCALING

Yoshifumi Harada

Department of Applied Physics
Faculty of Engineering
Fukui University
3-9-1 Bunkyo, Fukui 910 Japan

INTRODUCTION

The basic concept of dynamic scaling and dynamic universality for cri-tical dynamics has been reawakened largely because of the remarkable prog-ress on the recent studies of a new type of phase transition in a wide sense. In general critical phenomena, various physical quantities show ano-malies near the 2nd-order phase transition point which is a thermodynamic singular point, and response functions such as transport coefficient di-verge at this point.

On the other hand, quite recently another type of critical phenomena has been found, in which response functions diverge even near the 1st-order phase transition point that is not a thermodynamic singular point. These phenomena are now referred to as "pseudo-critical phenomena" (PCP), which are for the first introduced by us from the studies of sound absorption in binary fluids(Harada,1975,1980,1981). That is, PCP is defined as the crit-ical phenomena, in which "various physical quantities diverge at the point point where "spinodal", or the second derivatives of free energy with re-spect to order parameter vanish". Recently remarkable progress has been achieved on these phenomena, (Ikeda,1979; Saito,1978; Chu et al.,1969; de Gennes,1973; Bosio et al.,1975; Collins et al.,1973; Hashimoto et al.,1976,1978) which constitute a new branch of critical dynamics. Similar phenomena have been observed in a variety of material system, such as liq-uid crystal, binary intermetallic compound, ferroelectrics, metal-nonmetal transitions, heavy fermion systems, etc.

Theoretical studies on static PCP are at first presented by Saito(1978) using renormalization-group theories, and then by Ikeda(1979) and Boiko et al(1984) in terms of Landau mean-field theory. In their continued studies on "pseudo-critical exponent and the spinodal in metastable fluids", Mogel and Chalyj(1986) obtained the stable spinodal fixed points within the in-terval $4 > d > 2$ for the d-dimensional metastable fluid using the renor-malization group method.

The primary purpose of this article is to report some results of ultra-sound attenuation by 3-methylpentane and nitroethane(3M/N) system at vari-ous equilibrium states in the critical neighborhood of the temperature-con-centration phase diagram. By extrapolation of these data to ultrasound attenuation near phase separation and then to metastable state below the phase separation temperature one may determine a common "pseudo-spinodal curve" Tsp(c) described by Tc-Tsp(c) $\propto |c-c_c|^{1/\beta}$ where c is the concentration.

The secondary purpose of this report is to demonstrate that dynamical scaling theory for sound propagation, developed by Ferrell and Bhattachajee (F-B), is also applicable to the case of this pseudo-spinodal in critical binary mixtures at various concentrations, temperatures and frequencies. In the following section we shall first show typical example of critical sound attenuation results together with the analysis by using the (F-B) dynamical scaling theory. Finally, our ultrasound absorption studies on PCP in binary fluids will be shown in some detail.

THEORETICAL PREDICTION FOR PSEUDO-CRITICAL PHENOMENA AND F-B DYNAMICAL SCALING THEORY FOR SOUND PROPAGATION

[A] F-B DYNAMICAL SCALING THEORY FOR USUAL CRITICAL PHENOMENA AND CRITICAL SOUND ATTENUATION

It is well known that the principle of critical point universality asserts that onecomponent fluids near the gas-liquid critical point and binary mixtures near the consolute point belong to the same universality class. Light scattering data for fluid near the gas-lquid critical mixing point are consistent with the asymptotic behavior predicted for one component Landau Ginzuburg-Wilson model on the basis of the renormalization-group equations(Sengers,1982).

On the other hand, critical ultrasound propagation in fluid occurs in a different way compared with those in simple liquid and critical binary mixtures. The most striking difference has been known for a long time as follows. 1) In the case of liquid-gas phase transition : The ultrasonic velocity near Tc is typically of the order of 100m/sec and decreases on approaching the critical point, accompaning the anomalous increase of the heat capacity at constant volume.

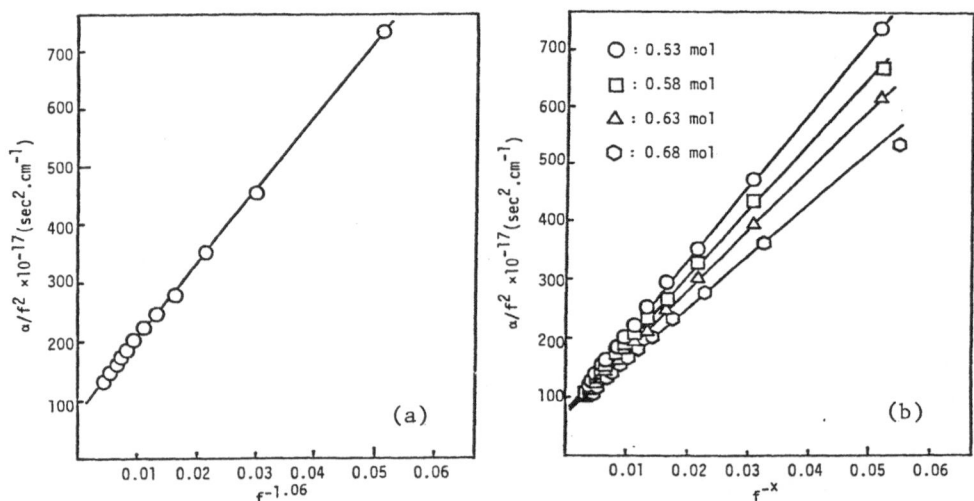

Fig. 1-(a)   Linear plot of α/f^2 values at Tc vs $f^{-1.06}$ for a critical binary mixture at consolute point
(b)   Linear plot of α/f^2 values at Tsp vs f^{-x} for critical binary mixtures of 3M/N at pseudo-spinodal.

2) In the case of critical binary mixture: The ultrasonic velocity is typically of the order of 1000m/sec, and no anomalous behavior is observed on approaching the critical point. Recently, these puzzlements have been solved quantitatively by Ferrell et al.(1981), who have pointed out that critical behavior of sound velocity must exist in binary liquids at sufficiently low accessible frequencies. From the low frequency experiments on nitrobenzen and n-hexane system, the adiabatic sound velocity was determined indirectly by Tanaka et al.(1982).

The mode coupling theory for sound propagation in a critical fluid has been compared successfully with sound absorption data in both simple liquid and critical binary mixtures(Kawasaki, 1976). Since we have pointed out some discrepancies between the experiments and the theories on the critical ultrasound propagation in a critical binary fluid (Harada,1980), several theoretical papers have been published in the same year (Ferrell et al., 1981; Kroll et al., 1981; Shiwa et al., 1981). Among the others, F-B theory gives the most useful information: a) Frequency dependent critical sound relaxation curve just above Tc can be obtained. b) Making use of the complex frequency dependent specific heat and renormalization group theory, reduced critical sound attenuation can be obtained. Furthermore, by using linear proportionality relation between α/f^2 and f^{-x}, we are able to obtain pseudospinodal point at each concentration. For these reasons, we shall analyze our data using the F-B dynamical scaling theory for critical sound propagation. The predicted sound absorption just above Tc including so called classical absorption (Navier-Stokes background term) is given by

$$\frac{\alpha_c}{f^2} = \frac{\pi^2 \alpha_c^0 g^2 U_c^2 a}{2 T_c b} (\omega_c^0/2\pi)^{\alpha_c^0/2} \cdot f^{-1-\alpha_c^0/2} + B \qquad (1)$$

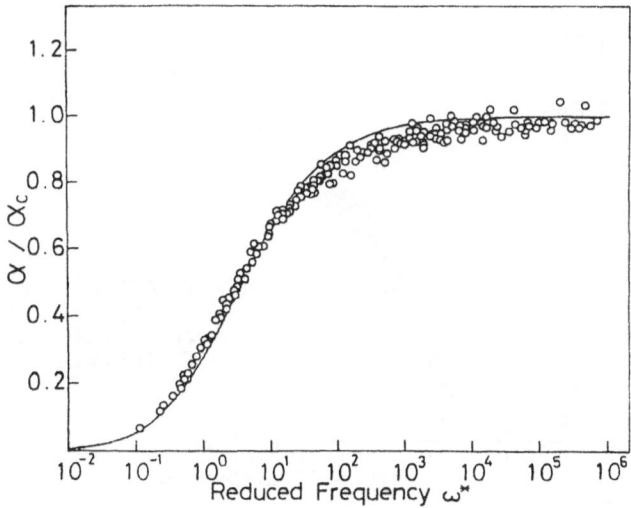

Fig. 2. Scaling plot, α/α_c vs ω^*, for 3-methylpentane + nitroethane. The solid line represents the F-B theory expressed by Eq. (2)

where a is the critical amplitude of specific heat, Uc the sound velocity
at Tc, g the coupling constant which can be obtained from the sudden
adiabatic pressure change(Clerk, 1982), ω_c^0 the critical amplitude of order
parameter relaxation frequency , b the background term of specific heat,
and B the classical absorption. Figure 1 shows that linear proportionality
relation between α/f^2 and f^{-x} based on the F-B dynamical scaling prediction given in Eq. (1) with ealier reported data by us (Harada et al,
1980).   This will also yield the classical background term from the intercept.   The good fit to the straight line is excellent confirmation of the
frequency dependence of α/f^2 predicted by Ferrell et al.(1981).

The scaled sound attenuation with respect to reduced frequencies ω^* can
be written in the form

$$\alpha/\alpha_c = \frac{2(1+p)\omega^*}{\pi} \int_0^\infty \frac{\ddot{u}du}{(1+u)^2} \cdot \frac{u(1+u)^p}{\omega^{*2} + u^2(1+u)^{2p}} \qquad (2)$$

where u=qξ represents the critical variable, p the parameter of the re-
laxationnal factor and in this case we use p=1/2, which is more appropriate for the three-dimensional fluids.   Figure 2 shows that the agreement
with our data at various frequencies ranging from 16.5 to 165 MHz is excellent with the F-B theory.

[B] PSEUDO-DYNAMICAL SCALING OF PSEUDO-SPINODAL AND PSEUDO-CRITICAL
SOUND ATTENUATION

In order to obtain a pseudospinodal curve, we have measured the ultrasound attenuation of 3M/N at various equilibrium states in the critical
neighborhood of temperature-concentration phase diagram.

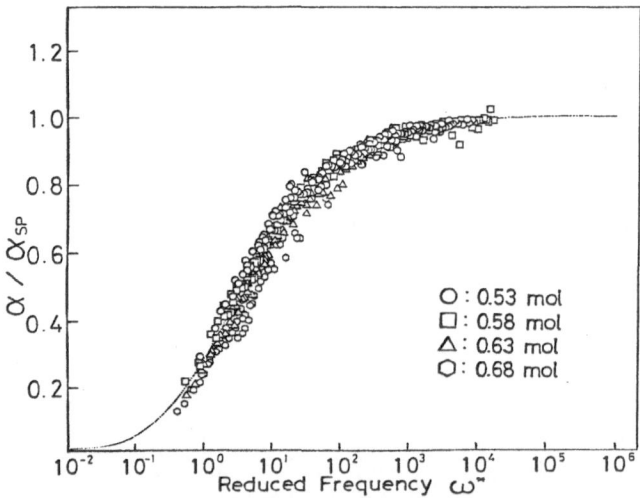

Fig. 3.   Scaling plot, α/α_{sp} vs ω^* , for 3-methylpentane + nitroethane
for various concentration of the phase diagram. The solid line
represents the F-B theory expressed by Eq.(4)

In the spirit of Chu's original approach for pseudo-spinodal, from the light scattering intensity and linewidth measurement of binary mixture, it has been observed that the diffusion coefficient D becomes smaller as $D \propto (T - Tsp(c))^{\gamma}$ when the temperature T approaches Tsp(c), where Tsp(c) represents the "pseudo-spinodal curve" in the temperature-concentration T-c plane(Chu et al., 1969). The pseudo-spinodal curve Tsp(c) is determined by extrapolation from the empirical form of D. The experiments were also performed for the binary mixture 3M/N in this pseudo-spinodal study. The co-existence curve of 3M/N was determined by observation of a meniscus in this laboratory and compared with other available experimental data(Wims et al., 1975). Based on the phase diagram of 3M/N system, we have obtained the pseudo-spinodal curve from the sound absorption measurements, in which we approached the phase separation temperatures from higher temperature at each fixed concentration. We selected five different samples: (a) c=0.50 mol %, (b) c=0.53 mol %, (c) c=0.58 mol %, (d) c=0.63 mol %, (e) c=0.68 mol %. 3-methylpentane of special grade was dehydrated by calcium chloride and purified by distillation. A sample of nitroethane of the same grade was purified repeatedly in a vacuum system by freezing and pumping. The standard pulse technique was used for the absorption measurments over the frequency range 16.5-165MHz. The temperature of the sample cell was regulated to within $\pm 0.00062\,^{\circ}C$ over periods of several ours using a modified apparatus of a lamp-radiation type thermostat.

We shall begin our analysis of the pseudocritical dynamic behavior of sound propagation with an examination of the frequency dependence of the attenuation at pseudo-spinodal points by extending the F-B theory for critical dynamics. The temperature and frequency dependent specific heat due to the adiabatic temperature oscillation at the thermodynamic limit is also written in the same form as F-B theory, and the pseudo-critical ultrasound attenuation just above Tsp can be written in the following form

$$\frac{\alpha_{sp}(c)}{f^2} = \frac{\pi^2 \alpha_s^0 g^2 U_{sp}^2(c) a}{2 T_{sp}(c) b} \cdot (\omega_s^0(c)/2\pi)^{\alpha_s^0/2} \cdot f^{-1-\alpha_s^0/2} + B \quad (3)$$

where a is the critical amplitude of specific heat, Usp the sound velocity at Tsp, g the coupling constant, ω_s^0 the pseudo-critical amplitude of pseudo-orderparameter relaxation frequency, b the background term of specific heat. Thus according to the renormalization-group theory and complex frequency dependent specific heat theory developed by Ferrell et al.(1981), we rewrite the following form for α/α_{sp} which is the same form as Eq.(2),

$$\alpha/\alpha_{sp}(c) = \frac{2(1+p)\omega^*}{\pi} \int_0^{\infty} \frac{x dx}{(1+x)^2} \cdot \frac{x(1+x)^p}{\omega^{*2} + x^2(1+x)^{2p}} \quad (4)$$

where α_{sp} represent sound attenuation at Tsp, ω^* the reduced frequency, $x=k\xi$ the pseudo-critical variable, p the parameter of the relaxation factor, c the concentration. As shown in Figure 1b, the linear relation between α/f^2 vs f^{-x} is satisfied at each concentration, but, the slope of these plots are decreased with increasing the concentration of pseudo-orderparameter. These results are the most significant feature compared with those at the usual consolute point. Figure 3 also shows that the agreement with our attenuation data near the pseudo-spinodal at various frequencies and concentrations is excellent with the F-B theory.

CONCLUSION

We have measured the ultrasound absorption in 3M/N at various equilibrium states in the critical neighborhood of temperature-concentration phase diagram over a wide range of frequency and temperature. In particular, we have applied the F-B frequency dependent specific heat theory for critical sound propagation, including dynamical scaling theory, to the pseudo-critical phenomena in 3M/N system, for which the ultrasound absorption data observed near phase separation limit at each several concentration gives excellent pseudo-spinodal curve for this system. In conclusion, from experimental and theoretical considerations, we have shown the excellent confirmation of the pseudo-dynamical scaling and pseudo-dynamic universality for sound propagation in 3M/N.

ACKNOWLEDGMENTS

The author expresses his sincere thanks to Professor T. Yasunaga for valuable discussion and suggestions. He is also grateful to Professor R. A. Ferrell for many valuable discussion and suggestions. The author is much indebted to Professor M. Inoue for his help in pre-paring the manu-script. Finally, Mr I. Kato is acknowledged for his collaboration in taking the experimental data and numerical analyses.

REFERENCES

Bhattacharjee, J. K., and Ferrell R. A., 1981, Phys. Rev., A24:1643.

Boiko, V. G., Chalyj, A. V., and Mogel H. J., 1984, Kiev., ITP-84-119E:38,. Academy of Science of the Ukrainian SSR, Institute for Theoretical Physics,.

Bosio, L., and Windsor, C. G., 1975, Phys. Rev. Lett., 35:1652.

Chu, B., Schoenes, F. J., and Fisher, M. E., 1969, Phys. Rev., 185:219.

Clerk, E. A., Sengers, J. V., Ferrell, R. A., and Bhattacharjee, J. K., 1983, Phys. Rev.., A27:2140.

Collins, M. F., and Teh, T. C., 1973, Phys. Rev. Lett.,30:781.

de Genne, P. G., 1969, Phys. Letters., 30:454.

Domb, C., and Green M. S., 1971-1976, "Phase Transitions and Critical Phenomena" Vol. 1-5A,B, Academic Press,.

Ferrell, R. A., and Bhattacharjee, J. K., 1980, Phys. Rev. Lett., 44:403.

Ferrell, R. A., and Bhattacharjee, J. K., 1981, Phys. Rev., A24:4095.

Garland, C. W., and Sanchez, G., 1983, J. Chem. Phys., 79:3090.

Harada, Y., 1975, Bussei Kenkyu., 25:167.

Harada, Y., 1978, "A General Discussion of Molecular Dynamics of Relaxation Processes" Research Group of GASRME,.

Harada, Y., Suzuki, Y., and Ishida, Y., 1980, J. Phys. Soc. Jpn., 48:705.

Harada, Y., 1980, J. Sci. Hiroshima Univ., Ser., A44:283.

Harada, Y., 1980, Phys. Rev., A21:928.

Harada, Y., Suzuki, Y., and Ishida, Y., 1980, J. Phys. Soc. Jpn., 48:703.

Hashimoto, T., Miyoshi, T., and Ohtuka, N., 1976, Phys. Rev., B 13:1119.

Ikeda, H., 1979, Prog. Theor. Phys. 61:1029.

Kawasaki, K., 1976, "Phase Transition and Critical Phenomena" ed. Domb, C., and Green, M. S.; 5A:165.

Kroll, D. M., and Ruhland J. M., 1981, Phys. Rev., A23:371.

Mogel, H. J., and Chalyj, A. V., 1986, Kiev., ITP-86-96E,. Academy of Science of the Ukrainian SSR, Institute for Theoretical Physics,.

Saito, Y., 1978, Prog. Theor. Phys., 59:375.

Sanchez, G., and Garland, C. W., 1983. J. Chem. Phys., 79:3100.

Sengers, J. V., 1980, "Proceedings of the 1980 Cargese Summer Institute on Phase Transitions" ed. Levy, M., Le Guillou, J. C., and Zinn-Justin, J., Plenum Publ. Corp.

Shiwa, Y., and Kawasaki K., 1981, Prog. Theor. Phys., 66:118.

Shiwa, Y., and Kawasaki K., 1981, Prog. Theor. Phys., 66:406.

Tanaka, H., Wada, Y., and Nakajima, H:, 1982, Chem. Phys., 68:223.

Wims, A. M., McIntyre, D., and Hynne, F., 1975, J. Chem. Phys., 37:5067.

NEUTRON AND X-RAY STUDIES ON TIME EVOLUTION OF THE ORDERED MULTI-LAMELLAR PHASE IN AQUEOUS SOLUTION OF DIPALMITOYL-PHOSPHATIDYLCHOLINE

Takayoshi Takeda, Satoru Ueno, Shigehiro Komura
Yoshinori Toyoshima and Kozo Akabori

Faculty of Integrated Arts and Sciences, Hiroshima
University, Hiroshima 730, Japan

INTRODUCTION

Bilayers of dipalmitoyl-phosphatidylcholine (DPPC) in water have a lamellar structure and exhibit phase transisions at the temperature T_{m1}=35.3 and T_{m2}=41.4 °C .[1,2)]   The transition at T_{m1} is evidenced by a small anomaly of specific heat and called a pretransition.   The transition at T_{m2} is associated with a pronounced anomaly of specific heat and is called a main transition, which is of the order-disorder type concerning the hydrocarbon chains.   The hydrocarbon chains are arranged in order and the bilayers of lipid have a wavy structure below T_{m2}, corresponding to P_β' phase.   The P_β' phase is transformed into L_β' phase below T_{m1}.   The disordered phase above T_{m2} is a liquid crystalline which is typically smectic, corresponding to $L\alpha$ phase[3)].

Small angle neutron scattering (SANS) and small angle X-ray scattering (SAXS) from DPPC in water were previously measured.[4)]   The SANS patterns in the DPPC-D_2O system changed drastically at the main transition temperature T_{m2}; the diffration peak which corresponds to the repeat distance of the lipid bilayers was observed above T_{m2}, but below T_{m2} it disappeared and a large diffuse scattering appeared.   The superficial difference between the SANS and SAXS patterns could be interpreted in terms of the difference of the structure function of the lipid for neutrons and X-rays.   In the present study, we found that this system is non-equilibrium system and has a large hysteresis in the temperature variation of the lamellar structure between heating and cooling process , and we studied time evolution of the ordered multi-lamellar phase in aqueous dispersions of DPPC by slow time-resolved SANS and SAXS.

EXPERIMENTAL PROCEDURE

L-α-DPPC was purchased from Sigma Chem. Co. and purified no more. D_2O was purchased from Merk Sharp and Dohme Canada Ltd.   The SANS experiments were performed using a spectrometer with a position-sensitive detector installed at the Kyoto University Reactor[5)].   The wave length of the neutrons λ was about 4 Å with a resolution of $\Delta\lambda/\lambda$ = 14 %(FWHM). The samples were placed in a silica glass cell with inner thickness of 2 mm and a beam slit of 0.5 x 2.0 cm. The temperature of the samples were controlled by circulating thermostated water (± 0.1 °C) and monitored by thermocouple dipped in the cell. The SAXS experiments were performed using CuKα radiation (λ = 1.542 Å) from the Rigaku Denki rotating anode point focus generator RU-200 and the Rigaku Denki mirror-monochromator (quartz ($10\bar{1}1$)) point focusing camera with PSPC. The samples were placed in a cell with inner thickness of 1 mm. The temperature of the samples was controlled in the same way as for SANS.

RESULTS

Figures 1 and 2 show the differential cross section per unit solid angle per unit volume of sample ($d\Sigma/d\Omega$) as a function of scattering vector Q obtained from SANS measurements for 500 s on 20 wt % DPPC-D$_2$O. The SANS patterns change drastically at the main transition temperature T_{m2}; the diffraction peaks at Q_e of about 0.095 Å^{-1} corresponding to the repeat distance d of about 66 Å of the lamellar structure appeared

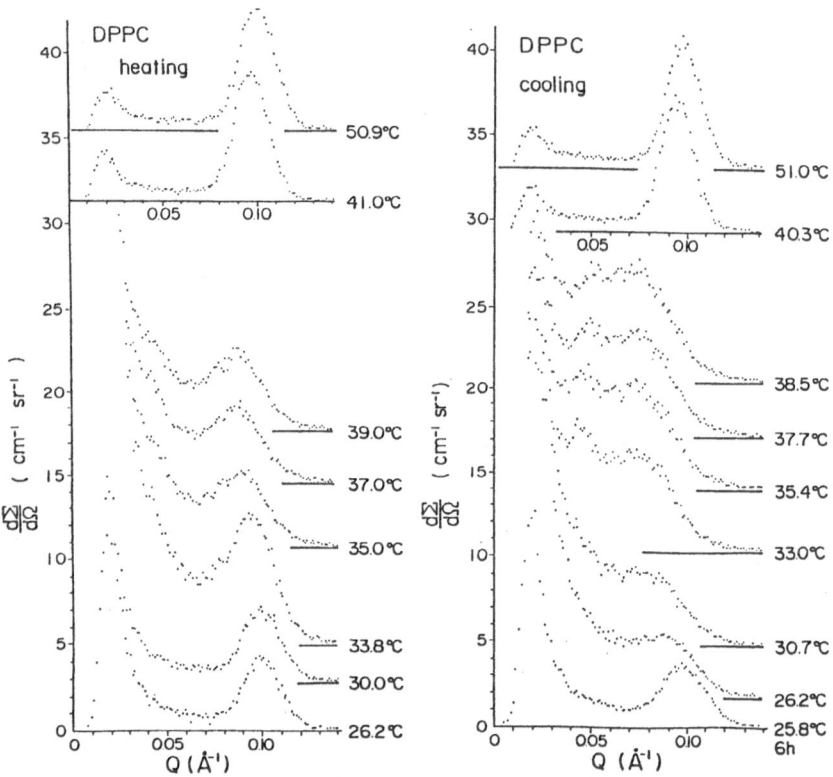

Fig.1　Temperature variation of macroscopic differential cross section ($d\Sigma/d\Omega$) as a function of scattering vector Q obtained from the SANS mesurements on 20 wt % DPPC-D$_2$O solution in heating and cooling process. The pattern at 26.2 °C changes to that at 25.8 °C being kept at room temperature for 6 h.

above T_{m2}. Below T_{m2}, a large diffuse scattering appeared in the range of Q less than 0.07 Å^{-1} and the diffraction peak disappears especially in cooling process. No appreciable time variation of the SANS patterns was observed above 35 °C in cooling process. On the other hand, the patterns vary with time below 35 °C. In Fig.1, the pattern at 26.2 °C changes to that at 25.8 °C being kept at room temperature at 6 h. Figure 2 shows the time variation of the SANS patterns at 24.2 °C. Figure 3 shows the time dependence of the SANS cross section at Q=0.03, 0.05 and 0.07 Å^{-1} at 24.2 °C. The cross section at 0.08 < Q < 0.1 Å^{-1} vary little though it seems to increase with time. Figure 4 shows the temperature variation of the scattering intensity as a function of Q obtained from SAXS measurements on 20 wt % DPPC-H$_2$O in heating and then cooling and the time variation of the SAXS pattern at 27.1 °C. Figure 5 shows the time variation obtained from the SAXS measurements at 23.6 °C after quenching from 47.1 °C. The time dependence of the repeat distance d of the lamellar structure obtained from Q at the peak in the SAXS pattern at 23.6 °C in Fig.5 is shown in Fig.6, though the peak in the early stage are not the diffraction peak but the reflection of the structure factor of one lipid bilayer.

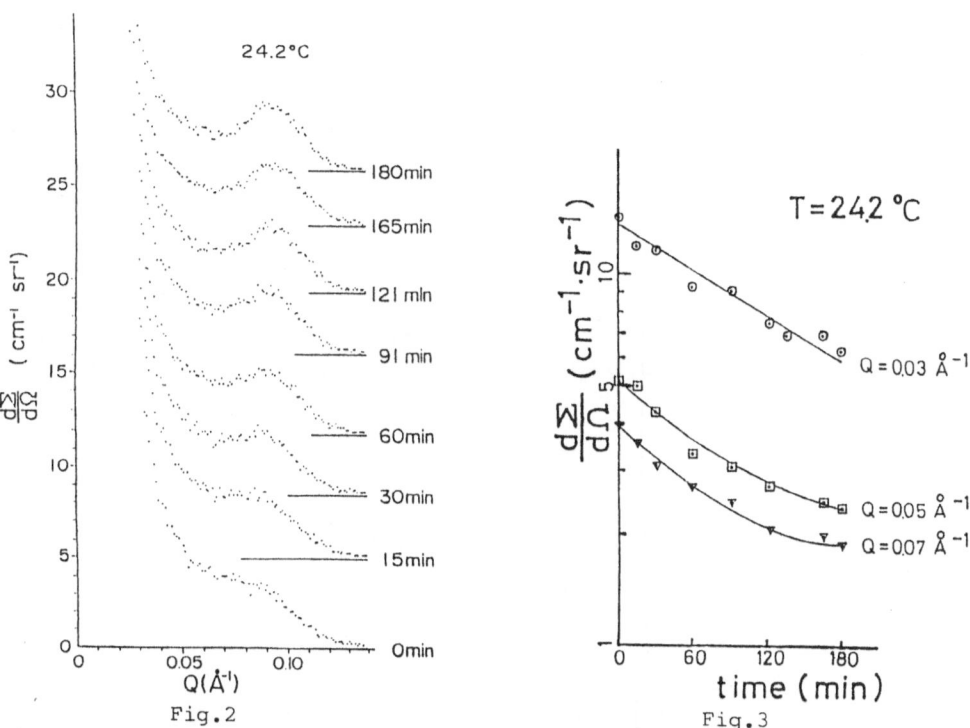

Fig.2

Fig.3

Fig.2 Time variation of macroscopic differential cross section (dΣ/dΩ) as a function of Q obtained from the SANS measurements on 20 wt % DPPC-D$_2$O solution at 24.2 °C after rapid cooling from 30 °C in cooling process.

Fig.3 Time dependence of the SANS cross section at Q=0.03, 0.05, and 0.07 Å^{-1} from the SANS measurements on 20 wt % DPPC-D$_2$O solution at 24.2 °C after rapid cooling from 30 °C in cooling process.

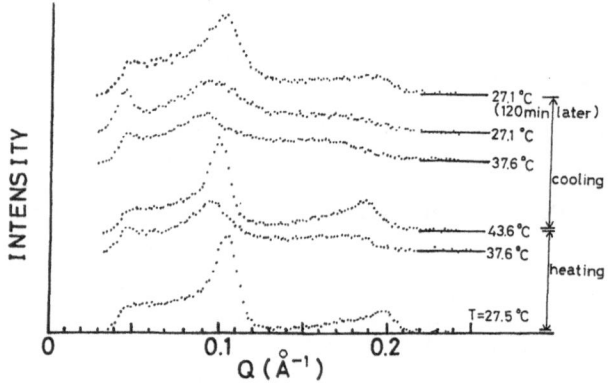

Fig.4   Temperature variation of the scattering intensity as a function
of Q obtained from the SAXS measurements on 20 wt % DPPC-H_2O in
heating and then cooling processes and the time variation of the SAXS
pattern at 27.1°C.

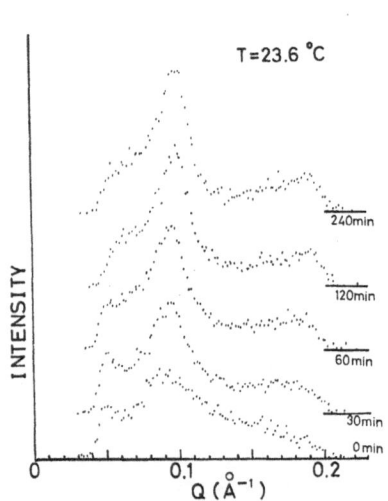

Fig.5   Time variation of intensity obtained from the SAXS measurements
on 20 wt % DPPC-H_2O at 23.6°C after quenching from 47.1°C.

DISCUSSION

We have calculated the modulated Patterson function P(r),

$$P(r) = \int_{Q_{min}}^{Q_{max}} [(d\Sigma(Q)/d\Omega)/f(Q)]\cos(Qr)dQ \qquad , \quad (1)$$

where f(Q) is the structure factor of one lipid bilayer and Q_{max} and Q_{min} are 0.1 and 0.02 A^{-1} respectively. Patterson function is equivalent to the space correlation function G(r),

$$G(r) = \int \rho(r')\rho(r'+r)dr' \qquad . \quad (2)$$

where $\rho(r)$ is the neutron scattering amplitude density. Fig.7 shows the example of P(r) calculated for the SANS patterns at room temperature in cooling process. The correlation between the lipid bilayers at the first neighbor at r=65 Å is seen but that beyond the first neighbor at r=130 and 180 Å disappears and appears again at about r=300 Å for t=0; this behavior is more remarkable at $T_{m1} < T < T_{m2}$. The short range correlation increases and the long range correlation decreases with time as shown in Fig. 7.

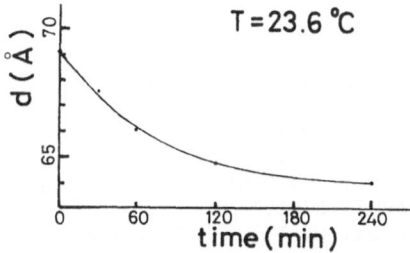

Fig.6 Time dependence of the repeat distance d of the lamellar structure obtained from the peak position in the SAXS pattern at 23.6 °C.

The results of SANS and SAXS indicate that the lamellar structure with the repeat distance of about 65 Å in Lα phase above T_{m2} is destroyed below T_{m2} in cooling process. It is considered that the hydrocarbon chains of DPPC are arranged in order within the bilayer below T_{m2} and the strains are accumulated in the bilayer, so that the bilayer ripples and the rippled bilayers below T_{m2} destroy the lamellar structure that has a single repeat distance. Below T_{m2}, however, it is considered that the lipid bilayers do not distribute at random but they correlate strongly with each other, which cause the very strong SANS diffuse scattering. This phase is considered to be a lamellar structure with many repeat units with various repeat distances. Below T_{m1}, the

fraction of the long repeat distance   decreases   and $L_\beta{}'$ phase evolves
with   time   as   shown   in   Fig.7.   The time dependence   of   d   in Fig.6
supports   this model.   From   Fig.3,   the   time   constant   was   estimated
to be about [6] h.
     Laggner[6]   also found   from time resolved   SAXS   that this system has
a strong   hysteretic effect   in the temperature   dependence.

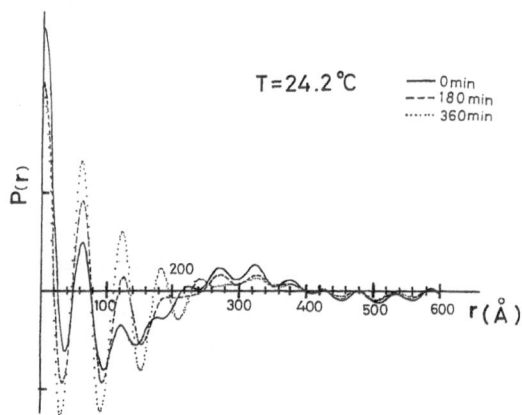

Fig.7   Time   variation   of   the   modulated   Patterson   function   (1)
calculated from the   SANS data   on 20 wt % $DPPC-D_2O$ solution at   24.2°C.
————— :t=0 h ,   ———— :t=3 h ,   •••••:t=6 h

REFERENCES

1) D. Chapman,   R. Williams and B. D. Ladbrooke :   Chem. Phys.   Lipids
   1 (1967) 445
2) H.-J. Hinz   and   J. M. Sturtevant :   J. biol. Chem. 247(1972)   6071.
3) T. Mitsui :   Adv. Biophy. 10(1978) 97.
4) T.   Takeda,   K. Akabori, Y. Toyoshima, S. Komura, and Y. Takebe: Jpn.
   J. Appl. Phys. 26(1987) 1791.
5) S.   Komura,   T. Takeda, H. Fujii, K. Osamura, Y. Toyoshima, K. Mochiki
   and K. Hasegawa :   Jpn. J. Appl. Phys.   22(1983) 351.
6) P. Laggner: to be published in   ' New Methods   in   X-ray   Absorption
   Scattering   and Diffraction' for Applications in   Structural   Biology,
   ed. B. Chance   and   M. D. Bartunik   (Academic Press,   London)

NEUTRON STUDY ON GROWTH PROCESS OF OXYGEN PRECIPITATES IN CZOCHRALSKI-

GROWN SILICON CRYSTALS

Takayoshi Takeda, Shigehiro Komura
Akira Ohsawa[+] and Koichiro Honda[+]

Faculty of Integrated Arts and Sciences
Hiroshima University, Hiroshima 730, Japan

[+] Fujitsu Laboratories Ltd., 1015 Kamikodanaka
Nakahara, Kawasaki 211, Japan

INTRODUCTION

The subject of the behavior of oxygen in Czochralski-grown silicon has been intensely studied in the last several years because of both beneficial and harmful effects caused by oxygen.[1] Silicon crystals used for manufacture of LSI are grown by the Czochralski method and contain interstitial oxygen of about 10^{18} atoms/cm^3. The oxygen concentration is much greater than the solubility limit at the temperature where the wafers are heat-treated for device fabrication of LSI and therefore oxygen precipitation takes place during the heat-treatment.

The macroscopic differential cross section per unit volume of the sample per unit solid angle $d\Sigma(Q)/d\Omega$ for small-angle neutron scattering (SANS) is given by,

$$\frac{d\Sigma}{d\Omega}(\vec{Q}) = \frac{1}{V_s} \left| \int_{V_s} (\rho(\vec{r}) - \bar{\rho}) \exp(i\vec{Q}\cdot\vec{r}) d^3\vec{r} \right|^2 \quad , \qquad (1)$$

where $\vec{Q}$ is the scattering vector, V_s is the sample volume, $\rho(\vec{r})$ is the scattering length density at the position $\vec{r}$ and $\bar{\rho}$ is the scattering length density of the matrix. The neutron scattering length densities of cristobalite SiO_2 and silicon are 0.344×10^{-11} and 0.207×10^{-11} cm^{-2}, respectively. Provided that the oxygen precipitates consist of cristobalite SiO_2, the contrast of the scattering length density $\Delta\rho = \rho_{SiO_2} - \rho_{Si}$ between precipitation (SiO_2) and matrix (Si) is fairly large. As precipitates are considered to be very dilute, we need not consider the interference among the precipitates and the cross section (1) can be described by Guinier's approximation for $QR_g \ll 1$,

$$\frac{d\Sigma}{d\Omega}(Q) = N(\Delta\rho)^2 V_p^2 \exp(-Q^2 R_g^2/3) \quad , \qquad (2)$$

where N is the number of cluster per unit volume, V_p is the volume of a precipitate and R_g is the radius of gyration. Since a silicon crystal is almost transparent against neutrons, the SANS experiments can be performed on a crystal annealed at a temperature from various periods without sample-destruction. The SANS study gives us the information of the oxygen precipitates integrated over the whole crystals. Thus we undertook SANS from a Czochralski grown silicon crystal annealed at different temperatures for various periods in order to investigate the growth mechanism of oxygen precipitates in the silicon crystal.[2] In the present paper growth process of oxygen precipitates is discussed.

EXPERIMENTAL

The dislocation-free Czochralski-grown silicon crystal used in this studies was 10.5 cm in diameter, being grown in the <100> direction. The dissolved oxygen concentration before annealing C_0 was 1.67 X 10^{18} atoms/cm^3 which was determined by an IR absorption method.[3] The sample for SANS were cut from the crystal such that two faces were normal to the growth direction [001] and that other faces were in the (110) and (1$\bar{1}$0) planes. Each sample was annealed in dry N_2 at the temperatures 600, 700, 750 or 800 °C. The fragments of the same crystal were annealed with the sample and used to determine the dissolved oxygen concentration C by the IR absorption method. The unprecipitated oxygen density $\bar{C}$ was estimated by $\bar{C} = C - C_E$, where C_E is the equilibrium oxygen density at the temperature. The values of $\bar{C}$ thus determined with increasing annealing time t at 750 and 800 °C are shown in Fig.1, where $C_E = 2\times10^{17}$ cm^{-3}. The small-angle neutron scattering experiment was performed using the spectrometer[4] installed at the Kyoto University Reactor(KUR). The incident beam was normal to the <110> direction of silicon crystals. The thickness of the samples along the neutron beam ranges from 5 to 10 cm.

RESULTS AND DISCUSSION

Figure 2 shows the macroscopic differential cross section $d\Sigma/d\Omega$ as a function of Q obtained from the SANS measurement on the samples annealed at 750 °C for different annealing times t. The solid lines are the gaussian curves with the radius of gyration R_g obtained by a least-squares fit of the cross section $d\Sigma/d\Omega$ to eq. (2) over the range of Q from 0.03 to 0.10 A^{-1}. The cross sections $(d\Sigma/d\Omega)_{<100>}$ obtained from the measurements at 700, 750, 800 °C with $\vec{Q}$ // <100> were greater than $(d\Sigma/d\Omega)_{<110>}$ with $\vec{Q}$ // <110> as shown in Fig.2. The former increased with increasing annealing time t. The radius of gyration R_g of $(d\Sigma/d\Omega)_{<100>}$ are about 20Å as shown in Fig.2 .

Figure 3 shows the

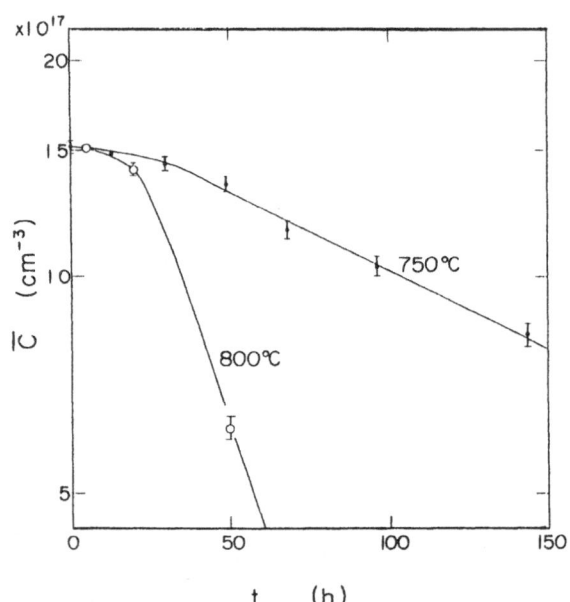

Fig.1 Logarithmic plot of unprecipitated oxygen density $\bar{C}$ in the Czochralski-grown silicon crystal as a function of annealing time t.

● ; at 750 ℃,   o : at 800 °C.

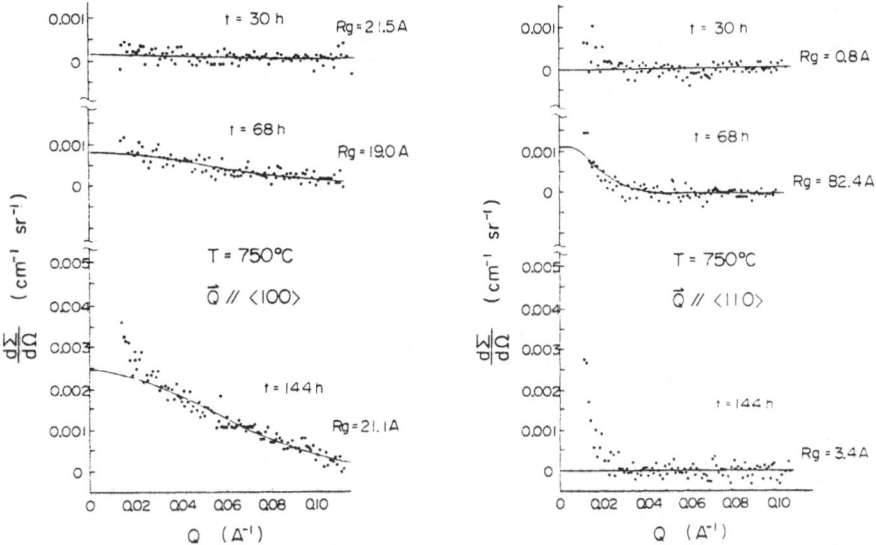

Fig.2   The macroscopic differential cross section $(d\Sigma/d\Omega)$ as a function of scattering vector Q obtained from the SANS measurement on the samples annealed at 750 °C for the annealing time t.   The solid lines show the gaussian curves with the radius of gyration R_g.

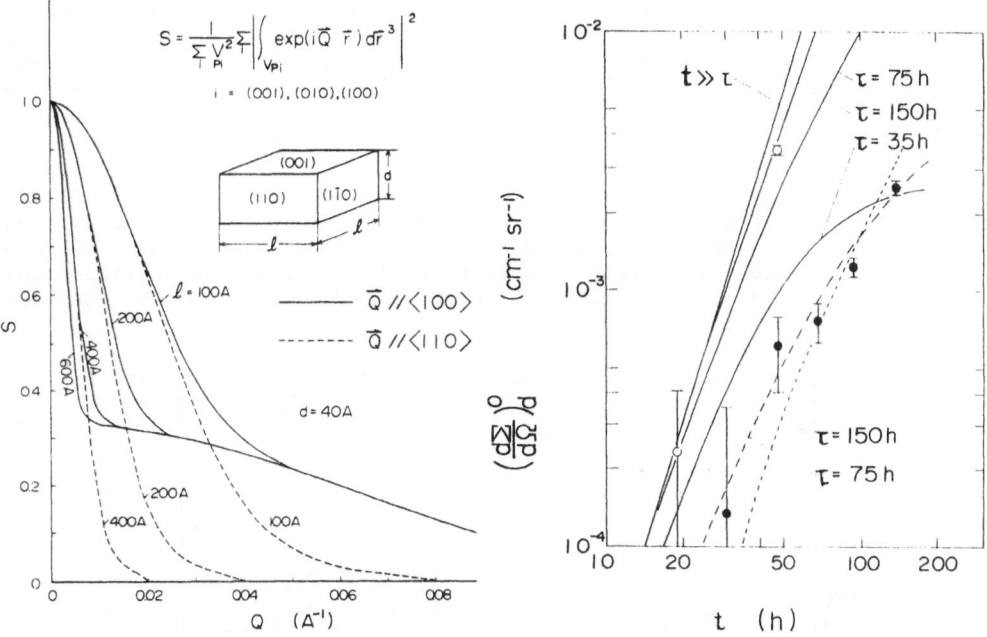

Fig.3   The caltulated scattering function S from eq. (3) the square-shaped platelet form with the plate surface parallel to the {100} plane and the peripherical edge in the <110> directions.   d is the thickness and l the edge length.

Fig.4   The logarithmic plot of the forward cross section $(d\Sigma/d\Omega)_0^d$ as a function of the annealing time t. ●;at 750 °C, ○;at 800 °C.
The solid line shows the cross section $(1/3)(d\Sigma(0)/d\Omega)^c$ calculated from eq. (12) for 750 °C. The dashed lines show $(1/3)(d\Sigma(0)/d\Omega)^c a$, where a is a reduction factor.   a=0.10 for τ =150 h and a=0.15 for τ =75 h.

calculated scattering function $S(\vec{Q})$ from the square-shaped platelet precipitates with the thickness d=40 Å and the edge length l, which have the plate surface parallel to the (100), (010) or (001) plane with equal probability. The function $S(\vec{Q})$ is given by,

$$S(Q) = \sum_i \left| \int_{V_{Pi}} \exp(iQ \cdot r) d^3 r / V_{pi} \right|^2 , \qquad (3)$$

where the i's represents the (100), (010) or (001) plane and V_{pi} is the volume of the precipitate which have the plate surface parallel to i-th plane. The scattering function $S(\vec{Q})$ with $\vec{Q}/\!/\langle 100\rangle$ decreased less rapidly with increasing Q than that with $\vec{Q}/\!/\langle 110\rangle$ and is almost equal to $|F_d|^2/3$ in the range of Q greater than $Q=2\pi/l$, where F_d is the structure function of a plate with the thickness of d and is given by, $F_d = (2/Qd)\sin(Qd/2)$. The factor 1/3 arises from the fact that only one of the three orientations of platelet contributes significantly to the scattering. The function $|F_d|^2$ is approximately equal to the gaussian function with $R_g = d/2$ for such small Q as $Qd < 1$. Figure 4 shows a logarithmic plot of the forward cross section $(d\Sigma/d\Omega)_d$ as a function of the annealing time t, where $(d\Sigma/d\Omega)_d$ is obtained by a least-square fit of the cross section $(d\Sigma/d\Omega)$ to the following equation

$$(d\Sigma/d\Omega) = (d\Sigma/d\Omega)_d^0 \, |F_d|^2$$

over the range of Q from 0.03 to 0.10 Å^{-1}, where d is fixed to 40 Å, corresponding to R_g of about 21 Å over the Q range. The mean value of R_g is 21.4 Å for 750 °C and 21.6 Å for 800 °C. These results indicate that the thickness of the oxygen precipitates d is about 41 Å. The forward cross section $(d\Sigma/d\Omega)_d^0$ increased as $t^{1.88 + 0.11}$ for the samples annealed at 750 °C and $t^{2.77 + 0.26}$ for the sample annealed at 800 °C. The cross section $(d\Sigma/d\Omega)_{\langle 110\rangle}$ decreases in the range of Q from 0.03 to 0.1 Å^{-1} with increasing l as shown in Fig.3. This behavior of $(d\Sigma/d\Omega)_{\langle 110\rangle}$ was observed as shown in Fig.2. These results agree with the earlier results obtained by TEM ; the oxygen precipitates have a square-shaped platelet form with the plate surface parallel to the 100 silicon lattice plane and with the peripherical edges along the $\langle 110\rangle$ directions and their thickness remains constant (about 40 Å) during the growth[5].

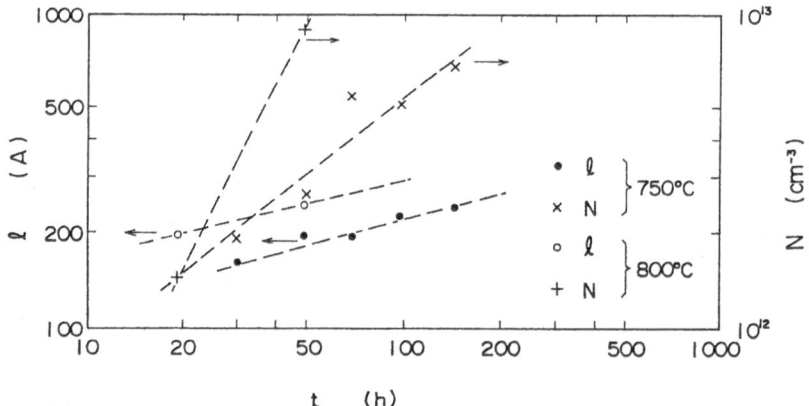

Fig.5 The edge length l and precipitate density N as a function of the annealing time t. ●; l at 750 °C, O; l at 800 °C, x; N at 750 °C, +; N at 800 °C. The dashed lines are guides for eyes for experimental data.

For the diffusion limited growth of randomly distributed particles, Ham[6] shows that

$$(dr_0(t)/dt) = [(C(t)-C_E)/(C_S-C_E)]D/r_0(t) \quad , \qquad (4)$$

where $r_0(t)$ is the radius of a spherical particle and for $(C(t)-C_E)/(C_I-C_E) < 0.5$,

$$(C(t)-C_E)/(C_I-C_E) = \exp(-t/\tau) \qquad\qquad (5)$$

where C_I is the initial oxygen density, C_E the equilibrium oxygen density at the temperature, C_S the oxygen density in the precipitate, and D the oxygen diffusion coefficient in silicon. Substituting eq.(5) to eq.(4) lead to

$$r_0^2(t) = 2D\tau[(C_I-C_E)/(C_S-C_E)][1-\exp(-t/\tau)] \quad , \quad (6)$$

and the particle volume is given by,

$$V_p(t) = (4\pi/3)r_0^3 \qquad . \qquad\qquad (7)$$

For $t \ll \tau$, eq.(6) reduces to

$$r_0^2(t) = 2D[(C_I-C_E)/(C_S-C_E)]t \qquad . \qquad (8)$$

If the square-shaped platelet precipitate grows in a spherical diffusion field since the precipitate is very small, the particle volume (7) can be replaced by the equivalent volume of the square-shaped platelet precipitate dl^2 and the growth law of the edge length $l(t)$ of a square-shape platelet is given by

$$l(t) = [8 \sqrt{2} \pi/(3d)]^{1/2}[[(C_I-C_E)/(C_S-C_E)]^{3/4}(Dt)^{3/4} \qquad . \quad (9)$$

a) Provided that all oxygen atoms which are released from interstitial sites of the silicon lattice coaleasce into the platelet precipitates and that each precipitate has the same size, we calculated the precipitate density N and l using the oxygen density released from the interstitial sites n $=C_0-C$ and the forward cross section $(d\Sigma/d\Omega)_d$.[6,2] These results are shown in Fig.5. The edge length l in Fig.5 increases as $l \propto t^{0.28}$. The results contradicts with eq.(9) which predicts $t^{0.75}$; Wada et al[5] reported that the precipitate grows according to eq.(9) and Newman et al[7] reported that it grows according to eq.(8). Presumably our assumption that each precipitate has the same size might be inappropriate. The nucleation of precipitates should occur during the annealing.

b) If we take into account the nucleation, the precipitate density N varies with time and can be approximated by,

$$N(t) = Jt \qquad , \qquad\qquad (10)$$

where J is the nucleation rate. We define the saturation time t_s after which the nucleation is suppressed.[2] Let us consider that the precipitates grow according to eq.(6). From eqs.(2) and (6), the forward neutron cross section at Q = 0 is estimated by,

$$\frac{d\Sigma}{d\Omega}(0) = (\Delta\rho)^2\int_0^{t_s} V_p^2(t-t')Jdt' \qquad , \qquad (12)$$

and was calculated for each annealing time t at 750 °C by substituting the following values; $J = 2 \times 10^8 \text{ cm}^{-3} \text{ s}^{-1}$, $C_I = 16.7 \times 10^{17}$ atoms/cm^3 [8) $C_s =$ 4.2×10^{22} atoms/cm^3, $C_E = 2 \times 10^{17}$ atoms/cm^3 and $D = 1.4 \times 10^{-13}$ cm^2 s^{-1} [8) and $\Delta \rho = 0.137 \times 10^{11}$ cm^{-2}. The calculated values of $(1/3)(d\Sigma(0)/d\Omega)^c$ were shown by the solid line in Fig.4. They increase more rapidly and much greater than the observed values of $(d\Sigma/d\Omega)_d$ at 750 °C. From eq.(5) and Fig.1, the time constant is estimated to be 150 h for 750 °C and 35 h for 800 °C. This result indicates that the oxygen precipitates grow more slowly than eq.(6), and that the contrast of the scattering density of oxygen precipitates against silicon is less than 0.137×10^{11} cm^{-2}. The latter means that the oxygen precipitates do not consist of cristobalite SiO_2 but a oxygen-deficient silicon oxide SiO_{2-x}. The former is considered to mean that the volume of the oxygen precipitates grows less rapidly than eq.(7), since the precipitates grow two-dimensionally. Another explanation for the less values of the observed cross section than the calculated values is that large oxygen precipitates which scatter neutrons at lower Q than 0.02 Å^{-1} grow at the expense of small precipitates.

In the case of samples annealed at 800 °C, the absolute values of the observed cross section were not compared with the calculated values for lack of the reported parameters. The cross section $(d\Sigma/d\Omega)_d^0$ of the sample annealed at 800 °C increased approximately according to eq.(12).

CONCLUSION

From the SANS study, it was confirmed that the oxygen precipitates in Czochoralski-grown silicon crystals grow two-dimensionally in a form of platelet with thickness of d = 40 Å, the plate surface being parallel to the {100} silicon lattice plane. The forward cross section for $\vec{Q}$ // <100> increased as $t^{1.88}$ at 750 °C and as $t^{2.77}$ at 800 °C. The edge length l and the density N of oxygen precipitates were estimated on the assumption that each precipitate has the same size. They are shown in Fig.5.

The cross sections were calculated by taking into account the nucleation of precipitates during the annealing and using the reported parameters at 750 °C. They were much greater than the observed values and increased more rapidly. From these analyses, it is considered that the oxygen precipitates consist of oxygen-deficient oxides (SiO_{2-x}). The mean size of oxygen precipitates grows according to less than 3/4-th power of annealing time t at 750 °C and to about 3/4-th power of t at 800 °C.

REFERENCES

1) J.R.Patel: Semiconductor Silicon(Electrochem. Soc., Princeton, N.J. 1981) p189.
2) T. Takeda, S. Komura, A. Ohsawa and K. Honda: Jpn. J. Appl. Phys. 26(1987)106.
3) ASTM F-121-79.
4) S. Komura, T. Takeda, H. Fujii, K. Osamura, Y. Toyoshima, K. Mochiki and K. Hasegawa: Jpn.J.Appl.Phys. 22(1983) 351.
5) K.Wada and N.Inoue: J.Cryst.Growth 49(1980) 794.
6) F. S. Ham: J. Phys. Chem. Solids 6(1958)335.
7) R. C. Newman, M. C. Claybourn, S. H. Kinder, S. Messoloras, A. S. Oates and R. J. Stewart: Semiconductor Silicon (Electrochem. Soc., Pennington, N.J. 1986) p766.
8) N.Inoue, K.Wada and J. Osaka: Semiconductor Silicon (Electrochem. Soc., Princeton, N.J. 1981) p282.

CONCLUDING REMARKS

S.Komura

Faculty of Integrated Arts and Sciences
Hiroshima University
Hiroshima 730, Japan

Now we are almost at the end of our symposium.

Here   I   should   like   to   provide   some   remarks   on   the   history
of our studies on "kinetic phase transitions."

The   study   of the topic in this symposium dates back   to   1687,   300
years ago, when Sir Isaac Newton published "Principia".   In this famous
book he showed not only that the motion of planets in the heavens is
governed by the law which also governs the motion of falling bodies on
earth, but also, in his third volume Principia, that the transformation
of substances are to be explained by essentially the same dynamics.

This year, 1987, we celebrate not only the 300th anniversary of
Principia but also the 30th anniversary of the monumental paper by Cahn &
Hilliard in 1957, which founded the subsequent development of researches
in this field.   We are very much honoured by the presence, here, of
Dr.Cahn.

On the logarithmic scale of time past, some 100 years ago, there
occurred a famous milestone, the equation of state proposed by Van der
Waals in 1873. This led to the important concepts of the critical point,
the binodal line, and the spinodal line.   Subsequently, the insight of
Gibbs, in 1878, enlarged our understanding of both the equilibrium and
non-equilibrium properties of inhomogeneous systems.

Our topic, namely, the "Dynamic Ordering Process" is certainly in the
direction that has been taken since the time of Newton, Van der Waals and
Gibbs. With such a thought in mind, we can proceed with confidence into
the future.

So far we have had many interesting lectures and enthusiastic
discussions.   Looking back to the posters a variety of phenomena and
different points of view have been presented to us. The demonstrations by
video players and personal computers have been so attractive.   We have
learned a lot and are very much impressed.   Some of the ideas are not
familiar to us and we need more time in which to digest them. Nevertheless I believe we are successfully promoting discussions among scientists
of different disciplines.

I emphasize here that it is the universal nature of the laws, such as the time evolution power law, the dynamical scaling law, the asymptotic power law of the structure function, that are applicable to various superficially quite different substances such as metals, polymers and liquids. This is one of the reasons why we are most excited and stimulated.

I should like to express my gratitude to all the participants who presented such nice lectures, posters, and discussions. Here you see the ingredients of our symposium: theoretical statistical physicists, computer physicists, experimental scattering physicists, metallurgists, polymer scientists and scientists in other fields.

These ingredients are mixed up to make a nice cocktail of this symposium. In particle physics language, experimental scattering physicists served as a gluon for several quarks to glue all the ingredients together.

I shall be very happy if this symposium serves as a step towards the next great successful jump forward.

We thank all the partipants, the members of the international advisory committee, the members of the organizing committee, the members of the local committee, the staff who helped to organize and carry out the symposium, and the many organizations and industrial companies, listed in the abstract book, who gave us donations.

I conclude my talk with the following three phrases,

more refined theories !!

more accurate experiments !!

more ample phenomena !!

SYMPOSIUM PARTICIPANTS

The numbers under the names refer to the photograph on page 567

Allen, Samuel M.
55
Department of Material Science and Engineering
Massachusetts Institute of Technology
Cambridge, Massachusetts 02139 U.S.A.

Beysens, Daniel
50
Service de Physique du Solide et de Resonance
Magnetique, Commissariat a l'Enenergie Atomique
Orme des Mersiers, 91191 Gif-Sur-Yvetle Cedex
France

Cahn, John W.
62
Institute for Materials Science and Engineering
National Bureau of Standards
Gaithersburg, Maryland 20899 U.S.A.

Chen, Haydn
22
Department of Metallurgy and Mining Engineering
University of Illinois at Urbana-Champaign
Urbana, Illinois 61801 U.S.A.

Chu, Benjamin
60
Department of Chemistry
State University of New York
Stony Brook, New York 11794-3400 U.S.A.

Descamps, Marc
66
Universite des Sciences et Techniques de Lille
"U.F.R. de Physiques" batiment P5 59655
Villeneuve d'Asq Cedex, France

Doi, Minoru
3
Department of Materials Science and Engineering
Nagoya Institute of Technology
Nagoya 466 Japan

Eguchi, Tetsuo
Department of Applied Physics
Fukuoka University
Fukuoka 814-01 Japan

Fratzl, Peter
47
Institut fur Festkorperphysik der Universitat
Wien, A-1090 Wien Strudlhofgasse 4, Austria

Fujikawa, Sin'ichirou
61
Faculty of Engineering
Tohoku University
Sendai 980 Japan

Fujita, Yuji
Toa Nenryo Kogyo K.K.
Ohi-cho, Irima-gun, Saitama 354 Japan

Furukawa, Hiroshi
28
Faculty of Education
Yamaguchi University
Yamaguchi 753 Japan

Goldburg, Walter
4
Physics Department
University of Pittsburgh
Pittsburgh, P.A. 15260 U.S.A.

Gunton, James D.
6
Department of Physics
Temple University
Philadelphia, Pa. 19122 U.S.A.

Guyot, Pierre
49
Institut National Polytechnique de Grenoble
ENSEEG, Domaine Universitaire BP75 38402
Saint Martin d'Heres France

Han, Charles
13
National Bureau of Standards
Polimers Division
Gaithersburg, MD 20899 U.S.A.

Hanamura, Toshihiro
93
Nippon Steel Corporation
R.& D. Labs.- I
Kawasaki, Kanagawa 211 Japan

Hara, Hiroaki
87
Faculty of Engineering
Tohoku University
Sendai 980 Japan

Harada, Yoshifumi
7
Faculty of Engineering
Fukui University
Fukui 910 Japan

Hasegawa, Hirokazu
45
Faculty of Engineering
Kyoto University
Kyoto 606 Japan

Hashimoto, Takeji
14
Faculty of Engineering
Kyoto University
Kyoto 606 Japan

Hayakawa, Hisao
57
Faculty of Science
Kobe University
Kobe 657 Japan

Heermann, Dieter W.
48
Institut fur physic
Johannes-Gutenberg Universitat
Standinger Weg 7, 6500 Maiz, Fed. Rep. Germany

Hirano, Ken'ichi
8
Faculty of Engineering
Tohoku University
Sendai 980 Japan

Hirotsu, Shunsuke
71
Faculty of Science
Tokyo Institute of Technology
Tokyo 152 Japan

Honda, Katsuya
29
Faculty of Engineering
Nagoya University
Nagoya 464 Japan

Ikeda, Hironobu
39
Faculty of Science
Ochanomizu University
Tokyo 112 Japan

Imai, Masayuki 85	Toyobo Co., Ltd., Research Center Ohtsu, Shiga 520-02 Japan
Inden, Gerhard 51	Max-Planck-Institut fur Eisenforschung GmbH Dusseldorf 14, Fed. Rep. Germany
Izumitani, Tatsuo 98	Daicel Chemical Industries, Ltd., Research Center Himeji, Hyogo 671-12 Japan
Izuyama, Takeo 80	College of Arts and Sciences Tokyo University Tokyo 153 Japan
Kaji, Keisuke	The Institute for Chemical Research Kyoto University Uji, Kyoto 611 Japan
Kanaya, Toshiji 46	The Institute for Chemical Research Kyoto University Uji, Kyoto 611 Japan
Karma, Alain 64	Department of Physics California Institute of Technology Pasadena, CA U.S.A.
Kashihara, Osamu 99	Polyplastics Co., Ltd. Research Center Fuji, Shizuoka 416 Japan
Katano, Susumu 23	Department of Physics Japan Atomic Energy Research Institute Ibaragi 319-11 Japan
Kawasaki, Kyozi 5	Faculty of Science Kyushu University Fukuoka 812 Japan
Kawasaki, Tatsuo 24	College of Liberal Arts and Sciences Kyoto University Kyoto 606 Japan
Kikuchi, Macoto 43	Faculty of Science Osaka University Osaka 560 Japan
Kikuchi, Ryoichi 11	Department of Materials, Sciences and Engineering University of Washington Seattle, WA 98195 U.S.A.
Kimura, Hatsuo 12	Faculty of Engineering Nagoya University Nagoya 464 Japan
Kitada, Masahiro	Hitachi Ltd. Central Research Laboratory Kokubunji, Tokyo 185 Japan
Komura, Shigehiro 10	Faculty of Integrated Arts & Sciences Hiroshima University Hiroshima 730 Japan

```
Komura, Shigeyuki,Jr.   College of Arts & Sciences
           72            Tokyo University
                         Tokyo 153 Japan

Konishi, Hiroyuki        Department of Physics
           35            Japan Atomic Energy Research Institute
                         Ibaragi 319-11 Japan

Kostorz, Gernot          Institut fur Angewandte Physik
           52            ETH-Honggerberg
                         CH-8093 Zurich Switzerland

Koyama, Yasumasa         Department of Metallurgy
           83            Tokyo Institute of Technology
                         Tokyo 152 Japan

Kozakai, Takao           Department of Materials Science and Engineering
           18            Nagoya Institute of Technology
                         Nagoya 466 Japan

Kozasu, Isao             Nippon Kokan K.K.
                         Steel Research Center
                         Kawasaki, Kanagawa 210 Japan

Kuroda, Toshio           Institute of Low Temperature Science
           36            Hokkaido University
                         Sapporo 060 Japan

Levine, Herbert          Schlumberger Doll Reseach
           75            Old Quarry Rd.
                         Ridgefield, CT 06877 U.S.A.

Liu, Yong                Faculty of Science
           17            Tokyo University
                         Tokyo 113 Japan

Mashiyama, Hiroyuki      Faculty of Science
           38            Yamaguchi University
                         Yamaguchi 753 Japan

Matsui, Mitsuhiko        Tokuyama Soda K.K.
           15            Fujisawa, Kanagawa 252 Japan

Matsumura, Syo           Department of Materials Science and Technology
           53            Graduate School of Engineering Science
                         Kyushu University
                         Fukuoka 816 Japan

Matsuura, Motohiro       Faculty of Engineering Science
           25            Osaka University
                         Osaka 560 Japan

Mazenko, Gene F.         The James Franck Institut and Department of
           63            Physics
                         The University of Chicago
                         Chicago, Illinois 60637 U.S.A.
```

Mitsui, Akitaka

Sumitomo 3M Ltd.
Corporate Development Laboratory
Sagamihara, Kanagawa 229 Japan

Miyazaki, Tohru
2

Department of Materials Science & Engineering
Nagoya Institute of Technology
Nagoya 466 Japan

Miyazaki Toshio
77

Faculty of Integrated Arts & Sciences
Hiroshima University
Hiroshima 730 Japan

Miyajima, Sasuke
58

Department of Engineering Physics
Chubu University
Aichi 487 Japan

Morii, Yukio
74

Department of Physics
Japan Atomic Energy Reseach Institute
Ibaragi 319-11 Japan

Mouritsen, Ole.G
65

Department of Structual Properties of Materials
The Technical University of Denmark
DK-2800 Lyngby, Denmark

Nagai, Tatsuo
33

Department of General Education
Kyushu Kyoritu University
Kitakyushu, Fukuoka 807 Japan

Nagler, Stephen
73

Department of Physics
University of Florida
Gainesville, FL 32611 U.S.A.

Nakao, Toshio
81

Sumitomo Bakelite Co., Ltd.
Yokohama, Kanagawa 245 Japan

Nakayama, Ryoji
68

Mitsubishi Metal Corporation
Central Research Laboratory
Ohmiya, Saitama 330 Japan

Nishioka, Kazumi
44

Department of Precision Mechanics
Tokushima University
Tokushima 770 Japan

Noda, Yukio
20

Faculty of Engineering Science
Osaka University
Osaka 560 Japan

Nose, Takuhei
67

Department of Polymer Chemistry
Tokyo Institute of Technology
Tokyo 152 Japan

Ochiai, Moyuru
34

Department of Electronics
North Shore College of SONY Institute
Kanagawa 243 Japan

Ohno, Makoto
84

Toyobo Co., Ltd.
Research Center
Ohtsu, Shiga 520-02 Japan

Ohta, Takao
79

Department of Physics
Kyushu University
Fukuoka 812 Japan

Okabe, Yutaka
42

Department of Physics
Tohoku University
Sendai 980 Japan

Okada, Mamoru
32

Department of Polymer Chemistry
Tokyo Institute of Technology
Tokyo 152 Japan

Oki, Kensuke
54

Department of Materials Science and Technology
Graduate School of Engineering Science
Kyushu University
Fukuoka 816 Japan

Onodera, Yukio
16

Department of Engineering Science
Tohoku University
Sendai 980 Japan

Onuki, Akira
27

Research Institute for Fundamental Physics
Kyoto University
Kyoto 606 Japan

Orihara, Hiroshi
95

Faculty of Engineering
Nagoya University
Nagoya 464 Japan

Osamura, Kozo
92

Department of Metallurgy
Kyoto University
Kyoto 606 Japan

Otsuki, Etsuo

Tokin Corporation
Kohriyama, Sendai 982 Japan

Sakuma, Akimasa
82

Hitachi Metals, Ltd.
Magnetic & Electronic Materials Research
Laboratry
Kumagaya, Saitama 360 Japan

Schwahn, Dietmar
59

Institut fur Festkorperforschung
der Kernforschungsanlage Julich GmbH
5170 Julich, Fed. Rep. Germany

Sekimoto, Ken
56

Faculty of Science
Kyushu University
Fukuoka 812 Japan

Senoo, Yoshiki
69

Toyota Central Research & Development Lab, Inc.
Nagakute, Aichi 480-11 Japan

Springer, Tasso
31

Institut fur Festkorperforschung
der Kernforschungsanlage Julich GmbH
5170 Julich, Fed. Rep. Germany

Srolovitz, David
76

Los Alamos National Laboratory
Los Alamos, NM 87545 U.S.A.

Suzuki, Masuo
9

Faculty of Science
Tokyo University
Tokyo 113 Japan

Takahashi, Masato

Faculty of Technology
Tokyo Metropolitan University
Tokyo 158 Japan

Takayasu, Hideki
90

Faculty of Science
Kobe University
Kobe 657 Japan

Takayasu, Mrs.
91

Faculty of Science
Kobe University
Kobe 657 Japan

Takebe, T
96

College of Liberal Arts & Sciences
Kyoto University
Kyoto 606 Japan

Takeda, Takayoshi
30

Faculty of Integrated Arts & Sciences
Hiroshima University
Hiroshima 730 Japan

Takenaka, M
100

College of Liberal Arts & Sciences
Kyoto University
Kyoto 606 Japan

Tanaka, Kohji
41

Nissin Steel Co.,Ltd.
Ichikawa, Chiba 272 Japan

Terao, Naoko
78

Faculty of Education
Hiroshima University
Hiroshima 730 Japan

Todoroki, Tsunehiko
19

Air-conditioner Division
Matsushita Electric Industrial Co. Ltd.
Kusatsu, Shiga 525 Japan

Tokuyama, Michio
37

Tohwa University
Fukuoka 815 Japan

Tomita, Hiroyuki
26

College of Liberal Arts & Sciences
Kyoto University
Kyoto 606 Japan

Tomokiyo, Yoshitsugu
1

Research Laboratory of High Voltage Electron
Microscope, Kyushu University
Fukuoka 812 Japan

Toyoki, Hiroyasu
94

Faculty of Education
Yamanashi University
Yamanashi 400 Japan

Uchida, Takashi
86

Faculty of Science
Hokkaido University
Sapporo 060 Japan

Ueno, Satoru
89

Faculty of Integrated Arts & Sciences
Hiroshima University
Hiroshima 730 Japan

Wada, Koh
88

Faculty of Science
Hokkaido University
Sapporo 060 Japan

Windsor, Colin G.
40

Materials Physics Division
Atomic Energy Research Establishment
Harwell, Didcot
Oxon, OX11 ORA, U.K.

Yagi, Toshirou
21

Department of Physics
Kyushu University
Fukuoka 812 Japan

Yamamoto, Naoki
70

Faculty of Engineering
Technological University of Nagaoka
Niigata 949-54 Japan

Yokoyama, Etsurou
97

Institute of Low Temperature Science
Hokkaido University
Sapporo 060 Japan

PHOTOGRAPH IDENTIFICATION

1. Tomokiyo, Yoshitsugu
2. Miyazaki, Tohru
3. Doi, Minoru
4. Goldburg, Walter
5. Kawasaki, Kyozi
6. Gunton, James D.
7. Harada, Yoshifumi
8. Hirano, Ken'ichi
9. Suzuki, Masuo
10. Komura, Shigehiro
11. Kikuchi, Ryoichi
12. Kimura, Hatsuo
13. Han, Charles
14. Hashimoto, Takeji
15. Matsui, Mitsuhiko
16. Onodera, Yukio
17. Liu, Yong
18. Kozakai, Takao
19. Todoroki, Tsunehiko
20. Noda, Yukio
21. Yagi, Toshirou
22. Chen, Haydn
23. Katano, Susumu
24. Kawasaki, Tatsuo
25. Matsuura, Motohiro

26. Tomita, Hiroyuki
27. Onuki, Akira
28. Furukawa, Hiroshi
29. Honda, Katsuya
30. Takeda, Takayoshi
31. Springer, Tasso
32. Okada, Mamoru
33. Nagai, Tatsuo
34. Ochiai, Moyuru
35. Konishi, Hiroyuki
36. Kuroda, Toshio
37. Tokuyama, Michio
38. Mashiyama, Hiroyuki
39. Ikeda, Hironobu
40. Windsor, Colin G.
41. Tanaka, Kohji
42. Okabe, Yutaka
43. Kikuchi, Macoto
44. Nisioka, Kazumi
45. Hasegawa, Hirokazu
46. Kanaya, Toshiji
47. Fratzl, Peter
48. Heermann, Dieter W.
49. Guyot, Pierre
50. Beysens, Daniel

51. Inden, Gerhard
52. Kostorz, Gernot
53. Matsumura, Syo
54. Oki, Kensuke
55. Allen, Samuel M.
56. Sekimoto, Ken
57. Hayakawa, Hisao
58. Miyajima, Sasuke
59. Schwahn, Dietmar
60. Chu, Benjamin
61. Fujikawa, Sin'ichirou
62. Cahn, John W.
63. Mazenko, Gene F.
64. Karma, Alain
65. Mouritsen, Ole G.
66. Descamps, Marc
67. Nose, Takuhei
68. Nakayama, Ryoji
69. Senoo, Yoshiki
70. Yamamoto, Naoki
71. Hirotsu, Shunsuke
72. Komura, Shigeyuki, Jr.
73. Nagler, Stephen
74. Morii, Yukio
75. Levine, Herbert

76. Srolovitz, David
77. Miyazaki, Toshio
78. Terao, Naoko
79. Ohta, Takao
80. Izuyama, Takeo
81. Nakao, Toshio
82. Sakuma, Akimasa
83. Koyama, Yasumasa
84. Ohno, Makoto
85. Imai, Masayuki
86. Uchida, Takashi
87. Hara, Hiroaki
88. Wada, Koh
89. Ueno, Satoru
90. Takayasu, Hideki
91. Takayasu, Mrs.
92. Osamura, Kozo
93. Hanamura, Toshihirc
94. Toyoki, Hiroyasu
95. Orihara, Hiroshi
96. Takebe, Tomoaki
97. Yokoyama, Etsurou
98. Izumitani, Tatsuo
99. Kashihara, Osamu
100. Takenaka, Mikito

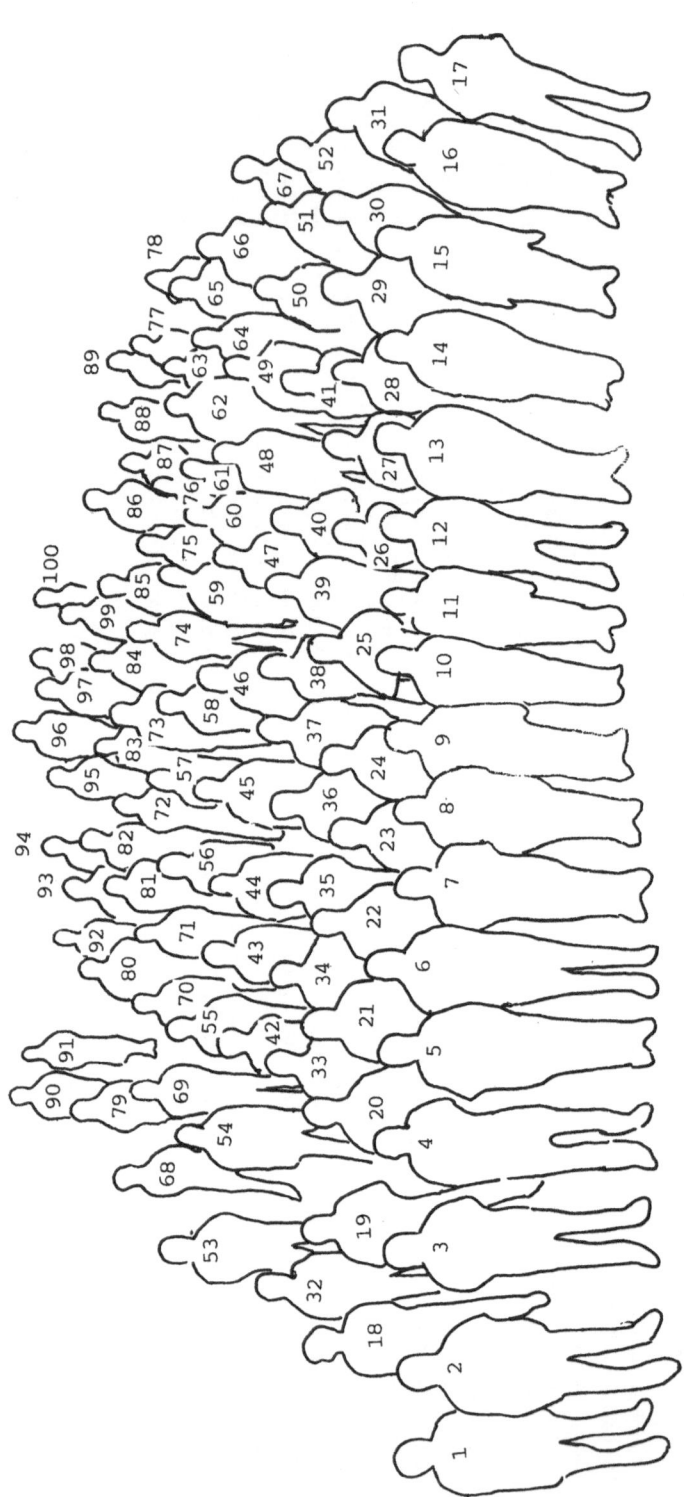

AUTHOR INDEX